Pharmacology of Ionic Channel Function: Activators and Inhibitors

Contributors

S. Adachi-Akahane, J. Barhanin, E.A. Barnard, J.B. Bergsman, H. Betz, J.M. Christie, M.-D. Drici, M. Endo, R.J. Evans, G. Giebisch, T. Gonoi, M. Gordey, A.O. Grant, M.E. Grunwald, R.J. Harvey, S.C. Hebert, F. Hofmann, M. Iino, T. Ikemoto, K. Imoto, A. Inanobe, D.E. Jane, N. Klugbauer, Y. Kubo, Y. Kurachi, E. Latorre, M. Lazdunski, C. Legros, F. Lesage, M. Mishina, D.T. Monaghan, T. Nagao, C.G. Nichols, R.W. Olsen, O. Pongs, S. Seino, D.A. Skifter, E. Stefani, J. Teulon, L. Toro, H.-W. Tse, R.W. Tsien, C. Vergara, W. Wang, D.B. Wheeler, J.L. Yakel, K.-W. Yau, H. Zhong

Editors:

M. Endo, Y. Kurachi, and M. Mishina

Springer

Professor Makato Endo, M.D., Ph.D.
Saitama Medical School
38 Morohongo, Moroyama
Iruma-Gun
Saitama 350-0495
Japan
e-mail: makoendo@saitama-med.ac.jp

Professor Dr. Yoshihisa Kurachi
Osaka University
Department of Pharmacology II
Graduate School of Medicine
and Faculty of Medicine
2-2 Yamada-oka, Suita
Osaka 565-0871
Japan
e-mail: ykurachi@pharma2.med.osaka-u.ac.jp

Professor Dr. Masayoshi Mishina
University of Tokyo
Department of Molecular
Neurobiology and Pharmacology
School of Medicine
Hongo 7-3-1, Bukyo-ku
Tokyo 113-0033
Japan
e-mail: mishina@m.u-tokyo.ac.jp

With 107 Figures and 25 Tables

ISBN 3-540-66127-1 Springer-Verlag Berlin Heidelberg New York

Library of Congress Cataloging-in-Publication Data
Pharmacology of ionic channel function : activators and inhibitors / editors, M. Endo,
Y. Kurachi, M. Mishina.
p. cm – (Handbook of experimental pharmacology; v. 147)
Includes bibliographical references and index.
ISBN 3540661271 (hardcover : alk. paper)
1. Ion channels. 2. Ion channels – Effect of drugs on. I. Endo, Makoto, 1933– .
II. Kurachi, Yoshihisa. III. Mishina, M. (Masayoshi), 1947– . IV. Series.
[DNLM: 1. Ion Channels – physiology. 2. Calcium Channel Blockers – pharmacology.
3. Cell Membrane – physiology. 4. Ion Channels – antagonists & inhibitors. 5. Membrane
Potentials – physiology. QH 603.I54 P536 2000]
QP905.H3 vol. 147
[QH603.I54]
615′.1s – dc21
[615′.7] 00-026585

Coverdesign: *design & production* GmbH, Heidelberg
Typesetting: Best-set Typesetter Ltd., Hong Kong
SPIN: 10647765 27/3020-5 4 3 2 1 0 – printed on acid-free paper

Preface

Life did not come into existence until living organisms developed the ability to establish an internal ionic environment which was quite different from that of the external world. Not only is the intracellular ionic composition suitable for various cellular housekeeping reactions, but its difference from the extracellular medium was ingeniously utilized by the cell for responding to stimuli given to the cell or to other changes in the environment. In the latter responses two major pathways are used: (1) alteration of the membrane potential that is formed by the difference in the ionic compositions across the boundary membrane combined with the different permeabilities of the membrane to each ion, and (2) alteration of the intracellular concentration of ions, particularly of calcium ions, which is minute under normal conditions and, therefore, easily altered. In both cases, ion channels in the boundary membrane play the key role, by changing the ionic permeability and by allowing ionic transport down the electrochemical potential gradient as a result of the permeability change. Since ion channels are thus vitally important in living organisms, they developed various kinds of ion channels, some of which, for example, are highly selective for a particular ion but others are rather nonselective. All the ion channels have special gating mechanisms of their own which are suitable for playing the given physiological role. The alteration of the functions of these ion channels by drugs or chemical agents, therefore, undoubtedly constitutes a very important field of pharmacology. Studies on ion channels have made great advances since the late 1980s, especially using molecular biology techniques. Many books have been published on ion channels, but we felt that we still need a comprehensive book which focuses on activatiors and inhibitors of ion channels. This book is meant to fulfill that need. Although some of the important ion channels, such as chloride channels or mechanically activated channels, have unfortunately not been included in this book, the editors of this volume believe that this book is still fairly comprehensive and very useful. We hope that the readers will agree that the authors have done an excellent job and that they will enjoy reading this stimulating volume.

Spring 2000

Makoto Endo
Saitama, Japan

List of Contributors

Adachi-Akahane, S., Laboratory of Pharmacology and Toxicology, Graduate School of Pharmaceutical Sciences, University of Tokyo, 7-3-1 Hongo, Bunkyo-ku, Tokyo 113-0033, Japan
e-mail: satomiaa@mol.f.u-tokyo.ac.jp

Barhanin, J., Institut de Pharmacologie Moléculaire et Cellulaire, CNRS-UPR 411, 660 Route des Lucioles, Sophia Antipolis, F-06560 Valbonne, France
e-mail: barhanin@ipmc.cnrs.fr

Barnard, E.A., Department of Pharmacology, University of Cambridge, Tennis Court Road, Cambridge CB2 1QJ, United Kingdom
e-mail: eb247@cam.ac.uk

Bergsman, J.B., Department of Molecular and Cellular Physiology, Beckman Center B105, Stanford University School of Medicine, Stanford, CA 94305-5345, USA

Betz, H., Max-Planck-Institut für Hirnforschung, Abteilung Neurochemie, Deutschordenstr. 46, D-60528 Frankfurt am Main, Germany
e-mail: Betz@mpi-frankfurt.mpg.de

Christie, J.M., Department of Pharmacology, University of Nebraska Medical Center, 600 S. 42nd Street, Box 986260, Omaha, NE 68198-6260, USA

Drici, M.-D., Institut de Pharmacologie Moléculaire et Cellulaire, CNRS-UPR 411, 660 Route des Lucioles, Sophia Antipolis, F-06560 Valbonne, France

Endo, M., Saitama Medical School, 38 Morohongo, Moroyama, Iruma-Gun, Saitama 350-0495, Japan
e-mail: makoendo@saitama-med.ac.jp

EVANS, R.J., Department of Cell Physiology and Pharmacology, University of Leicester, Medical Sciences Building, University Road, Leicester LE1 9HN, United Kingdom
e-mail: RJE6@le.ac.uk

GIEBISCH, G., Department of Cellular and Molecular Physiology, Yale University School of Medicine, 333 Cedar Street, New Haven, CT 06520-8026, USA
e-mail: gerhard.giebisch@yale.edu

GONOI, T., Research Center for Pathogenic Fungi and Microbial Toxicoses, Chiba University, 1-8-1, Inohana Chuo-ku, Chiba, Japan 260-8673, Japan
e-mail: gonoi@myco.pf.chiba-u.ac.jp

GORDEY, M., Department of Molecular and Medical Pharmacology, UCLA School of Medicine, 23-120 CHS, Box 951735, Los Angeles, CA 90024-1735, USA

GRANT. A.O., Cardiovascular Division, Department of Medicine, Duke University Medical Center, Durham, NC 27710-3504, USA
e-mail: aog@carlin.mc.duke.edu

GRUNWALD, M.E., Department of Molecular and Cell Biology, University of California at Berkeley, Berkeley, CA 94720, USA

HARVEY, R.J., Department of Pharmacology, The School of Pharmacy, 29-39 Brunswick Square, London WC1 N1AX, United Kingdom

HEBERT, S.C., Department of Cellular and Molecular Physiology, Yale University, School of Medicine, 333 Cedar Street, New Haven, CT 06520-8026, USA

HOFMANN, F., Institut für Pharmakologie und Toxikologie, Technische Universität München, Biedersteiner Str. 29, D-80802 München, Germany
e-mail: pharma@ipt.med.tu-muenchen.de

IINO, M., Department of Pharmacology, Graduate School of Medicine, The University of Tokyo; CREST, Japan Science and Technology Corporation, Bunkyo-ku, Tokyo 113-0033, Japan
e-mail: iino@m.u-tokyo.ac.jp

IKEMOTO, T., Department of Pharmacology, Saitama Medical School, 38 Morohongo, Moroyama, Iruma-Gun, Saitama 350-0495, Japan

IMOTO, K., National Institute for Physiological Sciences, Department of Information Physiology, Myodaiji, Okazaki 444-8585, Japan
e-mail: keiji@nips.ac.jp

INANOBE, A., Department of Pharmacology II, Graduate School of Medicine and Faculty of Medicine, Osaka University, 2-2 Yamada-oka, Suita, Osaka 565-0871, Japan

JANE, D.E., Department of Pharmacology, University of Bristol, Bristol, BS8 1TD United Kingdom
e-mail: david.jane@bristol.ac.uk

KLUGBAUER, N., Institut für Pharmakologie und Toxikologie, Technische Universität München, Biedersteiner Str. 29, D-80802 München, Germany

KUBO, Y., Department of Systems Physiology, Tokyo Medical and Dental University, Graduate School of Medicine and Dentistry, Yushima 1-5-45, Bunkyo-ku, Tokyo 113-8519, Japan
e-mail: ykubo@tmin.ac.jp

KURACHI, Y., Department of Pharmacology II, Graduate School of Medicine and Faculty of Medicine, Osaka University, 2-2 Yamada-oka, Suita, Osaka 565-0871, Japan
e-mail: ykurachi@pharma2.med.osaka-u.ac.jp

LATORRE, E., Avenida Arturo Prat 514, Casilla 1469, Valdivia, Chile
e-mail: ramon@cecs.cl

LAZDUNSKI, M., Institut de Pharmacologie Moléculaire et Cellulaire, CNRS-UPR 411, 660 Route des Lucioles, Sophia Antipolis, F.06560 Valbonne, France
e-mail: ipmc@ipmc.cnrs.fr

LEGROS, C., Institut für Neurale Signalverarbeitung, ZMNH, Martinistr. 52, D-20246 Hamburg, Germany

LESAGE, F., Institut de Pharmacologie Moléculaire et Cellulaire, CNRS-UPR 411, 660 Route des Lucioles, Sophia Antipolis, F.06560 Valbonne, France

MISHINA, M., Department of Molecular Neurobiology and Pharmacology, University of Tokyo, School of Medicine, Hongo 7-3-1, Tokyo 113-0033, Japan
e-mail: mishina@m.u-tokyo.ac.jp

MONAGHAN, D.T., Department of Pharmacology, Box 986260, University of Nebraska Medical Center, Omaha, NE 68198-6260, USA
e-mail: dtmonagh@unmc.edu

NAGAO, T., Laboratory of Pharmacology and Toxicology, Graduate School of Pharmaceutical Sciences, University of Tokyo, 7-3-1 Hongo, Bunkyo-ku, Tokyo 113-0033, Japan

NICHOLS, C.G., Department of Cell Biology and Physiology, Washington University School of Medicine, 660 South Euclid Ave., St. Louis, MO 63110, USA
e-mail: cnichols@cellbio.wustl.edu

OLSEN, R.W., Department of Molecular and Medical Pharmacology, UCLA School of Medicine, 23-120 CHS, Box 951735, Los Angeles, CA 90024-1735, USA
e-mail: ROlsen@mednet.ucla.edu

PONGS, O., Institut für Neurale Signalverarbeitung, ZMNH, Martinistr. 52, D-20246 Hamburg, Germany
e-mail: pointuri@uke.uni-hamburg.de

SEINO, S., Department of Molecular Medicine, Chiba University Graduate School of Medicine, 1-8-1, Inohana, Chuo-ku, Chiba 260-8673, Japan
e-mail: seino@molmed.m.chiba-u.ac.jp

SKIFTER, D.A., Department of Pharmacology, University of Nebraska Medical Center, 600 S. 42nd Street, Box 986260, Omaha, NE 68198-6260, USA

STEFANI, E., Facultad de Ciencias, Universidad de Chile, Centro de Estudios Cientificos de Santiago, Av. Presidente Errázuriz #3132, Casilla 16443, Santiago 9, Chile, and Departments of Anesthesiology, Physiology and Brain Research Institute, University of California, Los Angeles, CA, USA

TEULON, J., INSERM U-426, Faculté de Médecine Xavier Bichat, B.P. 416, 16 rue Henri Huchard, F-75870 Paris cedex, France
e-mail: teulon@bichat.inserm.fr

TORO, L., Departments of Anesthesiology, Physiology and Brain Research Institute, University of California, Los Angeles, CA, USA

TSE, H.-W., Department of Pharmacology, University of Bristol, Bristol, United Kingdom

TSIEN, R.W., Department of Molecular and Cellular Physiology, Beckman Center B105, Stanford University School of Medicine, Stanford, CA 94305-5345, USA
e-mail: rwtsien@leland.stanford.edu

VERGARA, C., Facultad de Ciencias, Universidad de Chile, Centro de Estudios Cientificos de Santiago, Av. Presidente Errázuriz #3132, Casilla 16443, Santiago 9, Chile

WANG, W., Department of Pharmacology, New York Medical College, New York, NY, USA

WHEELER, D.B., Department of Molecular and Cellular Physiology, Beckman Center B105, Stanford University School of Medicine, Stanford, CA 94305-5345, USA

YAKEL, J.L., Laboratory of Signal Transduction, National Institute of Environmental Health Sciences, National Institutes of Health, F2-08, P.O. Box 12233, 111 T.W. Alexander Drive, Research Triangle Park, NC 27709, USA
e-mail: yakel@niehs.nih.gov

YAU, K.-W., Department of Neuroscience and Ophthalmology, 9th Floor, Preclinical Teaching Building, Johns Hopkins University School of Medicine, 725 North Wolfe Street, Baltimore, MD 21205-2185, USA
e-mail: kwyau@mail.jhmi.edu

ZHONG, H., Department of Johns Hopkins University School of Medicine, 725 North Wolfe Street, Baltimore, MD 21205-2185, USA

Contents

Section I: Voltage-Dependent Ion Channels
A. Voltage-Dependent Na Channels 1

CHAPTER 1

Structure and Functions of Voltage-Dependent Na^+ Channels
K. Imoto. With 3 Figures 3

A. Introduction 3
B. General Architecture 4
C. α Subunit 4
I. Brain Types I, II, and III 7
1. Brain Type II/IIA 7
2. Brain Type I 8
3. Brain Type III 9
II. Skeletal Muscle μI/SkM1/SCN4A 9
III. Heart I/SkM2/hH1/SCN5A 9
IV. NaCh6 (Rat)/Scn8a (Mouse)/PN4 10
V. PN1/Na_s/hNE-Na/Scn9a 12
1. hNE-Na 12
2. Na_s 12
3. PN1 12
VI. SNS/PN3/NaNG/Scn10a 13
1. SNS/PN3/Scn10a 13
2. NaNG 13
VII. NaN/SNS2 13
VIII. Atypical Sodium Channels 14
1. $hNa_v2.1$ 14
2. $mNa_v2.3$ 14
3. SCL-11 14
D. Accessory Subunits 15
I. β1 Subunit 15
II. β2 Subunit 15
III. Other Associated Proteins 16
1. TipE 16
2. $Ankyrin_G$ 16

3. AKAP15 ... 17
4. Syntrophins ... 17
5. Extracellular Matrix Molecules ... 17
E. Genomic Structure ... 17
F. Concluding Remarks ... 17
References ... 19

CHAPTER 2

Sodium Channel Blockers and Activators

A.O. GRANT. With 3 Figures ... 27

A. Introduction ... 27
B. Classification and Structure of Na^+ Channels ... 27
C. Mechanisms of Na^+ Channel Blockade by Antiarrhythmic drugs ... 30
D. Models of Antiarrhythmic Drug Interaction with the Sodium Channel ... 32
E. The Highly Specific Na^+ Channel Blockers TTX and STX ... 38
F. Peptide Na Channel Blockers: μ Conotoxins ... 41
G. Na Channel Activators ... 42
H. Conclusions ... 45
References ... 45

B. Voltage-Dependent Ca-Channels

CHAPTER 3

Classification and Function of Voltage-Gated Calcium Channels

J.B. BERGSMAN, D.B. WHEELER, R.W. TSIEN. With 2 Figures ... 55

A. Generic Properties of Voltage-Gated Ca^{2+} Channels ... 55
I. Basic Functional Properties ... 55
II. Subunit Composition ... 56
1. α_1 ... 57
2. β ... 57
3. α_2/δ ... 58
4. γ ... 58
B. Classification of Native Ca^{2+} Channels According to Biophysical, Pharmacological, and Molecular Biological Properties ... 58
I. Molecular Biological Nomenclature ... 59
II. Ca_V1/L-Type Ca^{2+} Channels ... 59
III. Ca_V2 ... 61
1. $Ca_V2.2$/N-Type Ca^{2+} Channels ... 61
2. $Ca_V2.1$/P- and Q-Type Ca^{2+} Channels ... 62
3. $Ca_V2.3$/R-Type Ca^{2+} Channels ... 63

IV. Ca_V3/T-Type Ca^{2+} Channels ... 64
V. Note on Pharmacology ... 65
VI. Evolutionary Conservation of Ca^{2+} Channel Families ... 65
C. Functional Roles of Ca^{2+} Channels ... 66
I. Introduction/Subcellular Localization ... 66
II. Excitation-Contraction Coupling ... 66
III. Rhythmic Activity ... 67
1. Pacemaker ... 67
2. Other ... 67
IV. Excitation-Secretion Coupling ... 68
1. Generic Properties ... 68
2. Peripheral ... 69
3. Central ... 70
V. Postsynaptic Ca^{2+} Influx ... 71
1. Dendritic Information Processing ... 71
2. Excitation-Expression Coupling and Changes in Gene Expression ... 72
D. Concluding Remarks ... 73
References ... 73

CHAPTER 4

Structure of the Voltage-Dependent L-Type Calcium Channel
F. Hofmann, N. Klugbauer. With 3 Figures ... 87

A. Introduction ... 87
B. Subunit Composition and Genes of the Calcium Channel Complex ... 87
I. Subunit Composition of L-Type Calcium Channels ... 87
II. Genes ... 87
1. The α_1 Subunit ... 87
a) The L-Type α_1 Channels ... 89
α) The Class S α_1 Gene ... 89
β) The Class C α_1 Gene ... 89
γ) The Class D α_1 Gene ... 89
δ) The Class F α_1 Gene ... 90
b) The None L-Type α_1 Channels ... 90
α) The Class A α_1 Gene ... 90
β) The Class B α_1 Gene ... 90
γ) The Class E α_1 Gene ... 90
c) The Low Voltage-Activated α_1 Channels ... 90
α) The Class G and H Gene ... 90
2. Auxiliary Subunits of the Calcium Channel ... 91
a) The $\alpha_2\delta$ Subunit ... 91
b) The β-Subunit ... 92
c) The γ Subunit ... 93

III. Functional Domains of the α_1 Subunit 94
1. The Pore and Ion Selectivity Filter 94
2. Channel Activation 95
3. Channel Inactivation 96
IV. Sites for Interaction with Other Proteins 98
1. Interaction of the α_1 Subunit with the Ryanodine Receptor 98
2. Interaction of the α_1 Subunit with the β Subunit 99
V. Binding Sites for L-Type Calcium Channel Agonists and Antagonists 100
1. The Dihydropyridine Binding Site 100
2. The Phenylalkylamine and Benzothiazepine Binding Site 103
3. Modulation of Expressed L-Type Calcium Channel by cAMP-Dependent Phosphorylation 104
4. Modulation of Expressed L-Type Calcium Channel by Protein Kinase C-Dependent Phosphorylation 106
References 107

CHAPTER 5

Ca^{2+} Channel Antagonists and Agonists
S. Adachi-Akahane, T, Nagao. With 9 Figures 119

A. Ca^{2+} Channel Antagonists 119
I. Historical Background 119
II. Allosteric Interaction Between Ca^{2+} Channel Antagonist Binding Sites 121
III. Biophysical and Pharmacological Properties of Ca^{2+} Channel Antagonists 127
1. Dihydropyridines 128
2. Phenylalkylamines 130
3. Benzothiazepines 131
4. Other Ca^{2+} Channel Antagonists 132
IV. Binding Sites 133
1. Electrophysiological Identification of Binding Sites for Ca^{2+} Channel Blockers 133
2. Biochemical Characterization of Drug-Ca^{2+} Channel Interaction: Photoaffinity Labeling of Ca^{2+} Channels 135
3. Molecular Biological Characterization of Drug-Ca^{2+} Channel Interaction: Studies with Experimental Ca^{2+} Channel Mutants 135
B. Inorganic Blockers 138
C. Natural Toxins and Alkaloids 139
D. Ca^{2+} Channel Agonists 142
I. DHPs 142
II. Non-DHPs 144

E. Concluding Remarks 144
References 145

C. Voltage-Dependent K-Channels

CHAPTER 6

Overview of Potassium Channel Families: Molecular Bases of the Functional Diversity

Y. KUBO. With 7 Figures 157

A. Introduction 157
B. Primary Structure of the Main Subunit 157
I. 6-Transmembrane (TM) Type 157
II. 2-TM Type 158
III. 1-TM Type 158
IV. 2-Repeat Type 159
C. Heteromultimeric Assembly: Bases of Further Diversity 159
I. Heteromultimer Formation with Other Members of the Same Subfamily 159
1. Kv Channels 159
2. GIRK1,2,4 159
II. Suppression of Functional Expression by Heteromultimeric Assembly 160
III. Heteromultimeric Assembly of Main Subunits of Different Families 160
IV. Assembly with β Subunit 160
V. Assembly with Regulatory Subunits 161
VI. Assembly with Anchoring Protein 162
D. Structural Bases of the Gating Mechanism 162
I. Activation of Kv Channels 162
II. N-Type Inactivation of Kv Channels 163
III. C-Type Inactivation of Kv Channels 164
IV. Activation of IsK 165
E. Structural Bases of the Ion Permeation and Block 165
I. H5 Pore Region 165
II. Re-evaluation 165
III. Inward Rectification Mechanism 166
IV. Direct Structure Analysis 168
F. Structural Bases of Various Regulation Mechanisms 168
I. $G\beta\gamma$ 168
II. Block by Cytoplasmic ATP 169
III. Regulation by Phosphorylation 169
IV. Mg^{2+} as a Cytoplasmic Second Messenger 170
V. Regulation by Extracellular K^+ 170
VI. Other Mechanisms 170

G. Perspectives 170
References 171

CHAPTER 7

Pharmacology of Voltage-Gated Potassium Channels

O. PONGS, C. LEGROS. With 8 Figures 177

A. Introduction 177
B. Molecular and Functional Organization of the Voltage-Gated Potassium Channels 178
I. Structural Domains in Kvα-Subunits 178
II. Modulatory Kvβ-Subunits 181
C. Peptide Toxin Binding Sites 182
I. Scorpion Toxins 182
II. Snake Toxins 186
III. Sea Anemone Toxins 188
IV. Snail Toxins 189
V. Spider Toxins 190
D. Conclusions 191
References 191

CHAPTER 8

Voltage-Gated Calcium-Modulated Potassium Channels of Large Unitary Conductance: Structure, Diversity, and Pharmacology

R. LATORRE, C. VERGARA, E. STEFANI, L. TORO. With 2 Figures 197

A. Introduction 197
B. Channel Structure 198
C. Auxiliary Subunits 204
D. Calcium Sensitivity and Diversity of BK_{Ca} Channels in Different Cells and Tissues 205
E. Ca^{2+} Sensing Domain(s): The Calcium Bowl 207
F. Origin of Voltage Dependence in BK_{Ca} Channels 208
G. Channel Inactivation 209
H. Metabolic Modulation 210
I. Pharmacology 211
I. BK_{Ca} Channels Blockers 211
1. Toxins 211
2. Organic Blockers 212
a. Tetraethylammonium 212
b. Indole Diterpenes 213
c. General Anesthetics 213
II. BK_{Ca} Channel Activators 213
1. Activators Isolated from *Desmodium adscendens*: A Medicinal Herb 213

2. Anti-Inflamatory Aromatic Compounds (Fenamates) ... 214
3. Benzimidazolones ... 214
4. Phloretin ... 214
5. Ethanol ... 214
J. Summary and Conclusions ... 215
References ... 215

CHAPTER 9

Classical Inward Rectifying Potassium Channels: Mechanisms of Inward Rectification
C.G. Nichols. With 3 Figures ... 225

A. The Nature of Inward Rectification: Classical Considerations ... 225
B. The Inward Rectifier Ion Channel Family: Two Transmembrane Domain Potassium Channels ... 227
I. Kir 1 Subfamily ... 227
II. Kir 2 Subfamily ... 228
III. Kir 3 Subfamily ... 228
IV. Kir 4 and 5 Subfamilies ... 228
V. Kir 6 Subfamily ... 229
VI. KirD – a New Family of Double-Pored Inward Rectifier Channels? ... 229
VII. Inward Rectification in Other K^+ Channels ... 229
C. The Mechanism of Inward Rectification: Pore Block and Intrinsic ... 230
D. The Structure of the Kir Channel Pore: Binding Sites for Polyamines ... 231
E. The Structural Requirements for Inward Rectification: The Blocking Particles ... 233
F. The Physiological Significance of Polyamine-Induced Rectification ... 236
References ... 236

CHAPTER 10

ATP-Dependent Potassium Channels in the Kidney
G. Giebisch, W. Wang, S.C. Hebert. With 13 Figures ... 243

A. Introduction ... 243
B. The Function of ATP-Sensitive K Channels in the Proximal Tubule ... 243
C. The Function of ATP-Sensitive K Channels in the Thick Ascending Limb (TAL) of Henle's Loop ... 245
D. The Function of ATP-Sensitive K Channels in the Cortical Collecting Duct (CCD) ... 247
E. The Regulation of ATP-Sensitive K Channels ... 248

I. Proximal Tubule 248
II. Thick Ascending Limb of Henle's Loop 249
III. Cortical Collecting Tubules – Apical Membrane of Principal Cells 250
F. Properties of Cloned ATP-Sensitive K Channels (ROMK) 252
I. Channel Structure 252
II. Channel Isoforms and Localization 254
III. Comparison of ROMK with the Native Secretory ATP-Sensitive K Channel 256
IV. The Channel Pore-Rectification 256
V. Regulation by Phosporylation: Protein Kinase A (PKA) 257
VI. Regulation by Phosphorylation: Protein Kinase C (PKC) 259
VII. Regulation by Nucleotides 259
VIII. Regulation by Interaction with Cystic Fibrosis Transmembrane Conductance Regulator (CFTR) 260
IX. Regulation by pH 260
X. Regulation by Phosphoinositides 262
XI. Regulation of ROMK Density in CCD 262
G. ROMK and Bartter's Syndrome 263
References 264

CHAPTER 11

Structure and Function of ATP-Sensitive K^+ Channels
T. Gonoi, S. Seino. With 5 Figures 271

A. Introduction 271
B. Properties of K_{ATP} Channels in Native Tissues 272
I. Heart 272
II. Skeletal Muscles 273
III. Pancreatic β-Cells 273
IV. Brain 274
V. Smooth Muscles 275
VI. Kidney 275
VII. Mitochondria 276
C. Structure and Functional Properties of Reconstituted K_{ATP} Channels 276
I. The Pancreatic β-Cell Type K_{ATP} Channel 276
1. The Inwardly Rectifying K^+ Channel Subfamily Kir6.0 276
2. The Sulfonylurea Receptor SUR1 277
3. Reconstitution of the Pancreatic β-Cell Type K_{ATP} Channel 281
II. The Cardiac and Skeletal Muscle Type K_{ATP} Channel 283
1. The Sulfonylurea Receptor SUR2A 283

2. Reconstitution of the Cardiac and Skeletal Muscle Type K_{ATP} Channel 283
III. The Smooth Muscle Type K_{ATP} Channel 283
1. The Sulfonylurea Receptor SUR2B 283
2. Reconstitution of the Smooth Muscle Type K_{ATP} Channel 283
IV. The Vascular Smooth Muscle Type K_{ATP} Channel 284
1. Reconstitution of the Vascular Smooth Muscle Type K_{ATP} Channel 284
D. Physical Interaction and Stoichiometry of the Pancreatic β-Cell Type K_{ATP} Channel Subunits 284
I. Physical Interaction Between the SUR1 Subunit and the Kir6.2 Subunit 284
II. Subunit Stoichiometry of the SUR1/Kir6.2 Channel 285
E. Domains Conferring Sensitivities to the Nucleotides and Pharmacological Agents 285
I. ATP-Sensitivity 285
II. Nucleotide Diphosphate (NDP)-Sensitivity 286
III. Diazoxide-Sensitivity 286
IV. Sulfonylurea-Sensitivity 287
V. Mg^{2+}- and Spermin-Sensitivity 287
VI. Phentolamine-Sensitivity 287
VII. G-Protein Sensitivity 287
F. Pathophysiology of the Pancreatic β-Cell K_{ATP} Channel 288
I. Persistent Hyperinsulinemic Hypoglycemia of Infancy 288
II. Transgenic Mice 288
G. Conclusions 288
References 289

CHAPTER 12

G Protein-Gated K^+ Channels

A. INANOBE, Y. KURACHI*. With 11 Figures 297

A. Introduction 297
B. Acetylcholine-Activation of Muscarinic K^+ Channels 298
I. G Protein's Cyclic Reaction 299
II. Positive Cooperative Effect of GTP on the Muscarinic K^+ Channel Activity 302
III. Incorporation of Receptor-G Protein Reaction to the Model of K_{ACh} Channel 304
C. Molecular Analyses of G Protein-Gated K^+ Channels 305
I. Cloning of Inwardly Rectifying K^+ Channels and Kir Subunits for G Protein-Gated K^+ Channels 305
II. GIRK Subfamily 307

III. Expression of GIRK Channels ... 309
IV. Tetrameric Structure of Kir Channels ... 312
V. Molecular Mechanism Underlying Activation of the G Protein-Gated K^+ Channels by $\beta\gamma$ Subunits of G Protein ... 313
1. The G Protein $\beta\gamma$ Subunit-Binding Domains in GIRK Subunits ... 313
2. Putative Mechanism Underlying the G Protein $\beta\gamma$ Subunit-Induced Activation of the G Protein-Gated K^+ Channels ... 316
3. PIP_2-Mediation of $G_{\beta\gamma}$-Activation of K_G Channels ... 316
VI. The Possible Role of G Protein α Subunits in the G Protein-Gated K^+ Channel Regulation ... 317
1. Possibility of Microdomain Composed of Receptor, G Protein and the G Protein-Gated K^+ Channel ... 317
2. Specificity of Signal Transduction Based on the Receptor/G Protein/ G Protein-Gated K^+ Channel Interaction ... 317
D. Localization of the G Protein-Gated K^+ Channel Systems in Various Organs ... 318
I. Cardiac Atrial Myocytes ... 319
II. Neurons ... 321
III. Endocrine Cells ... 322
E. *Weaver* Mutant Mice and GIRK2 Gene ... 323
F. Conclusions ... 323
References ... 324

CHAPTER 13

Potassium Channels with Two Pore Domains
F. LESAGE, M. LAZDUNSKI. With 3 Figures ... 333

A. K^+ Channels with One Pore Domain ... 333
B. K^+ Channels with Two Pore Domains ... 334
I. TWIK, the Archetype of a Novel Structural Class of K^+ Channel ... 334
1. Cloning and Gene Organization ... 334
2. Functional Expression ... 335
3. Structure of the Channel ... 337
II. Related K^+ Channels in Mammals ... 337
1. TREK is an Unusual Outward Rectifier K^+ Channel ... 338
2. TASK is an Open Rectifier Channel Highly Sensitive to External pH ... 339
3. TRAAK Forms K^+ Channels Activated by Unsaturated Fatty Acids ... 340
III. Related Channels in Worm, Fly, Yeast, and Plant ... 340

C. Concluding Remarks 341
References 343

CHAPTER 14

Cardiac K+ Channels and Inherited Long QT Syndrome
M.-D. DRICI, J. BARHANIN. With 3 Figures 347

A. Long QT Syndromes 347
B. *HERG* and LQT2 348
I. The *HERG* Gene 348
II. I_{Kr} Current and LQT2 348
C. KvLQT1/IsK, LQT1, and LQT5 352
I. *KVLQT1* and *ISK* Genes 352
II. I_{Ks} Current, LQT1, and LQT5 352
III. Physiological Role of I_{Ks} in Cardiac Repolarization 355
D. Pharmacological Considerations in the Acquired LQTS 356
I. Determinants of Cardiac Repolarization 356
II. Pharmacological Modulation of Cardiac Repolarization and Acquired Long QT Syndromes 357
E. Conclusion 358
References 359

Section II: Ligand Operated Ion Channels

CHAPTER 15

Gating of Ion Channels by Transmitters: The Range of Structures of the Transmitter-Gated Channels
E.A. BARNARD. With 4 Figures 365

A. Introduction: The Scope of the Transmitter-Gated Channel Class 365
B. Structural Elements of the Membrane Domains of the Transmitter-Gated Channels 366
I. Transmembrane Domains 366
II. The α-Helix in Channel Transmembrane Domains 367
III. Supporting Transmembrane Structures 371
IV. Pore Loops (P-Domains) 371
C. The Subclasses of the Transmitter-gated Channels 373
I. The TGCs are in Completely Diverse Superfamilies 373
II. The Cys-Loop Receptors 374
III. Glutamate-Gated Cation Channels 378
IV. Channels Structurally Related to Voltage-Gated Channels 379
1. Cyclic Nucleotide-Gated Channels 379

2. Inositol Trisphosphate (IP_3) Receptors ... 380
3. Ryanodine Receptors ... 380
4. Vanilloid Receptors and Store-Operated Channels ... 380
V. Channels Topologically Related to Epithelial Na^+ Channels ... 381
1. P2X Channels ... 381
2. Proton-Gated Channels ... 381
3. Peptide-Gated Channels ... 383
VI. Channels Related to Inward Rectifier K^+ Channels ... 383
1. Nucleotide-Sensitive K^+ Channels ... 383
2. Nucleotide-Dependent K^+ Channels ... 384
3. Channels Containing Bi-Functional Kir Subunits ... 384
VII. Channels Related to ATP-Binding Transporters ... 385
VIII. Channels Related to Neurotransmitter Transporters ... 385
D. Conclusion ... 386
References ... 386

CHAPTER 16

Molecular Diversity, Structure, and Function of Glutamate Receptor Channels
M. MISHINA. With 2 Figures ... 393

A. Introduction ... 393
B. Structure and Molecular Diversity of the GluR Channel ... 393
I. Subunit Families and Subtypes ... 393
II. Primary Structure and Transmembrane Topology Model ... 394
C. AMPA Subtype ... 395
I. AMPA-Type Subunits ... 395
II. GluR2 Subunit and Ca^{2+} Permeability ... 396
III. Q/R Site as a Determinant of Channel Properties ... 396
IV. Phosphorylation ... 397
V. Autoimmune Disease ... 397
VI. GRIP, an Associated Protein ... 397
D. Kainate Subtype ... 397
E. NMDA Subtype ... 398
I. Heteromeric Nature of NMDA Receptor Channels ... 398
II. Dynamic Variations of the Distribution of the Subunits ... 399
III. Splice Variants ... 400
IV. Channel Pore and Gating ... 401
V. Agonist Binding ... 402
VI. Phosphorylation ... 403
VII. Modulation ... 403
VIII. Synaptic Plasticity, Learning, and Neural Development ... 404
IX. Associated Post–Synaptic Proteins ... 404

F. Additional Members of the GluR Channel Family 405
I. GluRδ Subfamily 405
II. GluRχ Subfamily 406
References 406

CHAPTER 17

Glutamate Receptor Ion Channels: Activators and Inhibitors
D.E. JANE, H.-W. TSE, D.A. SKIFTER, J.M. CHRISTIE,
D.T. MONAGHAN. With 15 Figures 415

A. Introduction 415
I. Receptor Classification 415
II. Molecular Biology of AMPA, Kainate, and NMDA Receptors 416
B. Pharmacology of AMPA Receptors 417
I. AMPA Receptor Agonists 417
II. Competitive AMPA Receptor Antagonists 419
1. Quinoxalinediones and Related Compounds 419
2. Decahydroisoquinolines 423
3. Isoxazoles 425
4. Phenylglycine and Phenylalanine Analogues 426
III. Benzodiazepine Analogues as Non-Competitive AMPA Receptor Antagonists 427
IV. Positive Allosteric Modulators 428
V. Channel Blockers 429
C. Kainate Receptor Pharmacology 431
I. Kainate Receptor Agonists 431
II. Competitive Kainate Receptor Antagonists 433
1. Quinoxalinediones and Related Compounds 433
2. Decahydroisoquinolines 433
3. Positive Allosteric Modulators Acting on Kainate Receptors 434
D. Therapeutic Potential of AMPA and Kainate Receptor Ligands 434
E. Pharmacology of NMDA Receptors 436
I. Therapeutic Considerations 436
II. The NMDA Receptor Glutamate Recognition Site 438
1. Glutamate Recognition Site Radioligands 438
2. Glutamate Binding Site Agonists. 439
3. Glutamate Recognition Site Competitive Antagonists 440
4. Antagonist Specificity for Subtypes of Glutamate Recognition Sites 443
III. NMDA Receptor Channel Blockers 444
1. Channel Blocker Pharmacology 444

2. Channel Blocker Receptor Subtype Selectivity 446
IV. The NMDA Receptor Glycine Recognition Site 447
1. Radioligand Binding and Functional Characteristics of the Glycine Receptor 447
2. NMDA Receptor Glycine Site Agonists 449
3. NMDA Receptor Glycine Site Antagonists 450
V. Allosteric Modulatory Sites on the NMDA Receptor 452
1. Polyamines 452
2. Spider and Wasp Toxins 453
3. Ifenprodil and Other NR2B Selective Compounds 453
4. Proton Inhibition 455
5. Zinc 455
F. Conclusions 456
References 459

CHAPTER 18

Structure, Diversity, Pharmacology, and Pathology of Glycine Receptor Chloride Channels

R.J. HARVEY, H. BETZ. With 2 Figures 479

A. Introduction 479
I. The Neurotransmitter Glycine 479
B. Structure and Diversity of Glycine Receptor Channels 479
I. GlyRs are Ligand-Gated Ion Channels of the nAChR Superfamily 479
II. Glycine Receptor Heterogeneity 480
III. The GlyR Ligand-Binding Domain 482
IV. Determinants of Ion Channel Function 483
V. Clustering of GlyRs by the Anchoring Protein Gephyrin 484
C. Pharmacology of Glycine Receptors 484
I. Strychnine is a Selective GlyR Antagonist 484
II. Amino Acids and Piperidine Carboxylic Acid Compounds 486
III. Antagonism by Picrotoxinin, Cyanotriphenylborate, and Quinolinic Acid Compounds 487
IV. Potentiation of GlyR Function by Anesthetics, Alcohol and Zn^{2+} 488
D. Pathology of Glycine Receptors 489
I. Mouse Glycine Receptor Mutants: *Spastic*, *Spasmodic*, and *Oscillator* 489
II. Mutations in GLRA1 Underlie the Human Hereditary Disorder Hyperekplexia 490
E. Conclusions 491
References 492

CHAPTER 19

$GABA_A$ Receptor Chloride Ion Channels
R.W. OLSEN, M. GORDEY. With 4 Figures ... 499

A. $GABA_A$ Receptors: Physiological Function, Molecular Structure, Pharmacological Subtypes ... 499
B. Activators and Inhibitors of $GABA_A$ Receptors ... 502
I. GABA Site ... 502
1. Agonists ... 502
2. Antagonists ... 504
II. The Picrotoxin Site ... 505
III. Benzodiazepine Site Ligands ... 506
IV. Barbiturates and Related Drugs ... 509
V. Neuroactive Steroids ... 511
VI. General Anesthetics: Propofol, Volatile Agents, and Alcohols ... 511
VII. Miscellaneous Agents ... 512
C. Discussion ... 512
References ... 512

CHAPTER 20

P2X Receptors for ATP: Classification, Distribution, and Function
R.J. EVANS. With 1 Figure ... 519

A. Introduction ... 519
B. Molecular Biology of P2X Receptors ... 519
I. A New Structural Family of Ligand Gated Ion Channels ... 520
II. The Extracellular Loop/Ligand Binding Site ... 520
III. Transmembrane Domains; Location of the Ionic Pore ... 522
1. Intracellular N and C Termini ... 523
2. Genomic Organisation, Human P2X Receptors and Chromosomal Location ... 523
C. Distribution of P2X Receptors ... 523
I. $P2X_1$ Receptors ... 524
II. $P2X_2$ Receptors ... 524
III. $P2X_3$ Receptors ... 525
IV. $P2X_4$ Receptors ... 525
V. $P2X_5$ Receptors ... 526
VI. $P2X_6$ Receptors ... 526
VII. $P2X_7$ Receptors ... 526
D. Functional Properties of P2X Receptors ... 527
I. General Features of P2X Receptors ... 527
1. $P2X_1$ Receptors ... 528
2. $P2X_2$ Receptors ... 528
3. $P2X_3$ Receptors ... 529

4. $P2X_2/P2X_3$ Heteromeric Receptors ... 529
5. $P2X_4$ Receptors ... 529
6. $P2X_5$ Receptors ... 530
7. $P2X_6$ Receptors ... 530
8. $P2X_7$ Receptors ... 530
II. Modulation of P2X Receptors ... 531
III. Native P2X Receptor Phenotypes; Molecular Correlates ... 532
1. Smooth Muscle ... 532
2. Sensory Neurons ... 533
3. Peripheral Neurons ... 534
4. Brain ... 534
5. Immune/Blood Cells ... 534
6. Salivary Gland ... 535
E. Future Directions ... 535
References ... 535

CHAPTER 21

The 5-HT_3 Receptor Channel: Function, Activation and Regulation
J.L. YAKEL. With 1 Figure ... 541

A. Introduction ... 541
B. Receptor Distribution ... 542
C. Molecular Structure ... 542
I. Sequence, Assembly, and Splice Variants ... 542
II. Gene Structure ... 544
III. Developmental Regulation ... 544
IV. Homo-Oligomeric Vs Hetero-Oligomeric Assembly ... 544
D. Function in the Nervous System ... 545
I. Presynaptic Role and Neurotransmitter Release ... 545
II. Postsynaptic Role ... 546
III. Physiological Properties ... 547
1. Receptor Activation ... 547
2. Single-Channel Properties ... 547
3. Desensitization ... 548
4. Ion Permeation and Pore Structure ... 549
5. Rectification and Voltage-Dependence ... 550
IV. Modulation, Synaptic Plasticity, and Learning and Memory ... 551
E. Pharmacological Properties ... 552
I. 5-HT_3R Ligands: Agonists and Antagonists ... 552
II. 5-HT_3R Ligand Binding Site ... 553
F. Allosteric Regulation ... 554
I. Alcohols ... 554
II. Anesthetics ... 554

III. 5-Hydroxyindole ... 555
G. Conclusion ... 555
References ... 556

CHAPTER 22

Cyclic Nucleotide-Gated Channels:
Classification, Structure and Function, Activators and Inhibitors

M.E. GRUNWALD, H. ZHONG, K.-W. YAU. With 2 Figures ... 561

A. Introduction ... 561
B. Structure ... 562
C. Ion Permeation Properties ... 563
D. Cyclic-Nucleotide Binding and Channel Gating ... 565
E. Modulations ... 568
I. $Ca^{2\pm}$-Calmodulin ... 568
II. $Ca^{2\pm}$ ... 569
III. Phosphorylation ... 569
IV. Transition Metals ... 570
V. Sulfhydryl Reagents ... 570
VI. Protons ... 571
VII. Other Modulators ... 571
F. Blockers ... 571
G. Conclusions ... 572
References ... 573

Section III: Miscellaneous Ion Channels – Intracellular Ca Release Channels

CHAPTER 23

Regulation of Ryanodine Receptor Calcium Release Channels

M. ENDO, T. IKEMOTO ... 583

A. Introduction ... 583
B. Molecular Structure and Function of RyR ... 584
C. Different Modes of Opening of RyR1 Calcium Release Channel ... 585
D. Activators of RyRs ... 588
I. Calcium, Strontium, and Barium Ions ... 589
II. Adenine Compounds ... 590
III. Caffeine and Related Compounds ... 590
IV. Ryanodine and Ryanoid ... 591
V. Halothane and Other Inhalation Anesthetics ... 592
VI. Oxidizing Agents and Doxorubicin ... 593

VII. Cyclic ADP-Ribose 593
VIII. Calmodulin and Other Endogenous Modulatory Proteins 593
IX. Imperatoxin Activator 594
X. Clofibric Acid 594
XI. Miscellaneous Activators 595
E. Inhibitors of RyRs 595
I. Magnesium Ion 595
II. Procaine and Other Local Anesthetics 596
III. Ruthenium Red 596
IV. Dantrolene 596
F. Closing Remarks 596
References 597

CHAPTER 24

Regulation of IP_3 Receptor Ca^{2+} Release Channels
M. IINO. With 1 Figure 605

A. Introduction 605
B. Molecular Structure and Function of IP_3R 605
C. Physiological Agonists and Modulators of IP_3R 607
I. IP_3 607
II. Ca^{2+} 609
III. ATP 610
IV. Phosphorylation 610
D. Activators of IP_3R 611
I. IP_3 Analogues 611
II. Caged IP_3 612
III. Thimerosal 612
IV. Immunophilin Ligands 613
V. Mn^{2+} 613
E. Inhibitors of IP_3R 613
I. Heparin 613
II. Xestospongin 614
III. Caffeine 614
IV. Cyclic ADP-Ribose 614
F. Comparison of Pharmacology Between IP_3R and RyR 615
G. Spatio-Temporal Patterns of IP_3R-Mediated Ca^{2+} Signals 615
H. Perspectives 617
References 617

CHAPTER 25

Ca^{2+}-Activated Non-Selective Cation Channels
J. TEULON 625

A. Introduction 625
B. Tissue Distribution 625
C. Conductive Properties 628
I. Unit Conductance and Voltage Dependence 628
II. Ion Selectivity 629
III. Ca-Permeable, Ca-Dependent Cation Channels: A Subtype of the NSC_{Ca} Channel? 630
D. Blockers and Pharmacological Stimulators 630
I. Blockers 630
II. Pharmacological Stimulators 632
E. Intracellular Regulatory Elements 633
I. Calcium Sensitivity 633
II. Inhibition by Intracellular Nucleotides 633
III. Tonic Influence of Intracellular ATP 635
IV. Stimulatory Effects of Intracellular Cyclic Nucleotides 635
V. Other Regulators: Internal pH and Oxidation 635
F. Phosphorylation-Dependent Regulation 636
I. Regulation via Protein Kinase A 636
II. Effects of Other Protein Kinases 637
G. Dependence on Hypertonicity 637
H. Agonist-Mediated Control of NSC_{Ca} Channels 638
I. Physiological Role 639
I. Excitable Cells: "Voltage Signal" 640
II. Exocrine Glands: Participation in Cl^- Transport 641
III. Other Epithelia: Speculative Functions 642
References 643

Subject Index 651

Section I
Voltage-Dependent Ion Channels
A. Voltage-Dependent Na Channels

CHAPTER 1

Structure and Functions of Voltage-Dependent Na^+ Channels

K. Imoto

A. Introduction

Voltage-gated sodium channels are responsible for the depolarizing phase of action potentials in nerve and muscle, and are essential for nerve conduction, excitation of neurons and skeletal and cardiac muscles, and other physiological processes (Hille 1992). Recent molecular biological approaches, combined with electrophysiological techniques, have allowed us to gain insights into molecular mechanisms of the ion channel operation. Furthermore, enduring efforts to discover new types of sodium channels have revealed the presence of multiple genes encoding sodium channel isoforms.

For the family of voltage-gated calcium channels, which are molecularly akin to the sodium channels, relationship between the structural (molecular biological) classification and the functional (electrophysiological and pharmacological) classification has been relatively well established, mainly because many pharmacological agents, such as dihydropyridines and ω-conotoxins, are available to distinguish one isoform from the others (for reviews see Hofmann et al. 1994; Mori et al. 1996). For sodium channels, several lines of evidence indicated coexistence of a slow-inactivating component or a tetrodotoxin-resistant fraction in neurons (Taylor 1993), but molecular heterogeneity of voltage-gated sodium channels was not seriously appreciated until recently. Lack of pharmacological tools and similarities in functional properties among the sodium channel isoforms have made it difficult to understand the consequence of molecular heterogeneity. As demonstrated for the NaCh6/Scn8a sodium channel in the cerebellar Purkinje cells (Raman and Bean 1997a; see below) however, all sodium channels do not behave in the same manner, consequently playing different physiological roles. This review mostly deals with the molecular heterogeneity of the mammalian voltage-gated sodium channels. For a more comprehensive description of the canonical structure-function relationships of selectivity filter, voltage sensor, inactivation gate, and phosphorylation sites, and drug binding sites, other reviews should be consulted (Patlak 1991; Heinemann et al. 1994; Catterall 1995; Guy and Durell 1995; Fozzard and Hanck 1996; Roden and George 1997).

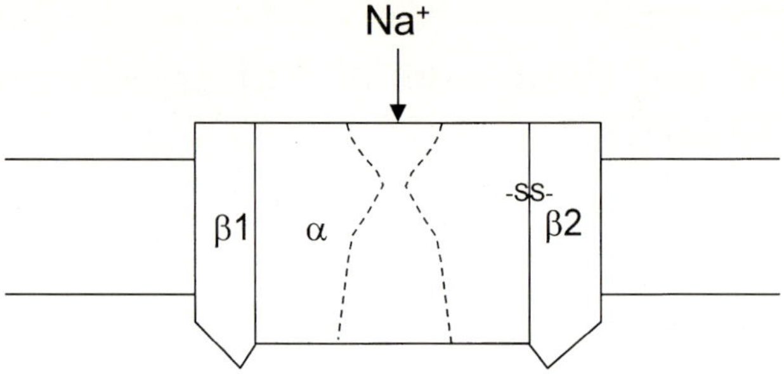

Fig. 1. The subunit structure of voltage-gated sodium channel. The rat brain sodium channel consists of the large α subunit, and smaller $\beta 1$ and $\beta 2$ subunits

B. General Architecture

The subunit composition of the voltage-gated sodium channels has been most thoroughly investigated for the rat brain sodium channel (CATTERALL 1995). The sodium channel complex contains the 260-kD α subunit, the 36-kD $\beta 1$ subunit, and the 33-kD $\beta 2$ subunit. The subunit stoichiometry is $\alpha:\beta 1:\beta 2 = 1:1:1$. The $\beta 2$ subunit is covalently linked to the α subunit, while the $\beta 1$ subunit is non-covalently associated. Both β subunits are transmembrane proteins. A heteromeric model for the subunit structure of the brain sodium channel is shown in Fig. 1.

Although the molecular structure of a potassium channel has been determined by X-ray crystallography (DOYLE et al. 1998; GULBIS et al. 1999), it has been unsuccessful with crystals of sodium channel proteins. Instead, electron microscopy has been used to study the tertiary structure, demonstrating that the sodium channel consists of four domains of different size and has a stain-filled pore in the center (SATO et al. 1998).

C. α Subunit

The α subunit is the main component of the voltage-gated sodium channels (Fig. 2). It consists of ~2000 amino acid residues. Analysis of amino acid sequences reveals four repeated units of homology (repeat I – repeat IV), each containing six hydrophobic, putative transmembrane segments (NODA et al. 1984, 1986a). Because there is no indication of signal peptide at the N-terminal end, and because there is a large C-terminal end, it is assumed that the N- and C-termini and the linking regions between repeats are exposed in the cytoplasmic side. The fourth hydrophobic segment, S4, of each repeat has a well-conserved motif of positively charged residues appearing every third residues. This motif contributes to sensing voltage changes (STÜHMER et al.

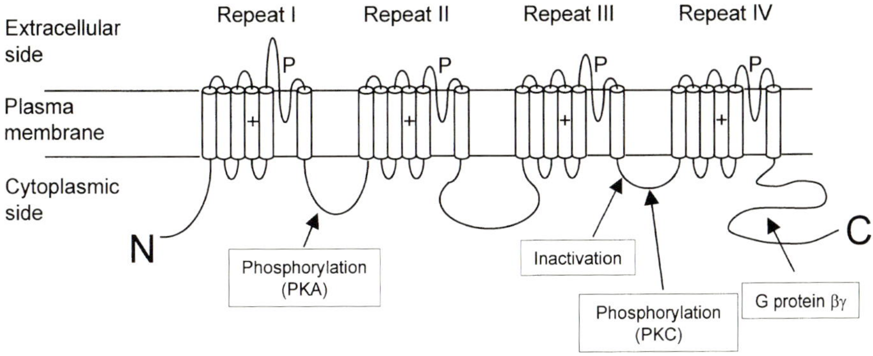

Fig. 2. Schematic structure of the sodium channel α subunit. The α subunit is unfolded and presented schematically. The α subunit forms channel pore (*P*), voltage sensor (*+*), inactivation gate, phosphorylation sites, and binding sites for various compounds

1989). S4 moves outward in response to depolarization and becomes accessible from the extracellular side (Yang and Horn 1995). The positive charges in S4 segments do not function equivalently. Neutralization of the fourth positive charge in repeats I or II produce the largest shifts in the voltage dependence of activation (Kontis et al. 1997). Whereas the hydrophobic regions show high homology among sodium channel isoforms, linker regions between repeats are less homologous, except for the linker connecting repeats III and IV.

The conserved III-IV linker is critical for fast inactivation. Cleavage of the linkage between repeats III and IV causes a strong reduction in the rate of inactivation (Stühmer et al. 1989). A cluster of three hydrophobic residues (IFM) in the linker is an essential component, possibly serving as a hydrophobic latch to stabilize the inactivated state (West et al. 1992; Kallenberger et al. 1996). The III-IV linker peptide can function as a fast inactivation gate even in a potassium channel (Patton et al. 1993). However, other parts of the α subunit are involved in fast inactivation. For example, alanine-scanning mutagenesis revealed that mutations in the putative transmembrane segment S6 of repeat IV substantially reduce fast inactivation (McPhee et al. 1995). A new technique of site-directed fluorescent labeling revealed that voltage sensors in repeats III and IV, but not I and II, are responsible for voltage-sensitive conformational changes linked to fast inactivation and are immobilized by fast inactivation (Cha et al. 1999).

The region between S5 and S6 of each repeat is now commonly called "P region" (P for "pore"), and is important for forming the channel pore and the selectivity filter. Search for the pore-forming region of the sodium channel was guided partly by the prediction by GUY (Guy and Conti 1990) and by the discovery of a mutation E387Q in the "P region" of repeat I, which abolishes tetrodotoxin sensitivity (Noda et al. 1989). Systematic mutagenesis studies around E387 and homologous positions of the other repeats identified the most critical amino acid residue for each repeat (TERLAU et al. 1991). They are D, E, K and A for repeats I–IV, respectively. Mutations K1422E in repeat III and A1714E in repeat IV dramatically change the ion-selectivity properties, to resemble those of calcium channels, suggesting that these amino acid residues form at least part of the selectivity filter (HEINEMANN et al. 1992a). The "P region" forms the binding site for tetrodotoxin and saxitoxin, which block the channel pore from the outer side. The difference in tetrodotoxin sensitivity among sodium channels is accounted for by an amino acid difference in the "P region" of repeat I (HEINEMANN et al. 1992b); the sensitive channels have aromatic amino acids (phenylalanine or tyrosine), while the resistant channels have cysteine or serine residue at the position.

The cAMP-dependent protein kinase (PKA) attenuates sodium current amplitude of the type IIA channel 20%–50% by phosphorylating serines located in the I-II linker. Among the five phosphorylation sites, the second site (S573) is necessary and sufficient to diminish sodium current amplitude (SMITH and GOLDIN 1997). Phosphatase 2A and calcineurin dephosphorylate sodium channels, counteract the effects of protein kinase A on sodium channel activity (CHEN et al. 1995). There is a consensus protein kinase C phosphorylation site in the III-IV linker (S1506 in type IIA, S1505 in heart I, S1321 in μI). Activation of protein kinase C decreases peak sodium current and slows its inactivation (NUMANN et al. 1991). Replacement of conserved serine residues reduces or abolishes the effect of protein kinase C on the type IIA and heart I channels (WEST et al. 1991; QU et al. 1996), but surprisingly it does not alter the effect on the μI channel (BENDAHHOU et al. 1995). Involvement of tyrosine kinases in regulation of neuronal sodium channels through *src* signaling pathway is also reported (HILBORN et al. 1998).

Sodium channels interact with G proteins. Coexpression of G protein $\beta\gamma$ subunits with the type IIA channel greatly enhances sodium currents, slows inactivation, and shifts the steady state inactivation curve to the depolarizing direction. Type IIA contains the proposed G$\beta\gamma$-binding motif, Q-X-X-E-R, in the C-terminal region, suggesting that type IIA channel is directly modulated by G$\beta\gamma$ subunits (MA et al. 1997). This motif is present in other isoforms of sodium channels, which include types I, III, NaCh6, Scn8a, hNE-Na, Na_s, and PN1, but not in heart I or μI.

Molecular cloning has detected multiple sodium channel genes, more than expected from electrophysiological and pharmacological measurements (Fig. 3). Multiple isoforms coexist, for example, at least brain types I, II, and III, and NaCh6/Scn8a are expressed in the rat central nervous system. Note that the primary transcript from a sodium channel gene undergoes a developmentally regulated complex pattern of alternative splicing that potentially generates as many as 100 different splice variants (THACHERAY and GANETZKY 1994). Moreover sodium channels that are expressed mainly in tissues outside of brain or muscles have been reported.

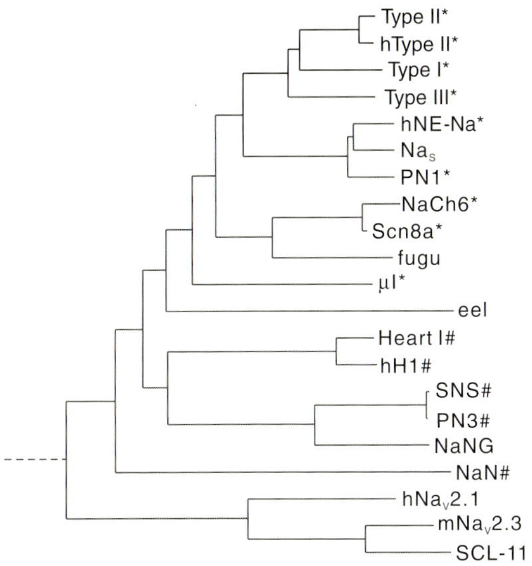

Fig. 3. Phylogenetic tree of mammalian sodium channel α subunit isoforms. The phyloginetic tree of voltage-gated soium channel family was generated using CLUSTAL W program (THOMPSON et al. 1994). The sequence of the T-type calcium channel was used to determine the root. For comparison, sequences from *Electrophorus electricus* and *Fugu rubripes* are included. *Asterisks* (*) and *sharps* (#) indicate tetrodotoxin-sensitive and tetrodotoxin-resistant channels, respectively, when functional channels are expressed from cDNAs in *Xenopus* oocytes or cultured cells. Sequences (with data base accession numbers in parentheses) are; Type II (X03639), hType II (M94055), Type I (X03638), Type III (Y00766), hNE-Na (X82835), Na$_s$ (U35238), PN1 (U79568), NaCh6 (L39018), Scn8a (U26707), FrSC (D37977), μI (M26643), EelNa (X01119), rHI (M27902), HH1(M77235), SNS (X92184), PN3 (U53833), NaNG (U60590), NaN (AF059030), hNa$_v$2.1 (M91556), mNa$_v$2.3 (L36179), SCL-11 (Y09164), and rat T-type Ca channel (AF027984)

I. Brain Types I, II, and III

1. Brain Type II/IIA

Molecular cloning of brain type II sodium channel was accomplished by NODA et al. (1986a) from rat brain. Its cDNA was the first to be functionally expressed successfully (NODA et al. 1986b). The type II channel has served as the archetypal sodium channel. The type II channel can be efficiently expressed in *Xenopus* oocytes. It exhibits classical tetrodotoxin-sensitivity with IC_{50} of ~10 nmol/l. Because expression level in *Xenopus* oocytes is so high, it has been used for detailed analysis of sodium channel, for example, quantal measurements of gating currents (CONTI et al. 1989). Type IIA, a variant of type II, was obtained independently (AULD et al. 1988). Type IIA differs at seven amino acid residues from type II (AULD et al. 1990). A difference N209D

(N in type II, D in type IIA) is caused by alternative splicing. Type II form is relatively abundant at birth, and gradually replaced by type IIA form as development proceeds (SARAO et al. 1991). Another difference F860L, which presumably resulted from reverse transcriptase error, caused slower inactivation and a shift of current-voltage relationship in the depolarizing direction (AULD et al. 1990).

Brain type II is a major sodium channel in the central nervous system (GORDON et al. 1987). It is preferentially expressed in the rostral areas, relatively dense in the forebrain, substantia nigra, hippocampus, and cerebellum (BECKH et al. 1989; WESTENBROEK et al. 1989). Immunohistochemistry revealed that, in hippocampus and cerebellum, type II is mainly localized in fibers, whereas type I is preferentially localized in cell bodies (WESTENBROEK et al. 1989).

The human counterpart of rat type II channel is HBA. Sequence identity is 97% at the amino acid level. HBA is successfully expressed transiently in CHO cells (AHMED et al. 1992).

2. Brain Type I

cDNA cloning of the rat brain sodium channel type I was reported by NODA et al. (1986a). Although initial attempts to characterize the functional properties in *Xenopus* oocytes were unsuccessful (NODA et al. 1986b), the same isoform was recloned recently (SMITH and GOLDIN 1998). The amino acid sequence of the recloned Rat I differs from the original sequence only at four positions. Three of them are located in putative cytoplasmic regions of the channel. The difference G979R (G in the newly reported sequence) is located in the S6 segment of the repeat II. Because the glycine residue in the S6 is conserved well in repeats I, II, and III of other isoforms of sodium channels, it is likely that the difference G979R caused the functional difference. However, to obtain a level of currents comparable to that of type II, a 500-fold greater amount of the type I mRNA must be injected into *Xenopus* oocytes, suggesting that other factors contribute to poorer functional expression of the type I channel.

The functional properties of the type I channel are generally similar to those of type II. The type I channel shows a high tetrodotoxin sensitivity with an apparent dissociation constant of 9.6 nmol/l. When type I is coexpressed with the $\beta1$ subunit, inactivation is accelerated as observed for type II. Coexpression of the $\beta2$ subunit results in only sight acceleration of inactivation (SMITH and GOLDIN 1998).

Voltage dependence of activation and inactivation for the type I channel is shifted to the positive direction, compared to that of the type II channel. The difference is more marked when the channels are coexpressed with the $\beta1$ and $\beta2$ subunits. At the membrane potential of –50 mV, more than two thirds of the type I channels are available, while more than two thirds of the type II channels are inactivated. Thus at resting potentials, the type I channels are more available for excitation. Type I recovers from inactivation more rapidly than type II channel (SMITH and GOLDIN 1998). These properties confer faster transmitting capability on type I, and may correspond to the observation of fast-spiking interneurons of rat hippocampus (MARTINA and JONAS 1997).

Expression of the type I mRNA rises postnatally with a stronger increase in caudal regions of the brain and in spinal cord (BECKH et al. 1989).

3. Brain Type III

The type III sodium channel was reported by KAYANO et al. (1988). Type III shows a high tetrodotoxin sensitivity when expressed in *Xenopus* oocytes (IC_{50} = 11 nmol/l; SUZUKI et al. 1988). The type III channel demonstrates a component of very slow decay (JOHO et al. 1990). Single channel analysis shows type III exhibits both fast gating and slow gating modes, switching between two gating modes (MOORMAN et al. 1990).

The type III mRNA is expressed predominantly at fetal and early postnatal stages in all regions of the brain (BECKH et al. 1989). It is also expressed in heart and skeletal muscle in minute quantities, but it may be attributable to coexisting neural tissues (SUZUKI et al. 1989). Because the β subunits are expressed in later stages of development, type III channel is assumed not to be associated with the β subunits.

II. Skeletal Muscle μI/SkM1/SCN4A

The μI cDNA was isolated from rat skeletal muscle library (TRIMMER et al. 1989). It is also called SkM1. The μI channel expressed in *Xenopus* oocytes is blocked by tetrodotoxin and μ-conotoxin at concentrations near 5 nmol/l. The μI channel is expressed in HEK (human embryonic kidney) cells transiently to give a large sodium current (up to 8 nA; UKOMADU et al. 1992). The μI channel exhibits slow inactivation kinetics in macroscopic currents and switching among slow, fast, and other additional modes at a single-channel level (ZHOU et al. 1991). However, the μI channel shows predominantly the faster component when coexpressed with the $\beta 1$ subunit (CANNON et al. 1993; WALLNER et al. 1993).

Mutations of the human skeletal muscle sodium channel gene, SCN4 A, cause various types of muscle diseases. They include hyperkalemic periodic paralysis, paramyotonia congenita, myotonia fluctuans, acetazolamide-sensitivie myotonia (see reviews: BARCHI 1995; CANNON 1996). Mutations disrupt inactivation and cause both myotonia (enhanced excitability) and attacks of paralysis (inexcitability resulting from depolarization).

III. Heart I/SkM2/hH1/SCN5A

The rat heart I was the first molecularly identified tetrodotoxin-resistant sodium channel (ROGART et al. 1989). It is also expressed in denervated and immature skeletal muscle (SkM2; Kallen et al. 1990). The SkM2 channel expressed in *Xenopus* oocytes is insensitive to low concentrations of tetrodotoxin but is ultimately blocked by this toxin with IC_{50} of 1.9 μmol/l. The human counterpart hH1 shows an even higher IC_{50} of 5.7 μmol/l (Gellens et al. 1992). Neither SkM2 nor hH1 is blocked by 100 nmol/l μ-conotoxin.

Mutations of the human heart sodium channel gene, SCN5A, result in the long QT syndrome 3 (LQT3) (Wang et al. 1995). Pathophysiological mechanism of LQT3 is not uniform. Channels with mutations in the III-IV linker (in-frame deletion of K1505-P1506-Q1507), autosomal dominant LTQ3 mutations, show a sustained inward current during long depolarizations. Single-channel recordings indicate that mutant channels fluctuate between normal and non-inactivating gating modes (Bennett et al. 1995). A sporadic mutation in S4 of repeat IV (R1623Q) increases probability of long opening and reopening (Kambouris et al. 1998).

Recently, an interesting observation that activation of PKA transforms the cardiac sodium channel into a calcium channel was reported (Santana et al. 1998). Note that molecular biological analyses have demonstrated the presence of other types of sodium channel in the heart, such as $hNa_v2.1$ and $mNa_v2.3$, whose function is unknown (see below). Furthermore, electrophysiological measurements have showed that a tetrodotoxin-sensitive sodium channel is present in sino-atrial node cells, exerting influence on heart rate (Baruscotti et al. 1996).

IV. NaCh6 (Rat)/Scn8a (Mouse)/PN4

The NaCh6 cDNA was isolated by RT-PCR using mRNAs prepared from rat brain, retina, and dorsal root ganglia, as well as from retrovirally transformed PC12 cells and primary cultures of neonatal cortical astrocytes. It was designated rat NaCh6 because it was the sixth rat full-length sodium channel sequence to be published (Schaller et al. 1995).

The mouse counterpart, Scn8a, was discovered independently in searching a causative mutation of "motor endplate disease" (*med*) of mouse (see below; Burgess et al. 1995). A new allele med^{tg} was made by non-targeted transgene insertion (Kohrman et al. 1995). Cosmid clones containing transgene junctions were isolated, and the transgenic insertion was found to disrupt a novel sodium channel gene, Scn8a. The complete cDNA was obtained by RT-PCR of cerebellar RNA and from mouse brain cDNA libraries. Scn8a is likely a mouse counterpart of NaCh6 (97% overall amino acid identity), although the I-II linker of Scn8a is much shorter. PN4 was isolated from rat DRG, and likely represents the same transcript as NaCh6 (Dietrich et al. 1998). Interestingly, NaCh6/Scn8a is most closely related to a brain cDNA from the pufferfish *Fugu*, with 83% overall sequence identity (Burgess et al. 1995).

Northern analysis shows that NaCh6/Scn8a is expressed in rat brain, cerebellum, spinal cord, but not in skeletal muscle, cardiac muscle, or uterus (Schaller et al. 1995; Burgess et al. 1995). Quantitative analysis of mRNA abundance using RNase protection assay revealed that NaCh6 mRNA is expressed in the brain as abundantly as types I, II, and III. Many neurons express NaCh6 mRNA. Those cells include motor neurons in the spinal cord and the brain stem, Purkinje cells and granular cells in the cerebellum, granule cells of the dentate gyrus, and CA1 and CA3 pyramidal cells. In situ hybridization analysis show that NaCh6 mRNA is expressed in cultured astrocytes as well as glia in the spinal cord white matter and Schwann cells (SCHALLER et al. 1995). More recently, single-cell RT-PCR analysis of cerebellar Purkinje cells detected mRNAs of brain I and NaCh6, but not of brain II (VEGA-SAENZ DE MIERA et al. 1997). Scn8a is the major contributor to the postnatal developmental increase of sodium current density in spinal motoneurons (GARCIA et al. 1998).

The neurological deficits of *med* mutant mice include lack of signal transmission at the neuromuscular junction, excess preterminal arborization, and degeneration of cerebellar Purkinje cells. There are three types of spontaneous mutation of Scn8a, *med*, med^J, and med^{jo}. The *med* and med^J mutations alter reading frames with premature stop codons close to the N-terminus of the protein (KOHRMAN et al. 1996a). The third allele med^{jo} (*jolting*) causes a milder form of disorder, exhibiting cerebellar ataxia only. The *jolting* mutation substitutes threonine for an evolutionary conserved alanine residue in the cytoplasmic S4-S5 linker of repeat II. Introduction this mutation into the brain IIA channel shifted the voltage dependence of activation by 14 mV in the depolarizing direction, without affecting the kinetics of fast inactivation or recovery from inactivation (KOHRMAN et al. 1996b).

Those lines of evidence described above suggest that NaCh6/Scn8a contributes to voltage-dependent sodium currents in cerebellar Purkinje cells. Purkinje cells are known for their unique electrical properties (LLINÁS and SUGIMORI 1980a,b). Purkinje cells show regular, spontaneous firing, and this distinctive firing pattern has been attributed to a persistent sodium conductance. In whole-cell patch clamp recording of dissociated rat Purkinje neurons, a tetrodotoxin-sensitive inward current was elicited when the membrane was repolarized to voltages between –60 mV and –20 mV after depolarization to +30 mV long enough to produce maximal inactivation (RAMAN and BEAN 1997a). This "resurgent" current likely contributes to repetitive firing. In *med* Scn8a mutant mice, peak sodium current of isolated Purkinje neurons is reduced to ~60% of normal control. The "resurgent" current is more drastically reduced to ~10% of normal. Furthermore, both spontaneous firing and evoked bursts of spikes are diminished (RAMAN et al. 1997b). The notion that NaCh/Scn8a is responsible for the "resurgent" subthreshold current and crucial for repetitive firing was confirmed recently by expressing Scna8 in *Xenopus* oocytes (SMITH et al. 1998). Scna8 channels coexpressed with the β subunits exhibited a persistent current that became larger with increasing depolarization. Interestingly, the "resurgent" currents are not observed in CA3 neurons where prominent Scn8a expression is demonstrated by in situ hybridization (VEGA-SAENZ DE MIERA et al. 1997).

V. PN1/Na$_s$/hNE-Na/Scn9a

The members of this group of sodium channels were discovered recently. They are tetrodotoxin-sensitive and similar to brain-type sodium channels in kinetic properties. Comparison of the amino acid sequences of PN1, hNE-Na, and Na$_s$ shows ~93% identity, suggesting that they may be counterparts of different species.

1. hNE-Na

This member of sodium channel genes was cloned from the human medullary thyroid carcinoma (hMTC) cell line (Klugbauer et al. 1995). It is expressed in hMTC cells, a C-cell carcinoma, and in thyroid and adrenal gland, but not in pituitary, brain, heart, liver, or kidney. The hNE-Na channel is successfully expressed in the absence and presence of the $\beta 1$ subunit in HEK cells. The hNE-Na α-subunit alone induce rapidly activating and inactivating inward currents. The threshold is –40 mV, and maximum amplitudes are reached at –10 mV. The inward current is tetrodotoxin-sensitive, with an IC_{50} value of 25 nmol/l. Coexpression of the $\beta 1$ subunit does not significantly affect the kinetic properties, except that the presence of $\beta 1$ subunit shifts the steady-state inactivation curve to the depolarizing direction by 20 mV (only in the absence of external calcium). The hNE-Na channel can elicit action potentials in HEK cells. It is likely that this type of sodium channel is responsible for tetrodotoxin-sensitive action potentials observed in adrenal chromaffine cells and in parafollicular C-cells in the thyroid.

2. Na$_s$

The Na$_s$ sodium channel was isolated from cultured rabbit Schwann cells (Belcher et al. 1995). Na$_s$ most closely resembles the hNE-Na channel in amino acid sequence, but its distribution is different. It is expressed not only in cultured Schwann cells but also in sciatic nerve, spinal cord, brain stem, cerebellum, and cortex. It is not determined in which cell types Na$_s$ is expressed in the brain. Schwann cells express brain type I and type II channels (Oh et al. 1994), and Na-G (Gautron et al. 1992) as well. Functional expression of Na$_s$ has not been reported.

3. PN1

PN1 is a sodium channel expressed principally in peripheral neurons, isolated from rat dorsal root ganglia (Toledo-Aral et al. 1997; Sangameswaran et al. 1997). The PN1 mRNA is detected in superior cervical, dorsal root, and trigeminal ganglia, and barely detectable in spinal cord. No transcripts are detected in skeletal muscle, cardiac muscle, or brain. PN1 gene expression seems confined to the neuronal population. Immunocytochemistry of cultured DRG (dorsal root ganglia) neurons and PC12 cells shows that the PN1 channel is targeted to neurite terminals (Toledo-Aral et al. 1997).

The sodium channel activity expressed in *Xenopus* oocytes by injecting PN1 mRNA is sensitive to tetrodotoxin with a half-maximal inhibitory concentration of 4.3 nmol/l. Inactivation kinetics is not accelerated by coinjection of the β1 or β2 subunit mRNAs (SANGAMESWARAN et al. 1997). PN1 is mapped very close to the brain types I–III in mouse chromosome 2 (KOZAK et al. 1996)

VI. SNS/PN3/NaNG/Scn10a

1. SNS/PN3/Scn10a

SNS (sensory neuron sodium channel) (AKOPIAN et al. 1996) and PN3 (peripheral nerve 3) (SANGAMESWARAN et al. 1996) are practically identical, differing at seven residues (99.6% identity). The SNS/PN3 isoform is expressed in small-diameter sensory neurons of dorsal root and trigeminal ganglia, but absent or detected very little in other peripheral or central neurons, glia, or non-neural tissues (AKOPIAN et al. 1996; SANGAMESWARAN et al. 1996). The SNS/PN3 channel is functionally expressed in *Xenopus* oocytes at a low level. The current is insensitive to tetrodotoxin, the estimated half-maximal inhibitory concentration being over 50 μmol/l. The voltage-dependence of activation of SNS/PN3 is shifted to the depolarizing direction (peak voltage at 10~20 mV), compared to that of type II channel and to sodium currents of native DRG neurons, suggesting that SNS/PN3 requires additional subunits to obtain proper properties. But slow inactivation is common to both native and recombinant sodium currents. The human ortholog, hPN3, exhibits similar properties of the shifted voltage dependence and the slow inactivation when expressed in *Xenopus* oocytes (RABERT et al. 1998). Insertion of an SNS-specific tetrapeptide, SLEN, in the S3-S4 linker of repeat IV into the corresponding position of the μ1 sodium channel does not alter kinetics of activation or inactivation, but accelerates recovery form inactivation (DIB-HAJJ et al. 1997).

Recently, generation of SNS knockout mice was reported (AKIPIAN et al. 1999). Null mutant mice are viable, fertile, and appear normal. They show a pronounced analgesia to noxious mechanical stimuli, small deficits in noxious thermoreception, and delayed development of inflammatory hyperalgesia (AKOPIAN et al. 1999).

2. NaNG

cDNA of NaNG was isolated from dog nodose ganglia (CHEN et al. 1997). The nodose ganglia contains most of the sensory cell bodies of the vagus neuron. NaNG most closely resembles SNS/PN3 (82% amino acid identity). NaNG is not expressed in CNS, heart or skeletal muscle. Experiments of functional expression have not been reported.

VII. NaN/SNS2

The NaN sodium channel is a new member of the family (DIB-HAJJ et al. 1998). The NaN retains all of the relevant landmark sequences of voltage-gated Na$^+$ channels, including the positively charged S4 and the P regions, but similarity to known Na^+ channels is only 42–50%. NaN is expressed preferentially in C type DRG and trigeminal ganglia neurons and down-regulated after axotomy (Dib-Hajj et al. 1998). SNS2, whose amino acid sequence is identical to that of NaN, is highly resistant to tetrodotoxin when expressed in HEK cells (Tate et al. 1998). SNS2 is activated at relatively negative potentials with a half activation potential of −45 mV.

VIII. Atypical Sodium Channels

The sodium channel subfamily of $hNa_v2.1$, $mNa_v2.3$, and SCL-11 have 40~50% identical and 60~70% homologous amino acid residues when compared with the classical sodium channels. But the amino acid sequences of $hNa_v2.1$, $mNa_v2.3$, and SCL-11 suggest the possibility that they do not function as voltage-gated sodium channels. Many positively charged amino acid residues of S4 segments are replaced with non-charged residues, and the III-IV linker essential for fast inactivation is significantly diverged from the consensus sequence. The amino acid residues in the "P region" critical for sodium selectivity are also altered; K–>S (repeat III) in $hNa_v2.1$, K–>N (repeat III) and A–>S (repeat IV) in SCL11.

1. $hNa_v2.1$

cDNA of $hNa_v2.1$ was obtained from both human adult heart and fetal skeletal muscle (George et al. 1992). The $hNa_v2.1$ mRNA is predominantly expressed in both heart and uterus. Faint signals are detected in brain, kidney and spleen.

2. $mNa_v2.3$

$mNa_v2.3$ cDNA was cloned from the mouse AT-1 atrial tumor cell line (Felipe et al. 1994). Northern blot analysis revealed that it is expressed in heart and uterus. Faint signals are also detected in brain, kidney, and skeletal muscle, as observed for $hNa_v2.1$. Immunohistochemistry and Western blot analysis showed $mNa_v2.3$ expression in the uterus is dramatically upregulated during pregnancy (Knittle et al. 1996).

3. SCL-11

The SCL-11 (sodium channel-like protein) cDNA was obtained from a rat dorsal root ganglion library. SCL-11 is expressed in dorsal root and trigeminal ganglia, sciatic nerve, pituitary, lung, urinary bladder, and vas deferens as well as PC12 and C6 glioma cells (Akopian et al. 1997). In situ hybridization of dorsal root ganglia shows signals from myelinating Schwann cells. The deduced amino acid sequence shows 98% identity to the rat partial clone Na-G (Gautron et al. 1992). It is unsuccessful to express voltage-dependent channel activity upon injection of mRNA into *Xenopus* oocytes.

D. Accessory Subunits

I. β1 Subunit

The β1 subunit is a membrane protein with a single transmembrane spanning domain (ISOM et al. 1992). The presence of a leader sequence indicates that the N-terminal region is located extracellularly, and the extracellular domain contains an immunoglobulin-like motif (ISOM and CATTERALL 1996). The β1 subunit is expressed in rat brain, spinal cord, heart, and skeletal muscle. There is a single gene encoding the β1 subunit (TONG et al. 1993; MAKITA et al. 1994a).

When expressed in *Xenopus* oocytes together with type IIA α subunit, the β1 subunit modulates channel function by accelerating the kinetics of inactivation and shifting its voltage dependence in the hyperpolarizing direction. Coexpression of the β1 subunit also increases the peak current amplitude approximately 2.5 times (ISOM et al. 1992). These effects of coexpression are also observed in a mammalian cell line (ISOM et al. 1995).

The β1 subunit has little or no effect on the gating of cardiac channels in recombinant expression systems (MAKITA et al. 1994a), although peak current amplitude is increased. But suppression of β1 subunit expression by antisense oligonucleotides prevents development of a mature (fast activating and fast inactivating) sodium current in mouse atrial tumor cells, suggesting that the gating of the cardiac sodium channel is modulated by the β1 subunit (KUPERSHMIDT et al. 1998). Molecular determinants of the β1 interaction were identified by analyzing chimeric sodium channels constructed from the human skeletal muscle (SkM1) and human heart sodium (hH1) channels. The S5-S6 loops of repeats I and IV of the α subunit and the N-terminal extracellular domain of the β1 subunit are responsible for interaction (MAKITA et al. 1996).

Recently, a subset of generalized epilepsy with febrile seizures has been reported to be associated with a mutation of the β1 subunit gene SCN1B (WALLACE et al. 1998). The mutation changes a conserved cysteine residue disrupting a putative disulfide bridge, and interferes with the ability of the β1 subunit to modulate the channel gating.

II. β2 Subunit

The β2 subunit is also a single-membrane spanning glycoprotein with a large N-terminal domain exposed in the extracellular side. The β2 subunit is covalently bound to the α subunit (ISOM et al. 1995). The amino acid sequence of the β2 subunit shows an interesting similarity with two separate segment of the neural cell adhesion molecule (CAM) contactin. One region contains an immunoglobulin-like motif. The other homologous region is the extracellular stalk portion. Because nearly all the immunoglobulin motifs interact with extracellular ligands, the β2 subunit probably also serves this function, possibly concentrating the sodium channels in specific locations.

The $\beta 2$ subunit is expressed in the brain and the spinal cord, but not outside of the nervous system (ISOM et al. 1995). Developmentally, the $\beta 2$ subunit mRNA is detectable at earlier stages than the $\beta 1$ mRNA. When expressed with the α subunit, the $\beta 2$ subunit increases sodium currents, but the augmenting effect is less prominent than that of the $\beta 1$ subunit (ISOM et al. 1995). A unique property of the $\beta 2$ subunit is expansion of the cell surface membrane. The $\beta 2$ subunit may stimulate fusion of intracellular transport vesicles with the plasma membrane.

The gene of the human counterpart is localized to human chromosome 11q3, close to the locus of Charcot-Marie-Tooth syndrome type 4B (CMT4B) (EUBANKS et al. 1997), but the SCN2B gene of patients with CMT4B was reported normal (BOLINO et al. 1998).

III. Other Associated Proteins

1. TipE

The *para* locus of *Drosophila* encodes the sodium channel (LOUGHNEY et al. 1989). A similar phenotypic mutant *tipE* (temperature-induced paralysis, locus E) was identified using genetic approach (FENG et al. 1995). TipE is a deduced protein of 452 amino acids, having two hydrophobic domains. The presumed transmembrane topology is that TipE has the N- and C-termini located in the cytoplasmic side, with the two transmembrane segments. TipE has no significant sequence homology to any other proteins. Functional expression of the *para* sodium channel in *Xenopus* oocytes is markedly augmented when TipE is coexpressed (FENG et al. 1995). TipE accelerates inactivation, as does the $\beta 1$ subunit for mammalian sodium channels (WARMKE et al. 1997). The mammalian counterpart of TipE has not been reported.

2. Ankyrin$_G$

It is generally believed that the maintenance of highly localized concentrations of the sodium channel at the axonal initial segments and nodes of Ranvier is important to the initiation and propagation of the saltatory action potential. Ankyrin links the sodium channel to the underlying cytoskeleton. The ankyrin present at the node corresponds to 480 kDa and 270 kDa alternatively spliced isoforms of ankyrin$_G$. The two brain-specific isoforms contain a unique stretch of sequence highly enriched in serine and threonine residues following the globular head domain (KORDELI et al. 1995). The β spectrin is precisely colocalized with both sodium channels and ankyrin$_G$ at the neuromuscular junctions (WOOD et al. 1998).

Cerebellum-specific knock-out of ankyrin$_G$ in mouse brain resulted in a progressive ataxia and subsequent loss of Purkinje neurons (ZHOU et al. 1998). In mutant cerebella, sodium channels were absent from axon initial segments of granule cell neurons, demonstrating that ankyrin$_G$ is essential for clustering sodium channels.

3. AKAP15

Phosphorylation of the α subunit by PKA reduces peak sodium current, with little change in the voltage dependence of activation or inactivation (see above). PKA is bound to brain sodium channels through interaction with a 15-kDa cAMP-dependent protein kinase anchoring protein (AKAP15) (Tibbs et al. 1998). AKAP15 also associates with skeletal muscle calcium channels (Gray et al. 1997).

4. Syntrophins

Syntrophins are modular proteins belonging to the dystrophin-associate glycoprotein complex and are thought to be involved in the maintenance of neuromuscular junction. Syntrophins contain one PDZ domain. This PDZ domain exhibits specific binding to the motif R/K/Q-E-S/T-X-V-COO^-. This motif is highly conserved in the sodium channel α subunits. This interaction is suggested to contribute to localization of the sodium channels (Schultz et al. 1998).

5. Extracellular Matrix Molecules

The $\beta 2$ subunits has an Ig-motif in its extracellular domain. The purified sodium channel and the extracellular domain of the $\beta 2$ subunit are shown to bind to tenascin-C and tenascin-R (Srinivasan et al. 1998). Tenascin-R knockout mice exhibited decreased conduction velocity of optic nerves, but the distribution of sodium channels at the nodes of Ranvier was not changed (Weber et al. 1999).

E. Genomic Structure

Structure of the sodium channel α subunit genes (Table 1) was extensively studied for SCN4A (George et al. 1993), SCN5A (Wang et al. 1996), SCN8A (Plummer et al. 1998), and Scn10a (Souslova et al. 1997). Each gene consists of 24–27 exons spanning 80–90 kb. The intron-exon structure is conserved well among genes (Souslova et al. 1997). SCN4A and SCN5A genes have atypical intron boundaries of AT-AC. Introns with this boundary are spliced by the U12-type splicesome (Sharp and Burge 1997).

Transcription of the type II sodium channel gene is regulated by binding of the repressor protein REST, a zing-finger protein, to the RE1 sequence (Tapia-Ramirez et al. 1997).

F. Concluding Remarks

Voltage-gated sodium channels are present not only at nodes of Ranvier and axon hillocks but also in soma and dendrites of neurons. A number of recent

Table 1. Sodium channel α and β subunit genes and their chromosomal localization

Locus	Chromosomal location (references[a])		Corresponding cDNA
	Human	Mouse	
α subunits			
SCN1A	2q24 (1)	2 (2)	Brain type I
SCN2A	2q23–24.3 (3)	2 (2)	Brain type II
SCN3A	2q24–31 (4)	2 (2)	Brain type III
SCN4A	17q23.1–25.3 (5)	11 (6)	μ1, SkM1
SCN5A	3p21 (7)	9 (6)	heart, SkM2
SCN6A	2q21–23 (8)	–	$hNa_v2.1$
SCN7A	2q36–37 (9)	2 (9)	Na-G, SCL-11
SCN8A	12q13 (10)	15 (10)	NaCh6, PN4
SCN9A	2q24	2 (11)	Na_s, hNE-Na, PN1
SCN10A	3p24.2–22 (12)	9 (13)	SNS, PN3
SCN11A	3p21 (14)	9	NaN, SNS2
β subunits			
SCN1B	19q13.1–2 (15)	7 (16)	β1
SCN2B	11q3 (17)	9 (18)	β2

[a] References: (1) MALO et al. 1994a; (2) MALO et al. 1991; (3) AHMED et al. 1992; (4) MALO et al. 1994b; (5) GEORGE et al. 1991, 1993; (6) KLOCKE et al. 1992; (7) GEORGE et al. 1995; WANG et al. 1996; (8) GEORGE et al. 1994; (9) POTTS et al. 1993; (10) BURGES et al. 1995; (11) BECKERS et al. 1996; (12) RABERT et al. 1998; (13) SOUSLOVA et al. 1997; (14) PLUMMER and MEISLER 1999; (15) MAKITA et al. 1994b; (16) TONG et al. 1993; (17) EUBANKS et al. 1997; (18) JONES et al. 1996. Data were also obtained from the LocusLink site (http://www.ncbi.nlm.nih.gov/LocusLink).

techniques, including high-speed fluorescence imaging and dendritic patch clamping, have provided new information on active involvement of sodium channels in dendritic propagation of action potentials (for reviews see JOHNSTON et al. 1996; STUART et al. 1997). Subtle differences in properties of sodium channels will influence the process of synaptic integration in important and complex ways.

Considerable progress in understanding sodium channel functions has been made in the field of medical genetics. It has been discovered that mutations of skeletal muscle and cardiac muscle sodium channels cause classically known disorders. For sodium channel isoforms predominantly expressing in the CNS, the *med* mutations was reported, and more recently, the mutation of the β1 subunit has been identified to be associated with a subset of generalized epilepsy with febrile seizures. It is likely that mutations of other isoforms can cause neurological disorders, which may include epilepsy and degenerative diseases. Analysis of functional differences of sodium channel isoforms will contribute to the elucidation of disease mechanisms and the development of new medical therapeutics.

References

Ahmed CMI, Ware DH, Lee SC, Patten CD, Ferrer-Montiel AV, Schinder AF, McPherson JD, Wagner-McPherson CB, Wasmuth JJ, Evans GA, Montal M (1992) Primary structure, chromosomal localization, and functional expression of a voltage-gated sodium channel from human brain. Proc Natl Acad Sci USA 89:8220–8224

Akopian AN, Sivilotti L, Wood JN (1996) A tetrodotoxin-resistant voltage-gated sodium channel expressed by sensory neurons. Nature 379:257–262

Akopian AN, Souslova V, Sivilotti L, Wood JN (1997) Structure and distribution of a broadly expressed atypical sodium channel. FEBS Lett 400:183–187

Akopian AN, Souslova V, England S, Okuse K, Ogata N, Ure J, Smith A, Kerr BJ, McMahon SB, Boyce S, Hill R, Stanfa LC, Dickenson AH, Wood JN (1999) The tetrodotoxin-resistant sodium channel SNS has a specialized function in pain pathways. Nature Neuroscience 2:541–548

Auld VJ, Goldin AL, Krafte DS, Marshall J, Dunn JM, Catterall WA, Lester HA, Davidson N, Dunn RJ (1988) A rat brain Na^+ channel α subunit with novel gating properties. Neuron 1:449–461

Auld VJ, Goldin AL, Krafte DS, Catterall WA, Lester HA, Davidson N, Dunn RJ (1990) A neutral amino acid change in segment IIS4 dramatically alters the gating properties of the voltage-dependent sodium channel. Proc Natl Acad Sci USA 87:323–327

Barchi RL (1995) Molecular pathology of the skeletal muscle sodium channel. Annu Rev Physiol 57:355–385

Baruscotti M, DiFrancesco D, Robinson RB (1996) A TTX-sensitive inward sodium current contributes to spontaneous activity in newborn rabbit sino-atrial cells. J Physiol (Lond) 492:21–30

Beckers MC, Ernst E, Belcher S, Howe J, Levenson R, Gros P (1996) A new sodium channel alpha-subunit gene (Scn9a) from Schwann cells maps to the Scn1a, Scn2a, Scn3a cluster of mouse chromosome 2. Genomics 36:202–205

Beckh S, Noda M, Lübbert H, Numa S (1989) Differential regulation of three sodium channel messenger RNAs in the rat central nervous system during development. EMBO J 8:3611–3616

Belcher SM, Zerillo CA, Levenson R, Ritchie JM, Howe JR (1995) Cloning of a sodium channel α subunit from rabbit Schwann cells. Proc Natl Acad Sci USA 92:11034–11038

Bendahhou S, Cummins TR, Potts JF, Tong J, Agnew WS (1995) Serine-1321-independent regulation of the μ1 adult skeletal muscle Na^+ channel by protein kinase C. Proc Natl Acad Sci USA 92:12003–12007

Bennett PB, Yazawa K, Makita N, George AL Jr (1995) Molecular mechanism for an inherited cardiac arrhythmia. Nature 376:683–685

Bolino A, Seri M, Caroli F, Eubanks J, Srinivasan J, Mandich P, Schenone A, Quattrone A, Romeo G, Catterall WA, Devot M (1998) Exclusion of the SCN2B gene as candidate for CMT4B. Eur J Hum Genet 6:629–634

Burgess DL, Kohrman DC, Galt J, Plummer NW, Jones JM, Spear B, Meisler MH (1995) Mutation of a new sodium channel gene, *Scn8a*, in the mouse mutant 'motor endplate disease'. Nature Genetics 10:461–465

Cannon SC, McClatchey AI, Gusella JF (1993) Modification of the Na^+ current conducted by the rat skeletal muscle α subunit by coexpression with a human brain β subunit. Pflügers Arch 423:155–157

Cannon SC (1996) Sodium channel defects in myotonia and periodic paralysis. Annu Rev Neurosci 19:141–164

Catterall W (1995) Structure and function of voltage-gated ion channels. Annu Rev Biochem 64:493–531

Cha A, Ruben PC, George AL Jr, Fujimoto E, Bezanilla F (1999) Voltage sensors in domains III and IV, but not I and II, are immobilized by Na^+ channel fast inactivation. Neuron 22:73–87
Chen J, Ikeda SR, Lang W, Isales CM, Wei X (1997) Molecular cloning of a putative tetrodotoxin-resistant sodium channel from dog nodose ganglion neurons. Gene 202:7–14
Chen TC, Law B, Kondratyuk T, Rossie S (1995) Identification of soluble protein phosphatases that dephosphorylate voltage-sensitive sodium channels in rat brain. J Biol Chem 270:7750–7756
Conti F, Stühmer W (1989) Quantal charge redistributions accompanying the structural transitions of sodium channels. Eur Biophys J 17:53–59
Dib-Hajj SD, Ishikawa K, Cummins TR, Waxman SG (1997) Insertion of a SNS-specific tetrapeptide in S3-S4 linker of D4 accelerates recovery from inactivation of skeletal muscle voltage-gated Na channel μ1 in HEK293 cells. FEBS Lett 416:11–14
Dib-Hajj SD, Tyrrell L, Black JA, Waxman SG (1998) NaN, a novel voltage-gated Na channel, is expressed preferentially in peripheral sensory neurons and down-regulated after axotomy. Proc Natl Acad Sci USA 95:8963–8968
Dietrich PS, McGivern JG, Delgado SG, Koch BD, Eglen RM, Hunter JC, Sangameswaran L (1998) Functional analysis of a voltage-gated sodium channel and its splice variant from rat dorsal root ganglia. J Neurochem 70:2262–2272
Doyle DA, Morais Cabral JH, Pfuetzner RA, Kuo A, Gulbis JM, Cohen SL, Chait BT, MacKinnon R (1998) The structure of the potassium channel: molecular basis of K^+ conduction and selectivity. Science 280:69–77
Eubanks J, Srinivasan J, Dinulos MB, Disteche CM, Catterall WA (1997) Structure and chromosomal localization of the β2 subunit of the human brain sodium channel. NeuroReport 8:2775–2779
Felipe A, Knittle TJ, Doyle KL, Tamkun MM (1994) Primary structure and differential expression during development and pregnancy of a novel voltage-gated sodium channel in the mouse. J Biol Chem 269:30125–30131
Feng G, Deák P, Chopra M, Hall LM (1995) Cloning and functional analysis of TipE, a novel membrane protein that enhances Drosophila *para* sodium channel function. Cell 82:1001–1011
Fozzard HA, Hanck DA (1996) Structure and function of voltage-dependent sodium channels: comparison of brain II and cardiac isoforms. Physiol Rev 76:887–926
Garcia KD, Sprunger LK, Meisler MH, Beam KG (1998) The sodium channel Scn8a is the major contributor to the postnatal developmental increase of sodium current density in spinal motoneurons. J Neurosci 18:5234–5239
Gautron S, Dos Santos G, Pinto-Henrique D, Koulakoff A, Gros F, Berwald-Netter Y (1992) The glial voltage-gated sodium channel: cell- and tissue-specific mRNA expression. Proc Natl Acad Sci USA 89:7272–7276
Gellens ME, George AL Jr, Chen L, Chahine M, Horn R, Barchi RL, Kallen RG (1992) Primary structure and functional expression of the human cardiac tetrodotoxin-insensitive voltage-dependent sodium channel. Proc Natl Acad Sci USA 89:554–558
George AL Jr, Ledbetter DH, Kalen RG, Barchi RL (1991) Assingment of a human skeletal muscle sodium channel alpha-subunit gene (SCN4A) to 17q23.1–25.3. Genomics 9:555–556
George AL Jr, Knittle TJ, Tamkun MM (1992) Molecular cloning of an atypical voltage-gated sodium channel expressed in human heart and uterus: evidence for a distinct gene family. Proc Natl Acad Sci USA 89:4893–4897
George AL Jr, Iyer GS, Kleinfield R, Kallen RG, Barchi RL (1993) Genomic organization of the human skeletal muscle sodium channel gene. Genomics 15:598–606
George AL Jr, Knops JF, Han J, Finley WH, Knittle TJ, Tamkun MM, Brown GB (1994) Assignment of the human voltage-dependent sodium channel alpha-subunit gene (SCN6A) to 2q21-q23. Genomics 19:395–397

George AL Jr, Varkony TA, Drabkin HA, Han J, Knops JF, Finley WH, Brown GB, Ward DC, Hass M (1995) Assignment of the human heart tetrodotoxin-resistant voltage-gated Na^+ channel alpha-subunit gene (SCN5 A) to band 3p21. Cytogenet Cell Genet 68:67–70

Gordon D, Merrick D, Auld V, Dunn R, Goldin AL, Davidson N, Catterall WA (1987) Tissue-specific expression of the R_I and R_{II} sodium channel subtypes. Proc Natl Acad Sci U S A 84:8682–8686

Gray PC, Tibbs VC, Catterall WA, Murphy BJ (1997) Identification of a 15-kDa cAMP-dependent protein kinase-anchoring protein associated with skeletal muscle L-type calcium channels. J Biol Chem 272: 6297–6302

Gulbis JM, Mann S, MacKinnon R (1999) Structure of a voltage-dependent K^+ channel β subunit. Cell 97:943–952

Guy HR, Conti F (1990) Pursuing the structure and function of voltage-gated channels. Trends Neurosci 13:201–206

Guy HR, Durell SR (1995) Structural models of Na^+, Ca^{2+}, and K^+ channels. Soc Gen Physiol Ser 50:1–16

Heinemann SH, Terlau H, Stühmer W, Imoto K, Numa S (1992a) Calcium channel characteristics conferred on the sodium channel by single mutations. Nature 356:441–443

Heinemann SH, Terlau H, Imoto K (1992b) Molecular basis for pharmacological differences between brain and cardiac sodium channels. Pflügers Arch 422:90–92

Heinemann SH, Schlief T, Mori Y, Imoto K (1994) Molecular pore structure of voltage-gated sodium and calcium channels. Brazilian J Med Biol Res 27:2781–2802

Hilborn MD, Vaillancourt RR, Rane SG (1998) Growth factor receptor tyrosine kinases acutely regulate neuronal sodium channels through the Src signaling pathway. J Neurosci 18:590–600

Hille B (1992) Ionic channels of excitable membrane, 2nd edn. SinauerAssociates, Sunderland, Mass

Hofmann F, Biel M, Flockerzi V (1994) Molecular basis for Ca^{2+} channel diversity. Annu Rev Neurosci 17:399–418

Isom LL, De Jongh KS, Patton DE, Reber BFX, Offord J, Charbonneau H, Walsh K, Goldin AL, Catterall WA (1992) Primary structure and functional expression of the β_1 subunit of the rat brain sodium channel. Science 256:839–842

Isom LL, Scheuer T, Brownstein AB, Ragsdale DS, Murphy BJ, Catterall WA (1995) Functional co-expression of the $\beta 1$ and Type IIA α subunits of sodium channels in a mammalian cell line. J Biol Chem 270:3306–3312

Isom LL, Ragsdale DS, De Jongh KS, Westenbroek RE, Reber BFX, Scheuer T, Catterall WA (1995) Structure and function of the $\beta 2$ subunit of brain sodium channels, a transmembrane glycoprotein with a CAM motif. Cell 83:433–442

Isom LL, Catterall WA (1996) Na^+ channel subunits and Ig domains. Nature 383:307–308

Jones JM, Meisler MH, Isom LL (1996) Scnb2, a voltage-gated sodium channel $\beta 2$ gene on mouse chromosome 9. Genomics 34:258–259

Johnston D, Magee JC, Colbert CM, Christie BR (1996) Active properties of neuronal dendrites. Annu Rev Neurosci 19:165–186

Joho RH, Moorman JR, VanDongen AMJ, Kirsh GE, Silberberg H, Schuster G, Brown AM (1990) Toxin and kinetic profile of rat brain type III sodium channels expressed in *Xenopus* oocytes. Brain Res Mol Brain Res 7:105–113

Kallen RG, Sheng Z-H, Yang J, Chen L, Rogert RB, Barchi RL (1990) Primary structure and expression of a sodium channel characteristic of denervated and immature rat skeleatal muscle. Neuron 4:233–242

Kallenberger S, Scheuer T, Catterall WA (1996) Movement of the Na^+ channel inactivation gate during inactivation. J Biol Chem 29:30971–30979

Kambouris NG, Nuss HB, Johns DC, Tomaselli GF, Marban E, Balser JR (1998) Phenotypic characterization of a novel long-QT syndrome mutation (R1623Q) in the cardiac sodium channel. Circulation 97:640–644

Kayano T, Noda M, Flockerzi V, Takahashi H, Numa S (1988) Primary structure of rat brain sodium channel III deduced from the cDNA sequence. FEBS Lett 228: 187–194

Klocke R, Kaupmann K, George AL Jr, Barchi RL, Jockusch H (1992) Chromosomal mapping of muscle-expressed sodium channel genes in the mouse. Mouse Genome 90:433–35

Klugbauer N, Lacinova L, Flockerzi V, Hofmann F (1995) Structure and functional expression of a new member of the tetrodotoxin-sensitive voltage-activated sodium channel family from human neuroendocrine cells. EMBO J 14:1084–1090

Knittle TJ, Doyle KL, Tamkun MM (1996) Immonolocalization of the $mNa_v2.3$ Na^+ channel in mouse heart: upregulation in myometrium during pregnancy. Am J Physiol 270:C688–C696

Kohrman DC, Plummer NW, Schuster T, Jones JM, Jang W, Burges DL, Galt J, Spear BT, Meisler MH (1995) Insertional mutation of the motor endplate disease (*med*) locus on mouse chromosome 15. Genomics 26:171–177

Kohrman DC, Harris JB, Meisler MH (1996a) Mutant detection in the *med* and *med*^{J} alleles of the sodium channel *Scna8a*. J Biol Chem 271:17576–17581

Kohrman DC, Smith MR, Goldin AL, Harris J, Meisler MH (1996b) A missense mutation in the sodium channel Scn8a is responsible for cerebellar ataxia in the mouse mutant *jolting*. J Neurosci 16:5993–5999

Kontis KJ, Rounaghi A, Goldin AL (1997) Sodium channel activation gating is affected by substitutions of voltage sensor positive charges in all four domains. J Gen Physiol 110:391–401

Kordeli E, Lambert S, Bennett V (1995) $Ankyrin_G$. J Biol Chem 270:2352–2359

Kozak CA, Sangameswaran L (1996) Genetic mapping of the peripheral sodium channel genes, Scn9a and Scn10a, in the mouse. Mamm Genome 7:787–788

Kupershmidt S, Yang T, Roden DM (1998) Modulation of cardiac Na^+ current phenotype by β1-subunit expression. Circ Res 24:441–447

Llinás R, Sugimori M (1980a) Electrophysiological properties of *in vitro* Purkinje cell somata in mammalian cerebellar slices. J Physiol (Lond) 305:171–195

Llinás R, Sugimori M (1980b) Electrophysiological properties of *in vitro* Purkinje cell dendrites in mammalian cerebellar slices. J Physiol (Lond) 305:197–213

Loughney K, Kreber R, Ganetzky B (1989) Molecular analysis of the *para* locus, a sodium channel gene in Drosophila. Cell 58:1143–1154

Ma JY, Catterall WA, Scheuer T (1997) Persistent sodium currents through brain sodium channels induced by G protein $\beta\gamma$ subunits. Neuron 19:443–453

Malo D, Schurr E, Dorfman J, Canfield V, Levenson R, Gros P (1991) Three brain sodium channel alpha-subunit genes are clustered on the proximal segment of mouse chromosome 2. Genomics 10:666–672

Malo MS, Blanchard BJ, Andresen JM, Srivastava K, Chen XN, Li X, Jabs EW, Korenberg JR, Ingram VM (1994a) Localization of a putative human brain sodium channel gene (SCN1 A) to chromosome band 2q24. Cytogenet Cell Genet 67:178–186

Malo MS, Srivastava K, Andresen JM, Chen XN, Korenberg JR, Ingram VM (1994b) Targeted gene walking by low stringency polymerase chain reaction: assignment of a putative human brain sodium channel gene (SCN3 A) to chromosome 2q24–31. Proc Natl Acad Sci USA 91:2975–2979

Makita N, Bennett PB, George AL Jr (1994a) Voltage-gated Na^+ channel β1 subunit mRNA expressed in adult human skeletal muscle, heart, and brain is encoded by a single gene. J Biol Chem 269:7571–7578

Makita N, Sloan-Brown K, Weghuis DO, Ropers HH, George AL Jr (1994b) Genomic organization and chromosomal assignment of the human voltage-gated Na^+ channel β 1 subunit gene (SCN1B). Genomics 23:628–634

Makita N, Bennett PB, George AL Jr (1996) Molecular determinants of β_1 subunit-induced gating modulation in voltage-dependent Na^+ channels. J Neurosci 16:7117–7127

Martina M, Jonas P (1997) Functional differences in Na^+ channel gating between fast-spiking interneurones and principal neurones of rat hippocampus. J Physiol (Lond) 505:593–603

McPhee JC, Ragsdale DS, Scheuer T, Catterall WA (1995) A critical role for transmembrane segment IVS6 of the sodium channel α subunit in fast inactivation. J Biol Chem 270:12025–12034

Moorman JR, Kirsh GE, VanDongen AMJ, Joho RH, Brown AM (1990) Fast and slow gating of sodium channels encoded by a single mRNA. Neuron 4:243–252

Mori Y, Mikala G, Varadi G, Kobayashi T, Koch S, Wakamori M, Schwartz A (1996) Molecular pharmacology of voltage-dependent calcium channels. Jpn J Pharmacol 72:83–109

Noda M, Shimizu S, Tanabe T, Takai T, Kayano T, Ikeda T, Takahashi H, Nakayama H, Kanaoka Y, Minamino N, Kangawa K, Matsuo H, Raftery MA, Hirose T, Inayama S, Hayashida H, Miyata T, Numa S (1984) Primary structure of *Electrophorus electricus* sodium channel deduced from cDNA sequence. Nature 312:121–127

Noda M, Ikeda T, Kayano T, Suzuki H, Takeshima H, Kurasaki M, Takahashi H, Numa S (1986a) Existence of distinct sodium channel messenger RNAs in rat brain. Nature 320:188–192

Noda M, Ikeda T, Suzuki H, Takeshima H, Takahashi T, Kuno M, Numa S (1986b) Expression of functional sodium channels from cloned cDNA. Nature 322:826–828

Noda M, Suzuki H, Numa S, Stühmer W (1989) A single point mutation confers tetrodotoxin and saxitoxin insensitivity on the sodium channel II. FEBS Lett 259:213–216

Numann R, Catterall WA, Scheuer T (1991) Functional modulation of brain sodium channels by protein kinase C phosphorylation. Science 254:115–118

Oh Y, Black JA, Waxman SG (1994) Rat brain Na^+ channel mRNAs in non-excitable Schwann cells. FEBS Lett 350:342–346

Patlak J (1991) Molecular kinetics of voltage-dependent Na^+ channels. Physiol Rev 71:1047–1080

Patton ED, West JW, Catterall WA, Golding AL (1993) A peptide segment critical for sodium channel inactivation functions as an inactivation gate in a potassium channel. Neuron 11:967–974

Plummer NW, Galt J, Jones JM, Burgess DL, Sprunger LK, Kohrman DC, Meisler MH (1998) Exon organization, coding sequence, physical mapping, and polymorphic intragenic markers for the human neuronal sodium channel gene SCN8A. Genomics 54:287–296

Plummer NW, Meisler MH (1999) Evolution and diversity of mammalian sodium channel genes. Genomics 57:323–331

Potts JF, Regan MR, Rochelle JM, Seldin MF, Agnew WS (1993) A glial-specific voltage-gated Na channel gene maps close to clustered genes for neuronal isoforms on mouse chromosome 2. Biochem Biophys Res Commun 197:100–104

Qu Y, Rogers JC, Tanada TN, Catterall WA, Scheuer T (1996) Phosphorylation of S1505 in the cardiac Na^+ channel inactivation gate is required for modulation by protein kinase C. J Gen Phyiol 108:375–379

Rabert DK, Koch BD, Ilnicka M, Obernolte RA, Naylor SL, Herman RC, Eglen RM, Hunter JC, Sangameswaran L (1998) A tetrodotoxin-resistant voltage-gated sodium channel from human dorsal root ganglia, hPN3/*SCN10A*. Pain 78:107–114

Raman IM, Bean BP (1997a) Resurgent sodium current and action potential formation in dissociated cerebellar Purkinje neurons. J Neurosci 17:4517–4526

Raman IM, Sprunger LK, Meisler MH, Bean BP (1997b) Altered subthreshold sodium currents and disrupted firing patterns in Purkinje neurons of *Scn8a* mutant mice. Neuron 19:881–891

Roden DM, George AL Jr (1997) Structure and function of cardiac sodium and potassium channels. Am J Physiol 273:H511–H525

Rogert RB, Cribbs LL, Muglia LK, Kephart DD, Kaiser MW (1989) Molecular cloning of a putative tetrodotoxin-resistant rat heart Na^+ channel isoform. Proc Natl Acad Sci USA 86:8170–8174

Sangameswaran L, Delgado SG, Fish LM, Koch BD, Jakeman LB, Stewart GR, Sze P, Hunter JC, Eglen RM, Herman RC (1996) Structure and function of a novel voltage-gated, tetrodotoxin-resistant sodium channel specific to sensory neurons. J Biol Chem 271:5953–5956

Sangameswaran L, Fish LM, Koch BD, Rabert DK, Delgado SG, Ilnicka M, Jakeman LB, Navakovic S, Wong K, Sze P, Tzoumaka E, Stewart GR, Herman RC, Chan H, Eglen RM, Hunter JC (1997) A novel tetrodotoxin-sensitive, voltage-gated sodium channel expressed in rat and human dorsal root ganglia. J Biol Chem 272: 14805–14809

Santana LF, Gómez AM, Lederer WJ (1998) Ca^{2+} flux through promiscuous cardiac Na^+ channels: Slip-mode conductance. Science 279:1027–1033

Sarao R, Gupta SK, Auld VJ, Dunn RJ (1991) Developmentally regulated alternative RNA splicing of rat brain sodium channel mRNAs. Nucl Acids Res 19:5673–5679

Sato C, Sato M, Iwasaki A, Doi T, Engel A (1998) The sodium channel has four domains surrounding a central pore. J Struct Biol 121:314–325

Schaller KL, Krzemien DM, Yarowsky PJ, Krueger BK, Caldwell JH (1995) A novel, abundant sodium channel expressed in neurons and glia. J Neurosci 15:3231–3242

Schultz J, Hoffmüller U, Krause G, Ashurst J, Macias MJ, Schmieder P, Schneider-Mergener J, Oschkinat H (1998) Specific interactions between the syntrophin PDZ domain and voltage-gated sodium channels. Nature Struct Biol 5:19–24

Sharp PA, Burge CB (1997) Classification of introns: U2-type or U12 type. Cell 91:875–879

Smith RD, Goldin AL (1997) Phosphorylation at a single site in the rat brain sodium channel is necessary and sufficient for current reduction by protein kinase A. J Neurosci 17:6086–6093

Smith RD, Goldin AL (1998) Functional analysis of the rat I sodium channel in *Xenopus* oocytes. J Neurosci 18:811–820

Smith MR, Smith RD, Plummer NW, Meisler MH, Goldin AL (1998) Functional analysis of the mouse Scn8a sodium channel. J Neurosci 18:6093–6102

Souslova VA, Fox M, Wood JN, Akopian AN (1997) Cloning and characterization of a mouse sensory neuron tetrodotoxin-resistant voltage-gated sodium channel gene, Scn10a. Genomics 41:201–209

Srinivasan J, Schachner M, Catterall WA (1998) Interaction of voltage-gated sodium channels with the extracellular matrix molecules tenascin-C and tenascin-R. Proc Natl Acad Sci USA 95:15753–15757

Stuart G, Spruston N, Sakmann B, Häusser M (1997) Action potential initiation and backpropagation in neurons of the mammalian CNS. Trends Neurosci 20:125–131

Stühmer W, Conti F, Suzuki H, Wang X, Noda M, Yahagi N, Kubo H, Numa S (1989) Structural parts involved in activation and inactivation of the sodium channel. Nature 339:597–603

Suzuki H, Beckh S, Kubo H, Yahagi N, Ishida H, Kayano T, Noda M, Numa S (1989) Functional expression of cloned cDNA encoding sodium channel III. FEBS Lett 228:195–200

Tapia-Ramirez J, Eggen BJ, Peral-Rubio MJ, Toledo-Aral JJ, Mandel G (1997) A single zinc finger motif in the silencing factor REST represses the neural-specific type II sodium channel promoter. Proc Natl Acad Sci USA 94:1177–1182

Tate S, Benn S, Hick C, Trezise D, John V, Mannion RJ, Costigan M, Plumpton C, Grose D, Gladwell Z, Kendall G, Dale K, Bountra C, Woolf CJ (1998) Two sodium channels contribute to the TTX-R sodium current in primary sensory neurons. Nature Neuroscience 1:653–655

Taylor CP (1993) Na+ currents that fail to inactivate. Trends Neurosci 16:455–460
Terlau H, Heinemann SH, Stuhmer W, Pusch M, Conti F, Imoto K, Numa S (1991) Mapping the site of block by tetrodotoxin and saxitoxin of sodium channel II. FEBS Lett 293:93–96
Thacheray JR, Ganetzky B (1994) Developmentally regulated alternative splicing generates a complex array of *Drosophila para* sodium channel isoforms. J Neurosci 14:2569–2578
Thompson JD, Higgins DG, Gibson TJ (1994) CLUSTAL W: improving the sensitivity of progressive multiple sequence alignment through sequence weighting, positions-specific gap penalties and weight matrix choice. Nucl Acids Res 22:4673–4680
Tibbs VC, Gray PC, Catterall WA, Murphy BJ (1998) AKAP15 anchors cAMP-dependent protein kinase to brain sodium channels. J Biol Chem 273:25783–25788
Toledo-Aral JJ, Moss BL, He Z-J, Koszowski AG, Whisenand T, Levinson SR, Wolf JJ, Silos-Santiago I, Halegoua S, Mandel G (1997) Identification of PN1, a predominant voltage-dependent sodium channel expressed principally in peripheral neurons. Proc Natl Acad Sci USA 94:1527–1532
Tong J, Potts JF, Rochelle JM, Seldin MF, Agnew WS (1993) A single β_1 subunit mapped to mouse chromosome 7 may be a common component of Na channel isoforms from brain, skeletal muscle and heart. Biochem Biophys Res Commun 195: 679–685
Trimmer JS, Cooperman SS, Tomiko SA, Zhou JY, Crean SM, Boyle MB, Kallen RG, Sheng Z, Barchi RL, Sigworth FJ, Goodman RH, Agnew WS, Mandel G (1989) Primary structure and functional expression of a mammalian skeletal muscle sodium channel. Neuron 3:33–49
Ukomadu C, Zhou J, Sigworth FJ, Agnew WS (1992) μI Na+ channels expressed transiently in human embryonic kidney cells: biochemical and biophysical properties. Neuron 8:663–676
Vega-Saenz de Miera E, Rudy B, Sugimori M, Llinás R (1997) Molecular characterization of the sodium channel subunits expressed in mammalian cerebellar Purkinje cells. Proc Natl Acad Sci USA 94:7059–7064
Wallace RH, Wang DW, Singh R, Scheffer IE, George AL Jr, Phillips HA, Saar K, Reis A, Johnson EW, Sutherland GR, Berkovic SF, Mulley JC (1998) Febrile seizures and generalized epilepsy associated with a mutation in the Na+-channel β1 subunit gene *SCN1B*. Nature Genetics 19:366–370
Wallner M, Weigl L, Meera P, Lotan I (1993) Modulation of the skeletal muscle sodium channel α-subunit by the β_1-subunit. FEBS Lett 336:535–539
Wang Q, Shen J, Splawski I, Atkinson D, Li Z, Robinson JL, Moss AJ, Towbin JA, Keating MT (1995) *SCN5A* mutations associated with an inherited cardiac arrhythmia, long QT syndrome. Cell 80:805–811
Wang Q, Li Z, Shen J, Keating MT (1996) Genomic organization of the human *SCN5A* gene encoding the cardiac sodium channel. Genomics 34:9–16
Warmke JW, Reenan RAG, Wang P, Qian S, Arena JP, Wang J, Wunderler D, Liu K, Kaczorowski GJ, Van der Ploeg LHT, Ganetzky B, Cohen CJ (1997) Functional expression of *Drosophila para* sodium channels. Modulation by the membrane protein TipE and toxin pharmacology. J Gen Physiol 110:119–133
Weber P, Bartsch U, Rasband MN, Czaniera R, Lang Y, Bluethmann H, Margolis RU, Levinson SR, Shrager P, Montag D, Schachner M (1999) Mice deficient for tenascin-R display alterations of the extracellular matrix and decreased axonal conduction velocities in the CNS. J Neurosci 19:4245–4262
West JW, Numann R, Murphy BJ, Scheuer T, Catterall WA (1991) A phosphorylation site in the Na+ channel required for modulation by protein kinase C. Science 254:866–868
West JW, Patton DE, Scheuer T, Wang Y, Goldin AL, Catterall WA (1992) A cluster of hydrophobic amino acid residues required for fast Na+ channel inactivation. Proc Natl Acad Sci USA 89:10905–10909

Westenbroek RE, Merrick DK, Catterall WA (1989) Differential subcellular localization of the R_I and R_{II} Na^+ channel subtypes in central neurons. Neuron 3:695–704
Wood SJ, Slater CR (1998) β spectrin is colocalized with both voltage-gated sodium channels and $ankyrin_G$ at the adult rat neuromuscular junction. J Biol Chem 140:675–684
Yang N, Horn R (1995) Evidence for voltage-dependent S4 movement in sodium channels. Neuron 15:213–218
Zhou J, Potts JF, Trimmer JS, Agnew WS, Sigworth FJ (1991) Multiple gating modes and the effect of modulating factors on the μI sodium channel. Neuron 7:775–785
Zhou D, Lambert S, Malen PL, Carpenter S, Boland LM, Bennett V (1998) $Ankyrin_G$ is required for clustering of voltage-gated Na channels at axon initial segments and for normal action potential firing. J Cell Biol 143:1295–1304

CHAPTER 2

Sodium Channel Blockers and Activators

A.O. GRANT

A. Introduction

Impulse conduction in brain and peripheral nerves, skeletal and cardiac muscle is sustained by transient increases in membrane permeability to sodium ions. This function resides in a family of integral membrane proteins, the voltage-gated Na^+ channels. Sodium channel blockers are an important class of therapeutic agents as anticonvulsants, local anesthetics, and antiarrhythmic drugs. The blockers have also proved to be important tools for structure-function studies of the Na^+ channel. The activators of Na^+ channels are potentially useful tools to study the mechanism of activation and inactivation. They may also form the basis for the development of novel positive inotropic agents and insecticides.

Over the past several years, the genes encoding the voltage-gated Na^+ channel in brain, peripheral nerve, skeletal and cardiac muscle have been cloned, sequenced, and expressed in heterologous systems (reviewed in CATTERALL 1992; FOZZARD and HANCK 1996). Site-directed mutagenesis and electrophysiologic studies have been combined to provide a wealth of new insight into the structure-function relationships of the Na^+ channel. The sites of action of channel blockers and activators with the Na^+ channel are under active investigation. The structure of the Na^+ channel has been reviewed in Chap. 1. Here I shall outline important features of the structure of the channel that relates to the action of Na^+ channel blockers and activators. The review will then focus on the recent studies that define sites and mechanisms of interaction of blockers and activators with the channel.

B. Classification and Structure of Na^+ Channels

The identification of different classes of Na^+ channels provides a basis for the pharmacodynamic distinctions between anticonvulsant, local anesthetic, and antiarrhythmic drug actions. Tissue distribution and susceptibility to block by the marine toxins tetrodotoxin (TTX) and μ-conotoxin (μCTX) have provided complementary bases for the classification of Na^+ channels. The major subtypes that have been defined include: brain (TTX-sensitive and μCTX-resistant), peripheral nerve (TTX-sensitive and TTX-resistant), skeletal muscle (TTX-sensitive and μCTX-sensitive), and cardiac muscle (TTX resistant and μCTX-resistant). The TTX-resistant subtypes are usually Cd^+ sensitive. The distribution of subtypes is not mutually exclusive. For example, denerved and embryonic skeletal muscle also expresses a TTX-resistant channel that may be identical to that expressed in the heart (Weiss and Horn 1986; White et al. 1991). Certain conduction system myocytes, e.g., those in the sinus node also express a TTX-sensitive Na^+ channel (Baruscotti et al. 1997).

The Na^+ channel isoforms are products of a multigene family. They consist of a major α-subunit and one or more auxiliary β-subunits. The α-subunit is sufficient for the expression of an ion-selective pore in frog oocytes and mammalian cells. The β_1-subunit increases the level of expression of functional brain, skeletal and cardiac Na^+ channels and accelerates the macroscopic inactivation rate (Isom et al. 1992; Makita et al. 1994; Nuss et al. 1995). The available data suggest that the α-subunit is the site of action of Na^+ channel blockers and activators. The remainder of the review will focus on the α-subunit.

I shall briefly recapitulate the structural organization of the Na^+ channel which has been discussed in considerable detail by Imoto. The α-subunit is organized as four homologous domains, DI-DIV (Fig. 1). Each domain con-

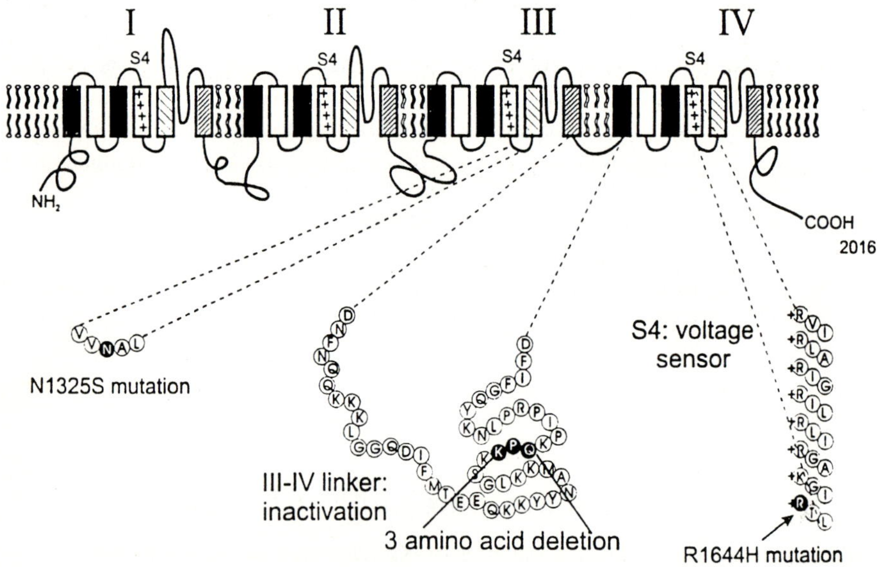

Fig. 1. Molecular organization of the human cardiac sodium channel (hH1). The SCN5 A gene encodes the sodium channel α-subunit, a protein 2016 amino acids long. The protein consists of four roughly homologous domains (*I–IV*) each containing six transmembrane spanning segments (S1–S6). The 52 amino acids linking domains III and IV are shown by their single letter codes; this region is known to be important for normal sodium channel inactivation. Three different mutations causing LQTS have been identified in SCN5 A. One is a deletion of three amino acids (KPQ) in the III-IV linker, and two are point mutations

sists of six transmembrane segments, S1–S6. The S4 segment of each domain is highly charged with lysine or arginine residues at every third position (STUEHMER et al. 1993; CATTERALL 1992). The outward rotation of the charged residues in S4 may account for the activation gating current. Neutralization of three of the four positive charges in S4 of DI reduced the valency of activation, without affecting inactivation (STUEHMER et al. 1993). The amino- and carboxy termini and the interdomain loops are intracellular. The DIII-DIV interdomain loop ($IDL_{III/IV}$) is short and highly conserved between Na^+ channel isoforms. The loop between transmembrane segments alternate between intra- and extracellular locations.

To date, two structural features of the Na^+ channel have proved important in understanding the action of blockers: the ion-permeation pathway and the inactivation gating mechanism. The extracellular loop between S5 and S6 of each domain is long and curves back into the membrane. The available data suggest that this loop forms the outer vestibule and selectivity filter of the ion-conducting pore. It may be divided into three regions: S5-P, the P segment, and P-S6 (FOZZARD and HANCK 1996). The ten residues that make up the P segment of rat brain 2, cardiac and skeletal muscle Na^+ channels are summarized in Table 1. Aspartate and glutamate residues in the P segment are conserved between isoforms. As we shall discuss below, mutations in the P segment affect ion conduction, selectivity, and TTX and STX binding. In fact, these small highly specific toxins with a rigid structure have proved pivotal in elucidating the nature of the selectivity filter and the process of permeation of the channel by Na^+ (FOZZARD and HANCK 1996).

The results of experiments using a variety of approaches have localized the inactivation gate to the $IDL_{III/IV}$ (CATTERALL 1992). Internal perfusion of

Table 1. P-loop sequences of Na channel isoforms and the heart Ca channel

Domain I	384
Br2	lslf**R**LMTQ**D**FW**E**Nlyq
Ht	lalf**R**LMTQ**D**CW**ER**lyq
Sk	lalf**R**LMTQ**D**YW**E**Nlfq
Domain II	942
Br2	ivfRVLCG**E**WI**E**Tmwd
Ht	iifRILCG**E**WI**E**Tmwd
Sk	ivfRILCG**E**WI**E**Tmwd
Domain III	1422
Br2	sllQVATF**K**GWM**D**imy
Ht	allQVATF**K**GWM**D**imy
Sk	sllQVATF**K**GWM**D**imy
Domain IV	1714
Br2	clfQITTSAGW**D**Glla
ht	clfQITTSAGW**D**Glls
Sk	clf**E**ITTSAGW**D**Glln

squid giant axon and cells with endopeptidases such as pronase and α-chymotrypsin remove Na^+ channel inactivation without affecting activation (ARMSTRONG et al. 1973). This suggests an intracellular location of the inactivation gate. Antibodies directed epitopes in $IDL_{III/IV}$ markedly slowed inactivation (VASSILEV et al. 1989). The injection of oocytes with separate cRNAs encoding DI-III and DIV resulted in Na^+ channels that fail to inactivate (STUEHMER et al. 1993). Patton and coworkers systematically deleted 10-residue segments of $IDL_{III/IV}$ and tentatively localized the inactivation gate to a 40-residue segment (PATTON et al. 1992). Subsequent experiments showed that the hydrophobic triplet IFM in $IDL_{III/IV}$ was critical for channel inactivation. The mutant channel IFM/QQQ was devoid of inactivation (WEST et al. 1992). A tentative model of the inactivation gate is that of a tilting disc that moves into the channel mouth to block Na^+ flow. Several mutations in this region are associated with incomplete Na^+ channel inactivation and cardiac arrhythmias (WANG et al. 1995a,b).

The process of inactivation is crucial to ion channel blockade. However, the agencies that remove inactivation are not equivalent. The sphere of changes that can affect Na^+ channel inactivation is far reaching. Peptide toxins can influence inactivation from extracellular site(s) (THOMSEN and CATTERALL 1989). Mutations in DIV S6 also produce channels that fail to inactivate (MCPHEE et al. 1994). These observations suggest any interaction between drug and inactivation cannot be interpreted as evidence for localization of the drug receptor to the inactivation gate.

C. Mechanisms of Na^+ Channel Blockade by Antiarrhythmic drugs

Blockade of Na^+ channels is an important mechanism of antiarrhythmic and local anesthetic drug action. Local anesthetic and antiarrhythmic drugs are principally small tertiary amines with an ionizable amino group and a hydrophobic tail. Antiarrhythmic drugs exert an anesthetic action on nerves. However, they block nerves at concentrations that are approximately 10–100 times greater than the antiarrhythmic concentrations. The isoform specific differences in susceptibility to marine neurotoxin block reflect differences in the structure of the channels (FOZZARD and HANCK 1996). However, the differences in isoform susceptibility to antiarrhythmic and local anesthetic drugs reflect differences in the characteristics of the nerve and cardiac action potential duration (APD) and the gating kinetics of the Na^+ channel (WRIGHT et al. 1997). The APD in nerve is about 10ms compared to 100–400ms in cardiac muscle (SMITH et al. 1996). The voltage dependence of activation and inactivation is ~10mV more negative in cardiac muscle (FOZZARD and HANCK 1996). These differences in APD and gating kinetics result in more prolonged occupancy in states susceptible to block in heart muscle and this is reflected in greater Na^+ channel blocking potency of drugs in the heart (WRIGHT et al.

1997). The available data suggest a common blocking mechanism of the Na^+ channel blockade in nerve and cardiac muscle.

Strichartz demonstrated that local anesthetic-class drugs effected two patterns of block: tonic block and phasic or frequency-dependent block (Strichartz 1973). Tonic block is the drug-induced current reduction during infrequent stimulation. Frequency-dependent block is the drug-induced current reduction during repetitive stimulation. Modeling by Starmer et al. (1990, 1991) suggests that tonic and frequency-dependent block are expressions of a common blocking mechanism. This conclusion is supported by structure-activity studies of Liu et al. (1994). The lidocaine derivative RAD-243 produced ~60% tonic block and little frequency-dependent block whereas the derivative L-30 produces 15% phasic block and 60% frequency-dependent block. There was a direct correlation between lipid solubility and tonic block.

Frequency-dependent block is the essence of antiarrhythmic drug action. The rapid succession of beats during a tachycardia are strongly suppressed, whereas the normal beats are little affected. The greater block during repeated excitation indicates the channel state(s) occupied during depolarization have a greater affinity for drug. During depolarization, the Na^+ channel passes through activated pre-open states followed by opening:

$$C_1 \rightleftharpoons C_2 \quad \cdots \quad C_N \rightleftharpoons O$$

Inactivation may occur from any of these pre-open or open states (Hodgkin and Huxley 1952). It is these activated pre-open, open, and inactivated states that have greater affinity for drug. Drugs dissociate from the Na^+ channels in the interval between depolarizations. If the intervals between depolarizations is less that five times the time constant for recovery, block accumulates. Eventually, a non-equilibrium steady-state is reached in which the rates of development and recovery from block are equal. The steady-state level of block is a function of the blocking and unblocking rate, and is characteristic for each drug. For most drugs, the blocking and unblocking rates are positively correlated. Subclasses of antiarrhythmic drugs can be identified based on their fast (class 1B), intermediate (class 1A) and slow kinetics of interaction with the Na^+ channel (Campbell 1983; Campbell and Vaughan Williams 1983). If the rate of stimulation is abruptly reduced, the rate at which the new steady-state level of block is achieved depends on the final rate of stimulation. Repetitive depolarization may actually enhance the rate at which equilibrium is achieved, a phenomenon termed use-dependent unblocking (Anno and Hondeghem 1990). Receptor-bound drug is trapped by the activation gate when the channel is in the rested state and is released at both threshold and subthreshold levels of depolarization.

The primary focus of the studies of antiarrhythmic drug in the 1970s and early 1980s was the characterization of the comparative kinetics of drug blockade using upstroke velocity and Na^+ current measurements. A number of models of Na^+ channel-drug receptor interaction was also proposed and critically examined. The recent focus has been the direct analysis of the relationship between channel gating and block, and the identification of the receptor site(s) for local anesthetic-class drugs with the Na^+ channel using site-directed mutagenesis and heterelogous expression in frog oocytes and mammalian cells. The earlier studies will be reviewed initially as they place the more contemporary studies in context. The studies of local anesthetic action using site-directed mutagenesis will be reviewed in detail.

D. Models of Antiarrhythmic Drug Interaction with the Sodium Channel

HILLE (1977b) and HONDEGHEM and KATZUNG (1977) independently proposed the modulated receptor model for drug interaction with the Na^+ channel. The basic postulates of the model are: (a) local anesthetic-class drugs bind to a specific receptor site on the Na^+ channel with affinities characteristic of each channel state; (b) the receptor is accessible through a hydrophilic pahway in the pore or a hydrophobic pathway through the membrane; (c) drug-associated Na^+ channels do not conduct Na^+, but make voltage-dependent gating transitions; (d) drug binding stabilizes the inactivated state. The inactivation curve of drug-associated channels is shifted to more negative potentials. Considerable effort has been applied to confirm the postulates of the modulated receptor model and to explore predictions that follow from the model.

Ligand-binding studies have provided support for the existence of a receptor site for local anesthetic-class drugs on the Na^+ channel. These drugs bind to a receptor on isolated myocytes with a rank order of potency and stereoselectivity that parallel their therapeutic action (SHELDON et al. 1987, 1991). Experiments by GRANT et al. (1993) demonstrated that it is the change in channel states rather than the membrane voltage that cause the change in the affinity of the receptor during repetitive depolarization. Studies with deltamathrin-modified Na^+ channels showed that disopyramide dissociated from open channels at normal rest potentials on a time frame of a few milliseconds compared with the hundreds of millisecond required for dissociation from rested channels at the same membrane potential. The dissociation of both anionic and cationic drug moieties is accelerated by membrane hyperpolarization (STRICHARTZ 1973; MATSUKI et al. 1984; KUO 1994). This suggests that it is the channel state occupied that determines dissociation rather than the field effect of hyperpolarization. The identification of the channel states that are blocked by drug presents a major challenge. Inasmuch as drug-associated channels do not conduct, their properties necessarily have to be inferred from the properties of the remaining drug-free channels. Those channels have to be activated to determine the fraction of channels that are available to conduct. Since block may occur during test activations, the measured current does not reflect the distribution of the channels in their various conformations prior to activation. The Na^+ channel exists in multiple conformational states at the potentials that result in channel activation. Therefore, it is a complex issue to determine which state(s) will be blocked with a given voltage-clamp paradigm. Single channel recording provides the best approach to define the states that are block as the open state of a channel can be unequivocally identified. Unfortunately, the long time required for drug studies makes this approach particularly challenging.

The modulated receptor model is the most comprehensive scheme that has been proposed and has received widespread acceptance. However, other significant models have been proposed. BALSER et al. (1996) proposed a model that is a departure from the modulated receptor model. The interaction of drug with its receptor alters the coupling between activation and inactivation such that macroscopic inactivation is accelerated. The model is based on the observation that lidocaine accelerates the macroscopic inactivation of the inactivation-deficient mutant channel IFM/QQQ. The simplest explanation for their result is that the increased relaxation reflected block. That hypothesis was rejected in part because of different apparent K_{Ds} for the peak and persistent current. Without other evidence that blockade of the peak current has reached equilibrium, a true K_D cannot really be calculated for the peak current. Starmer and colleagues proposed a guarded receptor model with the simplifying assumption that the receptor has single high and low affinity states (STARMER et al. 1984; STARMER and GRANT 1985). However, the channel gates control access to the receptor. Any channel gating model could be combined with simple binding kinetics to account for drug action. This permitted closed form solutions of the binding reactions and enabled binding parameters to be determined from the results of straightforward pulse train experiments (STARMER et al. 1990, 1991; CATTERALL and COPPERSMITH 1981; VALENZUELA et al. 1995). Data from a number of studies indicates that drugs bind to more than one channel state (GRANT et al. 1984). However, it is apparent that most drugs bind predominantly to one channel state, e.g., the open or the inactivated state. Therefore, application of the guarded receptor model remains a useful approach to determine association- and dissociation-rate constants of various drugs.

Convincing evidence of open state block of the Na^+ channel has been presented by a number of investigators (HORN et al. 1981; YAMAMOTO 1986; McDONALD et al. 1989; KOHLHARDT et al. 1989; KOHLHARDT and FICHTNER 1988; GRANT et al. 1993; BARBER et al. 1992; CARMELIET et al. 1989). The open time of the Na^+ channel is very brief, ~1 ms at room temperature and 0.04–0.07 msec at 35 °C (BENNDORF 1994). Therefore, to demonstrate open channel block, channel inactivation is usually slowed by chemical modification, e.g., *N*-bromocatemide, enzymes such as α-chymotrypsin, and the pyrethrin toxins (COHEN and BARCHI 1993; KOUMI et al. 1992; WASSERSTROM et al. 1993; GRANT et al. 1993). The mean open time is prolonged to several milliseconds or tens of milliseconds by these treatments. The block of these modified Na^+ channels can be described by Scheme 2

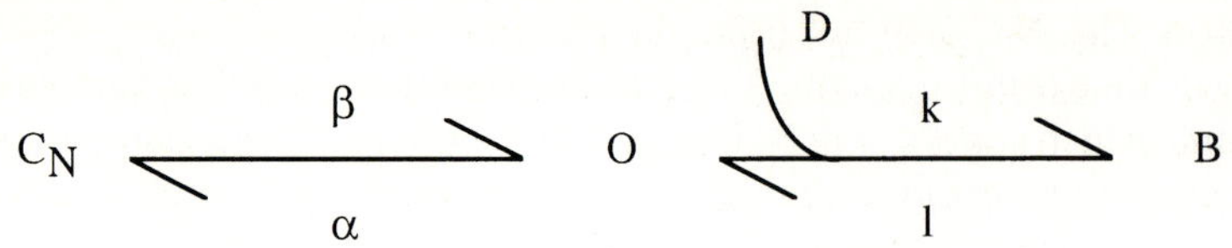

where C_N, O, and B refer to the closed, open, and blocked states of the Na^+ channel, D the drug, β and α the activation and deactivation rate constants, and k and l the drug association and dissociation rate constants. The pattern of block produced by an open channel blocker depends on the magnitudes of β and l. For blockers with fast dissociation rates (~10^5/s), the blocking events are not well resolved at the single channel level. Instead, blockade is evident as an increase in noise and a reduction of conductance of the open channel. This is the pattern of Na^+ channel blockade produced by the lidocaine derivative QX314 (GINGRICH et al. 1993). The dissociation constant, K_D for such open channels can be calculated from the following relationship:

$$\ln(i/i_O - 1) = \ln[D]/K_{Dv} \tag{2}$$

where i_o and i are the single channel currents in the absence and presence of a blocker (CORONADO and MILLER 1979). For QX314, i and i_o were examined over a range of concentrations and Eq. (2) applied. The K_{DV} was 4.4 mmol/l, indicating that QX314 was a weak blocker of the Na^+ channel. The single channel current can also be determined at a number of voltages and the apparent site of block determined from an extension of equation 2. Such an analysis suggested a blocking site subjected to 70% of the membrane field from the cytoplasmic side.

In the other patterns of open-channel block the residence time of the blocker on its Na^+ channel receptor is sufficiently long to record discrete blocking events. If $l >> \beta$ (Eq. 1), the blocking events are shorter than the normal shut periods and the openings are converted to bursts with a mean open time τo given by

$$\tau_O = 1/(\alpha + k[D]) \tag{3}$$

Disopyramide, penticainide, and propafenone produce this pattern of block (GRANT et al. 1993; CARMELIET et al. 1989; KOHLHARDT et al. 1989). All of these drugs have a similar association rate constant of 10^7/mol/l/s, suggesting that some fundamental process such as diffusion may determine block.

For the case in which $l << \beta$, the block states correspond to the long-lived shut states. The blocking events are intraburst gaps with mean 1/l. This pattern of block is produced by the specific neurotoxins TTX and STX, and quinidine (CRUZ et al. 1985; BENZ and KOHLHARDT 1991). For some local anesthetic-class antiarrhythmic drugs, e.g., lidocaine, the evidence of significant open channel block of any form is not convincing (GRANT et al. 1989; BENZ and KOHLHARDT 1992; BENNETT et al. 1995 – however, see NILIUS et al. 1987).

Many antiarrhythmic drugs, e.g., the class 1B drugs, lidocaine, mexiletine, and amiodarone produce progressive block as the duration of depolarization is increased beyond the initial transient inward current. This increased block may result from drug interaction with the small fraction of channels that remain open during prolonged depolarization. Alternatively, it could result from the interaction of drug with inactivated channels. Using single channel recordings, GRANT et al. (1989) showed progressive block of the Na^+ channel at late times in depolarizing trials without late opening. This is clear evidence for the occurrence of inactivated state block. The requirement for intact inactivation for block has been examined by slowing or removal of inactivation with endopeptidases such as pronase and chymotrypsin, amino acid modifying agents such as chloramine T, and the pyrethrin toxins. Following these treatments, the Na current assumes a compound wave form, with an initial transient followed by a persistent component. The results of these experiments have been inconclusive. Frequency-dependent block of the persistent component of neuronal Na^+ channel current lidocaine and tetraciane was diminished after chloramine-T modification whereas block by the open channel blockers *N*-propyl ajmaline and KC3791 persisted (ZABOROVSKAYA and KHODOROV 1994). Use-dependent block of neuronal Na^+ channel by etidocaine and QX 314 persisted after chloramine-T treatment whereas block is abolished after pronase treatment (WANG et al. 1987). Both modifying agents exert similar effects on inactivation. The non-specific nature of the modifying agents may account for the inconclusive results.

BENNETT et al. (1995) examined the block of wild-type cardiac Na^+ channels and channels with the inactivation disabling mutation IFM/QQQ in the $IDL_{III/IV}$. It was found that 25 μmol/l of lidocaine produced ~80% use-dependent block of wild-type Na^+ channels at a stimulus frequency of 5 Hz whereas 100 μmol/l of lidocaine produced less than 10% block in the IFM/QQQ mutant channel. The lack of use-dependent block by IFM/QQQ could be the result of slow association or rapid dissociation of drug from its receptor. When the onset of block was measured with a twin pulse protocol (conditioning pulse of increasing duration followed by a test pulse) no block developed with pulses up to 10 s in duration in the IFM/QQQ mutant channels. These data suggest that intact inactivation is required for the block of cardiac Na^+ channel by lidocaine. GRANT et al. (1996) examined the requirements of inactivation for *open* channel blockade with studies of the action of disopyramide in the mutant with inactivation partially removed (IFM/IQM). Open channel blockade with an association rate constant of 10^7/mol/l/s persisted in the mutant channel.

The use of congeners of lidocaine and variations of pH have provided insight into the location of the receptor site for local anesthetic-class drugs. The permanently charged derivatives of lidocaine QX314 and QX222 block most Na^+ channel types when applied from the cytoplasmic side of the membrane only (STRICHARTZ 1973; CAHALAN and ALMERS 1979). The cardiac isoform appears to be an exception in that it is blocked by externally applied

QX314 (Alpert et al. 1989). The receptor site can be accessed through the membrane phase. The clearest evidence for this has been obtained with single channel recordings. The mean open time of Na^+ channel currents is reduced when currents are recorded in the cell-attached configuration with drug-free micropippette solution and disopyramide applied to the superfusate (Grant et al. 1993). Lowering the external pH slows the rate of dissociation of local anesthetic-class drugs from the Na^+ channel (Hille 1977a,b; Grant et al. 1982). This suggests that the receptor-bound drug is accessible to external protons through the channel pore. Further studies on the local anesthetic receptor on the Na^+ channel have used by two approaches. Ragsdale et al. (1994) used the technique of scanning mutagenes in which residues within a given region of the Na^+ channel are systematically replaced by alanine. The F 1764 A mutation in the middle of the sixth transmembrane segment of the fourth domain, decreased block of rat brain II Na^+ channel by etidocaine to 1% of control. The mutation Y1771A also substantially reduced use-dependent block. Another mutation in the same region (N1769A) increased block. The quaternary derivative of lidocaine, QX314 blocks the neuronal Na^+ channel when applied from the cytoplasmic side of the membrane only. Another mutation in D4-S6, I1760A permitted block by *external* QX314, suggesting that residue I1760 controls access to the local anesthetic receptor site. Based on these experiments, Ragsdale et al. proposed the model illustrated in Fig. 2. F1764 and Y1771 are separated by two turns of an α-helix. The

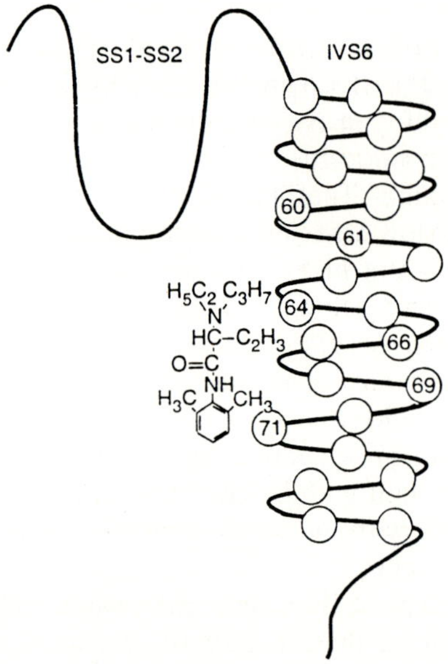

Fig. 2. Proposed orientation of amino acids in IVS6 with respect to a bound local anesthetic molecule in the ion-conducting pore. Segment SS1-SS2, which also contributes to the pore is shown as well. Amino acids at positions 1760, 1764, and 1771 are shown facing the pore lumen

aromatic nucleus of etidocaine binds to Y1771 and the tertiary amino end binds to F1764. Access to this binding site is controlled by I1769.

RAGSDALE et al. (1996) extended these studies with the examination of the class IA drug quinidine, the class IB drugs lidocaine and phenytoin, and the class IC drug flecainide. Prior studies had suggested that lidocaine and phenytoin are predominantly inactivated state blockers whereas flecainide and quinidine were open state blockers (GRANT et al. 1984). The F1764A mutation reduced lidocaine and phenytoin block 24.5- and 8.3-fold respectively. They provided supporting data that quinidine and flecainide block required activated channels whereas lidocaine and phenytoin block could be observed at threshold potential.

These important results should be interpreted with caution. These mutations produce only modest reduction in affinity when compared with the 100–1000-fold change in affinity with mutations that define the TTX receptor site (NODA et al. 1989; TERLAU et al. 1991). The F1764A mutation shifts the voltage dependence of availability +7mV on the voltage axis. Such a shift would predict an increase in the K_D several fold.

Single channel studies have conclusively demonstrated block of open Na^+ channels by some antiarrhythmic drugs such as penticainide, disopyramide, and propafenone. Therefore, the residues that line the Na^+ channel pore are candidate sites for local anesthetic interaction with the Na^+ channel. The ionizable amino group of the local anesthetic may bind to the negative residues in the selectivity filter of the pore. Sunami et al. (1997) examined the effects of mutations in the pore region of each domain of the skeletal muscle Na^+ channel, $\mu 1$ on block by lidocaine, its neutral derivatives, benzocaine and its permanently charged derivatives, QX222 and QX314. The largest effect was seen with the K1237E in domain III. However, the effect of the mutation on lidocaine affinity was modest, with a fourfold reduction in K_D. Mutation of selectivity residues in other domains D400A (domain I), E755A (domain II), and A1529D, (domain IV) enabled block by externally applied QX222 and QX314. Recovery from block by these drugs applied internally was markedly accelerated with the time constant of greater than 30min in the wild-type channel reduced to ~100s. A cautionary note is also appropriate when interpreting these results as some of the mutations shifted the voltage dependence of channel availability.

A significant value of models of drug interaction with the Na^+ channel is that they predict behavior that has not heretofore been examined. I shall examine the postulate of a single receptor site for local anesthetic-class drugs. If these drugs are interacting with a single receptor site, competition between blockers may occur at the receptor. Examples include the competition between a drug and its metabolites (BENNETT et al. 1988). In a conventional drug binding regimen in which competing ligands have continuous access to the receptor, block by the drugs should be additive. In the case of ion channels, access to the receptor is phasic and differences in the kinetics of binding of competing drugs can be amplified by repetitive depolarization. A drug with fast kinetics may compete with and displace a drug with slower kinetics. In the interval between depolarizations, the drug with fast kinetics also leaves the receptor site rapidly. The net result is that, over a limited range of stimulation frequencies and drug concentrations, less block may be observed with the combination of the two drugs than with the slow drug alone. This provides a basis for interpreting the observation that lidocaine can reverse the cardiac toxicity of a number of Na^+ channel blockers that have slow binding kinetics with the Na^+ channel (Wynn et al. 1986; von Dach and Streuli 1988; Whitcomb et al. 1989). The demonstration that some drugs block form sites within the pore provides a basis for reversing the toxic effect of open channel blockers with Na^+ salts (Bellet et al. 1959a,b; Pentel and Benowitz 1984; Kohlhardt et al. 1989; Barber et al. 1992).

E. The Highly Specific Na^+ Channel Blockers TTX and STX

TTX and STX are highly potent marine toxins that have proved very useful in the study of voltage-gated Na^+ channels. TTX is concentrated in the liver and gonads of some species of the *Spheroides* puffer fish and some species of *Taricha* newts. Saxitoxin is synthesized by the dinoflagellates *Gonyaulax catanella* and *Gonyaulux tamarensis* and concentrated by shellfish. TTX and STX are small rigid heterocyclic molecules that carry critical positive charges on one (TTX) or two (STX) guanidinium groups. Modification of the toxins close to the guanidinium group result in a dramatic loss of activity.

Narahashi et al. (1964) showed that TTX blocks Na^+ channels with high specificity. The stoichiometry of block is 1:1. Neuronal and skeletal muscle Na^+ channels are blocked by TTX with a K_D of ~10 nmol/l. In cardiac muscle, the K_D is in the μmol/l range (Cohen et al. 1981; Fozzard and Hanck 1996). Some Na^+ channels are resistant to TTX. STX and TTX compete for the same binding site (Henderson and Wang 1972; Henderson et al. 1973). Protons and mono- and divalent cations such as Ca^{2+} compete with TTX and STX for a binding site on the Na channel (Henderson et al. 1973, 1974). Carboxyl-modifying agents reduced TTX binding to the Na channel (Shrager and Profera 1973; Baker and Robinson 1975; Spalding 1980; Sigworth and Spaulding 1980). The data implicates negatively charged residues at the TTX/STX binding site. Schild and Moczydlowski (1991) used Zn^{2+} blockade as a tool to examine the possible structural basis of STX resistance in cardiac Na channels. They showed that Zn^{2+} was a direct competitive inhibitor of STX binding. As sulfhydryl groups are frequently coordinating ligands for Zn^{2+}, they examined the effect of iodoacetamide on Zn^{2+} and STX block of the cardiac Na channel. Zn^{2+} blockade was abolished and STX block was reduced 20-fold. This suggested that cysteine group(s) form a part of the STX binding site in cardiac Na channels.

Residues in the putative pore region (SS1-SS2) of the Na^+ channel were the logical sites for residues associated with TTX blockers (Table 1). NODA et al. (1989) showed that change of a single glutamate residue to glutamine (E387Q) in D1 SS2 of rat brain 2 Na^+ channel rendered the channel insensitive to TTX. TERLAU et al. (1991) reported a systematic analysis of mutations of the charged residues in each domain on TTX sensitivity and channel conductance. They identified two clusters of residues (D384, E942, K1422, and A 1714) and (E387, E945, M1425, and D1717) that were crucial to TTX binding. Charge altering mutations reduced TTX sensitivity by 100-fold or greater.

With the knowledge of the cluster of residues on the channel and the toxin groups that were critical for binding, LIPKIND and FOZZARD (1994) developed a model of the structural organization of the TTX and STX binding site, and the inferred structure of the channel pore. As illustrated in Fig. 3, each domain contributes a β-hairpin structure to pore. Carboxyl groups in the β-hairpins of domains I and III bind to the guanidinium group of TTX through salt bridges. The second guanidinium group of STX interacted with carboxyl groups of the domain IV β-hairpin. The domain III β-hairpin forms non-bonding interactions with the toxins. The picture that emerges is a funnel shaped toxin binding site with a width of ~12 Å at its outer vestibule and a narrow mouth, the selectivity filter of 3 × 5 Å. Energy calculations suggest that interaction of Na^+ with the pore is sufficient to allow dehydration of the ion, a process postulated to be necessary for ion permeation.

That this region of the α-subunit forms the channel vestibule and selectivity filter received considerable support from the work of HEINEMANN et al. (1992). They noted the striking similarity between SS1-SS2 region of Na^+ and Ca^{2+} channels. They showed that if the cluster of negative residues in SS1-SS2 of the brain Na^+ were increased by K1442E and/or A1714E, the channel was transformed to a Ca^{2+}-selective channel. The Lipkind-Fozzard model provided

Table 2. Neurotoxin receptor sites on the Na^+ channel[a]

Site	Toxin	Effects
1	Tetrodotoxin Saxitoxin μ-Conotoxins	Inhibitor of ionic conductance
2	Veratridine Batrachotoxin Aconitine Grayanotoxin	Persistent activation
3	α-Scorpion toxins Sea anemone toxins	Inhibit inactivation; enhance persistent activation
4	β-Scorpion toxins	Shift voltage dependence of activation

[a] Modified from CATTERALL 1992.

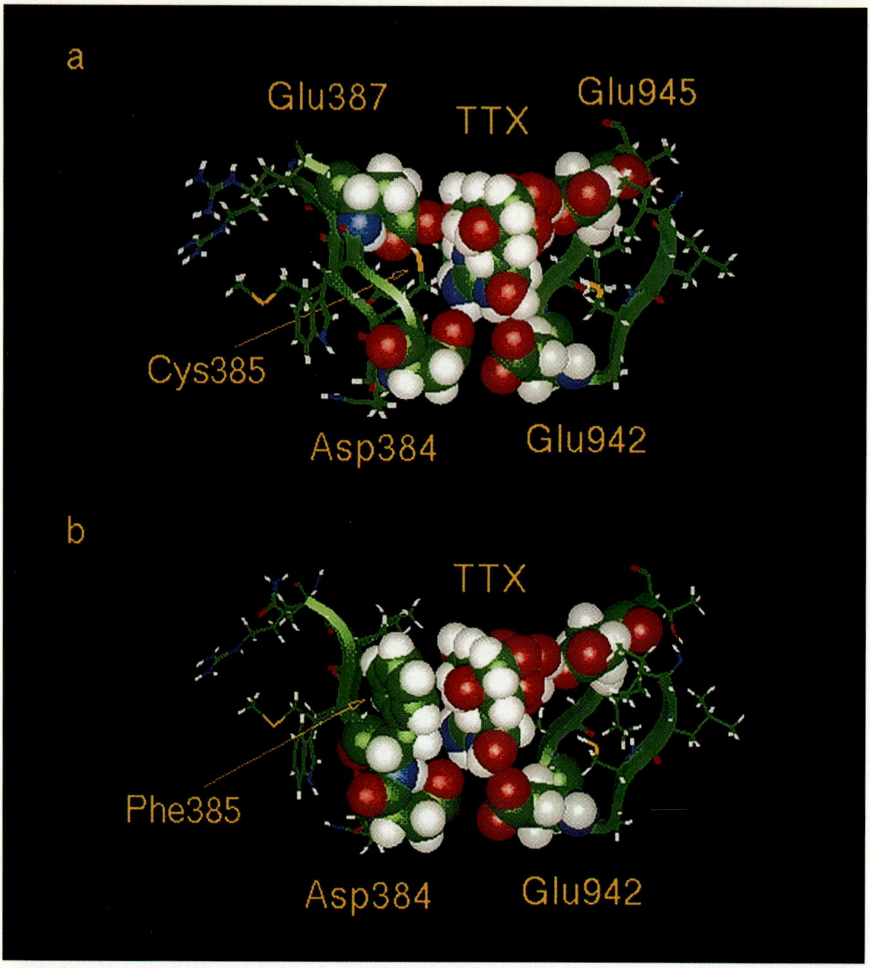

Fig. 3a,b. Lipkind-Fozzard tetrodotoxin (TTX) binding site model. TTX interacts with 4 residues on β-hairpin model (*green ribbon*) of domains I and II segments. Asp-384 and Glu-387 are shown as space-filling residues from domain I β-hairpin, and Glu-942 and Glu 945 are shown from domain II β-hairpin. In the model, guanidinium toxin interacts with Glu-387, Asp-384, and Glu-942. Hydroxyls of toxin interact with Glu-945. **a** Cardiac structure for domain I, with a Cys in position 385. **b** Substitution of Phe in position 385, with space-filling aromatic ring interacting with toxin hydrophobic surface. Image is rotated slightly from that in A to more clearly show relationship

a basis for interpretation of the observed isoform differences in TTX sensitivity in the Na^+ channel. Only two residues are different between the TTX- resistant cardiac isoform and the TTX-sensitive isoform. A cysteine at position 385 in the cardiac isoform replaces phenylalanine or tyrosine in the TTX-sensitive isoform and arginine 388 is replaced by asparagine. TTX block showed little sensitivity to mutations of the arginine residue. The C385

F mutation simultaneously increased TTX sensitivity ~1000-fold and reduced Cd^{2+} sensitivity (SATIN et al. 1992). The complementary experiment with the mutation Y385C converted the TTX sensitive skeletal muscle isoform to the TTX-resistant Cd^{2+} sensitive isoform (BACKX et al. 1992). The replacement of cysteine by phenylalanine or tyrosine permitted hydrophobic interactions with the toxin of ~5 kcal/mol, consistent with the greater TTX affinity. The recently cloned TTX-resistant sensory neuron Na^+ channel has a serine residue in the analogous position to the cysteine in the cardiac Na^+ channel. The S/F mutation decreased the IC_{50} for TTX blockade from 50 μmol/l to 2.8 nmol/l (SIVILOTTI et al. 1997).

F. Peptide Na Channel Blockers: μ Conotoxins

The μCTXs are a group of 22-amino acid peptides that have been purified from the venom of the marine snail Conus geographus (MOCZYDLOWSKI et al. 1986). The μCTXs block skeletal muscle, Na^+ channels, but have no effect on the neuronal, brain or cardiac Na channel (MOCZYDLOWSKI et al. 1986; KOBAYASHI et al. 1986). The structure of the major peptide μCTXS G IIIA shown below is very hydrophilic, with multiple charged residue and the uncommon amino acid trans-4-hydroxyproline.

R D C C T P Hyp KKC K D R Q C K Hyp Q R C C
. . . Hyp . . .

Structure-activity studies have shown that the guanidinium group on arginine 13 is essential for its blocking action (SATO et al. 1991). μCTX G IIIA produces reversible discrete block of bactrochotoxin-activated skeletal muscle Na^+ channels with a K_D of 100 nA. The dissociation rate constant is voltage dependent, decreasing e-fold for 43 mV of hyperpolarization; the association rate constant shows little voltage dependence. This pattern of block is similar to that produced by TTX.

μCTX G IIIA competitively inhibits the binding of ^3H-STX to Na^+ channels, suggesting overlap between their binding sites (KOBAYASHI et al. 1986). These binding studies, together with the importance of the guanidinium group for Na^+ channel blockade, suggested that μCTXs and TXX may bind at the same or overlapping binding sites. STEPHAN et al. (1994) examined the role of an E/Q mutation in the second cluster of residues that make up the selectivity filter. TTX sensitivity was reduced ~1000-fold whereas μCTX sensitivity was reduced ~4-fold. These results suggest that the μCTX and TTX binding sites are not identical. DUDLEY et al. (1995) performed additional mutations of the outer vestibule residues of the skeletal muscle Na^+ channel to further define the μCTX binding site. The mutation E758Q neutralizes one negative charge in SS1-SS2 of domain II and reduced μCTX binding affinity 48-fold. They developed a model for μCTX binding in which the guanidinium group of arginine 13 of μCTX interacted with two carboxyl groups in the selectivity filter. The charge neutralizing mutation E758Q decreased the association of μCTX with its receptor, suggesting that Glu-758 is involved in guiding the toxin to its receptor.

G. Na Channel Activators

A wide range of highly lipophilic molecules and neurotoxin peptides modify the gating and conductance of the Na^+ channel in a manner that generally promotes channel activation or failure of inactivation. Radioligand and voltage clamp studies suggest that they bind to one of five receptor sites on the Na^+ channel (CATTERALL 1992) (Table 2). The toxins within each group produce similar effects on channel gating. The toxins that act at receptor site 1 tetrodotoxin and saxitoxin have already been discussed. The functional effects of a representative member of each class of toxins acting at the other four sites will be presented in detail. Special features of the other toxins in each group will be discussed in brief.

The alkaloids veratridine, batrachotoxin (BTX), aconitine, and grayanotoxin bind to neurotoxin site 2. Competitive binding studies with [^{3}H]-batrachotoxin A 20-α benzoate indicate that they share a common receptor site (CATTERALL 1992). The mechanism of action and functional consequences of their interaction with the Na^+ channel have been investigated with voltage clamp of muscle fibers, whole-cell and single channel recordings. Repetitive depolarization is required for Na^+ channel modification and results in a reduction of the peak current and the simultaneous development of a persistent component of Na^+ current. The amplitude of the peak current lost is greater than that of the persistent current. Barnes and Hille demonstrated a reduction of single Na^+ channel conductance to 20–25% of normal following veratridine-induced channel modification. The apparent reversal potential of the modified channel was positive to that of the normal channel, suggesting a change in channel selectivity. The normal brief channel openings were converted to bursts of openings of duration ~ 1 s. Scheme 3 for channel modification has been proposed (SUTRO 1986; BARNES and HILLE 1988)

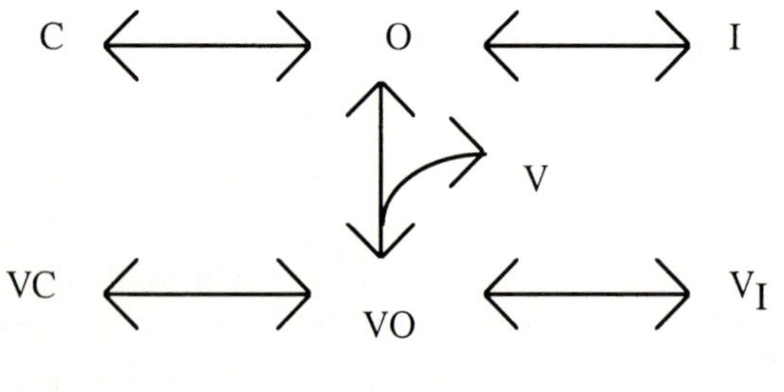

where C, O, and I are the closed, open, and inactivated states of the Na^+ channel, V veratridine, and VC, VO, and VI the corresponding veratradine-associated states.

Modification of the Na^+ channel occurs from the open state and is reversible at the normal resting potential. The changes in gating kinetics can be explained by a slowing of inactivation and a shift in activation to more negative potentials. The site 2 toxins may prove to be useful tools to clarify the coupling between activation and inactivation.

BTX shares many of the actions of veratridine. However, some of the action of the receptor site 2 toxins have been more thoroughly studied with BTX and some of its actions are also unique. QUANDT and NARAHASHI (1982) have shown that the distribution of open times is biexponential during BTX exposure. The brief open time of ~2 ms corresponds to the open time of the unmodified channel while there is a second open time of ~60 ms. This indicates that modification by BTX is an all or nothing phenomenon. Its interaction with the open Na^+ channel is irreversible. The voltage dependence of activation is shifted to more negative potentials by about 50 mV. As a result, channels are open at the normal resting potential. BTX increased the permeability of the Na^+ channel to divalent cations such as Ca^{2+} (KHODOROV 1985). BTX has proved to be a very useful tool to study the properties of the Na^+ channel. The permanent activation and marked prolongation of the open timees of BTX-modified Na^+ channels makes the drug particularly useful for studying the Na^+ channel properties in artificial lipid bilayers where the frequency response of the system is much slower than can be achieved with the patch clamp technique (MOCZYDLOWSKI et al. 1984a,b; KHODOROV 1985; ZAMPONI et al. 1993a,b).

The α-scorpion toxins (α-ScTX) isolated from the venom of *Tityus serrulatus*, *Leiurus quinquestriatus* and the toxins of nematocysts of the sea anemones, *Condylactis gigantea*, *Anemonia sulacta*, and *Anthopleura xanthogrammica* are polypeptides that interact with site 3, localized to the extracellular loops between S5 and S6 of domains I and IV on the Na^+ channel (THOMSEN and CATTERALL 1989). α-ScTX increases the peak amplitude of the Na^+ current at depolarized potentials and markedly slows the rate of macroscopic inactivation. At the single channel level, α-ScTX and anemone toxin prolong the single channel mean open time (KIRSCH et al. 1989; EL-SHERIF et al. 1992). The normal biphasic distribution of the voltage dependence of open times is converted to a monotonic increase. The burst duration is prolonged about 20-fold. These effects can be explained by a model in which the rate of transition from the open to the inactivated state is markedly slowed; activation is unaffected. Depolarization has been reported to promote or to have no effect on toxin dissociation from the Na^+ channel (KIRSCH et al. 1989; CAHALAN 1980). Since the site 3 polypeptide toxins act from an extracellulalr site yet influence inactivation, externally accessible residues may be important for the inactivation process. Alternatively, the modification of inactivation could be an allosteric effect.

The β-scorpion toxins (β-ScTX) interact with neurotoxin receptor site 4 (CATTERALL 1992). Their primary effect is to shift the voltage dependence of activation to more negative potentials. Their action is potentiated by membrane depolarization. The contrasting voltage dependence of the actions of the α- and β-ScTXs support the conclusion that these toxins are interacting at different sites.

The brevetoxins and ciguatoxin bind to neurotoxin receptor site 5 (Catterall 1992). Like the site 2 neurotoxins, they shift the voltage dependence of activation to more negative potentials and delay inactivation. While they have no effect on the binding of neurotoxins to sites 1 and 3, they enhance binding to sites 2 and 4. Their action at the single channel level and site of binding on the Na^+ channel have not been defined.

I shall discuss the action of two other Na^+ channels modifiers, the pyrethroids, and DPI 201–106 whose sites of action have not been localized to the neurotoxin sites 1–5, but are potentially important as biodegradable insecticides and a positive inotropic agent respectively. The pyrethroids are synthetic derivatives of pyrethrins. They are primarily esters or alcohols of chrysanthemic acid. Type 2 pyrethroids are distinguished from type 1 pyrethroids by the presence of an α-cyano group. There are differences in the actions of the type 1 and type 2 pyrethroids. In voltage clamp experiments, the peak Na^+ current is unaffected by pyrethroids (Narahashi 1998a). However, the peak transient is followed by a persistent component of current. Repolarization is followed by a prominent tail current. The time constant of relaxation of the tail current is much larger with the type II pyrethroids. At the single channel level, the pyrethroids prolong the mean open time markedly (Holloway et al. 1989). The monoexponential distribution of open times is converted to a biexponential with a second component with a mean open about ten times normal. The prolongation is even more marked with the type II pyrethroids. The single conductance is unchanged. The prominent tail current on repolarization permits the study of drug interaction with the Na^+ channel in the range of the normal resting potential (Narahashi 1998a; Holloway et al. 1989).

The change in whole-cell and single channel Na currents reflect multiple modifications of gating: (i) the activation-voltage relationship is shifted to more negative potentials; (ii) activation and deactivation are slowed; (iii) the inactivation-voltage relationship is shifted to more negative potentials; (iv) inactivation is slowed. TTX-resistant and invertebrate Na channels are more sensitive to the gating changes induced by the pyrethroids (Song and Narahashi 1996; Narahashi 1998b). The pyrethroids are also more potential at low temperature (Song and Narahashi 1996). This enhances the differential toxicity for insects (body temperature ~ambient of 25 °C) and most vertebrates (body temperature ≥37 °C) (Song and Narahashi 1996; Narahashi 1998b). The Na channel gating modification produced by the pyrethroids is reversed by vitamin E (Song and Narahashi 1995).

DPI 201–106 is a highly lipid soluble diphenyl piperazinyl indole derivative that prolongs the cardiac action potential and increases cardiac contractility (Scholtysik et al. 1985). The drug has an asymmetric center and the APD prolonging and positive inotropic effects reside with the S-enantiomer.

Buggisch et al. showed that these effects of DPI201–106 are blocked by tetrodotoxin (BUGGISCH et al. 1985). KOHLHARDT et al. (1986) showed that racemic DPI markedly slowed inactivation of a fraction of Na^+ channels. The open time of the modified channels is increased approximately tenfold. A fraction of the DPI201–106 modified channels had low and intermediate conductances of 5 and 8 pS compared with the normal conductance of 15 pS (NILIUS et al. 1989). The inactivation deficient channels resulted in a persistent component of Na^+ current and could account for the APD prolonging effect of DPI201–106. The R-enantiomer of DPI201–106 blocks the Na^+ channel. However, the disparate effects of the two enantiomers are effected from different receptors (ROMEY et al. 1987). DPI201–106 had promise as a positive inotropic agent. However, this may be limited by its arrhythmogenic potential.

H. Conclusions

The Na-channel blockers occupy an important place in the treatment of cardiac arrhythmias, seizures, and local anesthesia. Voltage clamp and site-directed mutagenesis have provided insight into their mechanism and site of action. Future studies should define the molecular organization of the blocker binding site(s) and the forces that control drug-channel interaction. Certain Na activators such as the pyrethroids are being reexamined as insecticides. Voltage clamp experiments have helped to clarify the basis of their low mammalian toxicity. Na channel activation as a strategy for increasing the force of contraction of the failing heart remain an area of active study.

Acknowledgement. This work was supported by grant HL32708 from the National Institutes of Health.

References

Alpert LA, Fozzard HA, Hanck DA, Makielski JC (1989) Is there a second external lidocaine binding site on mammalian cardiac cells? Am J Physiol 257:H79–H84

Anno T, Hondeghem LM (1990) Interaction of flecainide with guinea pig cardiac sodium channels. Importance of activation unblocking to the voltage dependence of recovery Circ Res 66:789–803

Armstrong CM, Bezanilla F, Rogas E (1973) Destruction of sodium conductance inactivation in squid axons perfused with pronase. J Gen Physiol 62:375–391

Backx PH, Yue DT, Lawrence JH, Marban E, Tomaselli GF (1992) Molecular localization of an ion-binding site within the pore of mammalian sodium channels. Science 257:248–251

Baker PF, Robinson KA (1975) Chemical modification of crab nerves can make them insensitive to the local anaesthetics tetrodotoxin and saxitoxin. Nature 257:412–414

Balser JR, Nuss HB, Orias DW, Johns DC, Marban E, Thomaselli GF, Lawrence JH (1996) Local anesthetics as effectors of allosteric gating. J Clin Invest 98:2874–2886

Barber MJ, Wendt DJ, Starmer CF, Grant AO (1992) Blockade of cardiac sodium channels. Competition between the permeant ion and antiarrhythmic drugs. J Clin Invest 90:368–381

Barnes S, Hille B (1988) Veratridine modifies open sodium channels. J Gen Physiol 91:421–443

Baruscotti M, Westenbroek R, Catterall WA, Difrancesco D, Robinson WA (1997) The newborn rabbit sino-atrial node expresses a neuronal type I-like Na+ channel. J Physiol 4983:641–648

Bellet S, Hamdan G, Somlyo A, Lara R (1959a) A reversal of cardiotoxic effects of procainamide. Am J Med Sci 237:177–189

Bellet S, Hamdan G, Somlyo A, Lara R (1959b) The reversal of cardiotoxic effects of quinidine by molar sodium lactate: an experimental study. Am J Med Sci 237:165–176

Benndorf K (1994) Properties of single cardiac Na channels at 35 °C. J Gen Physiol 104:801–820

Bennett PB, Woosley RL, Hondeghem LM (1988) Competition between lidocaine and one of its metabolites, glycylxylidide for cardiac sodium channels. Circulation 78:692–700

Bennett PB, Valenzuela C, Chen L-Q, Kallen RG (1995) On the molecular nature of the lidocaine receptor of cardiac Na+ channels. Circ Res 77:584–592

Benz I, Kohlhardt M (1991) Responsiveness of cardiac Na^+ channels to antiarrhythmic drugs: the role of inactivation. J Membrane Biol 122:267–278

Benz I, Kohlhardt M (1992) Differential response of DPI-modified cardiac Na^+ channels to antiarrhythmic drugs: no flicker blockade by lidocaine. J Membrane Biol 126:257–263

Buggisch D, Isenberg G, Ravens U, Scholtysik G (1985) The role of sodium channels in the effects of the cardiotonic compound DPI 201–106 on contractility and membrane potentials in isolated mammalian heart preparations. Europ J Pharm 118:303–311

Cahalan M (1980) Molecular properties of sodium channels in excitable membranes. In: Cotman CW, Poste G, Nicolson GL (eds) The cell surface and neuronal function. Elsevier/North-Holland Biomedical Press, pp 1–47

Cahalan MD, Almers W (1979) Interaction between quaternary lidocaine, the sodium channel gates and tetrodotoxin. Biophys J 27:39–56

Campbell TJ (1983) Kinetics of onset of rate-dependent effects of class 1 antiarrhythmic drugs are important in determining their effects on refractoriness in guinea pig ventricle, and provide a theoretical basis for their subclassification. Cardiovasc Res 17:344–352

Campbell TJ, Vaughan Williams EM (1983) Voltage- and time-dependent depression of maximum rate of depolarization of guinea-pig ventricular action potential by two new antiarrhythmic drugs, flecainide and lorcainide. Cardiovasc Res 17: 251–258

Carmeliet E, Nilius B, Vereecke J (1989) Properties of the block of single Na^+ channels in guinea-pig ventricular myocytes by the local anesthetic penticainide. J Physiol 409:241–262

Catterall WA, Coppersmith J (1981) Pharmacological properties of sodium channels in cultured rat heart cells. Mol Pharmacol 20:533–542

Catterall WA (1992) Cellular and molecular biology of voltage-gated sodium channels. Physiol Rev 72:515–548

Cohen CJ, Bean BP, Colatsky TJ, Tsien RW (1981) Tetrodotoxin block of sodium channels in rabbit purkinje fibres: interaction between toxin binding and channel gating. J Gen Physiol 78:383–411

Cohen SA, Barchi RL (1993) Voltage-dependent sodium channels. Internat Rev Cytol 137C:55–103

Coronado R, Miller C (1979) Voltage-dependent caesium blockade of a cation channel from fragmented sarcoplasmic reticulum. Nature 280:807–810

Cruz LJ, Gray WR, Olivera BM, Zeikus RD, Kerr L, Yoshikami D, Moczydlowski E (1985) Conus geographus toxins the discriminate between neuronal and muscle sodium channels. J Biol Chem 260:9280–9288

Dudley SC Jr, Todt H, Lipkind G, Fozzard HA (1995) A *m*-conotoxin-insensitive Na^+ channel mutant: possible localization of a binding site at the outer vestibule. Biophysical J 69:1657–1665

El-Sherif N, Fozzard HA, Hanck DA (1992) Dose-dependent modulation of the cardiac sodium channel by sea anemone toxin ATX11. Circ Res 70:285–301

Fozzard H, Hanck DA (1996) Structure and function of voltage-dependent sodium channels: comparison of brain II and cardiac isoforms. Physiolog Rev 76:887–926

Gingrich KJ, Beardsley D, Yue DT (1993) Ultra-deep blockade of Na^+ channels by a quaternary ammonium ion: catalysis by a transition-intermediate state? J Physiol 471:319–341

Grant AO, Strauss LJ, Wallace AG, Strauss HC (1982) The influence of pH on the electrophysiological effects of lidocaine in guinea pig ventricular myocardium. Circ Res 47:542–550

Grant AO, Starmer CF, Strauss HC (1984) Antiarrhythmic drug action. Blockade of the inward sodium current. Circ Res 55:427–439

Grant AO, Dietz MA, Gilliam FR III, Starmer CF (1989) Blockade of cardiac sodium channels by lidocaine: single channel analysis. Circ Res 65:1247–1262

Grant AO, Wendt DJ, Zilberter Y, Starmer CF (1993) Kinetics of interaction of disopyramide with the cardiac sodium channel: fast dissociation from open channels at normal rest potentials. J Membrane Biol 136:199–214

Grant AO, John JE, Nesterenko VV, Starmer CF (1996) The role of inactivation in open-channel block of the sodium channel: studies with inactivation-deficient mutant channels. Molec Pharmacol 50:1643–1650

Heinemann SH, Terlau H, Stuhmer W, Imoto K, Numa S (1992) Calcium channel characteristics conferred on the sodium channel by single mutations. Nature 356:441–443

Henderson R, Wang JH (1972) Solubilization of a specific tetrodotoxin-binding component from garfish olfactory nerve membrane. Biochemistry 11:4565–4569

Henderson R, Ritchie JM, Strichartz GR (1973) The binding of labelled saxitoxin to the sodium channels in nerve membranes. J Physiol 235:783–804

Henderson R, Ritchie JM, Strichartz GR (1974) Evidence that tetrodotoxin and saxitoxin act at a metal cation binding site in the sodium channels of nerve membrane. Proc Natl Acad Sci 71:3936–3940

Hille B (1977a) The pH-dependent rate of action of local anesthetics on the node of Ranvier. J Gen Physiol 69:475–496

Hille B (1977b) Local anesthetics: hydrophilic and hydrophobic pathways for the drug-receptor reaction. J Gen Physiol 69:497–515

Hodgkin AL, Huxley AF (1952) A quantitative description of membrane current and its application to conduction and excitation in nerve. J Physiol 117:500–544

Holloway SF, Salgado VL, Wu CH, Narahashi T (1989) Kinetic properties of single sodium channels modified by fenvalerate in mouse neuroblastoma cells. Pflugers Arch 414:613–621

Hondeghem LM, Katzung BG (1977) Time- and voltage-dependent interactions of antiarrhythmic drugs with cardiac sodium channels. Biochim et Biophys Acta 472:373–398

Horn R, Patlak J, Stevens CF (1981) The effect of tetramethylammonium on single sodium channel currents. Biophys J 36:321–327

Isom LL, De Jongh KS, Patton DE, Reber BFX, Oxford J, Charbonneau H, Walsh K, Goldin, AL, Catterall WA (1992) Primary structure and functional expression of the b_1 subunit of the rat brain sodium channel. Science 256:839–842

Khodorov BI (1985) Batrachotoxin as a tool to study voltage-sensitive sodium channels of excitable membranes. Prog Biophys Molec Biol 45:57–68

Kirsch GE, Skattebol A, Possani LD, Brown AM (1989) Modification of Na channel gating by an a scorpion toxin from *Tityus serrulatus*. J Gen Physiol 93:67–83
Kobayashi M, Wu CH, Yoshii M, Narahashi T, Nakamura H, Kobayashi J, Ohizumi Y (1986) Preferential block of skeletal muscle sodium channels by geographutoxin II, a new peptide toxin from *Conus geographus*. Pflugers Arch 407:241–243
Kohlhardt M, Froebe U, Herzig JW (1986) Modification of single cardiac Na^+ channels by DPI201–106. J Membrane Biol 89:163–172
Kohlhardt M, Fichtner H (1988) Block of single cardiac Na^+ channels by antiarrhythmic drugs: the effects of amiodarone, propafenone and diprafenone. J Membrane Biol 102:105–119
Kohlhardt M, Fichtner H, Froebe U, Herzig JW (1989) On the mechanism of drug-induced blockade of Na^+ current: interaction of antiarrhythmic compounds with DPI-modified single cardiac Na^+ channels. Circ Res 64:867–881
Koumi S, Sato R, Katori R, Hisatome I, Nagasawa K, Hayakawa H (1992) Sodium channel states control binding and unbinding behaviour of antiarrhythmic drugs in cardiac myocytes from the guinea pigCardiovasc Res 26:1199–1205
Kuo C-C (1994) Bean BP Na^+ channels must deactivate to recover from inactivation. Neuron 12:819–829
Lipkind GM, Fozzard HA (1994) A structural model of the tetrodotoxin and saxitoxin binding site of the Na^+ channel. Biophys J 66:1–13
Liu L, Wendt DJ, Grant AO (1994) Relationship between structure and sodium channel blockade by lidocaine and its amino-alkyl derivatives. J Cardiovasc Pharmacol 24:803–812
Makita N, Bennett PB Jr, George AL Jr (1994) Voltage-gated Na^+ channel b1 subunit mRNA expressed in adult human skeletal muscle, heart and brain is encoded by a single gene. J Biolog Chem 269:7571–7578
Matsuki N, Quandt FN, Ten Eick RE, Yeh JZ (1984) Characterization of the block of sodium channels by phenytoin in mouse neuroblastoma cells. J Pharmacol Exp Ther 228:523–530
McDonald TV, Courtney KR, Clusin WT (1989) Use-dependent block of single channels by lidocaine in guinea pig ventricular myocytes. Biophys J 55:1261–1266
McPhee JC, Ragsdale DS, Scheuer T, Catteral WA (1994) A mutation in segment IVS^ disrupts fast inactivation of sodium channels. Proc Natl Acad Sci 91:12346–12350
Moczydlowski E, Garber SS, Miller C (1984a) Batrachotoxin-activated Na^+ channels in planar lipid bilayers. J Gen Physiol 84:665–686
Moczydlowski E, Hall S, Garber SS, Strichartz GS, Miller C (1984b) Voltage-dependent blockade of muscle Na+ channels by guanidinium toxins. J Gen Physiol 84:687–704
Moczydlowski E, Olivera BM, Gray WR, Strichartz GR (1986) Discrimination of muscle and neuronal Na-channel subtypes by binding competition between [^{3}H]saxitoxin and *m*-conotoxins. Proc Natl Acad Sci 83:5321–5325
Narahashi T, Moore JW, Scott WR (1964) Tetrodotoxin blockage of sodium conductance increase in lobster giant axons. J Gen Physiol 47:965–974
Narahashi T (1998a) Toxins that modulate the sodium channel gating mechanism. Ann NY Acad Sci 479:133–151
Narahashi T (1998b) Chemical modulation of sodium channels. In: Soria B, Cea V (eds) Ion channel pharmacology. Oxford University Press pp 23–73
Nilius B, Benndorf K, Markwardt F (1987) Effects of lidocaine on single cardiac sodium channels. J Mol Cell Cardiol 19:865–874
Nilius B, Vereecke J, Carmeliet E (1989) Different conductance states of the bursting Na channel in guinea-pig ventricular myocytes. Pflugers Arch 413:242–248
Noda M, Suzuki S, Numa S, Stuhmer WA (1989) A single point mutation confers tetrodotoxin and saxitoxin insensitivity on the sodium channel II. FEBS Lett 259:213–216
Nuss HB, Chiamvimonvat N, Perez-Garcia MT, Tomaselli GF, Marban E (1995) Functional association of the b1 subunit with human cardiac (hH1) and rat skeletal muscle (ml) sodium channel a subunits expressed in *Xenopus oocytes*. J Gen Physiol 106:1171–1191

Patton DE, West JW, Catterall WA, Goldin AL (1992) Amino acid residues required for fast Na+-channel inactivation: charge neutralizations and deletions in the III-IV linker. Proc Natl Acad Sci 89:10905–10909
Pentel P, Benowitz N (1984) Efficacy and mechanism of action of sodium bicarbonate in the treatment of desipramine toxicity in rats. J Pharmacol Exp Ther 230:12–19
Quandt FN, Narahashi T (1982) Modification of single Na^+ channels by batrachotoxin. Proc Natl Acad Sci 79:6732–6736
Ragsdale DS, McPhee JC, Scheuer T, Catteral WA (1994) Molecular determinants of state-dependent block of Na^+ channels by local anesthetics. Science 265:1724–1728
Ragsdale DS, Mephee JC, Scheuer T, Catterall WA (1996) Common molecular determinants of local anesthetic, antiarrhythmic and anticonvulsant block of voltage-gated Na^+ channels. Proc Natl Acad Sci USA 93:9270–9275
Romey G, Quast U, Pauron D, Frelin C, Renaud JF, Lazdunski M (1987) Na^+ channels as sites of action of the cardioactive agent DPI 201–106 with agonist and antagonist enantiomers. Proc Natl Acad Sci 84:896–900
Satin J, Kyle JW, Chen M, Bell P, Cribbs LL, Fozzard HA, Rogart R B (1992) A mutant of TTX-resistant cardiac sodium channels with TTX-sensitive properties. Science 256:1202–1205
Sato K, Ishida Y, Wakamatsu K, Kato R, Honda H, Ohizumi Y, Nakamura H, Ohya M, Lancelin J-M, Kohda D, Inagaki F (1991) Active site of *m*-Conotoxin GIIIA, a peptide blocker of muscle sodium channels. J Biol Chem 266:16,989–16,991
Schild L, Moczydlowski E (1991) Competitive binding interaction between Zn^{2+} and saxitoxin in cardiac Na^+ channels. Biophys J 59:523–5370
Scholtysik G, Saltzmann R, Berthold R, Herzig JW, Quast U, Markstein R (1985) DPI 201–106, a novel cardioactive agent. Combination of cAMP-independent positive inotropic, negative chronotropic, action potential prolonging and coronary dilatory properties. Naunyn-Schmied Arch Pharmacol 985:329–325
Sheldon RS, Cannon NJ, Duff HJ (1987) A receptor for type 1 antiarrhythmic drugs associated with rat cardiac sodium channels. Circ Res 61:492–497
Sheldon RS, Hill RJ, Taouis M, Wilson LM (1991) Aminoalkyl structural requirements for interaction of lidocaine with the class I antiarrhythmic drug receptor on rat cardiac myocytes. Molec Pharmacol 39:609–614
Shrager P, Profera C (1973) Inhibition of the receptor for tetrodotoxin in nerve membranes by reagents modifying carboxyl groups. Biochimica et Biophysica Acta 318:141–146
Sigworth FJ, Spaulding BC (1980) Chemical modification reduces the conductance of sodium channels in nerve. Nature 283:293–295
Sivilotti L, Okuse K, Akopian AN, Moss S, Wood JN (1997) A single serine residue confers tetrodotoxin insensitivity on the rat sensory-neuron-specific sodium channel SNS. FEBS Lett 409:49–52
Smith PL, Baukrowitz T, Yellen G (1996) The inward rectification mechanism of the HERG cardiac potassium channel. Nature 379:833–836
Song J-H, Narahashi T (1995) Selective block of tetramethrim-modified sodium channels by (±)-α-tocopherol (vitamin E). J Pharm Exp Ther 275:1402–1411
Song J-H, Narahashi T (1996) Modulation of sodium channels of rat cerebellar purkinje neurons by the pyrethroid tetramethrin. J Pharm Exp Ther 277:445–453
Spalding BC (1980) Properties of toxin-resistant sodium channels produced by chemical modification in frog skeletal muscle. J Gen Physiol 305:485–500
Starmer CF, Grant AO, Strauss HC (1984) Mechanisms of use-dependent block of sodium channels in excitable membranes by local anesthetics. Biophys J 46:15–27
Starmer CF, Grant AO (1985) Phasic ion channel blockade: a kinetic and parameter estimation procedure. Mol Pharmacol 28:348–356
Starmer CF, Nesterenko VV, Gilliam FR, Grant AO (1990) Use of ionic currents to identify and estimate parameters in models of channel blockade. Am J Physiol 259:H626-H634
Starmer CF, Nesterenko VV, Undrovinas AI, Grant AO, Rosenshtraukh LV (1991) J Mol Cell Cardiol 23 Suppl 1:73–83

Stephan MM, Potts JF, Agnew WS (1994) The mI skeletal muscle sodium channel: mutation E403Q eliminates sensitivity to tetrodotoxin but not to *m*-conotoxins GIIIA and GIIIB. J Membrane Biol 137:1–8
Strichartz GR (1973) The inhibition of sodium currents in myelinated nerve by quaternary derivatives of lidocaine. J Gen Physiol 62:37–57
Stuehmer W, Conti F, Suzuki H, Wang X, Noda M, Yahagi N, Kubo H, Numa S (1993) Structural parts involved in activation and inactivation of the sodium channel. Nature 339:597–603
Sunami A, Dudley SC Jr, Fozzard HA (1997) Sodium channel selectivity filter regulates antiarrhythmic drug binding. Proc Natl Acad Sci 94:14,126–14,131
Sutro JB (1986) Kinetics of veratridine action on Na channels of skeletal muscle. J Gen Physiol 87:1–24
Terlau H, Heinemann SH, Stuehmer W, Pusch M, Conti F, Imato K, Numa S (1991) Mapping the site of block by tetrodotoxin and saxitoxin of sodium channel II. FEBS Lett 293:93–96
Thomsen WJ, Catterall WA (1989) Localization of the receptor site for a-scorpion toxins by antibody mapping: Implications for sodium channel topology. Proc Natl Acad Sci 86:10161–10165
Valenzuela C, Snyders DJ, Bennett PB, Tamargo J, Hondeghem LM (1995) Stereoselective block of cardiac sodium channels by bupivaciaine in guinea pig ventricular myocytes. Circulation 92:3014–3024
Vassilev P, Scheuer T, Catterall WA (1989) Inhibition of single sodium channels by a site-directed antibody. Proc Natl Acad Sci USA 86:8147–8151
Von Dach B, Streuli RA (1988) Lidocainbehandlung einer Vergiftung mit eibennadeln (Taxus baccata I) Schweiz Med Wochenschript 118:1113–1116
Wang GK, Brodwick MS, Eaton DC (1987) Inhibition of sodium currents by local anesthetics in Chloramine-T-treated squid axons. J Gen Physiol 89:645–667
Wang Q, Shen J, Li Z, Timothy K, Vincent GM, Priori SG, Schwartz PJ, Keating MT (1995a) Cardiac sodium channel mutations in patients with long QT syndrome, an inherited cardiac arrhythmia. Human Molec Genetics 4:1603–1607
Wang Q, Shen J, Splawski I, Atkinson D, Li A, Robinson JL, Moss AJ, Towbin JA, Keating MT (1995b) SCN5A mutations associated with an inherited cardiac arrhythmia, long QT syndrome. Cell 80:805–811
Wasserstrom JA, Liberty K, Kelly J, Santucci P, Myers M (1993) Modification of cardiac Na^+ channels by batrachotoxin: effects on gating, kinetics and local anesthetic binding. Biophys J 65:386–395
Weiss RE, Horn R (1986) Functional differences between two classes of sodium channels in developing rat skeletal muscle. Science 233:361–364
West JW, Patton DE, Scheuer T, Wang Y, Goldin AL, Catterall WA (1992) A cluster of hydrophobic aminoacid residues required for fast Na^+-channel inactivation. Proc Natl Acad Sci 89:10910–10914
Whitcomb DC, Gilliam FR III, Starmer CF, Grant AO (1989) Marked QRS complex abnormalities and sodium channel blockade by propoxyphene reversed with lidocaine. J Clin Invest 84:L1629–1643
White MM, Chen L, Kleinfield R, Kallen RG, Barchi RL (1991) SkM2, a Na^+ channel cDNA clone from denervated skeletal muscle, encodes a tetrodotoxin-sensitive Na^+ channel. Molec Pharmacol 39:604–608
Wright SN, Wang S-Y, Kallen RG, Wang GK (1997) Differences in steady-state inactivation between channel isoforms affect local anesthetic binding affinity. Biophysical J 73:779–788
Wynn J, Fingerhood M, Keefe D, Maza S, Miura D, Somberg JC (1986) Refractory ventricular tachycardia with flecainide. Am Heart J 112:174–175
Yamamoto D (1986) Dynamics of strychnine block of single sodium channels in bovine chromaffin cells. J Physiol 370:395–407
Zaborovskaya LD, Khodorov BI (1984) The role of inactivation in the cumulative blockage of voltage-dependent sodium channels by local anesthetics and antiarrhythmics. Gen Physiol Biophys 3:517–520

Zamponi GW, Doyle DD, French RJ (1993a) Fast lidocaine block of cardiac and skeletal muscle sodium channels: one site with two routes of access. Biophys J 65:80–90
Zamponi GW, Doyle DD, French RJ (1993b) State-dependent block underlies the tissue specificity of lidocaine action on batrachotoxin-activated cardiac sodium channels. Biophys J 65:91–100

B. Voltage-Dependent Ca-Channels

CHAPTER 3

Classification and Function of Voltage-Gated Calcium Channels

J.B. BERGSMAN, D.B. WHEELER, and R.W. TSIEN

A. Generic Properties of Voltage-Gated Ca^{2+} Channels

Voltage-gated Ca^{2+} channels are members of a superfamily of voltage-gated ion channels which also includes Na^+ channels and K^+ channels. Ca^{2+} channels transduce membrane potential changes to intracellular Ca^{2+}-signals in a wide variety of cell types, including nerve, endocrine, and muscle cells. Many types of Ca^{2+} channels have been characterized by pharmacological and biophysical criteria in various cell types. More recently, molecular cloning has revealed a wealth of genes encoding the subunits of native channels. Following a brief introduction to the basic properties and subunit composition of Ca^{2+} channels, we will proceed to an overview of their classification, molecular composition, and specialization for various functional roles. Details about structure-function appear in another chapter in this volume 147 (Chap. 4) and we will confine our structural comments here to those that pertain to classification of the channels. Likewise modulation of Ca^{2+} channels is left to other authors. In addition we will not touch on several other important aspects of regulation of $[Ca^{2+}]_i$, such as Ca^{2+} channels not gated by depolarization (PUTNEY 1997), Ca^{2+} sequestration and extrusion, and neuropathological conditions such as stroke, epilepsy, and migraine, some involving mutations of the Ca^{2+} channels themselves.

I. Basic Functional Properties

Our present-day understanding of Ca^{2+} channels began with their electrophysiological isolation and description. *Gating* describes the opening and closing of channels. Typically, Ca^{2+} channels open (or *activate*) within one or a few milliseconds after the membrane is depolarized from rest, and close (*deactivate*) within a fraction of a millisecond following repolarization. Activation of Ca^{2+} channels is steeply voltage-dependent: channels open more quickly and with higher likelihood with larger depolarizations. *Inactivation*, the closing of channels during maintained or repeated depolarizations, strongly influences the cytosolic Ca^{2+} signal that arises from cellular electrical activity. While inactivation is a general property of Ca^{2+} channels, the speed of entry into and recovery from inactivation varies widely.

In addition to gating we consider two properties concerning the conduction of Ca^{2+} through the channel. *Selectivity* of voltage-gated Ca^{2+} channels for

Ca^{2+} ions is remarkably high, so that Ca^{2+} is the main charge carrier even when Ca^{2+} is greatly outnumbered by other ions, as under normal physiological conditions. *Permeation* of Ca^{2+} through a single open Ca^{2+} channel can achieve rates of millions of ions per second when the electrochemical gradient is large. At driving forces reached physiologically, the flux rate is more modest, but sufficient to cause a large increase in $[Ca^{2+}]_i$ (>1 μmol/l) in a very localized domain (~1 μmol/l) near the mouth of the open channel.

II. Subunit Composition

The powerful functional capabilities of Ca^{2+} channels are rooted in their molecular architecture. Voltage-gated Ca^{2+} channels contain at their core a protein known as α_1, which is a large (200–260 kDa) transmembrane protein that contains the channel pore, the voltage-sensor, and the gating machinery. Most, or possibly all, channel types additionally contain subunits known as β, α_2, δ, and γ (Fig. 1), that come together with the α_1 subunit to form a large macromolecular complex. The first examples of each of these subunits were originally isolated from skeletal muscle transverse tubules by biochemical techniques more than a decade ago (CATTERALL and CURTIS 1987; CAMPBELL et al. 1988;

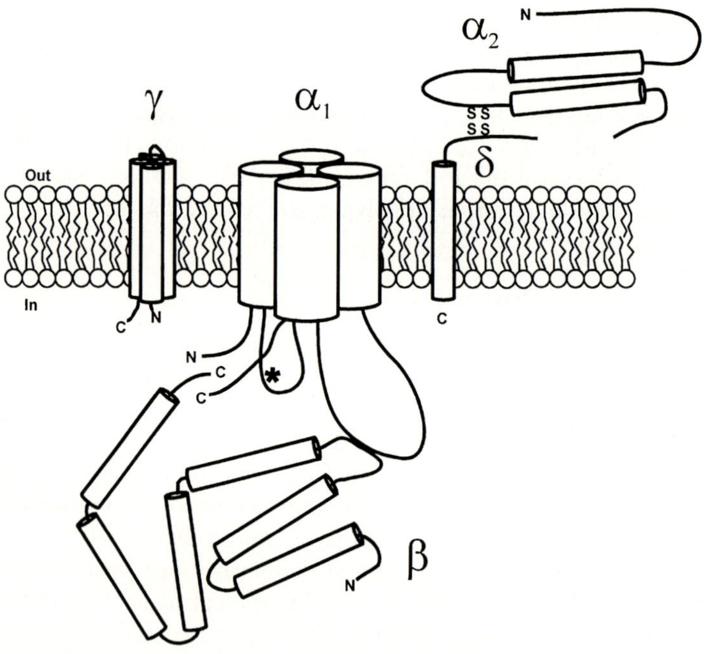

Fig. 1. Structural organization of the subunits comprising a generic voltage-gated Ca^{2+} channel. *Small cylinders* represent α helices, *large cylinders* in the α_1 subunit represent 6 α helices. *Asterisk* marks the II–III loop of the α_1 subunit

Catterall et al. 1988; Glossmann and Striessnig 1990). Each subunit has since been cloned in several forms.

Because the α_1 subunit appears to be able to form a functional Ca^{2+} channel on its own, the other subunits are sometimes referred to as auxiliary or ancillary subunits although they may dramatically affect channel gating, modulation, pharmacology, and expression. In the last few years our understanding of the relationship between the α_1 subunits and the native channel classes has become increasingly clear. While the α_1 subunit is the major determinant of channel properties, the high level of promiscuity in the association of the α_1 subunit with the various forms of the auxiliary subunits, combined with alternative splicing, can likely produce an incredible diversity of properties.

1. α_1

Much of the diversity of Ca^{2+} channel types seems to arise from the expression of multiple forms of the α_1 subunit, isolated by molecular cloning (e.g., Tanabe et al. 1987; Mikami et al. 1989; Mori et al. 1991; Starr et al. 1991; Dubel et al. 1992; Williams et al. 1992a; Williams et al. 1992b; Soong et al. 1993; Fisher et al. 1997; Cribbs et al. 1998; Perez-Reyes et al. 1998; Lee et al. 1999). Details of the various α_1 subunits will be examined thoroughly below.

2. β

All high voltage activated Ca^{2+} channels (see below) in their native state appear to contain β subunits – peripheral membrane proteins associated with the cytoplasmic aspect of the surface membrane with an apparent molecular weight of ~55–60 kDa (Glossmann et al. 1987; Takahashi et al. 1987). The β subunit of Ca^{2+} channels is not homologous to the $\beta1$ and $\beta2$ subunits of Na^{+} channels, which contain putative transmembrane spanning domains and are significantly glycosylated (Isom et al. 1994). β subunits serve several important and intriguing functions:

1. They play a key role in the proper targeting of the complex of Ca^{2+} channel subunits.
2. They are subject to regulation by protein kinases.
3. They act as modulators of the gating and pharmacological properties of α_1 subunits.

In the present work we concern ourselves only with the last function. For more information on the other functions see recent reviews (Hofmann et al. 1994; Isom et al. 1994; De Waard et al. 1996; Walker and De Waard 1998).

Four different types of β subunit are known to exist in mammals and are now known as β_1–β_4 (Birnbaumer et al. 1994). Diversity of these proteins is increased by alternative splicing (designated by lower case letters, β_{2a}, β_{2b}, etc.). In general, β subunits are not found in one organ or tissue exclusively. Whereas β_1 transcripts are expressed primarily in skeletal muscle, they also appear in brain. β_2 is predominantly expressed in heart, aorta, and brain, while β_3 is most abundant in brain but also present in aorta, trachea, lung, heart, and skeletal muscle. β_4 mRNA is expressed almost exclusively in neuronal tissues, with the highest levels being found in the cerebellum. Because each of the β subunits appears able to partner with each of the α_1 subunits, β subunit heterogeneity may contribute to the diversity of Ca^{2+} channels in a multiplicative manner; however it seems unlikely that β subunit differences are responsible for the differences between the major classes delineated below (L-, N-, P/Q-type, etc.).

3. α_2/δ

The $\alpha_2\delta$ subunit (175 kDa) is a dimer, consisting of glycosylated α_2 and δ proteins linked together by disulfide bonds, derived by posttranslational processing of a single parent polypeptide (ELLIS et al. 1988; DE JONGH et al. 1990; WILLIAMS et al. 1992b; KLUGBAUER et al. 1999). This pair of subunits has been shown to affect channel gating. The δ subunit is a transmembrane protein anchor and α_2 is entirely extracellular (JAY et al. 1991; HOFMANN et al. 1994). Three α_2/δ genes have been isolated: α_2/δ–1 and α_2/δ–2 have wide tissue distribution while α_2/δ–3 is brain specific (ANGELOTTI and HOFMANN 1996; KLUGBAUER et al. 1999). As with other Ca^{2+} channel subunits, α_2/δ diversity is increased by alternative splicing. The diversity of the α_2/δ genes has only recently begun to be characterized, and less is known about this subunit's effect on channel properties than that of the β subunit.

4. γ

A fifth subunit, known as γ (25–38 kDa) (BOSSE et al. 1990; JAY et al. 1990; EBERST et al. 1997; LETTS et al. 1998; BLACK and LENNON 1999), has four transmembrane domains. Like the α_2/δ subunits, the γ subunit is now starting to receive widespread attention and little is known about its effect on channel properties, although it has been shown to promote inactivation (EBERST et al. 1997; LETTS et al. 1998).

B. Classification of Native Ca^{2+} Channels According to Biophysical, Pharmacological, and Molecular Biological Properties

Multiple types of voltage-gated Ca^{2+} channels were first distinguished by voltage- and time-dependence of channel gating, single channel conductance and pharmacology (e.g., CARBONE and LUX 1984; NOWYCKY et al. 1985). One physiologically relevant characteristic which varies considerably among the different Ca^{2+} channel types is the degree of depolarization required to cause significant opening. Based on this criterion, voltage-gated Ca^{2+} channels are sometimes divided into two groups, low voltage-activated (LVA) and high voltage-activated (HVA). Use of all the criteria listed above has led to a more specific classification of native Ca^{2+} channels as T-, L-, N-, P/Q-, and R-type (TSIEN et al. 1987; LLINÁS et al. 1992; RANDALL and TSIEN 1995).

While this classification makes good sense in view of the varied biophysical properties and functional roles of the channel types in different organ systems, the relationship of these classes to the various cloned subunits has only recently been clarified. The recent findings from molecular cloning of Ca^{2+} channel subunits have greatly increased our understanding of Ca^{2+} channel diversity. This has allowed new perspective on the familial relationships between various channel types and a more precise characterization of the pharmacological properties of individual channel types.

I. Molecular Biological Nomenclature

Nine different Ca^{2+} channel α_1 subunit genes have been distinguished in mammalian brain and one in skeletal muscle and have been labeled classes A through I and S (SNUTCH et al. 1990; SNUTCH and REINER 1992; BIRNBAUMER et al. 1994). α_{1S} refers to the original Ca^{2+} channel clone from skeletal muscle, first isolated by the group of the late Shosaku Numa (TANABE et al. 1987) and the letters A–I refer to subsequently cloned channels. Based on sequence homology, the ten α_1 subunits can be assigned to various branches of a family tree as reviewed in Fig. 2. This sequence homology seems to follow channel properties and functional roles quite well. Following our newfound structural and functional understanding of the Ca^{2+} channels a new naming scheme similar to that used for voltage-gated K^+ channels has been proposed (W.A. Catterall et al., personal communication). In the following discussion we will adopt this scheme in which *v*oltage-gated *Ca*$^{2+}$ channels are designated Ca_V S.Tx, where *S* and *T* are numbers which refer to the subfamily and type respectively, and *x* is a letter which corresponds to any splice variants. The α_1 subunits are named correspondingly as α_1S.Tx. The numbers and letters are assigned in order of discovery, thus α_1S becomes $\alpha_1$1.1 and so on.

II. Ca_V1/L-Type Ca^{2+} Channels

L-type channels are generally categorized with the HVA group of channels, along with N-, P/Q-, and R-type channels. However, it is important to note that L-type channels may exhibit LVA properties under certain circumstances (AVERY and JOHNSTON 1996). L-type channels in vertebrate sensory neurons and heart cells were initially labeled as a *l*arge Ba^{2+} conductance contributing to a *l*ong-lasting current, with characteristic sensitivity to DHPs such as nifedipine or Bay K 8444 (BEAN 1985; NILIUS et al. 1985; NOWYCKY et al. 1985). Members of this group were subsequently identified in other excitable cells such as vascular smooth muscle, uterus, and pancreatic β cells. Later, the designation of L-type was extended to refer to all channels with strong sensitivity to DHPs, including those found in skeletal muscle (HOFMANN et al. 1988),

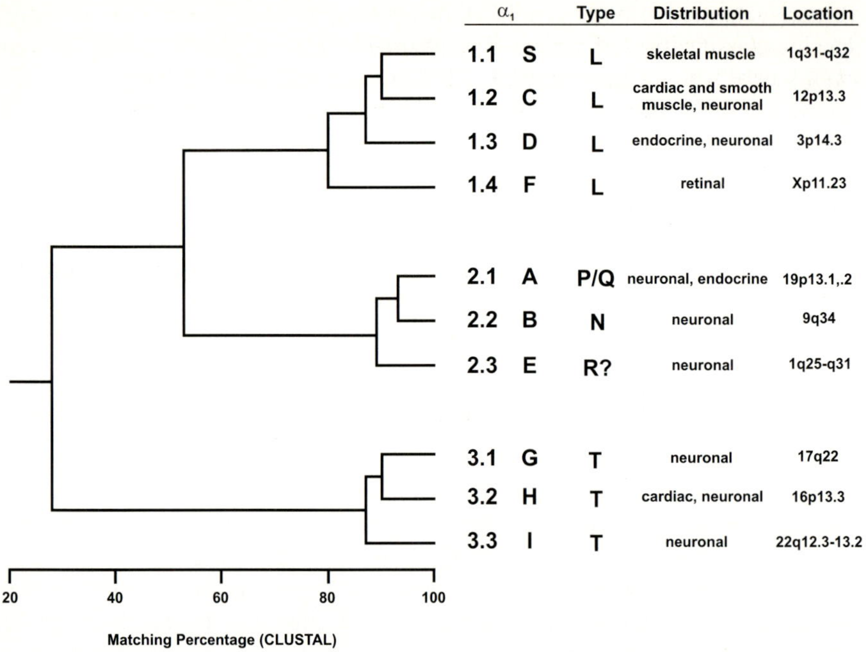

Fig. 2. Ca^{2+} channel α_1 subunit family tree. Sequences of membrane spanning and P loop regions were aligned and matching percentages determined using CLUSTAL. Corresponding current type supported by each α_1 subunit is given, as well as tissue distribution and chromosome location of the human gene. Sequence data provided by Dr. Perez-Reyes, Department of Pharmacology, University of Virginia

even though clear-cut biophysical distinctions between skeletal and cardiac L-type channels were already known (ROSENBERG et al. 1986). Thus, the category of L-type channels contains individual subtypes of considerable diversity. For example, three subtypes of L-type channel appear to co-exist in cerebellar granule neurons, two subtypes that resemble those found in heart and a third that shows prominent voltage-dependent potentiation (FORTI and PIETROBON 1993).

Three major subfamilies of α_1 subunits clearly emerge on the basis of sequence homology. The first subfamily ($\alpha_1$1) consists of four α_1 members. Along with the $\alpha_1$1.1 (α_{1S}) subunit from skeletal muscle, these include subunits first derived from heart muscle [$\alpha_1$1.2 (α_{1C})] (MIKAMI et al. 1989), neuroendocrine tissue [$\alpha_1$1.3 (α_{1D})] (WILLIAMS et al. 1992b), and retina [$\alpha_1$1.4 (α_{1F})] (FISHER et al. 1997; BECH-HANSEN et al. 1998; STROM et al. 1998). These cDNAs encode HVA channels classified as "L-type" because they are responsive to DHPs. The existence of four α_1 subunits, each capable of supporting L-type channel activity, provides an obvious starting point for attempts at understanding how L-type Ca^{2+} channel diversity might be generated from specific molecular structures. However, little information is yet available to link functionally distinct forms of L-type channel activity (e.g., FORTI and PIETROBON 1993; KAVALALI and PLUMMER 1994) to individual α_1 isoforms. While the $\alpha_1$1.1 subunit appears to be largely excluded from neurons according to Northern analysis and electrophysiological criteria, no sharp distinction has been made between currents generated by $\alpha_1$1.2 and $\alpha_1$1.3. Single channel recordings of expressed $\alpha_1$1.3 channels are lacking and analysis of the functional impact of various β subunits on $\alpha_1$1.2 and $\alpha_1$1.3 is not extensive.

Most of the attention to date has been focused on splice variations of $\alpha_1$1.2. These have a marked impact on channel behavior in several cases, producing:

1. Differences in sensitivity to DHPs in $\alpha_1$1.2 variants found in cardiac or smooth muscle (WELLING et al. 1993)
2. Differences in the voltage-dependence of DHP binding (SOLDATOV et al. 1995)
3. Differences in susceptibility to cyclic AMP-dependent phosphorylation (HELL et al. 1993b)

Further analysis will be greatly facilitated by knowledge of the genomic structure of the human $\alpha_1$1.2 gene, which spans an estimated 150kb of the human genome and is composed of 44 invariant and 6 alternative exons (SOLDATOV 1994). The L-type channel in chick hair cells incorporates an $\alpha_1$1.3 subunit that differs from the $\alpha_1$1.3 subunit in brain due to expression of distinct exons at three locations (KOLLMAR et al. 1997). It will be interesting to see if additional splice variations can account for L-type channel activity found at the resting potential of hippocampal neurons, possibly important for setting the resting $[Ca^{2+}]_i$ (AVERY and JOHNSTON 1996).

III. Ca_V2

The second α_1 subfamily consists of cDNAs which, when expressed, result in HVA channels which lack the characteristic DHP-response of L-type channels. These clones [$\alpha_1$2.1 (α_{1A}) (MORI et al. 1991), $\alpha_1$2.2 (α_{1B})(DUBEL et al. 1992), and $\alpha_1$2.3 (α_{1E}) (SOONG et al. 1993)] were derived from nervous tissue. Individual genes within this subfamily show ~89% identity with each other in the membrane spanning and pore forming regions but only ~53% or less with members of the $\alpha_1$1 subfamily.

1. $Ca_V2.2$/N-Type Ca^{2+} Channels

The most extensively characterized non-L-type Ca^{2+} channel was named N-type since it appeared to be largely specific to *n*eurons as opposed to muscle cells and was clearly *n*either T- nor L-type (NOWYCKY et al. 1985). It requires relatively negative resting potentials to be available for opening, somewhat like T-type, but is high voltage-activated, like L-type. This Ca^{2+} channel is potently and specifically blocked by a peptide toxin derived from the venom of the marine snail, *Conus geographus*, ω-conotoxin GVIA (ω-CTx-GVIA). The N-type channel is found primarily in presynaptic nerve terminals and neuronal dendrites in addition to cell bodies (WESTENBROEK et al. 1992). The N-type current can be assigned with a fairly high degree of certainty to $Ca_V2.2$ (α_{1B}), which, when expressed, conducts ω-CTx-GVIA-sensitive currents with characteristics that match those of native N-type channels (DUBEL et al. 1992; WILLIAMS et al. 1992a; FUJITA et al. 1993).

As discussed earlier, an important source of channel heterogeneity is the association of α_1 subunits with different ancillary subunits. A good example of this is provided by the N-type Ca^{2+} channel in brain. Biochemical analysis has shown that the $\alpha_1 2.2$ subunit associates with three different isoforms of β subunit in rabbit brain (SCOTT et al. 1996). Antibodies against individual β subunits were each able to immunoprecipitate ω-CTx-GVIA binding activity (a marker of $Ca_V2.2$), while immunoprecipitation of $\alpha_1 2.2$ showed its association with β_{1b}, β_3 and β_4.

Different isoforms of the N-type Ca^{2+} channel subunit $\alpha_1 2.2$ have been isolated from rat sympathetic ganglia and brain by LIN et al. (1997). Alternative splicing determines the presence or absence of small inserts in the S3–S4 regions of domains III and IV (SFMG and ET respectively). Different combinations of inserts in these putative extracellular loop regions are dominant in central (+SFMG, ΔET) vs peripheral (ΔSFMG, +ET) nervous tissue. Most interestingly, the gating kinetics of ΔET-containing clones (as found in the central form) are significantly faster than the +ET form (LIN et al. 1999). This work provides a clear example of how alternative splicing contributes to diverse functional properties.

2. $Ca_V2.1$/P- and Q-Type Ca^{2+} Channels

Currents carried by P-type channels were originally recorded from cell bodies of cerebellar *P*urkinje cells (LLINÁS et al. 1989, 1992). These channels are not blocked by DHPs or ω-CTx-GVIA, but are exquisitely sensitive to block by ω-Aga-IVA or ω-Aga-IVB, components of the venom of the funnel-web spider, *Agelenopsis aperta* (MINTZ et al. 1992a,b), with an IC50 of <1 nmol/l for ω-Aga-IVA (MINTZ and BEAN 1993). These channels support a current that hardly inactivates during depolarizations lasting for several seconds. They are seen in virtual isolation from other voltage-gated Ca^{2+} channels in cerebellar Purkinje neuron cell bodies, but also contribute substantially to somatic currents in many other central neurons (MINTZ et al. 1992a).

Initial observations of current supported by $\alpha_1 2.1$ (α_{1A}) suggested that it corresponded to the P-type channel (LLINÁS et al. 1992), consistent with the strong expression of this subunit in cerebellar Purkinje cells (MORI et al. 1991; STEA et al. 1994; MINTZ et al. 1992b). Closer comparison of the properties of $Ca_V2.1$ expressed in *Xenopus* oocytes and those of P-type channels in Purkinje cells, however, revealed clear differences. P-type channels activate at relatively negative potentials and support a sustained, non-inactivating current during depolarizing pulses longer than 1 s (LLINÁS et al. 1992; USOWICZ et al. 1992), whereas $\alpha_1 2.1$ subunits expressed in *Xenopus* oocytes activate at less negative potentials and exhibit marked inactivation within 100 ms (SATHER et al. 1993). Furthermore, the IC_{50} for ω-Aga-IVA block of $Ca_V 2.1$ expressed in oocytes (SATHER et al. 1993; STEA et al. 1994) or baby hamster kidney cells (NIIDOME et al. 1994) is 100–200 nmol/l. A current with these properties was characterized in the cell bodies of cerebellar granule neurons and named Q-type (ZHANG et al. 1993; RANDALL and TSIEN 1995) since it differed from the previously defined P-type current (which was also present in the granule neurons).

Subsequently channels of intermediate type have been found in several preparations (TOTTENE et al. 1996; FORSYTHE et al. 1998; MERMELSTEIN et al. 1999), indicating that instead of two discrete channel types, P and Q may represent points on a spectrum of channel properties. Additionally, evidence has been mounting that both channels are encoded by the same α_1 subunit (GILLARD et al. 1997; PIEDRAS-RENTERÍA and TSIEN 1998; PINTO et al. 1998; JUN et al. 1999), and it has been shown that differences in inactivation and toxin affinity, the basis for distinctions between these two types, can be explained in part by splice variants or subunit composition (LIU et al. 1996; BOURINET et al. 1999; MERMELSTEIN et al. 1999). With these facts in mind, the designation P-type or P/Q-type would be appropriate to indicate current through $Ca_V 2.1$ or ω-Aga-IVA/B- or ω-CTx-MVIIC-sensitive current, regardless of inactivation characteristics. P/Q-type channels have a similar distribution to N-type channels.

3. $Ca_V 2.3$/R-Type Ca^{2+} Channels

R-type Ca^{2+} channel currents were identified in cerebellar granule cells as a current that *r*emained in the presence of nimodipine, ω-CTx-GVIA, and ω-Aga-IVA, inhibitors of the L-, N-, and P/Q-type channels respectively (ELLINOR et al. 1993; ZHANG et al. 1993; RANDALL and TSIEN 1995). R-type currents have since been found in several other central nerve terminals (MEDER et al. 1997; NEWCOMB et al. 1998; WU et al. 1998). This predominantly HVA current decays rapidly and is at least partially responsive to low doses of Ni^{2+} and, in some preparations, SNX-482, a toxin derived from tarantula venom (NEWCOMB et al. 1998). Less is known about the molecular basis of R-type currents than for any of the other channel types. Of all the known α_1 subunits, $\alpha_1 2.3$ (α_1E) comes the closest. Expressed $Ca_V 2.3$ currents display certain attributes of R-type channels: they are readily blocked by Ni^{2+} (SOONG et al. 1993; WAKAMORI et al. 1994; WILLIAMS et al. 1994) and the spider toxin ω-Aga-IIIA (RANDALL and TSIEN 1998; ROCK et al. 1998), and display a single channel conductance of ~12–14 pS in 100 mmol/l Ca^{2+}, Ba^{2+}, or Sr^{2+} (SCHNEIDER et al. 1994; WAKAMORI et al. 1994; BOURINET et al. 1996; TOTTENE et al. 1996, 1999). In addition $Ca_V 2.3$ antisense treatment has been shown to reduce native R-type current (PIEDRAS-RENTERÍA and TSIEN 1998). Some studies have found reasons to question assignment of R-type currents to $Ca_V2.3$ (Soong et al. 1993; Bourinet et al. 1996; Tottene et al. 1996; Piedras-Rentería et al. 1997; Meir and Dolphin 1998); however some of these may be explained by diversity in R-type currents caused by splice variants and/or auxiliary subunit differences as seen for P- vs Q-type channels. Support for the possibility of R-type diversity comes from studies that show that SNX-482, a synthetic peptide neurotoxin, blocks R-type currents in some cell types but spares them in others (Newcomb et al. 1998) and differences in Ni^{2+} block and activation voltage in R-type current in the same cell type (Tottene et al. 1996).

IV. Ca_V3/T-Type Ca^{2+} Channels

LVA Ca^{2+} channels are exemplified by T-type channels, so named because they carry *t*iny unitary Ba^{2+} currents (6–8 pS with ~100 mmol/l Ba^{2+} or Ca^{2+} as charge carrier) that occur soon after the depolarizing step, giving rise to a *t*ransient average current (Carbone and Lux 1984; Nilius et al. 1985; Nowycky et al. 1985). Another defining characteristic of classical T-type channels is their slow deactivation following a sudden repolarization (Matteson and Armstrong 1986). T-type channel current records also exhibit a distinctive kinetic fingerprint: the superimposed current responses cross over each other in a pattern not found with other rapidly inactivating Ca^{2+} channels such as R-type (Randall and Tsien 1998). The kinetic properties are dominated by a strikingly voltage-dependent delay between the depolarizing step and the channel's first opening (Droogmans and Nilius 1989). In addition to these properties, T-type channels have a unique pharmacological profile, characterized by only mild sensitivity to 1,4-dihydropyridines (DHPs), such as nifedipine or nimodipine (Cohen and McCarthy 1987), but acute sensitivity to mibefradil (Ertel and Ertel 1997). A newly identified antagonist, kurtoxin, has recently been shown to affect $Ca_V3.1$ (Chuang et al. 1998). Kurtoxin is an α-scorpion toxin which also affects voltage-gated sodium channels and is currently the most specific antagonist with respect to T-type vs other Ca^{2+} channels. Within the overall category of T-type Ca^{2+} channel, further diversity has been found, particularly with respect to kinetic characteristics and pharmacology (Akaike et al. 1989; Kostyuk and Shirokov 1989; Huguenard and Prince 1992). Various subtypes of T-type Ca^{2+} channel may co-exist in the same cell type and show rates of inactivation differing by as much as fivefold, while sharing similar voltage-dependence of inactivation (Huguenard and Prince 1992). T-type channels are found in a wide variety of central and peripheral neurons.

The Ca_V3 subfamily of T-type channels is more distantly related to the two HVA subfamilies Ca_V1 and Ca_V2 than they are to each other (Fig. 2). Three genes in Ca_V3 have recently been identified, $Ca_V3.1$ (α_{1G}), $Ca_V3.2$ (α_{1H}), and $Ca_V3.3$ (α_{1I}) (Cribbs et al. 1998; Perez-Reyes et al. 1998; Lee et al. 1999). These genes encode LVA T-type channels when expressed without auxiliary subunits (Cribbs et al. 1998; Perez-Reyes et al. 1998; Lacinová et al. 1999; Lee et al.

1999). This is consistent with findings that native T-type currents are not dependent on auxiliary subunits (LAMBERT et al. 1997; LEURANGUER et al. 1998); however there is a report that coexpression of $\alpha_2\delta$ can increase expression of native T-type current (WYATT et al. 1998).

V. Note on Pharmacology

Pharmacology is the most widely used criterion when distinguishing various types of calcium currents. It should therefore be noted that antagonists discussed above are not perfectly selective. The P/Q-type blockers ω-Aga-IVA/B and ω-CTx-MVIIC all partially antagonize N-type channels at higher doses (MINTZ and SIDACH 1998; HILLYARD et al. 1992; GRANTHAM et al. 1994) and ω-Aga-IVA has been shown to have some effect on expressed $Ca_V2.3$ channels (SOONG et al. 1993; WILLIAMS et al. 1994). In addition to the lack of complete specificity of these toxins, it should also be noted that there are occasional reports of currents that display pharmacological properties that do not fit any of the above categories. These include currents blocked by both ω-CTx-GVIA and moderate doses of ω-Aga-IVA in rat supraoptic neurons (FISHER and BOURQUE 1995) and chicken forebrain synaptosomes (LUNDY et al. 1994) and a current reversibly blocked by ω-CTx-GVIA (MERMELSTEIN and SURMEIER 1997).

VI. Evolutionary Conservation of Ca^{2+} Channel Families

The evolutionary divergence of Ca_V1 and Ca_V2 Ca^{2+} channels occurred relatively early, as would be expected from the fairly low sequence homology between genes encoding channels from the two subfamilies (Fig. 2). This deduction can be corroborated by an examination of the distribution of Ca^{2+} channel types in organisms spread across many phyla. Both subfamilies of HVA channels are present in vertebrate species ranging from marine rays (HORNE et al. 1993) to humans (WILLIAMS et al. 1992a,b), and in many cases both are expressed within the same cells (e.g., RANDALL and TSIEN 1995). Amongst invertebrates, both channel types have been observed in mollusks (EDMONDS et al. 1990), insects (GRABNER et al. 1994; SMITH et al. 1996), and nematodes (SCHAFER and KENYON 1995). Given the widespread distribution of L- and non-L-type HVA Ca^{2+} channels across the animal kingdom their bifurcation must have occurred quite early during the speciation of Animalia. Presumably LVA and HVA channels diverged even earlier. A possible descendent of an ancestral HVA channel which resembles L-type channels has been cloned from jellyfish (JEZIORSKI et al. 1998). A "T-like" channel has been observed in paramecium (e.g., EHRLICH et al. 1988). LVA and HVA currents have been identified in cockroaches (GROLLEAU and LAPIED 1996) and leech (LU et al. 1997). Whether the various LVA currents are carried by channels with a molecular structure similar to Ca_V3 is not known.

C. Functional Roles of Ca^{2+} Channels

I. Introduction/Subcellular Localization

The diversity of voltage-gated Ca^{2+} channels is indicative of the variety of functional roles they are called upon to serve. With the exception of $\alpha_1$1.1, which appears highly localized to skeletal muscle, α_1 subunits are broadly distributed across the spectrum of exocytotic cells. At the level of individual cells, however, the different channel types often show distinct patterns of localization to different parts of the cell.

Ca^{2+} channels of the Ca_V1 subfamily are widely distributed in muscle, nerve and endocrine cells. Their unique biophysical properties and subcellular localization put them in a good position to act as transducers linking membrane depolarization to intracellular signaling. In the brain, for example, Ca_V1 channels are found in the cell bodies and proximal dendrites of hippocampal pyramidal cells (Westenbroek et al. 1990). $\alpha_1$1.2-containing channels were concentrated in clusters at the base of major dendrites, while Ca_V1.3 channels were more generally distributed across cell surface membrane of cell bodies and proximal dendrites (Hell et al. 1993a).

The Ca_V2 subfamily of Ca^{2+} channels is widely distributed both pre- and postsynaptically in the central and peripheral nervous systems. In most regions of the brain, antibodies against $\alpha_1$2.2 bind primarily on dendrites and nerve terminals (Westenbroek et al. 1992) whereas $\alpha_1$2.1 subunits are concentrated in presynaptic terminals and are present at lower density in the surface membrane of dendrites of most major classes of neurons (Westenbroek et al. 1995). Ca_V2.3 epitopes are found mostly on cell bodies, and in some cases in dendrites, of a broad range of central neurons (Yokoyama et al. 1995). Thus, these classes of Ca^{2+} channels seem to be well positioned to support both presynaptic Ca^{2+} influx that triggers neurotransmitter release and postsynaptic Ca^{2+} entry that helps shape the response downstream to that release.

Little is known about the subcellular distribution of the recently cloned Ca_V3 subfamily of Ca^{2+} channels. The only systematic study so far (Talley et al. 1999) contains no information regarding subcellular distribution of these proteins. In many cell types T-type currents seem to be found primarily in the dendrites as compared to somata (Karst et al. 1993; Markram and Sakmann 1994; Magee et al. 1995; Magee and Johnston 1995; Kavalali et al. 1997; Mouginot et al. 1997; but see Schultz et al. 1999). This is consistent with theories about their functional roles (see below).

II. Excitation–Contraction Coupling

L-type Ca^{2+} channels play a central role in excitation–contraction coupling in skeletal, cardiac, and smooth muscle, although other channel types may play a supporting role in some of these cells (Zhou and January 1998). In skeletal muscle, L-type Ca^{2+} channels contain the $\alpha_1$1.1, β_{1a}, γ_1, and $\alpha_2\delta$-1 subunits and are largely localized to the transverse tubule system. Ca^{2+} entry through the L-type channel is not required for skeletal muscle contraction (reviewed in MILLER and FREEDMAN 1984), in contrast to cardiac muscle, where Ca^{2+} entry is essential for contractility (NÄBAUER et al. 1989). Interestingly, blockade of L-type channels in skeletal muscle by organic Ca^{2+} antagonists completely inhibits contraction (EISENBERG et al. 1983). The explanation of these findings centers on gating charge movement in the T-tubule membrane, which was known to be essential for intracellular Ca^{2+} release (SCHNEIDER and CHANDLER 1973). DHPs eliminate charge movement, thereby blocking skeletal muscle contraction (RÍOS and BRUM 1987). The implication of these findings was that DHP-sensitive L-type Ca^{2+} channels act as voltage sensors to link T-tubule depolarization to intracellular Ca^{2+} release.

This hypothesis was tested in elegant experiments by Tanabe, Numa, Beam, and their colleagues. The cloning of the DHP receptor protein from skeletal muscle led immediately to its identification as a voltage-gated channel (TANABE et al. 1987). Later, expression of the cloned DHP receptor in dysgenic skeletal muscle myotubes showed that it could restore electrically evoked contractility in these formerly non-responsive cells (TANABE et al. 1988), along with L-type Ca^{2+} current (TANABE et al. 1988; GARCIA et al. 1994) and gating charge movement (ADAMS et al. 1990). While the skeletal DHP receptor allowed contraction even in the absence of extracellular Ca^{2+}, the cardiac L-type Ca^{2+} channel restored contractility only if Ca^{2+} entry occurred (TANABE et al. 1990). The structural basis of the skeletal-type excitation–contraction coupling was investigated with molecular chimeras. By inserting pieces of the $\alpha_1 1.1$ gene into an $\alpha_1 1.2$ background, TANABE et al. (1990) showed that the key domain was the intracellular loop joining repeats II and III of $\alpha_1 1.1$ (see asterisk in Fig. 1). More recently, other groups have shown that purified II-III loop fragments can activate directly the ryanodine receptor (LU et al. 1994; EL-HAYEK et al. 1995) and that this region may contain phosphorylation sites for the regulation of excitation–contraction coupling (LU et al. 1995).

III. Rhythmic Activity

1. Pacemaker

In cardiac cells, T-type Ca^{2+} channels are generally present at much lower density than L-type channels, if at all. However, T-type channels supply a major fraction of the current recorded in cells from the sinoatrial node, the natural source of cardiac rhythms, and thus provide a significant contribution to the inward current that drives the last stages of the pacemaker depolarization (HAGIWARA et al. 1988; LEI et al. 1998).

2. Other

T-type channels also support oscillatory activity and repetitive activity in the thalamus (JAHNSEN and LLINÁS 1984; MCCORMICK and BAL 1997). Along with an apamin-sensitive Ca^{2+}-activated K current, T-type channels in the nucleus reticularis generate rhythmic action potential bursts. In thalamocortical neurons the overlapping activation and inactivation curves of T-type currents support rebound burst firing in which a hyperpolarization is followed by a Ca^{2+} spike and results in the generation of several action potentials. Interestingly, expression of T-type channels in smooth muscle fluctuates in synchrony with the cell cycle (KUGA et al. 1996), and may be associated with cell proliferation (SCHMITT et al. 1995).

Excitation–Secretion Coupling

1. Generic Properties

The most commonly studied role of Ca^{2+} is its ability to trigger neurotransmitter release. The importance of Ca^{2+} ions in the release of neurotransmitter has been appreciated for more than 60 years (FENG 1936). Seminal work by DOUGLAS (1963) and KATZ (1969) and their colleagues demonstrated that Ca^{2+} ions exert their influence at the nerve terminal where they control the amount of neurotransmitter that is released. The action of Ca^{2+} ions in the regulation of neurotransmission was shown to be cooperative, requiring about four Ca^{2+} ions to bind to their receptor in order to trigger release (DODGE and RAHAMIMOFF 1967). The importance of Ca^{2+} action in the nerve terminal was further supported by the observation that injection of Ca^{2+} into the terminal triggered the release of transmitter at the squid giant synapse (MILEDI 1973). Subsequently, the Ca^{2+}-sensitive protein, aequorin, was used to show that presynaptic $[Ca^{2+}]_i$ increases during neurotransmission (LLINÁS and NICHOLSON 1975).

Studies using simultaneous voltage-clamp of the presynaptic terminal and postsynaptic axon of the squid giant synapse provided direct measurements of the Ca^{2+} currents in the presynaptic membrane that trigger the release of neurotransmitter (LLINÁS et al. 1981; AUGUSTINE et al. 1985). Ongoing issues include the identification of presynaptic Ca^{2+} channels and clarification of the functional consequences of their diversity (for other recent reviews, see OLIVERA et al. 1994; DUNLAP et al. 1995; REUTER 1996).

Ca^{2+} channels from the Ca_V2 subfamily are the primary types responsible for excitation–secretion coupling. Interestingly just as the II-III loop of the Ca_V1 channel interacts with the Ca^{2+} channel's effector for contraction, the II-III loop of the Ca_V2 channel interacts with its effector: the secretory apparatus (SHENG et al. 1994) (asterisk in Fig. 1). The specific type of channel involved in secretion from various cell types is discussed in greater detail below.

While the vast majority of studies of neurotransmitter release have failed to identify a role for L-type Ca^{2+} channels (DUNLAP et al. 1995), this subtype has been implicated in a few specialized forms of exocytosis. For example, activation of L-type channels is required for zona pellucida-induced exocytosis from the acrosome of mammalian sperm (FLORMAN et al. 1992). L-type channels also seem to play a role in mediating hormone release from endocrine cells. Inhibition of L-type Ca^{2+} channels reduces insulin secretion from pancreatic β cells (Ashcroft et al. 1994; Bokvist et al. 1995), oxytocin and vasopressin release from the neurohypophysis (Lemos and Nowycky 1989), luteinizing hormone-releasing hormone release from the bovine infundibulum (Dippel et al. 1995), and catecholamine release from adrenal chromaffin cells (Lopez et al. 1994). L-type channels also seem to play an important role in supporting release of GABA from retinal bipolar cells (Maguire et al. 1989; Duarte et al. 1992), as well as dynorphin release from dendritic domains of hippocampal neurons (Simmons et al. 1995). In some cases L-type channels may function to release excitatory amino acid transmitters, in response to particular patterns of activity (Bonci et al. 1998), in cells that exhibit graded potentials (Schmitz and Witkovsky 1997), during extended depolarizations with high K^{+}, or under the experimental influence of the DHP agonist Bay K 8644 (e.g., see Sabria et al. 1995).

In addition to admitting the Ca^{2+} which directly triggers neurotransmitter release, Ca^{2+} channels regulate and are regulated by the state of the nerve terminal. Ca^{2+} entry though the same channels which trigger transmitter release, and most likely through other presynaptic channels more distant from the release site (possibly including L-type channels) affects the background level of Ca^{2+} in the terminal, which regulates endocytosis, release probability, various dynamic parameters of the vesicle pool, as well as the channels themselves (reviewed in Neher 1998). Ca^{2+} channels also receive direct feedback about the state of the release machinery (Bezprozvanny et al. 1995; Bergsman and Tsien 2000; Degtiar et al. 2000).

2. Peripheral

At the neuromuscular junction, the release of neurotransmitter is generally mediated by a single Ca^{2+} channel type, although there is variation in the type that predominates from species to species. Invertebrate motor end plates utilize primarily P/Q-type channels. In crayfish, for example, inhibitory and excitatory transmitter release onto the claw opener muscle was completely abolished by ω-Aga-IVA, while ω-CTx-GVIA and nifedipine were both ineffective (Araque et al. 1994). In locusts and houseflies, motor end plate potentials are blocked by type I and II Agatoxins, which inhibit P/Q-type channels, but not by type III Agatoxins, which potently block both L- and N-type channels (Bindokas et al. 1991). In non-mammalian vertebrates, unlike invertebrates, neurotransmitter release at the neuromuscular junction is completely blocked by ω-CTx-GVIA. This is true for frogs (Kerr and Yoshikami 1984; Katz et al. 1995), lizards (Lindgren and Moore 1989), and chicks (De Luca et al. 1991; Gray et al. 1992). In mammals on the other hand, ω-CTx-GVIA does not seem to have any effect on the evoked release of acetylcholine at the neuromuscular junction (Sano et al. 1987; Wessler et al. 1990; De Luca et al. 1991; Protti et al. 1991; Bowersox et al. 1995). In contrast, block of P/Q-type Ca^{2+} channels by ω-CTx-MVIIC, ω-Aga-IVA, or FTx completely abolishes transmission in mice (Protti and Uchitel 1993; Bowersox et al. 1995; Hong and Chang 1995) and humans (Protti et al. 1996). In all of these species, neuromuscular transmission seems to rely on a single type of channel from the Ca_V2 subfamily.

In general, sympathetic neurons contain both L- and N-type Ca^{2+} channels but not P/Q-type channels (Hirning et al. 1988; Mintz et al. 1992a; Zhu and Ikeda 1993; but see Namkung et al. 1998). However, only N-type Ca^{2+} channels seem to be important for the release of norepinephrine, inasmuch as ω-CTx-GVIA blocks NE secretion (Hirning et al. 1988; Fabi et al. 1993) but DHPs do not (Perney et al. 1986; Hirning et al. 1988; Koh and Hille 1996). Along similar lines, N- but not L-type Ca^{2+} channels in sympathetic nerve terminals are susceptible to modulation of Ca^{2+} current via autoreceptors for NE or neuropeptide Y (Toth et al. 1993). Thus, sympathetic nerve endings are like motor nerve terminals in relying on a single predominant type of Ca^{2+} channel, in this case N-type, despite the sizable contribution of L-type channels to the global Ca^{2+} current. Reliance on N-type channels cannot be generalized to all autonomic terminals since P/Q-type channels play a prominent role in transmitter release in rodent urinary bladder (Frew and Lundy 1995; Waterman 1996) and also participate in triggering release of exocytosis from mouse sympathetic and parasympathetic nerve terminals (Waterman 1997; Waterman et al. 1997)

3. Central

At central synapses, unlike synapses in the periphery, neurotransmitter release often involves more than one Ca^{2+} channel type. Central neurons appear to be richly endowed with Ca^{2+} channels, with as many as five or six different types of channels in an individual nerve cell (Mintz et al. 1992a; Randall and Tsien 1995). Several recent papers have reported that neurotransmission at specific synapses in the CNS depends upon the concerted actions of more than one type of Ca^{2+} channel (Luebke et al. 1993; Takahashi and Momiyama 1993; Castillo et al. 1994; Regehr and Mintz 1994; Wheeler et al. 1994; Mintz et al. 1995). The relative importance of N-, P/Q-, and R-type Ca^{2+} channels can vary from one synapse to another. Studies of synapses in hippocampal and cerebellar slices suggest that the vast majority of single release sites are in close proximity to a mixed population of Ca^{2+} channels that jointly contribute to the local Ca^{2+} transient that triggers vesicular fusion (e.g., Mintz et al. 1995; but see also Reuter 1995; Poncer et al. 1997; Reid et al. 1997). The synergistic effect of multiple Ca^{2+} channels arises because of limitations on the Ca^{2+} flux through individual channels under physiological conditions. Indeed, the reliance on multiple types of Ca^{2+} channels was not absolute but could be relieved by increasing the Ca^{2+} influx per channel, either by prolonging the presynaptic action potential or by increasing $[Ca^{2+}]_o$ (Wheeler et al. 1996). The reliance on more than a single Ca^{2+} channel type may offer the advantage of precise control over Ca^{2+} influx and transmitter release by allowing for differential modulation (TSIEN et al. 1988; MOGUL et al. 1993; SWARTZ et al. 1993; MYNLIEFF and BEAM 1994).

V. Postsynaptic Ca^{2+} Influx

1. Dendritic Information Processing

Much of the electrical and biochemical signal processing in central neurons takes place within their dendritic trees. Ca^{2+} entry through voltage-gated channels is critical for many of these events. The idea that voltage-gated Ca^{2+} channels may contribute to electrogenesis in dendrites first arose in the interpretation of intracellular recordings from hippocampal pyramidal neurons (SPENCER and KANDEL 1961). Initial intradendritic voltage recordings were conducted on the dendritic arbors of cerebellar Purkinje neurons (LLINÁS and NICHOLSON 1971; LLINÁS and HESS 1976; LLINÁS and SUGIMORI 1980) and apical dendrites of hippocampal pyramidal neurons (WONG et al. 1979). The ability of dendrites to support Ca^{2+}-dependent action potential firing was reinforced by experiments where apical dendrites of pyramidal neurons were surgically isolated from their cell bodies in a hippocampal slice preparation (BENARDO et al. 1982; MASUKAWA and PRINCE 1984). These experiments revealed a variety of Ca^{2+}-dependent active responses in the dendrites of central neurons that could be elicited by excitatory postsynaptic potentials or injection of depolarizing current pulses.

Recent studies of the electrical properties of dendrites have been facilitated by the ability to visualize dendrites in brain slices, thus rendering dendrites accessible to patch electrodes (STUART et al. 1993). These studies revealed that back-propagating Na^{+}-dependent action potentials can activate dendritic Ca^{2+} channels, thereby causing substantial increases in intradendritic free Ca^{2+} (JAFFE et al. 1992; STUART and SAKMANN 1994; MARKRAM et al. 1995; SCHILLER et al. 1995; SPRUSTON et al. 1995). Subthreshold excitatory postsynaptic potentials can also open Ca^{2+} channels and result in more localized changes in intradendritic Ca^{2+} concentration (MARKRAM and SAKMANN 1994; YUSTE et al. 1994; MAGEE et al. 1995). T-type Ca^{2+} channels play a prominent role in dendritic Ca^{2+} signaling in hippocampal and cortical neurons (MAGEE et al. 1995), presumably due to their ability to open at relatively negative membrane potentials.

The presence of multiple types of voltage-gated Ca^{2+} channels on dendrites has been demonstrated by several techniques, including Ca^{2+} imaging (MARKRAM et al. 1995; WATANABE et al. 1998), dendrite-attached patch clamp recordings (USOWICZ et al. 1992; MAGEE and JOHNSTON 1995), and immunocytochemistry (WESTENBROEK et al. 1990, 1992, 1995; HELL et al. 1993a; YOKOYAMA et al. 1995). Recordings from isolated dendritic segments of acutely dissociated hippocampal neurons indicated that T-, N-, P/Q-, and R-type channels all contribute to the overall Ca^{2+} current in dendrites, with T-type current particularly enhanced when compared to somata (Kavalali et al. 1997).

2. Excitation-Expression Coupling and Changes in Gene Expression

A number of extracellular factors that influence cell growth and activity depolarize the membranes of their target cells (Hill and Treisman 1995). Membrane depolarization opens voltage-gated Ca^{2+} channels and the resulting influx of Ca^{2+} can trigger gene transcription (for a review, see Morgan and Curran 1989). L-type Ca^{2+} channels are thought to play a role in this cascade because agonists of these channels can induce expression of several protooncogenes in the absence of other stimuli (Morgan and Curran 1988). Indeed the mode and location of Ca^{2+} entry may be important to how the Ca^{2+} signal is interpreted by the cell (Ghosh et al. 1994; Rosen and Greenberg 1994). Some recent studies have shed light on the cascade of events that follows influx of Ca^{2+} through L-type channels.

An example of a signal-transduction cascade where Ca^{2+} entry is important involves the cAMP and Ca^{2+} response element (CRE), and its nuclear binding protein (CREB) (Montminy and Bilezikjian 1987; Hoeffler et al. 1988). The interaction of CREB with the CRE is facilitated when CREB is phosphorylated on serine-133 (Gonzalez and Montminy 1989). The phosphorylation of CREB is catalyzed by several kinases including Ca^{2+}-calmodulin kinases II and IV, cAMP-dependent protein kinase (Greenberg et al. 1992), and others. Thus, rises in $[Ca^{2+}]_i$ can act either directly, via Ca^{2+}-calmodulin and its dependent kinases, or indirectly, by stimulating Ca^{2+}-calmodulin-sensitive adenylate cyclase leading to increased cAMP levels. Recent work has shown that Ca^{2+} entry through L-type channels can trigger CREB phosphorylation (Yoshida et al. 1995; Deisseroth et al. 1998; Rajadhyaksha et al. 1999), and that the Ca^{2+} probably binds to a target molecule within $1\,\mu m$ of the point of entry (Deisseroth et al. 1996).

In addition to Ca^{2+}, Zn^{2+} influx is interesting because it regulates a wide variety of enzymes and DNA binding proteins, provides an important developmental signal, and may be involved in excitotoxicity and responses to trauma (for a review, see Smart et al. 1994). Interestingly, L-type Ca^{2+} channels can support Zn^{2+} influx into heart cells, where it can induce transcription of genes driven by a metallothionein promoter (Atar et al. 1995). Morphological studies have revealed that Zn^{2+} is highly enriched in a number of nerve fiber pathways, especially in boutons where it appears to be contained within vesicles (Smart et al. 1994). Furthermore, Zn^{2+} can be released from brain tissue during electrical or chemical stimulation (Assaf and Chung 1984; Howell et al. 1984; Charton et al. 1985). Given that Zn^{2+} can be released by synaptic activity, and can enter cells via voltage-dependent Ca^{2+} channels, it seems likely that Zn^{2+} may play an important role in excitation–expression coupling.

D. Concluding Remarks

Understanding of the diversity of voltage-gated Ca^{2+} channels has greatly increased over the last decade or so as a result of several synergistic approaches. The identification of multiple types of Ca^{2+} channels on the basis of biophysical and pharmacological criteria has been complemented by studies of the biochemistry and molecular biology of their underlying subunit components. The most recent advances have been made in understanding the basis of P/Q-, R-, and T-type Ca^{2+} channel activity. Considerable progress has also been made in clarifying molecular mechanisms of the structural features that distinguish individual types of Ca^{2+} channels and enable them to perform specialized functional roles or to respond to type-selective drugs. The largest area of uncertainty concerns the three-dimensional structures of Ca^{2+} channels and the structural basis of differences among channel subtypes.

References

Adams BA, Tanabe T, Mikami A, Numa S, Beam KG (1990) Intramembrane charge movement restored in dysgenic skeletal muscle by injection of dihydropyridine receptor cDNAs. Nature 346:569–572

Akaike N, Kanaide H, Kuga T, Nakamura M, Sadoshima J, Tomoike H (1989) Low-voltage-activated calcium current in rat aorta smooth muscle cells in primary culture. J Physiol (Lond) 416:141–160

Angelotti T, Hofmann F (1996) Tissue-specific expression of splice variants of the mouse voltage-gated calcium channel alpha2/delta subunit. Febs Letters 397:331–337

Araque A, Clarac F, Buno W (1994) P-type Ca2+ channels mediate excitatory and inhibitory synaptic transmitter release in crayfish muscle. Proc Natl Acad Sci USA 91:4224–4228

Ashcroft FM, Proks P, Smith PA, Ammala C, Bokvist K, Rorsman P (1994) Stimulus–secretion coupling in pancreatic β cells. J Cell Biochem 55:54–65

Assaf SY, Chung SH (1984) Release of endogenous Zn^{2+} from brain tissue during activity. Nature 308:734–736

Atar D, Backx PH, Appel MM, Gao WD, Marban E (1995) Excitation–transcription coupling mediated by zinc influx through voltage-dependent calcium channels. J Biol Chem 270:2473–2477

Augustine GJ, Charlton MP, Smith SJ (1985) Calcium entry and transmitter release at voltage-clamped nerve terminals of squid. J Physiol (Lond) 367:163–181

Avery RA, Johnston D (1996) Multiple channel types contribute to the low-voltage-activated calcium current in hippocampal CA3 pyramidal neurons. J Neurosci 16:5567–5582

Bean BP (1985) Two kinds of calcium channels in canine atrial cells. Differences in kinetics, selectivity, and pharmacology. J Gen Physiol 86:1–30

Bech-Hansen NT, Naylor MJ, Maybaum TA, Pearce WG, Koop B, Fishman GA, Mets M, Musarella MA, Boycott KM (1998) Loss-of-function mutations in a calcium-channel alpha1-subunit gene in Xp11.23 cause incomplete X-linked congenital stationary night blindness. Nature Genetics 19:264–267

Benardo LS, Masukawa LM, Prince DA (1982) Electrophysiology of isolated hippocampal pyramidal dendrites. J Neurosci 2:1614–1622

Bergsman JB, Tsien RW (2000) Syntaxin Modulation of Calcium Channels in Cortical Synaptosomes as Revealed by Botulinum Toxin C1. J Neurosci (in press)

Bezprozvanny I, Scheller RH, Tsien RW (1995) Functional impact of syntaxin on gating of N-type and Q-type calcium channels. Nature 378:623–626

Bindokas VP, Venema VJ, Adams ME (1991) Differential antagonism of transmitter release by subtypes of ω-Agatoxins. J Neurophysiol 66:590–601

Birnbaumer L, Campbell KP, Catterall WA, Harpold MM, Hofmann F, Horne WA, Mori Y, Schwartz A, Snutch TP, Tanabe T, Tsien RW (1994) The naming of voltage-gated calcium channels. Neuron 13:505–506

Black JL, 3rd, Lennon VA (1999) Identification and cloning of putative human neuronal voltage-gated calcium channel gamma-2 and gamma-3 subunits: neurologic implications. Mayo Clinic Proceedings 74:357–361

Bokvist K, Eliasson L, Ammala C, Renstrom E, Rorsman P (1995) Co-localization of L-type Ca2+ channels and insulin-containing secretory granules and its significance for the initiation of exocytosis in mouse pancreatic β-cells. EMBO J 14:50–57

Bonci A, Grillner P, Mercuri NB, Bernardi G (1998) L-Type calcium channels mediate a slow excitatory synaptic transmission in rat midbrain dopaminergic neurons. J Neurosci 18:6693–6703

Bosse E, Regulla S, Biel M, Ruth P, Meyer HE, Flockerzi V, Hofmann F (1990) The cDNA and deduced amino acid sequence of the γ subunit of the L-type calcium channel from rabbit skeletal muscle. Febs Lett 267:153–156

Bourinet E, Soong TW, Sutton K, Slaymaker S, Mathews E, Monteil A, Zamponi GW, Nargeot J, Snutch TP (1999) Splicing of alpha 1A subunit gene generates phenotypic variants of P- and Q-type calcium channels. Nat Neurosci 2:407–415

Bourinet E, Zamponi GW, Stea A, Soong TW, Lewis BA, Jones LP, Yue DT, Snutch TP (1996) The alpha 1E calcium channel exhibits permeation properties similar to low-voltage-activated calcium channels. J Neurosci 16:4983–4993

Bowersox SS, Miljanich GP, Sugiura Y, Li C, Nadasdi L, Hoffman BB, Ramachandran J, Ko CP (1995) Differential blockade of voltage-sensitive calcium channels at the mouse neuromuscular junction by novel ω-Conopeptides and ω-Agatoxin-IVA. J Pharmacol Exp Ther 273:248–256

Campbell KP, Leung AT, Sharp AH (1988) The biochemistry and molecular biology of the dihydropyridine-sensitive calcium channel. Trends Neurosci 11:425–430

Carbone E, Lux HD (1984) A low voltage-activated, fully inactivating Ca2+ channel in vertebrate sensory neurones. Nature 310:501–502

Castillo PE, Weisskopf MG, Nicoll RA (1994) The role of Ca2+ channels in hippocampal mossy fiber synaptic transmission and long-term potentiation. Neuron 12:261–269

Catterall WA, Curtis BM (1987) Molecular properties of voltage-sensitive calcium channels. Soc Gen Physiol Ser 41:201–213

Catterall WA, Seagar MJ, Takahashi M (1988) Molecular properties of dihydropyridine-sensitive calcium channels in skeletal muscle. J Biol Chem 263:3535–3538

Charton G, Rovira C, Ben-Ari Y, Leviel V (1985) Spontaneous and evoked release of endogenous Zn^{2+} in the hippocampal mossy fiber zone of the rat *in situ*. Exp Brain Res 58:202–205

Chuang RS, Jaffe H, Cribbs L, Perez-Reyes E, Swartz KJ (1998) Inhibition of T-type voltage-gated calcium channels by a new scorpion toxin. Nat Neurosci 1:668–674

Cohen CJ, McCarthy RT (1987) Nimodipine block of calcium channels in rat anterior pituitary cells. J Physiol (Lond) 387:195–225

Cribbs LL, Lee JH, Yang J, Satin J, Zhang Y, Daud A, Barclay J, Williamson MP, Fox M, Rees M, Perez-Reyes E (1998) Cloning and characterization of alpha1H from human heart, a member of the T-type Ca2+ channel gene family. Circulation Research 83:103–109

De Jongh KS, Warner C, Catterall WA (1990) Subunits of purified calcium channels. α_2 and δ are encoded by the same gene. J Biol Chem 265:14738–14741

De Luca A, Rand MJ, Reid JJ, Story DF (1991) Differential sensitivities of avian and mammalian neuromuscular junctions to inhibition of cholinergic transmission by ω-Conotoxin GVIA. Toxicon 29:311–320

De Waard M, Gurnett CA, Campbell KP. (1996) Structural and functional diversity of voltage-activated calcium channels. In Ion Channels, T Narahashi, ed. (New York: Plenum Press), pp. 41–87
Degtiar VE, Scheller RH, Tsien RW (2000) Syntaxin Modulation of Slow Inactivation Of N-type Calcium Channels. J Neurosci (submitted)
Deisseroth K, Bito H, Tsien RW (1996) Signaling from synapse to nucleus: postsynaptic CREB phosphorylation during multiple forms of hippocampal synaptic plasticity. Neuron 16:89–101
Deisseroth K, Heist EK, Tsien RW (1998) Translocation of calmodulin to the nucleus supports CREB phosphorylation in hippocampal neurons. Nature 392:198–202
Dippel WW, Chen PL, McArthur NH, Harms PG (1995) Calcium involvement in luteinizing hormone-releasing hormone release from the bovine infundibulum. Domest Anim Endocrinol 12:349–354
Dodge F Jr., Rahamimoff R (1967) Co-operative action a calcium ions in transmitter release at the neuromuscular junction. J Physiol 193:419–432
Douglas WW, Rubin RP (1963) The Mechanism of Catecholamine Release From the Adrenal Medulla and the Role of Calcium in Stimulus–Secretion Coupling. J Physiol 167:288–310
Droogmans G, Nilius B (1989) Kinetic properties of the cardiac T-type calcium channel in the guinea-pig. J Physiol (Lond) 419:627–650
Duarte CB, Ferreira IL, Santos PF, Oliveira CR, Carvalho AP (1992) Ca2+-dependent release of [^{3}H]GABA in cultured chick retina cells. Brain Res 591:27–32
Dubel SJ, Starr TV, Hell J, Ahlijanian MK, Enyeart JJ, Catterall WA, Snutch TP (1992) Molecular cloning of the alpha-1 subunit of an omega-conotoxin-sensitive calcium channel. Proc Natl Acad Sci USA 89:5058–5062
Dunlap K, Luebke JI, Turner TJ (1995) Exocytotic Ca2+ channels in mammalian central neurons. Trends Neurosci 18:89–98
Eberst R, Dai S, Klugbauer N, Hofmann F (1997) Identification and functional characterization of a calcium channel gamma subunit. Pflugers Archiv European Journal of Physiology 433:633–637
Edmonds B, Klein M, Dale N, Kandel ER (1990) Contributions of two types of calcium channels to synaptic transmission and plasticity. Science 250:1142–1147
Ehrlich BE, Jacobson AR, Hinrichsen R, Sayre LM, Forte MA (1988) Paramecium calcium channels are blocked by a family of calmodulin antagonists. Proc Natl Acad Sci USA 85:5718–5722
Eisenberg RS, McCarthy RT, Milton RL (1983) Paralysis of frog skeletal muscle fibres by the calcium antagonist D-600. J Physiol (Lond) 341:495–505
el-Hayek R, Antoniu B, Wang J, Hamilton SL, Ikemoto N (1995) Identification of calcium release-triggering and blocking regions of the II-III loop of the skeletal muscle dihydropyridine receptor. J Biol Chem 270:22116–22118
Ellinor PT, Zhang J-F, Randall AD, Zhou M, Schwarz TL, Tsien RW, Horne WA (1993) Functional expression of a rapidly inactivating neuronal calcium channel. Nature 363:455–458
Ellis SB, Williams ME, Ways NR, Brenner R, Sharp AH, Leung AT, Campbell KP, McKenna, E, Koch, WJ, Hui, A (1988). Sequence and expression of mRNAs encoding the α_1 and α_2 subunits of a DHP-sensitive calcium channel. Science 241:1661–1664
Ertel SI, Ertel EA (1997) Low-voltage-activated T-type Ca2+ channels. Trend Pharmacol Sci 18:37–42
Fabi F, Chiavarelli M, Argiolas L, Chiavarelli R, del Basso P (1993) Evidence for sympathetic neurotransmission through presynaptic N-type calcium channels in human saphenous vein. Br J Pharmacol 110:338–342
Feng TP (1936) Studies on the neuromuscular junction II. The universal antagonism between calcium and curarizing agencies. Chin J Physiol 10:513–528
Fisher SE, Ciccodicola A, Tanaka K, Curci A, Desicato S, D'Urso M, Craig IW (1997) Sequence-based exon prediction around the synaptophysin locus reveals a

gene-rich area containing novel genes in human proximal Xp. Genomics 45:340–347
Fisher TE, Bourque CW (1995) Distinct omega-agatoxin-sensitive calcium currents in somata and axon terminals of rat supraoptic neurones. Journal of Physiology 489:383–388
Florman HM, Corron ME, Kim TD, Babcock DF (1992) Activation of voltage-dependent calcium channels of mammalian sperm is required for zona pellucida-induced acrosomal exocytosis. Dev Biol 152:304–314
Forsythe ID, Tsujimoto T, Barnes-Davies M, Cuttle MF, Takahashi T (1998) Inactivation of presynaptic calcium current contributes to synaptic depression at a fast central synapse. Neuron 20:797–807
Forti L, Pietrobon D (1993) Functional diversity of L-type calcium channels in rat cerebellar neurons. Neuron 10:437–450
Frew R, Lundy PM (1995) A role for Q type Ca2+ channels in neurotransmission in the rat urinary bladder. Br J Pharmacol 116:1595–1598
Fujita Y, Mynlieff M, Dirksen RT, Kim MS, Niidome T, Nakai J, Friedrich T, Iwabe N, Miyata T, Furuichi T, Furutama D, Mikoshiab K, Mori Y, Beam KG (1993) Primary structure and functional expression of the omega-Conotoxin-sensitive N-type calcium channel from rabbit brain. Neuron 10:585–598
Garcia J, Tanabe T, Beam KG (1994) Relationship of calcium transients to calcium currents and charge movements in myotubes expressing skeletal and cardiac dihydropyridine receptors. J Gen Physiol 103:125–147
Ghosh A, Ginty DD, Bading H, Greenberg ME (1994) Calcium regulation of gene expression in neuronal cells. J Neurobiol 25:294–303
Gillard SE, Volsen SG, Smith W, Beattie RE, Bleakman D, Lodge D (1997) Identification of pore-forming subunit of P-type calcium channels: an antisense study on rat cerebellar Purkinje cells in culture. Neuropharmacology 36:405–409
Glossmann H, Striessnig J (1990) Molecular properties of calcium channels. Rev Physiol Biochem Pharmacol 114:1–105
Glossmann H, Striessnig J, Hymel L, Schindler H (1987) Purified L-type calcium channels: only one single polypeptide (α_1-subunit) carries the drug receptor domains and is regulated by protein kinases. Biomed Biochim Acta 46:S351–356
Gonzalez GA, Montminy MR (1989) Cyclic AMP stimulates somatostatin gene transcription by phosphorylation of CREB at serine 133. Cell 59:675–680
Grabner M, Bachmann A, Rosenthal F, Striessnig J, Schultz C, Tautz D, Glossmann H (1994) Insect calcium channels. Molecular cloning of an α_1-subunit from housefly (*Musca domestica*) muscle. FEBS Lett 339:189–194
Grantham CJ, Bowman D, Bath CP, Bell DC, Bleakman D (1994) Omega-conotoxin MVIIC reversibly inhibits a human N-type calcium channel and calcium influx into chick synaptosomes. Neuropharmacology 33:255–258
Gray DB, Bruses JL, Pilar GR (1992) Developmental switch in the pharmacology of Ca2+ channels coupled to acetylcholine release. Neuron 8:715–724
Greenberg ME, Thompson MA, Sheng M (1992) Calcium regulation of immediate early gene transcription. J Physiol (Paris) 86:99–108
Grolleau F, Lapied B (1996) Two distinct low-voltage-activated Ca2+ currents contribute to the pacemaker mechanism in cockroach dorsal unpaired median neurons. Journal of Neurophysiology 76:963–976
Hagiwara N, Irisawa H, Kameyama M (1988) Contribution of two types of calcium currents to the pacemaker potentials of rabbit sino-atrial node cells. J Physiol 395:233–253
Hell JW, Westenbroek RE, Warner C, Ahlijanian MK, Prystay W, Gilbert MM, Snutch TP, Catterall WA (1993a) Identification and differential subcellular localization of the neuronal class C and class D L-type calcium channel α_1 subunits. J Cell Biol 123:949–962
Hell JW, Yokoyama CT, Wong ST, Warner C, Snutch TP, Catterall WA (1993b) Differential phosphorylation of two size forms of the neuronal class C L-type calcium channel α_1 subunit. J Biol Chem 268:19451–19457

Hill CS, Treisman R (1995) Transcriptional regulation by extracellular signals: mechanisms and specificity. Cell 80:199–211

Hillyard DR, Monje VD, Mintz IM, Bean BP, Nadasdi L, Ramachandran J, Miljanich G, Azimi-Zoonooz A, McIntosh JM, Cruz LJ, et al (1992) A new Conus peptide ligand for mammalian presynaptic Ca2+ channels. Neuron 9:69–77

Hirning LD, Fox AP, McCleskey EW, Olivera BM, Thayer SA, Miller RJ, Tsien RW (1988) Dominant role of N-type Ca2+ channels in evoked release of norepinephrine from sympathetic neurons. Science 239:57–61

Hoeffler JP, Meyer TE, Yun Y, Jameson JL, Habener JF (1988) Cyclic AMP-responsive DNA-binding protein: structure based on a cloned placental cDNA. Science 242:1430–1433

Hofmann F, Biel M, Flockerzi V (1994) Molecular basis for Ca2+ channel diversity. Annu Rev Neurosci 17:399–418

Hofmann F, Oeken HJ, Schneider T, Sieber M (1988) The biochemical properties of L-type calcium channels. J Cardiovasc Pharmacol 12:S25–30

Hong SJ, Chang CC (1995) Inhibition of acetylcholine release from mouse motor nerve by a P-type calcium channel blocker, ω-Agatoxin IVA. J Physiol (Lond) 482:283–290

Horne WA, Ellinor PT, Inman I, Zhou M, Tsien RW, Schwarz TL (1993) Molecular diversity of Ca2+ channel α_1 subunits from the marine ray *Discopyge ommata*. Proc Natl Acad Sci USA 90:3787–3791

Howell GA, Welch MG, Frederickson CJ (1984) Stimulation-induced uptake and release of zinc in hippocampal slices. Nature 308:736–738

Huguenard JR, Prince DA (1992) A novel T-type current underlies prolonged Ca2+-dependent burst firing in GABAergic neurons of rat thalamic reticular nucleus. J Neurosci 12:3804–3817

Isom LL, De Jongh KS, Catterall WA (1994) Auxiliary subunits of voltage-gated ion channels. Neuron 12:1183–1194

Jaffe DB, Johnston D, Lasser-Ross N, Lisman JE, Miyakawa H, Ross WN (1992) The spread of Na^+ spikes determines the pattern of dendritic Ca2+ entry into hippocampal neurons. Nature 357:244–246

Jahnsen H, Llinás R (1984) Ionic basis for the electro-responsiveness and oscillatory properties of guinea-pig thalamic neurones *in vitro*. J Physiol 349:227–247

Jay SD, Ellis SB, McCue AF, Williams ME, Vedvick TS, Harpold MM, Campbell KP (1990) Primary structure of the γ subunit of the DHP-sensitive calcium channel from skeletal muscle. Science 248:490–492

Jay SD, Sharp AH, Kahl SD, Vedvick TS, Harpold MM, Campbell KP (1991) Structural characterization of the dihydropyridine-sensitive calcium channel α_2-subunit and the associated δ peptides. J Biol Chem 266:3287–3293

Jeziorski MC, Greenberg RM, Clark KS, Anderson PA (1998) Cloning and functional expression of a voltage-gated calcium channel alpha1 subunit from jellyfish. J Biol Chem 273:22792–22799

Jun K-S, Piedras-Rentería ES, Smith SM, Wheeler DB, Lee SB, Lee TG, Chin H, Adams ME, Scheller RH, Tsien RW, Shin H-S (1999) Ablation of P/Q-type Ca2+ channel currents and progressive, fatal ataxia in mice lacking the a1 A subunit. Nature Neuroscience (submitted)

Karst H, Joëls M, Wadman WJ (1993) Low-threshold calcium current in dendrites of the adult rat hippocampus. Neuroscience Letters 164:154–158

Katz B (1969) The Release of Neural Transmitter Substances, Edition (Liverpool: Liverpool University Press)

Katz E, Ferro PA, Cherksey BD, Sugimori M, Llinas R, Uchitel OD (1995) Effects of Ca2+ channel blockers on transmitter release and presynaptic currents at the frog neuromuscular junction. J Physiol (Lond) 486:695–706

Kavalali ET, Plummer MR (1994) Selective potentiation of a novel calcium channel in rat hippocampal neurones. J Physiol 480:475–484

Kavalali ET, Zhuo M, Bito H, Tsien RW (1997) Dendritic Ca2+ channels characterized by recordings from isolated hippocampal dendritic segments. Neuron 18:651–663

Kerr LM, Yoshikami D (1984) A venom peptide with a novel presynaptic blocking action. Nature 308:282–284
Klugbauer N, Lacinová L, Marais E, Hobom M, Hofmann F (1999) Molecular diversity of the calcium channel alpha2delta subunit. J Neurosci 19:684–691
Koh DS, Hille B (1996) Modulation by neurotransmitters of norepinephrine secretion from sympathetic ganglion neurons detected by amperometry. Soc Neurosci Abstr 22:507
Kollmar R, Fak J, Montgomery LG, Hudspeth AJ (1997) Hair cell-specific splicing of mRNA for the alpha1D subunit of voltage-gated Ca2+ channels in the chicken's cochlea. Proc Natl Acad Sci USA 94:14889–14893
Kostyuk PG, Shirokov RE (1989) Deactivation kinetics of different components of calcium inward current in the membrane of mice sensory neurones. J Physiol 409:343–355
Kuga T, Kobayashi S, Hirakawa Y, Kanaide H, Takeshita A (1996) Cell cycle-dependent expression of L- and T-type Ca2+ currents in rat aortic smooth muscle cells in primary culture. Circ Res 79:14–19
Lacinová L, Klugbauer N, Hofmann F (1999) Absence of modulation of the expressed calcium channel alpha1G subunit by alpha2delta subunits. Journal of Physiology 516:639–645
Lambert RC, Maulet Y, Mouton J, Beattie R, Volsen S, De Waard M, Feltz A (1997) T-type Ca2+ current properties are not modified by Ca2+ channel beta subunit depletion in nodosus ganglion neurons. J Neurosci 17:6621–6628
Lee JH, Daud AN, Cribbs LL, Lacerda AE, Pereverzev A, Klöckner U, Schneider T, Perez-Reyes E (1999) Cloning and expression of a novel member of the low voltage-activated T-type calcium channel family. J Neurosci 19:1912–1921
Lei M, Brown H, Noble D. Low-Voltage-Activated T-Type Calcium Channels. International Electrophysiology Meeting, Vol p103–109, 1998
Lemos JR, Nowycky MC (1989) Two types of calcium channels coexist in peptide-releasing vertebrate nerve terminals. Neuron 2:1419–1426
Letts VA, Felix R, Biddlecome GH, Arikkath J, Mahaffey CL, Valenzuela A, Bartlett FS, 2nd, Mori Y, Campbell KP, Frankel WN (1998) The mouse stargazer gene encodes a neuronal Ca2+-channel gamma subunit [see comments]. Nature Genetics 19:340–347
Leuranguer V, Bourinet E, Lory P, Nargeot J (1998) Antisense depletion of beta-subunits fails to affect T-type calcium channels properties in a neuroblastoma cell line. Neuropharmacology 37:701–708
Lin Z, Haus S, Edgerton J, Lipscombe D (1997) Identification of functionally distinct isoforms of the N-type Ca2+ channel in rat sympathetic ganglia and brain. Neuron 18:153–166
Lin Z, Lin Y, Schorge S, Pan JQ, Beierlein M, Lipscombe D (1999) Alternative splicing of a short cassette exon in alpha1B generates functionally distinct N-type calcium channels in central and peripheral neurons. J Neurosci 19:5322–5331
Lindgren CA, Moore JW (1989) Identification of ionic currents at presynaptic nerve endings of the lizard. J Physiol (Lond) 414:201–222
Liu H, De Waard M, Scott VES, Gurnet CA, Lennon VA, Campbell KP (1996) Indentification of three subnuits of the high affinity omega-conotoxin MVIIC-sensitive Ca2+ channel. J Biol Chem 271:13804–13810
Llinás R, Hess R (1976) Tetrodotoxin-resistant dendritic spikes in avian Purkinje cells. Proc Natl Acad Sci USA 73:2520–2523
Llinás R, Nicholson C (1971) Electrophysiological properties of dendrites and somata in alligator Purkinje cells. J Neurophysiol 34:532–551
Llinás R, Nicholson C (1975) Calcium role in depolarization–secretion coupling: an aequorin study in squid giant synapse. Proc Natl Acad Sci USA 72:187–190
Llinás R, Steinberg IZ, Walton K (1981) Relationship between presynaptic calcium current and postsynaptic potential in squid giant synapse. Biophys J 33:323–351

Llinás R, Sugimori M (1980) Electrophysiological properties of *in vitro* Purkinje cell dendrites in mammalian cerebellar slices. J Physiol 305:197–213
Llinás R, Sugimori M, Hillman DE, Cherksey B (1992) Distribution and functional significance of the P-type, voltage-dependent Ca2+ channels in the mammalian central nervous system. Trends Neurosci 15:351–355
Llinás RR, Sugimori M, Cherksey B (1989) Voltage-dependent calcium conductances in mammalian neurons: the P channel. Ann N Y Acad Sci 560:103–111
Lopez MG, Albillos A, de la Fuente MT, Borges R, Gandia L, Carbone E, Garcia AG, Artalejo AR (1994) Localized L-type calcium channels control exocytosis in cat chromaffin cells. Pflügers Arch 427:348–354
Lu J, Dalton JFt, Stokes DR, Calabrese RL (1997) Functional role of Ca2+ currents in graded and spike-mediated synaptic transmission between leech heart interneurons. Journal of Neurophysiology 77:1779–1794
Lu X, Xu L, Meissner G (1994) Activation of the skeletal muscle calcium release channel by a cytoplasmic loop of the dihydropyridine receptor. J Biol Chem 269:6511–6516
Lu X, Xu L, Meissner G (1995) Phosphorylation of dihydropyridine receptor II-III loop peptide regulates skeletal muscle calcium release channel function. Evidence for an essential role of the β-OH group of Ser687. J Biol Chem 270:18459–18464
Luebke JI, Dunlap K, Turner TJ (1993) Multiple calcium channel types control glutamatergic synaptic transmission in the hippocampus. Neuron 11:895–902
Lundy PM, Hamilton MG, Frew R (1994) Pharmacological identification of a novel Ca2+ channel in chicken brain synaptosomes. Brain Research 643:204–210
Magee JC, Christofi G, Miyakawa H, Christie B, Lasser-Ross N, Johnston D (1995) Subthreshold synaptic activation of voltage-gated Ca2+ channels mediates a localized Ca2+ influx into the dendrites of hippocampal pyramidal neurons. J Neurophysiol 74:1335–1342
Magee JC, Johnston D (1995) Characterization of single voltage-gated Na^+ and Ca2+ channels in apical dendrites of rat CA1 pyramidal neurons. J Physiol 487:67–90
Maguire G, Maple B, Lukasiewicz P, Werblin F (1989) γ-Aminobutyrate type B receptor modulation of L-type calcium channel current at bipolar cell terminals in the retina of the tiger salamander. Proc Natl Acad Sci USA 86:10144–10147
Markram H, Helm PJ, Sakmann B (1995) Dendritic calcium transients evoked by single back-propagating action potentials in rat neocortical pyramidal neurons. J Physiol (Lond) 485:1–20
Markram H, Sakmann B (1994) Calcium transients in dendrites of neocortical neurons evoked by single subthreshold excitatory postsynaptic potentials via low-voltage-activated calcium channels. Proc Natl Acad Sci USA 91:5207–5211
Masukawa LM, Prince DA (1984) Synaptic control of excitability in isolated dendrites of hippocampal neurons. J Neurosci 4:217–227
Matteson DR, Armstrong CM (1986) Properties of two types of calcium channels in clonal pituitary cells. J Gen Physiol 87:161–182
McCormick DA, Bal T (1997) Sleep and arousal: thalamocortical mechanisms. Annual Review of Neuroscience 20:185–215
Meder W, Fink K, Göthert M (1997) Involvement of different calcium channels in K+- and veratridine-induced increases of cytosolic calcium concentration in rat cerebral cortical synaptosomes. Naunyn-Schmiedebergs Arch Pharmacol 356:797–805
Meir A, Dolphin AC (1998) Known calcium channel alpha1 subunits can form low threshold small conductance channels with similarities to native T-type channels. Neuron 20:341–351
Mermelstein PG, Foehring R, Tkatch T, Song W-J, Baranauskas G, Surmeier D (1999) Properties of Q-Type Calcium Channels in Neostriatal and Cortical Neurons are Correlated with Beta Subunit Expression. J Neurosci 19:7268–7277
Mermelstein PG, Surmeier DJ (1997) A calcium channel reversibly blocked by omega-conotoxin GVIA lacking the class D alpha 1 subunit. Neuroreport 8:485–489

Mikami A, Imoto K, Tanabe T, Niidome T, Mori Y, Takeshima H, Narumiya S, Numa S (1989) Primary structure and functional expression of the cardiac dihydropyridine-sensitive calcium channel. Nature 340:230–233
Miledi R (1973) Transmitter release induced by injection of calcium ions into nerve terminals. Proc R Soc Lond (Biol) 183:421–425
Miller RJ, Freedman SB (1984) Are dihydropyridine binding sites voltage sensitive calcium channels? Life Sci 34:1205–1221
Mintz IM, Adams ME, Bean BP (1992a) P-type calcium channels in rat central and peripheral neurons. Neuron 9:85–95
Mintz IM, Bean BP (1993) Block of calcium channels in rat neurons by synthetic ω-Aga-IVA. Neuropharmacology 32:1161–1169
Mintz IM, Sidach S (1998) The Society for Neuroscience Abstract 24:1021
Mintz IM, Sabatini BL, Regehr WG (1995) Calcium control of transmitter release at a cerebellar synapse. Neuron 15:675–688
Mintz IM, Venema VJ, Swiderek KM, Lee TD, Bean BP, Adams ME (1992b) P-type calcium channels blocked by the spider toxin omega-Aga-IVA. Nature 355:827–829
Mogul DJ, Adams ME, Fox AP (1993) Differential activation of adenosine receptors decreases N-type but potentiates P-type Ca2+ current in hippocampal CA3 neurons. Neuron 10:327–334
Montminy MR, Bilezikjian LM (1987) Binding of a nuclear protein to the cyclic-AMP response element of the somatostatin gene. Nature 328:175–178
Morgan JI, Curran T (1988) Calcium as a modulator of the immediate-early gene cascade in neurons. Cell Calcium 9:303–311
Morgan JI, Curran T (1989) Stimulus–transcription coupling in neurons: role of cellular immediate-early genes. Trends Neurosci 12:459–462
Mori Y, Friedrich T, Kim MS, Mikami A, Nakai J, Ruth P, Bosse E, Hofmann F, Flockerzi V, Furuichi T, Mikoshiba K, Imoto K, Tanabe T, Numa S (1991) Primary structure and functional expression from complementary DNA of a brain calcium channel. Nature 350:398–402
Mouginot D, Bossu JL, Gähwiler BH (1997) Low-threshold Ca2+ currents in dendritic recordings from Purkinje cells in rat cerebellar slice cultures. J Neurosci 17:160–170
Mynlieff M, Beam KG (1994) Adenosine acting at an A_1 receptor decreases N-type calcium current in mouse motoneurons. J Neurosci 14:3628–3634
Näbauer M, Callewaert G, Cleemann L, Morad M (1989) Regulation of calcium release is gated by calcium current, not gating charge, in cardiac myocytes. Science 244:800–803
Namkung Y, Smith SM, Lee SB, Skrypnyk NV, Kim HL, Chin H, Scheller RH, Tsien RW, Shin HS (1998) Targeted disruption of the Ca2+ channel beta3 subunit reduces N- and L-type Ca2+ channel activity and alters the voltage-dependent activation of P/Q-type Ca2+ channels in neurons. Proc Natl Acad Sci USA 95:12010–12015
Neher E (1998) Vesicle pools and Ca2+ microdomains: new tools for understanding their roles in neurotransmitter release. Neuron 20:389–399
Newcomb R, Szoke B, Palma A, Wang G, Chen X, Hopkins W, Cong R, Miller J, Urge L, Tarczy-Hornoch K, Loo JA, Dooley DJ, Nadasdi L, Tsien RW, Lemos J, Miljanich G (1998) Selective peptide antagonist of the class E calcium channel from the venom of the tarantula Hysterocrates gigas. Biochemistry 37:15353–15362
Niidome T, Teramoto T, Murata Y, Tanaka I, Seto T, Sawada K, Mori Y, Katayama K (1994) Stable expression of the neuronal BI (class A) calcium channel in baby hamster kidney cells. Biochem Biophys Res Commun 203:1821–1827
Nilius B, Hess P, Lansman JB, Tsien RW (1985) A novel type of cardiac calcium channel in ventricular cells. Nature 316:443–446
Nowycky MC, Fox AP, Tsien RW (1985) Three types of neuronal calcium channel with different calcium agonist sensitivity. Nature 316:440–443

Olivera BM, Miljanich GP, Ramachandran J, Adams ME (1994) Calcium channel diversity and neurotransmitter release: the ω-Conotoxins and ω-Agatoxins. Annu Rev Biochem 63:823–867

Perez-Reyes E, Cribbs LL, Daud A, Lacerda AE, Barclay J, Williamson MP, Fox M, Rees M, Lee JH (1998) Molecular characterization of a neuronal low-voltage-activated T-type calcium channel [see comments]. Nature 391:896–900

Perney TM, Hirning LD, Leeman SE, Miller RJ (1986) Multiple calcium channels mediate neurotransmitter release from peripheral neurons. Proc Natl Acad Sci USA 83:6656–6659

Piedras-Rentería ES, Chen CC, Best PM (1997) Antisense oligonucleotides against rat brain alpha1E DNA and its atrial homologue decrease T-type calcium current in atrial myocytes. Proc Natl Acad Sci USA 94:14936–14941

Piedras-Rentería ES, Tsien RW (1998) Antisense oligonucleotides against alpha1E reduce R-type calcium currents in cerebellar granule cells. Proc Natl Acad Sci USA 95:7760–7765

Pinto A, Gillard S, Moss F, Whyte K, Brust P, Williams M, Stauderman K, Harpold M, Lang B, Newsom-Davis J, Bleakman D, Lodge D, Boot J (1998) Human autoantibodies specific for the alpha1A calcium channel subunit reduce both P-type and Q-type calcium currents in cerebellar neurons. Proc Natl Acad Sci USA 95:8328–8333

Poncer JC, McKinney RA, Gähwiler BH, Thompson SM (1997) Either N- or P-type calcium channels mediate GABA release at distinct hippocampal inhibitory synapses. Neuron 18:463–472

Protti DA, Reisin R, Mackinley TA, Uchitel OD (1996) Calcium channel blockers and transmitter release at the normal human neuromuscular junction. Neurology 46:1391–1396

Protti DA, Szczupak L, Scornik FS, Uchitel OD (1991) Effect of ω-Conotoxin GVIA on neurotransmitter release at the mouse neuromuscular junction. Brain Res 557:336–339

Protti DA, Uchitel OD (1993) Transmitter release and presynaptic Ca2+ currents blocked by the spider toxin ω-Aga-IVA. Neuroreport 5:333–336

Putney JW. (1997) Capacitative Calcium Entry, Edition (Austin, TX: R.G. Landes Company)

Rajadhyaksha A, Barczak A, Macías W, Leveque JC, Lewis SE, Konradi C (1999) L-Type Ca(2+) channels are essential for glutamate-mediated CREB phosphorylation and c-fos gene expression in striatal neurons. J Neurosci 19:6348–6359

Randall A, Tsien RW (1995) Pharmacological dissection of multiple types of Ca2+ channel currents in rat cerebellar granule neurons. J Neurosci 15:2995–3012

Randall A, Tsien RW (1998) Distinctive Biophysical and Pharmacological Features of T-Type Calcium Channels. In Low-Voltage-Activated T-Type Calcium Channels, RW Tsien, J-P Clozel J Nargeot, eds. (Basel, Switzerland: Adis International), pp. 29–43

Regehr WG, Mintz IM (1994) Participation of multiple calcium channel types in transmission at single climbing fiber to Purkinje cell synapses. Neuron 12:605–613

Reid CA, Clements JD, Bekkers JM (1997) Nonuniform distribution of Ca2+ channel subtypes on presynaptic terminals of excitatory synapses in hippocampal cultures. J Neurosci 17:2738–2745

Reuter H (1995) Measurements of exocytosis from single presynaptic nerve terminals reveal heterogeneous inhibition by Ca2+ channel blockers. Neuron 14:773–779

Reuter H (1996) Diversity and function of presynaptic calcium channels in the brain. Curr Opin Neurobiol 6:331–337

Ríos E, Brum G (1987) Involvement of dihydropyridine receptors in excitation–contraction coupling in skeletal muscle. Nature 325:717–720

Rock DM, Horne WA, Stoehr SJ, Hashimoto C, Cong RZ, M., Palma A, Hidayetoglu D, Offord J (1998) Does α_{1E} code for T-type Ca2+ channels? A comparison of recombinant α_{1E} Ca2+ channels with GH3 pituitary T-type and recombinant α_{1B}

Ca2+ channels. In Low-Voltage-Activated T-type Calcium Channels, J Nargeot, JP Clozel RW Tsien, eds. (Chester, England: Aidis Press), pp. 279–289
Rosen LB, Greenberg ME (1994) Regulation of c-*fos* and other immediate-early genes in PC12 cells as a model for studying specificity in neuronal signaling. Molec Neurobiol 7:203–216
Rosenberg RL, Hess P, Reeves JP, Smilowitz H, Tsien RW (1986) Calcium channels in planar lipid bilayers: insights into mechanisms of ion permeation and gating. Science 231:1564–1566
Sabria J, Pastor C, Clos MV, Garcia A, Badia A (1995) Involvement of different types of voltage-sensitive calcium channels in the presynaptic regulation of noradrenaline release in rat brain cortex and hippocampus. J Neurochem 64:2567–2571
Sano K, Enomoto K, Maeno T (1987) Effects of synthetic ω-Conotoxin, a new type Ca2+ antagonist, on frog and mouse neuromuscular transmission. Eur J Pharmacol 141:235–241
Sather WA, Tanabe T, Zhang J-F, Mori Y, Adams ME, Tsien RW (1993) Distinctive biophysical and pharmacological properties of class A (BI) calcium channel α_1 subunits. Neuron 11:291–303
Schafer WR, Kenyon CJ (1995) A calcium-channel homologue required for adaptation to dopamine and serotonin in *Caenorhabditis elegans*. Nature 375:73–78
Schiller J, Helmchen F, Sakmann B (1995) Spatial profile of dendritic calcium transients evoked by action potentials in rat neocortical pyramidal neurones. J Physiol (Lond) 487:583–600
Schmitt R, Clozel JP, Iberg N, Buhler FR (1995) Mibefradil prevents neointima formation after vascular injury in rats. Possible role of the blockade of the T-type voltage-operated calcium channel. Arterioscler Thromb Vasc Biol 15:1161–1165
Schmitz Y, Witkovsky P (1997) Dependence of photoreceptor glutamate release on a dihydropyridine-sensitive calcium channel. Neuroscience 78:1209–1216
Schneider MF, Chandler WK (1973) Voltage dependent charge movement of skeletal muscle: a possible step in excitation–contraction coupling. Nature 242:244–246
Schneider T, Wei X, Olcese R, Costantin JL, Neely A, Palade P, Perez-Reyes E, Qin N, Zhou J, Crawford GD, et, a (1994) Molecular analysis and functional expression of the human type E neuronal Ca2+ channel alpha 1 subunit. Receptors and Channels 255–270
Schultz LM, Christie RB, Sejnowski TJ. Distribution of T-Type Calcium Channels in CA1 Stratum Oriens Interneurons. Society for Neuroscience, Vol 1, p79.79, 1999
Scott VE, De Waard M, Liu H, Gurnett CA, Venzke, DP, Lennon, VA, Campbell, KP (1996) β Subunit heterogeneity in N-type Ca2+ channels. J Biol Chem 271:3207–3212
Sheng ZH, Rettig J, Takahashi M, Catterall WA (1994) Identification of a syntaxin-binding site on N-type calcium channels. Neuron 13:1303–1313
Simmons ML, Terman GW, Gibbs SM, Chavkin C (1995) L-type calcium channels mediate dynorphin neuropeptide release from dendrites but not axons of hippocampal granule cells. Neuron 14:1265–1272
Smart TG, Xie X, Krishek BJ (1994) Modulation of inhibitory and excitatory amino acid receptor ion channels by zinc. Prog Neurobiol 42:393–341
Smith LA, Wang XJ, Peixoto AA, Neumann EK, Hall LM, Hall JC (1996) A drosophila calcium channel α_1 subunit gene maps to a genetic locus associated with behavioral and visual defects. J Neurosci 16:7868–7879
Snutch TP, Leonard JP, Gilbert MM, Lester HA, Davidson N (1990) Rat brain expresses a heterogeneous family of calcium channels. Proc Natl Acad Sci USA 87:3391–3395
Snutch TP, Reiner PB (1992) Ca2+ channels: diversity of form and function. Curr Opin Neurobiol 2:247–253
Soldatov NM (1994) Genomic structure of human L-type Ca2+ channel. Genomics 22:77–87

Soldatov NM, Bouron A, Reuter H (1995) Different voltage-dependent inhibition by dihydropyridines of human Ca2+ channel splice variants. J Biol Chem 270:10540–10543

Soong TW, Stea A, Hodson CD, Dubel SJ, Vincent SR, Snutch TP (1993) Structure and functional expression of a member of the low voltage-activated calcium channel family. Science 260:1133–1136

Spencer WA, Kandel ER (1961) Electrophysiology of hippocampal neurons IV: fast potentials. J Neurophysiol 24:272–285

Spruston N, Schiller Y, Stuart G, Sakmann B (1995) Activity-dependent action potential invasion and calcium influx into hippocampal CA1 dendrites. Science 268:297–300

Starr TV, Prystay W, Snutch TP (1991) Primary structure of a calcium channel that is highly expressed in the rat cerebellum. Proc Natl Acad Sci USA 88:5621–5625

Stea A, Tomlinson WJ, Soong TW, Bourinet E, Dubel SJ, Vincent SR, Snutch TP (1994) Localization and functional properties of a rat brain α_{1A} calcium channel reflect similarities to neuronal Q- and P-type channels. Proc Natl Acad Sci USA 91:10576–10580

Strom TM, Nyakatura G, Apfelstedt-Sylla E, Hellebrand H, Lorenz B, Weber BH, Wutz K, Gutwillinger N, Rüther K, Drescher B, Sauer C, Zrenner E, Meitinger T, Rosenthal A, Meindl A (1998) An L-type calcium-channel gene mutated in incomplete X-linked congenital stationary night blindness. Nature Genetics 19:260–263

Stuart GJ, Dodt HU, Sakmann B (1993) Patch-clamp recordings from the soma and dendrites of neurons in brain slices using infrared video microscopy. Pflügers Arch 423:511–518

Stuart GJ, Sakmann B (1994) Active propagation of somatic action potentials into neocortical pyramidal cell dendrites. Nature 367:69–72

Swartz KJ, Merritt A, Bean BP, Lovinger DM (1993) Protein kinase C modulates glutamate receptor inhibition of Ca2+ channels and synaptic transmission. Nature 361:165–168

Takahashi M, Seagar MJ, Jones JF, Reber BF, Catterall WA (1987) Subunit structure of dihydropyridine-sensitive calcium channels from skeletal muscle. Proc Natl Acad Sci USA 84:5478–5482

Takahashi T, Momiyama A (1993) Different types of calcium channels mediate central synaptic transmission. Nature 366:156–158

Talley EM, Cribbs LL, Lee JH, Daud A, Perez-Reyes E, Bayliss DA (1999) Differential distribution of three members of a gene family encoding low voltage-activated (T-type) calcium channels. J Neurosci 19:1895–1911

Tanabe T, Beam KG, Adams BA, Niidome T, Numa S (1990) Regions of the skeletal muscle dihydropyridine receptor critical for excitation–contraction coupling. Nature 346:567–569

Tanabe T, Beam KG, Powell JA, Numa S (1988) Restoration of excitation–contraction coupling and slow calcium current in dysgenic muscle by dihydropyridine receptor complementary DNA. Nature 336:134–139

Tanabe T, Takeshima H, Mikami A, Flockerzi V, Takahashi H, Kangawa K, Kojima M, Matsuo H, Hirose T, Numa S (1987) Primary structure of the receptor for calcium channel blockers from skeletal muscle. Nature 328:313–318

Toth PT, Bindokas VP, Bleakman D, Colmers WF, Miller RJ (1993) Mechanism of presynaptic inhibition by neuropeptide Y at sympathetic nerve terminals. Nature 364:635–639

Tottene A, Moretti A, Pietrobon D (1996) Functional diversity of P-type and R-type calcium channels in rat cerebellar neurons. J Neurosci 16:6353–6363

Tottene A, Volsen S, Pietrobon D. The R-Type Calcium Current of Rat Cerebellar Granule Cells Comprises Three Components With Distinct Biophysical and Pharmacological Properties. Society for Neuroscience, Vol 1, p431.433, 1999

Tsien RW, Fox AP, Hess P, McCleskey EW, Nilius B, Nowycky MC, Rosenberg RL (1987) Multiple types of calcium channel in excitable cells. Soc Gen Physiol Ser 41:167–187

Tsien RW, Lipscombe D, Madison DV, Bley KR, Fox AP (1988) Multiple types of neuronal calcium channels and their selective modulation. Trends Neurosci 11:431–438

Usowicz MM, Sugimori M, Cherksey B, Llinás R (1992) P-type calcium channels in the somata and dendrites of adult cerebellar Purkinje cells. Neuron 9:1185–1199

Wakamori M, Niidome T, Furutama D, Furuichi T, Mikoshiba K, Fujita Y, Tanaka I, Katayama K, Yatani A, Schwartz A (1994) Distinctive functional properties of the neuronal BII (class E) calcium channel. Receptors Channels 2:303–314

Walker D, De Waard M (1998) Subunit interaction sites in voltage-dependent Ca2+ channels: role in channel function. Trends in Neurosciences 21:148–154

Watanabe S, Takagi H, Miyasho T, Inoue M, Kirino Y, Kudo Y, Miyakawa H (1998) Differential roles of two types of voltage-gated Ca2+ channels in the dendrites of rat cerebellar Purkinje neurons. Brain Research 791:43–55

Waterman SA (1996) Multiple subtypes of voltage-gated calcium channel mediate transmitter release from parasympathetic neurons in the mouse bladder. J Neurosci 16:4155–4161

Waterman SA (1997) Role of N-, P- and Q-type voltage-gated calcium channels in transmitter release from sympathetic neurones in the mouse isolated vas deferens. Br J Pharmacol 120:393–398

Waterman SA, Lang B, Newsom-Davis J (1997) Effect of Lambert-Eaton myasthenic syndrome antibodies on autonomic neurons in the mouse. Annals of Neurology 42:147–156

Welling A, Kwan YW, Bosse E, Flockerzi V, Hofmann F, Kass RS (1993) Subunit-dependent modulation of recombinant L-type calcium channels. Molecular basis for dihydropyridine tissue selectivity. Circ Res 73:974–980

Wessler I, Dooley DJ, Osswald H, Schlemmer F (1990) Differential blockade by nifedipine and ω-Conotoxin GVIA of α_1- and β_1-adrenoceptor-controlled calcium channels on motor nerve terminals of the rat. Neurosci Lett 108:173–178

Westenbroek RE, Ahlijanian MK, Catterall WA (1990) Clustering of L-type Ca2+ channels at the base of major dendrites in hippocampal pyramidal neurons. Nature 347:281–284

Westenbroek RE, Hell JW, Warner C, Dubel SJ, Snutch TP, Catterall WA (1992) Biochemical properties and subcellular distribution of an N-type calcium channel α_1 subunit. Neuron 9:1099–1115

Westenbroek RE, Sakurai T, Elliott EM, Hell JW, Starr TV, Snutch TP, Catterall WA (1995) Immunochemical identification and subcellular distribution of the α_{1A} subunits of brain calcium channels. J Neurosci 15:6403–6418

Wheeler DB, Randall A, Tsien RW (1994) Roles of N-type and Q-type Ca2+ channels in supporting hippocampal synaptic transmission. Science 264:107–111

Wheeler DB, Randall A, Tsien RW (1996) Changes in action potential duration alter reliance of excitatory synaptic transmission on multiple types of Ca2+ channels in rat hippocampus. J Neurosci 16:2226–2237

Williams ME, Brust PF, Feldman DH, Patthi S, Simerson S, Maroufi A, McCue AF, Velicelebi G, Ellis SB, Harpold MM (1992a) Structure and functional expression of an omega-conotoxin-sensitive human N-type calcium channel. Science 257:389–395

Williams ME, Feldman DH, McCue AF, Brenner R, Veliÿelebi G, Ellis SB, Harpold MM (1992b) Structure and functional expression of α_1, α_2, and β subunits of a novel human neuronal calcium channel subtype. Neuron 8:71–84

Williams ME, Marubio LM, Deal CR, Hans M, Brust PF, Philipson LH, Miller RJ, Johnson EC, Harpold MM, Ellis SB (1994) Structure and functional characterization of neuronal α_{1E} calcium channel subtypes. J Biol Chem 269:22347–22357

Wong RK, Prince DA, Basbaum AI (1979) Intradendritic recordings from hippocampal neurons. Proc Natl Acad Sci USA 76:986–990

Wu LG, Borst JG, Sakmann B (1998) R-type Ca2+ currents evoke transmitter release at a rat central synapse. Proc Natl Acad Sci USA 95:4720–4725

Wyatt CN, Page KM, Berrow NS, Brice NL, Dolphin AC (1998) The effect of overexpression of auxiliary Ca2+ channel subunits on native Ca2+ channel currents in undifferentiated mammalian NG108–15 cells. Journal of Physiology 510:347–360

Yokoyama CT, Westenbroek RE, Hell JW, Soong TW, Snutch TP, Catterall WA (1995) Biochemical properties and subcellular distribution of the neuronal class E calcium channel α_1 subunit. J Neurosci 15:6419–6432

Yoshida K, Imaki J, Matsuda H, Hagiwara M (1995) Light-induced CREB phosphorylation and gene expression in rat retinal cells. J Neurochem 65:1499–1504

Yuste R, Gutnick MJ, Saar D, Delaney KR, Tank DW (1994) Ca2+ accumulations in dendrites of neocortical pyramidal neurons: an apical band and evidence for two functional compartments. Neuron 13:23–43

Zhang J-F, Randall AD, Ellinor PT, Horne WA, Sather WA, Tanabe T, Schwarz TL, Tsien RW (1993) Distinctive pharmacology and kinetics of cloned neuronal Ca2+ channels and their possible counterparts in mammalian CNS neurons. Neuropharmacology 32:1075–1088

Zhou Z, January CT (1998) Both T- and L-type Ca2+ channels can contribute to excitation–contraction coupling in cardiac Purkinje cells. Biophys J 74:1830–1839

Zhu Y, Ikeda SR (1993) Adenosine modulates voltage-gated Ca^{2+} channels in adult rat sympathetic neurons. J Neurophysiol 70:610–620

CHAPTER 4

Structure of the Voltage-Dependent L-Type Calcium Channel

F. HOFMANN and N. KLUGBAUER

A. Introduction

Voltage-activated L-type calcium channels regulate the intracellular concentration of calcium and contribute thereby to calcium signaling in numerous cells. These channels are widely distributed in the animal kingdom and are an essential part of many excitatory and non-excitatory mammalian cells. The opening of these channels is primarily regulated by the membrane potential, but is also modulated by a wide variety of hormones, protein kinases, protein phosphatases, toxins, and drugs. Site directed mutagenesis has identified sites on these channels which interact specifically with other proteins, inhibitors, and ions. This chapter will focus on these recent developments. The older findings have been summarized in several excellent reviews (STRIESSNIG et al. 1993; HOFMANN et al. 1994; CATTERALL 1995; DE WAARD et al. 1996a).

B. Subunit Composition and Genes of the Calcium Channel Complex

I. Subunit Composition of L-Type Calcium Channels

Calcium channels are heterooligomeric complexes of five proteins (Fig. 1): (a) the α_1 subunit, which contains the binding sites for all known calcium channel blockers, the voltage-sensor, the selectivity filter and the ion-conducting pore; (b) the intracellularly located β subunit; (c + d) the $\alpha_2\delta$ subunit, a disulfide linked dimer; and (e) the transmembrane γ subunit (HOFMANN et al. 1994).

II. Genes

1. The α_1 Subunit

Most of the prominent features of the calcium channel complex can be assigned to the α_1 subunit. The α_1 subunit contains the ion-conducting pore, the selectivity filter of the pore, the voltage sensor and the interaction sites for the β subunits, the $\beta\gamma$ subunits of the G proteins, the $\alpha_2\delta$ subunit, the calcium channel blockers and activators. Nine individual genes have been identified for the α_1 subunit, which are homologous to each other and encode proteins of predicted molecular masses of 212–273 kDa. They belong to the same multi-

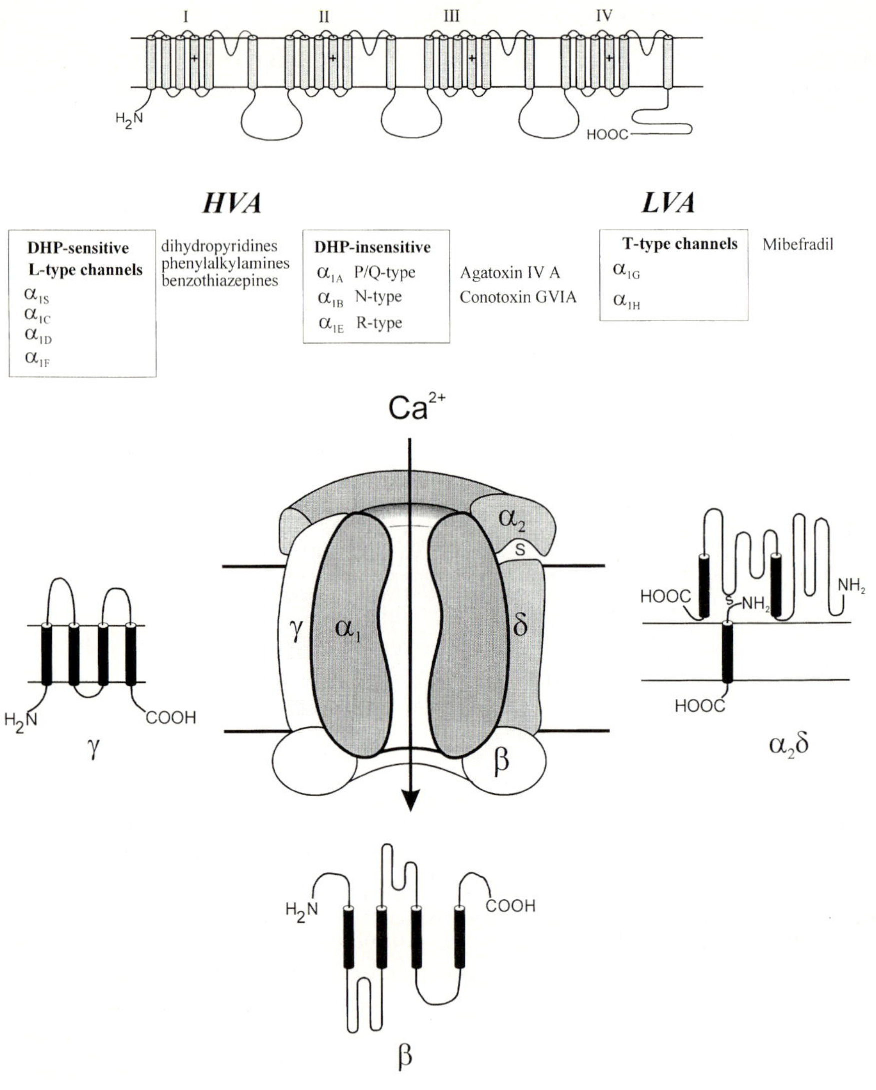

Fig. 1. Putative structure of the calcium channel complex. Proposed structures of the α_1 subunit (*top*) and the accessory β, $\alpha_2\delta$ and γ subunits are indicated. A disulfide bridge (s) connects the transmembrane δ and the extracellular α_2 subunit. The molecular diversity of the α_1 subunit and pharmacological properties are indicated. *HVA*, high voltage activated; *LVA*, low voltage activated

gene family as voltage-activated sodium and potassium channels and share a common ancestral protein with them. Hydrophobicity analysis of the α_1 subunits predicts a transmembrane topology with four homologous repeats, each containing five hydrophobic putative α helices and one amphiphatic segment (Fig. 1).

An early evolutionary event separated the α_1 subunits into the electrophysiologically distinct low voltage-activated (LVA) and high voltage-activated (HVA) calcium channels, which share less than 30% sequence identity. The two LVA genes G and H induce T-type current in the absence of additional subunits (PEREZ-REYES et al. 1998; CRIBBS et al. 1998). A later occurring event separated the HVA-channels again in two subfamilies, the four (C, D, F, S) dihydropyridine (DHP)-sensitive and the three (A, B, E) DHP-insensitive calcium channels. The A, B, and E genes are expressed almost exclusively in neuronal tissues. Both groups share about 50% identical amino acids, whereas the amino acid identity of the individual members of each subfamily is generally over 60%.

a) The L-Type α_1 Channels

α) The Class S α_1 Gene

The complete cDNA sequence of the class S gene was originally cloned from rabbit skeletal muscle (TANABE et al. 1987). Two isoforms of this calcium channel type can be identified in rabbit skeletal muscle: a 212 kDa polypeptide equivalent to the full length calcium channel transcript and a smaller 190 kDa protein, which is derived from the full length product by posttranslational proteolysis. This short form represents about 95% of the total α_{1S} calcium channel protein (DE JONGH et al. 1991).

β) The Class C α_1 Gene

The class C gene is expressed in heart and smooth muscle, in endocrine and neuronal cells. The human gene for the α_{1C} subunit is localized to the distal region of chromosome 12p13 (SCHULTZ et al. 1993). The gene spans about 150 kb and is composed of 44 invariant and over 6 alternative exons (SOLDATOV 1994). The α_1 subunit of the cardiac ($\alpha_{1C\text{-}a}$) (MIKAMI et al. 1989) and smooth muscle ($\alpha_{1C\text{-}b}$) (BIEL et al. 1990) calcium channel differ only at four sites and share 95% identical amino acids. Molecular analysis showed that the alternatively spliced exon 8, which codes for the IS6 segment, is differentially expressed in cardiac and vascular smooth muscle and is responsible in part for the different DHP sensitivity of the cardiac and vascular smooth muscle L-type current (WELLING et al. 1997); further details are discussed in Sect. C.III.1.

γ) The Class D α_1 Gene

The cDNA of the class D was isolated from neuronal and endocrine tissues and represents a neuroendocrine specific L-type calcium channel (WILLIAMS et al. 1992b; SEINO et al. 1992). Expression of α_{1D} cDNA in different host cells demonstrated only a small dihydropyridine sensitive inward current indicating that the native channel may contain an additional, so far unknown subunit.

δ) The Class F α_1 Gene

Analysis of the locus for the incomplete form of X-linked congenital stationary night blindness (CSNB2) identified mutations in a new L-type calcium channel α_1 subunit as cause of the disease (Strom et al. 1998; Bech-Hansen et al. 1998). The gene for the α_{1F} subunit is localized at Xp11.23. The F channel shows a 55–62% overall amino acid sequence identity with other L-type calcium channel α_1 subunits. Apparently, this channel is expressed specifically in the retina and required for optimal night vision.

b) *The None L-Type α_1 Channels*

α) The Class A α_1 Gene

Transcripts of the class A channel are present at high levels in the mammalian brain and peripheral nervous system (Mori et al. 1991; Starr et al. 1991). Because the α_{1A} transcripts are expressed in many neurons shown to possess P- and Q-type channels and because the properties of α_{1A} exhibits similarities with both of these channels (Stea et al. 1994), the class A cDNA is refered to as P/Q-type calcium channel.

β) The Class B α_1 Gene

The class B gene has been cloned exclusively from brain (Williams et al. 1992a; Dubel et al. 1992; Fujita et al. 1993). Expression studies using dysgenic myotubes or *Xenopus* oocytes revealed that α_{1B} induced a barium current which is inhibited by low concentrations of ω-conotoxin GVIA (Fujita et al. 1993; Williams et al. 1992a). The α_{1B} subunit also binds ω-conotoxin GVIA with high affinity (Dubel et al. 1992). These results identify the α_{1B} channel as the neuronal N-type calcium channel.

γ) The Class E α_1 Gene

The sixth gene has been cloned from rat, rabbit, and human brain libaries (Niidome et al. 1992; Soong et al. 1993; Williams et al. 1994; Schneider et al. 1994). Initially, this channel was characterized as an LVA T-type channel (Soong et al. 1993). However, later studies (Williams et al. 1994; Schneider et al. 1994) showed that the expressed α_{1E} channel has the activation and inactivation kinetics of a HVA neuronal channel. The human and rat α_{1E} currents have some properties in common with the R-type currents observed in cerebellar granule cells (Ellinor et al. 1993).

c) *The Low Voltage-Activated α_1 Channels*

α) The Class G and H Gene

The recently cloned class G and H α_1 subunits are LVA calcium channels, which have the basic electrophysiological characteristics of T-type channels (Perez-Reyes et al. 1998; Cribbs et al. 1998). The G gene localizes to human chromosome 17q22 and is expressed strongly in brain and less abundantly in heart. The expressed channel has a single channel conductance of 7.7 pS in 115 mmol/l Ba^{2+}. The current is blocked half maximally by Ni^{2+} at 1.1 mmol/l. The mibefradil block is slightly voltage dependent with IC_{50} values of 0.4 μmol/l and 0.1 μmol/l at a holding potential of –100 mV and –60 mV, respectively (KLUGBAUER et al. 1999b). The H gene localizes to the human chromosome 16p13.3 and is expressed strongly in kidney, at intermediate levels in heart, and at low abundance in brain. The expressed channel has a single channel conductance of 5.5 pS and is blocked by Ni^{2+} at micromolar concentrations and by mibefradil with an IC_{50} of 1.4 μmol/l at HP –90 mV (CRIBBS et al. 1998).

2. Auxiliary Subunits of the Calcium Channel

a) The $\alpha_2\delta$ Subunit

The skeletal muscle $\alpha_2\delta$-1 subunit is a highly glycosylated membrane protein of 125 kDa (ELLIS et al. 1988). The protein is posttranslationally cleaved to yield a disulfide-linked α_2 and δ protein (for older literature see HOFMANN et al. 1994; CATTERALL 1995; DE WAARD et al. 1996a). The δ part anchors the α_2 protein to the α_1 subunit via a single transmembrane segment, whereas the α_2 protein is localized extracellularly. This membrane topology of the $\alpha_2\delta$ subunit was confirmed and further refined (WISER at al. 1996; GURNETT et al. 1996, 1997; FELIX et al. 1997). Extensive splicing of this subunit results in at least five different isoforms, which are expressed in a tissue specific manner (ANGELOTTI and HOFMANN 1996). Two additional $\alpha_2\delta$ genes – $\alpha_2\delta$-2 and $\alpha_2\delta$-3 – have been identified recently (KLUGBAUER et al. 1999a). The primary structure of the novel $\alpha_2\delta$-2 and $\alpha_2\delta$-3 subunits is about 50% and 30% identical with the $\alpha_2\delta$-1 subunit, respectively. Northern blot analysis indicates that $\alpha_2\delta$-3 is expressed exclusively in brain, whereas $\alpha_2\delta$-2 is found in several tissues and $\alpha_2\delta$-1 is expressed ubiquitously. In situ hybridization of mouse brain sections showed mRNA expression of $\alpha_2\delta$-1 and $\alpha_2\delta$-3 in the hippocampus, cerebellum, and cortex, with $\alpha_2\delta$-1 strongly detected in the olfactory bulb and $\alpha_2\delta$-3 in the caudate putamen. The number of putative glycosylation sites and cysteine residues, hydropathicity profiles, and electrophysiological character of the $\alpha_2\delta$-3 subunit is similar to that of the $\alpha_2\delta$-1 subunit if expressed together with the α_{1C} and cardiac β_{2a} subunit (KLUGBAUER et al. 1999a). In general, coexpression of an $\alpha_2\delta$-1 subunit with α_1 and β subunits shifts the voltage-dependence of channel activation and inactivation in a hyperpolarizing direction, accelerates the kinetics of current inactivation, and increases the current amplitude (SINGER et al. 1991; DE WAARD et al. 1995a; GURNETT et al. 1996, 1997; BANGALORE et al. 1996; FELIX et al. 1997; QUIN et al. 1998b; KLUGBAUER et al. 1999a). Some inconsistencies in reported results can be accounted for by the experimental conditions, as various expression systems (*Xenopus* oocytes or mammalian cell lines), different charge carriers (Ba^{2+} or Ca^{2+}), different splice variants of the $\alpha_2\delta$-1 subunit, and different α_1 (α_{1C}, α_{1A}, α_{1E}) and β (β_1, β_2, β_3, or β_4) subunits were used. Detailed analysis of the effects of the α_2 and

δ proteins (GURNETT et al. 1996, 1997; FELIX et al. 1997) suggests that the extracellular α_2 protein enhances current density and the affinity for the DHP isradipine, whereas the transmembrane segment of the δ protein interacts with repeat III and some additional parts of the channel (GURNETT et al. 1997). Changes in the channel kinetics are associated with the expression of the δ protein.

The mechanism whereby $\alpha_2\delta$ modulates the conductance of α_1 is not clearly understood. The increase in current density can be partly accounted for by improved targeting of expressed α_1 subunit to the cell membrane (SHISTIK et al. 1995). The effects of the coexpression of $\alpha_2\delta$ subunit on time course and/or voltage dependence on current activation and inactivation also suggests a specific modulation of channel gating. In the presence of the $\alpha_2\delta$-1 subunit, the open probability of the channel is enhanced without a change in the mean open time (SHISTIK et al. 1995) and the amount of charge moved during channel activation increases (BANGALORE et al. 1996; QIN et al. 1998b). This increase in charge movement was coupled with an increased and unchanged maximal conductance, when the L-type α_{1C} calcium channel (BANGALORE et al. 1996) and neuronal α_{1E} channel (QIN et al. 1998b) were used, respectively. SHIROKOV (1998) reported that $\alpha_2\delta$-1 speeds up the transfer of the α_{1C} channel into a slow inactivated state and slows down its recovery. These changes in channel gating may underlie the observed effects on the inactivation of whole cell current.

b) The β-Subunit

The β subunits are intracellularly located proteins ranging from 50 to 72 kDa. Four genes – β_1, β_2, β_3, and β_4 – have been identified (RUTH et al. 1989; HULLIN et al. 1992; PEREZ-REYES et al. 1992; CASTELLANO et al. 1993) which give rise to several splice variants. A primary structure alignment of β subunits revealed that all share a common central core, whereas their N- and C-termini and a part of the central region differ significantly. Coexpression of a β subunit with various α_1 subunits increases peak current (SINGER et al. 1991) most likely by increasing the number of functional surface membrane channels and by facilitating channel pore opening (NEELY et al. 1993; JOSEPHSON and VARADI 1996; KAMP et al. 1996). With the exception of the rat brain β_{2a}, all other β subunits accelerate channel activation and inactivation and shift the steady state inactivation curve to hyperpolarized potential (SINGER et al. 1991; WEI et al. 1991; HULLIN et al. 1992; CASTELLANO et al. 1993). All four β subunits combine with the neuronal α_1 subunits (SCOTT et al. 1996; LIU et al. 1996; LUDWIG et al. 1997; PICHLER et al.1997; VOLSEN et al. 1997; VANCE et al. 1998). The brain expression of the β_4 subunit increases about tenfold between postnatal day 2 and maturity, in which time it associates with N- and P-type channels (VANCE et al. 1998). Mutation of the β_4 subunit in lethargic mice is associated with ataxia and seizures (BURGESS et al. 1997). The lethargic phenotype could be caused by the persistence of an immature N-type calcium channel coassembled with the β_{1b} subunit (McEnery et al. 1998). In contrast to neuronal calcium channels, the skeletal and cardiac muscle calcium channel are associated apparently exclusively with the β_{1a} and cardiac β_{2a} subunit (Ruth et al. 1989; Ludwig et al. 1997; Qin et al. 1998a).

Differential splicing of the primary transcripts of β_1 results in the expression of at least three isoforms (Ruth et al. 1989; Pragnell et al. 1991; Williams et al. 1992b). β_{1a} is exclusively expressed in skeletal muscle together with the α_{1S}, α_2/δ_A and γ_1 subunit, whereas the other two isoforms of β_1 were identified in brain and spleen (Powers et al. 1992). Deletion of the β_1 gene in mice leads to perinatal lethality (Gregg et al. 1996). The absence of the β_1 subunit lowers the concentration of the α_{1S} subunit in skeletal muscle and impairs thereby excitation-contraction coupling. Coexpression of the brain splice variant β_{1b} – but not that of the skeletal muscle β_{1a} variant – together with the α_{1S}, α_2/δ_A and γ_1 subunit has been reported to induce measurable inward current in oocytes suggesting that this specific splice variant has significant effects on the property of the skeletal muscle calcium channel (Ren and Hall 1997).

The β_2 gene is expressed abundantly in heart and to a lower degree in aorta, trachea, lung, and brain (Biel et al. 1991), whereas the β_3 specific mRNA is detectable in brain and different smooth muscle tissues (Hullin et al. 1992; Ludwig et al. 1997). The β_2 transcript is extensively spliced resulting in at least four different isoforms (Perez-Reyes et al. 1992; Hullin et al. 1992). The rabbit cardiac β_{2a} (Hullin at al. 1992) and the rat brain β_{2a} (Perez-Reyes et al. 1992) are N-terminal splice variants of the same gene. The rat brain β_{2a} has two cysteines at position 3 and 4 which are palmitoylated in vivo (Chien et al. 1996; Qin et al. 1998a). The β_{2a} expressed in rabbit heart does not contain the aminoterminal cysteines (Qin et al. 1998a) and is identical with the cloned cardiac β_{2a} (Hullin et al. 1992). Coexpressed with α_{1E}, the brain β_{2a} reduces the rate at which α_{1E} inactivates in response to depolarization, causes a right shift in steady-state inactivation curve, does not support facilitation of the α_{1C} current (Qin et al. 1998a), and prevents prepulse potentiation caused by G protein $\beta\gamma$subunit interaction with neuronal α_1 subunits (Herlitze et al. 1996). Prevention of the palmitoylation of the brain β_{2a} by mutation of the two cysteines to serines changes its properties to that of the cardiac β_{2a}, i.e., the mutated β_{2a} subunit accelerates channel activation and inactivation, shifts the steady-state inactivation curve to hyperpolarized potential, supports facilitation of the α_{1C}, current and interferes poorly with prepulse potentiation (Qin et al. 1998a). The extent of palmitoylation is affected by mutation in other regions of the neuronal β subunit, i.e., in a src homology 3 motif and in the β subunit interaction domain (Chien et al. 1998) (see also Sect. C.II.2).

c) The γ Subunit

The γ_1 subunit is an integral membrane protein consisting of 222 amino acids with a predicted molecular mass of 25 kDa (Bosse et al. 1990; Jay et al. 1990), which is exclusively expressed in skeletal muscle (Eberst et al. 1997). Recently, a second γ_2 subunit has been identified in brain which has 25% identity with γ_1 and is most highly expressed in cerebellum, olfactory bulb, cerebral cortex, thalamus and CA3, and dentate gyrus of the hippocampus (LETTS et al. 1998). The human γ_1 and γ_2 subunits are encoded on chromosome 17q23 and 22q12–13, respectively (POWERS et al. 1993; LETTS et al. 1998). Hydrophobicity analysis reveals the existence of four putative transmembrane helices with intracellular located amino- and carboxy-termini. The presence of two extracellular potential N-glycosylation sites is consistent with the observed strong glycosylation of these subunits. Coexpression of each γ subunit together with α_1, α_2/δ, and β subunits in oocytes induces a left shift in the steady-state inactivation curves (SINGER et al. 1991; LETTS et al. 1998). The γ_2 gene is mutated in stargazer mice leading to spike-wave seizures characteristic of absence epilepsy with accompanying defects in the cerebellum and inner ear (LETTS et al. 1998).

III. Functional Domains of the α_1 Subunit

1. The Pore and Ion Selectivity Filter

Part of the pore structure of the calcium channel is formed by the linker connecting the S5 and S6 transmembrane segments in repeat I to IV (GUY and CONTI 1990). This P region is thought to contribute to the outer vestibule of the channel pore and to span the outer half of the membrane. In analogy to the recently obtained crystal structure of the *Streptomyces lividans* potassium channel (DOYLE et al. 1998), the calcium channel pore can be envisioned to have the structure of an inverted teepee with the vertex inside the cell. The helices of the four S6 segments would form the poles of this teepee, which are widely separated near the outer membrane surface and converging towards a narrow zone at the inner surface. This outer structure would stabilize an inner ring formed by the four P-regions, which control the speed of permeation and the ion selectivity.

Mutational analysis of the α_{1C} (TANG et al. 1993; YANG et al. 1993) and α_{1A} (KIM et al. 1993) channel has shown that the four glutamic acid residues E413, E731, E1140, and E1441 (amino acid numbering is according to the $\alpha_{1C\text{-}b}$ sequence (BIEL et al. 1990)) in the P region of repeat I, II, III, and IV are critical in determining the ion selectivity of the calcium channel. Equivalent glutamates are present in all HVA calcium channels. Mutation of these glutamates decreased dramatically the affinity for Ca^{2+} or Cd^{2+} to block monovalent ion permeation (YANG et al. 1993; KIM et al. 1993; YATANI et al. 1994; ELLINOR et al. 1995; PARENT and GOPALAKRISHNAN 1995). The studies showed that these glutamates form the high affinity Ca^{2+} binding site within the pore that is responsible for the Ca^{2+} selectivity. The glutamic acid residues of each repeat contribute differently to the Ca^{2+} affinity, selectivity, and speed of permeation (TANG et al. 1993; PARENT and GOPALAKRISHNAN 1995; ELLINOR et al. 1995). Mutation of E1140 in repeat III has a much greater effect on ion selectivity and permeation than comparable mutations in the other three repeats.

LVA channels have aspartates instead of glutamates in the pore of repeat III and IV, which difference may be the cause of their distinct ion selectivity (PEREZ-REYES et al. 1998; CRIBBS et al. 1998; KLUGBAUER et al. 1999b).

To explain rapid permeation of calcium ions, different models have been discussed with one or two – high and low affinity – site(s) for Ca^{2+} (HESS and TSIEN 1984; TSIEN et al. 1987; ROSENBERG and CHEN 1991; KUO and HESS 1993; ARMSTRONG and NEYTON 1991). In a recent study, ELLINOR et al. (1995) demonstrated that these glutamates form a single high affinity Ca^{2+} site within the pore. This site may be accessed by two Ca^{2+} ions at the same time, thereby allowing rapid permeation. The cloned smooth muscle $\alpha_{1C\text{-}b}$ channel permeates rapidly Ca^{2+} at physiological pH and voltages and has a high unitary conductance (GOLLASCH et al. 1996), whereas the unitary conductance of the skeletal muscle α_{1S} subunit is half of that of the cardiac α_{1C} subunit (DIRKSEN et al. 1997). Unitary conductance was reduced from cardiac to skeletal muscle size, when the skeletal muscle IS5-IS6 linker was introduced in to the cardiac α_{1C} subunit (DIRKSEN et al. 1997). The net charge of the vestibule part of the cardiac and skeletal muscle IS5-IS6 linker is –5 and –2, respectively. It is plausible that the more negatively charged vestibule of the cardiac compared to skeletal muscle channel increases conduction by electrostatic attraction of Ca^{2+} ions to the channel pore.

Increased extracellular proton (H^+) concentrations that occur during episodes of intense neuronal activity or with ischemia in heart strongly inhibit ion permeation through open calcium channels (KUO and HESS 1993). A single H^+ binding site has been invoked. Analysis of the mutated α_{1C} subunit localized this site to the glutamates of the pore region. Controversial data have been published suggesting that H^+ binding requires either only E1140 in repeat III (KLÖCKNER et al. 1996) or E413 and E1140 in repeat I and III (CHEN and TSIEN 1997). The two glutamate model may explain better the unusual high pKa (pH>8) of the protonated site than the single glutamate model. The interpretation of these results is further complicated by the observation, that removal of protons increases L-type current only, when the α_{1C} subunit is expressed together with the cardiac β_{2a} subunit (SCHUHMANN et al. 1997).

2. Channel Activation

Mutational analysis in K^+ (PAPAZIAN et al. 1991; LIMAN et al. 1991) and Na^+ (STÜHMER et al. 1989) channels suggested that the positive charges of the S4 segments in each repeat function as voltage sensor. Mutation of individual S4 arginines in repeat I and III of a skeletal/cardiac α_1 chimera affected midpoint and time constant of activation, whereas those of repeat II and IV were without effect (GARCIA et al. 1997). Mutation of the leucine heptad motif present in the region of S4-S5 in repeat I and III yielded inconclusive results. The speed of calcium channel activation is also a property of the α_1 subunit and is modulated by the $\alpha_2\delta$ (see Sect. B.II.2.a) and β (see Sect. B.II.2.b) subunits. More than a fivefold difference in the speed of activation was observed

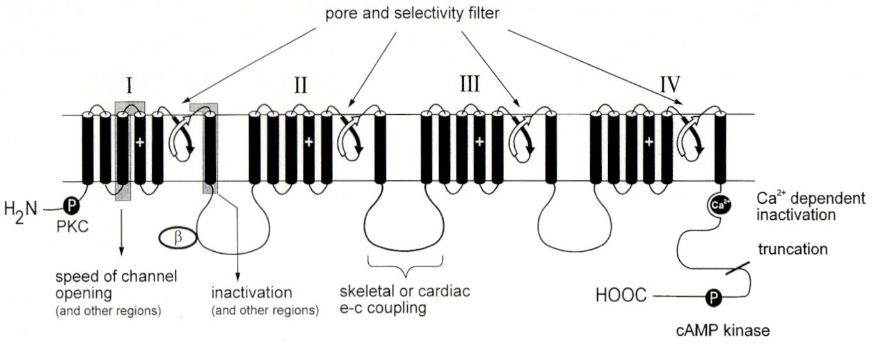

Fig. 2. Suggested topology of the L-type calcium channel α_1 subunit. The putative transmembrane configuration is based on the hydrophobicity analysis of the primary structure. The α_1 subunit consists of four homologous repeats (I, II, III, IV) each containing six membrane-spanning segments. The amphipathic segment which forms the voltage sensor of the channel is indicated by a +. *Black and white arrows* are part of the channel pore and contain the selectivity filter. *Grey boxes* indicate regions involved in activation or inactivation kinetics. *P* indicates sites for cAMP kinase or protein kinase C (*PKC*). *e–c coupling*, excitation – contraction coupling; β, binding site for β subunit; Ca^{2+}, interaction site for Ca^{2+} dependent inactivation

between the skeletal (slow) and cardiac (fast) α_1 subunits. Functional expression of chimeric calcium channels showed that repeat I determines the speed of activation (Fig. 2) (TANABE et al. 1991). Initially, the S3 segment and the linker IS3-IS4 was shown to control slow and fast activation (NAKAI et al. 1994). Analysis of several skeletal/cardiac chimeras suggests that, although unitary conductance and speed of activation are encoded in different parts of repeat I, the linker IS5-IS6 affects not only unitary conductance but also the speed of activation (DIRKSEN et al. 1997). In addition, the sequence between IIIS5 and IVS6 contributes also to the speed of channel activation (WANG et al. 1995).

3. Channel Inactivation

HVA-calcium channels show two types of inactivation: slow and fast inactivation. The slow inactivation is voltage-dependent, whereas the fast inactivation is caused by the permeating calcium ion. The kinetics of slow/voltage-dependent inactivation, which is observed with all HVA calcium channels, differ considerably between the various types of calcium channels and are important in determining the amount of calcium entry during electrical activity. The IS6 segment and its flanking regions are critical for the inactivation properties of the channel (ZHANG et al. 1994) as determined with chimeric α_1 subunits of channels with different inactivation rates, i.e., the α_1 subunits of the class C, class A and doe-1, an α_1 subunit cloned from the marine ray *Discopyge ommata*. Chimeras between the α_{1C} and α_{1S} calcium channels confirmed this conclusion (PARENT et al. 1995). However, inactivation of the α_{1C}

channel is also controlled by the intracellular carboxyterminal sequences (WEI et al. 1994). Removal of the carboxyterminus of the $\alpha_{1C\text{-a}}$ or $\alpha_{1C\text{-b}}$ subunit up to aa 1733 or 1728, respectively, increases the expressed current (WEI et al. 1994; KLÖCKNER et al. 1995; SEISENBERGER et al. 1995) without increasing the charge moved or the density of DHP binding sites (WEI et al. 1994). Therefore, truncation of the channel up to aa 1733 does not increase the number of channels but removes an inhibitory action of the carboxyterminus. Similar results have been obtained in vivo by perfusion of cardiac myocytes with trypsin (HESCHELER and TRAUTWEIN 1988). However, the trypsinated channel had lost its calcium sensitivity, whereas the truncated channel still showed calcium-dependent inactivation.

Calcium-sensitive inactivation of α_{1C} channels is a negative biological feedback mechanism, by which the increase of intracellular calcium speeds up channel inactivation and prevents a calcium overload of the cell. Using the L-type calcium current of guinea pig cardiac myocytes, HESCHELER and TRAUTWEIN (1988) showed that intracellular application of trypsin or carboxypeptidase increased the amplitude of calcium or barium current and decreased calcium-dependent inactivation. The trypsin-dependent increase in current amplitude was confirmed by others (SCHMID et al. 1995; YOU et al. 1995), whereas the loss of calcium-dependent inhibition was seen by YOU et al. (1995) but not by SCHMID et al. (1995). These discrepant results were clarified by the use of the cloned α_1 subunits (Fig. 2). Fast/Ca^{2+}-dependent inactivation is especially prominent in the cardiac and the smooth muscle channel and requires only the α_{1C} subunit (WELLING et al. 1993b; NEELY et al. 1994; ZONG and HOFMANN 1996). Intracellular Ca^{2+} inactivates calcium current by binding to a single site with an IC_{50} of 4 μmol/l Ca^{2+} (HÖFER et al. 1997) supporting the hypothesis of the presence of a single EF hand (BABITCH 1990). Exchange of amino acids between residues 1572 and 1651 by exons only found so far in the α_{1C} gene increases the speed of inactivation and, depending on the substitution, removes calcium-dependent inactivation (SOLDATOV et al. 1998; ZÜHLKE and REUTER 1998). Exchange of the same region of α_{1C} sequence for those of α_{1E} – a calcium insensitive channel – also results in a loss of calcium-dependent inhibition (DE LEON et al. 1995; ZHOU et al. 1997). However, no agreement exists on the importance of the EF hand binding motif, since exchange or removal of it did effect calcium sensitivity (SOLDATOV et al. 1998; ZÜHLKE and REUTER 1998) or had no effect (ZHOU et al. 1997). Further complication comes from the work of ADAMS and TANABE (1997). An α_{1C}/α_{1S} chimera, in which the carboxyterminal α_{1C} sequence 1633 to 2166 was replaced by the skeletal muscle sequence 1510 to 1873, had lost calcium-dependent inactivation. However, the same chimera, in which the last 211 amino acids from the skeletal muscle (sequence used 1510 to 1662) were removed, again showed Ca^{2+}-dependent inactivation. It is quite likely that these very different sequence modifications affected either the Ca^{2+} binding site, or the conformation of the carboxyterminus, that mediates channel inhibition or both. Agreement exist only insofar that Ca^{2+}-dependent inactivation requires only the α_{1C} subunit and binding of Ca^{2+} to the intracellular amino acid stretch between residues 1513 and approximately 1700.

IV. Sites for Interaction with Other Proteins

The α_1 subunit interacts with a number of proteins such as its auxiliary subunits $\alpha_2\delta$, β, and γ and proteins such as the ryanodine receptor and proteins necessary for fusion of a neurosecretory vesicle with the presynaptic membrane. The potential interaction sites for the γ subunit and the $\alpha_2\delta$ are unknown or have been outlined above (see Sect. B.II.2.a). Here we will consider only those interactions relevant to the α_{1S} and α_{1C} subunits.

1. Interaction of the α_1 Subunit with the Ryanodine Receptor

In cardiac muscle, excitation-contraction (e-c) coupling does not require a direct contact between the calcium channel and the ryanodine receptor type 2 (RyR-2). Calcium release from the sarcoplasmatic reticulum (SR) is triggered by the calcium flowing through the open L-type α_{1C} calcium channel into a restricted space between the plasma membrane and the SR (SHAM et al. 1995). In contrast in skeletal muscle, e-c coupling requires direct coupling between the α_{1S} subunit and the ryanodine receptor type 1 (RyR-1). The cytoplasmic loop between repeat II and III of the α_{1S} subunit, but not that of the α_{1C} subunit, affects ryanodine binding to skeletal muscle RyR-1 and induces calcium release from skeletal muscle SR (TANABE et al. 1990). The α_{1S} subunit can be replaced by a peptide containing the skeletal sequence E666 to L791 (LU et al. 1994). Later refinement of this peptide showed: (i) that phosphorylation of S687 (RÖHRKASTEN et al. 1988) in the peptide E666-E726 prevents activation of calcium release from the SR (LU et al. 1995); (ii) that activation of RyR-1 requires only the sequence T671-L690 (EL-HAYEK et al. 1995) which contains the essential basic cluster RKRRK (EL-HAYEK and IKEMOTO 1998); (iii) that activation of the RyR-1 by the peptide T671-L690 is prevented by the peptide E724-P760 which is localized in the carboxyterminal part of the II-III loop of α_{1S} (EL-HAYEK et al. 1995). Using α_{1S}/α_{1C} chimeras expressed in the dysgenic myotubes, NAKAI et al. (1998b) have slightly revised the site which interact with the RYR-1. Transfer of the skeletal muscle sequence between residues 711–765 to a cardiac α_{1C} subunit yields skeletal muscle type e-c coupling. The core region between residues 725–742 is necessary for e-c coupling but gives only a weak response (NAKAI et al. 1998b).

Activation of the RyR-1 is not affected by truncation of the intracellular tail of the α_{1S} sequence at N1662, suggesting that this part of the tail is not necessary for normal e-c coupling in skeletal muscle (BEAM et al. 1992). RyR-1 expression is not only necessary for normal e-c coupling, but also for a high density of the DHP receptor complex in skeletal muscle (NAKAI et al. 1996) and neurons (CHAVIS et al. 1996). Work with chimeric RyR-1/RyR-2 showed that the sequence from aa 1635 to 2636 of the RyR-1 couples to the α_{1S} subunit of the DHP-receptor, increases the density of the DHP receptor complex, and is necessary for calcium release from the SR (NAKAI et al. 1998a). In addition, the carboxyterminal sequence aa 2659–3720 couples to the DHP-receptor complex as evidenced by an increase in calcium current, but does not allow calcium release from the SR (NAKAI et al. 1998a) suggesting multiple contact sites between the skeletal muscle calcium channel complex and the cytosolic part of the RyR-1.

2. Interaction of the α_1 Subunit with the β Subunit

Coexpression of a β subunit with α_1 subunits alters the voltage-dependence, kinetics, and magnitude of the calcium channel current. The differences in reported effects most likely depend on the particular combination of both subunits and splice variants. These modulatory effects are the consequence of conformational changes in the quaternary structure resulting from the specific interaction of subunit surfaces (NEELY et al. 1993). To identify the β subunit interaction site on the α_1 subunit, an epitope library of the α_{1S} subunit was screened with a labeled β_{1b} subunit probe (PRAGNELL et al. 1994). The β subunit probe binds to the cytoplasmic linker between domain I and II of the α_1 subunit (Fig. 2). A detailed analysis of different α_1 subunits revealed that a highly conserved sequence motif, called AID for alpha subunit interaction domain, is reponsible for this specific interaction, i. e., 428QQ-E–L-GY–WI–E445 (amino acid numbering is according to the $\alpha_{1C\text{-}b}$ sequence (BIEL et al. 1990)) positioned 24 amino acids from the IS6 transmembrane domain in each α_1 subunit. Further mutations showed that only the sequence –437Y–WI441- is essential for high affinity binding of the β subunits, whereas the sequence -Q-ER- is necessary for binding of the $\beta\gamma$ subunit of G proteins to the neuronal α_{1A}, α_{1B}, and α_{1E} channels (DE WAARD et al. 1996b). The L-type calcium channels α_{1C}, α_{1D}, α_{1F}, and α_{1S}, which do not have the R in the -Q-ER sequence, do not bind the $\beta\gamma$ subunit of G proteins and their current amplitudes are not modified by G proteins. Mutation of the tyrosine to a serine (-Y–WI- to -S–WI-) reduces the affinity of the AID for β subunits dramatically (WITCHER et al. 1995). This mutation abolishes the stimulation of peak currents, the change in the inactivation kinetics, and the voltage-dependence of activation by the β subunit (DE WAARD et al. 1996b). In a biochemical assay, DE WAARD et al. (1995b) showed that the AID of the α_{1A} subunit binds $\beta 4$ with a K_D of 5 nmol/l. The relative affinities for the various β subunits to the AID_A were $\beta 4 > \beta 2a > \beta 1b >> \beta 3$. A second low affinity binding site (K_D about 100 nmol/l) for the β_4 and β_3 subunit has been detected in the carboxyterminal sequence of the α_{1A} subunit between residues 2090 and 2424 (WALKER et al. 1998) and the α_{1E} subunit (TAREILUS et al. 1997).

Since all four β subunits can modulate the kinetics and voltage dependence of the α_1 subunit and bind to the AID, it was likely that β subunits contain a conserved motif, which binds to AIDs. To identify this structural domain, a series of truncated and mutated β_{1b} subunits was constructed and tested to interact with α_{1A} in vitro (DE WAARD et al. 1994). A 30 amino acid domain of the β subunit (aa 215–245 of β_{1b}) is sufficient to induce all the modulatory effects of this subunit. This sequence stretch is located at the amino terminus of the second region of high conservation among all four β subunits. Modifications in this region changed or abolished the stimulation of calcium currents by the β subunit and the binding to the α_1 subunit.

Deletion of the β_1 subunit gene showed that a proper targeting of the α_{1S} subunit in skeletal muscle depends on the coexpression of the β_{1a} subunit (GREGG et al. 1996). Transient transfection of the β_1 cDNA in the deficient myotubes restored Ca^{2+} current, charge movement and Ca^{2+} transients (BEURG et al. 1997). Slightly different results were reported when the homozygous dysgenic (mdg/mdg) cell line GLT was used (NEUHUBER et al. 1998a). This cell line does not express the α_{1S} subunit. Proper targeting of the β_{1a} subunit required coexpression of the α_{1S} subunit, in which the AID subunit was not mutated (NEUHUBER et al. 1998a). Further experiments on the interaction and targeting of the α_{1S} subunit by the β_{1a} or neuronal β_{2a} in tsA201 cells yielded similar results (NEUHUBER et al. 1998b). The biological significance of these findings is not clear since: (i) the β_{1a} subunit is expressed in the absence of the α_{1S} subunit in mdg/mdg myotubes; (ii) the neuronal β_{2a} subunit is targeted by palmitoylation of the two amino terminal cysteines to the plasma membrane; (iii) palmitoylation of the β_{2a} subunit is affected significantly by mutations in the BID and other domains (CHIEN et al. 1998): (iv) it is difficult to understand how the β subunits affect barium current without colocalizing with the α_{1S} subunit (NEUHUBER et al. 1998a,b).

V. Binding Sites for L-Type Calcium Channel Agonists and Antagonists

1. The Dihydropyridine Binding Site

The L-type calcium channel ligands represent a clinically and experimentally important set of blockers and agonists. The major classes of these drugs are the dihydropyridines (DHP), phenylalkylamines (PAA), and benzothiazepines. Different techniques have been used to localize potential binding sites of these drugs on the calcium channel complex. Earlier experimental observations from photoaffinity labeling and peptide mapping studies on the skeletal muscle channel revealed that all three classes bind to the transmembrane region of repeat IV of the α_1 subunit (REGULLA et al. 1991; CATTERALL and STRIESSNIG 1992; KUNIYASU et al. 1998) with additional sites on repeat III (CATTERALL and STRIESSNIG 1992; KALASZ et al. 1993) and repeat I (KALASZ et al. 1993) for the DHPs. These localizations were refined by the use of chimeric α_{1C}/α_{1A} and α_{1C}/α_{1E} channels and site directed mutagenesis of single amino acids in the α_{1S} or α_{1C} subunit (Fig. 3). High affinity block of α_{1C} mediated barium current (I_{Ba}) by the DHP antagonist isradipine or (–)R-202–791 is prevented by mutation of the L-type specific amino acids (amino acid numbering

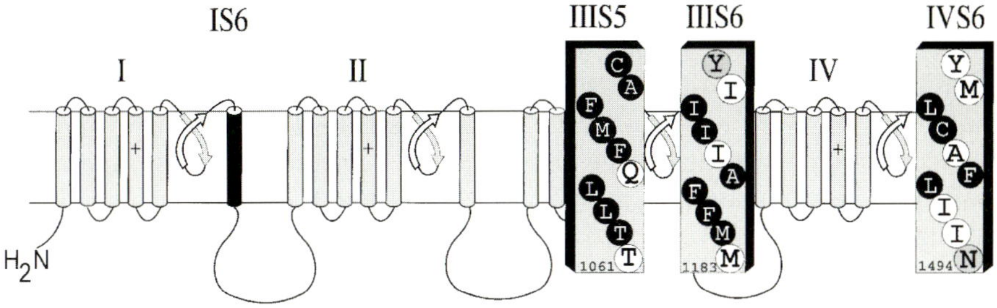

Fig. 3. Localization of interaction sites for calcium channel antagonists and agonists on the transmembrane IIIS5, IIIS6, and IVS6 segments. *Letters on white background* indicate residues that are different between dihydropyridine sensitive and insensitive calcium channels. *Letters on grey background* are residues that are conserved in all calcium channel sequences, but which participate also in the interaction with different ligands. IS6 indicates the transmembrane segment which is differentially spliced in cardiac and smooth muscle a_{1C} calcium channels and which accounts for the different sensitivity to dihydropyridines in these tissues

is according to the $\alpha_{1C\text{-}b}$ sequence (BIEL et al. 1990)) Thr1061 and Gln1065 in IIIS5 (ITO et al. 1997; HE et al. 1997), Ile 1175, Ile 1178, Met 1183, and the conserved Tyr1174 of IIIS6 (BODI et al. 1997; PETERSON et al. 1997) and Tyr1485, Met1486, Ile1493, and the conserved Asn1494 in IVS6 (SCHUSTER et al. 1996; PETERSON et al. 1997) (Fig. 3). The stimulation of I_{Ba} by the DHP agonists Bay K 8644 or (+)S-202–791 required mutation of less amino acids: Thr1061 in IIIS5 (ITO et al. 1997), Tyr1174 in IIIS6 (BODI et al. 1997), and Tyr1485, Met1486 in IVS6 (SCHUSTER et al. 1996). The largest effects were observed with mutation of Thr1061 to Tyr, which mutation lowered the affinity for isradipine more than 1000-fold (ITO et al. 1997). In contrast to these mutations, the replacement of the L-type specific Phe1484 in IVS6 by Ala decreased the IC_{50} for the DHP antagonists isradipine from 6.8 nmol/l to 0.014 nmol/l (PETERSON et al. 1997). More or less identical results were obtained, when the binding affinity of the mutated α_{1C} or α_{1S} subunit for isradipine was determined (HE et al. 1997; PETERSON et al. 1996). High affinity binding of DHPs requires Ca^{2+} (SCHNEIDER et al. 1991), which is coordinated by the glutamates in the pore region I, II, III, IV (MITTERDORFER et al. 1995). Mutation of the respective Glu to Gln in the α_{1S} pore region III and IV decreased the affinity for isradipine 10- to 40-fold (PETERSON and CATTERALL 1995). Although not completely excluded, it is unlikely that the high affinity binding of DHPs involves direct binding to the pore region glutamates. Most likely, the coordination of Ca^{2+} is required to allow the optimal conformation for high affinity binding. In contrast, isradipine binds with low affinity (IC_{50} about 2 μmol/l) to the open state of an α_{1C} subunit as revealed by the use of a channel, in which Tyr1485, Met1486, Ile1493 of IVS6 were mutated (LACINOVA and HOFMANN 1998). Possibly, binding to the pore region is involved in this low affinity block.

The transfer of parts of the α_{1C} sequence to the DHP insensitive neuronal α_{1A} subunit (GRABNER et al. 1996) confirmed the above concept. Detailed analysis using the α_{1A} subunit (SINNEGGER et al. 1997; HOCKERMAN et al. 1997b) or the α_{1E} subunit (ITO et al. 1997) showed that the L-type specific and the non-conserved amino acids (see above) had to be present to allow high affinity block and stimulation of these channels by the DHP antagonist isradipine and agonist Bay K 8644, respectively. The IC_{50} values for block of the chimeric channels was in the range of 10 nmol/l to 100 nmol/l. A similar value is obtained with the wild type α_{1C} channel at a holding potential of –80 mV, suggesting that these amino acids transfer the affinity for a "resting block." The high affinity block for DHPs requires inactivation of the L-type Ca^{2+} channel, which state results in IC_{50} values of 0.1 nmol/l or less. At the present it is not clear if this high affinity state requires the transfer of additional amino acids or cannot be obtained with the α_{1A} and α_{1E} subunit, since these channels inactivate at different membrane potentials leading to a different conformation of the binding site. Testing of the different mutations of the α_{1C} channel with charged and noncharged DHPs (BANGALORE et al. 1994) indicated that inactivation of the mutated channel affected the channel block differently. The noncharged DHP behaved like the usually used isradipine (LACINOVA et al. 1999). In contrast, the charged DHP blocked wild type and mutated α_{1C} channel with similar affinity, indicating that charged DHPs might bind to a different conformation of the channel and interact with different amino acids than the neutral DHPs.

The work of several groups suggested that the coexpression of a β and $\alpha_2\delta$ subunit is required for high affinity binding of DHPs (MITTERDORFER et al. 1994; LACINOVA et al. 1995; SUH-KIM et al. 1996; WEI et al. 1995). However, at the present time it cannot be decided, if these subunits help to localize the α_1 subunit in the membrane to obtain a correctly folded α_1 subunit or influence directly the binding site. It was reported that high affinity binding of DHPs was already observed when only the α_{1C} subunit was expressed alone (WELLING et al. 1993a). Investigation of several splice variants of the α_{1C} subunit showed that additional sequences affect the DHP sensitivity (WELLING et al. 1993b). In-depth analysis of the $\alpha_{1C\text{-}a}$ (cardiac) and $\alpha_{1C\text{-}b}$ (smooth muscle) sequence showed, that the alternative exon 8a or 8b, which codes for the IS6 segment, affects the affinity for neutral DHPs (WELLING et al. 1997). The cardiac $\alpha_{1C\text{-}a}$ channel, which contains the segment IS6a and is expressed in cardiac muscle, is blocked at higher concentrations of nisoldipine than the smooth muscle $\alpha_{1C\text{-}b}$ channel, which is expressed in vascular smooth muscle (WELLING et al. 1997). IC_{50} values for isradipine were 32 nmol/l and 8 nmol/l at a holding potential of –80 mV and 10 nmol/l and 1.3 nmol/l at a holding potential of –50 mV for the $\alpha_{1C\text{-}a}$ and $\alpha_{1C\text{-}b}$, respectively (L. Lacinová, unpublished results). Similar results were reported by ZÜHLKE et al. (1998), proving that the IS6 segment affects significantly the DHP block. It was possible that the change in affinity was caused by different inactivation kinetics of the two splice variants, since the IS6 segment strongly affects the inactivation kinetics of the channel (ZHANG et al. 1994). However, the inactivation kinetics of the two channels are either identical or opposite to expectation, i.e., the steady state inactivation of the cardiac $\alpha_{1C\text{-a}}$ channel occurred at more negative membrane potentials than that of the smooth muscle $\alpha_{1C\text{-b}}$ channel (HU and MARBAN 1998). Together with the earlier photoaffinity results (KALASZ et al. 1993), it is obvious that the increased affinity of the smooth muscle L-type calcium channel for DHPs is caused by structural differences in the IS6 segment, which contribute directly to the DHP binding pocket and not to the inactivation kinetics. Additional splice variations at the IIIS2 segment and in the intracellular carboxyterminal sequences could contribute as well to an altered DHP affinity (ZÜHLKE et al. 1998).

2. The Phenylalkylamine and Benzothiazepine Binding Site

Phenylalkylamines (PAA) such as verapamil, gallopamil, or devapamil block L-type calcium current use-dependent from the intracellular side of the membrane (HESCHELER et al. 1982) and affect the binding of DHPs by allosteric interaction (STRIESSNIG et al. 1993). In addition, benzothiazepines (BTZ) such as diltiazem interact allosterically with the binding of DHPs (STRIESSNIG et al. 1993). In contrast to PAAs, benzothiazepines label extracellular sites in the linker sequence between IVS5 and IVS6 in the α_{1S} subunit (WATANABE et al. 1993), in agreement with a recent report that the quaternary 1,5 BZT DTZ417 blocks the cardiac L-type channel only when applied from the extracellular site (KUROKAWA et al. 1997). More recently it was shown that similar to the PAA devapamil (CATTERALL and STRIESSNIG 1992), the 1,4-BZT semotiadil labels a short sequence of the IVS6 segment (KUNIYASU et al. 1998). The PAA verapamil blocks the L-type α_{1C} Ca^{2+} channel and the non-L-type α_{1A} and α_{1E} Ca^{2+} channels at similar concentrations in a state-dependent manner (CAI et al. 1997), whereas diltiazem blocked all three channels at similar concentrations, but only the α_{1C} Ca^{2+} channel in a state-dependent manner.

Molecular analysis of the α_{1C} subunit (SCHUSTER et al. 1996; HOCKERMAN et al. 1995, 1997a) showed that the L-type channel specific Ile1175 and the conserved Tyr1174, Phe1186, and Val1187 in IIIS6 and the L-type specific Tyr1485, Ala1489, and Ile1492 in IVS6 are necessary to form a high affinity PAA site (Fig. 3). In addition, the two glutamates (E1140 and E1441) in the pore region of repeat III and IV are necessary (amino acid numbering is according to the $\alpha_{1C\text{-b}}$ sequence (BIEL et al. 1990)) (Hockerman et al. 1997a). The effect of the mutation of the conserved Tyr1174 depends on the replacing amino acid. Substitution by phenylalanine decreased the affinity for devapamil 18-fold whereas substitution by an alanine increased the affinity 7-fold (HOCKERMAN et al. 1997a). The increased affinity of the Y1174 A mutant is most likely caused by a shift of −11 mV for the steady state inactivation curve. Transfer of the three IVS6 amino acids Y1485, A1489, and I1492 from the α_{1C} to the α_{1A} subunit introduced PAA and BZT sensitivity, when measured in a use-dependent protocol (HERING et al. 1996). Furthermore, it was shown that the triple mutation Y1485A, A1489S, and I1492A in IVS6 of the α_{1C} channel reduced use-dependent block of the three PAAs, devapamil, verapamil, and gallopamil, reduced the resting and depolarized block of devapamil, but affected poorly the resting and depolarized block of verapamil and gallopamil (Johnson et al. 1996).

Together these results show that the IVS6 segment is interacting with various PAAs and BZT. State-dependent block of the L-type channel is mediated by the same three amino acid residues in IVS6 for diltiazem and devapamil. However, different amino acids are required to allow high affinity interaction at resting state for diltiazem, verapamil, and gallopamil. A further problem arises from the finding that DHPs, PAAs, and BZTs interact with the same (Y1485) or with adjacent (I1492and I1493) amino acid side chains. It is difficult to reconcile this close location of interacting site chains with the previously described allosteric modulation of DHP binding by diltiazem or phenylalkylamines (Striessnig et al. 1993).

3. Modulation of Expressed L-Type Calcium Channel by cAMP-Dependent Phosphorylation

In the heart, the positive inotropic action of catecholamines is mainly caused by an increased calcium influx through L-type calcium channels. cAMP-dependent phosphorylation of the α_1 subunit or a closely associated protein increases the current three- to sevenfold (Osterrieder et al. 1982; Kameyama et al. 1985; Hartzell and Fischmeister 1992). Phosphorylation increases the availability of the channel to open upon depolarization by modulation of channel gating. Cardiac calcium channels also show facilitation of current amplitude during high frequency stimulation (Lee 1987) or after strong depolarization (Pietrobon and Hess 1990). Depolarization induced facilitation was supposed to require voltage-dependent phosphorylation of the channel by cAMP kinase (Artalejo et al. 1992). However, these results of Artalejo and colleagues were probably caused by the removal of secreted substances from the external solution and not by channel phosphorylation (Garcia and Carbone 1996). The adult skeletal muscle calcium channel is apparently not regulated by phosphorylation to a large extent. In contrast, the calcium channel of embryonic rat skeletal muscle myoballs shows voltage- and cAMP kinase-dependent facilitation (Sculptoreanu et al. 1993b). Facilitation depending on a strong depolarizing prepulse requires membrane localization of cAMP kinase (Johnson et al. 1994) by a 15kDa cAMP kinase anchoring protein (Gray et al. 1998).

In adult skeletal muscle, two forms of the α_{1S} subunit are present – a large 212kDa form, containing the complete sequence of the cloned α_{1S} cDNA, and a small 190kDa form, which is truncated between amino acid 1685 and 1699 (De Jongh et al. 1991). About 5% of the α_{1S} subunit are the large 212kDa form and over 90% belongs to the small 190kDa form (De Jongh et al. 1991). In intact rabbit skeletal muscle myotubes, cAMP kinase phosphorylates rapidly Ser1757 and Ser1854 in the large 212 kDa form and slowly Ser687 in the small 190 kDa form, which does not contain the cAMP kinase sites at Ser1757 and Ser1854 (ROTMAN et al. 1995). Expression of an α_{1S} cDNA, which is truncated at Asn1662 and encodes the small form, fully restored both excitation-contraction coupling and calcium current in dysgenic myotubes, consistent with the idea that the small form of the α_{1S} subunit performs both functions in adult muscle without cAMP-dependent phosphorylation (BEAM et al. 1992). These results are in line with the conclusion that the long form of the skeletal muscle α_{1S} channel is modulated by cAMP kinase in myoballs, but that this modulation is attenuated or not present in adult skeletal muscle, in which the short form prevails.

In contrast to the skeletal muscle L-type calcium channel, the precise mechanism of phosphorylation of the cardiac α_{1C} calcium channel is less clear. The fact that cAMP kinase-dependent phosphorylation affects significantly the function of the channel in vivo is undisputed. However, the mechanism causing the channel modulation is controversial. Rabbit heart sarcolemma contains a large 240 kDa and a small 210 kDa form of the α_{1C} subunit (DE JONGH et al. 1996). The small 210 kDa form is truncated at residue 1870 in the carboxy terminal sequence. The 240 kDa form is phosphorylated by cAMP kinase at Ser1928 (DE JONGH et al. 1996). The expressed full length 250 kDa $\alpha_{1C\text{-}a}$ subunit is phosphorylated in vivo in CHO cells (YOSHIDA et al. 1992) and HEK 293 cells (GAO et al. 1997). Phosphorylation of the α_{1C} subunit is prevented by the mutation S1928 A (GAO et al. 1997). The mutation S1928 A prevents also a decrease in barium current induced by the cAMP kinase inhibitor H-89 in X. oocytes (PERETS et al. 1996). However, a direct effect of cAMP kinase on current amplitude was not observed in X. oocytes (SINGER-LAHAT et al. 1994; BOURON et al. 1995; PERETS et al. 1996). In contrast to studies in oocytes, a cAMP-dependent increase in current amplitude was reported by several groups, who used either CHO or HEK cells as expression system (HAASE et al. 1993; PEREZ-REYES et al. 1994). Dialysis of the CHO cells with active cAMP kinase facilitated the peak barium inward current following a prepulse to positive membrane potentials (SCULPTOREANU et al. 1993a). cAMP kinase-dependent facilitation was also reported by BOURINET and coworkers (1994), who used the neuronal $\alpha_{1C\text{-}c}$ splice variant and the oocyte expression system. In a recent report these authors observed facilitation of barium currents in the absence of cAMP-dependent phosphorylation and showed that facilitation was observed only in the presence of the β_1, β_3, and β_4 subunit and was not supported by the neuronal β_{2a} subunit (CENS et al. 1998). Identical results were reported by QUIN et al. (1998a), which used a N-terminal truncated $\alpha_{1C\text{-}a}$ (expressed residues 60–2171) subunit. These recent results are in agreement with the earlier reports by KLEPPISCH et al. (1994) and BOURON et al. (1995) that facilitation of the α_{1C} current is independent of cAMP kinase-dependent phosphorylation. In a careful study, which used the $\alpha_{1C\text{-}a}$ and $\alpha_{1C\text{-}b}$ splice variants stably expressed in CHO and HEK 293 cells and transient expression of $\alpha_{1C\text{-}b}$, cardiac β_{2a} and $\alpha_2\delta$-1 Zong et al. (1995) showed, that the current amplitude of these cells was not affected significantly by internal dialysis with cAMP kinase inhibitor peptide, catalytic subunit of the cAMP kinase, or a combination of cAMP kinase and okadaic acid. Similar results were obtained by the coexpression of all subunits of the calcium channel complex, whereas the calcium current of cardiac myocytes was increased threefold during internal dialysis with active cAMP kinase or external superfusion with isoproterenol. Furthermore, dialysis of cardiac myocytes with the phosphatase inhibitor microcystin stimulated the calcium inward current more than twofold, whereas the current of the expressed calcium channel was not affected. These conflicting results were apparently solved, when Gao et al. (1997) reported that cAMP kinase-dependent stimulation of barium current required the coexpression of the cAMP kinase anchoring protein AKAP 79, $\alpha_{1C\text{-}a}$ and neuronal β_{2a} subunit in HEK 293 cells. AKAP 79 anchors the kinase at the plasma membrane. These authors reported that phosphorylation of Ser1928 was required for cAMP-dependent stimulation of barium currents. However, a careful reexamination of these results using overexpression of AKAP79 – cloned from the HEK 293 cells and identical to that used by Gao and coworker – failed to reproduce a cAMP kinase-dependent increase in current amplitude or facilitation of the current by strong depolarization (DAI et al. 1998). In contrast, cAMP-independent facilitation was observed, when $\alpha_{1C\text{-}a}$ and cardiac β_{2a}, or $\alpha_{1C\text{-}a}$ truncated at residue 1733 were used. Prepulse facilitation was prevented by expressing the $\alpha_{1C\text{-}a}$ and cardiac β_{2a} subunits together with the $\alpha_2\delta$-1 or $\alpha_2\delta$-3 subunit, in line with the known effect of the $\alpha_2\delta$ subunit on the gating of the channel. These results demonstrate clearly that facilitation of the cardiac L-type current can be observed with channels which do not contain the established cAMP kinase phosphorylation site at Ser1928.

4. Modulation of Expressed L-Type Calcium Channel by Protein Kinase C-Dependent Phosphorylation

L-type calcium channels are tightly regulated by hormonal and neuronal signals. Protein kinase C (PKC) is one such regulator, which increases cardiac, smooth muscle, and neuronal L-type current (LACERDA et al 1988; SCHUHMANN and GROSCHNER 1994; YANG and TSIEN 1993) by an increase in the open probability of the channel (YANG and TSIEN 1993). The response to PKC activators is usually biphasic, with an increase followed by a later decrease (LACERDA et al 1988; SCHUHMANN and GROSCHNER 1994). The biphasic response to PKC stimulators was fully reconstituted, when the $\alpha_{1C\text{-}a}$ subunit was expressed in X. oocytes (SINGER-LAHAT et al 1992). BOURON et al. (1995), who used a human α_{1C} splice form that has the same amino terminus as the $\alpha_{1C\text{-}b}$ subunit, observed only a decrease in current, suggesting that PKC-dependent regulation may be controlled by the different amino termini of the two splice variants. This prediction was confirmed (SHISTIK et al 1998). Deletion of amino acids 2–46 in the amino terminus of the $\alpha_{1C\text{-}a}$ subunit prevented PKC-dependent current increase. The effects of PKC activation were larger in the presence of the $\alpha_{1C\text{-}a}$ and $\alpha_2\delta$-1 subunit and were decreased by the coexpression of the cardiac β_{2a} subunit. Upregulation of the current was not affected by truncation of the $\alpha_{1C\text{-}a}$ subunit at residue 1665, or mutation of the proposed PKC phosphorylation site Ser533 in the I-II linker. Upregulation depended on the splice variation of the amino terminus and was not observed with the amino terminus of the $\alpha_{1C\text{-}b}$ subunit. In agreement with WEI et al. (1996), these studies show that, depending on the splice variant, the amino terminus affects channel gating and mediates PKC-dependent upregulation.

Acknowledgement. The work in the author's laboratory was supported by DFG, BMBF, and Fond der Chemie.

References

Adams B, Tanabe T (1997) Structural regions of the cardiac Ca channel alpha subunit involved in Ca-dependent inactivation. J Gen Physiol 110:379–389
Angelotti T, Hofmann F (1996) Tissue-specific expression of splice variants of the mouse voltage-gated calcium channel $\alpha_2\delta$ subunit. FEBS Lett 392:331–337
Armstrong CM, Neyton J (1991) Ion permeation through calcium channels. A one-site model. Ann N Y Acad Sci 635:18–25
Artalejo CR, Rossie S, Perlman RL, Fox AP (1992) Voltage-dependent phosphorylation may recruit Ca^{2+} current facilitation in chromaffin cells. Nature 358:63–66
Babitch J (1990) Channel hands. Nature 346:321–322 [letter]
Bangalore R, Baindur N, Rutledge A, Triggle DJ, Kass RS (1994) L-type calcium channels: asymmetrical intramembrane binding domain revealed by variable length, permanently charged 1,4-dihydropyridines. Mol Pharmacol 46:660–666
Bangalore R, Mehrke G, Gingrich K, Hofmann F & Kass, RS (1996) Influence of the L-type Ca-channel α_2/δ subunit on ionic and gating current in transiently transfected HEK 293 cells. Am J Physiol 270:H1521–1528
Beam KG, Adams BA, Niidome T, Numa S, Tanabe T (1992) Function of a truncated dihydropyridine receptor as both voltage sensor and calcium channel. Nature 360:169–171
Bech-Hansen NT, Naylor MJ, Maybaum TA, Pearce WG, Koop B, Fishman GA, Mets M, Musarella MA, Boycott KM (1998) Loss-of-function mutations in a calcium-channel alpha1-subunit gene in Xp11.23 cause incomplete X-linked congenital stationary night blindness. Nat Genet 19:264–267
Beurg M, Sukhareva M, Strube C, Powers PA, Gregg RG, Coronado R (1997) Recovery of Ca^{2+} current, charge movements, and Ca^{2+} transients in myotubes deficient in dihydropyridine receptor beta 1 subunit transfected with beta 1 cDNA. Biophys J 73:807–818
Biel M, Hullin R, Freundner S, Singer D, Dascal N, Flockerzi V, Hofmann F (1991) Tissue-specific expression of high-voltage-activated dihydropyridine-sensitive L-type calcium channels. Eur J Biochem 200:81–88
Biel M, Ruth P, Bosse E, Hullin R, Stühmer W, Flockerzi V, Hofmann F (1990) Primary structure and functional expression of a high voltage activated calcium channel from rabbit lung. FEBS Lett 269:409–412
Bodi I, Yamaguchi H, Hara M, He-M, Schwartz A, Varadi G (1997) Molecular studies on the voltage dependence of dihydropyridine action on L-type Ca^{2+} channels. Critical involvement of tyrosine residues in motif IIIS6 and IVS6. J Biol Chem 272:24952–24960

Bosse E, Regulla S, Biel M, Ruth P, Meyer HE, Flockerzi V, Hofmann F (1990) The cDNA and deduced amino acid sequence of the γ subunit of the L-type calcium channel from rabbit skeletal muscle. FEBS Lett 267:153–156

Bourinet E, Charnet P, Tomlinson WJ, Stea A, Snutch TP, Nargeot J (1994) Voltage-dependent facilitation of a neuronal alpha 1C L-type calcium channel. EMBO J 13:5032–5039

Bouron A, Soldatov NM, Reuter H (1995) The beta 1-subunit is essential for modulation by protein kinase C of an human and a non-human L-type Ca^{2+} channel. FEBS Lett 377:159–162

Burgess DL, Jones JM, Meisler MH, NoebelsJL (1997) Mutation of the Ca^{2+} channel beta subunit gene Cchb4 is associated with ataxia and seizures in the lethargic (lh) mouse. Cell 88:385–392

Cai D, Mulle JG, Yue DT (1997) Inhibition of recombinant Ca^{2+} channels by benzothiazepines and phenylalkylamines: class-specific pharmacology and underlying molecular determinants. Mol Pharmacol 51:872–881

Castellano A, Wei X, Birnbaumer L, Perez-Reyes E (1993) Cloning and expression of a neuronal calcium channel beta subunit. J Biol Chem 268:12359–12366

Catterall WA (1995) Structure and function of voltage-gated ion channels. Annu Rev Biochem 64:493–531

Catterall WA, Striessnig J (1992) Receptor sites for Ca^{2+} channel antagonists. Trends Pharmacol Sci 13:256–262

Cens T, Restituito S, Vallentin A, Charnet P (1998) Promotion and inhibition of L-type Ca2+ channel facilitation by distinct domains of the subunit. J Biol Chem 273:18308–18315

Chavis P, Fagni L, Lansman JB, Bockaert J (1996) Functional coupling between ryanodine receptors and L-type calcium channels in neurons. Nature 382:719–722

Chen XH, Tsien RW (1997) Aspartate substitutions establish the concerted action of P-region glutamates in repeats I and III in forming the protonation site of L-type Ca^{2+} channels. J Biol Chem 272:30002–30008

Chien AJ, Carr KM, Shirokov RE, Rios E, Hosey MM (1996) Identification of palmitoylation sites within the L-type calcium channel beta2a subunit and effects on channel function. J Biol Chem 271:26465–26468

Chien J, Gao TY, Perez-Reyes E, Hosey MM (1998) Membrane targeting of L-type calcium channels – role of palmitoylation in the subcellular localization of the beta(2a) subunit. J Biol Chem 273:23590–23597

Cribbs LL, Lee JH, Yang J, Satin J, Zhang Y, Daud A, Barclay J, Williamson MP, Fox M, Rees M, PerezReyes E (1998) Cloning and characterization of alpha 1H from human heart, a member of the T-type Ca^{2+} channel gene family. Circ Res 83:103–109

Dai S, Klugbauer N, Zong X, Seisenberger S, Hofmann F (1999) The role of subunit composition on prepulse facilitation of the cardiac L-type calcium channel. FEBS Lett 442:70–74

De Jongh KS, Murphy BJ, Colvin AA, Hell JW, Takahashi M, Catterall WA (1996) Specific phosphorylation of a site in the full-length form of the alpha 1 subunit of the cardiac L-type calcium channel by adenosine 3′,5′-cyclic monophosphate-dependent protein kinase. Biochemistry 35:10392–10402

De Jongh KS, Warner C, Colvin AA, Catterall WA (1991) Characterization of two size forms of the α_1 subunit of skeletal muscle L-type calcium channels. Proc Natl Acad Sci USA 88:10778–10782

de Leon M, Wang Y, Jones L, Perez-Reyes E, Wie X, Soong TW, Snutch TP, Yue DT (1995) Essential Ca^{2+}-binding motif for Ca^{2+}-sensitive inactivation of L-type Ca^{2+} channels. Science 270:1502–1506

De Waard M, Campbell KP (1995a) Subunit regulation of the neuronal α_{1A} Ca^{2+} channel expressed in *Xenopus* oocytes. J Physiol 485:619–634

De Waard M, Gurnett CA, Campbell KP (1996a) Structural and functional diversity of voltage-activated calcium channels. Ion Channels 4:41–87

De Waard M, Pragnell M, Campbell, KP (1994) Ca^{2+} channel regulation by a conserved β subunit domain. Neuron 13:495–503

De Waard M, Scott VE, Pragnell M, Campbell KP (1996b) Identification of critical amino acids involved in alpha1-beta interaction in voltage-dependent Ca^{2+} channels. FEBS Lett 380:272–276

De Waard M, Witcher DR, Pragnell M, Liu H, Campbell KP (1995b) Properties of the α_1-β anchoring site in voltage-dependent Ca^{2+} channels. J Biol Chem 270: 12056–12064

Dirksen RT, Nakai J, Gonzalez A, Imoto K, Beam KG (1997) The S5-S6 linker of repeat I is a critical determinant of L-type Ca^{2+} channel conductance. Biophys J 73:1402–1409

Doyle DA, Cabral JM, Pfuetzner RA, Kuo A, Gulbis JM, Cohen SL, Chait BT, MacKinnon R (1998) The structure of the potassium channel: molecular basis of K^+ conduction and selectivity. Science 280:69–77

Dubel SJ, Starr TVB, Hell J, Ahlijanian MK, Enyart JJ, Catterall WA, Snutch TP (1992) Molecular cloning of the α_1 subunit of an ω-conotoxin-sensitive calcium channel. Proc Natl Acad Sci USA 89:5058–5062

Eberst R, Dai S, Klugbauer N, Hofmann F (1997) Identification and functional characterization of a calcium channel γ subunit. Pflüger's Archiv 433:633–637

El-Hayek R, Antoniu B, Wang J, Hamilton SL, Ikemoto N (1995) Identification of calcium release-triggering and blocking regions of the II-III loop of the skeletal muscle dihydropyridine receptor. J Biol Chem 270:22116–22118

El-Hayek R, Ikemoto N (1998) Identification of the minimum essential region in the II-III loop of the dihydropyridine receptor alpha 1 subunit required for activation of skeletal muscle-type excitation-contraction coupling. Biochemistry 37:7015–7020

Ellinor PT, Yang J, Sather WA, Zhang JF, Tsien, RW (1995) Ca^{2+} channel selectivity at a single locus for high-affinity Ca^{2+} interactions. Neuron 15:1121–1132

Ellinor PT, Zhang JF, Randall AD, Zhou M, Schwarz TL, Tsien RW, Horne WA (1993) Functional expression of a rapidly inactivating neuronal calcium channel. Nature 363:455–458

Ellis SB, Williams ME, Ways NR, Ellis SB, Brenner R, Sharp AH, Leung AT, Campbell KP McKenna E, Koch WJ, Hui A, et al. (1988) Sequence and expression of mRNAs encoding the α_1 and α_2 subunits of a DHP-sensitive calcium channel. Science 241:1661–1664

Felix R, Gurnett CA, De Waard M, Campbell KP (1997) Dissection of functional domains of the voltage-dependent Ca2+ channel alpha2delta subunit. J Neuroscience 17:6884–6891

Fujita Y, Mynlieff M, Dirksen RT, Kim MS, Niidome T, Nakai J, Friedrich T, Iwabe N, Miyata T, Furuichi T, Furutama D, Mikoshiba K, Mori Y, Beam KG (1993) Primary structure and functional expression of the ω-conotoxin-sensitive N-type channel from rabbit brain. Neuron 10:585–598

Gao T, Yatani A, Dell'Acqua ML, Sako H, Green SA, Dascal N, Scott JD, Hosey MM (1997) cAMP-dependent regulation of cardiac L-type Ca^{2+} channels requires membrane targeting of PKA and phosphorylation of channel subunits. Neuron 19:185–196

Garcia AG, Carbone E (1996) Calcium-current facilitation in chromaffin cells. Trends Neurosci 19:383–384

Garcia J, Nakai J, Imoto K, Beam KG (1997) Role of S4 segments and the leucine heptad motif in the activation of an L-type calcium channel. Biophys J 72:2515–2523

Gollasch M, Ried C, Liebold M, Haller H, Hofmann F, Luft FC (1996) High permeation of L-type calcium channels at physiological calcium concentrations: homogeity and dependence on the α_1 subunit. Am J Physiol 271:C842–C850

Grabner M, Wang Z, Hering S, Striessnig J, Glossmann H (1996) Transfer of 1,4-dihydropyridine sensitivity from L-type to class A (BI) calcium channels. Neuron 16:207–218

Gray PC, Johnson BD, WestenbroekRE, Hays LG, Yates JR III, Scheuer T, Catterall WA, Murphy BJ (1998) Primary structure and function of an A kinase anchoring protein associated with calcium channels. Neuron 20:1017–1026

Gregg RG, Messing A, Strube C, Beurg M, Moss R, Behan M, Sukhareva M, Haynes S, Powell JA, Coronado R, Powers PA (1996) Absence of the beta subunit (cchb1) of the skeletal muscle dihydropyridine receptor alters expression of the alpha 1 subunit and eliminates excitation-contraction coupling. Proc Natl Acad Sci USA 93:13961–13966

Gurnett CA, De Waard M, Campbell KP (1996) Dual function of the voltage-dependent Ca^{2+} channel a2/δ subunit in current stimulation and subunit interaction. Neuron 16:431–440

Gurnett CA, Felix R, Campbell KP (1997) Extracellular interaction of the voltage-dependent Ca^{2+} channel alpha2delta and alpha1 subunits. J Biol Chem 272:18,508–18,512

Guy HR, Conti F (1990) Pursuing the structure and function of voltage-gated channels. Trends Neurosci 13:201–206

Haase H, Karczewski P, Beckert R, Krause EG (1993) Phosphorylation of the L-type calcium channel beta subunit is involved in beta-adrenergic signal transduction in canine myocardium. FEBS Lett 335:217–222

Hartzell HC, Fischmeister R (1992) Direct regulation of cardiac Ca^{2+} channels by G proteins: neither proven nor necessary? Trends Pharmacol Sci 13:380–385

He M, Bodi I, Mikala G, Schwartz A (1997) Motif III S5 of L-type calcium channels is involved in the dihydropyridine binding site. A combined radioligand binding and electrophysiological study. J Biol Chem 272:2629–2633

Hering S, Aczel S, Grabner M, Doring F, Berjukow S, Mitterdorfer J, Sinnegger MJ, Striessnig J, Degtiar VE, Wang Z, Glossmann H (1996) Transfer of high sensitivity for benzothiazepines from L-type to class A (BI) calcium channels. J Biol Chem 271:24471–24475

Herlitze S, Garcia DE, Mackie K, Hille B, Scheuer T, Catterall WA (1996) Modulation of Ca^{2+} channels by G-protein $\beta\gamma$ subunits. Nature 380:258–262

Hescheler J, Pelzer D, Trube G, Trautwein W (1982) Does the organic calcium channel blocker D600 act from inside or outside on the cardiac cell membrane? Pflügers Arch 393:287–291

Hescheler J, Trautwein W (1988) Modification of L-type calcium current by intracellularly applied trypsin in guinea-pig ventricular myocytes. J Physiol 404:259–274

Hess P, Tsien RW (1984) Mechanism of ion permeation through calcium channels. Nature 309:453–456

Hockerman GH, Johnson BD, Abbott MR, Scheuer T, Catterall WA (1997a) Molecular determinants of high affinity phenylalkylamine block of L-type calcium channels in transmembrane segment IIIS6 and the pore region of the alpha1 subunit. J Biol Chem 272:18759–18765

Hockerman GH, Johnson BD, Scheuer T, Catterall (1995) Molecular determinants of high affinity phenylalkylamine block of L-type calcium channels. J Biol Chem 270:22119–22122

Hockerman GH, Peterson BZ, Sharp E, Tanada TN, Scheuer T, Catterall WA (1997b) Construction of a high-affinity receptor site for dihydropyridine agonists and antagonists by single amino acid substitutions in a non-L-type Ca^{2+} channel. Proc Natl Acad Sci USA 94:14906–14911

Höfer G, Hohenthanner K, Baumgartner W, Groschner K, Klugbauer N, Hofmann F, Romanin C (1996) Intracellular Ca^{2+} inactivates L-type Ca^{2+} channels with a Hill coefficient of approximately 1 and a K_i of approximately 4 μM by reducing channels's open probability. Biophys J 73:1857–1865

Hofmann F, Biel M, Flockerzi V (1994) Molecular basis for Ca^{2+} channel diversity. Annu Rev Neurosci 17:399–418

Hu H, Marban E (1998) Isoform-specific inhibition of L-type calcium channels by dihydropyridines is independent of isoform-specific gating properties. Mol Pharmacol 53:902–907

Hullin R, Singer-Lahat D, Freichel M, Biel M, Dascal N, Hofmann F, Flockerzi V (1992) Calcium channel β subunit heterogeneity: functional expression of cloned cDNA from heart, aorta and brain. EMBO J 11:885–890

Ito H, Klugbauer N, Hofmann F (1997) Transfer of the high affinity dihydropyridine sensitivity from L-type to non-L-type calcium channel. Mol Pharmacol 52:735–740

Jay SD, Ellis SB, McCue AF, Williams ME, Vedvick TS, Harpold MM, Campbell KP (1990) Primary structure of the γ subunit of the DHP-sensitive calcium channel from skeletal muscle. Science 248:490–492

Johnson BD, Hockerman GH, Scheuer T, Catterall WA (1996) Distinct effects of mutations in transmembrane segment IVS6 on block of L-type calcium channels by structurally similar phenylalkylamines. Mol Pharm 50:1388–1400

Johnson BD, Scheuer T, Catterall WA (1994) Voltage-dependent potentiation of L-type Ca^{2+} channels in skeletal muscle cells requires anchored cAMP-dependent protein kinase. Proc Natl Acad Sci USA 91:11492–11496

Josephson IR, Varadi G (1996) The beta subunit increases Ca^{2+} currents and gating charge movements of human cardiac L-type Ca^{2+} channels. Biophys J 70:1285–1293

Kalasz H, Watanabe T, Yabana H, Itagaki K, Naito K, Nakayama H, Schwartz A, Vaghy PL (1993) Identification of 1,4-dihydropyridine binding domains within the primary structure of the alpha 1 subunit of the skeletal muscle L-type calcium channel. FEBS Lett 331:177–181

Kamejama M, Hofmann F, Trautwein W (1985) On the mechanism of β-adrenergic regulation of the Ca channel on the guinea pig heart. Pflügers Arch 405:285–293

Kamp TJ, Perez-Garcia MT, Marban E (1996) Enhancement of ionic current and charge movement by coexpression of calcium channel beta 1A subunit with alpha 1C subunit in a human embryonic kidney cell line. J Physiol 492:89–96

Kim MS, Morii T, Sun LX, Imoto K, Mori Y (1993) Structural determinants of ion selectivity in brain calcium channel. FEBS. 318:145–148

Kleppisch T, Pedersen K, Strübing C, Bosse-Doenecke E, Flockerzi V, Hofmann F, Hescheler J (1994) Double-pulse facilitation of smooth muscle α_1 subunit Ca^{2+} channels expressed in CHO cells. EMBO J 13:2502–2507

Klöckner U, Mikala G, Schwartz A, Varadi G (1995) Involvement of the carboxyl-terminal region of the alpha 1 subunit in voltage-dependent inactivation of cardiac calcium channels. J Biol Chem 270:17,306–17,310

Klöckner U, Mikala G, Schwartz A, Varadi G (1996) Molecular studies of the asymmetric pore structure of the human cardiac voltage- dependent Ca^{2+} channel. Conserved residue, Glu-1086, regulates proton-dependent ion permeation. J Biol Chem 271:22,293–22,296

Klugbauer N, Lacinová L, Marais E, Hobom M, Hofmann F (1999a) Molecular diversity of the calcium channel $\alpha_2\delta$ subunit. J Neurosci 19:684–691

Klugbauer N, Marais E, Lacinová L, Hofmann F (1999b) A T-type calcium channel from brain. Pflügers Arch 437:710–715

Kuniyasu A, Itagaki K, Shibano T, Iino M, Kraft G, Schwartz A, Nakayama H (1998) Photochemical identification of transmembrane segment IVS6 as the binding region of semotiadil, a new modulator for the L-type voltage-dependent Ca^{2+} channel. J Biol Chem 273:4635–4641

Kuo CC, Hess P (1993) Ion permeation in through the L-type Ca^{2+} channel in rat PC12 cells: two sets of ion binding sites in the pore. J Physiol 466:629–655

Kurokawa J, Adachi-Akahane S, Nagao T (1997) 1,5-benzothiazepine binding domain is located on the extracellular side of the cardiac L-type Ca2+ channel. Mol Pharmacol 51:262–268

Lacerda AE, Rampe D, Brown AM (1988) Effects of protein kinase C activators on cardiac Ca^{2+} channels. Nature 335:249–251

Lacinová L, An RH, Xia J, Ito H, Klugbauer N, Triggle D, Hofmann F, Kass RS (1999) Distinction in the molecular determinants of charged and neutral dihydropyridine block of L-type calcium channels. J Pharmacol Exp Ther 289:1472–1479

Lacinová L, Hofmann F (1998) Isradipine interacts with the open state of the L-type calcium channel at high concentrations. Receptor and Channels 6:153–164

Lacinová L, Ludwig A, Bosse E, Flockerzi V, Hofmann F (1995) The block of the expressed L-type calcium channel is modulated by the β_3 subunit FEBS Lett 373:103–107

Lee KS (1987) Potentiation of the calcium-channel currents of internally perfused mammalian heart cells by repetitive depolarization. Proc Natl Acad Sci USA 84:3941–3945

Letts VA, Felix R, Biddlecome GH, Arikkath J, Mahaffey CL, Valenzuela A, Bartlett FS, Mori Y, Campbell KP, Frankel WN (1998) The mouse stargazer gene encodes a neuronal Ca^{2+}-channel gamma subunit. Nature Gen 19:340–347

Liman ER, Hess P, Weaver F, Koren G (1991) Voltage-sensing residues in the S4 region of a mammalian K^+ channel. Nature 35:752–756

Liu H, De Waard M, Scott VES, Gurnett CA, Lennon VA, Campbell KP (1996) Identification of three subunits of the high affinity omega-conotoxin MVIIC-sensitive Ca^{2+} channel. J Biol Chem 271:13804–13810

Lu X, Xu L, Meissner G (1994) Activation of the skeletal muscle calcium release channel by a cytoplasmic loop of the dihydropyridine receptor. J Biol Chem 269:6511–6516

Lu X, Xu L, Meissner G (1995) Phosphorylation of dihydropyridine receptor II-III loop peptide regulates skeletal muscle calcium release channel function. Evidence for an essential role of the beta-OH group of Ser687. J Biol Chem 270:18459–18464

Ludwig A, Flockerzi V, Hofmann F (1997) Regional expression and cellular localization of the α_1 and β subunit of high voltage-activated calcium channels in rat brain J Neurosci 17:1339–1349

McEnery MW, Copeland TD, Vance CL (1998) Altered expression and assembly of N-type calcium channel α_{1B} and β subunits in epileptic *lethargic* (*lh/lh*) mouse. J Biol Chem 273:21435–21438

Mikami A, Imoto K, Tanabe T, Niidome T, Mori Y, Takeshima H, Narumiya S, Numa S (1989) Primary structure and functional expression of the cardiac dihydropyridine-sensitive calcium channel. Nature 340:230–233

Mitterdorfer J, Froschmayr M, Grabner M, Striessnig J, Glossmann H (1994) Calcium channels: the beta-subunit increases the affinity of dihydropyridine and Ca^{2+} binding sites of the alpha 1-subunit. FEBS Lett 352:141–145

Mitterdorfer J, Sinnegger MJ, Grabner M, Striessnig J, Glossmann H (1995) Coordination of Ca^{2+} by the pore region glutamates is essential for high-affinity dihydropyridine binding to the cardiac Ca^{2+} channel alpha 1 subunit. Biochemistry 34:9350–9355

Mori Y, Friedrich T, Kim MS, Mikami A, Nakai J, Ruth P, Bosse E, Hofmann F, Flockerzi V, Furuichi T, Mikoshiba K, Imoto K, Tanabe T, Numa S (1991) Primary structure and functional expression from complementary DNA of a brain calcium channel. Nature 350:398–402

Nakai J, Adams BA, Imoto K, Beam KG (1994) Critical roles of the S3 segment and S3-S4 linker of repeat I in activation of L-type calcium channels. Proc Natl Acad Sci USA 91:1014–1018

Nakai J, Dirksen RT, Nguyen HT, Pessah IN, Beam KG, Allen PD (1996) Enhanced dihydropyridine receptor channel activity in the presence of ryanodine receptor. Nature 380:72–75

Nakai J, Sekiguchi N, Rando TA, Allen PD, Beam KG (1998a) Two regions of the ryanodine receptor involved in coupling with L-type Ca^{2+} channels. J Biol Chem 273:13403–13406

Nakai J, Tanabe T, Konno T, Adams B, Beam KG (1998b) Localization in the II-II loop of the dihydropyridine receptor of a sequence critical for excitation-contraction coupling. J Biol Chem 273:24983–24986

Neely A, Olcese R, Wei X, Birnbaumer L, Stefani E (1994) Ca^{2+}-dependent inactivation of a cloned cardiac Ca^{2+} channel alpha 1 subunit (alpha 1C) expressed in Xenopus oocytes. Biophys J 66:1895–1903

Neely A, Wei X, Olcese R, Birnbaumer L, Stefani E (1993) Potentiation by the beta subunit of the ratio of the ionic current to the charge movement in the cardiac calcium channel. Science 262:575–578

Neuhuber B, Gerster U, Doring F, Glossmann H, Tanabe T, Flucher BE (1998a) Association of calcium channel alpha1S and beta1a subunits is required for the targeting of beta1a but not of alpha1S into skeletal muscle triads. Proc Natl Acad Sci USA 95:5015–5020

Neuhuber B, Gerster U, Mitterdorfer J, Glossmann H, Flucher BE (1998b) Differential effects of Ca^{2+} channel beta1a and beta2a subunits on complex formation with alpha1S and on current expression in tsA201 cells. J Biol Chem 273:9110–9118

Niidome T, Kim MS, Friedrich T, Mori Y (1992) Molecular cloning and characterization of a novel calcium channel from rabbit brain. FEBS Lett 308:7–13

Osterrieder W, Brum G, Hescheler J, Trautwein W, Flockerzi V, Hofmann F (1982) Injection of subunits of cyclic AMP-dependent protein kinase into cardiac myocytes modulates Ca^{2+} current. Nature 298:576–578

Papazian DM, Timpe LC, Jan YN, Jan LY (1991) Alteration of voltage-dependence of Shaker potassium channel by mutations in the S4 sequence. Nature 349:305–310

Parent L, Gopalakrishnan M (1995) Glutamate substitution in repeat IV alters divalent and monovalent cation permeation in the heart Ca^{2+} channel. Biophys J 69:1801–1813

Perets T, Blumenstein Y, Shistik E, Lotan I, Dascal N (1996) A potential site of functional modulation by protein kinase A in the cardiac Ca^{2+} channel alpha 1C subunit. FEBS Lett 384:189–192

Perez-Reyes E, Castellano A, Kim HS Bertrand P, Baggstrom E, Lacerda AE, Wei XY, Birnbaumer L (1992) Cloning and expression of a cardiac/brain beta subunit of the L-type calcium channel. J Biol Chem 267:1792–1797

Perez-Reyes E, Cribbs LL, Daud A, Lacerda AE, Barclay J, Williamson MP, Fox M, Rees M, Lee JH (1998) Molecular characterization of a neuronal low-voltage-activated T-type calcium channel. Nature 391:896–900

Perez-Reyes E, Yuan W, Wei X, Bers DM (1994) Regulation of the cloned L-type cardiac calcium channel by cyclic AMP-dependent protein kinase. FEBS Lett 342:119–123

Peterson BZ, Catterall WA (1995) Calcium binding in the pore of L-type calcium channels modulates high affinity dihydropyridine binding. J Biol Chem 270: 18201–18204

Peterson BZ, Johnson BD, Hockerman GH, Acheson M, Scheuer T, Catterall WA (1997) Analysis of the dihydropyridine receptor site of L-type calcium channels by alanine-scanning mutagenesis. J Biol Chem 272:18752–18758

Peterson BZ, Tanada TN, Catterall WA (1996) Molecular determinants of high affinity dihydropyridine binding in L-type calcium channels. JBiolChem 271:5293–5286

Pichler M, Cassidy TN, Reimer D, Haase H, Kraus R, Ostler D, Striessnig J (1997) Beta subunit heterogeneity in neuronal L-type Ca^{2+} channels. J Biol Chem 272:13877–13882

Pietrobon D, Hess P (1990) Novel mechanism of voltage-dependent gating in L-type calcium channels. Nature 346:651–655

Powers PA, Liu S, Hogan K, Gregg RG (1992) Skeletal muscle and brain isoforms of a β-subunit of human voltage-dependent calcium channels are encoded by a single gene. J Biol Chem 267:22967–22972

Powers PA, Liu S, Hogan K, Gregg RG (1993) Molecular characterization of the gene encoding the γ subunit of the human skeletal muscle 1,4-dihydropyridine-sensitive

Ca^{2+} channel (CACNLG), cDNA sequence, gene structure, and chromosomal location. J Biol Chem 268:9275–9279
Pragnell M, De Waard M, Mori Y, Tanabe T, Snutch TP, Campbell KP (1994) Calcium channel β-subunit binds to a conserved motif in the I-II cytoplasmic linker of the α_1 subunit. Nature 368:67–70
Pragnell M, Sakamoto J, Jay SD, Campbell KP (1991) Cloning and tissue-specific expression of the brain calcium channel β-subunit. FEBS Lett 291:253–258
Quin N, Platano D, Olcese R, Costantin JL, Stefani E, Birnbaumer L (1998a) Unique regulatory properties of the type 2a Ca^{2+} channel beta subunit caused by palmitoylation. Proc Natl Acad Sci USA 95:4690–4695
Quin N, Olcese R, Stefani E, Birnbaumer L (1998b) Modulation of human neuronal alpha 1E-type calcium channel by alpha 2 delta-subunit. Am J Physiol 274:C1324–C1331
Regulla S, Schneider T, Nastainczyk W, Meyer HE, Hofmann F (1991) Identification of the site of interaction of the dihydropyridine channel blockers nitrendipine and azidopine with the calcium-channel α_1 subunit. EMBO J 10:45–49
Ren D, Hall LM (1997) Functional expression and characterization of skeletal muscle dihydropyridine receptors in Xenopus oocytes. J Biol Chem 272:22393–22396
Röhrkasten A, Meyer H, Nastainczyk W, Sieber M, Hofmann F (1988) cAMP-dependent protein kinase rapidly phosphorylates Ser 687 of the rabbit skeletal muscle receptor for calcium channel blockers. J Biol Chem 263:15325–15329
Rosenberg RL, Chen XH (1991) Characterization and localization of two ion binding sites within the pore of cardiac L-type calcium channels. J Gen Physiol 97:1207–1225
Rotman EI, Murphy BJ, Catterall WA (1995) Sites of selective cAMP-dependent phosphorylation of the L-type calcium channel alpha 1 subunit from intact rabbit skeletal muscle myotubes. J Biol Chem 270:16371–16377
Ruth P, Röhrkasten A, Biel M, Bosse E, Regulla S, Meyer HE, Flockerzi V, Hofmann F (1989) Primary structure of the β subunit of the DHP-sensitive calcium channel from skeletal muscle. Science 245:1115–1118
Schmid R, Seydl K, Baumgartner W, Groschner K, Romanin C (1995) Trypsin increases availability and open probability of cardiac L-type Ca^{2+} channels without affecting inactivation induced by Ca^{2+}. Biophys J 69:1847–1857
Schneider T, Regulla S, Hofmann F (1991) The Devapamil binding site of the purified skeletal muscle CaCB-receptor is modulated by micromolar and millimolar calcium. Eur J Biochem 200:245–253
Schneider T, Wei X, Olcese R, Costantin JL, Neely A, Palade P, Perez-Reyes E, Qin N, Zhou J, Crawford GD, Smith RG, Appel SH, Stefani E, Birnbaumer L (1994) Molecular analysis and functional expression of the human type E neuronal Ca^{2+} channel α_1 subunit. Receptors Channels 2:255–270
Schuhmann K, Groschner K (1994) Protein kinase-C mediates dual modulation of L-type Ca^{2+} channels in human vascular smooth muscle. FEBS Lett 341:208–212
Schuhmann K, Voelker C, Höfer GF, Plügelmeier H, Klugbauer N, Hofmann F, Romanin C, Groschner K (1997) Essential role of the beta subunit in modulation of C class L-type Ca^{2+} channels by intracellular pH. FEBS Lett 408:75–80
Schultz D, Mikala G, Yatani A, Engle DB, Iles DE, Segers B, Sinke RJ, Weghuis DO, Klockner U, Wakamori M et-al. (1993) Cloning, chromosomal localization, and functional expression of the alpha 1 subunit of the L-type voltage-dependent calcium channel from normal human heart. Proc Natl Acad Sci USA 90:6228–6232
Schuster A, Lacinova L, Klugbauer N, Ito H, Birnbaumer L, Hofmann F (1996) The IVS6 segment of the L-type calcium channel is critical for the action of dihydropyridines and phenylalkylamines. EMBO J 15:2365–2370
Scott VE, De Waard M, Liu H, Gurnett CA, Venzke DP, Lennon VA, Campbell KP (1996) Beta subunit heterogeneity in N-type Ca^{2+} channels. J Biol Chem 271:3207–3212

Sculptoreanu A, Rotman E, Takahashi M, Scheuer T, Catterall WA (1993a). Voltage-dependent potentiation of the activity of cardiac L-type calcium channel α_1 subunit due to phosphorylation by cAMP-dependent protein kinase. Proc Natl Acad Sci USA 90:10135–10139

Sculptoreanu A, Scheuer T, Catterall WA (1993b). Voltage-dependent potentiation of L-type Ca^{2+} channels due to phosphorylation by cAMP-dependent protein kinase. Nature 364:240–243

Seino S, Chen L, Seino M, Blondel D, Takeda J, Johnson JH, Bell, GI (1992). Cloning of the α_1 subunit of a voltage-dependent calcium channel expressed in pancreatic β-cells. Proc Natl Acad Sci USA 89:584–588

Seisenberger C. Welling A, Schuster A, Hofmann F (1995) Two stable cell lines for screening of calcium channel blockers. N-Sch Arch Pharmacol 352:662–669

Sham JS, Cleemann L, Morad M (1995) Functional coupling of Ca^{2+} channels and ryanodine receptors in cardiac myocytes. Proc Natl Acad Sci USA 92:121–125

Shirokov R, Ferreira G, Yi J. Ríos E (1998) Inactivation of gating currents of L-type calcium channels. Specific role of the $\alpha_2\delta$ subunit. J Gen Physiol 111:807–823

Shistik E, Ivanina T, Blumenstein Y, Dascal N (1998) Crucial role of N terminus in function of cardiac L-type Ca^{2+} channel and its modulation by protein kinase C. J Biol Chem 273:17901–17909

Shistik E, Ivanina T, Puri T, Hosey M, Dascal N (1995) Ca^{2+} current enhancement by $\alpha 2/\delta$ and β subunits in *Xenopus* oocytes: contribution of changes in channel gating and $\alpha 1$ protein level. J Physiol 489:55–62

Singer D, Biel M, Lotan I, Flockerzi V, Hofmann F, Dascal N (1991) The roles of the subunits in the function of the calcium channel. Science 253:1553–1557

Singer-Lahat D, Gershon E, Lotan I, Hullin R, Biel M, Flockerzi V, Hofmann F, Dascal N (1992) Modulation of cardiac Ca^{2+} channels in *Xenopus* oocytes by protein kinase C. FEBS Lett 306:113–118

Singer-Lahat D, Lotan I, Biel M, Flockerzi V, Hofmann F, Dascal N (1994) Cardiac calcium channels expressed in *Xenopus* oocytes are modulated by dephosphorylation but not by cAMP-dependent phosphorylation. Receptors Channels 2:215–226

Sinnegger MJ, Wang Z, Grabner M, Hering S, Striessnig J, Glossmann H, Mitterdorfer J (1997) Nine L-type amino acid residues confer full 1,4-dihydropyridine sensitivity to the neuronal calcium channel alpha1 A subunit. Role of L-type Met1188. J Biol Chem 272:27686–27693

Soldatov NM (1994) Genomic structure of human L-type Ca^{2+} channel. Genomics 22:77–87

Soldatov NM, Oz M, O'Brien KA, Abernethy DR, Morad M (1998) Molecular determinants of L-type Ca^{2+} channel inactivation. Segment exchange analysis of the carboxyl-terminal cytoplasmic motif encoded by exons 40–42 of the human alpha1 C subunit gene. J Biol Chem 273:957–963

Soong TW, Stea A, Hodson CD, Dubel SJ, Vincent SR, Snutch TP (1993) Structure and functional expression of a member of the low voltage-activated calcium channel family. Science 260:1133–1136

Starr TVB, Prystay W, Snutch TP (1991) Primary structure of a calcium channel that is highly expressed in the rat cerebellum. Proc Natl Acad Sci USA 88:5621–5625

Stea A, Tomlinson WJ, Soong TW, Bourinet E, Dubel SJ, Vincent SR, Snutch TP (1994) Localization and functional properties of a rat brain α_{1A} calcium channel reflect similarities to neuronal Q- and P-type channels. Proc Natl Acad Sci USA 91:10576–10580

Striessnig J, Berger W, Glossmann H (1993) Molecular properties of voltage-dependent Ca^{2+} channels in excitable tissues. Cell Physiol Biochem 3:295–317

Strom TM, Nyakatura G, Apfelstedt-Sylla E,; Hellebrand H, Lorenz B, Weber BH, Wutz K, Gutwillinger N, Ruther K, Drescher B, Sauer C, Zrenner E, Meitinger T, Rosenthal A, Meindl A (1998) An L-type calcium-channel gene mutated in incomplete X-linked congenital stationary night blindness. Nat Genet 19:260–263

Stühmer W, Conti F, Suzuki H, Wang XD, Noda M, Yahagi N, Kubo H, Numa S (1989) Structural parts involved in activation and inactivation of the sodium channel. Nature 339:597–603

Suh-Kim H, Wei X, Klos A, Pan S, Ruth P, Flockerzi V, Hofmann F, Perez-Reyes E, Birnbaumer L (1996) Reconstitution of the skeletal muscle dihydropyridine receptor. Functional interaction among α_1, β, γ and $\alpha_2\delta$ subunits Receptors and Channels 4:217–225

Tanabe T, Adams BA, Numa S, Beam KG (1991) Repeat I of the dihydropyridine receptor is critical in determining calcium channel activation kinetics. Nature 352:800–803

Tanabe T, Beam KG, Adams BA, Niidome T, Numa S (1990) Regions of the skeletal muscle dihydropyridine receptor critical for excitation-contraction coupling. Nature 346:567–569

Tanabe T, Takeshima H, Mikami A, Flockerzi V, Takahashi H, Kangawa K, Kojima M, Matsuo H, Hirose T, Numa S (1987) Primary structure of the receptor for calcium channel blockers from skeletal muscle. Nature 328:313–318

Tang S, Mikala G, Bahinski A, Yatani A, Varadi G, Schwartz A (1993) Molecular localization of ion selectivity sites within the pore of a human L-type cardiac calcium channel. J Biol Chem 268:13026–13029

Tareilus E, Roux M, Qin N, Olcese R, Zhou J, Stefani E, Birnbaumer L (1997) A Xenopus oocyte beta subunit: evidence for a role in the assembly/expression of voltage-gated calcium channels that is separate from its role as a regulatory subunit. Proc Natl Acad Sci USA 94:1703–1708

Tsien RW, Hess P, McCleskey EW, Rosenberg RL (1987) Calcium channels: mechanisms of selectivity, permeation, and block. Annu Rev Biophys Biophys Chem 16:265–290

Vance CL, Begg CM, Lee WL, Haase H, Copeland TD, McEnery MW (1998) Differential expression and association of calcium channel alpha1B and beta subunits during rat brain ontogeny. J Biol Chem 273:14495–14502

Volsen SG, Day NC, McCormack AL, Smith W, Craig PJ, Beattie RE, Smith D, Ince PG, Shaw PJ, Ellis SB, Mayne N, Burnett JP, Gillespie A, Harpold MM (1997) The expression of voltage-dependent calcium channel beta subunits in human cerebellum. Neuroscience 8:161–174

Walker D, Bichet D, Campbell KP, De Waard M (1998) A beta 4 isoform-specific interaction site in the carboxyl-terminal region of the voltage-dependent Ca^{2+} channel alpha 1A subunit. J Biol Chem 273:2361–2367

Wang Z, Grabner M, Berjukow S, Savchenko A, Glossmann H, Hering S (1995) Chimeric L-type Ca^{2+} channels expressed in Xenopus laevis oocytes reveal role of repeats III and IV in activation gating. J Physiol 486:131–137

Watanabe T, Kalasz H, Yabana H, Kuniyasu A, Mershon J, Itagaki K, Vaghy PL, Naito K, Nakayama H, Schwartz A (1993) Azidobutyryl clentiazem, a new photoactivatable diltiazem analog, labels benzothiazepine binding sites in the alpha 1 subunit of the skeletal muscle calcium channel. FEBS Lett 334:261–264

Wei X, Neely A, Lacerda AE, Olcese R, Stefani E, Perez-Reyes E, Birnbaumer L (1994) Modification of Ca^{2+} channel activity by deletions at the carboxyl terminus of the cardiac alpha 1 subunit. J Biol Chem 269:1635–1640

Wei X, Neely A, Olcese R, Lang W, Stefani E, Birnbaumer L (1996) Increase in Ca^{2+} channel expression by deletions at the amino terminus of the cardiac alpha 1C subunit. Receptors Channels. 4:205–215

Wei X, Pan S, Lang W, Kim H, Schneider T, Perez-Reyes E, Birnbaumer L (1995) Molecular determinants of cardiac Ca^{2+} channel pharmacology. Subunit requirement for the high affinity and allosteric regulation of dihydropyridine binding. J Biol Chem 270:27106–27111

Wei XY, Perez-Reyes E, Lacerda AE, Schuster G, Brown AM, Birnbaumer, L (1991) Heterologous regulation of the cardiac Ca^{2+} channel α_1 subunit by skeletal muscle β and γ subunits. J Biol Chem 266:21943–21947

Welling A, Bosse E, Cavalié A, Bottlender R, Ludwig A, Nastainczyk W, Flockerzi V, Hofmann F (1993a) Stable coexpression of calcium channel α_1, β and α_2/δ subunits in a somatic cell line. J Physiol 471:749–765
Welling A, Kwan YW, Bosse E, Flockerzi V, Hofmann F, Kass RS (1993b) Subunit-dependent modulation of recombinant L-type calcium channels: molecular basis for dihydropyridine tissue selectivity. Circ Res 73:974–980
Welling A, Ludwig A, Zimmer S, Klugbauer N, Flockerzi V, Hofmann F (1997) Alternatively spliced IS6 segments of the α_{1C} gene determine the tissue-specific dihydropyridine sensitivity of cardiac and vascular smooth muscle L-type calcium channels. Circ Res 81:526–532
Williams ME, Brust PF, Feldman DH, Saraswathi P, Simerson S, Maroufi A, McCue AF, Velicelebi G, Ellis SB, Harpold MM (1992a) Structure and functional expression of an ω-conotoxin-sensitive human N-type calcium channel. Science 257:389–395
Williams ME, Feldman DH, McCue AF, Brenner R, Velicelebi G, Ellis SB, Harpold MM (1992b) Structure and functional expression of α_1, α_2/δ and β subunits of a novel human neuronal calcium channel subtype. Neuron 8:71–84
Williams ME, Marubio LM, Deal CR, Hans M, Brust PF, Philipson LH, Miller RJ, Johnson EC, Harpold MM, Ellis SB (1994) Structure and functional characterization of neuronal α_{1E} calcium channel subtypes. J Biol Chem 269:22347–22357
Wiser O, Trus M, Tobi D, Halevi S, Giladi E, Atlas D (1996) The α_2/δ subunit of voltage sensitive Ca^{2+} channels is a single transmembrane extracellular protein which is involved in regulated secretion. FEBS Lett 379:15–20
Witcher DR, De Waard M, Liu-H, Pragnell M, Campbell KP (1995) Association of native Ca^{2+} channel beta subunits with the alpha 1 subunit interaction domain. J Biol Chem 270:18088–18093
Yang J, Ellinor PT, Sather WA, Zhang JF, Tsien RW (1993) Molecular determinants of Ca^{2+} selectivity and ion permeation in L-type Ca^{2+} channels. Nature 366:158–161
Yang J, Tsien RW (1993) Enhancement of N- and L-type calcium channel currents by protein kinase C in frog sympathetic neurons. Neuron. 10:127–136
Yatani A, Bahinski A, Mikala G, Yamamoto S, Schwartz A (1994) Single amino acid substitutions within the ion permeation pathway alter single-channel conductance of the human L-type cardiac Ca^{2+} channel. Circ Res 75:315–323
Yoshida A, Takahashi M, Nishimura S, Takeshima H, Kokubun S (1992) Cyclic AMP-dependent phosphorylation and regulation of the cardiac dihydropyridine-sensitive Ca channel. FEBS Lett 309:343–349
You Y, Pelzer DJ, Pelzer S (1995) Trypsin and forskolin decrease the sensitivity of L-type calcium current to inhibition by cytoplasmic free calcium in guinea pig heart muscle cells. Biophys J 69:1838–1846
Zhang JF, Ellinor PT, Aldrich RW, Tsien RW (1994) Molecular determinants of voltage-dependent inactivation in calcium channels. Nature 372:97–100
Zhou J, Olcese R, Qin N, Noceti F, Birnbaumer L, Stefani E (1997) Feedback inhibition of Ca^{2+} channels by Ca^{2+} depends on a short sequence of the C terminus that does not include the Ca^{2+} -binding function of a motif with similarity to Ca^{2+} -binding domains. Proc Natl Acad Sci USA 94:2301–2305
Zong X, Hofmann F (1996) Ca^{2+}-dependent inactivation of the class C L-type Ca^{2+} channel is a property of the α_1 subunit. FEBS Letters 378:121–125
Zong X, Schreieck J, Mehrke G, Welling A, Schuster A, Bosse E, Flockerzi V, Hofmann F (1995) On the regulation of the expressed L-type calcium channel by cAMP-dependent phosphorylation. Pflügers Arch 430:340–347
Zühlke RD, Bouron A, Soldatov NM, Reuter H (1998) Ca^{2+} channel sensitivity towards the blocker isradipine is affected by alternative splicing of the human alpha1C subunit gene. FEBS Lett 427:220–224
Zühlke RD, Reuter H (1998) Ca^{2+} sensitive inactivation of L-type Ca^{2+} channels depends on multiple cytoplasmic amino acid sequences of the alpha1C subunit. Proc Natl Acad Sci USA 95:3287–3294

CHAPTER 5

Ca^{2+} Channel Antagonists and Agonists

S. ADACHI-AKAHANE and T. NAGAO

A. Ca^{2+} Channel Antagonists

I. Historical Background

Ca^{2+} channel antagonists were originally developed as coronary vasodilators. Ca^{2+} antagonism, as a new principle of pharmacological action of coronary drugs, was reported by Albert Fleckenstein in 1964 (FLECKENSTEIN 1964). Shortly after that, verapamil, gallopamil (D600), nifedipine, and diltiazem were shown to suppress cardiac E-C coupling in that they abolished contractile force without a major change in the action potential. These drugs were termed Ca^{2+} antagonists, because the inhibitory actions of these drugs were antagonized by increasing the extracellular Ca^{2+} concentration (FLECKENSTEIN 1983). The vascular smooth muscle E-C coupling also turned out to be susceptible to Ca^{2+} antagonism (FLECKENSTEIN 1977). In the 1970s, the voltage-clamp technique made it possible to demonstrate the specific suppression of the voltage-dependent slow Ca^{2+}-influx by verapamil, D600, nifedipine, and diltiazem. These studies opened up a new concept of "Ca^{2+} antagonism" as a new therapeutic principle in the treatment of cardiovascular diseases such as hypertension, angina pectoris, cerebral, and peripheral vascular disorders. The use of Ca^{2+} channel antagonists as a pharmacological tool helped clarify the biophysical and molecular properties of voltage-dependent L-type Ca^{2+} channels.

The molecular basis of Ca^{2+} antagonism is the block of plasmalemmal L-type Ca^{2+} channels. The voltage-dependent Ca^{2+} channels are classified as summarized in Table 1. However, as we discuss later, many of the organic Ca^{2+} channel blockers exert their effects not by simply occluding the channel pore but rather by modifying the channel gating in a manner similar to the allosteric inhibition of enzymes. Later generation of Ca^{2+} channel antagonists block not only L-type but also other channels, such as N-type Ca^{2+} channels, T-type Ca^{2+} channels, Na^+ channels, or K^+ channels, which turned out to be clinically beneficial.

Table 1. Voltage-dependent Ca^{2+} channel subtypes

Type	Threshold for activation	Inactivation voltage (mV)	Single channel conductance* (pS)	Blocker	Subunit composition α_1 subunit	Accessory subunits	Distribution	Function
L	>–30 mV HVA	–60 ~ –10	11 ~ 25	DHPs	α_{1S}	$\beta, \alpha_2/\delta, \gamma$	Skeletal muscle	Excitation-contraction coupling
				PAAs	α_{1Ca}	$\beta, \alpha_2/\delta$	Heart	
				BTZs	α_{1Cb}	$\beta, \alpha_2/\delta$	Smooth muscle	Excitation-secretion coupling (?)
				Calciseptine	α_{1Cc}	$\beta, \alpha_2/\delta$	Brain, Heart, Pituitary, Adrenal	Neurotransmission (?)
					α_{1D}	$\beta, \alpha_2/\delta$	Brain, Heart, Cochlea, Pancreas, Kidney, Ovary	
					α_{1F}	$\beta, \alpha_2/\delta$ (?)	Retina	
N	>–30 mV HVA	–120 ~ –30	10 ~ 22	ω-CgTxGVIA	α_{1B}, $\beta, \alpha_2/\delta$		Neuron	Neurotransmission
P/Q	>–40 mV HVA	?	9 ~ 19	ω-Aga IV_A FTX	α_{1A}, $\beta, \alpha_2/\delta, \gamma$ (?)		Brain, Neuron Pituitary, Cochlea	Neurotransmission
R	>–40 mV HVA	–100 ~ –40	14	Ni^{2+} Mibefradil	α_{1E}, $\beta, \alpha_2/\delta$		Brain, Retina, Cochlea, Heart	Repetitive firing
T	>–70 mV LVA	–110 ~ –50	7 ~ 10	Ni^{2+}	α_{1G}	(?)	Brain	Pacemaker potential
				Mibefradil	α_{1H}	(?)	Brain, Heart, Kidney	Repetitive firing
				Kurtoxin	α_{1I}	(?)	Brain	
				Octanol				
				Flunaridine				

* Measured with Ba^{2+} at 80–110 mM as a charge carrier.

II. Allosteric Interaction Between Ca^{2+} Channel Antagonist Binding Sites

In the early 1980s, the saturable high- and low-affinity binding of tritiated DHPs to membranes from heart muscle or brain were reported (GLOSSMANN et al. 1982). The high-affinity binding component represented a stereoselective binding of DHPs to L-type Ca^{2+} channels. The high density expression of L-type Ca^{2+} channels in skeletal muscle allowed the purification of Ca^{2+} channels (GLOSSMANN et al. 1983a; CURTIS and CATTERALL 1984) and subsequent cloning of the DHP receptor (TANABE et al. 1987). The low-affinity binding components are not related to L-type Ca^{2+} channels (GLOSSMANN et al. 1985).

The three major chemical classes of Ca^{2+} channel antagonists, 1,4-dihydropyridines (DHPs, Fig. 1), phenylalkylamines (PAAs, Fig. 2), and 1,5-benzothiazepines (BTZs, Fig. 3) have chemically different structures. Equilibrium and kinetic binding studies, by use of high-affinity tritiated probes of the three classes of Ca^{2+} channel antagonists, indicated that DHPs, PAAs, and BTZs bind to distinct sites on the Ca^{2+} channel. Moreover, the drug binding to the respective site affect other binding sites in a reciprocal manner. Such allosteric interaction between the DHP-, PAA-, and BTZ-binding domains was summarized in "allosteric model" (Fig. 4). The DHP-binding was noncompetitively inhibited by verapamil, but stimulated by D-*cis*-diltiazem. The PAA-binding was noncompetitively inhibited by DHPs and BTZs. The specific binding of [^{3}H]diltiazem was potentiated by DHPs, but inhibited by PAAs (GLOSSMANN et al. 1983b; BALWIERCZAK et al. 1987). The positive allosteric interaction between DHP- and BTZ-binding sites takes place in a temperature-dependent manner. For instance, diltiazem stimulated the binding of [^{3}H]isradipine at 37°C, but incompletely inhibited it at 2°C (GLOSSMANN et al. 1985). Diltiazem increased the affinity of [^{3}H]isradipine-binding to rabbit skeletal muscle T-tubular membranes through modulation of both association and dissociation rates of [^{3}H]isradipine-binding at 37°C (IKEDA et al. 1991). Electrophysiological study using guinea pig ventricular myocytes demonstrated that nitrendipine potentiates the blocking action of diltiazem on L-type Ca^{2+} channel currents in a temperature-dependent manner (KANDA et al. 1998). Some diltiazem analogs, such as azidobutyryl diltiazem and DTZ323, however, rather inhibited the DHP-binding at both 37°C and 2°C as a result of modulation of both association and dissociation rates of the DHP-binding (NAITO et al. 1989; HAGIWARA et al. 1997). Whether binding sites for PAAs and BTZs are identical or distinct has been questioned, because PAAs completely inhibit the BTZ-binding with an increase of K_d values in an apparently competitive manner. However, the two receptors appear to be distinct, because the dissociation rate of diltiazem is markedly increased in the presence of PAAs, indicating the negative allosteric modulation of BTZ-binding by PAA. In contrast, BTZs do not change the dissociation rate of diltiazem (GARCIA et al. 1986; IKEDA et al. 1991; HAGIWARA et al. 1997). The binding properties of PAA, devapamil ((–)D888) is unique: it resembles those of

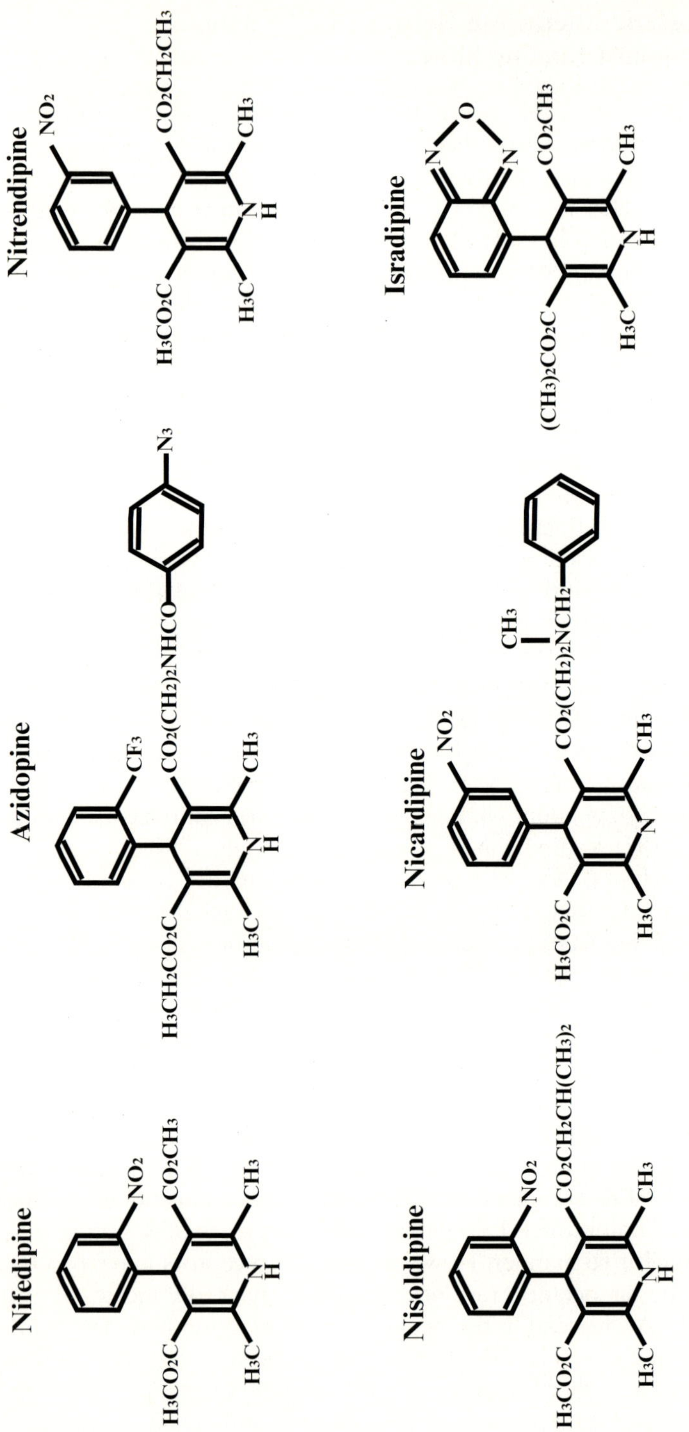

Fig. 1. Structure of 1,4-dihydropyridine (DHP) Ca^{2+} channel antagonists

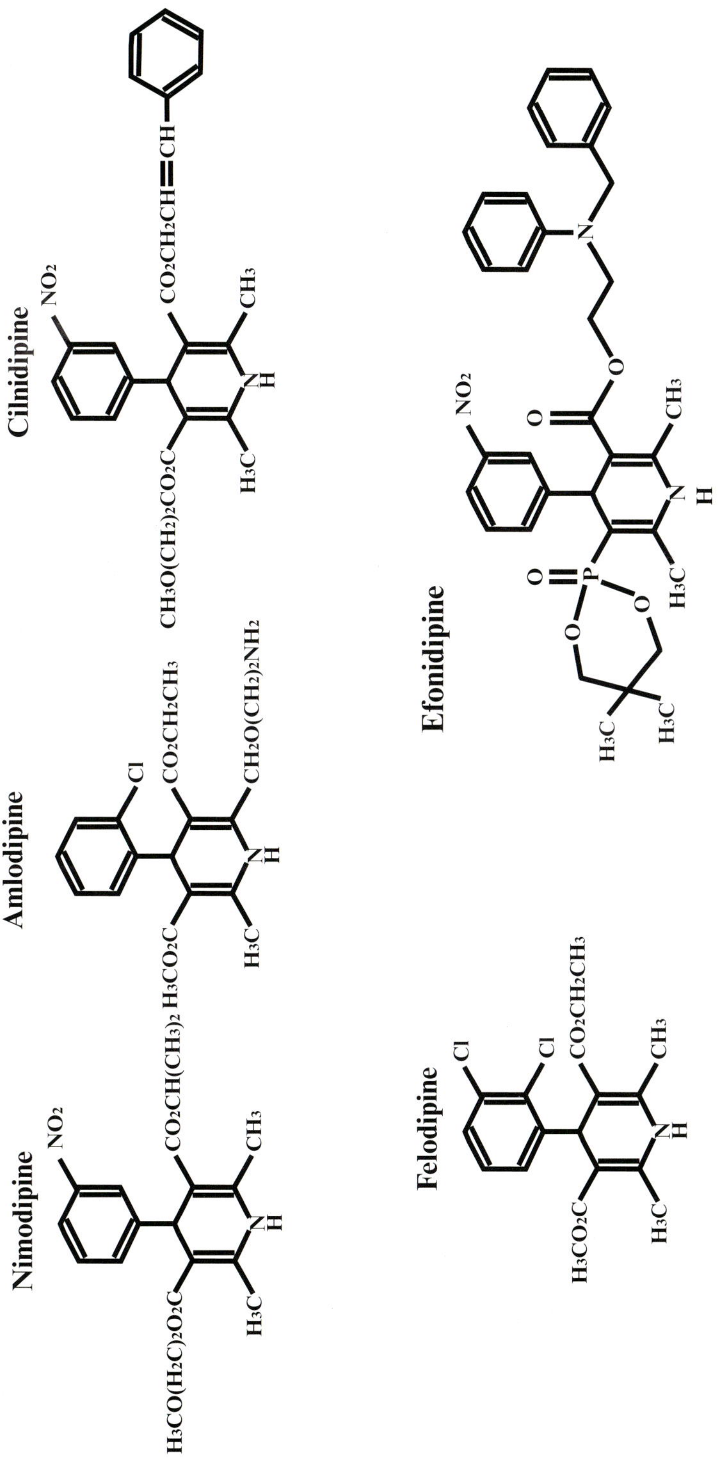

Fig. 1. *Continued*

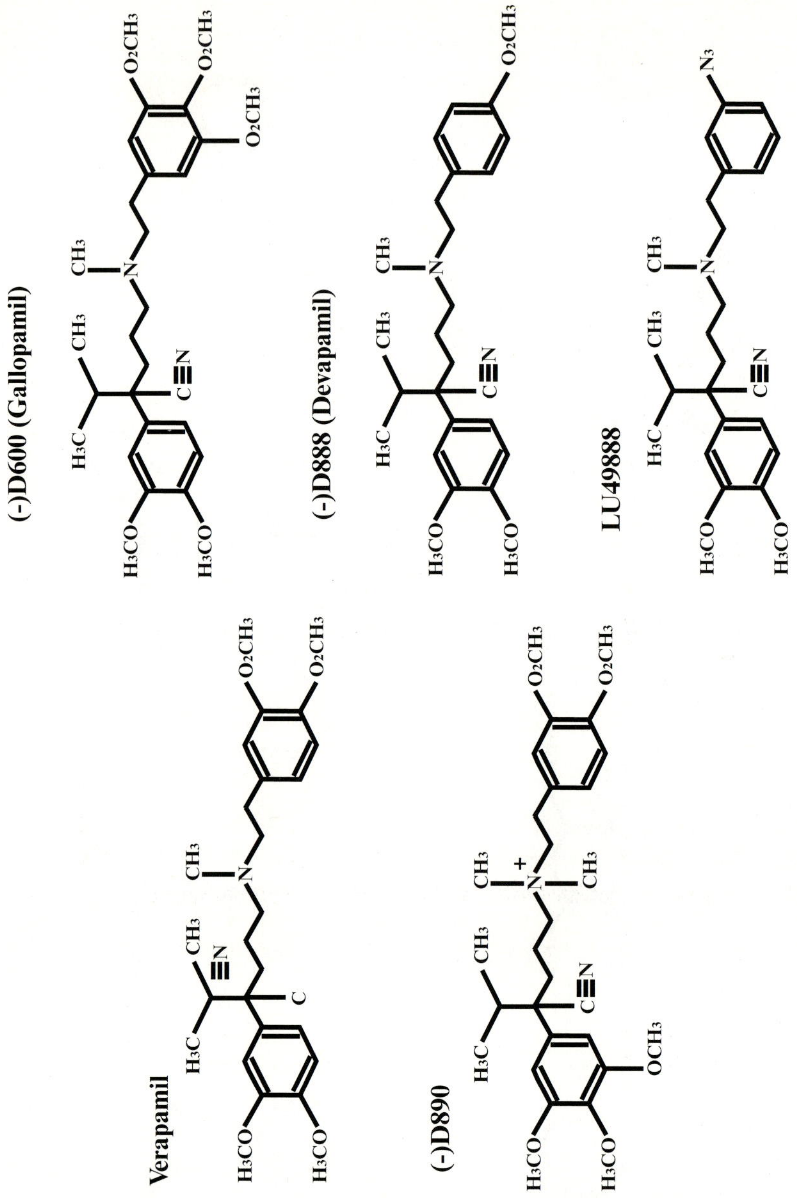

Fig. 2. Structure of phenylalkylamine (PAA) Ca^{2+} channel antagonists

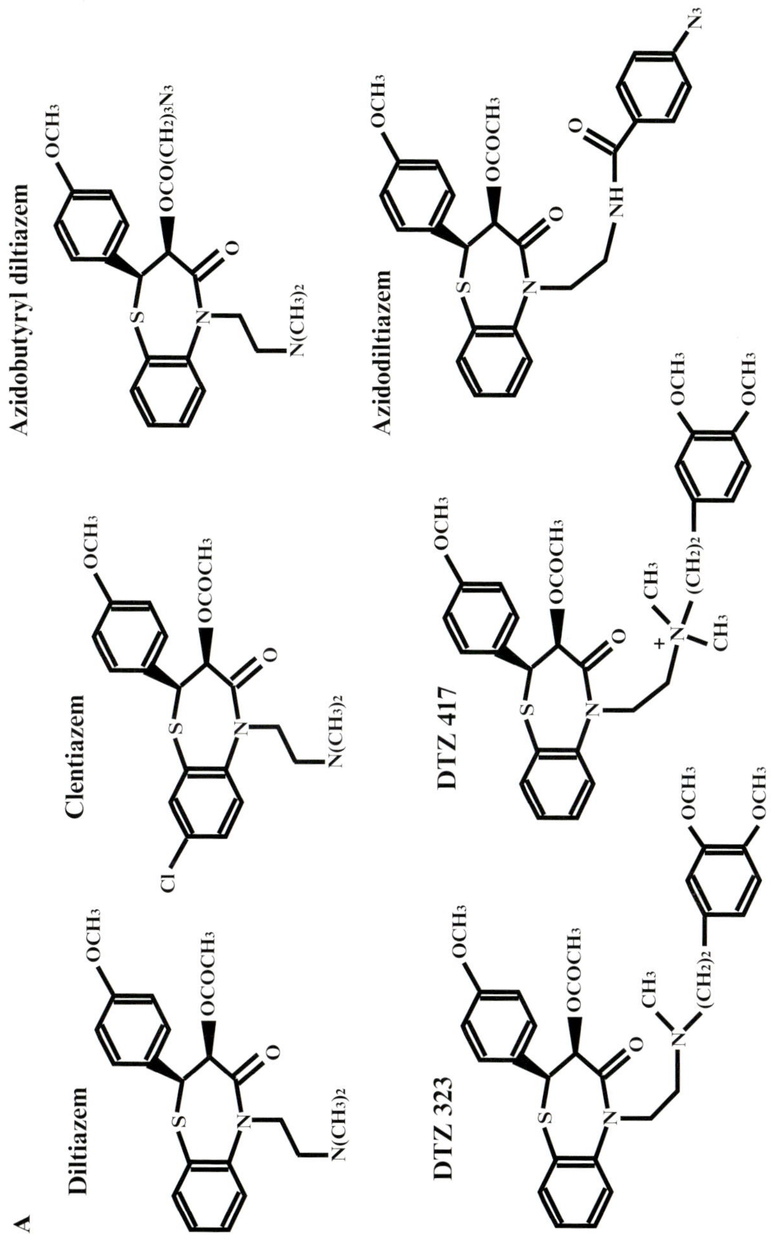

Fig. 3. A Structure of 1,5-benzothiazepine (BTZ) Ca^{2+} channel antagonists. **B** Benzothiazepine analogs

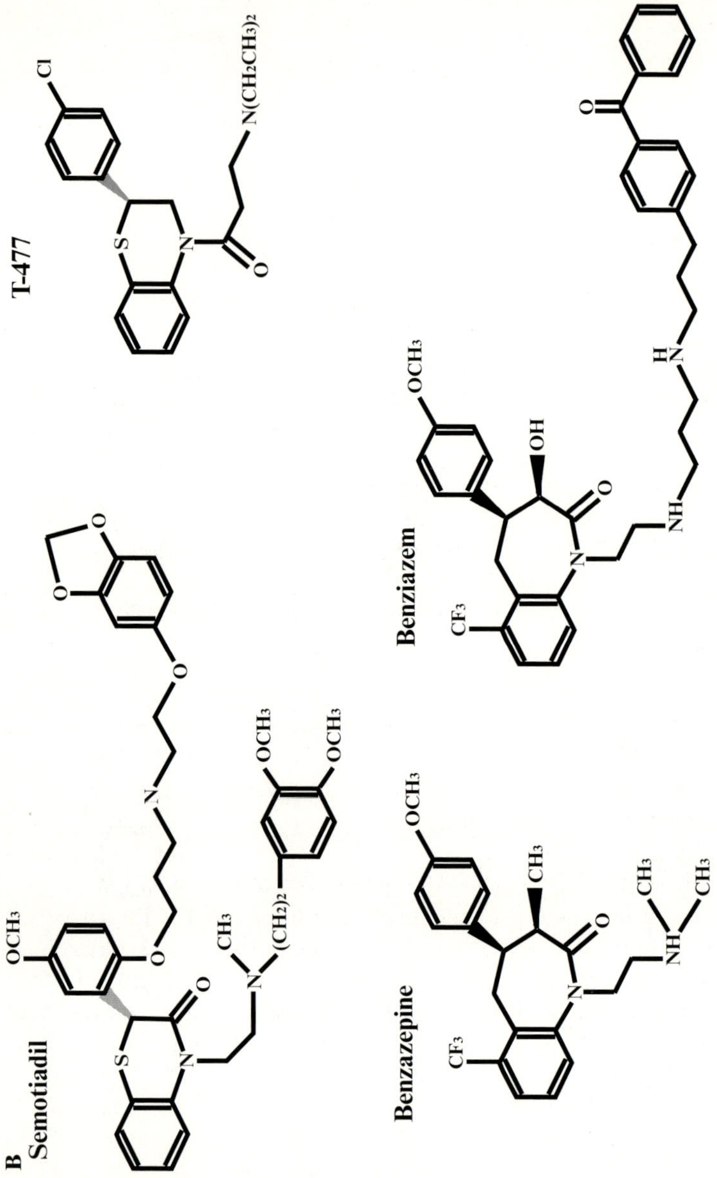

Fig. 3. *Continued*

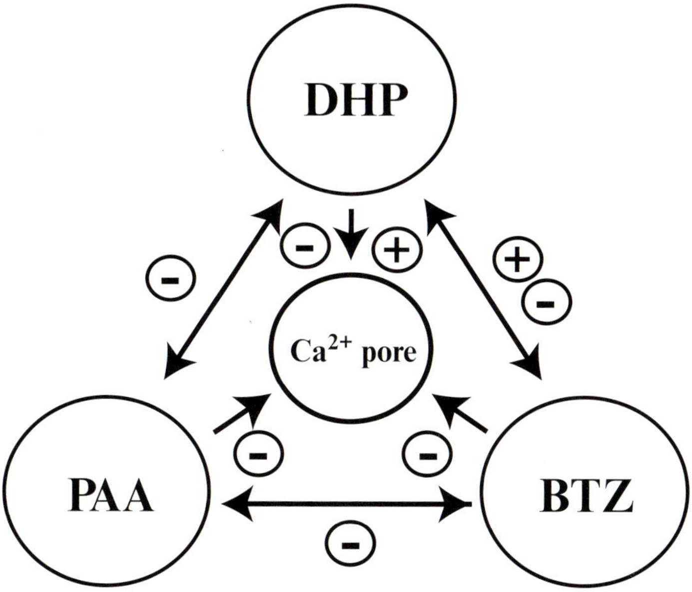

Fig. 4. Schematic representation of reciprocal allosteric modulation between binding sites for Ca^{2+} channel antagonists

D-*cis*-diltiazem, because it stimulates equilibrium binding of DHPs and vice versa (STRIESSNIG et al. 1986; REYNOLDS et al. 1986). As we discuss later, this compound appears to bind to both PAA and BTZ sites.

Drug binding to specific binding domains of Ca^{2+} channel antagonists is also affected by divalent cations, such as Ca^{2+}. Ca^{2+} binds to the Ca^{2+} channel and modulates, positively or negatively, the binding of Ca^{2+} channel antagonists. Treatment of membrane preparation from brain with EDTA abolished the high-affinity binding of DHP, which was restored by the addition of micromolar concentration of Ca^{2+} (GLOSSMANN et al. 1985). Similar Ca^{2+}-dependence was observed in PAA binding to skeletal muscle L-type Ca^{2+} channels (KNAUS et al. 1992). The binding of Ca^{2+} channel antagonists to the respective binding domains was inhibited by the addition of higher concentrations of (millimolar) divalent cations.

III. Biophysical and Pharmacological Properties of Ca^{2+} Channel Antagonists

Ca^{2+} channel antagonists are clinically useful because of their tissue selectivity, determined by the unique biophysical properties of Ca^{2+} channel antagonists such as voltage-dependence, use-dependence, ion channel selectivity, and by the dependence of cell function on the Ca^{2+} influx.

1. Dihydropyridines

DHPs are highly selective for vascular smooth muscles. They lack antiarrhythmic properties and usually do not have a depressant effect on myocardial contractility, because their direct depressant effects are offset by a reflex increase in sympathetic tone induced by vasodilation (see Table 2). Nifedipine reduces peripheral resistance and prevents coronary artery spasm. It has antihypertensive and antianginal properties. Nicardipine, isradipine, amlodipine, and felodipine are more specific for vascular smooth muscle than for cardiac muscle. Nimodipine is selective for cerebral vasculature (Fig. 1). In electrophysiological studies, the first generation of hydrophobic DHPs, such as nifedipine and nitrendipine, showed resting block (tonic block), which was augmented by membrane depolarization (LEE and TSIEN 1983; BEAN 1984; see Fig. 5). Such voltage-dependence of DHP effects accounts for the tissue-selectivity of DHPs for vascular smooth muscle cells (membrane potentials: around –50 mV) vs ventricular myocytes (membrane potentials: around –80 mV). Further analysis demonstrated that DHPs in neutral molecular form dissociate from the Ca^{2+} channel extremely rapidly, thus masking the state-dependent (use-dependent) block (SANGUINETTI and KASS 1984). In contrast, DHPs in the ionized form and amlodipine show the use-dependent block, mainly due to their slow dissociation from the channel through the hydrophilic pathway (KASS et al. 1989). Amlodipine is positively charged at physiological pH, because it possesses a basic amino side chain (Fig. 1). In addition to the hydrophobic interaction of DHP structure with phospholipid acyl chains of the membrane bilayer, amlodipine's protonated amino side chain serves ionic interaction with the charged anionic oxygen of the phosphate head group of the membrane (MASON et al. 1989). Such ionic interaction is believed to be responsible for its slow onset and long half-life. DHPs stabilizes the inactivated state of L-type Ca^{2+} channels, and thus shift the steady-state inactivation curve toward hyperpolarized potentials. The single channel analysis demonstrated that DHP antagonists, such as nitrendipine, exert mixed effects: first, DHPs increase the blank sweeps and speeds up the inactivation of L-type Ca^{2+} channel currents due to the reduction of late reopenings, which results in the reduction of average current amplitude. However, in non-blank sweeps, open

Table 2. Pharmacological effects of Ca^{2+} channel antagonists

	Verapamil	Nifedipine	Nimodipine	Diltiazem
Vasodilation				
Perpheral	++	+++	+	+
Coronary	++	+++	+	+++
Cerebral	+	+	+++	+
Heart rate	↓	+	–	↓
SA node	↓	–	–	↓↓
AV node	↓↓	–	–	↓
Contractility	↓↓	+	–	↓

time distribution was rather prolonged, and the latencies-to-first opening were shortened (HESS et al. 1984; MCDONALD et al. 1994). The balance between the two seemingly opposite effects of DHPs determines whether the compound behaves as a Ca^{2+} channel antagonist or as an agonist.

The later generation of DHPs block multiple ion channels in addition to the L-type Ca^{2+} channel. High concentration of DHPs have been reported to block the N-type Ca^{2+} channel (IC_{50} value with nicardipine, 10 μmol/l, DIOCHOT et al. 1995), the T-type Ca^{2+} channel (IC_{50} value with nifedipine, 5 μmol/l, AKAIKE et al. 1989a), cardiac voltage-dependent Na^+ channel (IC_{50} value with nitrendipine, 3 μmol/l, YATANI and BROWN 1985), and voltage-dependent K^+ channels (IC_{50} value with nicardipine, 1 μmol/l, FAGNI et al. 1994). The manner of block of K^+ channels by DHPs, however, is different from that of L-type Ca^{2+} channels (AVDONIN et al. 1997).

Among the third generation of Ca^{2+} channel antagonists, amlodipine and cilnidipine have been shown to block not only L-type but also N-type Ca^{2+} channels (IC_{50} values with cilnidipine in DRG neurons, 100 nmol/l for L-type vs 200 nmol/l for N-type, FUJII et al. 1997; UNEYAMA et al. 1997; FURUKAWA et al. 1997). Such N-type effect appears to be responsible for their inhibitory effect on sympathetic neurotransmission (HOSONO et al. 1995), and thus the prevention of reflex tachycatdia. 1,4-DHP structure may be a useful starting compound for developing Ca^{2+} channel antagonists specific for neuronal Ca^{2+} channels (TRIGGLE 1999).

Efonidipine shows negative chronotropic effect with greater potency compared to its negative inotropic potency (MASUDA et al. 1995). Recently it has been reported that efonidipine selectively suppresses the phase IV pacemaker

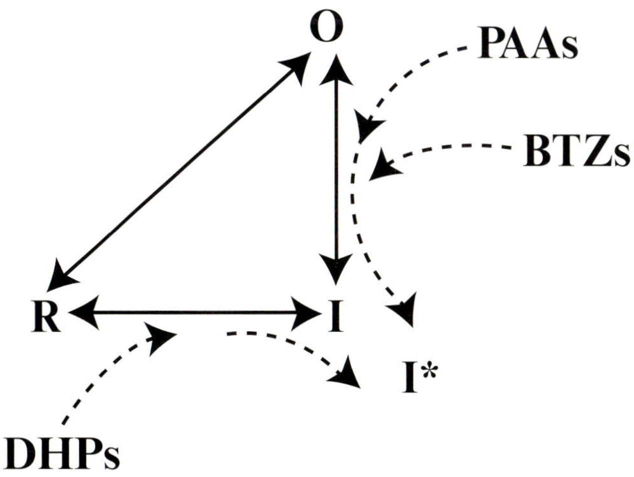

Fig. 5. Simplified diagram of resting block and use-dependent block by Ca^{2+} channel antagonists. I* represents the drug-bound inactivated state. Binding of DHPs, PAAs, or BTZs may lead to distinct inactivated conformation

depolarization in sino-atrial nodal cells through inhibition of both L-type and T-type Ca^{2+} channel currents (MASUMIYA et al. 1997). Felodipine has also been reported to block not only L-type but also T-type Ca^{2+} channel currents (COHEN et al. 1992).

Nimodipine, nicardipine, and isradipine show selectivity for L-type Ca^{2+} channels in cerebral artery. Nimodipine, shows high permeability through the blood-brain barrier. However, the vasodilating effect, rather than the block of neuronal L-type Ca^{2+} channels, appear to contribute to the neuroprotective effects of Ca^{2+} channel antagonists (KOBAYASHI et al. 1998).

DHPs have been shown to inhibit phosphodiesterases (PDEs) (IIJIMA et al. 1984; SHARMA 1997). The PDE-inhibitor activity may contribute to the potent vasodilating effect and the merely marginal cardio-depressive effect of DHPs. The PDE-inhibition, however, appears to potentiate the reflex tachycardia (Table 2).

2. Phenylalkylamines

Verapamil is less potent, compared to DHPs, as a vasodilator in vivo. With doses sufficient to produce vasodilation, it shows more direct negative chronotropic, dromotropic, and inotropic effects than with dihydropyridines. The intrinsic negative inotropic effect of verapamil is partially neutralized by both a decrease in afterload and the subsequent reflex increase in adrenergic tone. Verapamil is used for angina pectoris, hypertension, paroxysmal supraventricular tachycardia, atrial flutter, and fibrillation. Verapamil interrupts reentrant tachycardias and slows the ventricular response to rapid atrial rates by prolonging AV conduction time and the refractory period of the AV node.

Verapamil has been shown to block L-type Ca^{2+} channels in the open state in a manner similar to block of cardiac voltage-dependent Na^{+} channels by local anesthetics (MCDONALD et al. 1984). Verapamil blocks Ca^{2+} channels in a use-dependent manner (LEE and TSIEN 1983, see Fig. 5). Studies with very high concentration of D600 indicated that, during depolarization, the decay of Ba^{2+} currents through L-type Ca^{2+} channels was speeded up by the drug (TIMIN and HERING 1992). Single channel analysis have shown that D600 markedly shortens openings of Ca^{2+} channels and prolongs closed times in cardiac myocytes (PELZER et al. 1985; MCDONALD et al. 1989). Ca^{2+} channel antagonists that act in a use-dependent manner require the activation of Ca^{2+} channels to gain access to their binding sites either in the open state or in the inactivated state following the opening of Ca^{2+} channels. Since Ca^{2+} channel antagonists delay the recovery of Ca^{2+} channels from the inactivated state, high-frequency of depolarizing pulses accelerate the shift of Ca^{2+} channel gating to the inactivated state. The use-dependent block develops more rapidly and more strongly at depolarized membrane potentials, which is explained by the voltage-dependence of dissociation rates of the drugs (MCDONALD et al.

1994). Recovery from the block of L-type Ca^{2+} channel currents at –70 mV was investigated for DHPs and PAAs in cardiac tissue, which was fast for nisoldipine (τ_1 = 1.5 s and 12 = 30 s; SANGUINETTI and KASS 1984) and slow for D600 (τ_2 = 2.4 min; MCDONALD et al. 1984).

Verapamil has been shown to be useful in certain forms of ventricular tachycardia triggered by delayed after depolarizations. Verapamil causes high-affinity block of HERG K^+ channel (I_{Kr}), in native cardiac myocytes with IC_{50} value close to those reported for the block of L-type Ca^{2+} channels (IC_{50} = 1.43×10^{-7} mol/l for HERG current vs $IC_{50} = 1.64 \times 10^{-7}$ mol/l for L-type Ca^{2+} channel current). Verapamil block of HERG channels was use- and frequency-dependent. A quaternary verapamil, *N*-methyl-verapamil, blocked HERG channel only from the intracellular side, indicating that verapamil enters the cell membrane in the neutral form to act at a site within the pore in a manner similar to its block of the L-type Ca^{2+} channel (ZHANG et al. 1999).

Verapamil produced a potent use-dependent block of type IIA Na^+ channels expressed in a mammalian cell line. The drug bound to open and inactivated Na^+ channels during the depolarizing pulses and slowed repriming of drug-bound channels during the interpulse intervals (RAGSDALE et al. 1991).

Verapamil has also been reported to reverse multiple drug-resistance via a mechanism distinct from the block of L-type Ca^{2+} channels (HUET and ROBERT 1988; PEREIRA et al. 1995).

3. Benzothiazepines

Diltiazem exerts peripheral vasodilating effect and mild negative chronotropic effect. Despite the fact that diltiazem and verapamil produce similar effects on the SA node and AV node, the negative inotropic effect of diltiazem has been reported to be modest. Diltiazem is used for angina pectoris, hypertension, supra-ventricular tachycardia, atrial flutter, and fibrillation. Diltiazem (D-*cis*-diltiazem) has been shown to block the L-type Ca^{2+} channel partially at the resting state (tonic block) and mainly in a use-dependent manner (LEE and TSIEN 1983; TUNG and MORAD 1983; KANAYA and KATZUNG 1984, see Fig. 5). Single channel study demonstrated that diltiazem decreases open probability of L-type Ca^{2+} channels in ventricular myocytes (ZAHARADNIKOVA 1992). The negative inotropic effect of diltiazem was markedly augmented by slight depolarization of the resting membrane potential from –80 mV to –60 mV, which is very close to the change of the resting membrane potential during ischemia. Such augmentation of the diltiazem effect was explained by its voltage-dependence of the use-dependent block of the L-type Ca^{2+} channel current (OKUYAMA et al. 1994). Further kinetic analysis has shown that the dissociation rate of diltiazem is greatly dependent on the membrane potential of this range (–90 mV to –60 mV) (KUROKAWA et al. 1997a; YAMAGUCHI et al. 1999; see Fig. 5). Such voltage-dependence of the use-dependent block of L-type Ca^{2+} channels by diltiazem may contribute to its negative chronotropic effect and cardioprotection during ischemia. When the Ca^{2+} channel blocking effect was compared among the Ca^{2+} channel subtypes in the expression system, diltiazem blocked these Ca^{2+} channels at similar concentrations but only the L-type α_{1C} Ca^{2+} channel in a state-dependent manner (CAI et al. 1997).

A diltiazem analog, clentiazem is more selective for cerebral arteries than diltiazem (KIKKAWA et al. 1994). A diltiazem analog, T-477, has little selectivity among voltage-dependent Ca^{2+} channel subtypes and showed protective action in animal stroke model, which was in contrast to diltiazem that is selective for L-type Ca^{2+} channels but lacks neuroprotection (KOBAYASHI et al. 1997; KOBAYASHI and MORI 1998). Ca^{2+} channel antagonists with multiple action appear to protect neurons from ischemic damage more efficiently than those ones highly selective for L-type Ca^{2+} channels (SPEDDING et al. 1995).

The affinity of other stereoisomers of diltiazem, L-*cis*-diltiazem, D-*trans*-diltiazem, L-*trans*-diltiazem, is almost 100-fold lower than that of D-*cis*-dilaizem (IKEDA et al. 1991). However, both D-*cis*-dilaizem and L-*cis*-diltiazem block cardiac voltage-dependent Na^+ channels in a voltage-dependent manner with IC_{50} values close to that for L-type Ca^{2+} channel (IC_{50} values at –70 mV, 2.8×10^{-5} mol/l for L-type Ca^{2+} channel vs 10^{-5} mol/l for Na^+ channel, YAMAGUCHI et al. 1999; personal communication by Tomida et al.). Such Na^+ channel blocking effect appears to prevent the ischemia-reperfusion injury through inhibition of Na^+ accumulation during ischemia and thus preventing the Ca^{2+} influx via reverse mode Na^+-Ca^{2+} exchange activity (ITOGAWA et al. 1996; NISHIDA et al. 1999a, 1999b).

Na^+-Ca^{2+} L-*cis*-diltiazem has been used as a pharmacological tool for selective block of cGMP-activated cation channel of photoreceptors (STERN et al. 1986; HAYNES 1992).

4. Other Ca^{2+} Channel Antagonists

In addition to the three major classes of Ca^{2+} antagonists, there is a vast number of organic molecules that modulate activity of L-type Ca^{2+} channels (Fig. 6). For instance, benzolactam (HOE 166) appears to share the binding site with DHPs. Diphenylbutylpiperidines (fluspirilene, pimozide) are among the most potent inhibitor of skeletal muscle Ca^{2+} channels.

Diphenylalkylamines (DPAAs), such as bepridil, flunarizine, cinnarizine, and fendiline block voltage-dependent Ca^{2+}, Na^+, and K^+ channels. Bepridil has been used as an antianginal agent with multiple therapeutic actions. It blocks L-type Ca^{2+} channels, exerts fast block of cardiac Na^+ channel similar to lidocaine, inhibits Na^+-Ca^{2+} exchanger, prolongs the QT-interval as a result of a blockade of I_{Kr} and I_{Ks}, and inhibits calmodulin (GILL et al. 1992; CHOUABE et al. 1998). Flunarizine, a potent non-selective Ca^{2+} channel blocker showed neuroprotection in in vitro and in vivo animal models. Flunarizine blocks both L-type and T-type Ca^{2+} channels (AKAIKE et al. 1989a,b), and Na^+-influx that may participate in the induction of the ischemic neuronal damage (KOBAYASHI and MORI 1998).

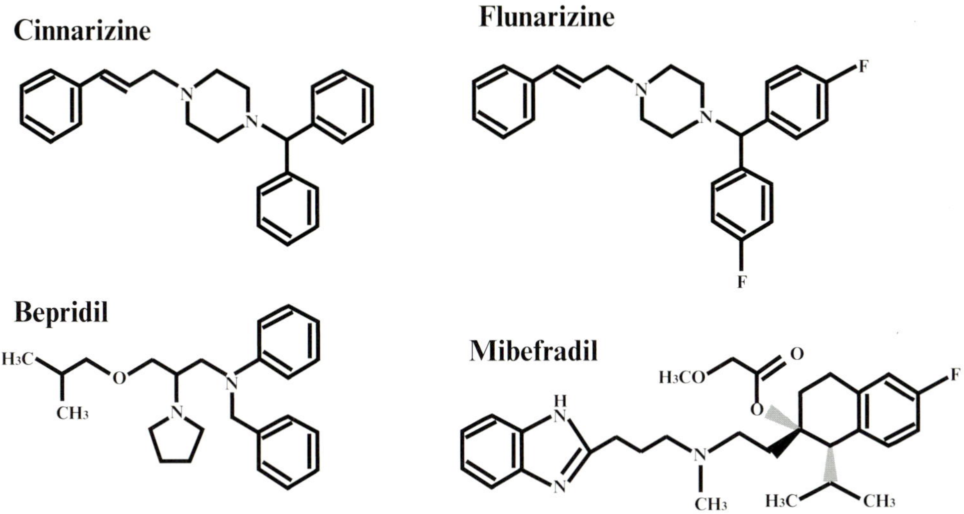

Fig. 6. Ca^{2+} channel antagonists other than the classic three groups

Mibefradil is the most selective organic blocker of T-type Ca^{2+} channels (Mishra and Hermsmeyer 1994; Clozel et al. 1997). However, it has multiple effects including the voltage-dependent block of L-type Ca^{2+} channels (Bezprozvanny and Tsien 1995) and I_{Kr} as well as I_{Ks} (Chouabe et al. 1998). Different class of organic compounds such as tetramethrine, octanol, amiloride, diphenylhidantoin have been suggested to block T-type Ca^{2+} channel, although they are not absolutely selective for T-type Ca^{2+} channel.

IV. Binding Sites

1. Electrophysiological Identification of Binding Sites for Ca^{2+} Channel Blockers

The classic Ca^{2+} antagonists have a secondary or tertiary amino group. Thus the fractional ratio of the drug in membrane permeable neutral form vs in the membrane impermeable ionized form depends on the pH condition. The use of permanently charged quaternary derivatives of Ca^{2+} channel antagonists helped characterize the sidedness of their action. It also introduced a view that the binding sites for DHPs, BTZs, and PAAs are distinct entities.

DHP derivatives with pKa values less than 3.5, such as isradipine, nitrendipine, and nifedipine, are in neutral molecular form at physiological pH so that they are permeable through the lipid bilayer membrane. An examination with a quaternary DHP, SDZ 207–180, localized the DHP interaction site on the extracellular side (Kass et al. 1991; Bangalore et al. 1994). Extracellular application of SDZ 207–180 caused voltage-dependent block of Ca^{2+}

channel currents, whereas SDZ 207–180 as well as amlodipine was ineffective when applied intracellularly.

Methoxyverapamil ((–)D600) is in the ionized form at physiological pH (95% at pH 7.3). It was originally reported that (–)D600 acts on the L-type Ca^{2+} channel from the extracellular side in isolated rabbit vascular and ileac smooth muscle cells (OHYA et al. 1987). This hypothesis was reinvestigated, by use of D890, a quaternary form of (–)D600, comparing the effects of the external application with the intracellular application through a patch-pipette in vascular smooth muscle cells. These studies indicated that the quaternary PAAs block Ca^{2+} channel currents only from the intracellular side as was reported in cardiac myocytes (HESCHELER et al. 1982; LEBLANC et al. 1989). However, in contrast to D890, (–)D600 blocked Ica only from the extracellular side, which was opposite to the results found in cardiac myocytes.

Early study with quaternary diltiazem showed that the compound produces the use-dependent block of the L-type Ca^{2+} channel current from both the extracellular and the intracellular sides of the membrane in guinea pig ventricular myocytes (ADACHI-AKAHANE et al. 1993). However, D-*cis*-Diltiazem blocked the L-type Ca^{2+} channel current preferentially from the extracellular side, because the intracellular application required more than 1000-fold higher concentration than the extracellular application for producing the use-dependent block of the Ca^{2+} channel (ADACHI-AKAHANE and NAGAO 1993). Quaternary diltiazem, given to the extracellular solution, potentiated the [^{3}H]israripine binding to intact rat ventricular myocytes, indicating that the diltiazem-binding site is accessible from the extracellular side of the L-type Ca^{2+} channel (KANDA et al. 1997). A novel 1,5-benzothiazepine analog, DTZ323, blocks the L-type Ca^{2+} channel current through high affinity binding to the BTZ site (KUROKAWA et al. 1997a; HAGIWARA et al. 1997). The sidedness of the BTZ site was determined by use of a quaternary derivative of DTZ323, DTZ417, in guinea pig ventricular myocytes (KUROKAWA et al. 1997b). This compound produced the use-dependent block of L-type Ca^{2+} channel currents only from the extracellular side but not from the intracellular side. These studies confirmed that the BTZ site is accessible from the extracellular side of the L-type Ca^{2+} channel.

Benzazepines were also reported as a competitive inhibitor of diltiazem binding, and a permanently charged benzazepine, SQ 32.428, was used to localize the approximate binding site of BTZ/benzazepine antagonists near the extracellular side of L-type Ca^{2+} channel (HERING et al. 1993; SEYDL et al. 1993).

Radioligand binding studies and the structure-activity relationship studies of a PAA analog, devapamil ((–)D888), suggested that this compound binds not only to the DHP site but also to the BTZ site (REYNOLDS et al. 1986; KIMBALL et al. 1992, 1993). Recent study using quaternary devapamil (qD888) revealed that this compound acts on the extracellular BTZ site as well as the intracellular PAA site within the L-type Ca^{2+} channel in A7r5 cells (BERJUKOV et al. 1996).

2. Biochemical Characterization of Drug-Ca^{2+} Channel Interaction: Photoaffinity Labeling of Ca^{2+} Channels

The development of the photoaffinity ligands of Ca^{2+} channel antagonists helped identify the drug-binding sites within the L-type Ca^{2+} channel. The drugs bind to their receptor sites within Ca^{2+} channels. Upon photo-activation, they label a region that is in close proximity to the high-affinity binding site on the Ca^{2+} channel molecule in a covalent manner. Earlier experimental observations derived from photoaffinity-labeling studies suggested that the three major classes of Ca^{2+} antagonists bind to α_1 subunit of the L-type Ca^{2+} channel complex (Ferry et al. 1984, 1987; Striessnig et al. 1991; Naito et al. 1989). Specifically photolabeled and purified α_1-subunit polypeptides were subjected to limited proteolysis with various proteases and then the mapping of the polypeptide subfragments (Catterall and Striessnig 1992).

A photoreactive verapamil, LU49888, was used to determine the binding site for PAAs on the rabbit skeletal muscle α_{1S} subunit solubilized and partially purified by affinity chromatography. The covalent label for PAA was localized on a 42 amino acid segment that includes transmembrane segment S6 in motif IV and short adjacent carboxyl tail of the α_{1S} subunit (Striessnig et al. 1990).

Similar photoaffinity-labeling strategies were employed to identify the binding site of DHPs using azidopine, diazepine, and isradipine (Regulla et al. 1991; Nakayama et al. 1991; Striessnig et al. 1991). The sites comprised part of S6 and the extracellular loop in motif III as the primary site, and S6 in motif IV as the "secondary or peripheral" site. Somewhat different sites were identified when photoaffinity labeling of skeletal muscle membrane preparations was followed by purification and trypsin digestion (Kalasz et al. 1993). The specific labels were found in the extracellular loops between S5 and S6 segments of motifs I, III, and IV.

A 1,5-benzothiazepine ligand, azidobutyryl clentiazem, specifically labeled the S5-S6 loop of motif IV (Watanabe et al. 1993). Benziazem, a benzazepine derivative that had been reported to bind to BTZ site, labeled S6 of motifs III and IV (Kraus et al. 1996). 1,4-Benzothiazepine, semotiadil ([^{3}H]D51–4700), labeled a short sequence of IVS6 of α_{1S} (Kuniyasu et al. 1998). However, semotiadil may bind to a site distinct from BTZ site, since that compound produced negative allosteric modulation of diltiazem-binding (Nakayama et al. 1994).

Such inconsistency of the results may derive from pitfalls of photoaffinity labeling methods. Photoreactive groups on the side chain are at a distance of 10–15 Å from the core structure of the derivatives. Another problem may be the mapping resolution limited by antibodies used for immunoprecipitation and by the fragment size of proteolytic digestion (typically 5–10 kDa).

3. Molecular Biological Characterization of Drug-Ca^{2+} Channel Interaction: Studies with Experimental Ca^{2+} Channel Mutants

To overcome the limitation of the photoaffinity labeling approach, chimeric α_1 subunits containing α_{1C} and brain α_{1A} or α_{1E} were constructed and expressed in *Xenopus* oocytes to test the sensitivity to Ca^{2+} channel antagonists.

In chimeric Ca^{2+} channels, sensitivity to DHPs was lost when a region from S3 to S6 of motif IV of α_{1C} was replaced by the corresponding region of the DHP-insensitive α_{1A} subunit (TANG et al. 1993).

The gain-of-function experiments using DHP-insensitive α_{1A} subunit revealed that IIIS5, that had not been phtotolabeled by photoreactive DHPs, is also important for restoring the high-affinity DHP-binding. A minimum component required for transferring DHP sensitivity comprised IIIS5, IIIS6, including the connecting linker, as well as the IVS5-IVS6 linker and IVS6 (GRABNER et al. 1996; SINNEGGER et al. 1997; HOCKERMANN et al. 1997c). Different regions in repeat IV appeared to be responsible for the sensitivity to DHP agonists and antagonists. Determinants for agonist action of DHP are located within S6, whereas those for DHP antagonists are in the N-terminal portion of the S5-S6 linker.

The systematic single mutation analysis was carried out to refine the amino acid sequences responsible for DHP-binding. Two amino acids in IIIS5, seven amino acids in IIIS6, and four in IVS6 appear to form the pocket for DHP-binding. Among them, Tyr1120 (IIIS6), Ile1124 (IIIS6), Met 1129 (IIIS6), and Asn 1429 (IVS6) showed strong contributions (MITTERDORFER et al. 1996; PETERSON et al. 1996, 1997; SCHUSTER et al. 1996; ITO et al. 1997; HE et al. 1997). Four among the 13 amino acids are common between L-type and non-L-type α_1 subunits. Introduction of nine amino acid residues in IIIS5, IIIS6, and IVS6 into α_{1A} transferred the pharmacological action of DHP antagonists and agonists, such as leftward shift of the steady-state inactivation curve, slowing of deactivation, and shift of voltage-dependent activation curve (SINNEGGER et al. 1997). Results of the single amino acid analysis of the DHP-binding pockets are summarized in Fig. 7.

Investigation of several splice variants of the α_{1C} subunit showed that additional sequences affect the DHP sensitivity. Detailed analysis of the α_{1Ca} (cardiac type) and α_{1Cb} (smooth muscle type) sequences showed that the alternative splicing at exon 8a/8b that codes for the IS6 segment, affects the affinity for DHPs. Nisoldipine produced significantly larger block of the α_{1Cb} channel compared to the α_{1Ca} Ca^{2+} channel (WELLING et al. 1993, 1997; ZÜHLKE et al. 1998; MOREL et al. 1998). However, such difference of DHP sensitivity could not be explained by the difference of the voltage-dependence of the inactivation kinetics between α_{1Ca} and α_{1Cb} (HU and MARBAN 1998). Thus the splice variant may form different conformation of the DHP-binding pocket. In addition, splice variations at the IIIS2 segment and at the carboxy terminal sequences appear to contribute to the affinity of DHPs (ZÜHLKE et al. 1998).

Analogous α_{1A} chimeras containing IVS6 of α_{1C} confirmed that this segment confers the PAA sensitivity (DÖRING et al. 1996). Site-directed mutations at Tyr1420Ala, Ala1424Ser, or Ile1427Ala in motif IVS6 decreased the binding affinity of devapamil by approximately 10 times, and those of paired combinations of these mutations reduced the affinity by more than 100 times,

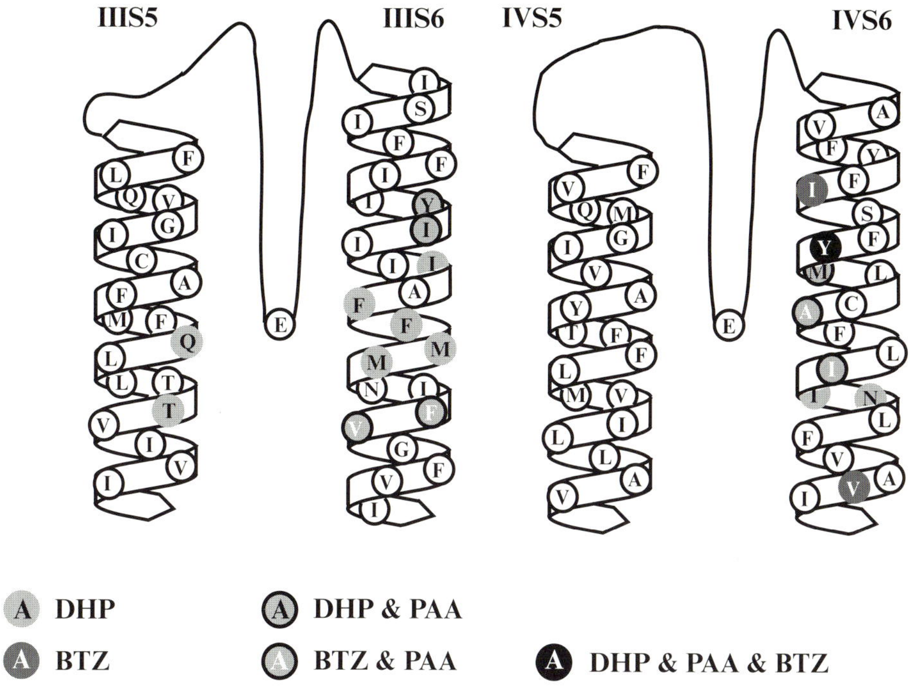

Fig. 7. Amino acid residues that contribute to the formation of Ca^{2+} channel antagonist binding domain. The amino acid sequences and numbering of IIIS5, IIIS6, IVS5, and IVS6 are according to those of human cardiac α_{1C} subunit (Genebank accession number L04569)

supporting the previous findings that IVS6 is responsible for the high-affinity PAA block of α_{1C} channels (Hockerman et al. 1995; Hering et al. 1996). In addition, Tyr1120, Ile1121, Phe1132, and Val1133 in IIIS6 turned out to contribute to PAA binding. The replacement of the conserved Tyr1120 by Phe decreased the affinity for devapamil by 20-fold, whereas replacement by Ala rather increased the affinity by 7-fold (Hockerman et al. 1997a). Two conserved Glu in the Ca^{2+} selectivity filter also affected the PAA binding affinity (Hockerman et al. 1997b).

The triple mutation Tyr1420Ala, Ala1424Ser, and Ile1427Ala in IVS6 of the α_{1C} channel reduced the use-dependent block by devapamil, verapamil, or gallopamil, and reduced the resting block by devapamil, but not by verapamil or gallopamil (Johnson et al. 1996). Transfer of IVS6 amino acids Tyr1420, Ala1424, and Ile1427 from the α_{1C} subunit to the α_{1A} subunit introduced the use-dependent block of Ca^{2+} channel currents by PAA as well as by BTZ (Hering et al. 1996). These results indicate that the state-dependent block by PAA or BTZ of the L-type Ca^{2+} channel is mediated by the same three amino acid residues in IVS6. The homologous amino acids of S6 segments that are localized to the inner mouth of the pore affects the use-dependent block of

Ca^{2+} channels by PAAs and BTZs. Such amino acids may determine the structure of the Ca^{2+} channel pore in the inactivated state and thus determine the kinetics of the drug dissociation from its receptor site in the pore (HERING et al. 1997, 1998). However, different amino acids are required to reproduce the voltage-dependence of the use-dependent block by PAAs and BTZs. Comparison of carp α_{1S} and rat α_{1C} demonstrated that the additional amino acid residues, Ile1417 and Val1434, are also important determinants of the use-dependent block of Ca^{2+} channel currents by diltiazem (BERJUKOW et al. 1999).

The subunit composition also affects the gating kinetics, which appear to determine the susceptibility to the use-dependent block by PAAs and BTZs (SOKOLOV et al. 1999).

As summarized in Fig. 7, the amino acids that are critical for transferring the blocking action of DHPs, PAAs, and BTZs are largely overlapped, which seems to contradict the findings from radio-ligand binding experiments, i.e., distinct binding sites and the allosteric interaction between DHP-, PAA-, and BTZ-sites. Another discrepancy is that the affinity of mutant Ca^{2+} channels for Ca^{2+} channel blockers is generally lower than that of native Ca^{2+} channels by 10–100-fold. For instance, the IC_{50} values of mutant Ca^{2+} channels are 10^{-8} mol/l to 10^{-7} mol/l. Such IC_{50} values are close to those for the block of native Ca^{2+} channels at a resting membrane potential of around –80 mV, suggesting that those amino acids reproduce the DHP-binding at resting state. The additional amino acids may be involved in the high affinity binding at more depolarized membrane potentials. The involvement of additional amino acids in the high affinity block of the L-type Ca^{2+} channel by DHPs is implicated by the recently cloned L-type Ca^{2+} channel α_1 subunits from jellyfish (JEZIORSKI et al. 1998) and sea squirt (OKAMURA et al. 1999). Phylogenetic analysis indicated that these α_1 channels are close to the mammalian L-type Ca^{2+} channel α_1 subunit. These α_1 subunits contained all amino acids identified in IIIS5, IIIS6, and IVS6 which have been shown to be necessary for producing block by Ca^{2+} channel antagonists of the mammalian α_{1C} channel. However, interestingly enough, when expressed in *Xenopus* oocytes, these channels were poorly sensitive to DHP antagonists and insensitive to agonists.

It has been reported that high-affinity binding of Ca^{2+} to L-type Ca^{2+} channels, with a Kd value of less than 1 μmol/l, stabilizes the DHP-binding (GLOSSMANN et al. 1985). Mutation analysis also proved that high affinity binding of DHPs requires Ca^{2+} and the coordination of Glu in the pore regions of repeat I, II, III, and IV (MITTERDORFER et al. 1995; PETERSON and CATTERALL 1995). The binding of Ca^{2+} to the pore region may be necessary for maintaining the optimal conformation of α_{1C} subunit for the high affinity binding of DHPs.

B. Inorganic Blockers

Three divalent cations, Ca^{2+}, Ba^{2+}, or Sr^{2+} pass readily through all known voltage-dependent Ca^{2+} channels. Most of other divalent and trivalent cations such as La^{3+}, Cd^{2+}, Co^{2+}, Mn^{2+}, Ni^{2+}, and Mg^{2+} act as Ca^{2+} channel blockers.

However, it has been demonstrated that those cations, such as Mg^{2+}, Mn^{2+}, Cd^{2+}, Zn^{2+}, and Be^{2+} carry inward currents (HAGIWARA and BYERLY 1981; HESS et al. 1986). Ca^{2+} channels pass monovalent cations when the concentration of external divalent cation is in submicromolar level. Large currents can be carried by organic cations, such as hydrazinium, hydroxylammonium, and methylammonium. These currents are blocked by Ca^{2+} channel blockers and by micromolar concentration of Ca^{2+} added to the extracellular solution. Single channel recording of Ca^{2+} channel current using Ba^{2+} as a charge carrier demonstrated that Cd^{2+} chops the unitary Ba^{2+} current into bursts that appeared to arise from discrete blocking and unblocking transitions. Such kinetic features suggested a simple reaction between a blocking ion and an open channel, and that Cd^{2+} lodges within the pore. Ca^{2+} is both an effective permeator and a potent blocker because it dehydrates rapidly (unlike Mg^{2+}) and binds to the pore with appropriate affinity (unlike Cd^{2+}) (LANSMAN et al. 1986). There is no absolute distinction between "blocking" and "permeant" ions, but only quantitative differences in the rates at which they enter and leave the pore. Ca^{2+} entry and exit rates could be resolved when micromolar Ca^{2+} blocked unitary Li^{+} fluxes through the Ca^{2+} channel. The blocking rate was independent of voltage, but varied linearly with Ca^{2+} concentration, thus suggesting that the initial Ca^{2+}-pore interaction takes place at the outer side of the membrane field and much faster than the overall process of Ca^{2+} ion transfer. The unblocking rate did not vary with the extracellular Ca^{2+} concentration, but increased steeply with hyperpolarization, as if blocking Ca^{2+} ion was electrically driven from the pore into the cytoplasm (BYERLY et al. 1985).

High-voltage activated Ca^{2+} channels are inhibited by micromolar concentrations of Cd^{2+} but are resistant to this concentration range of Ni^{2+}, whereas Low-voltage activated Ca^{2+} channels have similar sensitivities to Cd^{2+} and Ni^{2+} (see Table 1).

Intracellular Mg^{2+} in milimolar concentration blocks the L-type, N-type, and P/Q-type Ca^{2+} channels.

C. Natural Toxins and Alkaloids

Tetrandrine, a bis-benzylisoquinoline alkaloid purified from the Chinese medical herb *Radix stephania tetrandrae*, has been used for the treatment of angina and hypertension in traditional Chinese medicine (Fig. 8). Tetrandrine appears to bind to the diltiazem-binding site of the L-type Ca^{2+} channel, because it competitively inhibits $[^3H]$diltiazem binding, enhances $[^3H]$nitrendipine-binding, and incompletely inhibits $[^3H]$D600 binding (KING et al. 1988). In electrophysiological studies, tetrandrine has been shown to block both T- and L-type Ca^{2+} channel currents with higher potency for the L-type Ca^{2+} channel, which is in contrast to diltiazem that is highly specific for L-type Ca^{2+} channels over T-type Ca^{2+} channels. Another difference between tetrandrine and diltiazem is that tetrandrine exerts mostly tonic block on $I_{Ca(L)}$,

Alkaloids

Tetrandrine

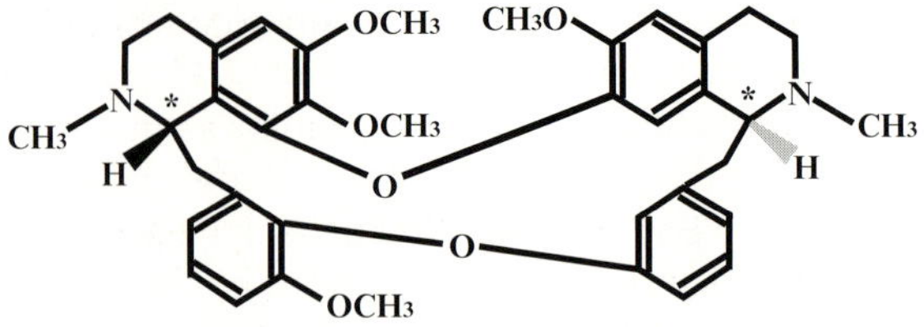

Peptide toxins

Calciseptine

RICYHKASLPRATKTCVENT
CYKMFIRTQREYISERGCGC
PTAMWPYQTGCCKGDRCNK

ω-Conotoxin GVIA

CKSOGSSCSOTSYNCCRSCNOYTKRCY

ω-Agatoxin IVA

KKKCIAKDYGRCKWGGTPCCRGRG
CICSIMGTNCECKPRLIMEGLGLA

Fig. 8. Alkaloid and peptide toxins that block voltage-dependent Ca^{2+} channels

whereas diltiazem blocks $I_{Ca(L)}$ in a use-dependent manner (RUBIO et al. 1993; WU et al. 1997). Tetrandrine blocks the voltage-dependent Na^{+} channel currents in a voltage-dependent manner with relatively high potency close to that for L-type Ca^{2+} channel block, which may contribute to the inhibitory effect of this alkaloid on the supraventricular tachycardia (RUBIO et al. 1993). Tetrandrin has also been suspected of inhibiting the intracellular Ca^{2+} handling system, such as SR Ca-ATPase and the store-operated Ca^{2+} entry (LEUNG et al. 1994; LIU et al. 1995; WANG et al. 1997).

FTX, isolated from *Agelenopsis aperta* venom with a molecular mass in the range 200–400 Da, was originally reported as a specific blocker of P-type Ca^{2+} channel in Purkinje cells (HILLMANN et al. 1991). The active substance in FTX is a nonaromatic polyamine (LLINÁS et al. 1992). FTX was conjugated to an affinity gel and used for the purification of P-type Ca^{2+} channel.

Calciseptine is the first peptide that was reported for L-type selective blockade (DE WEILLE et al. 1991, Fig. 8). This 60-amino acid peptide toxin with eight cysteines forming four disulfide bridges, isolated from the venom of black mamba *Dendroaspis polylepis polulepis*, abolished both cardiac cell contraction and smooth muscle contraction. Similar activity was reported for the major 60-amino acid component of the same venom, FS2, that differs from calciseptine by only two amino acid residues (ALBRAND et al. 1995). In rat brain membrane preparation, calciseptine inhibited the [^{3}H]nitrendipine binding in a competitive manner and potentiated the [^{3}H]diltiazem binding with an increase of affinity, strongly indicating that calciseptine binds to the DHP-site of L-type Ca^{2+} channel. Interestingly, calciseptine did not affect either [^{3}H]verapamil binding or ω-[^{125}I]conotoxin binding (YASUDA et al. 1993). Action potentials and L-type Ca^{2+} channel currents in aortic smooth muscle cells (A7r5 cell line) were potently blocked by calciseptine, although T-type Ca^{2+} channel currents in neuroblastoma cells and N-type Ca^{2+} channel currents in insulinoma cells were not affected (DE WEILLE et al. 1991). Calciseptine inhibited L-type Ca^{2+} channel currents in guinea pig portal vein cells by reducing the open probability of unitary currents (TERAMOTO et al. 1996). NMR study of the solution structure of FS2, having three loops similar to *angusticeps*-type toxins such as fasciculin 1 (choline esterase inhibitor), suggested that the characteristic strongly hydrophobic domain in loop III may be responsible for the specific binding of FS2 to the L-type Ca^{2+} channel. The region between the flanking proline residues (PTAMWP) has been proposed as the critical domain of calciseptine and FS2 for their specific binding to Ca^{2+} channel (KINI et al. 1998).

Interestingly, both calciseptine and DHPs have been reported to inhibit the store-depletion-induced Ca^{2+} influx (WILLMOTT et al. 1996), although diltiazem and verapamil do not have such effect.

Calcicludine is another 60-amino acid mamba toxin, isolated from the venom of *Deudroaspis angusticeps* (SCHWEITZ et al. 1994). This toxin is unique in that it blocks all major types of Ca^{2+} channels with very high affinity (Kd = 15 pmol/l), except for the skeletal muscle L-type Ca^{2+} channel, but its binding does not affect DHP binding or ω-CgTx-GVIA binding.

Polypeptide toxins, ω-CgTx isolated from different species of *Conus* snails and ω-Aga from the funnel-web spider *Agelenopsis aperta*, have been used as powerful pharmacological probes for exploring the diversity of Ca^{2+} channels (OHIZUMI et al. 1997). Marine snails of the genus *Conus* produce disulfide-rich Ca^{2+} channel-blocking peptides, the ω-conotoxins (OLIVERA et al. 1985, 1994). A prepropeptide precursor of ω-CgTx, approximately 70 amino acids in length, is posttranslationally processed to the mature form of 24–29 amino acids by cleavage of the conserved N-terminal regions. "Four-loop Cys scaffold" is the characteristic arrangement of cysteine (Cys) residues common among the ω-conotoxins (HILLYARD et al. 1989). The native configuration of disulfide bonds is apparently crucial for the ω-conotoxins to exert the blocking effect on Ca^{2+} channels. On the other hand, the amino acids in loops between Cys residues are hypervariable. The only conserved non-Cys residue among ω-conotoxins is Gly5. Considerable divergence in amino acid sequences in the loop regions of the toxin does not attenuate the Ca^{2+} channel target specificity, although the loop regions are responsible for the binding specificity to Ca^{2+} channel subtypes.

The venom of the American funnel-web spider *Agelenopsius aperta* is the source of blocker toxins such as ω-agatoxins, α-agatoxins, and μ-agatoxins. The agatoxins are heterogeneous group of polypeptides with a molecular mass of 5–10 kDa classified by three different bioassays, block of neuromuscular transmission in housefly body muscle (ω-Aga-I and ω-Aga-II, OLIVERA et al. 1994), inhibition of ω-CgTx-binding to chick synaptosomal membrane (ω-Aga-III, VENEMA et al. 1992), and inhibition of $^{45}Ca^{2+}$ entry into both chick and rat synaptosomes (ω-Aga-IV, MINTZ et al. 1992). cDNA cloning studies and amino acid sequencing of the purified preparations of the toxins have provided structural information about the ω-agatoxins. ω-Agatoxins are more diverse than the ω-conotoxins in their primary structures, although they are similar in the abundance in Cys residues (8–12 residues). Peptide toxins with 48-amino acids, such as ω-Aga-IVA and ω-Aga-IVB, block P/Q type Ca^{2+} channels.

The two-dimensional ^{1}H-NMR technique has been used to elucidate the three-dimensional structure for ω–CgTx-GVIA, ω-Aga-IVA, and w-Aga-IVB (DAVIS et al. 1993; PALLAGHY et al. 1993; ADAMS et al. 1993; KIM et al. 1995). Irrespective of differences in the number of disulfide bonds and relatively low homology in primary structure, the toxins share structural characteristics, being composed of a short triple-stranded β-sheet and several reverse turns. In both group of toxins, patches of positively charged residues or, in ω–CgTx-GVIA, tyrosine residues with hydroxy residues are distributed on the molecular surface. These residues may contribute to the target specificity of these toxins.

Kurtoxin, a new peptide toxin isolated from the venom of a South African scorpion (*Parabuthus transvaalicus*), has recently been shown to bind to the α_{1G} subunit of T-type Ca^{2+} channel with high affinity (K_d = 15 nmol/l) and distinguishes between T-type Ca^{2+} channels and other voltage-dependent Ca^{2+} channels such as α_{1A}, α_{1B}, α_{1C}, or α_{1E} (CHUANG et al. 1998).

D. Ca^{2+} Channel Agonists

I. DHPs

Some racemic DHPs, such as Bay K 8644, exert dual action on L-type Ca^{2+} channels: one enantiomer ((+) Bay K 8644) works as a Ca^{2+} channel antagonist and its stereoisomer ((–) Bay K 8644) acts as a Ca^{2+} channel agonist (Fig. 9). As discussed in A.III.1, DHPs produce mixed effects of antagonist and agonist. The positive inotropic effect of (–)Bay K 8644 and its enhancement of Ca^{2+} channel currents have been reported in cardiac myocardium (SCHRAMM

1. Dihydropyridines

Bay k 8644

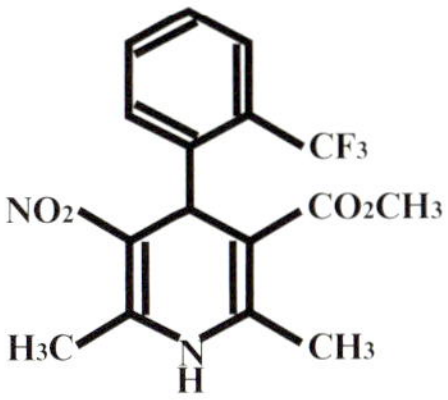

Sdz (+) (S) 202 791

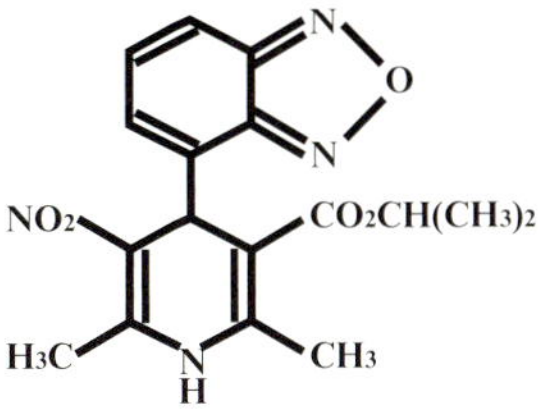

CGP 28392

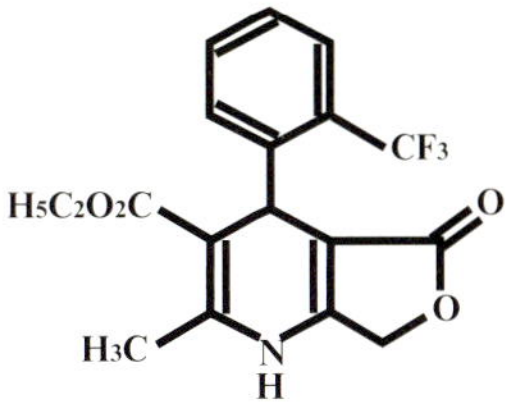

2. Others

FPL 64176

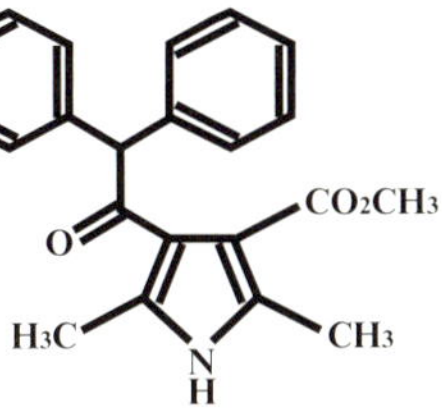

Fig. 9. Ca^{2+} channel agonists

et al. 1983; Thomas et al. 1985). The single channel studies of the L-type Ca^{2+} channels demonstrated that (–)Bay K 8644 increases P_o coupled with induction of long openings and reduction of blank sweeps (Hess et al. 1984; Ochi et al. 1984; McDonald et al. 1994). (–)Bay K 8644 (at 10^{-7} mol/l to 10^{-6} mol/l) enhances the peak Ca^{2+} channel currents with negative shift of both activation and inactivation curves by 10–15 mV. (–)Bay K 8644 accelerates the gating charge movement of L-type Ca^{2+} channel on depolarization (Josephson and Sperelakis 1990). Thus both activation and inactivation rates of Ca^{2+} channel currents are accelerated, and the deactivation rate is significantly reduced (Reuter et al. 1988). In contrast, when the L-type Ca^{2+} channels are phosphorylated by cAMP-dependent protein kinase, (–)Bay K 8644 significantly slows the inactivation of the Ca^{2+} channel current (Tsien et al. 1986; Tiaho et al. 1990). Like other DHP derivatives, the binding site for (–)Bay K 8644 has also been localized to the extracellular side of the L-type Ca^{2+} channels (Strübing et al. 1993). However, the exact binding site could be different from that of dihydropyridine antagonists (Grabner et al. 1996; Mitterdorfer et al. 1996). Interestingly, it has been reported that (–)Bay K 8644 depletes the SR Ca^{2+} by gating the Ca^{2+}-release from the SR in ventricular myocytes even in the absence of the extracellular Ca^{2+}, presumably via its interaction with the L-type Ca^{2+} channel, but not via direct interaction with ryanodine receptors

(SATOH et al. 1998). (–)Bay K 8644 also modifies Ca^{2+}-dependent inactivation of L-type Ca^{2+} channel and the Ca^{2+}-induced Ca^{2+} release process in ventricular myocytes (ADACHI-AKAHANE et al. 1999).

1,4-difydropyridine (–)-(R) SDZ 202–791 works as a Ca^{2+} channel antagonist, while its enantiomer (+)-(S) SDZ 202–791 activates the L-type Ca^{2+} channels by prolongation of open time and reduction of the number of blank sweeps (KOKUBUN et al. 1986; REUTER et al. 1988). Both compounds have been reported to shift steady-state inactivation curve toward more negative potentials. Their binding to DHP receptors was potentiated by membrane depolarization. At potentials positive to –20 mV, the Ca^{2+} channel activator effect of (+)-(S) SDZ 202–791 turned over into a blocking effect.

The DHP agonist CGP 28392, like (–)Bay K 8644, exerts a positive inotropic effect by increasing Ca^{2+} influx through the L-type Ca^{2+} channel, and thus increasing the SR Ca^{2+} content. The single Ca^{2+} channel analysis revealed that CGP 28392 prolongs the mean open time (KOKUBUN and REUTER 1984).

(–)Bay K 8644 as well as CGP 28392 have been shown to enhance the cardiac Na^{+} channel with an increase of P_o and availability (KOHLHEART et al. 1989).

II. Non-DHPs

Benzoylpyrrole-type Ca^{2+} activator, FPL 64176, appears to act on the L-type Ca^{2+} channel through interaction with the site distinct from those for classical DHP Ca^{2+} channel antagonists (Fig. 9). Unlike (–)Bay K 8644, FPL 64176 did not show inhibitory effect up to 10^{-5} mol/l, which may be the reason for the activity as a Ca^{2+} channel agonist of approximately two fold higher than (–)Bay K 8644 (ZHENG et al. 1992). FPL 64176 prolongs the mean open time of the L-type Ca^{2+} channel in a voltage-dependent manner, more at hyperpolarized potentials and less at depolarized potentials, thus producing dramatic prolongation of tail current. The Ca^{2+} channel activity developes with some delay upon depolarization, which produces a markedly slow activation of whole cell Ca^{2+} channel current unlike (–)Bay K 8644 (KUNZE and RAMPE, 1992). The negative allosteric interaction between the binding sites of FPL 64176 and (–)Bay K 8644 has also been suggested (RAMPE and DAGE 1992). The intracellular application of FPL 64176 (10^{-6} mol/l) had no effect, whereas the subsequent extracellular application of the same concentration strongly enhanced the Ca^{2+} channel current (>10-fold), thus implying that the binding site is near the extracellular side of L-type Ca^{2+} channel (RAMPE and LACERDA 1991).

E. Concluding Remarks

Ca^{2+} channel antagonists are clinically useful because of their unique tissue selectivity. The most important determinant of the tissue selectivity is the voltage-dependent interaction between Ca^{2+} channels and Ca^{2+} channel antagonists. Existence of Ca^{2+} channel antagonists that block both L-type and N-type Ca^{2+} channels, L-type and T-type Ca^{2+} channels, or L-type Ca^{2+} channels and Na^{+} channels implies the common structure around the drug-binding pocket in those ion channels. Studies on the structure-function relationship of Ca^{2+} channels, the identification of drug-binding pockets, should clarify the molecular basis for voltage-dependence and the tissue selectivity of Ca^{2+} channel antagonists and agonists.

References

Adachi-Akahane S, Amano Y, Okuyama R, Nagao T (1993) Quaternary diltiazem can act from both sides of the membrane in ventricular myocytes. Jpn J Pharmacol 61:263–266

Adachi-Akahane S, Nagao T (1993) Binding site for diltiazem is on the extracellular side of the L-type Ca^{2+} channel. Circulation 88:I-230

Adachi-Akahane S, Cleemann L, Morad M (1999) Bay K 8644 modifies Ca^{2+} cross signaling between DHP and ryanodine receptors in rat ventricular myocytes. Am J Physiol 276:H1178–H1189

Adams ME, Mintz IM, Reily MD, Thanabal V, Bean BP (1993) Structure and properties of ω–AgatoxinIVB, a new antagonist of P-type calcium channels. Mol Pharmacol 44:681–688

Akaike N, Kostyuk PG, Osipchuk YV (1989a) Dihydropyridine-sensitive low-threshold calcium channels in isolated rat hypothalamic neurons. J Physiol (Lond) 412:181–195

Akaike N, Kanaide H, Kuga T, Nakayama M, Sadoshima J, Tomoike H (1989b) Low-voltage-activated Ca^{2+} current in rat aorta smooth muscle cells in primary culture. J Physiol 416:141–160

Albrand J-P, Blackledge MJ, Pascaud F, Hollecker M, Marion D (1995) NMR and restrained molecular dynamics study of the three-dimensional solution structure of toxin FS2, a specific blocker of the L-type calcium channel, isolated from black mamba venom. Biochemistry 34:5923–5937

Avdonin V, Shibata EF, Hoshi T (1997) Dihydropyridine action on voltage-dependent potassium channels expressed in xenopus oocytes. J Gen Physiol 109: 169–180

Bangalore R, Baindur N, Rutledge A, Triggle DJ, Kass RS (1994) L-type calcium channels: asymmetrical intramembrane binding domain revealed by variable length, permanently charged 1,4-dihydropyridines. Mol Pharmacol 46:660–666

Balwierczak JL, Johnson CL, Schwartz A (1987) The relationship between the binding site of $[^3H]$-*d-cis*-diltiazem and that of other non-dihydropyridine calcium entry blockers in cardiac sarcolemma. Mol Pharmacol 31:175–179

Bean BP (1984) Nitrendipine block of cardiac calcium channels: high-affinity binding to the inactivated state. Proc Natl Acad Sci USA 81:6388–6392

Berjukov S, Aczel S, Beyer B, Kimball SD, Dichtl M, Hering S, Striessnig J (1996) Extra- and intracellular action of quaternary devapamil on muscle L-type Ca^{2+}-channels. Br J Pharmacol 119:1197–1202

Berjukow S, Gapp F, Aczel S, Sinnegger MJ, Mitterdorfer J, Glossmann H, Hering S (1999) Sequence differences between α_{1C} and α_{1S} Ca^{2+} channel subunits reveal structural determinants of a guarded and modulated benzothiazepine receptor. J Biol Chem 274:6154–6160

Bezprozvanny I, Tsien RW (1995) Voltage-dependent blockade of diverse types of voltage-gated Ca^{2+} channels expressed in Xenopus oocytes by the Ca^{2+} channel antagonist mibefradil (Ro 40–5967). Mol Pharmacol 48:540–549

Byerly L, Chase PB, Stimers JR (1985) Permeation and interaction of divalent cations in calcium channels of snail neurons. J Gen Physiol 85:491–518
Cai D, Mulle JG, Yue DT (1997) Inhibition of recombinant Ca^{2+} channels by benzothiazepines and phenylalkylamines: class-specific pharmacology and underlying molecular determinants. Mol Pharmacol 51:872–881
Catterall WA, Striessnig J (1992) Receptor sites for Ca^{2+} channel antagonists. Trends Pharmacol Sci 13:256–262
Chouabe C, Drici MD, Romey G, Brafanin J, Lazdunski M (1998) HERG and KvLQT1/IsK, the cardiac K^+ channels involved in long QT syndoromes, are targets for calcium channel blockers. Mol Pharmacol 54:695–703
Chuang RS-I, Jaffe H, Cribbs L, Perez-Reyes E, Swartz KJ (1998) Inhibition of T-type voltage-gated calcium channels by a new scorpion toxin. Nature Neurosci 1:668–674
Clozel JP, Ertel EA, Ertel SI (1997) Discovery and main pharmacological properties of mibefradil (Ro 40–5967), the first selective T-type calcium channel blocker. J Hypertens 15:S17–25
Cohen CJ, Ertel EA, Smith MM, Venema VJ, Adams ME, Leibowitz MD (1992) High affinity block of myocardial L-type calcium channels by the spider toxin omega-Aga-toxin IIIA: advantage over 1,4-dihydropyridines. Mol Pharmacol 42:947–951
Curtis BM, Catterall WA (1984) Purification of the calcium antagonist receptor of thevoltage-sensitive calcium channel from skeletal muscle transverse tubules. Biochemistry 23:2113–2218
Davis JH, Bradley EK, Miljanich GP, Nadasdi L, Ramachandran J, Basus VJ (1993) Solution structure of ω-conotoxin GVIA using 2-D NMR spectroscopy and relaxation matrix analysis. Biochemistry 32:7396–7405
de Weille JR, Schweitz H, Maes P, Tartar A, Lazdunski M (1991) Calciseptine, a peptide isolated from black mamba venom, is a specific blocker of the L-type calcium channel. Proc Natl Acad Sci 88:2437–2440
Diochot S, Richard S, Baldy-Moulinier M, Nargeot J, Valmier J (1995) Dihydropyridines, phenylalkylamines and benzothiazepines block N-, P/Q- and R-type calcium currents. Pflügers Arch 431:10–19
Döring F, Degtiar VE, Grabner M, Striessnig J, Hering S, Glossmann H (1996) Transfer of L-type calcium channel IVS6 segment increases phenylalkylamine sensitivity of alpha 1 A. J Biol Chem 271:11745–11749
Fagni L, Bossu J-L, Bockaert J (1994) Inhibitory effects of dihydropyridines on macroscopic K^+ currents and on the large-conductance Ca^{2+}-activated K^+ channel in cultured cerebellar granule cells. Pflügers Arch 429:176–182
Ferry DR, Rombusch M, Goll A, Glossmann H (1984) Photoaffinity labelling of Ca^{2+} channels with [^{3}H]azidopine. FEBS Lett 169:112–122
Ferry DR, Goll A, Glossmann H (1987) Photoaffinity labelling of the cardiac calcium channel. Biochem J 243:127–135
Fleckenstein A (1964) Die bedeutung der energiereichen phosphate für kontraktilität und tonus des myokards. Verh Dtsch Ges Inn Med 70:81–99
Fleckenstein A (1977) Specific pharmacology of calcium in myocardium, cardiac pacemakers, and vascular smooth muscle. Ann Rev Pharmacol Toxicol 17:149–166
Fleckenstein A (1983) History of calcium antagonists. Circ Res 52(Suppl I):3–16
Fujii S, Kameyama K, Hosono M, Hayashi Y, Kitamura K (1997) Effect of cilnidipine, a novel dihydropyridine Ca^{++}-channel antagonist, on N-type Ca^{++} channel in rat dorsal root ganglion neurons. J Pharmacol Exp Ther 280:1184–1191
Furukawa T, Nukada T, Suzuki K, Fujita Y, Mori Y, Nishimura M, Tamanaka M (1997) Voltage and pH dependent block of cloned N-type Ca^{2+} channels by amlodipine. Br J Pharmacol 122:1136–1140
Garcia ML, King VF, Siegl PKS, Reuben JP, Kaczorowski GJ (1986) Binding of Ca^{2+} entry blockers to cardiac sarcolemmal membrane vesicles: Characterization of diltiazem-binding sites and their interaction with dihydropyridine and aralkylamine receptors. J Biol Chem 261:8146–8157

Gill A, Flain SF, Damiano BP, Sit SP, Brannan MD (1992) Pharmacology of bepridil. Am J Cardiol 69:11D–16D
Glossmann H, Ferry DR, Luebbecke F, Mewes R, Hofmann F (1982) Calcium channels: direct identification with radioligand binding studies. Trends Pharmacol Sci 3:431–437
Glossmann H, Striessnig J (1983a) Purification of the putative calcium channel from skeletal muscle with the aid of [^{3}H]-nimodipine binding. Naunun-Schmiedeberg's Arch Pharmacol 323:1–11
Glossmann H, Linn T, Rombusch M, Ferry DR (1983b) Temperature-dependent regulation of d-cis-[^{3}H]diltiazem binding to Ca^{2+} channels by 1,4-dihydropyridine channel agonists and antagonists. FEBS Lett 160:226–232
Glossmann H, Ferry DR, Goll A, Striessnig J, Zering G (1985) Calcium channels and calcium channel drugs: Recent biochemical and biophysical findings. Arzneim Forsch 35:1917–1935
Grabner M, Wang Z, Hering S, Striessnig J, Glossmann H (1996) Transfer of 1,4-dihydropyridine sensitivity from L-type to class A (BI) calcium channels. Neuron 16:207–218
Hagiwara M, Adachi-Akahane S, Nagao T (1997) High-affinity binding of DTZ323, a novel derivative of diltiazem, to rabbit skeletal muscle L-type Ca^{2+} channels. J Pharmacol Exp Ther 281:173–179
Hagiwara N, Byerly L (1981) Calcium channel. Ann Rev Neurosci 4:69–125
Haynes LW (1992) Block of the cyclic GMP-gated channel of vertebrate rod and cone photoreceptors by l-cis-diltiazem. J Gen Physiol 100:783–801
He M, Bodi I, Mikala G, Schwartz A (1997) Motif III S5 of L-type calcium channles is involved in the dihydropyridine binding site. A combined radioligand binding and electrophysiological study. J Biol Chem 272:2629–2633
Hering S, Savchenko A, Strubing C, Lakitsch M, Striessnig J (1993) Extracellular localization of the benzothiazepine binding domain of L-type calcium channels. Mol Pharmacol 43:820–826
Hering S, Aczel S, Grabner M, Doring F, Berjukov S, Miterdorfer J, Sinnegger MJ, Striessnig J, Degitar VE, Wang Z, Glossmann H (1996) Transfer of high sensitivity of benzothiazepines from L-type to class A (BI) calcium channels. J Biol Chem 271:24471–24475
Hering S, Aczel S, Kraus RL, Berjukov S, Striessnig J, Timin EN (1997) Molecular mechanism of use-dependent calcium channel block by phenylalkylamines: role of inactivation. Proc. Natl Acad Sci USA 94:13323–13328
Hering S, Berjukow S, Aczel S, Timin EN (1998) Ca^{2+} channel block and inactivation: common molecular determinants. Trends Pharmacol Sci 19:439–443
Hescheler J, Pelzer D, Trube G, Trautwein W (1982) Does the organic calcium channel blocker D600 act from inside or outside on the cardiac cell membrane? Pflügers Arch 393:287–291
Hess P, Tsiwn RW (1984) Mechanism of ion permeation through calcium channels. Nature 309:453–456
Hess P, Lansmann JB, Tsien RW (1986) Calcium channel selectivity for divalent and monovalent cations. Voltage and concentration dependence of suingle channel current in vetricular heart cells. J Gen Physiol 88:293–319
Hillmann D, Chen S, Aung TT, Cherksey B, Sugimori M, Linas RR (1991) Localization of P-ytpe calcium channels in the central nervous system. Proc Natl Acad Sci (USA) 88:7076–7080
Hillyard DR, Olivera BM, Woodward S, Corpuz GP, Gray WE, Ramilo CA, Cruz LJ (1989) A molluscivorous cone toxin: Conserved frameworks in conotoxins. Biochemistry 28:358–361
Hockerman GH, Johnson BD, Scheuer T, Catterall WA (1995) Molecular determinants of high-affinity phenylalkylamine block of L-type calcium channels. J Biol Chem 270:22119–22122
Hockerman GH, Johnson BD, Abbott MR, Scheuer T, Catterall WA (1997a) Molecular determinants of high-affinity phenylalkylamine block of L-type calcium chan-

nels in transmembrane segment IIIS6 and the pore region of the α_1 subunit. J Biol Chem 272:18759–18765

Hockerman GH, Peterson BZ, Johnson BD, Catterall WA (1997b) Molecular determinants of drug binding and action on L-type calcium channels. Ann Rev Pharmacol Toxicol 37:361–396

Hockerman GH, Johnson BD, Sharp E, Tanada TN, Scheuer T, Catterall WA (1997c) Construction of a high-affinity receptor site for dihydropyridine agonists and antagonists by single amino acid substitutions in a non-L-type calcium channel. Proc Natl Acad Sci (USA) 94:14906–14911

Hosono M, Fujii S, Hiruma T, Watanabe K, Hayashi Y, Ohnishi H, Tanaka Y, Kato H (1995) Inhibitory effect of cilnidipine on vascular sympathetic neurotransmission and subsequent vasoconstriction in spontaneously hypertensive rats. Jpn J Pharmacol 69:127–134

Hu H, Marban E (1998) Isoform-specific inhibition of L-type calcium channels by dihydropyridine is independent of isoform-specific gating properties. Mol Pharmacol 53:902–907

Huet S, Robert J (1988) The reversal of doxorubicin resistance by verapamil is not due to an effect on calcium channels. Int J Cancer 41:283–286

Iijima T, Yanagisawa T, Taira N (1984) Increase in the slow inward current by intracellularly applied nifedipine and nicardipine in single ventricular cells of the gionea-pig heart. J Mol Cell Cardiol 16:1173–1177

Ikeda S, Oka J-I, Nagao T (1991) Effects of diltiazem stereoisomers on binding of *d-cis*-[^{3}H]diltiazem and (+)-[^{3}H]PN200–110 to rabbit T-tubule calcium channels. Eur J Pharmacol 208:199–205

Ito H, Klugbauer N, Hofmann F (1997) Transfer of the high affinity dihydropyridine sensitivity from L-type To non-L-type calcium channel. Mol Pharmacol 52:735–740

Itogawa E, Kurosawa H, Yabanam H, Murata S (1996) Protective effect of *l-cis* diltiazem on hypercontracture of rat myocytes induced by veratridine. Eur J Pharmacol 317:401–406

Jeziorski MC, Greenberg RM, Clark KS, Anderson PAV (1998) Cloning and functional expression of a voltage-gated calcium channel α_1 subunit from Jellyfish. J Biol Chem 273:22792–22799

Johnson BD, Hockerman GH, Scheuer T, Catterall WA (1996) Distinct effects of mutations in transmembrane segment IVS6 on block of L-type calcium channels by structurally similar phenylalkylamines. Mol Pharmacol 50:1388–1400

Josephson R, Sperelakis N (1990) Fast activation of cardiac Ca^{2+} channel gating charge by the dihydrophyridine agonist, BAY K 8644. Biophys J 58:1307–1311

Kalasz H, Watanabe T, Yabana H, Itagaki K, Naito K, Nakayama H, Schwartz A, Vaghy PL (1993) Identification of 1,4-dihydropyridine binding domains within the primary structure of the α_1 subunit of the skeletal muscle L-type calcium channel. FEBS Lett 331:177–181

Kanaya S, Katzung BG (1984) Effects of diltiazem on transmembrane potential and current of right ventricular papillary muscle of ferrets. J Pharmacol Exp Ther 228:245–251

Kanda S, Kurokawa J, Adachi-Akahane S, Nagao T (1997) Ditliazem derivatives modulate the dihydropyridine-binding to intact rat ventricular myocytes. Eur J Pharmacol 319:101–107

Kanda S, Adachi-Akahane S, Nagao T (1998) Functional interaction between benzothiazepine- and dihydropyridine binding sites of cardiac L-type Ca^{2+} channels. Eur J Pharmacol 358:277–287

Kass RS, Arena JP (1989) Influence of pHo on calcium channel block by amlodipine, a charged dihydropyridine compound. Implications for location of the dihydropyridine receptor. J Gen Physiol 93:1109–1127

Kass RS, Arena JP, Chin S (1991) Block of L-type calcium channels by charged dihydropyridines. Sensitivity to side of application and calcium. J Gen Physiol 98: 63–75

Kikkawa K, Yamauchi R, Suzuki T, Banno K, Murata S, Tezuka T, Nagao T (1994) Clentiazem improves cerebral ischemia induced by carotid artery occlusion of stroke-prone spontaneously hypertensive rats. Stroke 25:474–480

Kim JI, Konishi S, Iwai H, Kohno T, Gouda H, Shimada I, Sato K, Arata Y (1995) Three-dimensional solution structure of the calcium channel antagonist ω-agatoxin IVA: consensus molecular folding of calcium channel blockers. J Mol Biol 250:659–671

Kimball SD, Floyd DM, Das J, Hunt JT, Krapcho J, Rovnyak G, Duff KJ, Lee VG, Moquin RV, Turk CF (1992) Benzazepinone calcium channel blockers. 4. Structure-activity overview and intracellular binding site. J Med Chem 35:780–793

Kimball SD, Hunt JT, Barrish JC, Das J, Floyd DM, Lago MW, Lee VG, Spergel SH, Moreland S, Hedberg SA (1993) 1-Benzazepin-2-one calcium channel blockers–VI. Receptor-binding model and possible relationship to desmethoxyverapamil. Bioorganic & Medicinal Chem 1:285–307

King VF, Garcia ML, Himmel D, Reuben JP, Lam YK, Pan JX, Han GQ, Kaczorowski GJ (1988) Interaction of tetrandrine with slowly inactivating calcium channels. Characterization of calcium channel modulation by an alkaloid of Chinese medicinal herb origin. J Biol Chem 263:2238–2244.

Kini RM, Caldwell RA, Wu QY, Baumgarten CM, Feher JJ, Evans HJ (1998) Flanking proline residues identify the L-type Ca^{2+} channel binding site of calciseptine and FS2. *Biochemistry* 37:9058–9063

Knaus HG, Moshammer T, Kang HC, Haugland RP, Glossmann H (1992) A unique fluorescent phenylalkylamine probe for L-type Ca^{2+} channels. J Biol Chem 267: 2179–2189

Kobayashi T, Strobeck M, Schwartz A, Mori Y (1997) Inhibitory effects of a new neuroprotective diltiazem analogue, T-477, on cloned brain Ca^{2+} channels expressed in Xenopus oocytes. Eur J Pharmacol 332:313–320

Kobayashi T, Mori Y (1998) Ca^{2+} channel antagonists and neuroprotection from cerebral ischemia. Eur J Pharmacol 363:1–15

Kohlhardt M, Fichtner H, Herzig JW (1989) The response of single cardiac sodium channels in neonatal rats to the dihydropyridines CGP 28392 and (-)-Bay K 8644. Naunyn-Schmiedeberg's Arch Pharmacol 340:210–218

Kokubun S, Reuter H (1984) Dihydropyridine derivatives prolong the open state of Ca channels in cultured cardiac cells. Proc Natl Acad Sci USA 81:4824–4827

Kokubun S, Prod'hom B, Becker C, Porzig H, Reuter H (1986) Studies on Ca channels in intact cardiac cells: voltage-dependent effects and cooperative interactions of dihydropyridine enantiomers. Mol Pharmacol 30:571–584

Kraus R, Reishl B, Kimball SD, Grabner M, Murphy BJ, Catterall WA, Striessnig J (1996) Identification of benz(othi)azepine-binding regions within L-type calcium channel α_1 subunits. J Biol Chem 271:20113–20118

Kuniyasu A, Itagaki K, Shibano T, Iino M, Kraft G, Schwartz A, Nakayama H (1998) Photochemical identification of transmembrane segment IVS6 as the binding region of semotiadil, a new modulator for the L-type voltage-dependent Ca^{2+} channel. J Biol Chem 273:4635–4641

Kunze DL, Rampe D (1992) Characterization of the effects of a new Ca^{2+} channel activator, FPL 64176, in GH3 cells. Mol Pharmacol 42:666–670

Kurokawa J, Adachi-Akahane S, Nagao T (1997a) Effects of a novel, potent benzothiazepine Ca^{2+} channel antagonist, DTZ323, on guinea-pig ventricular myocytes. Eur J Pharmacol 325:229–236

Kurokawa J, Adachi-Akahane S, Nagao T (1997b) 1,5-Benzothiazepine binding domain is located on the extracellular side of the cardiac L-type Ca^{2+} channel. Mol Pharmacol 51:262–268

Lansman JB, Hess P, Tsien RW (1986) Blockade of current through single calcium channels by Cd^{2+}, Mg^{2+}, and Ca^{2+}. Voltage and concentration dependence of calcium entry into the pore. J Gen Physiol 88:321–347

Leblanc N, Hume JR (1989) D600 block of L-type Ca^{2+} channel in vascular smooth muscle cells: comparison with permanently charged derivatives, D890. Am J Physiol 257:C689–695
Lee KS, Tsien RW (1983) Mechanism of calcium channel blockade by verapamil, D600, diltiazem, and nitrendipine in single dialysed heart cells. Nature Lond 302:709–794
Leung YM, Kwan CY, Loh TT (1994) Dual effects of tetrandrine on cytosolic calcium in human leukemic HL-60 cells: intracellular calcium release and calcium entry blockade. Br J Pharmacol 113:767–774
Liu QY, Li B, Gang JM, Karpinski E, Pang PKT (1995) Tetrandrine, a Ca^{2+} antagonist: effects and mechanisms of action in vascular smooth muscle cells. J Pharmacol Exp Ther 273:32–39
Llinás R, Sugimori M, Hillman DE, Cherksey B (1992) Distribution and functional significance of the P-type, voltage-dependent Ca^{2+} channels in the mammalian central nervous system. Trends Neurosci 15:351–355
Mason RP, Campbell SF, Wang SD, Herbette LG (1989) Comparison of location and binding for the positively charged 1,4-dihydropyridine calcium channel antagonist amlodipine with uncharged drugs of this class in cardiac membranes. Mol Pharmacol 36:634–640
Masuda Y, Miyajima M, Shudo C, Tanaka S, Shigenobu K, Kasuya Y (1995) Cardiovascular selectivity of 1,4-dihydropyridine derivatives, efonidipine (NZ-105), nicardipine and structure related compounds in idolated guinea-pig tissues. Gen Pharmacol 26:339–345
Masumiya H, Tanaka H, Shigenobu K (1997) Effects of Ca^{2+} channel antgonists on sinus node: prolongation of late phase 4 depolarization by efonidipine. Eur J Pharmacol 335:15–21
McDonald TF, Pelzer D, Trautwein W (1984) Cat ventricular muscle treated with D600: characteristics of calcium channel block and unblock. J Physiol Lond 352:217–241
McDonald T, Pelzer D, Trautwein W (1989) Dual action (stimulation, inhibition) of D600 on contractility and calcium channels in guinea-pig and cat heart cells. J Physiol Lond 414:569–586
McDonald TF, Pelzer S, Trautwein W, Pelzer DJ (1994) Regulation and modulation of calcium channels in cardiac, skeletal, and smooth muscle cells. Physiol Rev 74: 365–507
Mintz IM, Venema VJ, Swiderek KM, Lee TD, Bean BP, Adams ME (1992) P-type calcium channels blocked by the spider toxin w-Aga-IVA. Nature 355:827–829
Mishra SK, Hermsmeyer K (1994) Inhibition of signal Ca^{2+} in dog coronary arterial vascular muscle cells by Ro 40–5967 J Cardiovasc Pharmacol 24:1–7
Mitterdorfer J, Froschmayr M, Grabner M, Striessnig J, Glossmann H (1995) Coordination of Ca^{2+} by the pore region glutamates is essential for high-affinity dihydropyridine binding to the cardiac Ca^{2+} channels α_1 subunit. Biochemistry 34:9350–9355
Mitterdorfer J, Wang Z, Sinneger MJ, Hering S, Striessnig J, Grabner M, Glossmann H (1996) Two amino acid residues in the IIIS5 segment of L-type calcium channels differentially contribute to 1,4-dihydropyridine sensitivity. J Biol Chem 271:30330–30335
Morel N, Buryi V, Feron O, Gomez JP, Christen MO, Godfraind T (1998) The action of caocium channel blockers on recombinant L-type calcium channel α_1-subunits. Br J Pharmacol 125:1005–1012
Mori Y, Mikala G, Varadi G, Kobayashi T, Koch S, Wakamori M, Schwartz A (1996) Molecular pharmacology of voltage-dependent calcium channels. Jpn J Pharmacol 72:83–109
Naito K, McKenna M, Schwartz A, Vaghy PL (1989) Photoaffinity labelling of the purified skeletal muscle calcium antagonist receptor by a novel benzothiazepine, [^{3}H]azidobutyryl diltiazem. J Biol Chem 264:21211–21214

Nakayama H, Taki M, Striessnig J, Glossmann H, Catterall WA, Kanaoka Y (1991) Identification of 1,4-dihydropyridine binding regions within the α_1 subunit of skeletal muscle Ca^{2+} channels by photoaffinity labeling with diazipine. Proc Natl Acad Sci USA 88:9203–9207

Nakayama K, Nozawa K, Fukuda Y (1994) Allosteric interaction of semotiadil fumarate, a novel benzothiazine, with 1,4-dihydropyridines, phenylalkylamines, and 1,5-benzothiazepines at the Ca^{2+}-channel antagonist binding sites in canine skeletal muscle membranes. J Cardiovasc Pharmacol 23:731–740

Nishida M, Sakamoto K, Urushidani T, Nagao T (1999a) Treatment with l-cis diltiazem before reperfusion reduces infarct size in the ischemic rat heart in vivo. Jpn J Pharmacol 80:319–325

Nishida M, Urushidani T, Sakamoto K, Nagao T (1999b) L-cis diltiazem attenuates intracellular Ca^{2+} overload by metabolic inhibition in guinea pig myocytes. Eur J Pharmacol 385:225–230

Ochi R, Hino N, Niimi Y (1984) Prolongation of calcium channel open time by the dihydropyridine derivative BAY K 8644 in cardiac myocytes. Proc Jpn Acad 60:153–156

Ohizumi Y (1997) Application of phyriologically active substances isolated from natural resources to pharmacological studies. Jpn J Pharmacol 73:263–289

Ohya Y, Terada K, Kitamura K, Kuriyama H (1987) D600 blocks the Ca^{2+} channel from the outer surface of smooth muscle cell membrane of the rabbit intestine and portal vein. Pflügers Arch 408:80–82

Okamura Y (1999) Functional expression of a protochordate L-type Ca^{2+} channel. Biophys J 76:A340

Okuyama R, Adachi-Akahane S, Nagao T (1994) Differential potentiation by depolarization of the effects of calcium antagonists on contraction and Ca^{2+} current in guinea-pig heart. Br J Pharmacol 113:451–456

Olivera BM, Gray WR, Zeikus R, McIntosh JM, Varga J, Rivier J, de Santos V, Cruz LJ (1985) Peptide neurotoxin from fish-hunting cone snails. Science 230:1338–1343

Olivera BM, Milanich GP, Ramachandran J, Adams ME (1994) Calcium channel diversity and neurotransmitter release: The ω-conotoxins and ω-agatoxins. Annu Rev Biochem 63:823–867

Pallaghy PK, Duggan BM, Pennington MW, Norton RS (1993) Three-dimensional structure in solution of the calcium channel blocker omega-conotoxin. J Mol Biol 234:405–420

Pelzer D, Cavalie A, Trautwein W (1985) Guinea-pig ventricular myocytes treated with D600: mechanism of calcium channel blockade at the level of single channels. In: *Recent Aspects in Calcium Antagonism*, edited by Litchtlen PR, Stuttgart, Germany: Schattauer, 3–26

Pereira E, Teodori E, Dei S, Gualtieri F, Garnier-Suillerot A (1995) Reversal of multidrug resistance by verapamil analogues. Biochem Pharmacol 50:451–457

Peterson BZ, Catterall WA (1995) Calcium binding in the pore of L-type calcium channels modulates high affinity dihydropyridine binding. J Biol Chem 270:18201–18204

Peterson BZ, Tanada TN, Catterall WA (1996) Molecular determinants of high affinity dihydropyridine binding in L-type calcium channels. J Biol Chem 271:5293–5286

Peterson BZ, Johnson BD, Hockerman GH, Acheson M, Scheuer T, Catterall WA (1997) Analysis of the dihydropyridine receptor site of L-type calcium channels by alanine-scanning mutagenesis. J Biol Chem 272:18752–18758

Ragsdale DS, Scheuer T, Catterall WA (1991) Frequency and voltage-dependent inhibition of type IIA Na^+ channels, expressed in a mammalian cell line, by local anesthetic, antiarrhythmic, and anticonvulsant drugs. Mol Pharmacol 40:756–765

Ramp D, Dage RC (1992) Functional interactions between two Ca^{2+} channel activators (-)Bay K 8644 and FPL 64176, in smooth muscle. Mol Pharmacol 41:599–602

Rampe D, Lacerda AE (1991) A new site for the activation of cardiac calcium channels defined by the nondihydropyridine FPL 64176. J Pharmacol Exp Ther 259:982–987

Regulla S, Schneider T, Nastainczyk W, Meyer HE, Hofmann F (1991) Identification of the site of interaction of the dihydropyridine cahnnel blockers nitrendipine and azidopine with the calcium-channel a1 subunit. EMBO J 10:45–49

Reuter H, Porzig H, Kokubun S, Prod'hom B (1988) Voltage-dependent effect of 1,4-dihydropyridine enantiomers on Ca channels in cardiac cells. Annals N.Y. Acad Sci 522:189–199

Reynolds IJ, Snowman AM, Snyder SH (1986) (-)-[^{3}H]desmethoxyverapamil labels multiple calcium channel modulator receptors in brain and skeletal muscle membranes: differentiation by temperature and dihydropyridines. J Pharmacol Exp Ther 237:731–738

Rubio LS, Garrido G, Llanes L, Alvarez JL (1993) Effects of tetrandrine on Ca^{2+}- and Na^+-currents of single bullfrog cardiomyocytes. J Mol Cell Cardiol 25:801–813

Sanguinetti MC, Kass RS (1984) Voltage-dependent block of calcium channel current in the calf cardiac Purkinje fiber by dihydropyridine calcium channel antagonists. Circ Res 55:336–348

Satoh H, Katoh H, Velez P, Fill M, Bers DM (1998) Bay K 8644 increases resting Ca^{2+} spark frequency in ferret ventricular myocytes independent of Ca influx: contrast with caffeine and ryanodine effects. Circ Res 83:1192–1204

Schramm M, Thomas G, Towart R, Franckowiak G (1983) Novel dihydropyridines with positive inotropic action through activation of Ca^{2+} channels. Nature Lond 303:535–537

Schuster A, Lacinova L, Klugbauer N, Ito H, Birnbaumer L, Hofmann F (1996) The IV6 segment of the L-type calcium channel is critical for the action of dihydropyridines and phenylalkylamines. EMBO J 15:2365–2370

Schweitz H, Heurteaux C, Bios P, Moinier D, Romey G, Lazdunski M (1994) Calciclu-dine, a venom peptide of tje Kunitz-type protease inhibitor family, is a potent blocker of high-threshold Ca^{2+} channels with a high affinity for L-type chanels in cerebellar granule neurons. Proc Natl Acad Sci USA 91:878–882

Seydl K, Kimball D, Schindler H, Romanin C (1993) The benzazepine/benzothiazepine binding domain of the cardiac L-type Ca^{2+} channel is accessible only from the extracellular side. Pflügers Arch 424:552–554

Sharma RK, Wang JH, Wu Z (1997) Mechanisms of inhibition of calmodulin-stimulated cyclic nucleotide phosphodiesterase by dihydropyridine calcium antagonists. J Neurochem 69:845–850

Sinnegger MJ, Wang Z, Grabner M, Hering S, Striessnig J, Glossmann H, Mitterdorfer J (1997) Nine L-type amino acid residues confer full 1,4-dihydropyridine sensitivity to the neuronal calcium channel alpha 1 A subunit: role of L-type methionine-1188. J Biol Chem 272:27686–27693

Sokolov S, Weiß RG, Kurka B, Gapp F, Hering S (1999) inactivation determinant in the I-II loop of the Ca^{2+} channel α_1-subunit and β-subunit interaction affect sensitivity for the phenylalkylamine (-)gallopamil. J Physiol 519:315–322

Spedding M, Kenny B, Chatelain P (1995) New drug binding sites in Ca^{2+} channels. Trends Pharmacol Sci 16:139–142

Stern JH, Kaupp UB, MacLeish PR (1986) Control of the light-regulated current in rod photoreceptors by cyclic GMP, calcium, and *l-cis*-diltiazem. Proc Natl Acad Sci USA 83:1163–1167

Striessnig J, Moosburger K, Goll A, Ferry DR, Glossmann H (1986) Stereoselective photoaffinity labelling of the purified 1,4-dihydropyridine receptor of the voltage-dependent calcium channel. Eur J Biochem 161:603–609

Striessnig J, Glossmann H, Catterall WA (1990) Identification of a phenylalkylamine binding region within the alpha 1 subunit of skeletal muscle Ca^{2+} channels. Proc Natl Acad Sci USA 87:9108–9112

Striessnig J, Murphy BJ, Catterall WA (1991) Dihydropyridine receptor of L-type Ca^{2+} channels: identification of binding domains for [^{3}H](+)-PN200–110 and [^{3}H]azidopine within the alpha 1 subunit. Proc Natl Acad Sci USA 88:10769–10773

Strübing C, Hering S, Glossmann H (1993) Evidence for an external location of the dihydropyridine agonist receptor site on smooth muscle and skeletal muscle calcium channels. Br J Pharmacol 108:884–891

Tanabe T, Takeshima H, Mikami A, Flockerzi V, Takahashi H, Kangawa K, Kojima M, Matsuo M, Hirose T, Numa S (1987) Primary structure for receptor for calcium channel blockers from skeletal muscle. Nature Lond 328:313–318

Tang CM, Presser F, Morad M (1988) Amiloride selectively blocks the low threshold (T) calcium channel. Science 240:213–215

Tang S, Mikala G, Bahinski A, Yatani A, Varadi G, Schwarts A (1993) Molecular localization of ion selectivity sites within the pore of a human L-type cardiac calcium channel. J Biol Chem 268:13026–13029

Teramoto N, Ogata, R, Okuba K, Kameyama A, Kameyama M, Watanabe T, Kuriyama H, Kitamura K (1996) Effects of calciseptine on unitary barium channel currents in guinea-pig portal vein. Pflügers Arch. Eur J Physiol 432:462–470

Thomas G, Chung M, Cohen CJ (1985) A dihydropyridine (Bay k 8644) that enhances calcium currents in guinea pig and calf myocardial cells. A new type of positive inotropic agent. Circ Res 56:87–96

Tiaho F, Richard S, Lory P, Nerbonne JM, Nargeot J (1990) Cyclic-AMP-dependent phosphorylation modulates the stereospecific activation of cardiac Ca channels by BayK8644. Pflügers Archive 417:58–66

Timin EN, Hering S (1992) A method for estimation of drug affinity constants to the opne conformational state of calcium channels. Biophys J 63:808–814

Triggle DJ (1999) The pharmacology of ion channles: with particular reference to voltage-gated Ca^{2+} channels. Eur J Pharmacol 375:311–325

Tsien RW, Bean BP, Hess P, Lansmann JB, Nilius B, Nowycky MC (1986) Mechanisms of calcium channel modulation by b-adrenergic agents and dihydropyridine calcium agonists. J Mol Cell Cardiol 18:691–710

Tung L, Morad M (1983) Voltage- and frequency-dependent block of diltiazem on the slow inward current and generation of tension in frog ventricular muscle. Pflügers Arch 189–198

Uneyama H, Takahara A, Dohmoto H, Yoshimoto R, Inoue K, Akaike N (1997) Blockade of N-type Ca^{2+} current by cilnidipine (FRC-8653) in acutely dissociated rat sympathetic neurones. Br J Pharmacol 122:37–42

Venema VJ, Swiderek KM, Lee TD, Hathway GM, Adams ME (1992) Antagonism of synaptosomal calcium channels by subtypes of ω-agatoxins. J Biol Chem 267:2610–2615

Wang G, Lemos JR (1995) Tetrandrine: a new ligand to block voltage-dependent Ca^{2+} and Ca^{+}-activated K^{+} channels. Life Sci 56:295–306

Watanabe T, Kalasz H, Yabana H, Kuniyasu A, Mershon J, Itagaki K, Vaghy PL, Naito K, Nakayama H, Schwartz A (1993) Azidobutyryl clentiazem, a new photoactivatable diltiazem analog, labels benzothiazepine binding sites in the α_1 subunit of the skeletal muscle calcium channel. FEBS Lett 334:261–264

Welling A, Kwan YW, Bosse E, Flockerzi V, Hofmann F, Kass RS (1993) Subunit-dependent modulation of recombinant L-type calcium channels: molecular basis for dihydropyridine tissue selectivity. Circ Res 73:974–980

Welling A, Ludwig A, Zimmer S, Klugbauer N, Flockerzi V, Hofmann F (1997) Alternatively spliced IS6segments of the a1 C gene determine the tissue-specific dihydropyridine sensitivity of cardiac and vascular smooth muscle L-type calcium channels. Circ Res 81:526–532

Willmott NJ, Choundhury Q, Flowerm RJ (1996) Functional importance of the dihydropyridine-sensitive, yet voltage-insensitive store-operated Ca^{2+} influx of U937 cells. FEBS Lett 394:159–164

Wu SN, Hwang TL, Jan CR, Tseng CJ (1997) Ionic mechanisms of tetrandrine in cultured rat aortic smooth muscle cells. Eur J Pharmacol 327:233–238

Yamaguchi S, Adachi-Akahane S, Nagao T (1999) The mechanism for the voltage-dependent block of diltiazem on L-type Ca^{2+} channels in guinea-pig ventricular myocytes. Jpn J Pharmacol 79(Suppl-I):40P

Yasuda O, Morimoto S, Chen Y, Jiang B, Kimura T, Sakakibara S, Koh E, Fukuo K, Kitano S, Ogihara T (1993) Calciseptine binding to a 1,4-dihydropyridine recognition site of the L-type calcium channel of rat synaptosomal membrnaes. Biochem. Biophys. Res Comm 194:587–594

Yatani A, Brown AM (1985) The calcium channel blocker nitrendipine blocks sodium channels in neonatal rat cardiac myocytes. Circ Res 56:868–875

Zahradnikova A, Zahradnik I (1992) Interaction of diltiazem with single L-type calcium channels in guinea-pig ventricular myocytes. Gen Physiol Biophys 11:535–543

Zhang S, Zhou Z, Gong Q, Makielski JC, January CT (1999) Mechanism of block and identification of the verapamil binding domain to HERG potassium channels. Circ Res 84:989–998

Zheng W, Rampe D, Triggle DJ (1992) Pharmacological, radioligand binding, and electrophysiological characteristics of FPL 64176, a new nondihydropyridine Ca^{2+} channel activator, in cardiac and vascular preparations. Mol Pharmacol 40:734–741

Zühlke RD, Bourbon A, Soldatov NM, Reuter H (1998) Ca^{2+} channel sensitivity towards the blocker isradipine is affected by alternative splicing of the human α_{1C} subunit gene. FEBS Lett 427:220–224

C. Voltage-Dependent K-Channels

CHAPTER 6

Overview of Potassium Channel Families: Molecular Bases of the Functional Diversity

Y. KUBO

A. Introduction

The structural and functional diversity is one of the most characteristic features of K^+ channels among other ion channels. In the K^+ channel superfamily there are some distinctly different families, and each family consists of many subfamilies. As the number of members in each subfamily is also large in number, total numbers of genes for K^+ channels are vast. The electrophysiological properties of each member are different and the function is also highly diversified. This functional diversity enables fine regulation of membrane potential and electrical excitability of cells.

The other characteristic feature of K^+ channels is that they are structurally small. Na^+ channels and Ca^{2+} channels consist of four repeats, but K^+ channels have only one repeat in one molecule. As four subunits assemble to form a tetramer in the case of K^+ channels, the functional unit is similar in size to Na^+ and Ca^{2+} channels. However, the difference is significant from the practical aspects of the experiments. Due to the small size of a subunit, it is feasible to manipulate the gene, and the expression experiments using heterologous expression system such as *Xenopus* oocyte are also easy in general. For these reasons, the structure-function study of K^+ channels has made remarkable progress.

In this chapter, the diversity of the structure, function, and regulation of K^+ channels and the progress of the structure-function study will be reviewed. There are some informative reviews of the related content which should also be referred to (JAN and JAN 1992, 1997; KUBO 1994; WEI et al. 1996; DASCAL 1997; ISOMOTO et al. 1997; NICHOLS and LOPATIN 1997; ARUILAR-BRYAN et al. 1998).

B. Primary Structure of the Main Subunit

I. 6-Transmembrane (TM) Type

The first voltage-gated K^+ channel (Kv) gene was isolated from a Drosophila mutant, *Shaker*, which has an abnormality of the K^+ channel gene (PAPAZIAN et al. 1987). In contrast with Na^+ channels which have four tandem repeats, the

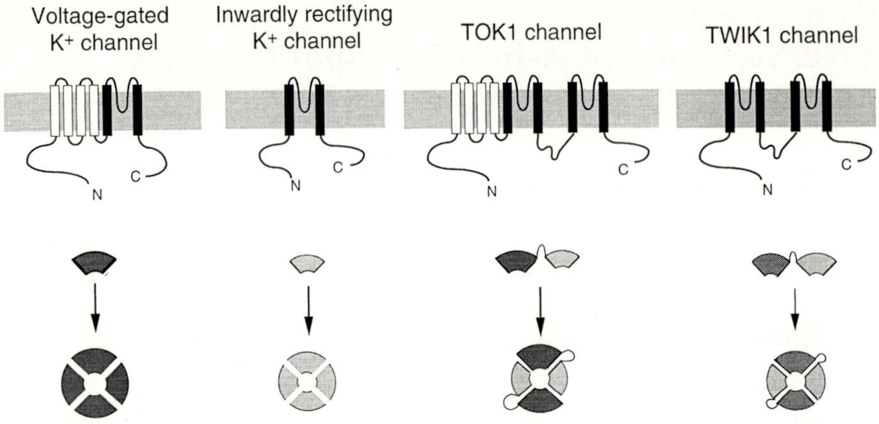

Fig. 1. Diversity of the structure of K^+ channel family. Schematic drawing of the topology (*upper*) and the top view (*lower*)

Shaker channel had only one repeat. In the one repeat there were six transmembrane regions and H5 pore region which comprises a channel pore (JAN and JAN 1997). The fourth transmembrane region, S4, has several positively charged amino acids, and functions as a voltage-sensor. By the extent of fast inactivation, the voltage-gated K^+ channels are classified into delayed rectifier type and A-type, but the structural difference is minor (Fig. 1).

II. 2-TM Type

The presence of inward rectifying K^+ channels (Kir), which have distinctly different properties from Kv, has been known physiologically since 1949. In 1993, cDNAs for inwardly rectifying K^+ channels, ROMK1 (Ho et al. 1993) and IRK1 (KUBO et al. 1993) were isolated by the expression cloning method. It was revealed that the inwardly rectifying K^+ channels have only two transmembrane regions and H5 pore region, and that the structure corresponds to the latter half of Kv channels (KUBO et al. 1993). As expected from the fact that the inward rectifier does not show voltage-dependence as Kv does, there was no S4-like voltage-sensor region (Fig. 1).

III. 1-TM Type

cDNA for a very unique channel, Isk or minK, was isolated by the expression cloning method (TAKUMI et al. 1988). This channel had only one transmembrane region, and it did not even have the K^+ selective filter region conserved in all other K^+ channels. Recently it was reported that Isk formed a functional heteromultimer with a member of Kv, KvLQT1, and expressed a slowly activating current observed in the heart (BARHANIN et al. 1996; SANGUINETTI et al. 1996).

IV. 2-Repeat Type

Channels which are composed of tandemly ligated 6-TM type and 2-TM type (Ketchum et al. 1995; Lesage et al. 1996a), and channels composed of two tandemly ligated 2-TM type (Lesage et al. 1996b; Fink et al. 1996) were isolated. This family can be understood as a 2-repeat type, in contrast with Na^+ channels which have four repeats and K^+ channels which have only one repeat. It is expected that two subunits of this family assemble to form a functional channel (Fig. 1).

C. Heteromultimeric Assembly: Bases of Further Diversity

I. Heteromultimer Formation with Other Members of the Same Subfamily

It is known that different members of the same family assemble to form functionally different heteromultimeric channels. This is a further basis for the diversity of the K^+ channel function, on top of the diversity of the genes. In the following some representative examples are shown.

1. Kv Channels

It was shown in heterologous expression systems that a member in the *Shaker* family and a member in the *drk* family co-assemble, and the property of the channel is different from either *Shaker* or *drk*. The structural basis for the assembly was identified to lie in the N-terminus cytoplasmic region (Li et al. 1992). It was also shown that this heteromultimeric channel actually exists in the brain by analyzing co-immuno-precipitating molecules (Sheng et al. 1993).

2. GIRK1,2,4

GIRK1, an inwardly rectifying K^+ channel which is activated by direct interaction with GTP binding protein, was isolated by sequence homology with IRK1 (Kubo et al. 1993b) and by expression cloning (Dascal et al. 1993). As the functional expression of GIRK1 was not sufficiently high, Krapivinsky et al. (1995) expected that there should be a molecule which assembles to form highly expressable channel. They purified a protein which co-immunoprecipitates with GIRK1 from the heart, and determined the partial amino acid sequence. Using the obtained information, they isolated a cDNA designated CIR. It was confirmed that coexpression of GIRK1 and CIR induces large inward K^+ current. It was concluded that the hetero-multimer of GIRK1 and CIR is the muscarinic K^+ channel in the heart (Krapivinsky et al. 1995; Iizuka et al. 1995). GIRK2, which was isolated by sequence homology with GIRK1, was also showed to form a heteromultimer with GIRK1, and GIRK1/2 channel is thought to be the major form of the G-protein coupled inward rectifier K^+ channel in the brain (Lesage et al. 1995; Velimirovic et al. 1996).

II. Suppression of Functional Expression by Heteromultimeric Assembly

There are some members in the Kv family which do not show functional expression of themselves. The Kv 8.1 channel, which was non-functional of itself, assembled with Kv2.1 or with Kv3.4 and suppressed their functional expression (Hugnot et al. 1996). A variant of a member of inwardly rectifying K^+ channel (Kir) family, Kir2.2v, was shown to function as a negative regulator for Kir2.2 (Namba et al. 1996). As Kir2.2v is not a cDNA but a genomic clone, the actual expression and physiological role of Kir2.2 remains to be tested.

III. Heteromultimeric Assembly of Main Subunits of Different Families

Isk expressed a extremely slowly activating current when injected alone into *Xenopus* oocytes (Takumi et al. 1988). As the point mutation changed the ion selectivity and the permeation, it was thought that the channel actually form the permeation pathway (Goldstein and Miller 1991). By cys scan mutagenesis study, it was confirmed that the sites of the transmembrane region faces to the aqueous pore, and forms a permeation pathway (Wang et al. 1996). However, the extremely slow current was not reported in cells in the animal organs, and the physiological function of the Isk molecule was not known. It was reported recently that coexpression of KvLQT1 voltage-gated K^+ channel with Isk enhances the expression of KvLQT1 current, and the expressed current was similar to the slowly activating current in the cardiac muscle (Barhanin et al. 1996; Sanguinetti et al. 1996) (Fig. 2). Furthermore, the assembly of Isk with HERG, a member of Kv family, was also reported (MacDonald et al. 1997). Thus, the physiological significance of Isk appeared not to make a very unique channel, but to form a functional heteromultimer by assembling with KvLQT1 or with HERG. The mode and stoichiometry of the assembly of 1-TM subunit and 6-TM subunit is a very interesting question for future studies.

IV. Assembly with β Subunit

The $\beta\gamma$ subunit of Na^+ channels has been known to accelerate inactivation. The β subunit for K^+ channels was not known until recently. In 1994, the β subunit of voltage-gated K^+ channels was isolated and functionally characterized to accelerate inactivation (Rettig et al. 1994). After that, the presence of three types of β subunits were reported, implying the complexity of the regulation (Heinemann et al. 1996). It was also reported that the β subunits not only reg-

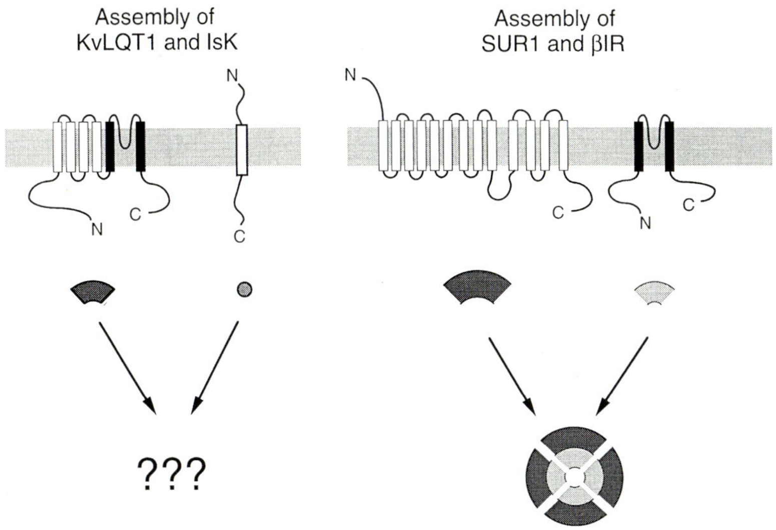

Fig. 2. Schematic drawing of heteromultimerization of main subunits and accessory subunits

ulate the inactivation kinetics but also promote the expression of the main subunit to the cytoplasmic membrane (SHI et al. 1996). As for the inward rectifying K^+ channels, the presence of β subunit is not known yet.

V. Assembly with Regulatory Subunits

The inward rectifying K^+ channel which is blocked by cytoplasmic ATP was characterized physiologically by NOMA (1983), but the cDNA had not been isolated until recently. The break-through was the purification of the protein which binds to sulfonylurea, a inhibitor of the channel, and the succeeding cDNA (SUR1) cloning (AGUILAR-BRYAN et al. 1995). INAGAKI et al. (1995a) isolated a uKATP (Kir6.1) by sequence homology with GIRK1, and then isolated βIR (Kir6.2), a β cell specific inward rectifier, by sequence homology with uKATP (INAGAKI et al. 1995b) (Fig. 2). When βIR was coexpressed with SUR1, channels were expressed which highly resemble with the ATP-sensitive K^+ channels of the pancreatic β cell in terms of the ATP sensitivity, the sulfonylurea sensitivity and the channel pore properties (INAGAKI et al. 1995b). By sequence homology with SUR1, SUR2A (INAGAKI et al. 1996) and SUR2B (ISOMOTO et al. 1996) were isolated succeedingly, and it was reported that SUR2A and βIR form a cardiac type ATP sensitive K^+ channel (INAGAKI et al. 1996), and that SUR2B and Kir 6.1 form a smooth muscle type ATP sensitive K^+ channels (ISOMOTO et al. 1996). The stoichiometry of βIR and SUR was studied, and it was suggested that 4 and 4 subunits assemble together (CLEMENT et al. 1997; INAGAKI et al. 1997; SHYNG and NICHOLS 1997) (Fig. 2).

It is an interesting but open question how the information of the binding of ATP or sulfonylurea is transmitted to the gate. Recently ROMK2 channel was reported to form functional heteromultimer with a ATP binding protein named CFTR (cystic fibrosis transmembrane regulator) (McNicholas et al. 1996). It was also reported that ROMK1 channel assembles with SUR1 (Ammala et al. 1996).

VI. Assembly with Anchoring Protein

For the physiological function of channels it is obviously important to make channels clustered at a hot spot. Kim et al. (1995) identified such a molecule, PSD-95. This molecule has a domain called PDZ, and PDZ domain interacts with a motif of Kv channel in the cytoplasmic chain to make the channels clustered. Similar observations were reported on Kir members such as IRK3 or KAB2. In the case of IRK3, the channels dispersed when the channels were phosphorylated by protein kinase A (Cohen et al. 1996). KAB2 is a weakly inward rectifying K^+ channel, which shows a distinct subcellular localization in the inner ear (Hibino et al. 1997) and in the renal tubules (Ito et al. 1996). KAB2 was also shown to be clustered by a protein with PDZ domain (Horio et al. 1997). The functional significance of channels in neurons is very different depending on the location where it appears, such as soma, axon, and dendrites. Thus, the localization is expected to be finely regulated both by the transporting system and by the anchoring molecules. More molecules which enable fine regulation of the subcellular localization and clustering will be found from now.

D. Structural Bases of the Gating Mechanism

I. Activation of Kv Channels

The hypothesis that S4 region which is abundant in positively charged amino acid functions as a voltage sensor, was proved by the results that the voltage-dependency decreased when the number of positive charge was reduced by mutagenesis (Papazian et al. 1991; Liman et al. 1991; Logothetis et al. 1992) (Fig. 3). It was also shown that residues outside of the S4 region play a role (Mathur et al. 1997; Planells et al. 1995), and that mutations of non-charged amino acids in S4 also affect the voltage dependency (Lopez et al. 1991).

It has been unknown whether conformational change actually occur in S4 region or not. Some models, such as a sliding helix model, were advocated, but it was not proved that the S4 actually moves upon depolarization. Isacoff's group developed a new challenging method to monitor the actual movement (Mannuzzu et al. 1996). They introduced cys residues to various positions of S4 by mutagenesis, and labeled the cys residues with a fluorescent probe, whose fluorescence intensity changes depending on the environment, i.e., the intensity differs in the lipid and in the water. When the probe was introduced

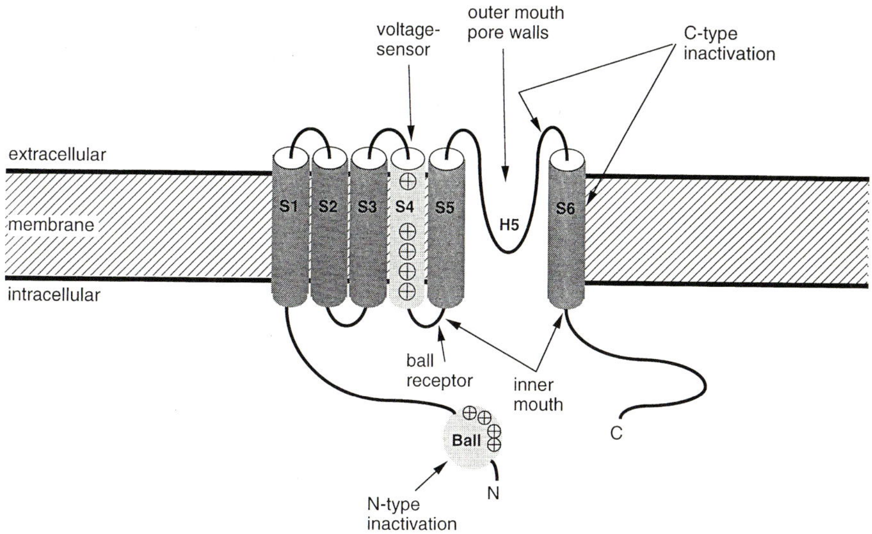

Fig. 3. Location of functional devices of voltage-gated K^+ channels on the molecule

in the center of S4 region, changes in the fluorescence intensity was observed upon depolarization. This result demonstrates that this region comes out from the membrane to the outside when depolarized. They also showed that the time course of the change of the fluorescent signal correlated well to the time course of the gating current, confirming that the change of the fluorescence intensity truly reflects conformational change during gating (MANNUZZU et al. 1996). They also compared the accessibility of the cys reacting reagents to the introduced cys residues in the S4 region at depolarized and hyperpolarized states, and concluded that the S4 region translocates by six amino acid residues length towards the extracellular direction when the membrane potential was depolarized (LARSSON et al. 1996) (Fig. 4).

II. N-Type Inactivation of Kv Channels

Some Kv channels show fast inactivating current which is called A-current. As a mechanism for the fast inactivation, a model in which a ball plugs a channel pore was postulated. Hoshi et al. proved this model experimentally. The fast inactivation disappeared by deleting the N-terminus region (N-type inactivation) and application of a peptide of the deleted region restored the fast inactivation in a dose-dependent manner (HOSHI et al. 1990) (Fig. 3). A domain called NIP, which inhibits the N-type inactivation by the ball, was identified recently also in the adjacent region of the ball domain (ROEPER et al. 1998).

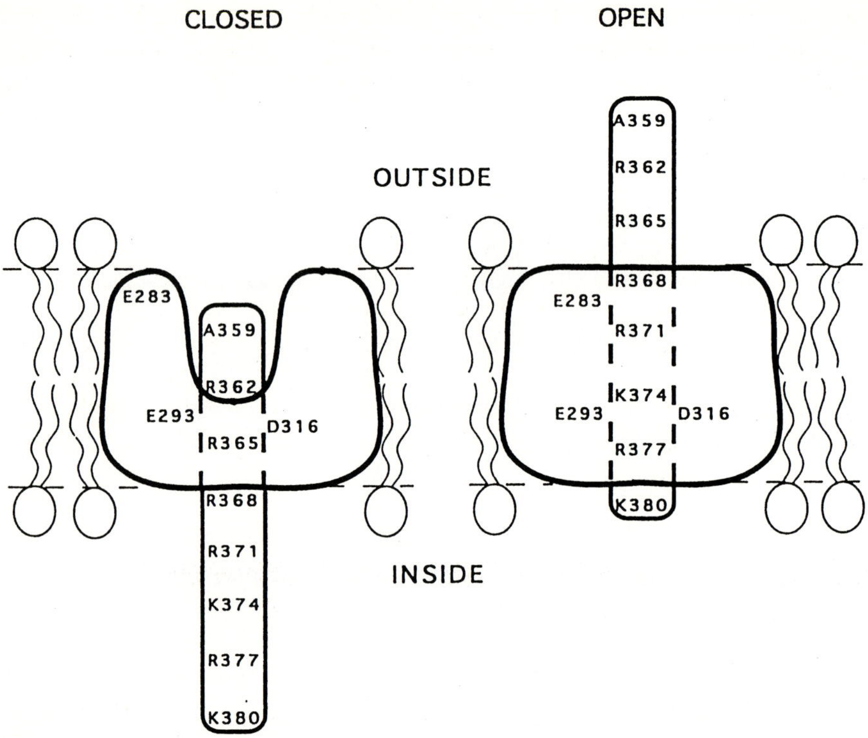

Fig. 4. Dynamic translocation of the S4 region of voltage-gated K^+ channel demonstrated by Larsson et al. (1996)

III. C-Type Inactivation of Kv Channels

Shaker K^+ channels show slow inactivation even after removing the N-type inactivation. Hoshi et al. (1991) observed that the speed of the slow inactivation differs in the splice variants which have difference in the region from S6 to C-terminus, and designated this inactivation to be C-type inactivation. They identified the structural basis at amino acid residues in the S6 region. Later on, an amino acid residue at the external mouth of the H5 pore region, T449, was reported to be critically important (Lopez-Barneo et al. 1993) (Fig. 3). Lopez-Barneo et al. (1993) observed that this inactivation proceeds faster when extracellular K^+ is lower. It was demonstrated that depletion of K^+ ion at the external mouth of the pore leads the channel to the C-inactivated state (Baukrowitz and Yellen 1995). Furthermore, Liu et al. (1996) demonstrated a dynamic conformational change at the external mouth of the pore during C-type inactivation by monitoring changes of the accessibility (i.e., modification speed) of the cysteine modifiers. The C-type inactivation, which was thought to be a static change in contrast with activation, was also proved to accompany a dynamic structural change of the pore. Supporting the dynamic rearrangement of the pore, the ion selectivity of Kv channels was also shown to change during C-type inactivation (STARKUS et al. 1997).

IV. Activation of IsK

When Isk was expressed alone it showed extremely slowly activating current. It was reported that the activation speed depends on the channel density. By cross-linking the assembled channels, it was shown that instantaneously activating current appeared from the second depolarizing pulse but not by the first one. From this result, the slowly activating step was concluded to reflect a step of the assembly of channel subunits (VARNUM et al. 1993). The recently reported heteromultimer channel of Isk and KvLQT1 might have different mechanisms of activation.

E. Structural Bases of the Ion Permeation and Block

I. H5 Pore Region

When *Shaker* was first isolated, H5 pore region was not paid too much attention. Later on, the various mutations of this region were shown to cause changes of the pore properties, such as the conductance, the ionic selectivity, and the sensitivities to blockers applied from internal side or external side of the membrane. Thus, it was postulated that the H5 region form the channel pore and that the middle of the H5 region faces to the internal side of the membrane (YOOL and SCHWARTZ 1991; YELLEN et al. 1991) (Fig. 3). The result of the cys scan mutagenesis study on H5 region of Kv channels was also compatible with this scheme (PASCUAL et al. 1995).

II. Re-evaluation

In addition to H5 region, some regions other than H5 were also shown to form part of the pore. Mutations in S4-S5 linker region (SLESINGER et al. 1993) and part of S6 region (LOPEZ et al. 1994) also affected the channel conductance and the ionic selectivity, and it is accepted that they also form part of the pore (Fig. 3).

In the case of inward rectifier K^+ channels, it also had a highly conserved motif for K^+ selective channels, and the pore structure was postulated to be highly similar (KUBO et al. 1993a). However, it was shown recently that the H5 region of ROMK1 (Kir1.1) does not face to the internal side, by the glycosylation site scan study (SCHWALBE et al. 1996) and by the cys scan study (Y. KUBO et al. 1998). Thus, the pore structure of Kir might be quite different from those of Kv channels.

III. Inward Rectification Mechanism

Inwardly rectifying K^+ channels do not sense the membrane potential as clearly as Kv channels do. Under various extracellular K^+ conditions they allow inward flow of K^+ below E_K and little outward current above E_K. The channel looks as if to sense the shift from E_K in various extracellular K^+ conditions (Hagiwara et al. 1976) (Fig. 5). The inward rectification property of inward rectifier in the cardiac muscle cells was reported to be due to a block by cytoplasmic Mg^{2+} (Matsuda et al. 1987; Vandenberg 1987) and an intrinsic gating which remains in the absence of Mg^{2+} (Matsuda 1988; Ishihara et al. 1989). Using isolated cDNA clones, the intrinsic gating was discovered to be due to a block by cytoplasmic polyamines, such as spermine (Lopatin et al. 1994; Ishihara et al. 1996) (Fig. 6). One of the structural bases which defines the high sensitivity to the blockers was identified to be Asp residue in the center of M2 region (Stanfield et al. 1994; Wible et al. 1994). In addition, contributions of other parts of the channel to the extent of rectification was reported (Taglialatela et al. 1994; Kubo et al. 1996). Yang et al. (1995) reported that Glu at 224 in the putative C-terminal cytoplasmic chain is also critical. It is not known how the chain affects the sensitivity to the cytoplasmic pore blockers, but a possibility that this region is folded into the inner part of the pore could be postulated (Nichols and Lopatin 1997).

Matsuda (1988) reported that three blocking ions block the channels independently to form 1/3, 2/3 sublevels, from the single channel analysis of the outward current in low internal Mg^{2+}, and the inward current in low external Cs^+ (Matsuda et al. 1989). They discussed that the result can be

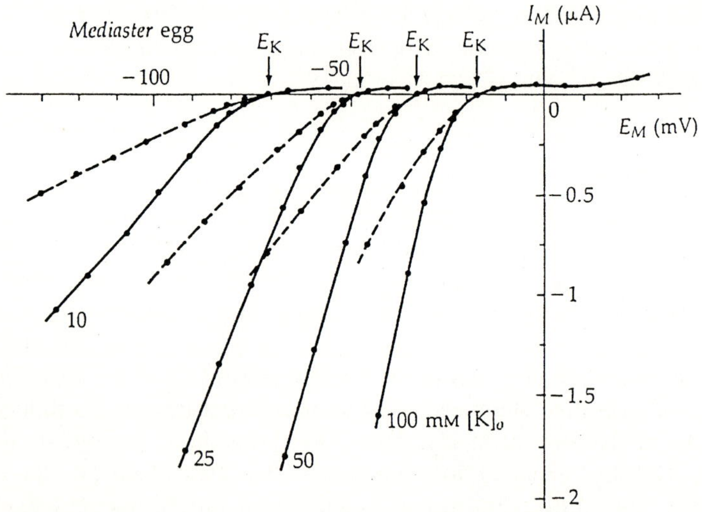

Fig. 5. Current-voltage relationship of the inward rectifier K^+ current in various extracellular K^+ condition shown by Hagiwara et al. (1976)

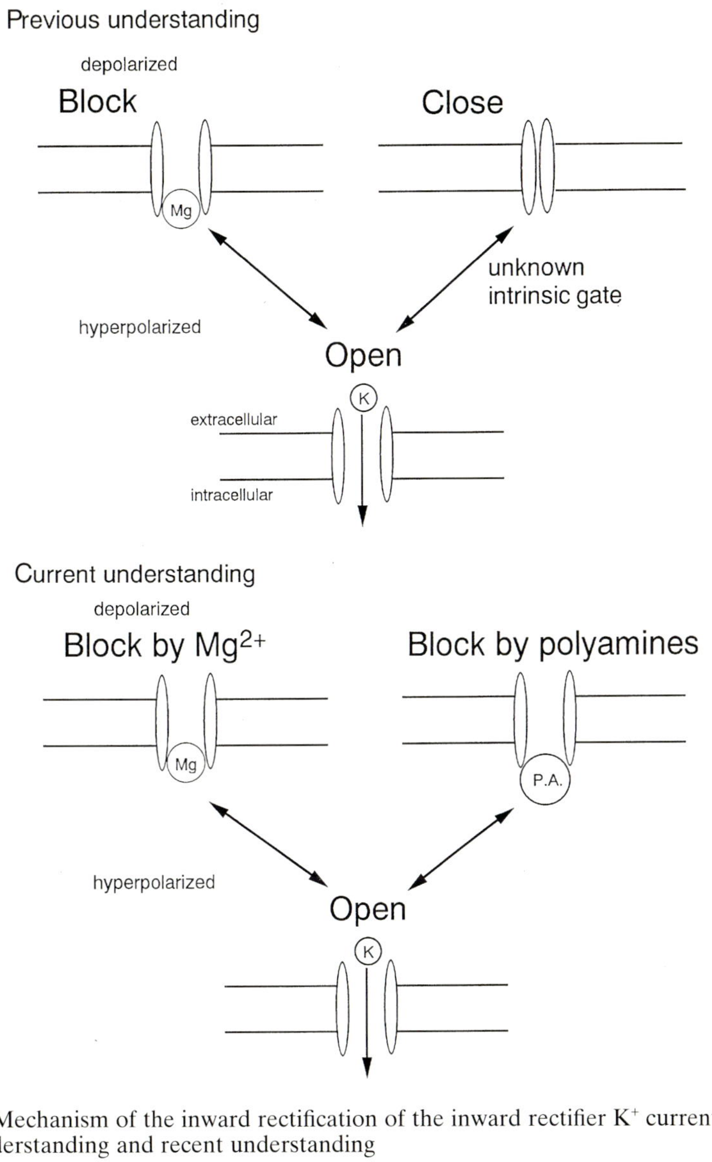

Fig. 6. Mechanism of the inward rectification of the inward rectifier K^+ current. Previous understanding and recent understanding

explained most simply by a model in which three blocking ions block three parallel permeation pathways independently. On the other hand, the stoichiometry of IRK1 was also shown to be a tetramer (Yang et al. 1995) similarly with Kv channels (Liman et al. 1992; Li et al. 1994), suggesting that there is only one pore in the center. It is necessary to obtain a detailed image of the block by cytoplasmic Mg^{2+} to find the structural basis which can coordinate independent interaction of three blocking ions.

IV. Direct Structure Analysis

The structure analysis of membrane proteins by X-ray crystallography is not easy in general, due to the difficulty of the preparation of crystals, and the number of studies is very limited. Unwin (1995) resolved the structure of acetylcholine receptor using tubular crystals in the closed state and in the open state at 9Å resolution. Recently, Doyle et al. (1998) resolved the structure of a bacterial K^+ channel, KcsA at 3.2Å resolution. This channel belongs to 2TM type but the amino acid sequence of the H5 region is closer to Kv channels. On the basis of this structure, for the first time they showed experimentally the detailed image of the permeation pathway and the selective filter, and discussed the mechanism of selectivity and permeation of K^+ ions. Further structural analysis at high resolution of channel proteins in various states is awaited to show the dynamic conformational changes during gating.

F. Structural Bases of Various Regulation Mechanisms

I. $G\beta\gamma$

There was a debate over whether the G-protein coupled inwardly rectifying K^+ channel of the cardiac muscle cells is activated by α subunit or by $\beta\gamma$ subunit of G protein. As shown consistently by Kurachi and his colleagues (Logothetis et al. 1987; Kurachi 1995), the channel was concluded to be activated by $G\beta\gamma$ subunits from the results using GIRK cDNA and recombinant G protein (Reuveny et al. 1994). The α subunits rather inhibited the channel activity by absorbing the free $G\beta\gamma$ subunits (Schreibmayer et al. 1996). For the functional interaction with $G\beta\gamma$ subunits, the C-terminus cytoplasmic region was reported to be important (Slesinger et al. 1993; Huang et al. 1995; Kubo and Iizuka 1996). On the other hand, biochemical binding was observed at both N and C-terminus cytoplasmic regions (Fig. 7). It was also reported that the α subunit binds to the N-terminus, and the mutation of this region causes slowing down of the activation kinetics upon receptor stimulation (Slesinger et al. 1995) (Fig. 7). Thus, the heterotrimeric G protein might be prepared at the N-terminus chain so that free $G\beta\gamma$ subunits can be supplied without delay upon receptor stimulation.

Luchian et al. (1997) reported that a C-terminal peptide of GIRK1 blocks the GIRK channels. It was speculated that the C-terminus is like a blocking ball, and the interaction of $G\beta\gamma$ releases the block (Kubo 1994; Luchian et al. 1997). Huang et al. (1998) reported that PIP_2 activates GIRK channel directly, and that $G\beta\gamma$ stabilizes the effect of PIP_2. This study suggests that the interaction of PIP_2 with GIRK is most critically and directly important. The mechanistic link between $G\beta\gamma$ binding and pore opening will be elucidated in the near future.

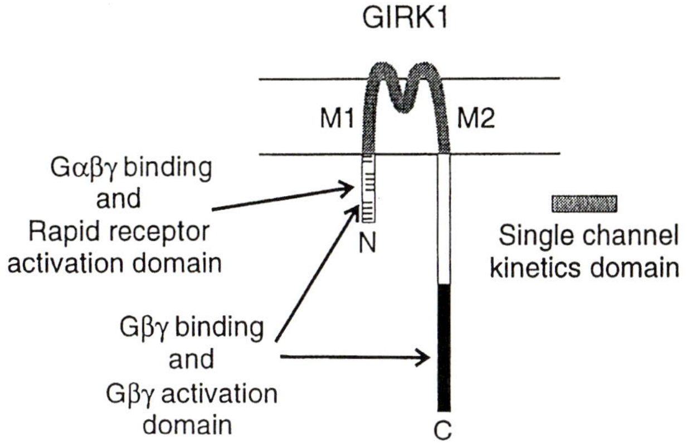

Fig. 7. The domains of G-protein coupled K^+ channels for the interaction with GTP binding proteins shown by SLESINGER et al. (1995)

II. Block by Cytoplasmic ATP

*β*IR expresses ATP-sensitive K^+ current only when cointroduced with SUR1 (INAGAKI et al. 1995b). However, *β*IR whose C-terminus is truncated expressed ATP sensitive K^+ current alone (TUCKER et al. 1997). It was found that the ATP sensitivity is determined by *β*IR, and that SUR1 adds sensitivity to sulfonylurea (TUCKER et al. 1997). A clear image, how the information of ATP binding causes opening of the channel, has not yet been obtained.

III. Regulation by Phosphorylation

Regulation of channel activities by phosphorylation is reported in various channels such as IRK1 (FAKLER et al. 1994; WISCHMEYER and KARSCHIN 1996). In the case of HERG (human ether-go-go related gene) channels, it is known that phosphorylation changes the gating property (R. Schonherr et al., in preparation). HERG channel structurally belongs to the Kv channels, but the electrophysiological properties look like inward rectifier. It was reported that the electrophysiological property is due to fast inactivation (O-I transition) and slow deactivation (O-C transition) (SMITH et al. 1996). Upon depolarization, outward current is not clearly seen due to a fast inactivation, and upon repolarization the channels transit from I to O, then O to C states. As this O-C transition is slow, significant inward K^+ current can be observed. It was reported that the structural basis for the slow O-C transition lies in the N-terminal cytoplasmic region, and that for fast O-I transition lies in the H5 pore region by SCHONHERR and HEINEMANN (1996). They reported these gating properties change by phosphorylation.

IV. Mg^{2+} as a Cytoplasmic Second Messenger

A novel regulation mechanism of IRK3 channel was advocated recently (Chuang et al. 1997). IRK3 current decreased upon stimulation of the coexpressed m1 muscarinic receptor. However, the second messenger involved in this regulation was not Ca^{2+} or protein kinase C, which are generally expected to be downstream of m1 receptor. By thorough analysis, it was concluded that cytoplasmic Mg^{2+} works as a second messenger, and downregulates the activity of IRK3 channels (Chuang et al. 1997), although the route from the m1 receptor to the increase in the cytoplasmic Mg^{2+} is not established. This down regulation is independent of the highly sensitive block of the outward current which causes inward rectification.

V. Regulation by Extracellular K^+

C-type inactivation of Kv channels is a phenomenon which depends on K^+_o, and the activity regulation of inward rectifier K^+ channels has similar aspects (Kubo 1996). Inward rectifier behaves as if it senses the shift from E_K (Hagiwara et al. 1976), but it is confined to the case in which K^+_o was changed. When K^+_i concentration was changed, the conductance-voltage curve was not clearly shifted (Hagiwara et al. 1979; Matsuda 1991; Kubo 1996). Thus, the inward rectifier senses not the shift from E_K but the combination factor of the $[K^+]_o$ and the membrane potential. In addition, the effect of K^+_o to IRK1 channel was severely changed when an amino acid residue, which corresponds to T449 of *Shaker* which plays a critical role for C-type inactivation, was mutated (Kubo 1996). Thus, the regulation mechanism of IRK1 by K^+_o might be somehow relevant to the C-type inactivation of Kv channels.

VI. Other Mechanisms

Cyclosporin A is a blocker of cyclophilin, a class of peptidyl-prolyl isomerase. Chen et al. (1998) reported that cyclosporin selectively blocks the expression of IRK1 channels. They speculated that the isomerization of prolyne residues might be a novel regulation mechanism of channel activities of IRK1.

Cohen et al. (1997) observed that channel activity of a splice variant of *Shaker* was regulated by the status of oxidation/reduction of a methionine residue in the N-terminus cytoplasmic region, and said that the regulation of the channel activity by oxidation/reduction states have an important meaning in the physiological and pathophysiological conditions.

G. Perspectives

In this chapter, recent progress in the molecular biological studies of K^+ channels have been reviewed. What are the problems to be solved from now on? In terms of the isolation of cDNA clones, m-current channels, Ih current channels, and Na^+ activated K^+ channels could be the main targets in the near future. As to the structure-function relationship, it is thought that there are many more problems which could be solved by the experimental approach at present. At the same time, it is desirable to take a direct approach by developing new techniques. Needless to say, we also await a solution of the structure of channels proteins at various states at high resolution. The structure analysis of membrane protein is difficult in general, but the door has already been opened as seen in the cases of the acetylcholine receptor (Unwin 1995) and the bacteria K^+ channel (Doyle et al. 1998). It is expected that structure analysis will make a steady and remarkable progress in the near future.

References

Aguilar-Bryan L, Clement JP, Gonzalez G, Kunjilwar K, Babenko A, Bryan J (1998) Towards understanding the assembly and structure of KATP channels. Physiol Rev 78:227–245

Aguilar-Bryan L, Nichols CG, Wechsler SW, et al. (1995) Cloning of the beta cell high-affinity sulfonylurea receptor: a regulator of insulin secretion. Science 268:423–426

Ammala C, Moorhouse A, Gribble F, et al. (1996) Promiscuous coupling between the sulphonylurea receptor and inwardly rectifying potassium channels. Nature 379:545–548

Barhanin J, Lesage F, Guillemare E, Fink M, Lazdunski M, Romey G (1996) K(V)LQT1 and lsK (minK) proteins associate to form the I(Ks) cardiac potassium current. Nature 384:78–80

Baukrowitz T, Yellen G (1995) Modulation of K^+ current by frequency and external $[K^+]$: a tale of two inactivation mechanisms. Neuron 15:951–960

Chen H, Kubo Y, Hoshi T, Heinemann SH (1998) Cyclosporin A selectively reduces the functional expression of Kir2.1 potassium channels in Xenopus oocytes. FEBS letters 422:307–310

Chuang H, Jan YN, Jan LY (1997) Regulation of IRK3 inward rectifier K^+ channel by m1 acetylcholine receptor and intracellular magnesium. Cell 89:1121–1132

Ciorba MA, Heinemann SH, Weissbach H, Brot H, Hoshi T. (1997) Modulation of potassium channel function by methionine oxidation and reduction. Proc Natl Acad Sci USA 94:9932–9937

Clement JP, Kunjilwar K, Gonzalez G, et al. (1997) Association and stoichiometry of K(ATP) channel subunits. Neuron 18:827–838

Cohen NA, Brenman JE, Snyder SH, Bredt DS (1996) Binding of the inward rectifier K^+ channel Kir 2.3 to PSD-95 is regulated by protein kinase A phosphorylation. Neuron 17:759–767

Dascal N (1997) Signaling via the G-protein-activated K^+ channels. Cell Signal. 9:551–573

Doyle DA, Cabral JM, Pfuetzner RA, Kuo A, Gulbis JM, Cohen SL, Chait BT, MacKinnon R (1998) The structure of the potassium channel: Molecular basis of K^+ conduction and selectivity. Science 280:69–77

Fakler B, Brandle U, Glowatzki E, Zenner HP, Ruppersberg JP (1994) Kir2.1 inward rectifier K^+ channels are regulated independently by protein kinases and ATP hydrolysis. Neuron 13:1413–1420

Fink M, Duprat F, Lesage F, et al. (1996) Cloning, functional expression and brain localization of a novel unconventional outward rectifier K^+ channel. EMBO J 15:6854–6862

Goldstein SA, Miller C (1991) Site-specific mutations in a minimal voltage-dependent K^+ channel alter ion selectivity and open-channel block. Neuron 7:403–408

Hagiwara S, Miyazaki S, Rosenthal NP (1976) Potassium current and the effect of cesium on this current during anomalous rectification of the egg cell membrane of a starfish. J Gen Physiol 67:621–638

Hagiwara S, Yoshii M (1979) Effects of internal potassium and sodium on the anomalous rectification of the starfish egg as examined by internal perfusion. J Physiol 292:251–265

Heinemann SH, Rettig J, Graack HR, Pongs O (1996) Functional characterization of Kv channel beta-subunits from rat brain. J Physiol 493:625–633

Hibino H, Horio Y, Inanobe A, et al. (1997) An ATP-dependent inwardly rectifying potassium channel, KAB-2 (Kir4. 1), in cochlear stria vascularis of inner ear: its specific subcellular localization and correlation with the formation of endocochlear potential. J Neurosci 17:4711–4721

Ho K, Nichols CG, Lederer WJ, et al. (1993) Cloning and expression of an inwardly rectifying ATP-regulated potassium channel. Nature 362:31–38

Horio Y, Hibino H, Inanobe A, et al. (1997) Clustering and enhanced activity of an inwardly rectifying potassium channel, Kir4.1, by an anchoring protein, PSD-95/SAP90. J Biol Chem 272:12885–12888

Hoshi T, Zagotta WN, Aldrich RW (1990) Biophysical and molecular mechanisms of Shaker potassium channel inactivation. Science 250:533–53

Hoshi T, Zagotta WN, Aldrich RW (1991) Two types of inactivation in Shaker K^+ channels: Effects of alterations in the carboxy-terminal region. Neuron 7:547–556

Huang CL, Slesinger PA, Casey PJ, Jan YN, Jan LY (1995) Evidence that direct binding of G beta gamma to the GIRK1 G protein-gated inwardly rectifying K^+ channel is important for channel activation. Neuron 15:1133–1143

Huang CL, Feng S, Hilgemann DW (1998) Direct activation of inward rectifier potassium channels by PIP2 and its stabilization by $G\beta\gamma$. Nature 391:803–806

Hugnot JP, Salinas M, Lesage F, et al. (1996) Kv8.1, a new neuronal potassium channel subunit with specific inhibitory properties towards Shab and Shaw channels. EMBO J 15:3322–3331

Iizuka M, Kubo Y, Tsunenari I, Pan CX, Akiba I, Kono T (1995) Functional characterization and localization of a cardiac-type inwardly rectifying K^+ channel. Receptors Channels 3:299–315

Inagaki N, Gonoi T, Clement JP, et al. (1996) A family of sulfonylurea receptors determines the pharmacological properties of ATP-sensitive K^+ channels. Neuron 16:1011–1017

Inagaki N, Gonoi T, Clement JP, et al. (1995) Reconstitution of IKATP: an inward rectifier subunit plus the sulfonylurea receptor. Science 270:1166–1170

Inagaki N, Gonoi T, Seino S (1997) Subunit stoichiometry of the pancreatic beta-cell ATP-sensitive K^+ channel. FEBS Lett 409:232–236

Inagaki N, Tsuura Y, Namba N, et al. (1995) Cloning and functional characterization of a novel ATP-sensitive potassium channel ubiquitously expressed in rat tissues, including pancreatic islets, pituitary, skeletal muscle, and heart. J Biol Chem 270:5691–5694

Ishihara K, Hiraoka M, Ochi R (1996) The tetravalent organic cation spermine causes the gating of the IRK1 channel expressed in murine fibroblast cells. J Physiol 491:367–381

Ishihara K, Mitsuie T, Noma A, Takano M (1989) The Mg2+ block and intrinsic gating underlying inward rectification of the K^+ current in guinea-pig cardiac myocytes. J Physiol 419:297–320

Isomoto S, Kondo C, Kurachi Y (1997) Inwardly rectifying potassium channels: their molecular heterogeneity and function. Jpn J Physiol 47:11–39

Isomoto S, Kondo C, Yamada M, et al. (1996) A novel sulfonylurea receptor forms with BIR (Kir6.2) a smooth muscle type ATP-sensitive K^+ channel. J Biol Chem 271:24321–24324

Ito M, Inanobe A, Horio Y, et al. (1996) Immunolocalization of an inwardly rectifying K^+ channel, K(AB)-2 (Kir4.1), in the basolateral membrane of renal distal tubular epithelia. FEBS Lett 388:11–15

Jan LY, Jan YN (1992) Structural elements involved in specific K^+ channel functions. Annu Rev Physiol 54:537–555

Jan LY, Jan YN (1997) Cloned potassium channels from eukaryotes and prokaryotes. Annu Rev Neurosci 20:91–123

Ketchum KA, Joiner WJ, Sellers AJ, Kaczmarek LK, Goldstein SA (1995) A new family of outwardly rectifying potassium channel proteins with two pore domains in tandem. Nature 376:690–695

Kim E, Niethammer M, Rothschild A, Jan YN, Sheng M (1995) Clustering of Shaker-type K^+ channels by interaction with a family of membrane-associated guanylate kinases. Nature 378:85–88

Krapivinsky G, Gordon EA, Wickman K, Velimirovic B, Krapivinsky L, Clapham DE (1995) The G-protein-gated atrial K^+ channel IKACh is a heteromultimer of two inwardly rectifying K^+-channel proteins. Nature 374:135–141

Kubo Y (1994) Towards the elucidation of the structural-functional relationship of inward rectifying K^+ channel family. Neurosci-Res 21:109–117

Kubo Y (1996) Effects of extracellular cations and mutations in the pore region on the inward rectifier K^+ channel IRK1. Receptors Channels 4:73–83

Kubo Y, Baldwin TJ, Jan YN, Jan LY (1993a) Primary structure and functional expression of a mouse inward rectifier potassium channel. Nature 362:127–133

Kubo Y, Iizuka M (1996) Identification of domains of the cardiac inward rectifying K^+ channel, CIR, involved in the heteromultimer formation and in the G-protein gating. Biochem Biophys Res Commun 227:240–247

Kubo Y, Miyashita T, Kubokawa K (1996) A weakly inward rectifying potassium channel of the salmon brain. Glutamate 179 in the second transmembrane domain is insufficient for strong rectification. J Biol Chem 271:15,729–15,735

Kubo Y, Reuveny E, Slesinger PA, Jan YN, Jan LY (1993b) Primary structure and functional expression of a rat G-protein-coupled muscarinic potassium channel. Nature 364:802–806

Kurachi Y (1995) G protein regulation of cardiac muscarinic potassium channel. Am J Physiol 269:C821–830

Larsson HP, Baker OS, Dhillon DS, Isacoff EY (1996) Transmembrane movement of the Shaker K^+ channel S4. Neuron 16:387–397

Larsson O, Ammala C, Bokvist K, Fredholm B, Rorsman P (1993) Stimulation of the KATP channel by ADP and diazoxide requires nucleotide hydrolysis in mouse pancreatic beta-cells. J Physiol 463:349–365

Lesage F, Guillemare E, Fink M, et al. (1995) Molecular properties of neuronal G-protein-activated inwardly rectifying K^+ channels. J Biol Chem 270:28,660–28,667

Lesage F, Guillemare E, Fink M, et al. (1996a) A pH-sensitive yeast outward rectifier K^+ channel with two pore domains and novel gating properties. J Biol Chem 271:4183–4187

Lesage F, Guillemare E, Fink M, et al. (1996b) TWIK-1, a ubiquitous human weakly inward rectifying K^+ channel with a novel structure. EMBO J 15:1004–1011

Li M, Jan YN, Jan LY (1992) Specification of subunit assembly by the hydrophilic amino-terminal domain of the Shaker potassium channel. Science 257:1225–1230

Li M, Unwin N, Stauffer KA, Jan YN, Jan LY (1994) Images of purified Shaker potassium channels. Curr Biol 4:110–115

Liman ER, Hess P, Weaver F, Koren G (1991) Voltage-sensing residues in the S4 region of a mammalian K^+ channel. Nature 353:752–756

Liman ER, Tytgat J, Hess P (1992) Subunit stoichiometry of a mammalian K^+ channel determined by construction of multimeric cDNAs. Neuron 9:861–871

Liu Y, Jurman ME, Yellen G (1996) Dynamic rearrangement of the outer mouth of a K^+ channel during gating. Neuron 16:859–867

Logothetis DE, Kurachi Y, Galper J, Neer EJ, Clapham DE (1987) The beta gamma subunits of GTP-binding proteins activate the muscarinic K^+ channel in heart. Nature 325:321–326

Logothetis DE, Movahedi S, Satler C, Lindpaintner K, Nadal Ginard B (1992) Incremental reductions of positive charge within the S4 region of a voltage-gated K^+ channel result in corresponding decreases in gating charge. Neuron 8:531–540

Lopatin AN, Makhina EN, Nichols CG (1994) Potassium channel block by cytoplasmic polyamines as the mechanism of intrinsic rectification. Nature 372:366–369

Lopez GA, Jan YN, Jan LY (1991) Hydrophobic substitution mutations in the S4 sequence alter voltage-dependent gating in Shaker K^+ channels. Neuron 7:327–336

Lopez GA, Jan YN, Jan LY (1994) Evidence that the S6 segment of the Shaker voltage-gated K^+ channel comprises part of the pore. Nature 367:179–182

Lopez-Barneo J, Hoshi T, Heinemann SH, Aldrich RW (1993) Effects of external cations and mutations in the pore rgion on C-type inactivation of Shaker potassium channels. Receptors and Channels 1:61–71

Luchian T, Dascal N, Sessauer C, Platzer D, Davidson N, Lester HA, Schreibmayer W (1997) A C-terminal peptide of the GIRK1 subunits directly blocks the G protein-activated K^+ channel (GIRK) expressed in Xenopus oocytes. J Physiol 505:13–22

Mannuzzu LM, Moronne MM, Isacoff EY (1996) Direct physical measure of conformational rearrangement underlying potassium channel gating. Science 271: 213–216

Mathur R, Zheng J, Yan Y, Sigworth FJ (1997) Role of the S3-S4 linker in Shaker potassium channel activation. J Gen Physiol 109:191–199

Matsuda H (1988) Open-state substructure of inwardly rectifying potassium channels revealed by magnesium block in guinea-pig heart cells. J Physiol 397:237–258

Matsuda H (1991) Effects of external and internal K^+ ions on magnesium block of inwardly rectifying K^+ channels in guinea-pig heart cells. J Physiol 435:83–99

Matsuda H, Matsuura H, Noma A (1989) Triple-barrel structure of inwardly rectifying K^+ channels revealed by Cs^+ and Rb^+ block in guinea-pig heart cells. J Physiol 413:139–157

Matsuda H, Saigusa A, Irisawa H (1987) Ohmic conductance through the inwardly rectifying K channel and blocking by internal Mg^{2+}. Nature 325:156–159

McDonald TV, Yu Z, Ming Z, et al. (1997) A minK-HERG complex regulates the cardiac potassium current I(Kr). Nature 388:289–292

McNicholas CM, Guggino WB, Schwiebert EM, Hebert SC, Giebisch G, Egan ME (1996) Sensitivity of a renal K^+ channel (ROMK2) to the inhibitory sulfonylurea compound glibenclamide is enhanced by coexpression with the ATP-binding cassette transporter cystic fibrosis transmembrane regulator. Proc Natl Acad Sci USA 93:8083–8088

Namba N, Inagaki N, Gonoi T, Seino Y, Seino S (1996) Kir2.2v: a possible negative regulator of the inwardly rectifying K^+ channel Kir2.2. FEBS Lett 386:211–214

Nichols CG, Lopatin AN (1997) Inward rectifier potassium channels. Annu Rev Physiol 59:171–191

Noma A (1983) ATP-regulated K^+ channels in cardiac muscle. Nature 305:147–148

Papazian DM, Schwarz TL, Tempel BL, Jan YN, Jan LY (1987) Cloning of genomic and complementary DNA from Shaker, a putative potassium channel gene from Drosophila. Science 237:749–753

Papazian DM, Timpe LC, Jan YN, Jan LY (1991) Alteration of voltage-dependence of Shaker potassium channel by mutations in the S4 sequence. Nature 349:305–310

Pascual JM, Shieh C-C, Kirsch GE, Brown AM (1995) K^+ pore structure revealed by reporter cysteins at inner and outer surfaces. Neuron 14:1055–1063

Planells Cases R, Ferrer Montiel AV, Patten CD, Montal M (1995) Mutation of conserved negatively charged residues in the S2 and S3 transmembrane segments of a mammalian K^+ channel selectively modulates channel gating. Proc Natl Acad Sci USA 92:9422–9426

Rettig J, Heinemann SH, Wunder F, et al. (1994) Inactivation properties of voltage-gated K^+ channels altered by presence of beta-subunit. Nature 369:289–294

Reuveny E, Slesinger PA, Inglese J, et al. (1994) Activation of the cloned muscarinic potassium channel by G protein beta gamma subunits. Nature 370:143–146
Roeper J, Sewing S, Zhang Y, Sommer T, Wanner SG, Pongs O (1998) NIP domain prevents N-type inactivation in voltage-gated potassium channels. Nature 391: 390–393
Sanguinetti MC, Curran ME, Zou A, et al. (1996) Coassembly of K(V)LQT1 and minK (IsK) proteins to form cardiac I(Ks) potassium channel. Nature 384:80–83
Schonherr R, Heinemann SH (1996) Molecular determinants for activation and inactivation of HERG, a human inward rectifier potassium channel. J Physiol 493: 635–642
Schreibmayer W, Dessauer CW, Vorobiov D, et al. (1996) Inhibition of an inwardly rectifying K^+ channel by G-protein alpha-subunits. Nature 380:624–627
Schwalbe RA, Wang Z, Bianchi L, Brown AM (1996) Novel sites of N-glycosylation in ROMK1 reveal potative pore-forming segment H5 as extracellular. J Biol Chem 271:24201–24206
Sheng M, Liao YJ, Jan YN, Jan LY (1993) Presynaptic A-current based on heteromultimeric K^+ channels detected in vivo. Nature 365:72–75
Shi G, Nakahira K, Hammond S, Rhodes KJ, Schechter LE, Trimmer JS (1996) Beta subunits promote K^+ channel surface expression through effects early in biosynthesis. Neuron 16:843–852
Shyng SL, Nichols CG ((1997) Octameric stoichiometry of the KATP channel complex. J Gen Physiol 110:655–664
Slesinger PA, Jan YN, Jan LY (1993) The S4-S5 loop contributes to the ion-selective pore of potassium channels. Neuron 11:739–749
Slesinger PA, Reuveny E, Jan YN, Jan L Y (1995) Identification of structural elements involved in G protein gating of the GIRK1 potassium channel. Neuron 15:1145–1156
Smith PL, Baukrowitz T, Yellen G (1996) The inward rectification mechanism of the HERG cardiac potassium channel. Nature 379:833–836
Stanfield PR, Davies NW, Shelton PA, et al. (1994) A single aspartate residue is involved in both intrinsic gating and blockage by Mg^{2+} of the inward rectifier, IRK1. J Physiol 478:1–6
Starkus JG, Kuschel L, Rayer MD, Heinemann SH (1997) Ion conduction through C-type inactivated Shaker channels. J Gen Physiol 110:539–550
Taglialatela M, Wible BA, Caporaso R, Brown AM (1994) Specification of pore properties by the carboxyl terminus of inwardly rectifying K^+ channels. Science 264:844–847
Takumi T, Ohkubo H, Nakanishi S (1988) Cloning of a membrane protein that induces a slow voltage-gated potassium current. Science 242:1042–1045
Tucker SJ, Gribble FM, Zhao C, Trapp S, Ashcroft FM (1997) Truncation of Kir6.2 produces ATP-sensitive K^+ channels in the absence of the sulphonylurea receptor. Nature 387:179–183
Unwin, N. (1995) Acetylcholine receptor channel imaged in the open state. Nature 373:37–43
Varnum MD, Busch AE, Bond CT, Maylie J, Adelman JP (1993) The min K channel underlies the cardiac potassium current IKs and mediates species-specific responses to protein kinase C. Proc Natl Acad Sci USA 90:11,528–11,532
Vandenberg CA (1987) Inward rectification of a potassium channel in cardiac ventricular cells depends on internal magnesium ions. Proc Natl Acad Sci USA 84:2560–2564
Velimirovic BM, Gordon EA, Lim NF, Navarro B, Clapham DE (1996) The K^+ channel inward rectifier subunits form a channel similar to neuronal G protein-gated K^+ channel. FEBS Lett 379:31–37
Wang KW, Tai KK, Goldstein SA (1996) MinK residues line a potassium channel pore. Neuron 16:571–577
Wei A, Jegla T, Salkoff L (1996) Eight potassium channel families revealed by the C. elegans genome project. Neurophermacology 35:805–829

Wible BA, Taglialatela M, Ficker E, Brown AM (1994) Gating of inwardly rectifying K^+ channels localized to a single negatively charged residue. Nature 371:246–249

Wischmeyer E, Karschin A (1996) Receptor stimulation causes slow inhibition of IRK1 inwardly rectifying K^+ channels by direct protein kinase A-mediated phosphorylation. Proc Natl Acad Sci USA 93:5819–5823

Yang J, Jan YN, Jan LY (1995) Control of rectification and permeation by residues in two distinct domains in an inward rectifier K^+ channel. Neuron 14:1047–1054

Yang J, Jan YN, Jan LY (1995) Determination of the subunit stoichiometry of an inwardly rectifying potassium channel. Neuron 15:1441–1447

Yellen G, Jurman ME, Abramson T, MacKinnon R (1991) Mutations affecting internal TEA blockade identify the probable pore-forming region of a K^+ channel. Science 251:939–942

Yool AJ, Schwarz TL (1991) Alteration of ionic selectivity of a K^+ channel by mutation of the H5 region. Nature 349:700–704

CHAPTER 7

Pharmacology of Voltage-Gated Potassium Channels

O. Pongs and C. Legros

A. Introduction

Voltage-gated potassium (Kv) channels play an important role in many cellular functions correlated with changes in excitability. Their functions range from the setting of basal levels of membrane potential to shaping action potentials in excitable cells (Papazian et al. 1987). About a decade ago, the first Kv channel cDNAs were cloned from *Drosophila* (Kamb et al. 1987; Pongs et al. 1988; Tempel et al. 1988) and mammals (Stühmer et al. 1988, 1989). The cDNAs encoded -in comparison to the then known Ca- and Na-channel α-subunits- considerably smaller protein sequences. Their analysis showed that Kv α-subunits contain a membrane-spanning core region with six hydrophobic transmembrane segments (S1–S6) flanked by hydrophilic amino- and carboxyterminal sequences. They are exposed to the cytoplasmic side of the plasma membrane. Subsequently, it was shown that four Kvα-subunits make up a functional Kv channel (see Fig. 1) (MacKinnon 1991). Assembly of the four subunits may occur in the form of oligo- or heteromultimers (Isacoff et al. 1990; Ruppersberg et al. 1990; Christie et al. 1990). In fact, assembly of Kvα-subunits to heteromultimers appears to be a wide spread phenomenon in eukaryotic cells, making it difficult to correlate native potassium outward-currents with cloned Kv channel subunits. Nevertheless, heterologous expression studies with cloned Kv channel subunits have shown that the various homo- and heteromultimeric Kv channel assemblies may mediate the whole spectrum of rapidly- to non-inactivating outward currents observed in physiological studies.

Certain Kv channels may also contain accessory Kvβ subunits (Rettig et al. 1994). Kvβ subunits are tightly associated with *Shaker* like Kvα subunits (Rehm et al. 1988; Parcej et al. 1989). Most likely, the corresponding native Kv channels are composed of four Kvα and four Kvβ subunits. It is, however, not known if all Kv channels contain both Kvα and Kvβ-subunits. The presence of Kvβ subunits can modulate Kv channel gating (Rettig et al. 1994; Heinemann et al. 1996); in particular it may confer rapid inactivation on otherwise non-inactivating Kv channels. Thus, the possible combinations of Kvα- and Kvβ-subunits are potentially a rich source to generate a highly diverse Kv channel population in excitable and non-excitable cells. Most likely, the diverse Kv channels exhibit not only diverse electrophysiological properties, but also

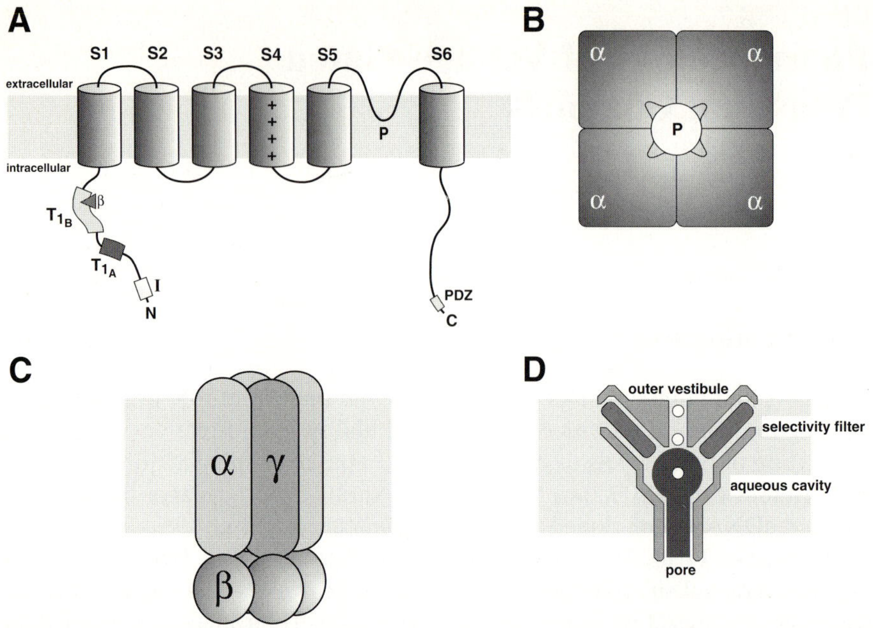

Fig. 1. A Putative topology of one α-subunit of voltage-gated potassium channels. The membrane inserted core region comprises six hydrophobic segments depicted as cylinders and a P-domain which enters and exits the membrane from the extracellular side. The amino-terminus (*N*) may contain an inactivating domain (*I*), a tetramerization or assembly domain (T_1A and T_1B), a Kvβ-subunit binding domain (*arrow*); the carboxy-terminus (*C*) may contain a PDZ-binding domain (*PDZ*). **B** Four Kvα subunits assemble as tetramer. Most likely, four P-domains (*loops*) contribute to the conducting path and the outer mouth of the pore (*P*) which is viewed from top. **C,D** Profile view of the organisation of the *Streptomyces lividans* K^+ channel. Major functional aspect deduced from the 3.2 Å resolution crystal structure of the *S. lividans* K channel. Schematic diagram modified from Doyle et al. (1998)

distinct pharmacologies. Unfortunately, however, only a few peptide toxins are known to target a few distinct Kv channels. Therefore, the finding of new Kv channel toxins with different specificities would greatly aid the study of Kv channel diversity and its physiological implications.

B. Molecular and Functional Organization of the Voltage-Gated Potassium Channels

I. Structural Domains in Kvα-Subunits

All Kvα subunits appear to have a common secondary structure (Chandy and Gutman 1994). It comprises a cytoplasmic amino-terminus that can vary in length from a few tens to a few hundred amino acids. Also, the cytoplasmic carboxy terminus of Kvα subunits is of similarly variable length. By contrast, the membrane-integrated core domain has a quite constant length of ~200 amino acids comprising six membrane-spanning segments (S1–S6) and a hydrophobic pore (P) forming loop between S5 and S6. The P-loop enters and exits the lipid bilayer from the extracellular side (see Fig. 1A). Apparently, this membrane topology is a hallmark of Kv α-subunits. Domains involved in Kvα subunit assembly and in Kvβ subunit binding have been allocated to cytoplasmic domains. In the *Shaker* Kv family a cytoplasmic amino-terminal domain directing Kvα-subunit tetramerization (T-domain) has been well defined (Li et al. 1992; Shen and Pfaffinger 1995). T-domains may specify the specificity of Kvα subunit binding. Also, certain T-domains specify the association of Kvα- and Kvβ-subunits (Yu et al. 1996; Sewing et al. 1996). For *eag*-type Kv channels a carboxy-terminal domain has been shown to be important for assembly of functional channels. The core domain contains voltage-sensor and pore-forming sequences/amino acids, respectively. The occurrence of positively charged amino acids with a characteristic arrangement, e.g., (Lys/Arg-X-Y)$_n$ in segment S4 may contribute to the gating charge movements underlying the voltage-sensitive gating of Kv channels (Papazian et al. 1991). It has been proposed that the voltage-sensing gating machinery may involve additional amino acids of the core region, in particular some in segment S2 (Seoh et al. 1996). Note that many more Kv channel mutants have been found which suggest the involvement of additional structural elements in Kv channel gating.

The pore of Kv channels appears to be lined by amino acids of the P-region (Yool et al. 1991; Hartmann et al. 1991), the S4-S5 linker (Durell and Guy 1996; Gómez et al. 1997), and S6 (Lopez et al. 1994). The data were obtained by a tremendous amount of elegant work combining biochemical, structural, and electrophysiological experiments. Most of this work has been obtained with cloned *Shaker* Kv channels. It is believed that other Kv channels contain structurally homologous domains. Typically, the P-region contains a K channel signature sequence (GYGD or GFGN) (Heginbotham et al. 1994). This sequence is a hallmark of K channels and probably forms an essential part of the Kv channel pore (Durell and Guy 1996). It is likely that the signature sequence contributes to an important potassium ion binding site in the Kv channel pore and may contribute to the selectivity filter of Kv channels (Heginbotham et al. 1994).

Recently, the crystal structure of a bacterial K channel protein (KcsA) has been solved to a 3.2 Å resolution (Doyle et al. 1998). The results of this landmark work are in excellent agreement with the predictions made by the in vitro mutagenesis studies on cloned Kv channel subunits. The crystal structure shows that the outer mouth of the KcsA-channel and the selectivity filter is indeed formed by residues and backbone of the P-loop. Then the KcsA channels pore widens to an inner "lake" before it narrows down again to a funnel-like structure which is made up by KcsA residues that are equivalent to S6 residues in Kv channel subunits (Fig. 1D). As the original paper, which describes the KcsA crystal structure, also gives an excellent discussion of the implications of this structure for potassium ion permeation and K channel selectivity, we refer the reader to this landmark article for further details.

Following activation, many Kv channels inactivate with widely differing inactivation time courses ranging from a few ms to seconds. Two major types of inactivation have been discerned: C-type and N-type (HOSHI et al. 1990; CHOI et al. 1991). C-type inactivation has been correlated with carboxy (C) terminal amino acids, in particular in S6; N-type inactivation has been correlated with an amino (N) terminal domain present in some Kvα and Kvβ subunits. The N-type inactivating domain is referred to as the *ball* domain as its function is reminiscent of a ball tethered to a chain. The ball domain can bind to a receptor near or at the pore and thereby rapidly close the open Kv channel pore. In addition to inactivating domains, an additional domain at the aminoterminus of Kv1.6 channels has been discovered recently (ROEPER et al. 1998). This domain is able to neutralize the activity of the ball domain by an as yet unknown mechanism.

Kv α-subunits are members of a large gene superfamily (Fig. 2). To date, three major Kv channel branches can be discerned in this superfamily: (i) the *Shaker* branch related to the *Drosophila Shaker* gene; (ii) the KCN branch related to the *Drosophila eag* gene; (iii) the KCNQ branch related to the human KvLQT1 (KCNQ1) gene. The *Shaker* gene family presently consists of nine subfamilies. Each subfamily comprises Kvα-subunits encoded in three to nine different genes. Members of the *Shaker* superfamily are designated as Kvm.n, where m refers to the subfamily and n to the member within the given subfamily. For example, Kv1.1 would be member 1 in the *Shaker*-related Kv channel subfamily 1. The KCNB-branch does not have yet a formalized nomenclature. Lately, it has become clear that this branch includes genes related to the Drosophila *ether-à-go-go* gene (eag, erg, elk), but also ones encoding pacemaker and cyclic-nucleotide gated channels (LUDWIG et al. 1998). Finally, the relatively new KCNQ-branch has four members so far, which appear to belong to the same subfamily (CHARLIER et al. 1998). I will only discuss the *Shaker*-related Kv channels, since their structure, function and pharmacology has been studied in considerable detail. (SALINAS et al.).

It has been shown in vitro and in vivo that many Kvα subunits (ISACOFF et al. 1990; RUPPERSBERG et al. 1990; CHRISTIE et al. 1990) are able to assemble not only with themselves, but also with homologous subunits. Thus, Kv channels may be heteromultimers comprising various combinations of Kvα subunits. In fact, the theoretically possible combinations of Kvα subunits in Kv channels would make it possible that each neuron expresses an individual set of distinct Kv channels according to its needs. Extensive biochemical and immunocytochemical work has shown for the Kv1α subunit family that members of this family heteromultimerize with each other extensively, but not with members of other Kv subfamilies (e.g., Kv2, Kv3). How general this kind of restriction in the assembly of Kvα-subunits is, is not known. Recent data have shown that Kv2 α-subunits may assembly with members of Kv5-Kv9 subfamilies (SALINAS et al. 1997; PATEL et al. 1997; POST et al. 1996). Kv5-Kv9

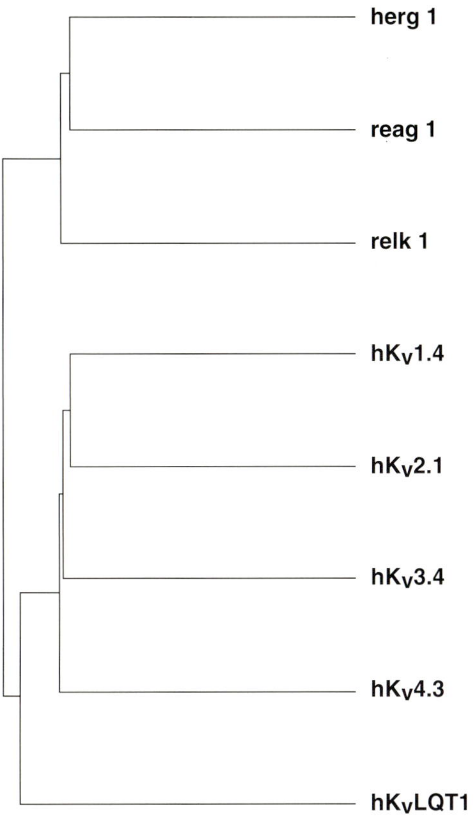

Fig. 2. Dendrogram of representative members of the superfamily of voltage-sensitive ion channels. The horizontal branch lengths are inversely proportional to the similarity between the sequences. The dendrogram was derived from an unweighted pair-group method using arithmetic averages comparison of the S1-S6 domains of the channels indicated at right using PILEUP multiple sequence analysis program (GCG)

subunits may be referred to as Kvγsubunits. Kvγsubunits do not express functional Kv channels by themselves, but only in conjunction with Kvα-subunits, e.g., Kv2.1 and Kv2.2 subunits.

II. Modulatory Kvβ-Subunits

K-channel α-subunits are often coassembled with auxiliary β-subunits (Rehm and Lazdunski 1988; Parcej and Dolly 1989; Rettig et al. 1994; Heinemann et al.). Two types of Kvβ-subunits have been discovered – membrane-integrated β-subunits (Hanner et al. 1997) and cytoplasmic β-subunits (Fig. 1C) (Rettig et al. 1994). Kvβ-subunits are apparently not integral membrane proteins. The Kvβ-subunits may function as chaperons aiding assembly and/or transport to the plasma membrane (Shi et al. 1996). They may facilitate activation. Kvβ1 and Kvβ3-subunits contain N-type inactivating domains which confer rapid inactivation to otherwise non-inactivating Kv channels. Thus, the modal gating of Kv channels may not only depend on α-subunit composition, but also on the β-subunits. It is not known whether Kvβ subunits influence the pharmacology of Kv channels. In the cases which we have studied, we did not find a significant influence of Kvβ subunits on toxin binding to *Shaker* type Kv channels (O. Pongs, unpublished experiments). However, this may need further investigation. Also, whether Kvβ subunits are influential or not may depend on the Kv channel structures which are involved in toxin binding.

C. Peptide Toxin Binding Sites

Although a three-dimensional structure of Kv channels is not known, it is safe to assume that only a few sequences of Kvα subunits are exposed to the extracellular space (see Fig. 1). Sequences between segments S1 and S2, S3 and S4, and S5 and S6 are most likely those that face the extracellular space. Most Kvα-subunits are glycosylated between S1 and S2. This may further restrict potential peptide toxin binding sites. The remaining S3/S4 and S5/S6 loop sequences contribute to the voltage-activated gating machinery (TANG and PAPAZIAN 1997) and, respectively, to the Kv channel pore (YOOL and SCHWARZ 1991; HARTMANN et al. 1991; HEGINBOTHAM et al. 1994). Therefore, it may be expected that the binding of peptide toxins, which are not membrane permeable but bind to extracellular Kvα-subunit domains, may have two effects. The toxin may bind to S5/S6 loop sequences and thereby block the Kv channel pore from the outside. Alternatively, the toxin may bind to the S3/S4 loop sequence and modify the voltage-dependent activation of Kv channels.

I. Scorpion Toxins

Polypeptides from scorpion venoms constitute a large family of basic globular miniproteins sharing a common scaffold called α/β, which consist in an α-helix cross-linked to a double stranded β-sheet by two disulfide bridges (Fig. 3) (BONTEMS et al. 1991a,b). More than 40 sequences of scorpion toxin which act on voltage-gated and calcium-activated K channels have been described (Fig. 4). These peptides contain 29–39 amino acid residues and 3 or 4 disulfide bridges. Despite the nomenclature suggested by MILLER (1995), the classification of the K channel-blocking scorpion peptides remains discussed. They contain a highly conserved motif corresponding to the interacting surface with voltage-gated K channels, G_{26}-$[*]_{26a}$-K_{27}-C_{28}-$(M/I)_{29}$- $(N/G)_{30}$-X_{31}-K_{32}-C_{33}-$(n)_{34}$-C_{35} ([*] corresponds to a deletion or to small residue, ñ is a charged residue) (MILLER 1995).

Scorpion toxins, e.g., agitoxin (GROSS and MACKINNON 1996), charybdotoxin (GOLSTEIN et al. 1994), and kaliotoxin (AIYAR et al. 1995), block the Kv channel pore. The interaction of peptide toxins with the entry of the vestibule

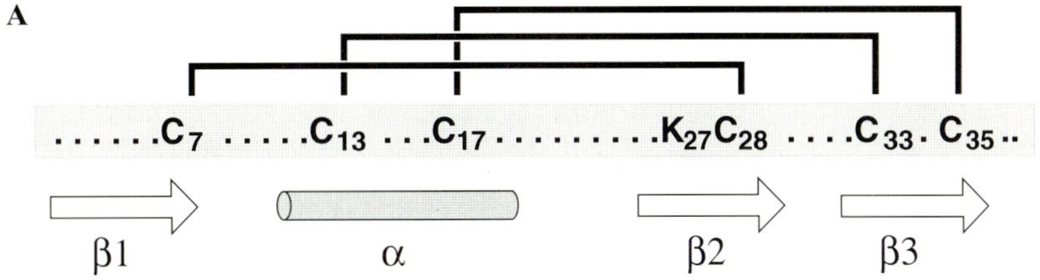

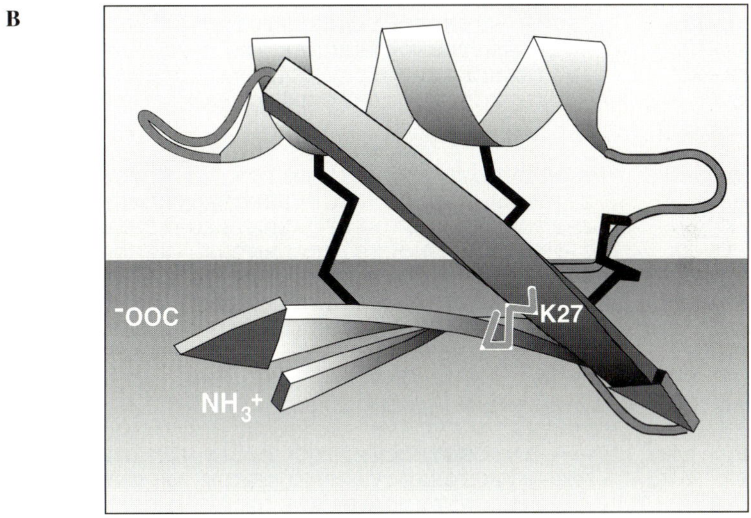

Fig. 3. **A** Schematic consensus sequence of charybdotoxin-like potassium-channel blocking peptide family (*C*, cysteine; *K*, Lysine). Disulfide bridges are indicated by *lines*. Conserved areas of secondary structure are indicated under the sequence by *arrows* for β-sheet and a *grey cylinder* for α-helix (α). **B** Schematic representation of α-KTx1.1 backbone fold. Disulfide bridges are sandwiched between an α-helix on top and a β-sheet at the bottom. NH^{+}_{3} and COO^{-} indicates amino- and carboxy-terminal end, respectively. Critical K27, which most likely projects into the potassium channel pore, is explicitly indicated. The structure has been adopted from Miller (1995) and Bontems et al. (1991a,b)

of *Shaker* type Kv channel pores has been studied in great detail. Mutational studies have mapped the binding sites to amino acids between S5 and S6, i.e., at or near the extracellular entrance of the Kv channel pore (Goldstein et al. 1993, 1994; Aiyar et al. 1995). Conversely, widespread point mutations of the toxins have identified Lys-27 as a residue which is very important for toxin affinity (Goldstein et al. 1993, 1994; Aiyar et al. 1995). This residue behaves as if 20% of the transmembrane potential field affect its interaction with the Kv channel pore. This suggested that Lys-27 comprises a crucial positively

A

```
KTX        GVEINVKCSGSPQCLKPCKDA-GMRFG-KCMNRKCHCTPK
KTX2       -VRIPVSCKHSGQCLKPCKDA-GMRFG-KCMNGKCDCTPK
BmKTX      -VGINKSCKHSGQCLKPCKDA-GMRFG-KCINGKCDCTPK*
AgTX1      GVPINVKCTGSPQCLKPCKDA-GMRFG-KCINGKCHCTPK
AgTX2      GVPINVSCTGSPQCIKPCKDA-GMRFG-KCMNRKCHCTPK
AgTX3      GVPINVPCTGSPQCIKPCKDA-GMRFG-KCMNRKCHCTPK
OsK-1      GVIINVKCKISRQCLEPCKKA-GMRFG-KCMNRKCHCTPK

MgTX       -TIINVKCTSPKQCLPPCKAQFGQSAGAKCMNGKCKCYPH
HgTX1      -TVIDVKCTSPKQCLPPCKAQFGIRAGAKCMNGKCKCYPH
NTX        -TIINVKCTSPKQCSKPCKELYGSSAGAKCMNGKCKCYNN*
PiTX-Kα    -TI---SCTNPKQCYPHCKKETGYPN-AKCMNRKCKCFGR
PiTX-Kβ    -TI---SCTNEKQCYPHCKKETGYPN-AKCMNRKCKCFGR
TsTX-Kα    -VFINAKCRGSPECLPKCKEAIGKAA-GKCMNGKCKCYP
ClTX       -ITINVKCTSPQQCLRPCKDRFGQHAGGKCINGKCKCYP

IbTX       Z-FTDVDCSVSKECWSVCKDLFGVDRG-KCMGKKCRCYQ
Lq2        Z-FTQESCTASNQCWSICKRLHNTNRG-KCMNKKCRCYS
ChTX       Z-FTNVSCTTSKECWSVCQRLHNTSRG-KCMNKKCRCYS
Lq 18-2    Z-FTQESCTASNQCWSICKRLHNTNRG-KCMNKKCRCYS
Lq 15-1    G-LIDVRCYDSRQCWIACKKVTGSTQG-KCQNKQCRCY
BmTX1      Z-FTDVKCTGSKQCWPVCKQMFGKPNG-KCMNGKCRCYS
BmTX2      Z-FTNVKCTASKQCWPVCKKLFGTYRG-KCMNSKCRCYS
```

B

```
Pi1        ----LVKCRGTSDCGRPCQQQTGCPN-SKCINRMCKCYGC
MTX        -----VSCTGSKDCYAPCRKQTGCPN-AKCINKSCKCYGC*
HsTX1      -----ASCRTPKDCADPCRKETGCPYG-KCMNRKCKCNRC
```

Fig. 4. A Shorter toxins cross-linked by three disulfide bridges: KTX, Kaliotoxin from *Androctonus mauretanicus mauretanicus* (CREST et al. 1992); KTX2, Kaliotoxin 2 from *Androctonus australis* (LARABA-DJEBARI et al. 1994); BmKTX, Kaliotoxin from *Buthus martenzi* (ROMI-LEBRUN et al. 1997); AgTX1, AgTX2, AgTX3, Agitoxin, 2 and 3 from *Leiurus quinquestriatus hebraeus* (GARCIA et al. 1994); OsK-1 from *Orthochirus scrobiculosus* (GRISHIN et al. 1996); Ts TX-Kα from *Tityus serrulatus* (ROGOWSKI et al. 1994); MgTX, Margatoxin from *Centruroides margaritus* (GARCIA-CALVO et al. 1993); HgTX1. Hongotoxin-1 from *Centruroides limbatus* (KOSCHAK et al. 1998); NTX, Noxiustoxin from *Centruroides noxius* (POSSANI et al. 1982); PiTX-Kα and PiTX-Kβ from *Pandinus imperator* (ROGOWKI et al. 1996); CITX from *Centruroides limpidus limpidus* (MARTIN et al.); IbTX, Iberiotoxin from *Buthus tamulus* (GALVEZ et al. 1990); ChTX, Charybdotoxin, Lq2, and Lq18–2 or Lq15–1 from *Leiurus quinquestriatus hebraeus*, respectively (GIMENEZ-GALLEGO et al. 1988; LUCCHESI et al. 1989; HARVEY et al. 1995); BmTX1 and BmTX2 from *Buthus martenzi* (ROMI-LEBRUN et al. 1997). *, amidated C-terminal natural toxin. **B** Shorter toxin cross-linked by four disulfide bridges: Pi 1 from *Pandinus imperator* (OLAMENDI-PORTUGAL et al. 1996); MTX, Maurotoxin from *Scorpio maurus palmatus* (KHARRAT et al. 1997); HsTX1 from *Heterometrus spinnifer* (LEBRUN et al. 1997). Extra half-cystine are *underlined* charged amino-acid chain, protruding from the toxin's surface. Thereby, Lys-27 may plug the Kv channel pore as a tethered surrogate potassium ion (Fig. 5), which enters, but cannot pass through the conduction path. Several experimental observations support this idea. Replacement of Lys-27 with Gln or Asn weakens charybdotoxin and kaliotoxin affinities by several orders of magnitude (GOLDSTEIN et al. 1994; AIYAR et al. 1995); neutral substitutions of Lys-27 make the toxin block insensitive to applied voltage and to internal K^+ (GOLDSTEIN et al. 1993; HIDALGO and MACKINNON 1995).

The specificity of toxin-Kv channel pore interactions could be provided by additional interactions between the surface of the toxin and the amino-acid

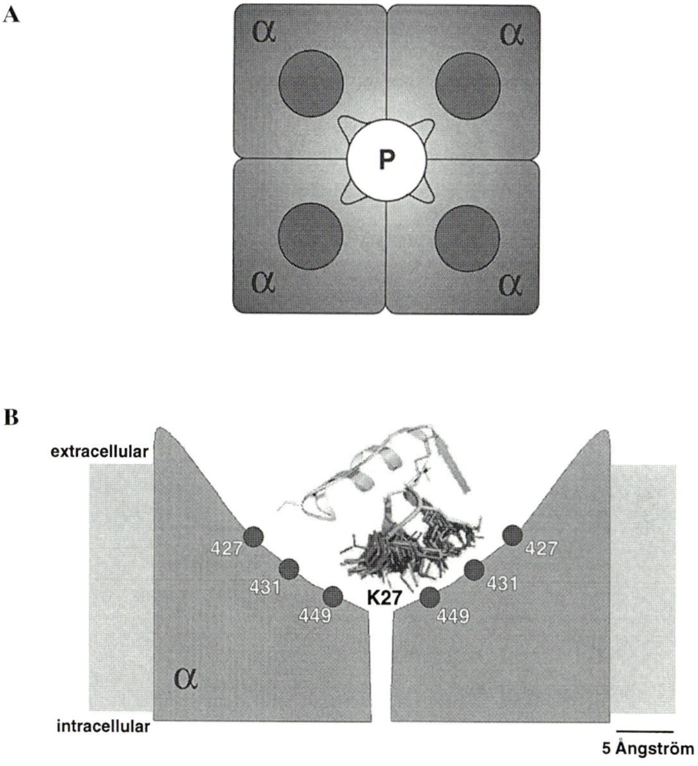

Fig. 5. A Diagram illustrating feature of pore blocking toxins. Shown is a side view with two Kvα subunits forming half a pore. The toxin binds to the surface of the outer mouth of the pore diagrammed as a wide vestibule. A positively charged lysine residue (K27) of respective peptide toxins interacts with the conducting path and thereby occludes the pore. **B** Diagram illustrating the topology of the extracellular vestibule of the *Shaker* probe by ChTX. Modified from NAINI and MILLER (1996). Individual structures of ChTX with certain residues are shown on the top. Note the position of the K27 in the entryway of the pore. Closed contact of certain residues of the *Shaker* (*427*, *431*, and *449*) are pointed out and positioned according to the electrostatic compliance analysis (NAINI and MILLER 1996)

residue neighboring the pore (Fig. 5). Interacting surfaces have been studied through thermodynamic mutant cycles in order to establish the toxin spatial arrangement with respect to the Kv channel (AIYAR et al. 1995; HIDALGO and KINNON 1995; NARANJO and MILLER 1996). The extensive mutational data that has been accumulated on binding of toxins to Kv channel pores has led to propose detailed molecular models for the spatial arrangement of interacting residues of toxin and Kv channel pore (LIPKIND and FOZZARD 1997). Note, however, that alternative spatial models and interaction schemes have been proposed. In part, this reflects some discrepancies between available experimental data and/or different assumptions and estimates of atomic distances in the outer vestibule of Kv channel pores. In particular, the 1:1 stoichiometry of toxin binding to the outer vestibule does not necessarily infer a symmetrical interaction (GROSS and MACKINNON 1996). Also, this interaction may induce conformational changes comparable to the ones observed during channel opening and closing (KROVETZ et al. 1991). Yet there is a fair agreement that the vestibule is ~10 Å deep and 25–30 Å wide and that this rather wide vestibule serves as an interaction surface for toxin binding.

Recently, it has been shown that replacement of only three residues in the pore vestibule of KcsA channels, which are located in homologous positions to the ones important for scorpion toxin binding in *Shaker* channels, enables the KcsA channel to bind agitoxin 2 (MACKINNONN et al. 1998). This result is in good agreement with the previous topological studies on *Shaker* channel-toxin interaction.

II. Snake Toxins

Two families of snake toxins have been characterized as blocking peptides of K channels, both from mamba venoms: the dendrotoxin family, which facilitate transmitter release from nerve terminals and a second family composed of polypeptides related to phospholipase A2, which are responsible for muscle paralysis (HARVEY and ANDERSON 1985). The dendrotoxins constitute a family of basic peptides of 57–61 amino acid residues, reticulated with 3 disulfide bridges, and structurally related to the Kunitz family of protease inhibitors (Fig. 6) (FORAY et al. 1993; LANCELIN et al. 1994). α-Dendrotoxins inhibit specifically Kv1.1, Kv1.2, and Kv1.6 with nanomolar affinity (GRUPPE et al. 1990; GRISSMER et al. 1994). α-Dendrotoxin was instrumental for the purification of mammalian *Shaker* type Kv channels (PARCEJ et al. 1992; REID et al. 1992). The use of dendrotoxin as a Kv channel ligand has aided the characterization of the heterooligomeric composition of native Kv channels (SCOTT et al. 1994).

The dendrotoxin binding site on Kv channels is mainly composed of amino acid residues located between transmembrane segments S5 and S6 (Ala352, Glu353, and Tyr379 of the Kv1.1) (HURST et al. 1991). It has been proposed that dendrotoxin binds near the vestibule of the Kv channel pore and then occludes the pore by steric hindrance. In addition, through-space electrostatic interactions may stabilize toxin binding (HURST et al. 1991; TYTGAT et al. 1995).

A

DTX_I	ZPLRKLCILHRNPGRCYQKIPAFYYNQKKKQCEGFTWSGCGGNSNRFKTIEECRRTCIRK
DTX_K	-GAAKYCKLPLRIGPCKRKIPSFYYKWKAKQCLPFDYSGCGGNANRFKTIEECRRTCVG
DTX_E	LQHRTFCKLPAEPGPCKASIPAFYYNWAAKKCQLFHYGGCKGNANRFSTIEKCRHACVG
DTXα	ZPRRKLCILHRNPGRCYDKIPAFYYNQKKKQCERFDWSGCGGNSNRFKTIEECRRTCIG
DTXβ	--RPYACELIVAAGPCMFFISAFYYSKGANKCYPFTYSGCRGNANRFKTIEECRRTCVV
DTXγ	-----------------LPAEFGRQFNSFYXCLPFLFSGCGGXAXXFQTIGECR-----
DTXδ	--AAKYCKLPVRYGPCKKKIPSFYYKWKAKQCLPFDYSGCGGNANRFKTIEECRRTCVG

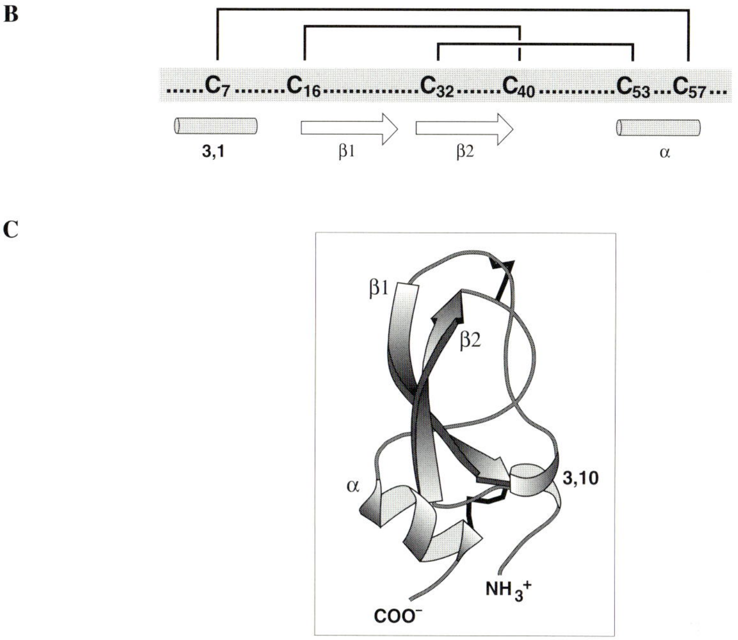

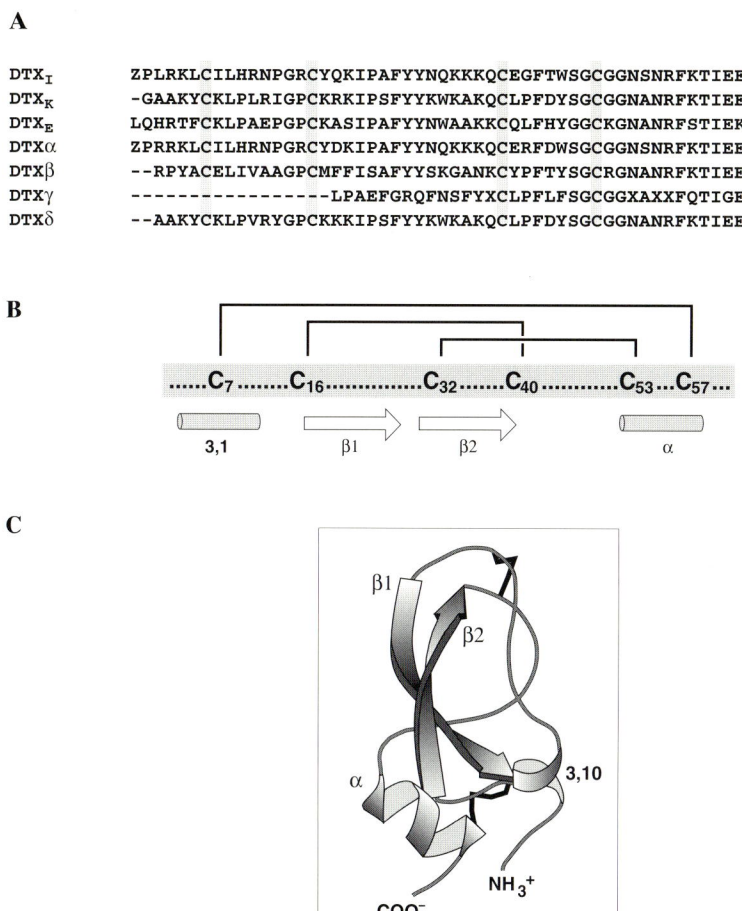

Fig. 6. **A** Alignment of amino acid sequences of snake toxins which block voltage-gated potassium channels. DTX_I, DTX_K and DTX_E, from *Dendroaspis polylepis polylepis* (black mamba) (Anderson and Harvey 1985). DTXα, DTXβ, DTXγ, and DTXδ from *Dendroaspis angusticeps* (green mamba) (Benishin et al. 1988). *X* indicates an identified residue and – indicates that no sequence is available. **B** Schematic representation of dendrotoxin I backbone fold. Secondary structures are marked: *β1*, *β2*, *β*-sheet; *α*, *α*-helix, *3, 10*, 3, 10-helix. NH_3^+ and COO^- indicate amino- and carboxy-terminal ends, respectively. The structures has been adopted from Lancelin et al. (1994)

Most likely, all four Kvα-subunits contribute to the toxin binding site with an energy additivity feature (Tytgat et al. 1995). Point mutagenesis studies showed that β-turn and 3_{10}-helix domains constitute the interaction surface of dendrotoxin with its Kv channel receptor (Fig. 6B) (Smith et al. 1997). Further studies may provide a detailed picture of the dendrotoxin residues that play a key role for binding to Kv channels.

III. Sea Anemone Toxins

Two families of sea anemone peptides that are able to bind with high affinities to Kv channels have been described (Fig. 7A,B). K channels blocking peptides from sea anemone venoms have been identified on the basis of competition experiments with dendrotoxin (KARLSSON et al. 1991). Kalicludines form a class of peptides structurally homologous both to dendrotoxins and to Kunitz inhibitor (SCHWEITZ et al. 1995). Interestingly, they block Kv1.2 channels and also inhibit trypsin. Therefore, it was suggested that Kalicludines may constitute an evolutionary link between Kv channel

A

```
BgK     -VCRDWFKETACRHAKSLGNCRTSQKYRAN-CAKTCELC
ShK     RSCIDTIPKSRC----TAFQCKHSMKYRLSFCRKTCGTC
AsKS    -ACKDNFAAATCKHTKENKNC-GSQKYATN-CAKTCGKC
HmK     RTCKDLIPVSEC----TDIRCRTSMKYRLNLCRKTCGSC
```

B

```
DTX-I   ZPLRKLCILHRNPGRCYDKIPAFYYNQKKKQCEGFTWSGCGGNSNRFKTIEECRRTCIRK
AsKC1   --INKDCLLPMDVGRCRASHPRYYYNSSSKRCEKFIYGGCRGNANNFHTLEECEKVCGVR
AsKC2   --INKDCLLPMDVGRCRARHPRYYYNSSSRRCEKFIYGGCRGNANNFITKKECEKVCGVR
AsKC3   --INGDCELPKVVGRCRARFPRYYYNLSSRRCEKFIYGGCGGNANNFHTLEECEKVCGVRS
BPTI    --RPDFCLEPPYTGPCKARIIRYFYNAKAGLCQTFVYGGCRKRNNKSAEDCMRTCGCA
```

C

```
BDS-I   AAPCFCSGKPGRGDLWILRGTCPGGYGYTSNCYKWPNICCYPH
BDS-II  AAPCFCPGKPDRGDLWILRGTCPGGYGYTSNCYKWPNICCYPH
```

D

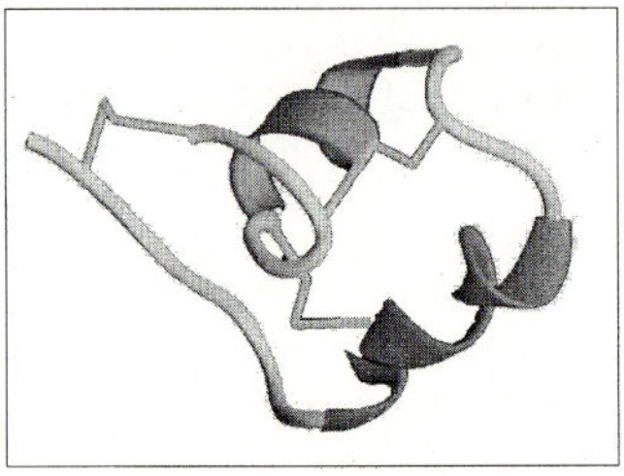

Fig. 7A–D. Alignment of amino acid sequences from Sea anemone toxins which block voltage-gated K channels. **A** BgK, from *Bunodosoma granulifera*, ShK, from *Stichodactyla helianthus*, (TUDOR et al. 1996) AsKS from *Anemonia sulcata* (SCHWEITZ et al. 1995), HmK from *Heteractis magnifica* (GENDEH et al. 1997). **B** DTXI from the black mamba, *Dendrosapis polylepis polylepis* (HARVEY and ANDERSON 1985), AsKC1, AsKC2 and AsKC3, kalicludines from *Anemonia sulcata* (SCHWEITZ et al. 1995), BPTI, bovin peptide trypsin inhibitor. **C** BDS-I and BDS-II, blood depressing substance, from *Anemonia sulcata* (DIOCHOT et al. 1998). **D** Ribbon representation of the structure of BgK resolved by NMR. The globular architecture contains two perpendicular α-helix stabilized with three disulfide bridges (DAUPLAIS et al. 1997)

toxins and protease inhibitors of the Kunitz type inhibitor (Schweitz et al. 1995).

The second family of sea anemone toxins is composed of short basic peptides of 35–37 amino acid residues reticulated with 3 disulfide bridges (Fig. 7B,D) (Karlsson et al. 1991; Schweitz et al. 1995; Tudor et al. 1996; Gendeh et al. 1997). All of them are able to compete with dendrotoxin in binding assays with rat brain synaptosomes. In vitro they block Kv1.1 and Kv1.2 channels (Schweitz et al. 1995; Gendeh et al. 1997). The toxin ShK presents a high affinity blocker for Kv1.3 channels (Pennington et al. 1995). Mapping studies of the residues important for ShK toxicity showed that the crucial ShK residues are functionally equivalent to those shown to be important for charybdotoxin binding (Dauplais et al. 1997). Again, a critical lysine residue in a protruding position on a flat surface, associated to the aromatic ring of a tyrosine residue, is correlated with the pore blocking toxin activity. This may suggest a convergent evolution of toxins with a tethered lysine residue for plugging the vestibule of Kv channels.

Lately, toxins have been isolated from *Anemonia sulcata* venom. Strikingly, the new toxins block specifically Kv3 channels (Diochot et al. 1998). In vitro expressed Kv3.4. channels are blocked with an IC_{50} of 47 nmol/l. The toxins BDS-I and BDS-II are composed of 43 amino acid residues and are reticulated with three disulfide bridges (Fig. 7C). They form a new group of K channels blocking peptides, structurally close to sea anemone toxins that block Na channel (Driscoll et al. 1989).

IV. Snail Toxins

The predatory marine snails of the *Conus* genus present a source of Kv channel toxins (Olivera et al. 1990). The κ-conotoxin PVIIA, contained in the "fin-popping" fraction of the venom from *C. purpurascens*, is a 27 amino acid residue basic peptide reticulated with three disulfide bridges, which inhibits *Shaker* type Kv channels (Fig. 8A) (Terlau et al. 1996). Two-dimensional NMR studies have shown that this conotoxin comprises two large parallel loops stabilized by a triple-stranded anti-parallel β-sheet and three disulfide bridges. This structural fold is similar to the ones found with the other conotoxins (Scanlon et al. 1997; Savarin et al. 1998). Also, κ-conotoxin PVIIA occludes the Kv channel conduction pore by binding to the external vestibule (Scanlon et al. 1997; Shon et al. 1998; Savarin et al. 1998). Probably, the toxin utilizes a lysine key for Kv channel block in the surface of interaction, as described for the scorpion toxins (Scanlon et al. 1997; Goldstein et al. 1994). Mutagenesis studies showed that the S5-S6 loop contains the binding site of the κ-conotoxin PVIIA, as described for charybdotoxin (Shon et al. 1998). It remains to be shown, however, whether the presence of a functional diad such as reported for snake, scorpion and sea anemone toxins, may account of the activity of the κ-conotoxin PVIIA.

A

```
CRIONQKCFQHLDDCCSRKCNRFNKCV
```

B

```
HaTx1   -ECRYLFGGCKTT-SDCCKHLGC-KFRDKYC--AWDFTFS
HaTx2   -ECRYLFGGCKTT-ADCCKHLGC-KFRDKYC--AWDFTFS
HpTx1   -DCGTIWHYCGTDQSECCEGWKCSR---QLCKYVIDW-
HpTx2   DDCGKLFSGCDTN-ADCCEGYVC-R---LWCKL--DW-
HpTx3   -ECGTLFSGCSTH-ADCCEGFIC-K---LWCRYERTW-
```

Fig. 8. A Amino acid sequence of κ-conotoxin PVIIA (TERLEAU et al. 1996). **B** Spider toxins which act on K^+ channels. HaTx1 and 2, hanatoxins from *Grammostola spatulata* (SWARTZ and MACKINNON 1995) HpTx1, HpTx2 and HpTx3, heteropodatoxins from *Heteropoda venatoria* (SANGUINETTI et al. 1997)

V. Spider Toxins

The first spider peptides able to act on K channel were purified from the venom of a Chilean tarantula and were called hanatoxin1 (HaTx1) and hanatoxin2 (HaTx2) (SWARTZ and MACKINNON 1995). Hanatoxins are basic peptides containing 35 amino acid residues reticulated by three disulfide bridges (see Fig. 8B), which are able to bind specifically to the surface of Kv2.1 channels. However, unlike the other toxins, hanatoxin does not interfere with the conducting path. Accordingly, residues between segments S5 and S6 are not critical for hanatoxin binding (SWARTZ and MACKINNON 1995, 1997a). Also, multiple hanatoxin molecules can simultaneously bind to Kv2.1 channels in marked contrast to the 1:1 stoichiometry of pore-blocking toxins. The receptor site for hanatoxin has been mapped to the S3-S4 linker region (SWARTZ and MACKINNON 1997a). Thus, hanatoxin may bind to the surface of Kv2.1 channels at four equivalent sites. Upon binding, hanatoxin is modified through the gating of Kv2.1 channels. Probably, hanatoxin binds more tightly to the closed state of the Kv2.1 channel and thereby, shifts channel opening to more depolarized voltages (SWARTZ and MACKINNON 1997b). Thus, the voltage activation relation for Kv2.1 channels is shifted to more depolarized membrane voltages, when hanatoxin is bound to its receptor at or near the S3-S4 linker region (SWARTZ and MACKINNON 1997a,b). Pore blocking toxins like agitoxin 2 and hanatoxin can simultaneously bind to chimeric *Shaker*/Kv2.1 channels (SWARTZ and MACKINNON 1997a). This made it possible to estimate 15 Å as minimal distance of the hanatoxin receptor site from the central pore axis on the surface of Kv channels (Fig. 5A).

Three new toxins isolated from the venom of the malaysian spider *Heteropoda venatoria* share sequence homology with hanatoxins (Fig. 8B) (SANGUINETTI et al. 1997). These peptides, called heteropodatoxins (HpTx1, HpTx2, and HpTx3) are able to block selectively voltage-gated K channel from rat myocytes identified as Kv4.2 in a voltage-dependent manner. In contrast with the other K channel blocking peptides already described, these peptides present a global negative charge, suggesting that the toxin-channel interactions is supported by contact between negative residues of the toxins and positive charged residues of the channel.

D. Conclusions

Kv channels may contain four Kvα- and four Kvβ-subunits. The Kvβ-subunits are located on the cytoplasmic side of the channel and may not contribute to peptide toxin binding sites localized to the extracellular surface of Kvα subunits. There, the outer mouth of Kv channels comprises amino acids of the S5/S6 linker region including the P-domain. It may be modeled as a wide vestibule and serves as receptor for pore blocking peptide toxins. These toxins bind to Kv channels in a 1:1 stoichiometry. Alternatively, peptide toxins like hanatoxin may not block the Kv channel pore, but may bind to another surface receptor which comprises amino acid residues of the S3/S4 linker region. In this case, Kv channel gating is modified.

References

Aiyar J, Withka JM, Rizzi JP, Singleton DH, Andrews GC, Lin W, Boyd J, Hanson DC, Simon M, Dethlefs B, Lee CL, Hall JE, Gutman GA, Chandy KG (1995) Topology of the pore-region of a K^+ channel revealed by the NMR-derived structures of scorpion toxins, Neuron 15:1169–1181

Benishin CG, Sorensen RG, Brown WE, Kreuger BK, Blaustein MP (1988) Four polypeptide components of green mamba venom selectively block certain potassium channels in rat brain synaptosomes, Mol Pharmacol 34:152–159

Bontems F, Roumestand C, Boyot P, Gilquin B, Doljanski Y, Menez A, Toma F (1991a) Three-dimensional structure of natural charybdotoxin in aqueous solution by ^{1}N-NMR, Eur J Biochem 196:119–128

Bontems F, Roumestand C, Gilquin B, Menez A, Toma F (1991b) Refined structure of charybdotoxin: common motifs in scorpion toxins and insect defensins, Science 254: 1521–1523

Chandy G, Gutman GA (1994) Voltage-gated K^+ channels. In: North RA (ed) Handbook of Receptors and Channels. CRC Press, Boca Raton, Florida, pp 1–71

Charlier C, Singh NA, Ryan SG, Lewis TB, Reus BE, Leach RJ, Leppert M (1998) A pore mutation in a novel KQT-like potassium channel gene in an idiopathic epilepsy family, Nature Genet 18:53–55

Choi KL, Aldrich RW, Yellen G (1991) Tetraethylammonium blockade distinguishes two inactivation mechanisms in voltage-gated K^+ channels. Proc Natl Acad Sci USA 88:5092–5095

Christie MJ, North RA, Osborne PB, Douglass J, Adelman JP (1990) Heteropolymeric potassium channels expressed in *Xenopus* oocytes from cloned subunits, Neuron 4:405–411

Crest M, Jacquet G Gola M, Zerrouk H, Benslimane A, Rochat H, Mansuelle P, Martin-Eauclaire MF (1992) Kaliotoxin, a novel peptidyl inhibitor of neuronal BK-type Ca^{2+}-activated K^+ channels characterized from *Androctonus mauretanicus* venom. J Biol Chem 267:1640–1647

Dauplais M, Lecoq A, Song J, Cotton J, Jamin N, Gilquin B, Roumestand C, Vita C, de Medeiros CLC, Rowan EG, Harvey AL, Menez A (1997) On the convergent evolution of animal toxins. Conservation of a diad of functional residues in potassium channel-blocking toxins with unrelated structures. J Biol Chem 272:4302–4309

Diochot S, Schweitz H, Béress L, Lazdunski M (1998) Sea anemone peptides with a specific blocking activity against the fast inactivating potassium channel Kv3.4. J Biol Chem 273:6744–6749
Doyle DA, Cabral JM, Pfuetzner RA, Kuo A, Gulbis JM, Cohen SL, Chait BT, MacKinnon R (1998) The structure of the potassium channel: molecular basis of K^+ conduction and selectivity. Science 280:69–77
Driscoll PC, Gronenborn AM, Beress L, Clore GM (1989) Determination of the three-dimensional solution structure of the antihypertensive and antiviral protein BDS-I from the sea anemone *Anemonia sulcata*: a study using nuclear magnetic resonance and hybrid distance geometry-dynamical simulated annealing. Biochemistry 28:2188–2198
Durrell SR, Guy HR (1996) Structural model of the outer verstibule and selectivity filter of the *Shaker* voltage-gated K^+ channel. Neuropharmacology 35:761–773
Foray MF, Lancelin JM, Hollecker M, Marion D (1993) Sequence-specific 1H-NMR assignment and secondary structure of black mamba dendrotoxin I, a highly selective blocker of voltage-gated potassium channels. Eur J Biochem 211:813–820
Galvez A, Gimenez-Gallelgo G, Reuben JP, Roy-Contancin L, Feigenbaum P, Kaczorowski GJ, Garcia ML (1990) Purification and characzerization of a unique potent peptidyl probe for high conductance calcium-activated potassium channel from venom of the scorpion *Buthus tamulus*. J Biol Chem 265:11,083–11,090
Garcia ML, Garcia-Calvo M, Hidalgo P, Lee A, MacKinnon R (1994) Purification and characterization of three inhibitors of voltage-dependent K^+ channels from *Leiurus quinquestriatus* var *Hebraeus* venom. Biochemistry 33:6834–6839
Garcia-Calvo M, Leonard RJ, Novick J, Stevens SP, Schmalhofer W, Kaczorowski GJ, Garcia ML (1993) Purification, characterization, and biosynthesis of margatoxin, a component of *Centruroides margaritatus* venom that selectively inhibits voltage-dependent potassium channels. J Biol Chem 268:18866–18874
Gendeh GS, Young LC, de Medeiros CLC., Jeyaseelan K, Harvey AL, Chung MCM (1997) A new potassium channel from the sea anemone *Heteractis magnifica*: isolation, cDNA cloning, and functional expression. Biochemistry 36:11461–11471
Gimenez-Gallego G, Naiva MA, Reuben JP, Katz JP, Kaczorowski GJ, Garcia ML (1988) Purification, sequence, and model structure of charybdotoxin, a potent selective inhibitor of calcium-activated potassium channels. Proc Natl Acad Sci USA 85:3329–3333
Goldstein SAN, Miller C (1993) Mechanism of charybdotoxin block of a voltage-gated K channel. Biophys J 65:1613–1619
Goldstein SAN, Pheasant DJ, Miller C (1994) The charybdotoxin receptor of a *Shaker* K^+ channel: peptide and channel residues mediating molecular recognition. Neuron 12:1377–1388
Gómez JM, Lorra C, Pardo LA, Stühmer W, Pongs O, Heinemann SH, Elliott AA (1997) Molecular basis for different pore properties of potassium channels from the rat brain Kv1 gene family. Pflügers Archiv 434:661–668
Grishin EV, Yu V, Korolkova SA, Lipkin AV, Nosyreva ED, Pluzhnikov KA, Sukhanov, SV, Volkova TM (1996) Structure and function of the potassium channel inhibitor from black scorpion venom. Pure & Appl Chem 68:2105–2109
Grissmer S, Nguyen AN, Aiyar J, Hanson DDC, Mather RJ, Gutman GA, Karmilowicz MJ, Auperin DD, Chandy G (1994) Pharmacological characterization of five cloned voltage-gated K^+ channels, types Kv1.1, Kv1.2, Kv1.3, Kv1.5, and 3.1, stably expressed in mammalian cell lines. Mol Pharm 45:1227–1234
Gross A, MacKinnon R (1996) Agitoxin footprinting the *Shaker* potassium channel pore. Neuron 16:399–406
Gruppe A, Schroter KH, Ruppersberg JP, Stocker M, Drewes T, Beckh S, Pongs O (1990) Cloning and expression of a human voltage-gated potassium channel A novel member of the RCK potassium channel family. EMBO J 9:1749–1756
Hanner M, Schmalhofer WA, Munujos P, Knaus HG, Kaczorowski GJ, Garcia ML (1997) The β subunit of the high-conductance calcium-activated potassium channel

contributes to the high-affinity receptor for charybdotoxin. Proc. Natl Acad SciUSA 94:2853–2858
Hartmann HA, Kirsch GE, Drewe JA, Joho R, Brown AM (1991) Exchange of conduction pathways between two related K^+ channels. Science 251:942–944
Harvey AL, Anderson AJ (1985) Dendrotoxins: snake toxins that block potassium channels and facilitate neurotransmitter release. Pharmacol Ther 31:33–55
Harvey AL, Vantapour EG, Rowan EG, Pinkkasfeld S, Vita C, Menez A, Martin-Eauclaire MF (1995) Structure-activity studies on scorpion toxins that block potassium channels. Toxicon 33:425–436
Heginbotham L, Lu Z, Abramson T, MacKinnon R (1994) Mutations in the K^+ channel signature sequence. Biophys J 66:1061–1067
Heinemann SH, Rettig J, Graack HR, Pongs O (1996) Functional characterisation of Kv-channel β-subunit from rat brain. J Physiol 4933:625–633
Hidalgo P, MacKinnon R (1995) Revealing the architecture of a K^+ channel pore through mutant cycles with a peptide inhibitor. Science 268:307–310
Hille B (1992) Ionic channels of excitable membranes, 2nd edn. Sinauer Associates, Sunderland, Massachusetts
Hoshi T, Zagotta WN, Aldrich RW (1990) Biophysical and molecular mechanisms of *Shaker* potassium channel inactivation. Science 250:533–538
Hurst RS, Busch AE, Kavanaugh MP, Osborne PB, North RA, Adelman JP (1991) Identification of amino acid residues involved in dendrotoxin block of rat voltage-dependent potassium channels. Mol Pharmacol 40:572–576
Isacoff EY, Jan YN, Jan LY (1990) Evidence for the formation of heteromultimeric potassium channels in *Xenopus* oocytes. Nature 345:530–534
Kamb A, Iverson LE, Tanouye MA (1987) Molecular characterisation of *Shaker*, a *Drosophila* gene that encodes a potassium channel. Cell 50:405–413
Karlsson E, Adem A, Aneiros A, Castaneda O, Harvey AL Jolkkonen M, Sotolongo V (1991) New toxins from marine organism. Toxicon 29:1168
Kharrat R, Mansuelle F, Sampieri F, Crest M, Oughideni R, Van Rietschoten J, Martin-Eauclaire MF, El Ayeb M (1997) Maurotoxin, a four disulfide bridge toxin from *Scorpio maurus* venom: purification, structure and action on potassium channels. FEBS Lett 406:284–290
Koschak A, Bugianesi RM, Mitterdorfer J, Kaczorowski GJ, Garcia ML, Knaus HG (1998) Subunit composition of brain voltage-gated potassium channels determined by hongotoxin-1, a novel peptide derived from *Centruroides limbatus* venom. J Biol Chem 27:32639–32644
Krovetz HS, VanDongen HMA and VanDongen AMJ (1997) Atomic distance estimates from disulfides and high-affinity metal-binding sites in a K+ channel pore. Biophysic J 72:117–126
Lancelin JM, Foray MF, Poncin M, Hollecker M, Marion D (1994) Proteinase inhibitor homologues as potassium channel blockers. Nat Struct Biol 1:246–250
Laraba-Djebari F, Legros C, Crest M, Céard B, Romi R, Mansuelle P, Jacquet G, Van Rietschtoten J, Gola M, Rochat H, Bougis PE, Martin-Eauclaire MF (1994) The Kaliotoxin family enlarged Purification, characterization, and precursor nucleotide sequence of KTX2 from Androctonus australis venom. J Biol Chem 269: 32835–32843
Lebrun B, Romi-Lebrun R, Martin-Eauclaire MF, Yasuda A, Ishiguro M, Oyama Y, Pongs O, Nakajima T (1997) A four-disulfide-bridged toxin, with high affinity towards voltage-gated K^+ channels, isolated from *Heterometrus spinnifer* (*Scorpionidae*) venom. Biochem J 32:8321–8327
Li M, Jan YN, Jan LY (1992) Specification of subunit assembly by the hydrophilic amino-terminal domain of the *Shaker* potassium channel. Science 257:1225–1240
Lipkind GM, Fozzard HA (1997) A model of scorpion toxin binding to voltage-gated K^+ channels. J Membrane Biol 158:187–196
Lopez GA, Jan YN, Jan LY (1994) Evidence that the S6 segment of the *Shaker* voltage-gated K^+ channel comprises part of the pore. Nature 367:179–182

Lucchesi K, Ravindran A, Young H, Moczydlowski EJ (1989) Analysis of the blocking activity of charybdotoxin homologs and iodinated derivatives against Ca^{2+}-activated K^+ channels. Memb Biol 109:269–281
Ludwig A, Zong X, Jeglitsch M, Hofmann F, Biel M (1998) A family of hyperpolarization -activated mammalian cation channels. Nature 393:587–591
MacKinnon R, Cohen SL, Kuo A, Lee A, Chait BT (1998) Structural conservation in prokaryotic and eukaryotic potassium channels. Science 280:106–109
MacKinnon R (1991) Determination of the subunit stoichiometry of a voltage-activated potassium channel. Nature 350:232–235
Martin BM, Ramirez AN, Gurrola GB, Nobile M, Prestipino G, Possani L (1994) Novel K^+ channel-blocking toxins from the venom of the scorpion *Centruroides limpidus* Karsch. Biochem J 304:51–56
Miller C (1995) The charybdotoxine family of K^+ channel-blocking peptides. Neuron 15:5–10
Naini A, Miller C (1996) A symmetry-driven search for electrostatic interaction partners in Charybdotoxin and a voltage-gated K^+ channel. Biochemistry 35:6181–6187
Naranjo D, Miller C (1996) A strongly interacting pair of residues on the contact surface of Charybdotoxin and a *Shaker* K^+ Channel. Neuron 16:123–130
Olamendi-Portugal T, Gomez-Lagunas F, Gurrola GB, Possani LD (1996) A novel structural class of K^+-channel blocking toxin from the scorpion *Pandinus imperator.* Biochem J 315:977–981
Olivera BM, Rivier J, Clark C, Ramilo CA, Corpuz GP, Abogadie FC, Mena EE, Woodward SR, Hillyard DR, Cruz LJ (1990) Diversity of *Conus* neuropeptides. Science 249:257–263
Papazian DM, Schwarz TL, Tempel BL, Jan YN, Jan JY (1987) Cloning of genomic and complementary DNA from *Shaker*, a putative potassium channel gene from *Drosophila*. Science 237:749–753
Papazian DM, Timpe LC, Jan YN, Jan LY (1991) Alteration of voltage dependence of *Shaker* potassium channel by mutations in the S4 sequence. Nature 349:305–310
Parcej DN, Dolly JO (1989) Dendrotoxin receptor from bovine synaptic plasma membranes Binding properties, purification and subunit composition of a putative constituent of certain voltage-activated K^+ channels. Biochem J. 257:899–903
Parcej DN, Scott VES, Dolly JO (1992) Oligomeric properties of a α-dendrotoxin-sensitive potassium channels purified from bovine brain. Biochemistry 31:11084–11088
Patel AJ, Lazdunski M, Honoré E (1997) Kv2.1/Kv9.3, a novel ATP-dependent delayed-rectifier K^+ channel in oxygen-sensitive pulmonary artery myocytes. EMBO J 16:6615–6625
Pennington MW, Byrnes ME, Zaydenberg I, Khaytin I, De Chastonay J, Krafte DS Hill R, Mahnir VM, Volberg WA, Gorczyca W, Kem WR (1995) Chemical synthesis and characterization of ShK toxin: a potent potassium channel inhibitor from a sea anemone. Int J Peptide Protein Res 46:354–358
Pongs O, Kecskemethy N, Müller R, Krah-Jentgens I, Baumann A, Kiltz HH, Canal I, Lamazares S, Ferrus A (1988) *Shaker* encodes a family of putative potassium channel proteins in the nervous system of *Drosophila* EMBO J 7:1087–1096
Possani LD, Martin BM, Svendsen I (1982) The primary structure of noxiustoxin: a K^+ channel blocking peptide, purified from the venom of the scorpion *Centruroides noxius hoffmann.* Carlsberg Res Comm 47:285–289
Post MA, Kirsch GE, Brown AM (1996) Kv2.1 and electrically silent Kv6.1 potassium channel subunits combine and express a novel current. FEBS Letters 399:177–182
Rehm H, Lazdunski M (1988) Purification and subunit structure of a putative K^+ channel protein identified by its binding properties for dendrotoxin I. Proc Natl Acad Sci USA 85:4919–4923
Reid PF, Pongs O, Dolly JO (1992) Cloning of a bovine voltage-gated K^+ channel gene utilising partial amino acid sequence of a dendrotoxin-binding protein from brain cortex. FEBS Lett 302:31–34

Rettig R, Heinemann RH, Wunder F, Lorra C, Parcej DN, Dolly JO, Pongs O (1994) Inactivation properties of voltage-gated K^+ channels altered by presence of β-subunit. Nature 369:289–294

Roeper J, Sewing S, Zhang Y, Sommer T, Wanner SG, Pongs O, (1998) NIP domain prevents N-type inactivation in voltage-gated potassium channels. Nature 391:390–393

Rogowski RS, Collins JH, O'Neill TJ, Gustafson TA, Werkman TR, Rogowski MA, Tenenholz TC, Weber DJ, Blaustein MP (1996) Three new toxins from the scorpion *Pandinus imperator* selectively block certain voltage-gated K^+ channels. Mol Pharmacol 50:1167–1177

Rogowski RS, Krueger BK, Collins JH, Blaustein MP (1994) Tityus toxin-Kα blocks voltage-gated non-inactivating K^+ channels and unblocks inactivating K^+ channels blocked by α-dendrotoxin in synaptosomes. Proc Natl Acad Sci USA 91:1475–1479

Romi-Lebrun R, Lebrun B, Martin-Eauclaire MF, Ishiguro M, Escoubas P, Wu FQ, Hisada M, Pongs O, Nakajima T (1997) Purification, characterization, and synthesis of three novel toxins from the Chinese scorpion *Buthus martensi*, which act on K^+ channels. Biochemistry, 36:13473–13482

Ruppersberg JP, Schröter KH, Sakmann B, Stocker M, Sewing S, Pongs O (1990) Heteromultimeric channels formed by rat brain potassium channel proteins. Nature 345:535–537

Salinas M, de Weille J, Guillemare E, Lazdunski M, Hugnot JP (1997) Modes of regulation of *Shab* K^+ channel activity by the Kv8.1 subunit. J Biol Chem 272:8774–8780

Sanguinetti MC, Johnson JH, Hammerland LG, Kelbaugh PR, Volkmann RA, Saccomano NA, Mueller AL (1997) Heteropodatoxins: peptides isolated from spider venom that block Kv4.2 potassium channels. Mol Pharmacol 51:491–498

Savarin P, Guenneugues M, Gilquin B, Lamthanh H, Gasparini S, Zinn-Justin S, Ménez A (1998) Three-dimensional structure of K-Conotoxin PVIIA, a novel potassium channel-blocking toxin from cone snails. Biochemistry 37:5407–5416

Scanlon MJ, Naranjo D, Thomas L, Alewood PF, Lewis R, Craik DJ (1997) Solution structure and proposed binding mechanism of a novel potassium channel toxin K-conotoxin PVIIA. Structure 5:1585–1597

Schweitz H, Bruhn T, Guillemare E, Moinier D, Lancelin JM, Beress L, Lazdunski M (1995) Kalciludine and Kaliseptine. Two different classes of sea anemon toxins for voltage-sensitive K^+ channels. J Biol Chem 270:25121–25126

Scott VES, Parcej DN, Keen JN, Findlay JB, Pongs O, Dolly J.O. (1990) α-dendrotoxin acceptor from bovine brain is a K^+ channel protein: evidence from the N-terminal sequence of its larger subunit. J Biol Chem 265:20094–20097

Seoh SA, Sigg D, Papazian DM, Bezanilla F (1996) Voltage-sensing residues in the S2 and S4 segments of the *Shaker* K^+ channel. Neuron 16:1159–1167

Sewing S, Roeper J, Pongs O (1996) Kvβ1 subunit binding specific for *Shaker*-related potassium channel α-subunits. Neuron 16:455–463

Shen V, Pfaffinger P (1995) Molecular recognition and assembly sequences involved in the subfamily-specific assembly of voltage-gated K^+ channel subunit proteins. Neuron 14:625–633

Shi G, Nakahira K, Hammond S, Rhodes KJ, Schechter LE, Trimmer JS (1996) Beta subunits promote K+ channel surface expression through effects early in biosynthesis. Neuron 16:843–852

Shon KJ, Stocker M, Terlau H, Stühmer W, Jacobsen R, Walker C, Grilley M, Watkins M, Hillyard DR, Gray WR, Olivera BM (1998) kappa-Conotoxin PVIIA is a peptide inhibiting the *Shaker* K+ channel. J Biol Chem 273:33–38

Smith LA, Olson MA, Lafaye PJ, Dolly JO (1995) Cloning and expression of mamba toxins. Toxicon 33:459–474

Smith LA, Reid PF, Wang FC, Parcej DN, Schmidt JJ, Olson MA, Dolly JO (1997) Site-directed mutagenesis of Dendrotoxin K reveals amino acids critical for its interaction with neuronal K^+ channels. Biochemistry 36:7690–7696

Stühmer W, Ruppersberg JP, Schröter KH, Sakmann B, Stocker M, Giese KP, Perschke A, Baumann A, Pongs O (1989) Molecular basis of functional diversity of voltage-gated potassium channels in mammalian brain. EMBO J 8:3235–3244

Stühmer W, Stocker U, Sakmann B, Seeburg P, Baumann A, Gruppe A, Pongs O (1988) Potassium channels expressed from rat brain cDNA have delayed rectifier properties. FEBS Lett 241:199–206

Swartz KJ, MacKinnon R (1995) An inhibitor of the Kv2.1 potassium channel isolated from the venom of a Chilean tarantula. Neuron 15:941–949

Swartz KJ, MacKinnon R (1997a) Mapping the receptor site for Hanatoxin a gating modifier of voltage-dependent K^+ channels. Neuron 18:675–682

Swartz KJ, MacKinnon R (1997b) Hanatoxin modifies the gating of a voltage-dependent K^+ channel through multiple binding sites. Neuron 18:665–673

Tang CY, Papazian DM (1997) Transfer of voltage independence from a rat olfactory channel to the *Drosophila ether-à-go-go* K^+ channel. J Gen Physiol 109:301–311

Tempel BL, Jan YN, Jan LY (1988) Cloning of a probable potassium channel gene from mouse brain. Nature 332:837–839

Terlau H, Shon KJ, Grillez M, Stocker M, Stühmer W, Olivera BM (1996) Strategy for rapid immobilization of prey by a fish-hunting marine snail. Nature 381:148–151

Tudor JE, Pallaghy PK, Pennington MW, Norton RS (1996) Solution structure of ShK toxin, a novel potassium channel inhibitor from a sea anemone. Nat Struct Biol 33:317–320

Tytgat J, Debont T, Carmeliet E, Daenens P (1995) The α-Dendrotoxin footprint on a mammalian potassium channel. J Biol Chem 270:24776–24781

Yool AJ, Schwarz TL (1991) Alteration of ionic selectivity of a K^+ channel by mutations of the H5 region. Nature 349:700–704

Yu W, Xu J, Li M (1996) NAB domain is essential for the subunit assembly of both α-α and α-β complexes of *Shaker*-like potassium channels. Neuron 16:441–453

CHAPTER 8

Voltage-Gated Calcium-Modulated Potassium Channels of Large Unitary Conductance: Structure, Diversity, and Pharmacology

R. LATORRE, C. VERGARA, O. ALVAREZ, E. STEFANI, and L. TORO

A. Introduction

Calcium-sensitive and voltage-dependent channels of large unitary conductance (BK_{Ca}) are found ubiquitously distributed in different cells and tissues where they participate in regulating many cellular processes (LATORRE et al. 1989). Because cytosolic Ca^{2+} activates BK_{Ca} channels they play an important role in coupling chemical to electrical signaling. In neurons they contribute to action potential repolarization (e.g., SAH 1996) and in presynaptic terminals they appear to modulate transmitter release (ROBITAILLE et al. 1993; KNAUS et al. 1996; YAZEJIAN et al. 1997; but see WARBINGTON et al. 1996). BK_{Ca} channels are present abundantly in virtually all types of smooth muscle cells and they are crucial in controlling smooth muscle tone (ANWER et al. 1993; BRAYDEN and NELSON 1992; NELSON et al. 1995; for reviews see NELSON and QUAYLE 1995; SANDERS 1992). These channels also control fluid secretion (PETERSON 1986) and fluid reabsorption (GUGGINO et al. 1987). In chick cochleae different variants of the BK_{Ca} channel may help to determine the characteristic frequency of each hair cell helping to establish the tonotopic map (DHASAKUMAR et al. 1997; ROSENBLATT et al. 1997; RAMANATHAN et al. 1999). BK_{Ca} channels were cloned from *Drosophila* taking advantage of the existence of the mutant *slowpoke* (*Slo*) in which this potassium current is absent (ELKINS et al. 1986; GHO and MALLARD 1986). The primary sequence of the *Slo* protein showed that that BK_{Ca} channels belong to the S4 superfamily (ATKINSON et al. 1991; ADELMAN et al. 1992). The S4 superfamily encompass voltage-dependent Na^+, Ca^{2+}, and K^+ (K_V) channels. The channel-forming *Slo* protein (α subunit) is associated in some tissues with a smaller modulatory β subunit (KNAUS et al. 1994). Unlike K_V channels that possess six transmembrane (S1–S6) segments, BK_{Ca} channels are endowed with a seventh transmembrane (S0) segment that leads to an exoplasmic N-terminus (MEERA et al. 1997; WALLNER et al. 1997; for reviews see TORO et al. 1998; VERGARA et al. 1998). There are many functional subtypes of BK_{Ca} channels that differ in their Ca^{2+} sensitivity, toxin sensitivity, and single channel gating. Only one gene encoding the α subunit has been identified. Therefore, the molecular basis of this functional diversity may include alternative RNA splicing of a single transcript (ATKINSON et al. 1991; ADELMAN et al. 1992; BUTLER et al. 1993; TSENG-CRANK et al. 1994; PALLANK and GANETZKY 1994; WALLNER et al. 1995; MCCOBB et al. 1995; FERRER et al. 1996; ROSENBLATT et al. 1997; DHASAKUMAR et al. 1997; SAITO et al. 1997; XIE and MCCOBB 1998), modulation by auxiliary subunits and/or metabolic modulation. The function of BK_{Ca} channels can be modulated by a wide variety of intracellular and extracellular factors (TORO and STEFANI 1993; LEVITAN 1994). In particular, regulatory mechanisms such as protein phosphorylation have been studied in detail with the conclusion that BK_{Ca} channels form part of a regulatory complex tightly associated with protein kinases and phosphatases (CHUNG et al. 1991; REINHART and LEVITAN 1995; PREVARSKAYA et al. 1995; SCHUBERT et al. 1999). BK_{Ca} channels can also be modulated via endogenous or purified G-proteins (TORO et al. 1990; KUME et al. 1992; SCORNIK et al. 1993; WALSH et al. 1996). These channels possess a well studied pharmacology characterized by a fast blockade induced by micromolar concentrations of external tetraethylammonium (TEA) (VERGARA et al. 1984; YELLEN 1984); and a highly specific slow blockade induced by external iberiotoxin (CANDIA et al. 1992; GIANGIACOMO et al. 1992; GARCIA et al. 1995). The channel is insensitive to apamin and to 4-aminopyridine (WALLNER et al. 1995). In addition, a number of non-peptidyl compounds that act by increasing channel activity have been identified. These BK_{Ca} channel gating-modifiers include soyasaponins (MCMANUS et al. 1993; GIANGIACOMO et al. 1998), fenamates (OTTOLIA and TORO 1994), benzimidazolones (OLESEN 1994) and flavonoids (KOH et al. 1994).

The defining characteristics of BK_{Ca} channels are their high single channel conductance (~250 pS in symmetric 0.1 mol/l KCl), high K^+ selectivity and the fact that their open probability, P_o, is increased by membrane depolarization as well as by increases in $[Ca^{2+}]_i$ (LATORRE et al. 1989; MCMANNUS 1991; LATORRE 1994; TORO et al. 1998; VERGARA et al. 1988). These properties allow BK_{Ca} to act as feedback modulators of the activity of voltage- dependent Ca^{2+} channels with whom they coexist, particularly in neurons (ROBITAILLE et al. 1993; YAZEJIAN et al. 1997; MARRION and TRAVALIN 1998), and smooth muscle cells (NELSON et al. 1995). However, it is important to note here that BK_{Ca} channels can fully open and the maximal gating charge can be obtained in the absence of Ca^{2+} at strong depolarizations (MEERA et al. 1996; CUI et al. 1997; STEFANI et al. 1997; COX et al. 1997; DIAZ et al. 1998). These results strongly suggest the presence of an intrinsic voltage sensor in the BK_{Ca} protein whose displacement is induced by voltage and facilitated by cytosolic Ca^{2+}. In this review we discuss recent advances in structure, gating properties, diversity, modulation and pharmacology of BK_{Ca} channels.

B. Channel Structure

Primary sequence analysis of the pore-forming (α) subunit of BK_{Ca} channels from different species has revealed several interesting points. The primary sequences among different mammalian BK_{Ca} channels are almost identical (>97% amino acid identity), and they share a high degree of homology with the sequences of the six transmembrane segments S1–S6 of the family of voltage-gated K^+ (K_v) channels (K_v) (SHIH and GOLDIN 1997; JAN and JAN 1997). Particularly striking is the homology among positively charged amino acids in the S4 segment that is part of the voltage sensor in K_v channels (LARSSON et al. 1996; CHA and BEZANILLA 1997). From the four basic residues that determine the voltage dependence in *Shaker* K^+ channels (AGGARWAL and MACKINNON 1996; SHEO et al. 1996; BEZANILLA, 2000), three are conserved in *Slo*. Alignment of two evolutionary distant *Slo* channels, the *C. elegans* homologue (n*Slo*) and the mouse m*Slo*, shows that there is a high degree of sequence conservation between these two channels that extends into the carboxyl-terminal (Fig. 1) (WEI et al. 1996). Sequence conservation is interrupted by a short stretch of amino acid residues that divides the *Slo* protein into two functional domains, the "core" and the "tail"; these two domains do not produce functional channels by themselves, but do so when coexpressed (WEI et al. 1994). Hydrophobicity plots for the mammalian, *Drosophila* or *C. elegans* BK_{Ca} channels show besides the seven transmembrane regions (S0–S6), four other segments (S7–S10) with lower overall hydrophobicity when compared to S0, S1, or S6 (WALLNER et al. 1996; MEERA et al. 1997; SCHREIBER et al. 1998) (see Figs 1 and 2). From these four hydrophobic segments, which comprise almost 70% of the whole subunit, S9 and S10 segments are cytosolic. This was demonstrated by: (a) in vitro translation of the tail region (that includes S9 and S10) which resulted in a soluble protein; and (b) "cross-cramming" reconstitution experiments where a "silent" patch expressing core protein produced functional channels when introduced into the cytoplasm of an oocyte expressing tail protein (MEERA et al. 1997). Since S8 and S9 regions in d*Slo* and n*Slo* show a rather low hydrophobicity and S8 may be too short to span the membrane, it is likely that S7 and S8 are also cytosolic. However, the nature of S7 and S8 regions needs to be explored further.

With respect to the amino-terminal region, it was generally accepted that the BK_{Ca} channel had a topology similar to K_V channels with an intracellular N-terminus and six transmembrane domains. However, WALLNER et al. (1996) and MEERA et al. (1997) have postulated an N-terminal topology consisting of an extracellular N-terminus and seven transmembrane segments (S0–S6). Several lines of evidence indicated that this was the case: in vitro translation and glycosylation experiments, functional expression of signal sequence fusions, sequence alignments, and hydrophobicity plots. In the majority of sequence alignments and models, the first segment that WALLNER et al. (1996) named S0 was considered as S1 and the fourth hydrophobic segment, identified as S3 by the same authors, was considered to be extracellular. In both models, S4, S5, S6 regions are equivalent. Conclusive experiments showing that the amino end of the BK_{Ca} channel is indeed extracellular was obtained by introducing epitope tags in different regions of the *Slo* protein and analyzing intact vs permeabilized cells (MEERA et al. 1997). A c-myc tag at the NH2-terminus was readily labeled in *intact* cells using either antibody coated magnetic beads or fluorescent-labeled antibodies. Consistent with the

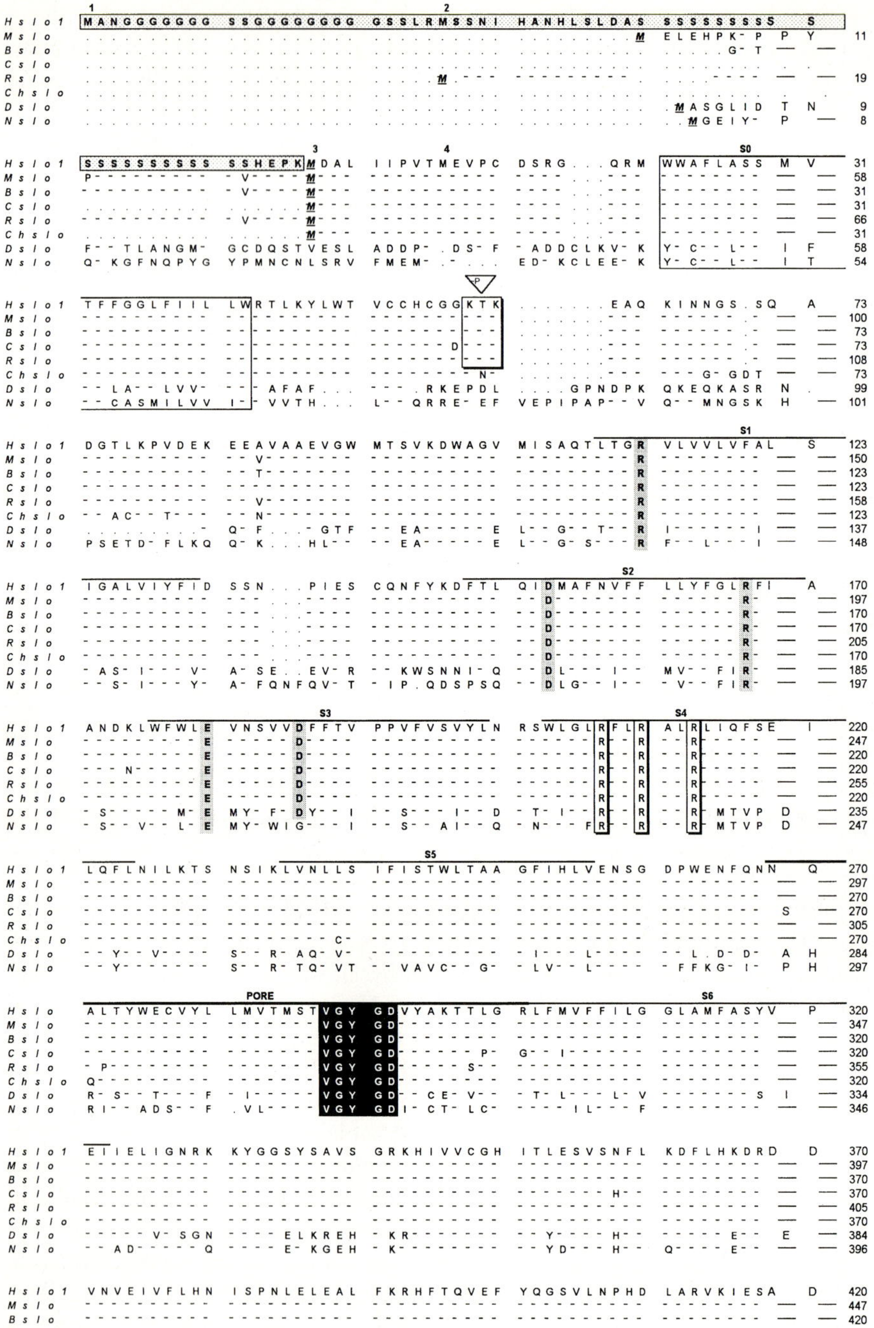

Hslo1
Mslo
Bslo
Cslo
Rslo
Chslo
Dslo
Nslo
S0
S1
S2
S3
S4
S5
PORE
S6

Fig. 1

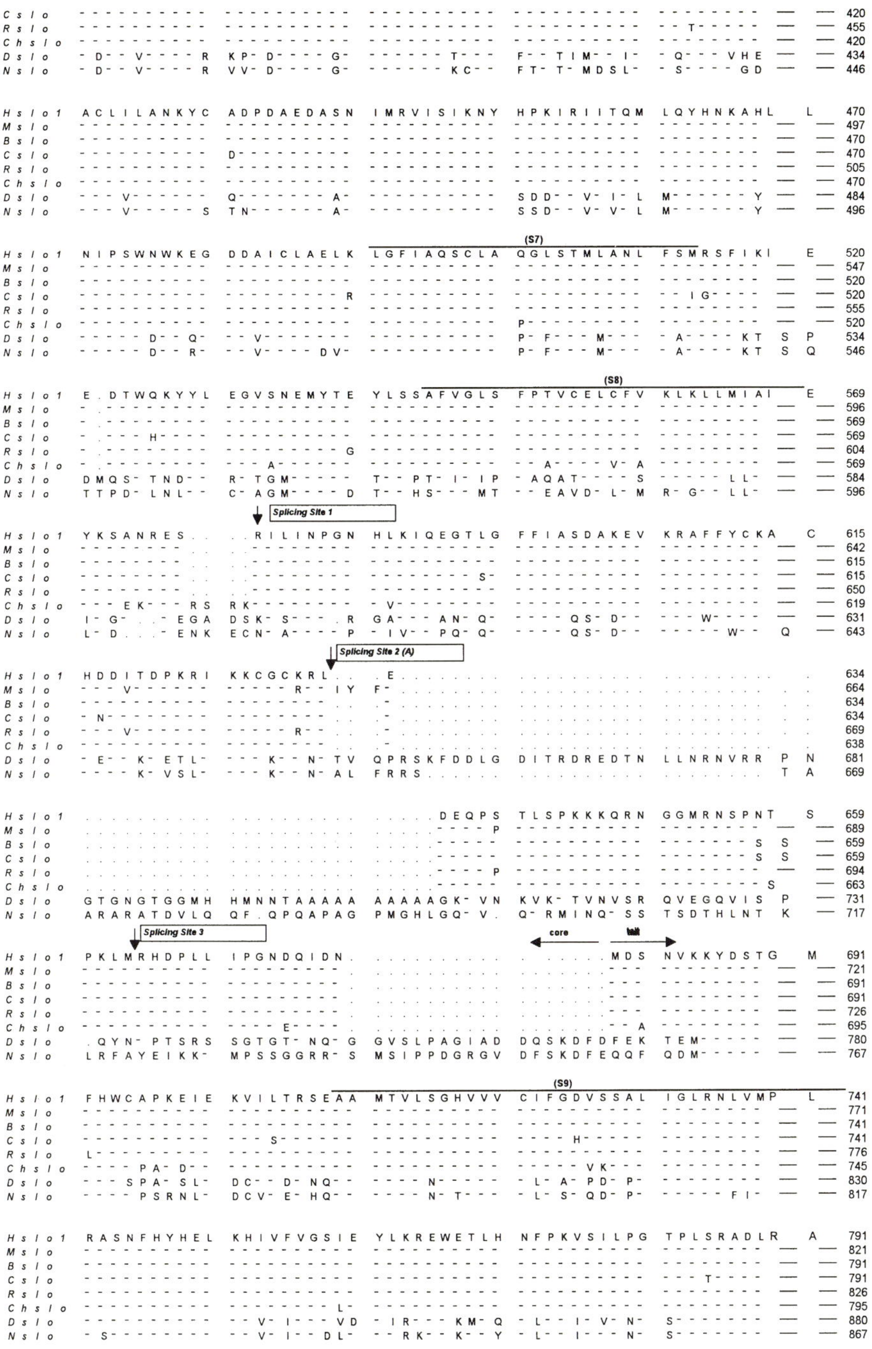

Fig. 1

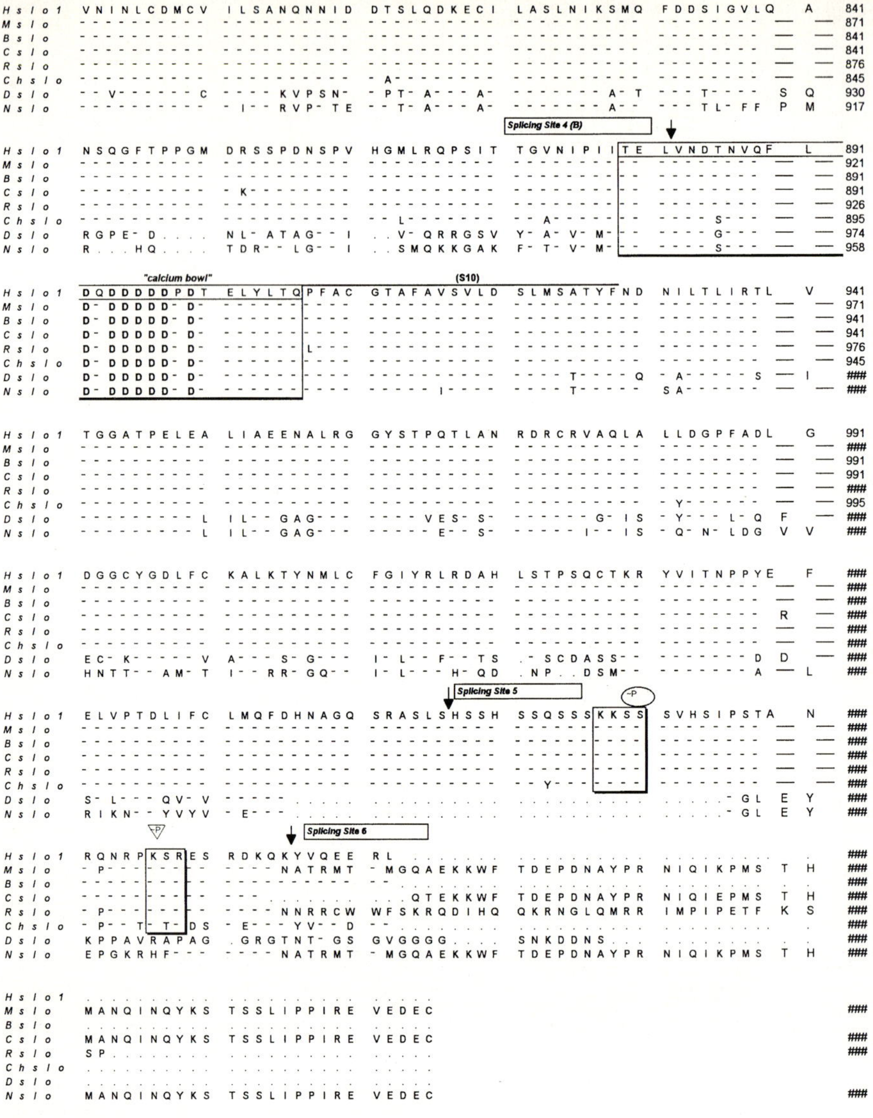

Fig. 1. Sequence alignments of *Slo* channels from different species. *H*, human; *M*, mouse; *B*, bovine; *C*, canine; *R*, rat; *Ch*, chicken; *D*, *Drosophila melanogaster*; *N*, *Caenorhabditis elegans*. All sequences are shown for the N- and C-termini. Numbering starts at M3 for *Hslo, Bslo, Cslo,* and *Chslo*. M1, M2, M3, and M4 denote four possible starting codons with Kozak consensus sequences in *Hslo*. Since there is a high degree of homology among species, the apparent lack of M1 and M2 in some of them may be due to failure of the reverse transcription. M3 has been used as starting codon in functional studies. *Dashes*, identical amino acids as in *Hslo*; *dots*, gaps introduced for better alignment; *~P*, strong putative phosphorylation site (consensus sequence is boxed); *triangles* for PKC; *circle* for PKG. Transmembrane segments S0–S6 and hydrophobic regions at the carboxyl terminus (S7–S10) are marked with a *line*. *Gray boxes*, conserved charged residues in S1–S3. Critical residues in voltage depending gating in S4 region are highlighted. Pore region is *double lined*. *Black box*, K^+ channel pore signature sequence. *Vertical arrows*, splicing sites. *Horizontal arrows*, borders of "core" and "tail" domains which form functional channels when coexpressed as separable domains (Wei et al. 1994). *Box double lines*, "calcium bowl" (Schreiber and Salkoff 1997)

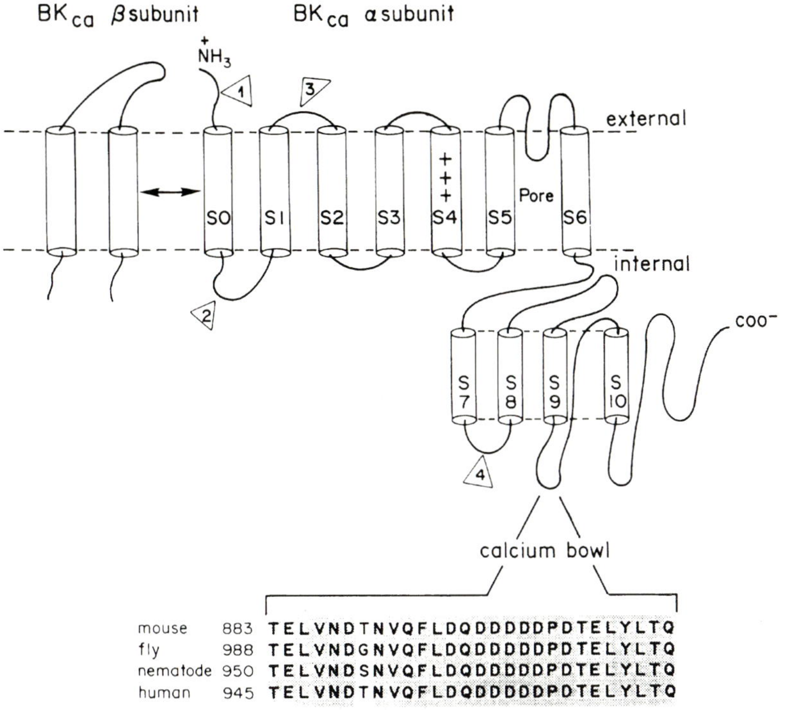

Fig. 2. Proposed BK_{Ca} membrane topology. Numbers in the *flags* indicate the positions of the epitopes used to determine the model shown here. The regulatory β subunit is also shown. A potential site involved in calcium activation, the "calcium-bawl", has been identified within the highly conserved region between segments S9 and S10

extracellular N-terminus, the loop between segments S1 and S2 was extracellular. Thus the BK_{Ca} channel has a unique topology consisting of an extracellular N-terminus, seven N-terminal transmembrane domains (S0–S6) and an intracellular C-terminus (MEERA et al. 1997) (Figs. 1 and 2).

The TEA blocking behavior of BK_{Ca} channels revealed the tetrameric formation of the BK_{Ca} channels. Tetraethylammonium binds with high affinity to the wild type *Slo* due to the presence of a tyrosine (Y) located in the pore region in positions 294 and 308 in the h*Slo* and d*Slo* channels, respectively. All K^+ channels containing a tyrosine in a corresponding position in the pore region have high affinities for TEA. Those K^+ channels that lack this particular tyrosine have much lower affinities (JAN and JAN 1992). SHEN et al. (1994) injected *Xenopus* oocytes with two different RNAs encoding the wild type d*Slo* channel and the mutant Y308 V having very distinct TEA binding affinities. The expressed *Slo* channels showed four different conductance levels in the presence of TEA. The amplitude of these conductance levels was that expected of channels containing 1, 2, 3, or 4 tyrosine residues assuming that the tyrosine of each channel subunit contributes equally to the TEA binding energy (HEIGENBOTHAM and MACKINNON 1992). Therefore, BK_{Ca} channels are tetrameres.

C. Auxiliary Subunits

BK_{Ca} channels purified from smooth muscle are tightly associated with an auxiliary β subunit (GARCIA-CALVO et al. 1994). The smooth muscle β subunit (named $\beta 1$, KCNB1) has a proposed topology of spanning the membrane twice, with the N-and C-termini residing in the cytoplasmic side (KNAUS et al. 1994; JIANG et al. 1999) (Fig. 2). The $\beta 1$ subunit causes a dramatic increase in the apparent Ca^{2+} sensitivity of channels from smooth muscle. It shifts the voltage range of activation by 60–100 mV in the hyperpolarizing direction when measured in symmetric 110 mmol/l K^+ (MCMANUS et al. 1995; WALLNER et al. 1995; DWORETZKY et al. 1996; MEERA et al. 1996). The hyperpolarizing shift along the voltage axis is switched on by micromolar Ca^{2+} (MEERA et al. 1996) and it seems to be regulated by external [K^+] (REIMAN et al. 1997). The $\beta 1$ subunit is found together with BK_{Ca} channels in trachea, aorta, coronary, and probably in smooth muscle of other tissues such as uterus and intestine (KNAUS et al. 1994; VOGALIS et al. 1996; TANAKA et al. 1997; JIANG et al. 1999). However, the $\beta 1$ subunit is not an obligatory component of all BK_{Ca} channels (TSENG-CRANCK et al. 1996; SAITO et al. 1997; CHANG et al. 1997; HANNER et al. 1997). The $\beta 1$ subunit increases the binding affinity of the scorpion toxin charybdotoxin (ChTX) by increasing the ChTX association rate and decreasing the dissociation rate (HANNER et al. 1997; but see DWORETZKY et al. 1996); it also increases the IC50 for iberiotoxin (DWORETZKI et al. 1996; L. Toro and P. Meera, unpublished observations). Channel activity is increased by nanomolar concentrations of the agonist dehydrosoyasaponin I only in oocytes expressing the β and α subunits (MCMANUS et al. 1995). Micromolar amounts of dehydrosoyasaponin I are needed to increase the activity of the α subunit alone (WALLNER et al. 1999). Thus, α and $\beta 1$ subunits together contribute to the properties of BK_{Ca} channels. The S0 segment of the α subunit is crucial for the functional interaction between α and $\beta 1$ subunits. The normally unresponsive d*Slo*, becomes modulated by the $\beta 1$ subunit when the 41 N-terminal amino acids, including the S0, are exchanged with the corresponding amino acids from the responsive h*Slo* (WALLNER et al. 1996).

TANAKA et al. (1997) compared the calcium and dehydrosoyasaponin I sensitivities of currents obtained after the controlled expression of α and β channel subunits in a heterologous system with the sensitivities of currents obtained from freshly dissociated human coronary smooth muscle. They found that most native channels were coupled to β subunits and, therefore, could be activated by the levels of calcium attained during calcium sparks (NELSON et al. 1995).

Drosophila BK_{Ca} channels can specifically interact with other proteins through their carboxyl terminus. Two such proteins have been identified: dSL1P1 (XIA et al. 1998) and Slob (SCHOPPERLE et al. 1998). These two proteins seem to modulate the *Drosophila* channel in different ways. Direct application of Slob to the intracellular side of *dSlo* channels increased channel activity; whereas dSL1P1 expression seems to decrease the number of BK_{Ca} channels in the plasma membrane. Interestingly, the distribution of dSL1P1 and d*Slo* transcripts coincide throughout the *Drosophila* nervous system. Recently, WALLNER et al. (1999) identified a novel β subunit homolog, dubbed $\beta 2$ (KCNMB2), whose main characteristic is to cause inactivation of the BK_{Ca} channel through a "ball and chain" mechanism. The $\beta 2$ subunit, similarly to the $\beta 1$ subunit, increases the Ca^{2+} sensitivity and modifies its pharmacology. Dehydrosoyasaponin I activates BK_{Ca} channels coexpressed with the $\beta 2$ subunit in the nanomolar range similar to the effect seen on $\alpha + \beta 1$ subunit channels.

D. Calcium Sensitivity and Diversity of BK_{Ca} Channels in Different Cells and Tissues

Data for the probability of opening, as a function of internal Ca^{2+} concentration are usually fit with the relationship

$$P_O = P_O^{max}[Ca^{2+}]^n / \left([Ca^{2+}]^n + Kd^n\right) \tag{1}$$

where n is the Hill coefficient and Kd is the apparent dissociation constant. Equation 1 is immediately obtained if we assume a highly cooperative scheme in which n Ca^{2+} ions must bind simultaneously to sites on a receptor in order to open a channel:

$$\text{closed channel} \quad \overset{nCa}{\rightleftharpoons} \quad \text{open channel} \qquad \text{(R-1)}$$

Scheme 1 does not account for by the behavior of BK_{Ca} channels since channels can open independently of internal Ca^{2+} when $[Ca^{2+}]i \leq 100$ nmol/l (MEERA et al. 1996) and in the virtual absence of internal Ca^{2+} (PALLOTA 1985; MEERA et al. 1996; CUI et al. 1997; COX et al. 1997; STEFANI et al. 1997). Therefore, fittings of P_o-$[Ca^{2+}]$ data to Eq. (1) should be taken as purely empirical and the meaning of *n* should be interpreted very cautiously. A more general way to compare Ca^{2+} sensitivity of BK_{Ca} channels from different tissues is (1) to plot the midpoint of the voltage activation curve against $[Ca^{2+}]i$. In this case the results can be confronted with a specific model for the channel gating kinetics (e.g., COX et al. 1997) or (2) using two-dimensional analysis of single-channel currents at different internal Ca^{2+} concentrations (e.g., ROTHBERG and MAGLEBY 1998).

Calcium sensitivities of BK_{Ca} channels in different cells and tissues were reviewed extensively by MCMANUS (1991). The apparent K_d of Eq. (1) is highly variable for different channels and variations in Ca^{2+} sensitivity in the same tissue are also found (MOCZYDLOWSKI and LATORRE 1983; TORO et al. 1991). The origin of the different Ca^{2+} sensitivities in BK_{Ca} channels may reside in the presence of: (a) different alternatively spliced variants (LAGRUTTA et al. 1994; TSENG-CRANK et al. 1994; DHASAKUMAR et al. 1997; ROSENBLATT et al. 1997; SAITO et al. 1997; XIE and MCCOBB 1998); (b) the relative expression of β subunit in a given tissue (RAMANATHAN et al. 1999) and/or (c) the formation of heteromultimers.

In *Drosophila*, LAGRUTTA et al. (1994) found two d*Slo* spliced located in the carboxyl terminal having different Ca^{2+} sensitivities. BK_{Ca} channel diversity in the brain is also generated by means of alternative RNA splicing. TSENG-CRANCK et al. (1994) characterized nine BK_{Ca} channel splice variants from human brain. Two channel variants with different exons between region S8 and S9 showed different Ca^{2+} sensitivities. SAITO et al. (1997) identified a rat *Slo* variant containing 59 amino acids between S8 and S9 regions that left-shifted the voltage activation curve of $BK_{Ca.}$ The work of ROSENBLATT et al. (1997), DHASAKUMAR et al. (1997) and JONES et al. (1998) is another example of how differential splicing of an RNA transcript is used as a mechanism for generating BK_{Ca} channel diversity. In the cochlea, the properties of BK_{Ca} channels play a major role in determining the electrical tuning of individual hair cells. Several spliced variants of the BK_{Ca} channel (c*Slo*) from the receptor epithelium of the chick cochlea were cloned (ROSENBLATT et al. 1997; DHASAKUMAR et al. 1997; JIANG et al. 1997; JONES et al. 1998). Seven RNA splice sites were located and if the formation of heterotetrameres is unrestricted 576 different BK_{Ca} channels could be expressed from the c*Slo* channel (ROSENBLATT et al. 1997). Splice variants were expressed differentially in the hair cells along the frequency axis of the epithelia. In particular, two c*Slo* isoforms show differences in their Ca^{2+} sensitivity pattern. This finding provides a possible molecular mechanism to account for one component of frequency tuning in hair cells (but see below and RAMANATHAN et al. 1999). XIE and MCCOBB (1988) described a hormonal control of *Slo* splice variants in rat adrenal chromaffin cells. One of the BK_{Ca} channel variants, strex-2, was found to decrease abruptly after hypophysectomy; this decrease was prevented by adrenocorticotropic hormone injections. BK_{Ca} channels having the strex amino acid sequence (strex-2 channel) activate at more hyperpolarized voltages than those that do not have this sequence ("zero" splice variant).

The molecular basis underlying functional diversity in BK_{Ca} Ca^{2+} sensitivity also includes the association of the α with the β subunit. For example, BK_{Ca} channels present in skeletal muscle, a tissue with a very low level of β subunit expression (TSENG-CRANK et al. 1996; HANNER et al. 1997) are less sensitive to Ca^{2+} than those BK_{Ca} channels of smooth muscle where the β subunit expresses abundantly (e.g., TANAKA et al. 1997; JIANG et al. 1999). Differential expression of the β subunit in the cochlea appears to be crucial in determining the electrical tuning of hair cells. Hair cell *Slo* β subunit decreases from lowest (apical) to higher frequencies regions of the epithelia. RAMANATHAN et al. (1999) argued that alternative splicing of the *Slo* gene is not enough to provide the functional heterogeneity of BK_{Ca} channels in hair cells. The *Slo* splice variant in hair cells revealed little or no difference in equilibrium or kinetic parameters. On the other hand, interaction between the β with α *Slo* splice variants may produce the necessary channel activation kinetic range needed for electrical tuning of the cochlear hair cells.

Functional diversity due to heterotetrameric formation of BK_{Ca} channels has not been shown in an heterologous system. However, as mentioned before, heterotetrameric BK_{Ca} channels are expressed when RNAs encoding channels with distinct TEA binding affinities are injected into oocytes (SHEN et al. 1994). WU et al. (1997) examined the Ca^{2+} activation characteristics of BK_{Ca} channels isolated from avian nasal glands reconstituted into lipid bilayers. They found that the Ca^{2+} sensitivity varied from channel to channel but it is possible to pool the Ca^{2+} activation curves into five clusters. One simple explanation to this finding is to assume that the tetrameric channels are formed by two distinct subunits possessing different Ca^{2+} sensitivities. These different subunits may derive, as discussed above, from alternatively spliced variants.

E. Ca^{2+} Sensing Domain(s): The Calcium Bowl

The work with chimeric m*Slo*-d*Slo* channels led WEI et al. (1994) to suggest that the α subunit has two functional domains: the core encompasses transmembrane segments S0 to S8 and the tail (Fig. 1). The tail has been associated with Ca sensitivity. On the other hand, the core domain was associated to single channel conductance, channel open time, and voltage dependence. The BK_{Ca} channel α subunit primary sequence does not show any of the consensus sequences for Ca^{2+} binding. In this case the strategy used to detect potential calcium binding sites was the scanning of 3 or more acidic residues within a moving frame of 12 amino acids, a method by which most Ca^{2+} binding domains can be identified (KRAUSE et al. 1997). This approach showed 12 potential Ca binding sites in h*Slo*, (6 of them in the tail domain) and mutations have been performed in these regions.

In order to differentiate between an effect upon a "putative" Ca^{2+} binding site and an indirect effect upon domains that participate in other events that lead to channel opening, the selectivity of the Ca^{2+} binding site was determined. Given that Sr^{2+}, Mn^{2+}, and Cd^{2+} can also activate BK_{Ca} channels (OBERHAUSER et al. 1988), a mutation affecting a Ca^{2+} binding site should also alter the selectivity of the site for these divalent cations. Mutations in m*Slo* (SCHREIBER and SALKOFF 1997) or h*Slo* (KRAUSE et al. 1997) channels indicate that out of the 12 potential sites, at least two different domains participate in channel activation. One of these domains is the "Ca bowl" (Figs. 1 and 2), a 28 amino acid stretch that is the most conserved region between different species throughout the complete protein and concentrates many negative charges, mostly aspartates (Figs 1 and 2). Several different mutations in this region cause a 50 mV positive shift in the voltage activation curve that is not observed when Cd^{2+} is used to activate the channels. This result indicates that this region is highly selective for calcium over cadmium and that there must be a second site to which Cd^{2+} binds (SCHREIBER and SALKOFF 1997). KRAUSE et al. (1996, 1997) also found that this region is involved in calcium binding associated to modulation of channel gating. Their h*Slo* mutant D886 N showed a decreased Ca^{2+} sensitivity and an altered selectivity sequence for channel activation between Ca^{2+}, Sr^{2+}, and Mn^{2+} when compared to the wild type channel. Moss et al. (1996) proposed that the calcium bowl is involved in Ca^{2+} binding based on their study of the similarity of part of the BK_{Ca} carboxyl-terminus with the Ca^{2+} binding loop of serine proteases.

The original proposal of WEI et al. (1994) that associates calcium binding exclusively to the tail region seems untenable. Using the same criteria mentioned above, KRAUSE et al. (1996) have identified a region outside the "tail" and in the S6-S7 linker (h*slo* mutant D358 N) as participating in Ca^{2+} binding. Also, WALLNER et al. (1996) found differences in Ca^{2+} sensitivity among wild type and different h*Slo*-d*Slo* chimeric channels where "core" regions were interchanged. These observations strongly suggest that calcium sensitivity of BK_{Ca} channels is determined not just by the carboxyl domain but by the whole protein.

F. Origin of Voltage Dependence in BK_{Ca} Channels

Since BK channels are activated by voltage *and* by cytoplasmic Ca^{2+}, their voltage sensing mechanism may not be the same as that used by purely voltage-dependent channels. A mechanism to explain the voltage dependence in BK_{Ca} channels was one in which the binding of Ca^{2+} is voltage-dependent implying that Ca^{2+} binding was a necessary step to open the channel (WONG and LECAR 1982; MOCZYLOWSKI and LATORRE 1983). However, it has become clear that for BK_{Ca} channels the ion gating hypothesis is untenable. First, the ion gating model demands a linear relationship between the half activation potential ($V_{1/2}$; voltage at which P_o=0.5) and the $[Ca^{2+}]_i$. However, for $[Ca^{2+}]$ 100 nmol/l the *hSlo* channel becomes $[Ca^{2+}]$-independent (MEERA et al. 1996). Second, using the *mSlo* channel WEI et al. (1994), and more recently, CUI et al. (1997) showed a marked decrease in the slope of the $V_{1/2}$ – $[Ca^{2+}]_i$ relation at $[Ca^{2+}]_i$>10^{-4} mol/l. Third, STEFANI et al. (1997) demonstrated that h*Slo* possesses an intrinsic voltage sensor by measuring gating currents. As shown for ionic currents, at low $[Ca^{2+}]_i$ these gating currents are purely voltage-dependent. Raising the $[Ca^{2+}]_i$ shifted the P_o-voltage and the charge-voltage curves towards the left along the voltage axis, but the limiting gating charge as well as the limiting open probability were found to be $[Ca^{2+}]$-independent. In contrast to other voltage-dependent channels where charge moves preferentially between closed states, in *hSlo* channels charge also moves between open states. The total charge per h*Slo* channel is 4–5 elementary charges. This value is smaller than the one obtained for other voltage-dependent K^+ channels of the S4 superfamily. In *Shaker* K^+ channels the charge per channel is 13 elementary charges. SCHOPPA et al. 1992; SEOH et al. 1996; AGGARWAL and MACKINNON 1996; NOCETI et al. 1996. In BK_{Ca} channels, the positively charged S4 segment (Fig. 1) is a good candidate to be or form part of the voltage sensor. In the voltage-dependent *Shaker* K^+ channels the distribution of accessible positively charged residues of the S4 segment to cysteine reactive species is a function of the gating state of the channel (MANNUZU et al. 1996; LARSSON et al. 1996; BAKER et al. 1998). Only four of the seven charged residues in S4 contribute significantly to the gating charge: arginines (R) in positions 362, 365, 368, and 371 (SEOH et al. 1996; AGGARWAL and MACKINNON 1996; BAZANILLA 1999). As determined from the P_o-V curves, in *hSlo* channels only two of the positively charged residues of the S4 segment contribute to the channel voltage dependence: arginine 210 and 213 (DIAZ et al. 1998). In *Shaker* K^+ channels these positions correspond to residues R368 and R371). Fewer charges in the S4 region of *hSlo* could explain the finding of less gating charges per channel in *hSlo* compared with *Shaker* K^+ channels.

The results from macroscopic current measurements have been explained by models in which the Ca^{2+}-binding steps are independent from the voltage-dependent conformational changes that the channel undergoes during activation (COX et al. 1997). The model assumes that Ca^{2+} binds to open and closed conformations and that the voltage-dependent steps reside in the open-closed channel transitions. However, it is important to note here that a gating kinetic model of BK_{Ca} channels should also consider their single channel and gating current properties: (1) the large and slow fluctuations in open probability with time ("wanderlust kinetics") (SILBERBERG et al. 1996); (2) the long closed Ca^{2+}-independent intervals that limit channel activation at high $[Ca^{2+}]$ (ROTHBERG et al. 1996) (3) the brief lifetime closed states described by Rothberg and MAGLEBY (1998); and (4) the purely voltage-dependent gating currents with charge displacements occurring between closed *and* open states (STEFANI et al. 1997). This last feature of h*Slo* gating currents is difficult to reconcile with the Cox et al. (1997) model since in this model charge moves only between closed to open transitions.

G. Channel Inactivation

Chromaffin, PC12, and pancreatic β cells express a fast inactivating BK channel apparently involved in modulating the pattern of neurosecretion (LINGLE et al. 1996; SOLARO et al. 1997; DING et al. 1998). Their inactivation process can be removed by internal application of trypsin, suggesting that a cytoplasmic portion of the channel-forming protein may be involved. However, this cytosolic domain does not behave as an open channel blocker (ball-and-chain mechanism) since occupancy of the internal vestibule of the channel by quaternary ammonium blockers does not slow inactivation (SOLARO et al. 1997). Moreover, the *Shaker* B ball peptide failed to slow down inactivation despite the fact that it is able to interact with BK channels and behaves as an internal open-pore blocker (FOSTER et al. 1992; TORO et al. 1992; KUKULJAN et al. 1995). Therefore, block of permeation by the inactivating protein domain does not takes place by interacting with a receptor located in the internal mouth of this BK channel. In rat adrenal chromaffin cells BK channel appear to be a heteromultimer composed of subunits carrying the inactivation domain and others deprived of it. Removal of inactivation by trypsin is best accounted for by an average of two to three inactivation domains per channel (DING et al. 1998).

Current induced by the α subunit and a new identified $\beta 2$ subunit (WALLNER et al. 1999) closely resembles the characteristics of the inactivating currents from chromaffin cells. The amino terminal of this β subunit contains a 19 amino acid "ball peptide" that behaves as an open channel blocker.

Ca^{2+} sensing domain(s): the calcium bowl. In spite of its calcium sensitivity, the α subunit primary sequence does not have any of the concensus sites for calcium binding. A mutation in the 'calcium bowl' a 28 amino acid stretch between segmeents S9 and S10 makes it possible to identify this region as one of the calcium binding sites (KRAUSE et al. 1996; SCHREIBER and SALFOFF 1997). This region concentrates many negative charges, mostly aspartates and it is the most conserved region between different species throughout the complete protein (Fig. A). SCHREIBER and SALKOFF (1997) showed that the calcium bowl is highly selective for calcium and that there must be at least one more region, that can be activated also by Cd^{2+}, that participates in channel activation.

H. Metabolic Modulation

Besides Ca^{2+} and associated regulatory subunits, BK_{Ca} channels are also metabolically modulated. Metabolic modulation has been extensively studied in smooth muscles and to a lesser extent in other tissues and includes a variety of agonists and intracellular pathways (for a review see TORO and STEFANI 1991). Some examples are potent vasoconstrictors such as angiotensin II (TORO et al. 1990; MINAMI et al. 1995) and thromboxane A_2 (SCORNIK et al. 1992; TANAKA and TORO 1996) that cause inhibition of BK_{Ca} channels. Others are vasorelaxants such as nitro compounds (WILLIAMS et al. 1988; ROBERTSON et al. 1993; PENG et al. 1996; STOCKAND and SANSOM 1996a; BYCHKOV et al. 1998; LI et al. 1998) and β-adrenergic agents (TORO et al. 1990; KUME et al. 1992) that induce BK_{Ca} channel activation. The mechanisms of action on BK_{Ca} channels may be summarized in: (1) a direct interaction with the channel such as the case of nitric oxide (BOLOTINA et al. 1994; SHIN et al. 1997), G proteins (TORO et al. 1990; KUME et al. 1992; SCORNIK et al. 1993; WALSH et al. 1996; LEE et al. 1997; LI and CAMPBELL 1997), arachidonic acid and metabolites (KIRBER et al. 1992; ZOU et al. 1996), carbon monoxide (WANG et al. 1997; WANG and WU 1997), and steroids (FARRUKH et al. 1998; VALVERDE et al. 1999); (2) through second messenger pathways such as phosphorylation/dephosphorylation cycles (reviews: TORO and STEFANI 1993; LEVITAN 1994) and changes in the redox state (LEE et al. 1994; THURINGER and FINDLAY 1997; WANG et al. 1997b); and (3) changes in "bulk" (YUAN et al. 1996) or "local" intracellular Ca^{2+} (PORTER et al. 1998) by Ca^{2+} release from intracellular stores (PORTER et al. 1998) or Ca^{2+} entry (LEMOS 1995; MOREAU et al. 1996).

It is becoming evident that neurotransmitters, neuropeptides, vasoactive substances, and widely used therapeutic agents modulate the activity of BK_{Ca} channels using more than one of the mechanisms mentioned above. Isoproterenol, a β-adrenergic agonist used to prevent smooth muscle contraction, activates BK_{Ca} channel activity via a direct G protein effect, but also via phosphorylation by a cAMP dependent protein kinase (SCORNIK et al.

1993; KUME et al. 1994). Nitric oxide releasing compounds, clinically used for their vasorelaxant effect, activate smooth muscle BK_{Ca} channels using all three mechanisms. Nitric oxide may act directly on BK_{Ca} channels (BOLOTINA et al. 1994; SHIN et al. 1997). Nitric oxide may also stimulate cGMP-dependent pathways and BK_{Ca} by direct cGMP-mediated phosphorylation (ROBERTSON et al. 1993; ARCHER et al. 1994; ALIOUA et al. 1995; STOCKAND and SANSOM 1996), and indirectly via phosphatase activation (ZHOU et al. 1996), or an increase of Ca^{2+} spark frequency (PORTER et al. 1998). In cortical neurons, neurotrophin-3 stimulates BK_{Ca} channels through a signaling pathway that includes tyrosine kinase, phospholipase C, and protein dephosphorylation (HOLM et al. 1997). In general, the multiplicity of mechanisms triggered by a metabolite or external drug should allow an exquisite fine-tuning of BK_{Ca} channel activity. Evidently, BK_{Ca} channel metabolic modulation will be governed by the relative expression and/or colocalization of receptors, BK_{Ca} channels, and intracellular proteins or organelles in a given cell type.

The molecular target of BK_{Ca} channel modulation by nitric oxide, G proteins, arachidonic acid and metabolites, carbon monoxide, steroids, redox state and phosphorylation/dephosphorylation may be its α and/or regulatory subunits. However, very little is known about the modulation of *Slo* channels by these agents. Although α and β proteins have consensus sequences for phosphorylation (TORO et al. 1998) and PKA and PKG modulate native BK_{Ca} channels to a large extent, phosphorylation by these kinases has not been possible to demonstrate using inside-out patches of cells expressing the canine *Slo* (*cSlo)* channel (VOGALIS et al. 1996). However, in *hSlo* channels PKA dependent phosphorylation activated the α and inhibited α/β channels (DWORETZKY et al. 1996), whereas PKG directly phosphorylates the α subunit in vivo (ALIOUA et al. 1998). Comparison of *cSlo* and *hSlo* sequences shows no obvious amino acid changes that could explain the lack of phosphorylation in *cSlo*. Experimental differences rather than sequence variations may explain the results. In fact, phosphorylation by PKG was observed in HEK cells expressing *cSlo* and a phosphorylation site identified at serine 1031 (Fig. 1) (TUKAO et al. 1999). Reducing agents increase *hSlo* channel activity while oxidizing agents reduce it (DI CHIARA and REINHART 1997; WANG Z-W et al. 1997). Whether intracellular redox couples like NAD/H or glutathion play a modulatory role on *hSlo* channel needs to be addressed. It is evident that investigations up to now have characterized the effects that modulators may have on mammalian *Slo* channels, but work needs to be done to identify the molecular determinants responsible for the modulatory responses.

I. Pharmacology

I. BK_{Ca} Channels Blockers

1. Toxins

Charybdotoxin was the first high affinity toxin discovered able to inhibit BK_{Ca} channel activity (MILLER et al. 1985). The toxin isolated from the venom of the scorpion *L. quinquestratus* is a 37-amino acid peptide and blocks BK_{Ca} channels at nanomolar concentrations according to a bimolecular reaction. The toxin occludes the pore and prevents ion conduction by binding to the extracellular entryway of the channel. Charybdotoxin has made it possible to isolate and to purify the BK_{Ca} channel as well as to identify the molecular nature of its β subunit (KNAUS et al. 1994). Scorpion toxins of the ChTX type contain six cysteine residues and they fall into three different subclasses (GARCIA et al. 1994). Within each subclass toxins exhibit an amino acid identity larger than 70%. The first subclass is composed of ChTX and iberiotoxin (IbTX); the second subclass consists of margatoxin (MgTX) and noxiustoxin (NxTX); and the third subclass contains the agitoxins (AgTX1–4). Once the structure of ChTX (e.g., BONTEMS et al. 1991) and analogues was elucidated, these toxins became an extremely useful tool to inquire about the arrangement of amino acid residues in the external vestibule of voltage-dependent potassium channels (e.g., MACKINNON 1991; HIDALGO and MACKINNON 1995; RANGANATHAN et al. 1996). Arrangements have been confirmed by the recent determination of the crystal structure of a bacterial K^+ channel (DOYLE et al. 1998). Moreover, the fact that one amino acid residue in AgTX2 (lysine 27) is in close proximity to a K^+ binding site located in the *Shaker* K^+ channel allowed RANGANATHAN et al. (1996) to locate the position of the amino acid residues that make the selectivity filter in this channel. The disadvantage of ChTX is its low selectivity for BK_{Ca} channels. Charybdotoxin inhibits with high affinity the voltage-dependent $K_V1.3$ channel and with a lower affinity the $K_V1.2$ channel (GRISSMER et al. 1994). Charybdotoxin also inhibits other Ca^{2+}-activated K^+ channels of intermediate and small conductance. Of particular interest for the present review is IbTX since is highly selective for BK_{Ca} channels (GARCIA et al. 1995). The mechanism for binding of this toxin to the BK_{Ca} channel is similar to that of ChTX (CANDIA et al. 1992; GIANGIACOMO et al. 1992) and binds with a K_d of about 1 nmol/l which is about ten times smaller than that for ChTX. KOSCHAK et al. (1997) engineered a double IbTX mutant in which aspartate (D) 19 was replaced by a tyrosine and tyrosine 36 was replaced by a phenylalanine (F). This mutant was subsequently radioiodinated to high specific activity with ^{125}I ([^{125}I]D19Y/Y36F IbTX). Since IbTX seems to be highly specific for BK_{Ca} channels, it is a powerful tool to determine the distribution of BK_{Ca} channels and in purifying BK_{Ca} channel complexes.

2. Organic Blockers

a) Tetraethylammonium

Two different binding sites for TEA have been located in BK_{Ca} channels: a high affinity external TEA binding site with a $K_d = 0.14–0.29$ mmol/l and a low affinity internal binding site, $K_d = 27–60$ mmol/l (LATORRE 1994; VERGARA et al. 1999). As discussed above the *Slo* protein has a tyrosine (Y) located in the pore region in position 308 in d*Slo* and all potassium channels containing a Y in this position show high affinity for external TEA. Tetraethylammonium and derivatives have been used as probes of the pore structure (VILLARROEL et al.

1988; VERGARA et al. 1999). The external binding site is specific for TEA and appears to select quaternary ammonium ions by size. On the other hand, the internal TEA site contains a hydrophobic pocket able to accommodate the long hydrophobic tail of compounds such as nonyltrimethylammonium.

b) Indole Diterpenes

Indole diterpenes are the most potent non-peptidyl compounds BK_{Ca} channel inhibitors and they were identified based on their ability to modulate ChTX binding (KNAUS et al. 1994). They are fungal metabolite and cause tremors in animal that consume contaminated grains. Some compounds as paxilline and verruculogen, stimulate ChTX binding, while others such as aflatrem and penitrem A inhibit the binding of the toxin to BK_{Ca} channels. Of these compounds the best characterized electrophysiologically is paxilline (GRIBKOFF et al. 1996; SANCHEZ and MCMANUS 1996). This drug inhibits by binding to a site located in the cytoplasmic side of the α subunit with a Hill coefficient of 1 and with a K_d of 2.2 nmol/l. The K_d is $[Ca^{2+}]$-dependent, increasing as the internal $[Ca^{2+}]$ is augmented (SANCHEZ and MACMANUS 1994). DRIFKOFF et al. (1996), on the other hand, have described a high affinity (K_d = 9 nmol/l) and a low affinity site (K_d = 530 nmol/l).

c) General Anesthetics

Three general anesthetics, isoflurane, enflurane, and halotahane inhibit Ca^{2+}-activated K^+ channels in chromaffin cells (PANCRAZIO et al. 1992). This is of importance regarding the mechanism of action of general anesthetics since at the synaptic level the BK_{Ca} channel modulate transmitter release (ROBITAILLE et al. 1993; KNAUS et al. 1996; YAZEJIAN et al. 1997). Ketamine, a general anesthetic different from inhalation anesthetics, blocks BK_{Ca} channels in GH_3 cells. In GH_3 cells ketamine decreases P_o with a K_d of about 20 μmol/l (DENSON et al. 1994). However, ketamine was ineffective in reducing BK_{Ca} currents induced by h*Slo* in *Xenopus* oocytes (GRIBCOFF et al. 1996).

II. BK_{Ca} Channel Activators

1. Activators Isolated from *Desmodium adscendens*: A Medicinal Herb

Three organic compounds present in a crude extract of a medical herb used in Ghana to treat ailments related to smooth muscle contraction have proved to be potent activators of BK_{Ca} channels (MCMANUS et al. 1993). The compounds were identified as triterpenoid glycosides: dehydrosoyasaponin (DHS-I), soyasaponin I, and soyasaponin III. The most potent of these compounds is DHS-I, acting at nanomolar concentrations and from the internal side only, increases P_o. DHS-I increases the rate of dissociation of ChTX and since the toxin binds to a site located in the external side of the pore, interaction between these two compounds is mediated by an allosteric mechanism. DHS-

I does not activate BK_{Ca} channels in the absence of Ca^{2+} and requires the presence of the β subunit to exert its activation effect in the nanomolar range (McMANUS et al. 1995; WALLNER et al. 1999). However, DHS-I can activate the α subunit alone at micromolar concentrations (WALLNER et al. 1999). GIANGIACOMO et al. (1998) propose a model where binding of four DHS-I molecules bind preferentially to the open channel for maximal activation.

2. Anti-Inflamatory Aromatic Compounds (Fenamates)

Several compounds that are commonly used as Cl^- channel blockers activate BK_{Ca} channels (OTTOLIA and TORO 1994). External 100 mmol/l flufenamic or niflumic acid activate BK_{Ca} channels by increasing P_o by about 40% whereas the same concentration of mefenamic acid increases P_o by only 10%. Internal niflumic acid also activates BK_{Ca} channels but less effectively. Externally applied niflumic acid does not interfere with channel block by charybdotoxin; conversely, partial blockade induced by external TEA does not hinder BK_{Ca} channel activation by niflumic acid. These results indicate that fenamates act at a site distinct for that for charybdotoxin or TEA. Activation of BK_{Ca} channels induced by fenamates has similar characteristics in oocytes expressing either m*Slo* or *hSlo* (WALNER et al. 1995; GRIBKOFF et al. 1996). These results are of importance since they strongly suggest that the fenamate binding site is located in the α subunit of *Slo*.

3. Benzimidazolones

Several benzamidazolones such as NS004 (OLESEN et al. 1994), NS1608 (STROBAEK et al. 1996), and NS1619 (GRIBKOFF et al. 1996) are highly effective in increasing BK_{Ca} currents aortic smooth muscle, in HEK 293 cells transfected with h*Slo* and in oocytes expressing m*Slo* or h*Slo* channels, respectively. The potency sequency is the following: NS1608 > NS1604 > NS1619. Like the fenamates, the bezimidazolones shift the the BK_{Ca} channel voltage activation curve towards the left along the voltage axis with a $K_d = 2.1\,\mu$mol/l in the case of NS 1608 (STROBAEK et al. 1996).

4. Phloretin

This flavonoid that is able to decrease voltage-dependent Na^+ and K^+ conductances in axons (KLUSEMAN and MEVES 1991; STRICHARTZ et al. 1980) *activates* BK_{Ca} channels by shifting to the left the voltage activation curve. At 80 mmol/l phloretin shifts to the left the P_o vs voltage curve by 64 mV (KOH et al. 1994). Given the differential influences of phloretin on K_V and BK_{Ca} channels, this flavonoid may be useful as a pharmacological tool to discriminate their gating properties.

5. Ethanol

This alcohol increases the P_o of skeletal muscle BK_{Ca} channels incorporated into lipid bilayers at clinically relevant concentrations (25–200 nM) (CHU et al.

1998). It is important to note here that at 50 mM ethanol increases P_o about eightfold. A ethanol concentration of 50 mM is equivalent to a 0.2% weight/volume ethanol solution and the experimenter should be very careful when testing the effect of compounds on BK_{Ca} channels that, due to their solubility, are dissolved in ethanol. Ethanol also affects BK_{Ca} channel activity in isolated neurohypophysial terminals (DOLPICO et al. 1996) and the activity of the m*Slo* channel expressed in oocytes of *Xenopus laevis* (DOLPICO et al. 1998).

J. Summary and Conclusions

The membrane topology of BK_{Ca} channel-forming protein (α subunit) was resolved. The protein spans the membrane seven times (S0–S7) leaving an external amino-terminus and a large cytoplasmic carboxyl-terminus. In some tissues, particularly in smooth muscle, BK_{Ca} channels are accompanied by a modulatory $\beta 1$ subunit. In chromaffin cells, a $\beta 2$ subunit may cause inactivation. The segment S0 is crucial for the functional interaction between the α and the β subunit. BK_{Ca} channels appear to originate from a single gene (*slowpoke*) and attain its great diversity on the basis of splicing and the formation of heteromultimers. Functional diversity of BK_{Ca} channels also originates by association with other proteins such as β subunits, dSL1P1 and Slob and/or metabolic regulation. BK_{Ca} channels and their great diversity play an important role in a number of physiological processes. For example, the smooth muscle tone, in determining the tonotopic map of the chicken cochlea and in controlling the excitable properties of epinephrine secreting cells. BK_{Ca} channels can be now considered as voltage-gated and calcium-modulated since they possess an intrinsic voltage-sensor, probably part of or the S4 transmembrane domain. The discovery and characterization of a number of peptidyl toxins, organic blokers, and channel openers have provided valuable tools in the study of BK_{Ca} channel function.

Acknowledgements. This work was supported by Chilean grant FNI 197–0739, FNI 100890 and Catedra Presidencial and a group of Chilean companies (CODELCO, CMPC, CGE, Minera Escondida, NOVAGAS, Bussiness Design Ass., and XEROX Chile) (to R. Latorre). Grant FNI 198–1053 (to C. Vergara). NIH grants HL54970 and HL47382 (to L. Toro) and GM52203 (to E. Stefani). L.T. is an Established Investigator of the American Heart Association. Human frontier in Science Program grant was given to R. Latorre and L. Toro. We thank Catherine Card for her assistance in preparing the manuscript.

References

Aggarwal SK, MacKinnon R (1996) Contribution of the S4 segment to gating charge in the Shaker K^+ Channel. 1996 Neuron 1996, 16:1169–1177

Alioua A, Huggins JP, Rousseau E (1995) PKG-Iα phosphorylates the α-subunit and upregulates reconstituted GK_{Ca} channels from tracheal smooth muscle. Am J Physiol Lung Cell Mol Physiol 268:L1057–L1066

Alioua A, Tanaka Y, Wallner M, Hofmann F, Ruth P, Meera P, Toro L (1998) The large conductance voltage-dependent and calcium-sensitive K+ channel, hSlo, is target of a GMP-dependent protein kinase phosphorilation in vivo. J Biol Chem 273:32950–32956

Anwer K, Oberti C, Pérez, GJ, Perez-Reyes N, McDougall JK, Monga M, Sanborn BM, Stefani E, Toro L (1993) Calcium-activated K^+ channels as modulators of human myometrial contractile activity. Am J Physiol Cell Physiol 265:C976–C985

Archer SL, Huang JMC, Hampl V, Nelson DP, Shultz PJ, Weir EK (1994) Nitric oxide and cGMP cause vasorelaxation by activation of a charybdotoxin-sensitive K channel by cGMP-dependent protein kinase. Proc Natl Acad Sci USA 91:7583–7587

Atkinson NS, Robertson GA, Ganetzky B (1991) A component of calcium-activated potassium channels encoded by the Drosophila slo locus. Science 253:551–555

Baker OS, Larsson HP, Mannuzzu LM, Isacoff EY (1998) Three transmembrane conformations and sequence-dependent displacement of the S4 domain in Shaker K^+ channel gating. Neuron 20:1283–1294.

Bezanilla F. (2000) The voltage sensor in voltage-dependention channels. Physiol Rev (In Press)

Bolotina VM, Najibi S, Palacino JJ, Pagano PJ, Cohen RA (1994) Nitric oxide directly activates calcium-dependent potassium channels in vascular smooth muscle. Nature 368:850–853

Bontems F, Roumestand C, Menez A, Toma F (1991) Refined structure of charybdotoxin:common motifs in scorpion toxins and insect defensins. Science 254:1521–1523

Butler A, Tsunoda S, McCobb DP, Wei A, Salkoff L (1993) mSlo, a complex mouse gene encoding "maxi" calcium-activated potassium channels. Science 261:221–224

Bychkov RM, Gollasch T, Steinke C, Ried FC, Luft J, Haller H (1998) Calcium-activated potassium channels and nitrate-induced vasodilation in human coronary arteries J Pharmacol Exp Ther 285:293–298

Candia S, Garcia ML, Latorre R (1992) Mode of action of iberiotoxin, a potent blocker of the large conductance Ca^{2+}-activated K^+ channel. Biophys J 63:583–590

Cha A, Bezanilla F (1997) Characterizing voltage-dependent conformational changes in the Shaker K^+ channel with fluorescence. Neuron 19:1127–1140

Chang CP, Dworetzky SI, Wang J, Goldstien ME (1997) Differential expression of the α and β subunits of the large-conductance calcium-activated potassium channel: implication for channel diversity. Brain Res Mol Brain Res 45:33–40

Chu B, Dopico AM, Lemos JR, Triestman SN (1998) Ethanol potentiation of calcium-activated potassium channels reconstituted into planar lipid bilayers. Mol Pharmacol 54:397–406

Chung S, Reinhart PH, Martin BL, Brautigan D, Levitan IB (1991) Protein kinase activity closely associated with a reconstituted calcium-activated potassium channel. Science 253:560–562

Cox DH, Cui J, Aldrich RW (1997) Allosteric gating of a large conductance Ca^{2+}-activated K^+ channel. J Gen Physiol 110:257–281

Cui J, Cox DH, Aldrich RW (1997) Intrinsic voltage dependence and Ca^{2+} regulation of mslo large conductance Ca-activated K^+ channels. J Gen Physiol 109:647–673

Denson DD, Duchatelle P, Eaton DC (1994) The effect of racemic ketamine on the large conductance Ca^{2+}-activated potassium channel (BK) channels in GH_3 cells. Brain Res 638:61–68

Dhasakumar S, Navaratnam S, Bell TJ, Dinh Tu T, Cohen EL, Oberholtzer JC (1997) Differential distribution of Ca^{2+}-activated K^+ channel splice variants among hair cells along the tonopic axis of the chick cochlea. Neuron 19:1077–1085

Diaz L, Meera P, Amigo J, Stefani E, Alvarez O, Toro L, Latorre R (1998) Role of the S4 segment in a voltage-dependent calcium-sensitive potassium (hSlo) channel. J Biol Chem 273:32430–32436

DiChiara T, Reinhart PH (1997) Redox modulation of hslo Ca^{2+}-activated K^+ channels. J Neurosci 17:4942–4955

Ding JP, Li ZW, Lingle CJ (1998) Inactivating BK channels in rat chromaffin cells may arise from heteromultimeric assembly of distinct inactivation-competent and noninactivating subunits. 1998 Biophys J 74:268–289

Dopico AM, Lemos JR, Treitman SN (1996) Ethanol increases the activity of large conductance Ca^{2+}-activated K^+ channels in isolated neurohypophysial terminals. Mol Pharmacol 49:40–48

Dopico AM, Anantharam V, Treitman SN (1998) Ethanol increases the activity of large conductance Ca^{2+}-dependent K^+ (mSlo) channels: functional interaction with cytosolic Ca^{2+}. J Pharmacol Exp Ther 284:258–268

Doyle DA, Cabral JM, Pfuetzner RA, Kuo A, Gulbis JM, Cohen SL, Chait BT, MavKinnon R (1998) The structure of the potassium channel: molecular basis of K^+ conduction and selectivity. Science 280:69–81

Dworetzky SI, Boissard CG, Lum-Ragan JT, Mckay MC, Post-Munson DJ, Trojnacki JY, Chang C-P, Gribkoff VK (1996) Phenotypic alteration of a human BK (hslo) channels by hSlo subunit coexpression: changes in blocker sensitivity, activation/relaxation and inactivation kinetics, and protein kinase A modulation. J Neurosci 16:4543–4550

Farrukh IS, Peng W, Orlinska U, Hoidal JR (1998) Effect of dehydroepiandrosterone on hypoxic pulmonary vasoconstriction: a Ca(2+)-activated K(+)-channel opener. Am J Physiol Lung Cell Mol Physiol 274:L186–L195

Ferrer J, Wasson J, Salkoff L, Permutt MA (1996) Cloning of human pancreatic islet large conductance Ca^{2+}-activated K^+ channel (Hslo) cDNA's: evidence for high levels of expression in pancreatic islets and identification of a flanking genetic marker. Diabetologia 39:891–898

Foster CD, Chung S, Zagotta WN, Aldrich RW, Levitan IB (1992) A peptide derived from the Shaker B K^+ channel produces short and long blocks of reconstituted Ca^{2+}-dependent K^+ channels. Neuron 9:229–236

Garcia ML, Garcia-Calvo M, Hidalgo P, Lee A, MacKinnon R (1994) Purification and characterization of three inhibitors of voltage-dependent potassium channels from Leirus quinquestriatus var. hebraeus venom. Biochemistry 33:6834–6839

Garcia ML, Knaus H, Munujos P, Slaughter RS, Kaczorowski GJ (1995) Charybdotoxin and its effects on potassium channels. Am J Physiol Cell Physiol 269:C1–C10

Garcia-Calvo M, Knaus HG, Mc Manus OB, Giangiacomo KM, Kaczorowski GJ, Garcia ML, Knaus HG (1994) Purification and reconstitution of the high-conductance, calcium activated potassium channel from smooth muscle, a representative of the mSlo and slowpoke family of potassium channels. J Biol Chem 269:3921–3924

Giangiacomo KM, Garcia ML, McManus OB (1992) Mechanism of iberiotoxin block of the large-conductance calcium-activated potassium channel from bovine aortic smooth muscle. Biochemistry 31:6719–6727

Giangiacomo KM, Kamassah A, McManus OB (1998) Mechanism of maxi-K channel activation by dehydrosaponin-I. J. Gen. Physiol. 112:485–501

Grissmer S, Nguyen AN, Aiyar J, Hanson DC, Mather RJ, Gutman GA, Karmilowicz MJ, Auperin DD, Chandy KG (1994) Pharmacological characterization of five cloned voltage-gated K^+ channels, types Kv1.1, 1.2, 1.3, 1.5 and 3.1, stably expressed in mammalian cell lines. Mol Pharmacol 45:1227–1234

Gribkoff VK, Lum-Ragan JT, Boisssard CG, Post-Munson DJ, Meanwell NA, Starret JE, Kowslowski ES, Romine JL, Trojnacki JT, Craig McKay M, Zhong J, Dworetzky SI (1996) Effect of channel modulators on cloned large-conductance calcium-activated potassium channels. Mol Pharmacol 50:206–217

Hanner M, Schmalhofer WA, Munujos P, Knaus HG, Kaczorowski GJ, Garcia ML (1997) The β subunit of the high-conductance calcium-activated potassium channel contributes to the high-affinity receptor for charybdotoxin. Proc Natl Acad Sci USA 94:2853–2858

Heginbotham L, MacKinnon R (1992) The aromatic binding site for tetraethylammonium ion on potassium channels. Neuron 8:483–491

Holm NR, Christophersen P, Olesen SP, Gammeltoft S (1997) Activation of calcium-dependent potassium channels in mouse brain neurons by neurothrophin-3 and nerve growth factor. Proc Natl Acad Sci USA 94:1002–1006

Jan LY, Jan YN (1992) Tracing the roots of ion channels. Cell 69:715–719

Jan LY, Jan YN (1997) Cloned potassium channels from eukaryotes and prokaryotes. Ann Rev Neurosci 20:91–123

Jiang G-J, Zidanic M, Michaels RL, Griguer C, Fuchs PA (1997) cSlo encodes calcium-activated potassium channels in the chick's cochlea. Proc R Soc Lond B 264:731–737

Jiang Z, Wallner M, Meera P (1999) Human and rodent MaxiK channel β subunit genes: cloning and characterization. Genomics 55:57–67

Jones EM, Laus C, Fettiplace R (1998) Identification of Ca^{2+}-activated K^+ channel splice variants and their distribution in the turtle cochlea. Proc R Soc Lond B 265:685–692

Knaus H-G, Folander K, Garcia-Calvo M, Garcia ML, Kaczorowski GJ, Smith M, Swanson R (1994) Primary sequence and immunological characterization of subunit of high conductance Ca^{2+}-activated K^+ channel from smooth muscle. J Biol Chem 269:17274–17278

Knaus H-G, McManus OB, Lee SH, Schmalhofer WA, Garcia-Calvo M, Hekms LMH, Sanchez M, Giangiacomo K, Reuben JP, Smith III AB, Kaczorowski GJ, Garcia ML (1994) Tremorgenic indole alkaloids potently inhibit smooth muscle high conductance Ca^{2+}-activated K^+ channels. Biochemistry 33:5819–5828

Koh D-S, Reid G, Vogel W (1994) Activating effect of the flavoid phloretin on Ca^{2+}-activated K^+ channels in myelinated nerve fibres of Xenopus laevis. Neurosci Lett 165:167–170

Koschak A, Koch RO, Liu J, Kaczorowski GJ, Rienhardt P, Garcia ML, Knaus H-G (1997) [125I]Iberiotoxin-D19Y/Y36F, the first selective, high specific activity radioligand for high-conductance calcium-activated potassium channels. Biochemistry 36:1943–1952

Krause JD, Foster CD, Reinhart PH (1996) Localization of Ca^{2+} domains in human (hslo) Ca^{2+}- activated K^+ channels. Soc Neurosci Abstract # 473.1, 22:1194

Krause JD, Gross JM, Foster CD, Reinhart PH (1997) Localization of Ca^{2+} domains in human (hslo) Ca^{2+}- activated K^+ channels. Soc Neurosci Abstract 680.3, 23:1737

Kukuljan M, Labarca P, Latorre R (1995) Molecular determinants of ion conduction and inactivation in K^+ channels. 1995 Am J Physiol Cell Physiol 37:C535–C556

Kirber MT, Ordway RW, Clapp LH, Walsh JV Jr, Singer JJ (1992) Both membrane stretch and fatty acids directly activate large conductance Ca^{2+}-activated K^+ channels in vascular smooth muscle cells. FEBS Lett 297:24–28

Kume H, Graziano MP, Kotlikoff MI (1992) Stimulatory and inhibitory regulation of calcium-activated potassium channels by guanine nucleotide-binding proteins. Proc Natl Acad Sci USA 89:11051–11055

Kume H, Hall IP, Washabau RJ, Takagi K, Kotlikoff MI (1994) β-Adrenergic agonists regulate K_{Ca} channels in airway smooth muscle by cAMP-dependent and -independent mechanisms. J Clin Invest 93:371–379

Lagrutta A, Shen KZ, North RA, Adelman JP (1994) Functional differences among alternatively spliced variants of Slowpoke, a Drosophila calcium-activate potassium channel. J Biol Chem 269:20347–20351

Larsson PH, Baker OS, Dhillon DS, Isacoff EY (1996) Transmembrane movement of the Shaker K^+ channel S4. Neuron 16:387–397

Latorre R, Oberhauser A, Labarca P, Alvarez O (1989) Varieties of calcium-activated potassium channels. Ann Rev Physiol 51:385–399

Latorre R (1994) Molecular workings of large conductance (Maxi) Ca^{2+}-activated K^+ channels. In: Peracchia C (ed) Handbook of Membrane Channels: Molecular and Cellular Physiology Academic Press, Inc. San Diego, Ca pp 79–102

Lee MY, Chung S, Bang HW, Baek BK, Uhm D (1997) Modulation of large conductance Ca^{2+}-activated K^+ channel by Galphah (transglutaminase II) in the vascular smooth muscle cell. Pflugers Arch 433:671–673

Lemos VS, Takeda K (1995) Neuropeptide Y_2-type receptor-mediated activation of large- conductance Ca^{2+}-sensitive K^+ channels in a human neuroblastoma cell line. Pflugers Arch 430:534–540

Levitan IB (1994) Modulation of ion channels by protein phosphorylation and dephosphorylation. Annu Rev Physiol 56:193–212

Li PL, Campbell WB (1997) Epoxyeicosatrienoic acids activate K^+ channels in coronary smooth muscle through a guanine nucleotide binding protein. Circ Res 80:877–884

Li PL, Jin MW, Campbell WB (1998) Effect of selective inhibition of soluble guanylyl cyclase on the K(Ca) channel activity in coronary artery smooth muscle. Hypertension 31:303–308

Lingle CJ, Solaro CR, Prakriya M, Ding JP (1996) Calcium-activated potassium channels in adrenal chromaffin cells. In Ion Channels. Edited by Narahashi T, Plenum Press, NY, 4:261–301

Mannuzzu LM, Moronne MM, Isacoff EY (1996) Direct physical measure of conformational rearrangement underlying potassium channel gating. 1996 Science 271:213–216

MacKinnon R (1991) Using mutagenesis to study potassium channel mechanisms. J Bioenerg Biomembr 23:647–663

Marrion NV, Tavalin SJ (1998) Selective activation of Ca^{2+}-activated K^+ channels by colocalized Ca^{2+} channels in hippocampal neurons. Nature 395:900–904

McManus OB (1991) Calcium-activated potassium channels: regulation by calcium. J Bioenerg Biomembr 23:537–560

McManus OB, Harris GH, Giangiacomo KM, Feigenbaum P, Reuben JP, Addy ME, Burka JF, Kaczorowski GJ, Garcia ML (1993) An activator of calcium-dependent potassium channels isolated from a medicinal herb. Biochemistry 32:6128–6133

McManus OB, Helms LM, Pallank M, Ganetzky B, Swanson R, Leonard RJ (1995) Functional role of the beta subunit of high conductance calcium-activated potassium channels. Neuron 14:645–650

Meera P, Wallner M, Jiang Z, Toro L (1996) A calcium switch for the functional coupling between α (hslo) and β subunits ($K_{V,Ca\beta}$) of maxi K channels. Febs Lett 382:84–88

Meera P, Wallner M, Song M, Toro L (1997) Large conductance voltage-and calcium-dependent K^+ channel, a distinct member of voltage-dependent ion channels with seven N-terminal transmembrane segments (S0-S6), an extracellular N terminus, and an intracellular (S9-S10) C terminus. Proc Natl Acad Sci USA 94:14066–14071

Minami KY, Hirata Y, Tokumura A, Nakaya Y, Fukuzawa K (1995) Protein kinase C-independent inhibition of the Ca(2+)-activated K+ channel by angiotensin II and endothelin-1. Biochem Pharmacol 49:1051–1056

Moczydlowski E, Latorre R (1983) Gating kinetics of Ca^{2+}-activated K^+ channels from rat muscle incorporated into planar lipid bilayers. Evidence for two voltage-dependent Ca^{2+} binding reactions. J Gen Physiol 82:511–542

Moreau R, Hurst AM, Lapointe JY, Lajeunesse D (1996) Activation of maxi-K channels by parathyroid hormone and prostaglandin E2 in human osteoblast bone cells. J Membr Biol 150:175–184

Moss GWJ, Marshall J, Moczydlowski E (1996) Hypothesis for a serine proteinase-like domain at the COOH terminus of slowpoke calcium-activated potassium channels. J Gen Physiol 108:473–484

Nelson MT, Cheng H, Rubart M, Santana LF, Bonev AD, Knot HJ, Lederer WJ (1995) Relaxation of arterial smooth muscle by calcium sparks. Science 270:633–637

Noceti F, Baldelli P, Wei X, Qin N, Toro L, Birnbaumer L, Stefani E (1996) Effective gating charges per channel in voltage dependent K^+ and Ca^{2+} channels. J Gen Physiol 108:143–155

Oberhauser A, Alvarez O, Latorre R (1988) Activation by divalent cations of a Ca^{2+}-activated K^+ channel from skeletal muscle membrane. J Gen Physiol 92:67–86

Olesen S-P, Munch E, Moldt P, Drejer J (1994) Selective activation of Ca^{2+}-dependent K^+ channels by novel benzimidazolone. Eur J Pharmacol 251:53–59

Ottolia M, Toro L (1994) Potentiation of large-conductance K_{Ca} channels by niflumic, flufenamic and mefenamic acids. Biophys J 67:2272–2279

Pancrazio JJ, Park WK, Lynch IIIC (1992) Effects of enflurane on the voltage-gated membrane currents of bovine adrenal chromaffin cells. Neurosci Lett 146:147–151

Peng W, Hoidal JR, Farrukh IS (1996) Regulation of Ca(2+)-activated K+ channels in pulmonary vascular smooth muscle cells: role of nitric oxide. J Appl Physiol 81:1264–1272

Porter VA, Bonev AD, Knot HJ, Heppner TJ, Stevenson AS, Kleppisch T, Lederer WJ, Nelson MT (1998) Frequency modulation of Ca^{2+} sparks is involved in regulation of arterial diameter by cyclic nucleotides. Am J Physiol Cell Physiol 274: C1346–1355

Ramanathan K, Michael TH, Jiang G-J, Hiel H, Fuchs PA (1999) A molecular mechanism for electrical tuning of cochlear hair cells. Science 283:215–217

Ranganathan R, Lewis JH, MacKinnon R (1996) Spatial localization of the K^+ channel selectivity filter by mutant cycle-based structure analysis. Neuron 16:131–139

Reinhard PH, Levitan IB (1995) Kinase and phosphatase activities intimately associated with a reconstituted calcium-dependent potassium channel. J Neurosci 15:4572–4579

Robertson BE, Schubert R, Hescheler J, Nelson MY (1993) cGMP-dependent protein kinase activates Ca-activated K channels in cerebral artery smooth muscle cells. Am J Physiol Cell Physiol 265:C299–C303

Robitaille R, Garcia ML, Kaczorowski GJ, Charlton MP (1993) Functional colocalization of calcium and calcium-gated potassium channels in control of transmitter release. Neuron 11:645–655

Rosenblatt KP, Sun Z-P, Heller S, Hudspeth AJ (1997) Distribution of Ca^{2+}-activated K^+ channel isoforms along the tonopic gradient of the chicken cochlea. Neuron 19:1061–1075

Rothberg BS, Magleby KL (1998) Kinetic structure of large conductance Ca^{2+}-activated channels suggest that the gating includes transitions through intermediate or secondary states. A mechanism for flickers. J Gen Physiol 111:751–780

Rothberg BS, Bello RA, Song L, Magleby KL (1996) High Ca^{2+} concentrations induce a low activity mode and reveal Ca^{2+}-independent long shut intervals in BK channels from rat muscle. J Physiol 493.3:673–689

Sah P (1996) Ca^{2+}-activated K^+ currents in neurones: types, physiological roles and modulation. TINS 19:150–154

Saito M, Nelson C, Salfkoff L, Lingle CJ (1997) A cysteine-rich domain defined by a novel exon in a Slo variant in rat adrenal chromaffin cells and PC12 cells. J Biol Chem 18:11710–11717

Sanchez M, McManus OB (1996) Paxilline inhibition of the alpha-subunit of the high conductance calcium-activated potassium channel. Neuropharmacology 35: 963–968

Schoppa NE, McCormack K, Tanouye MA, Sigworth FJ (1992) The size of gating charge in wild-type and mutant Shaker potassium channels. Science 255:1712–1715

Schopperle WM, Holquvist MH, Zhou Y, Wang J, Wang Z, Griffith LC, Keselman I, Kusinitz F, Dagan D, Levitan IB (1998) Slob, a novel protein that interacts with the Slowpoke calcium-dependent potassium channel. Neuron 20:565–573

Schreiber M, Salkoff L (1997) A novel calcium-sensing domain in the BK channel. Biophys J 73:1355–1363

Schreiber M, Wei A, Yuan A, Gaut J, Saito M, Salkoff L (1998) Slo3, a novel pH-dependent sensitive K^+ channel from mammalian spermatocytes. J Biol Chem 273:3509–3516

Schubert R, Noack T, Serebryakov VN (1999) Protein kinase C reduces the KCa current of rat tail artery smooth mucle cells. Am J Physiol 276:C648–C658

Scornik FS, Codina J, Birnbaumer L, Toro L (1993) Modulation of coronary smooth muscle K_{Ca} channels by $G_s\alpha$ independent of phosphorylation by protein kinase A. Am J Physiol Lung Cell Mol Physiol 265:H1460–H1465

Scornik FS, Toro L (1992) U46619, a thromboxane A2 agonist, inhibits KCa channel activity from pig coronary artery. Am J Physiol Cell Physiol 262:C708–C713

Seoh S-A, Sigg D, Papazian D, Bezanilla F (1996) Voltage-sensing residues in the S2 and S4 segments of the Shaker K channel. 1996 Neuron 16:1159–1167

Shen K-Z, Lagrutta A, Davies NW, Standen NB, Adelman JP, North RA (1994) Tetraethylammonium block of Slowpoke calcium-activated potassium channels expressed in Xenopus oocytes:Evidence for tetrameric channel formation. Pflugers Arch 426:440–445

Shih TM, Goldin al. (1997) Topology of the Shaker potassium channel probed with hydrophilic epitope insertions. J Cell Biol 136:1037–1045

Shin JH, Chung S, Park EJ, Uhm DY, Suh CK (1997) Nitric oxide directly activates calcium-activated potassium channels from rat brain reconstituted into planar lipid bilayer. FEBS Lett 415:299–302

Silberberg SD, Lagrutta A, Adelman JP, Magleby KL (1996) Wanderlust kinetics and variable Ca^{2+}-sensitivity of Drosophila, a large conductance Ca^{2+}-activated K^+ channel, expressed in oocytes. Biophys J 70:2640–2651

Solaro CR, Ding JP, Li ZW, Lingle CJ (1997) The cytosolic inactivation domains of BK_i channels in rat chromaffin cells do not behave like simple, open channel blockers. Biophys J 73:819–830

Stefani E, Ottolia M, Noceti F, Olcese R, Wallner M, Latorre R, Toro L (1997) Voltage-controlled gating in a large conductance Ca^{2+}-sensitive K^+ channel (hslo). Proc Natl Acad Sci USA 94:5427–5431

Stockand JD, Sansom SC (1996a) Role of large Ca(2+)-activated K+ channels in regulation of mesangial contraction by nitroprusside and ANP. Am J Physiol Cell Physiol 270:C1773–1779

Stockand JD, Sansom SC (1996b) Mechanism of activation by cGMP-dependent protein kinase of large Ca(2+)-activated K+ channels in mesangial cells. Am J Physiol Cell Physiol 271:C1669–1677

Tanaka Y, Meera P, Song M, Knaus HG, Toro L (1997) Molecular constituents of maxi K_{Ca} channels in human coronary smooth muscle: predominant $\alpha + \beta$ subunit complexes. J Physiol 502:545–557

Tanaka Y, Toro L (1996) Inhibitory effect by thromboxane A_2 agonist on K_{Ca} channel activity in human coronary artery smooth muscle cells. Biophys J 70:A396

Thuringer D, Findlay I (1997) Contrasting effects of intracellular redox couples on the regulation of maxi-K^+ channels in isolated myocytes from rabbit pulmonary artery. J Physiol 500:583–592

Toro L, Amador M, Stefani E (1990) ANG II inhibits calcium-activated potassium channels from coronary smooth muscle in lipid bilayers. Am J Physiol Heart Circul Physiol 258:H912–H915

Toro L, Ramos-Franco J, Stefani E (1990) GTP-dependent regulation of myometrial K_{Ca} channels incorporated into lipid bilayers. J Gen Physiol 96:373–394

Toro L, Stefani E (1993) Modulation of maxi calcium-activated K channels. Role of ligands, phosphorylation and G-proteins. In: Dickey B, Birnbaumer L (eds) Handbook of Experimental Pharmacology. Vol. 108 "GTPases in Biology". Springer-Verlag, New York pp. 561–579

Toro L, Stefani E, Latorre R (1992) Internal blockade of a Ca^{2+}-activated K^+ channel by Shaker B inactivating "ball" peptide. Neuron 9:237–245

Toro L, Vaca L, Stefani E (1991) Calcium-activated potassium channels from coronary smooth muscle reconstituted in lipid bilayers. Am J Physiol Heart Circul Physiol 260:H1779–H1789

Toro L, Wallner M, Meera P, Tanaka Y (1998) Maxi K_{Ca}, a unique member of the voltage-gated K channel superfamily. News Physiol Sci 13:112–117

Tseng-Crank J, Foster CD, Jrause JD, Mertz R, Godinot R, DiChiara TJ, Reinhart PH (1994) Cloning, expression, and distribution of functionally distinct Ca^{2+}-activated K^+ channel isoforms from human brain. Neuron 13:1315–1330

Tseng-Crank J, Godinot N, Johansen TE, Ahring PK, Strobaer D, Mertz R, Foster CD, Olesen S-P, Reinhart PH (1996) Cloning, expression and distribution of Ca^{2+}-activated K^+ channel from subunit from human brain. Proc Natl Acad Sci USA 92:9200–9205

Tukao M, Mason HS, Briton FC, Kenyon JL, Horowitz B, Keet KD (1999) Cyclic GMP protein-dependent kinase activates cloned BK_{Ca} channels expressed in mammalian cells by direct phosphorylation of serine 1072. J Biol Chem 74:10927–10935

Valverde MA, Rojas P, Amigo J, Cosmelli D, Orio P, Bahamonde MI, Mann GE, Vergara C, Latorre R (1999) Acute activation of Maxi-K channels (*hSlo*) by estradiol binding to the β subunit. Science 285:1929–1931

Vergara C, Latorre R, Marrion NV, Adelman JP (1988) Calcium-activated potassium channels. Curr Op Neurobiol (in press)

Vergara C, Alvarez O, Latorre R (1999) Localization of the K^+ lock-in and the Ba^{2+} binding sites in a voltage-gated calcium modulated channel: implications for survival of K^+ permeability. J Gen Physiol (In press)

Villarroel A, Alvarez O, Oberhauser A, Latorre R (1988) Probing a Ca^{2+} activated K^+ channel with quaternary ammonium ions. Pflüger Arch 413:118–126

Wallner M, Meera P, Ottolia M, Kaczorowski G, Latorrre R, Garcia ML, Stefani E, Toro L (1995) Cloning, expression and modulation by a β-subunit of a human maxi K_{Ca} channel cloned from human myometrium. Receptors and Channels 3:185–199

Wallner M, Meera P, Toro L (1996) Determinant for subunit regulation in high-conductance voltage-activated and Ca^{2+}-sensitive K^+ channels: An additional transmembrane region at the N terminus. Proc Natl Acad Sci USA 93:14922–14927

Wallner M, Meera P, Toro L (1999) Molecular basis of fast inactivation in voltage and Ca^{2+}-activated K^+ channels: a transmembrane β subunit homolog. Proc Natl Acad Sci USA 96:4137–4142

Walsh KB, Wilson SP, Long KJ, Lemon SC (1996) Stimulatory regulation of the large-conductance calcium-activated potassium channel by G proteins in bovine adrenal chromaffin cells. Mol Pharmacol 49:379–386

Wang R, Wu L (1997) The chemical modification of KCa channels by carbon monoxide in vascular smooth muscle cells. J Biol Chem 272:8222–8226

Wang R, Wu L, Wang Z (1997) The direct effect of carbon monoxide on KCa channels in vascular smooth muscle cells. Pflugers Arch 434:285–291

Wang Y-X, Fleischmann BK, Kotlikoff MI (1997) Modulation of maxi-K^+ channels by voltage dependent Ca^{2+} channels and methacholine in single airway myoccytes. Am J Phys 272:C1151–C1159

Wang Z-W, Nara M, Wang Y-X, Kotlikoff ML (1997) Redox regulation of large conductance Ca^{2+}-activated K^+ channels in smooth muscle. J Gen Physiol 110: 35–44

Warbington F, Hillman T, Adams C, Stern M (1996) Reduced transmitter release conferred by mutations in the slowpoke-encoded Ca^{2+} -activated K^+ channel gene of Drosophila. Invert Neurosci 2:51–60

Wei A, Solaro CR, Lingle CJ, Salkoff L (1994) Calcium sensitivity of BK-type K_{Ca} channels determined by a separable domain. 1994 Neuron 13:671–681

Wei A, Jegla T, Salkoff L (1996) Conserved classes of potassium channel genes identified from the Caenorhabditis elegans genome. Neuropharmacology 35:805–829

White RE, Lee AB, Shcherbatko AD, Lincoln TM, Schonbrunn A, Armstrong DL (1993) Potassium channel stimulation by natriuretic peptides through cGMP-dependent dephosphorylation. Nature 361:263–266

Williams DL Jr, Katz GM, Roy-Contancin L, Reuben JP (1988) Guanosine 5′-monophosphate modulates gating of high-conductance Ca^{2+}-activated K^+ channels in vascular smooth muscle cells. Proc Natl Acad Sci USA 85:9360–9364

Wu JV, Trevor J, Stampe P (1997) Clustered distribution of calcium sensitivities: an indication of heterotetrameric gating components in Ca^{2+}-activated K^+ channels reconstituted from avian nasal gland cells. J Membr Biol

Xia X, Hirschberg B, Smolik S, Forte M, Adelman JP (1998) dSlo interacting protein 1, a novel prorein that interacts with large-conductance calcium-activated potassium channels. J Neurosci 18:2360–2369

Xie J, McCobb DP (1998) Control of alternative splicing of potassium channels by stress hormones. Science 280:443–446

Yazejian B, DiGregorio DA, Vergara J, Poage RE, Meriney SD, Grinnel AD (1997) Direct measurement of presynaptic calcium and calcium-activated potassium currents regulating neurotransmitter release at cultured Xenopus nerve-muscle synapses. J Neurosci 17:2990–3001

Yellen G (1984) Ionic permeation and blockade in Ca-activated K-channels of bovine chromaffin cells. J Gen Physiol 84:157–186

Yuan XJ, Sugiyama T, Goldman WF, Rubin LJ, Blaustein MP (1996) A mitochondrial uncoupler increases KCa currents but decreases KV currents in pulmonary artery myocytes. Am J Physiol Cell Physiol 270:C321–C331

Zhou XB, Ruth P, Schlossmann J, Hofmann F, Korth M (1996) Protein phosphatase 2 A is essential for the activation of Ca^{2+}-activated K+ currents by cGMP-dependent protein kinase in tracheal smooth muscle and Chinese hamster ovary cells. J Biol Chem 271:19760–19767

CHAPTER 9

Classical Inward Rectifying Potassium Channels: Mechanisms of Inward Rectification

C.G. NICHOLS

A. The Nature of Inward Rectification: Classical Considerations

Potassium channels are highly selective for potassium ions over other cations. They have been broadly classified into two main families (HILLE 1992). So-called "voltage-gated" K channels are typically closed at negative membrane potentials and open following depolarization beyond about –40 mV. "Inward rectifier" K channels show an almost opposite dependence on membrane potential. They are open at negative membrane potentials and close following depolarization. The change of conductance with voltage is referred to as "rectification", and the term is used to indicate both voltage-dependent channel "gating" and voltage-dependence of the open channel current. Strong inward rectification (Fig. 1A) was first described in skeletal muscle (KATZ 1949), and is very prominent in cardiac myocytes, and in glial cells and neurons in the central nervous system (NAKAJIMA et al. 1988; NEWMAN 1993; BRISMAR and COLLINS 1989). Rectification of these channels is such that conductance declines to zero about 40 mV positive to the potassium reversal potential (NOBLE 1965; VANDENBERG 1994). The high conductance at negative voltages allows cells to maintain a stable resting potential, but the reduced conductance at positive potentials avoids short-circuiting the action potential. "Weak" (Fig. 1A) inward rectifier ATP-sensitive K^+ (K_{ATP}) channels allow substantial outward current to flow at positive potentials (NOMA 1983). Between these two channel types, K channels showing intermediate rectification properties are found throughout the nervous system, many of them being activated by G-proteins or other second messenger systems (KANDEL and TAUC 1966; CONSTANTI and GALVAN 1983; INOUE et al. 1988; WILLIAMS et al. 1988; NAKAJIMA et al. 1988; NEWMAN 1993; BRISMAR and COLLINS 1989).

HODGKIN and HUXLEY (1952) developed a common nomenclature to describe the opening (activation) of voltage-gated K^+ and Na^+ channels following depolarization, the subsequent closing of the channels (inactivation), the reversal of the activation process following hyperpolarization (deactivation), and the subsequent recovery of availability of channels at negative voltages (recovery from inactivation). In such channels, there is now much evidence to support the hypothesis that activation and deactivation result from the voltage-dependent movement of the highly charged S4 segment within the

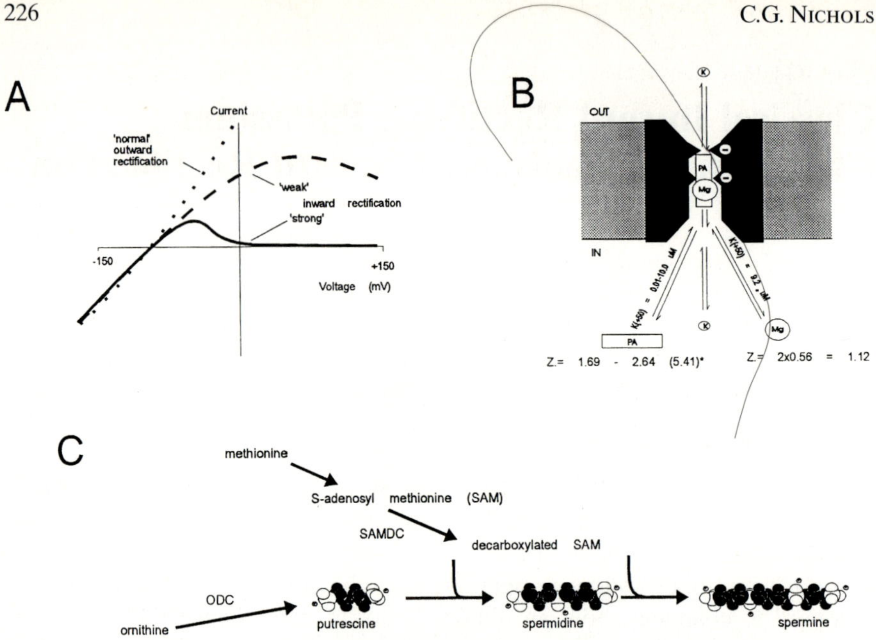

Fig. 1. A Idealized current-voltage relationship of strong and weak inward rectifier K channels. Both conduct significantly at diastolic potentials, but strong inward rectifiers pass little or no current during action potentials. **B** Schematic diagram of proposed pore blocking mechanisms causing inward rectification. Blocking ions enter the pore from the inside, and binding can be relieved by potassium ions entering the pore from the outside. Although the effective valency ($z\delta$) of Mg^{2+} block is consistent with one Mg^{2+} ion permeating about 50% of the voltage field, polyamines may enter more deeply, experimental data suggests that more than one polyamine molecule actually enters the field, giving an effective valency ($z\delta$) >4 as the block approaches saturation. **C** Schematic diagram of polyamine structure and outline of synthetic pathway in animal cells. Amines are shown in *white*, methyl groups are shown in *black*. All amines are charged at neutral pH

membrane (LIMAN et al. 1991; PAPAZIAN et al. 1991; TYTGAT et al. 1992, 1993). Rapid inactivation of many voltage-gated K (Kv) channels results from block of the open channel by a cytoplasmic "ball". In some Kv channels, this "ball" consists of the amino terminus (HOSHI et al. 1990; ZAGOTTA et al. 1990) of the channel protein. In some cases, the C-terminal and specific residues within the S6 domain are also shown to be involved in Kv inactivation (HOSHI et al. 1991; STOCKER et al. 1991).

As discussed below, recent experiments show that in inward rectifiers, the reduction of channel current at positive potentials results from block of the open channel by polyamines and Mg^{2+} ions. This is somewhat analogous to inactivation of voltage-gated channels, and some Kv channels also show a mild inward rectification resulting from voltage-dependent block by cytoplasmic Mg^{2+} (FORSYTHE et al. 1992; LOPATIN and NICHOLS 1994; RETTIG et al. 1992). Additionally, most inwardly rectifying channels also show some tendency to close at negative voltages, although, the mid-point voltage for such closure is typically around −80 mV to −100 mV, and the steepness is much less than for "deactivation" of Kv channels (Koumi et al. 1994; Lopatin et al. 1995; Nichols et al. 1994). The parallels between the voltage-dependent behavior of Kv and Kir channels suggests that the voltage-dependent behavior of each, although quantitatively different, might arise from fundamentally similar processes in channels which are actually of fundamentally similar structures, i.e., both Kir and Kv channels share the "inner core" of the Kir channel (Nichols 1993; Doyle et al. 1998). Hence, closure of both Kv and Kir channels at positive potentials can result from pore block by internal cations or inactivating particles. Inward rectification of potassium channels was first recognized by Bernard Katz (Katz 1949). Twenty years later, Clay Armstrong (Armstrong 1969) suggested that inward rectification might result from voltage dependent block by an intracellular cation. Twenty years on again, two groups (Vandenberg 1987; Matsuda et al. 1987) demonstrated that intracellular Mg^{2+} ions were indeed capable of causing inward rectification by just such a mechanism. In the last five years, inward rectifier K^+ channel subunits have been cloned, and expressed at high levels in recombinant systems. This has led to the realization that intracellular polyamines are in fact major determinants of inward rectification (Fakler et al. 1994, 1995; Ficker et al. 1994; Lopatin et al. 1994, 1995; Lopatin and Nichols 1996) also acting as cytoplasmic blocking particles. This chapter will consider mechanisms of inward rectification, and the structural basis for the phenomenon.

B. The Inward Rectifier Ion Channel Family: Two Transmembrane Domain Potassium Channels

Cloning of the first members of the new Kir channel family [Kir1.1a (ROMK1), Kir2.1 (IRK1), and Kir3.1 (GIRK1)], in 1993 (Ho et al. 1993; Kubo et al. 1993a,b; Dascal et al. 1993) ushered in a new era of research on the physiology of inward rectifiers. Kir channel subunits have only two transmembrane domains (Ho et al. 1993; Kubo et al. 1993a; Choe et al. 1995; Nichols 1993), but they retain the H5-loop that is responsible for K^+ selectivity (Heginbotham et al. 1992). Utilizing mutations that express channels with altered rectification properties, there is evidence that, like Kv channels (MacKinnon 1991), Kir channels form as tetramers (Glowatzki et al. 1995; Yang et al. 1995; Shyng and Nichols 1997). There are now at least six Kir channel sub-families (Doupnik et al. 1995), each sharing ~40% amino acid identity between one another, and ~60% identity between individual members within each sub-family.

I. Kir 1 Subfamily

Kir1.1 (Ho et al. 1993) encodes a "weak" inward rectifier, and is expressed predominantly in the kidney, but also in various brain tissues (Ho et al. 1993; Boim et al. 1995). Alternate splicing at the 5' end generates multiple Kir1.1 splice variants (SHUCK et al. 1994; YANO et al. 1994; ZHOU et al. 1994).

II. Kir 2 Subfamily

Three distinct Kir 2 subfamily members have been cloned to date, all encoding "strong" inward rectifiers that differ in single channel conductance (Kir2.1 ~20 pS, Kir2.2 ~35 pS, Kir2.3 ~10 pS, all in 140 mmol/l external $[K^+]$), and in sensitivity to phosphorylation and other second messengers (CHANG et al. 1996; FAKLER et al. 1994; HENRY et al. 1996; MAKHINA et al. 1994). Kir2 subfamily members are expressed in the heart and nervous system, (ISHII et al. 1994; KUBO et al. 1993a; PERIER et al. 1994; PESSIA et al. 1996; WIBLE et al. 1994), and the time- and voltage-dependent rectification of the expressed channels are virtually indistinguishable from native iK1 channels in the heart (ISHIHARA et al. 1989, 1994; KURACHI 1985; OLIVA et al. 1990; STANFIELD et al. 1994), or the inward rectifier K current in glilal cells (NEWMAN 1993).

III. Kir 3 Subfamily

Members of the Kir 3 family all express G-protein activated strong inward rectifier K channels (KUBO et al. 1993b; DASCAL et al. 1993. LESAGE et al. 1994), and there is now substantial evidence that they express G-protein coupled receptor activated currents in heart, brain, and endocrine tissues (KUBO et al. 1993b; KARSCHIN et al. 1994; FERRER et al. 1995). KRAPIVINSKY et al. (1995) demonstrated that Kir3.4 subunits co-assemble with Kir3.1 (GIRK1) to form the cardiac muscarinic receptor-activated iK,Ach. Additional studies have provided evidence for a promiscuous coupling between the various members of the Kir3 sub-family (DUPRAT et al. 1995; FERRER et al. 1995; ISOMOTO et al. 1996; KOFUJI et al. 1995; SPAUSCHUS et al. 1996).

IV. Kir 4 and 5 Subfamilies

Two more subfamilies of Kir channels have been discovered in brain and other tissues (Kir4 and Kir5, 12, 153). Kir4.1 forms weak inward rectifier K channels when expressed alone, but Kir5.1 does not form channels in homoeric expression in oocytes (BOND et al. 1994). These two subunits can actually co-express to form novel channels, and tandem dimers and tetramers in a specific 4–5-4–5 arrangement reproduces the characteristics of these channels (PESSIA et al. 1996). Intriguingly, a 4–4-5–5 tetrameric arrangement produces channels with the properties of homomeric Kir4.1 channels, providing evidence for the importance of subunit position in the properties of heterotetrameric Kir channel.

V. Kir 6 Subfamily

INAGAKI et al. (1995a) isolated a novel, ubiquitously expressed gene which they named uKATP1(Kir6.1 in the unified nomenclature). A pancreatic-specific isoform (Kir6.2), was subsequently found to encode a weak inward rectifier K_{ATP} channel (INAGAKI et al. 1995b), although expression of active channels required co-expression of Kir6.2 (or Kir6.1) with the high affinity sulfonylurea receptor (SUR). Mutation of homologous pore-lining residues in Kir1.1 (LU and MACKINNON 1994) and Kir6.2 (SHYNG et al. 1997) clearly demonstrate that Kir6.2 forms the channel pore in an analogous way to other Kir subunits, and the SUR subunit provides a regulatory subunit (NICHOLS et al. 1996).

VI. KirD – a New Family of Double-Pored Inward Rectifier Channels?

KETCHUM et al. (1995) described a novel yeast K channel subunit (TOK1) which appeared to be formed from a Kir subunit in tandem with a six-transmembrane domain Kv subunit, and expressed outwardly rectifying K currents in *Xenopus* oocytes. LESAGE et al. (1996) reported the cloning and expression of a similarly structured channel (which they called TWIK-1), consisting of two Kir subunits in tandem. Although only limited expression data is available, currents through TWIK-1 channels appear to be weakly inwardly rectifying, similar to those expressed by Kir1.1 channels. It seems likely that this TWIK-1 cDNA was formed from a gene duplication, and provides a whole new series of possibilities for the generation of novel Kir channels.

VII. Inward Rectification in Other K^+ Channels

Many, if not all, Kv channels also show weak inward rectification under physiological conditions (FORSYTHE et al. 1993; FRENCH and WELLS 1977; LOPATIN and NICHOLS 1994; RETTIG et al. 1992). Like the rectification of weak inward rectifiers in Kir1 and Kir6 sub-families, rectification of these channels involves a weakly voltage-dependent block by internal Na^+ and Mg^{2+} (see below). Other recently cloned Kv channels actually show quite strong inward rectification, superimposed on steep voltage-dependent activation typical of Kv channels (SANGUINETTI et al. 1995; TRUDEAU et al. 1995). These channels underlie the delayed rectifier current in human cardiac ventricular muscle (iKr), and mutations in these genes are responsible for certain inherited forms of long QT syndrome (CURRAN et al. 1995). SMITH et al. (1996) have examined the rectification properties of expressed HERG and concluded that rectification results from "C-type" voltage-dependent inactivation, an incompletely understood intrinsic process that is present in other Kv channels (HOSHI et al. 1991), but is distinguishable from strong inward rectification in Kir channels (see below).

C. The Mechanism of Inward Rectification: Pore Block and Intrinsic

Armstrong (1969) suggested that inward rectification might result from a voltage-dependent block of the channel pore by cytoplasmic cations, since application of tetraethyl ammonium ions to the cytoplasmic surface of Kv channels induces an inward rectification by blocking the channel pore. Subsequently, Mg^{2+} and Na^{+} ions were shown to cause inward rectification of weakly inward rectifying K_{ATP} channels (CIANI et al. 1988; HORIE et al. 1987), and of cardiac I_{K1} channels. However, a seemingly intrinsic voltage-dependence of the conductance was also clearly a dominant cause of inward rectification in strong inward rectifier channels (KELLY et al. 1992; KURACHI 1985; MATSUDA 1991; MATSUDA et al. 1987, 1989; OLIVA et al. 1990; SILVER and DECOURSEY 1990; VANDENBERG 1987). For both Mg^{2+}-induced, and "intrinsic", rectification, a strong dependence on external [K^{+}] (K_o) was demonstrated; increasing K_o relieves the rectification. For Mg^{2+} induced rectification, this effect is explained by K^{+} ion binding at external sites and "knocking-off" Mg^{2+} from sites deeper inside a multi-ion pore (ARMSTRONG 1971; HILLE and SCHWARZ 1978; YELLEN 1984). An intriguing observation made by MATSUDA (1988) was that "intrinsic" rectification of cardiac inward rectifier K^{+} channels gradually disappears with time after excision of a membrane patch into the inside-out configuration. Following the cloning of strong inward rectifier K^{+} channel genes (Kir2.x gene family members), it was possible to observe high levels of expressed inward rectifier currents. In macro-patch experiments on Kir2.3 channels expressed in *Xenopus* oocytes, we observed that rectification disappeared when patches were isolated (LOPATIN et al. 1994), but was restored when we moved the patch back towards the oocyte. This indicated that rectification disappeared because some factor, or factors, were being lost from the oocyte interior, and that these "intrinsic rectifying factors" were actually being released from intact oocytes. We conditioned solutions by exposure to intact oocytes, allowing us to make some rudimentary biochemical characterization of "intrinsic rectifying factors", sufficient to indicate that they are actually polyamines (spermine, spermidine, putrescine) (Fig. 1), metabolites of amino acids that are found in almost all cells (TABOR and TABOR 1984). Application of these polyamines to inside-out patches containing Kir2.x channels restores all the essential features of "intrinsic" rectification (LOPATIN et al. 1994, 1995). Less potent than spermine and spermidine, putrescine and cadaverine also cause rectification with similar efficacy to the rectification caused by Mg^{2+}. The voltage-dependence of spermine and spermidine block are steeper than Mg^{2+} block (LOPATIN et al. 1994, 1995; FAKLER et al. 1994; FICKER et al. 1994), explaining why inward rectification in endogenous cells is steeper than that produced by Mg^{2+} ions (HILLE 1992) (see Fig. 1).

The voltage dependence of spermine and spermidine unblock rates match the rate constants of channel activation in cell-attached patches (LOPATIN et al. 1995). Kir1.1 (ROMK1), Kir4.1 channels, Kir6.2 (K_{ATP}) channels, and delayed rectifier Kv2.1 (DRK1) channels all show only "weak" inward rectification. In contrast to Kir2.x channels, they are only blocked by millimolar concentrations of Mg^{2+} and polyamines (LOPATIN et al. 1994; FAKLER et al. 1994; NICHOLS et al. 1994; SHYNG et al. 1997), and the block is only weakly voltage-dependent. The steepness of the voltage dependence of channel block by polyamines increases as the charge on the polyamine increases (LOPATIN et al. 1994), and mutations that alter Mg^{2+} block sensitivity also alter polyamine blocking affinity (FAKLER et al. 1994; YANG et al. 1995). As expected for a channel blocker that interacts with permeant ions inside the pore, external potassium ions substantially relieve rectification (LOPATIN and NICHOLS 1996).

D. The Structure of the Kir Channel Pore: Binding Sites for Polyamines

As discussed above, there is now very strong evidence that polyamines and Mg^{2+} cause rectification by a voltage-dependent block of the channel pore. Mg^{2+} ions are spherical charges, with diameters similar to K^+ ions, and it is reasonable to suggest that they block the channel by occupying K^+ ion binding sites within the pore. On the other hand, spermine is a very long (almost 20 Å long) and thin molecule (diameter ~ 3 Å), with spatially distributed positive charges. It is a possibility that in blocking Kir channels, spermine lies in the long pore, each charge associating with a different site that would otherwise be occupied by K^+ ions (LOPATIN et al. 1995). YANG et al. (1995) examined steady-state polyamine block of Kir2.1 channels over a wide concentration range, and their data suggest that at least two polyamines bind in the channel, with different affinities. We also initially proposed that two polyamines independently enter the channel pore, partly in order to account for the very large charge movement (more than five elementary charges) that accompanies spermine block (LOPATIN et al. 1995).

All potassium channels contain a highly conserved region which includes an extracellular loop (H5- or P-loop) with a –Gly-X-Gly triplet that forms the K^+ selectivity filter (HARTMANN et al. 1991; MACKINNON and YELLEN 1990; YOOL and SCHWARTZ 1990) between two transmembrane domains. Mutagenesis followed by biophysical analysis demonstrates that the transmembrane region following the P-loop is also involved in forming the permeation pathway (AIYAR et al. 1994; LIU et al. 1997). Multiple studies have indicated that a specific residue in the second transmembrane domain M2 of Kir2.1 (IRK1) is a major determinant of the potency of Mg^{2+} or polyamine block, and hence whether a channel will show classical strong inward rectification. When this residue is a negatively charged glutamate or aspartate, high affinity block is observed, and neutralization of this residue reduces or abolishes both Mg^{2+} and polyamine blocking affinity (FAKLER et al. 1994; LOPATIN et al. 1994; LU and MACKINNON 1994; WIBLE et al. 1994; FICKER et al. 1994). A histidine residue at this site also leads to permanent rectification at low internal pH (LU and

MACKINNON 1995). The rectification is titrated at higher pH, as the histidine residue is neutralized, but is insensitive to external pH, indicating that internal, but not external, protons have free access to this site. This is consistent with the idea that a tight selectivity filter, formed by the H5 region, exists at the outer mouth of the channel and blocks access of ions other than K^+ to the long inner vestibule. Studies with chimeras between "weakly" rectifying Kir1.1 (ROMK1) and "strongly" rectifying Kir2.1 (IRK1) indicated that the C-terminal region, beyond M2, might contain the necessary structural elements for strong inward rectification and high affinity Mg^{2+} block. (PESSIA et al. 1995; TAGLIALATELA et al. 1994). YANG et al. (1995) demonstrated that E224 (in the C-terminal of Kir2.1) is also a determinant of both Mg^{2+} and polyamine sensitivity, and RUPPERSBERG et al. (1996) subsequently demonstrated that both absolute and relative off-rates of different polyamines and Mg^{2+} from the channel depend critically on the amino acid at residue 84 (in IRK1), which is positioned at the entrance to the M1 transmembrane domain. These latter results suggest that the region immediately before M1 (containing residue 84), and the region immediately after M2 (containing residue 224) contribute to forming the internal entrance to the pore.

Very recently, these predictions have been dramatically confirmed by determination of the crystal structure of KcsA, a K channel from *S. lividans* (DOYLE et al. 1998) (Fig. 2). Although there is presently little functional char-

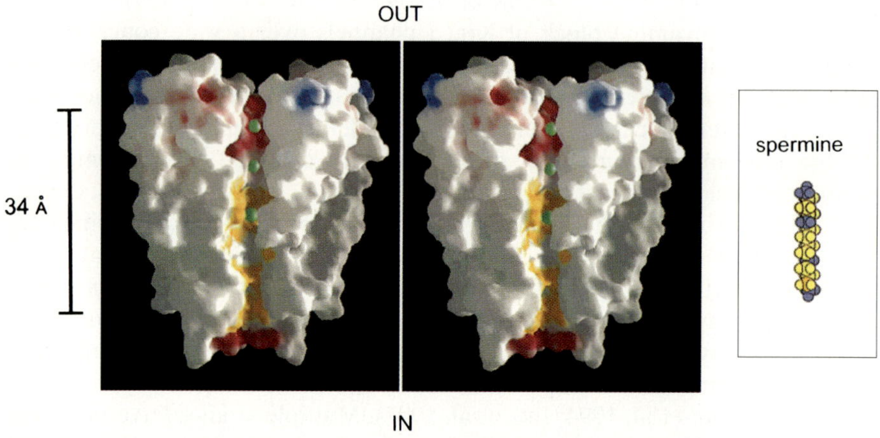

Fig. 2. The molecular surface of the KcsA potassium channel and contour of the channel pore. The two images are a stereoview of the solvent accessible surface of the K channel colored according to physical properties – *Blue* corresponds to highly positively charged, *red* corresponds to highly negatively charged. *Yellow* areas correspond to carbon atoms of hydrophobic side chains lining the inner vestibule. The *green* CPK spheres represent K ions in the conduction pathway. (Reproduced with author's permission from DOYLE et al. 1998.) To the *right* is a space filling model of spermine in extended linear conformation, approximately to the same scale. *Blue* represents positively charged amines, *yellow* represents hydrophobic methyl groups

acterization of this channel, it is structurally a member of the K^+ channel family, and contains two transmembrane domains with a H5, or P-loop, containing the K channel signature Gly-X-Gly motif. The crystal structure demonstrates that the P-loop region forms a shallow disc at the outer surface of the membrane with a long inner vestibule that extends at least 20 Å (long enough to accommodate a spermine molecule in extended form) into and through the membrane (Fig. 2). The width of the inner vestibule is variable, but with a maximum diameter of about 10 Å. It is a tantalizing possibility that the binding site for the blocking polyamine that causes inward rectification is physically in this vestibule, the exact structure of the narrow entrance determining the on- and off-rates for polyamines, which can vary by several orders of magnitude for different inward rectifiers (Ruppersberg et al. 1994; Lopatin et al. 1995; Shyng et al. 1997).

E. The Structural Requirements for Inward Rectification: The Blocking Particles

Although rectification is clearly conferred by positively charged ions binding within the channel pore, not all charged or polar molecules can cause rectification. Lopatin et al. (1994) showed that while polyamines conferred strong rectification, related bulkier, dipolar, or non-linear molecules (e.g., GABA, creatinine, lysine) failed to block Kir2.1 channels, suggesting that a molecule must possess both the correct structure and charge density or distribution to confer strong rectification. Because the most energetically favorable conformation of endogenous polyamines in free solution is an extended linear chain (Romano et al. 1992), it seems likely that these molecules enter the long pore of the channel and lie in the pore to block it. To examine the structural requirements of the blocking species more systematically, we have recently examined the ability of series of mono- and diamino alkanes to block Kir2.1 channels. Although short chain monoamines (MA1–MA4) were without obvious effect, compounds with longer alkyl chains (5–12 methylene groups) produced significant inward rectification at concentrations below 100 μmol/l (Pearson and Nichols 1998). The blocking potency increased with the alkyl chain length, $V_{1/2}$ increasing by ~–10 mV per additional methylene group, whereas the effective valence (i.e., voltage dependence) of monoamine block ($z\delta$) was relatively constant at a value ~2.2. The increase in blocking potency results primarily from a decreased off-rate as the chain length is increased, indicating a strong hydrophobic interaction in the binding site. Similarly, all diamines tested (DA2–DA12) blocked Kir 2.1 channels at micromolar concentrations and again, increasing the alkyl chain length increased the blocking affinity (Pearson and Nichols 1998). In contrast to the behavior of monoamines, the effective valence of diamine block increased steeply with increase in chain length. This monotonic increase of $z\delta$ with alkyl chain length, with constant valence (+2) of the blocking particle, is striking, and reminiscent of the effect of alkyl chain length on the blocking potency of bis-quaternary amines (i.e., alkyl backbones with trio-ethylamine groups at each end) in Ca-activated and sarcoplasmic reticular K^+ channels (e.g., FRENCH and SHOUKIMAS 1981; MILLER 1982), raising the possibility that block by these different compounds shares common features.

Original suggestions, based on biophysical analysis, that inward rectifier K channels consist of a long narrow pore (HILLE and SCHWARZ 1978), are now dramatically confirmed by the crystal structure of the bacterial KcsA K^+ channel (DOYLE et al. 1998). The model we originally put forward to account for polyamine-induced rectification was one in which the polyamines enter and block the pore in an extended linear conformation – "long pore plugging" (LOPATIN et al. 1995), such that the polyamines should lay "vertically" inside the long narrow pore, binding through electrostatic interaction with a negatively charged binding site. The systematic analysis of mono- and diamines demonstrates that block does not result from a purely electrostatic interaction with residues in the channel pore, and indicates that hydrophobic interaction of the alkyl chain must stabilize binding in the pore. Although somewhat of a surprise, the crystal structure of KcsA reveals that the inner vestibule of the channel is actually very hydrophobic (DOYLE et al. 1998), and this hydrophobic vestibule is indeed large enough to accommodate a molecule as large as spermine, in extended linear conformation (Fig. 2).

The very high effective valence of block by spermine (~5.4 LOPATIN et al. 1995) and by long diamines (>4) (PEARSON and NICHOLS 1998) might be explained by multiple molecules sequentially entering the channel pore. However, the strong interaction of polyamine block with external K^+ ions (LOPATIN and NICHOLS 1996) suggests an alternative probability, namely that the polyamine entry into, and binding in, the channel pore "sweeps" K^+ ions outwards, contributing extra charge movement to the binding process, as discussed by RUPPERSBERG et al. (1994). With increasing chain length of linear diamines, it is not likely that the increase in valence results from increase in the number of blocking particles, so the increase in $z\delta$ cannot result from more charge being contributed by the blocking ion. Instead, the increase in charge associated with the blocking process may result from progressively more charge (i.e., permeating K^+ ions) being swept out of the pore as the blocking particle increases in size (RUPPERSBERG et al. 1994). The maximum diameter of the inner vestibule of the KcsA channel is in the order of 10 Å (DOYLE et al. 1998), and is accessed from the cytoplasm through a "tunnel" that is 18 Å long and ~5 Å wide (Fig. 2). Assuming that the mammalian inward rectifier inner pore is of similar dimensions, then the following model (Fig. 3) presents itself to explain diamine and monoamine block. We hypothesize that the long pore of the inward rectifier is accessible to polyamines, and to even the longest diamines (PEARSON and NICHOLS 1998). Moreover, we suggest that, since the diamine charges cannot occupy a single point in space or in the electric field, they may occupy a balanced position relative to the negative charges at the

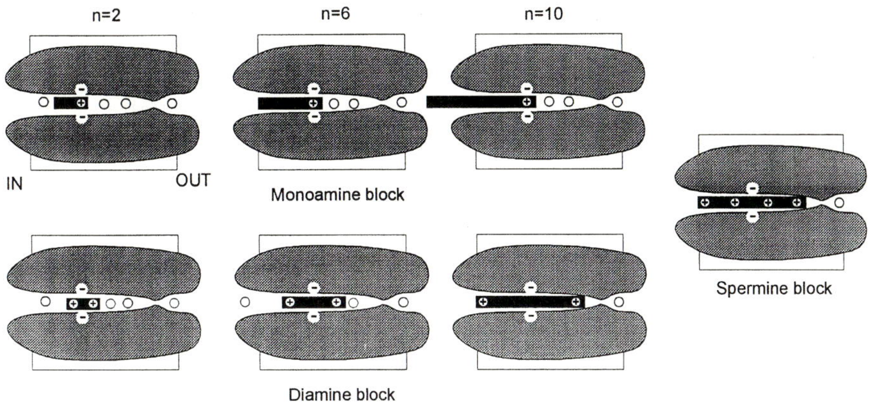

Fig. 3. Hypothetical accommodation of polyamines within the pore of Kir channels. Kir pores are hypothesized to have a long (>20 Å) inner vestibule with strong electronegativity provided by the ring of negative charges at the "rectification controller" position. Diamines (*below*) can occupy the pore, their two charges being equidistant from the center of the ring of negativity at the "rectification controller". As alkyl chain length increases, the "head" amine is pushed deeper into the pore, displacing K^+ ions to the outside of the cell. Monoamines (*above*) can also occupy the pore, but the "head" amine remains in the ring of electronegativity provided by the "rectification controller", the alkyl chain stretching back out of the pore. Spermine (*right*) can occupy the pore, its four charges being symmetrically arranged along the axis of the pore, perpendicular to the ring of negativity at the "rectification controller"

"rectification controller", i.e., for diamines, the two positive charges will be equidistant from the center of the ring of negative charges (Fig. 3). Thus, as the alkyl chain length increases, a single diamine will occupy more and more of the available space, with more and more K^+ ions being displaced to the outside of the cell, moving more charge outwards and thereby increasing the net charge movement associated with channel block (Fig. 3). A similar suggestion may account for the results of FAKLER et al. (1997) who observed that the apparent electrical distance for tetraalkylammonium block of Kir1.1 channels increased from 0.83 to 0.93 to 1.44 as the alkyl chain length increased from 2 to 3 to 4.

If this mechanism is indeed causing increased steepness of voltage dependence with increase in diamine length, then how can we account for monamine (MA) block having a steepness of voltage dependence that is independent of alkyl chain length? We hypothesize that the major determinants of blocker *depth* within the inner pore are the charged groups on the blocker. Thus, for MAs, the single charge stabilizes at essentially the same *depth* (i.e., at the level of the "rectification controller" (the ring of four negative charges in the M2 segments). The increasing alkyl chain then stretches further and further out of the pore, and the K^+ ion displacing effect of increasing chain length observed with the DAs is absent (Fig. 3).

F. The Physiological Significance of Polyamine-Induced Rectification

Polyamines are present in almost every cell, and have been the subject of interest as cellular metabolites since their discovery by van Leeuwenhoek (1678). They may have a role as stabilizing moieties for DNA (Tabor and Tabor 1984), and are essential for normal and neoplastic cell growth. In order to reproduce the degree of rectification seen in intact cells (Fakler et al. 1995; Ficker et al. 1994), only nanomolar to micromolar concentrations of free polyamines would be required. Induction of inward rectification may in fact be the most potent physiological property of polyamines. Although cellular levels are strongly buffered, total cellular polyamine concentrations (up to several millimolar) (Seiler 1994) are clearly in excess of those required to cause very strong rectification of Kir channels. Since there is a very steep voltage-dependence of polyamine block, it is likely that cytoplasmic polyamine levels will always be in the range whereby they will cause rectification at physiological voltages. Treatment of cells with inhibitors of polyamine synthetic enzymes has been shown to relieve inward rectification in RBL-1 cells (Bianchi et al. 1996) and in oocytes (Shyng et al. 1996). We also utilized a CHO cell line that is deficient in ornithine decarboxylase activity (ODC) (Steglich and Scheffler 1982), and requires putrescine in the medium for normal cell growth, to demonstrate the effects of polyamine depletion on the rectification of expressed Kir2.3 channels. In these cells, removal of putrescine leads to gradual decline in intracellular levels of putrescine, then spermidine, and finally spermine (Steglich and Scheffler 1982). These changes correlate with alterations in Kir2.3 kinetics predicted by excised-patch experiments (Lopatin et al. 1995). In native tissues, the effects of altered polyamine levels on inward rectification and excitability remain largely unexplored, but it remains an exciting possibility that changes in cellular polyamine levels will physiologically regulate excitability.

Acknowledgements. Our own experimental work has been supported by the N.I.H. and the American Heart Association

References

Aiyar J, Nguyen AN, Chandy KG, Grissmer S (1994) The P–region and S6 of Kv3.1 contribute to the formation of the ion conduction pathway. Biophysical Journal. 67:2261–2264

Armstrong CM (1969) Inactivation of the potassium conductance and related phenomena caused by quaternary ammonium ion injected in squid axons. J Gen Physiol 54:553–575

Armstrong CM (1971) Interaction of tetraethylammonium ion derivatives with the potassium channel of giant axons. J Gen Physiol 58:413–437

Bianchi L, Roy ML, Taglialatela M, Lundgren DW, Brown AM, Ficker E (1996) Regulation by spermine of native inward rectifier K^+ channels in RBL-1 cells. Journal of Biological Chemistry 271:6114–6121

Boim MA, Ho K, Shuck ME, Bienkowski MJ, Block JH, Slightom JL, Yang Y et al. (1995) ROMK inwardly rectifying ATP-sensitive K^+ channel II Cloning and distribution of alternative forms. American Journal of Physiology 268:F1132-F1140

Bond CT, Pessia M, Xia XM, Lagrutta A, Kavanaugh MP, Adelman JP (1994) Cloning and expression of a family of inward rectifier potassium channels. Receptors and Channels 2:183–191

Brismar T, Collins VP (1989) Inwardly rectifying potassium channels in human malignant glioma cells. Brain Research 480:249–258

Chuang H, Jan YN, Jan LY (1996) A strongly inwardly rectifying K channel modulated by M1 acetylcholine receptor. Biophysical Journal 70:A73

Choe S, Stevens CF, Sullivan JM (1995) Three distinct structural environments of a transmembrane domain in the inwardly rectifying potassium channel ROMK1 defined by perturbation. Proceedings of the National Academy of Sciences of the United States of America 92:12046–12049

Ciani S, Ribalet B (1988) Ion permeation and rectification in ATP-sensitive channels from insulin-secreting cells (RINm5F): effects of K^+, Na^+ and Mg^{2+}. Journal of Membrane Biology 103:171–180

Constanti A, Galvan M (1983) Fast inward-rectifying current accounts for anomalous rectification in olfactory cortex neurones. Journal of Physiology 335:153–178

Curran ME, Splawski I, Timothy KW, Vincent GM, Green ED, Keating MT (1995) A molecular basis for cardiac arrhythmia: HERG mutations cause long QT syndrome. Cell 80:795–804

Dascal N, Schreibmayer W, Lim NF, Wang W, Chavkin C, DiMagno L et al. (1993) Atrial G protein–activated K^+ channel: expression cloning and molecular properties. PNAS 90:10235–10239

Doupnik CA, Davidson N, Lester HA (1995) The inward rectifier potassium channel family. Current Opinion in Neurobiology 5:268–277

Doyle DA, Cabral JM, Pfuetzner RA, Kuo A, Gulbis JM, Cohen SL, Chait BT, MacKinnon R (1998) The structure of the potassium channel: Molecular basis of K^+ conduction and selectivity. Science 280:69–77

Duprat F, Lesage F, Guillemare E, Fink M, Hugnot JP, Bigay J et al. (1995) Heterologous multimeric assembly is essential for K+ channel activity of neuronal and cardiac G-protein-activated inward rectifiers. Biochemical and Biophysical Research Communications 212:657–663

Fakler B, Antz C, Ruppersberg JP (1997) Anomalous dependence of inward-rectifier K^+ channel block on the size of the blocking ion. Biophysical Journal 72:A232 (Abst)

Fakler B, Brandle U, Bond C, Glowatzki E, Konig C, Adelman JP et al. (1994) A structural determinant of differential sensitivity of cloned inward rectifier K^+ channels to intracellular spermine. FEBS Letters 356:199–203

Fakler B, Brandle U, Glowatzki E, Weidemann S, Zenner HP, Ruppersberg JP (1995) Strong voltage-dependent inward rectification of inward rectifier K^+ channels is caused by intracellular spermine. Cell 80:149–154

Fakler B, Brandle U, Glowatzki E, Zenner HP, Ruppersberg JP (1994) Kir21 inward rectifier K^+ channels are regulated independently by protein kinases and ATP hydrolysis. Neuron 13:1413–1420

Ferrer J, Nichols CG, Makhina EN, Salkoff L, Bernstein J, Gerhard D et al. (1995) Pancreatic islet cells express a family of inwardly rectifying K^+ channel subunits which interact to form G-protein-activated channels. Journal of Biological Chemistry 270:26086–26091

Ficker E, Taglialatela M, Wible BA et al. (1994) Spermine and spermidine as gating molecules for inward rectifier Kchannels. Science 266:1068–1072

Forsythe ID, Linsdell P, Stanfield PR (1992) Unitary A-currents of rat locus coeruleus neurones grown in cell culture: rectification caused by internal Mg^{2+} and Na^+. Journal of Physiology 451:553–583

French RJ, Shoukimas JJ (1981) Blockage of squid axon potassium conductance by internal tetra-N-alkylammonium ions of various sizes. Biophysical Journal 34: 271–291

French RJ, Wells JB (1977) Sodium ions as blocking agents and charge carriers in the potassium channel of the squid giant axon. Journal of General Physiology 70:707–724

Glowatzki E, Fakler G, Brandle U, Rexhausen U, Zenner HP, Ruppersberg JP, Fakler B (1995) Subunit-dependent assembly of inward-rectifier K^+ channels. Proceedings of the Royal Society of London B261:251–261

Hartmann HA, Kirsch GE, Drewe JA, Taglialatela M, Joho RH, Brown AM (1991) Exchange of conduction pathways between two related K^+ channels. Science 251:942–944

Heginbotham L, Abramson T, MacKinnon R (1992) A functional connection between the pores of distantly related ion channels as revealed by mutant K^+ channels. Science 258:1152–1155

Henry P, Pearson WL, Nichols CG (1996) Protein kinase C inhibition of cloned inward (HRK1/Kir23) K^+ channels. J Physiol 495:681–688

Hille B (1992) Ionic channels of excitable membranes. Sinauer Associates Inc, Sunderland, Massachusetts

Hille B, Schwarz W (1978) Potassium channels as multi-ion single-file pores. Journal of General Physiology 72:409–442

Ho K, Nichols CG, Lederer WJ, Lytton J, Vassilev PM, Kanazirska MV et al. (1993) Cloning and expression of an inwardly rectifying ATP-regulated potassium channel. Nature 362:31–38

Hodgkin AL, Huxley AM (1952) Currents carried by sodium and potassium ions through the membrane of the giant axon of Loligo. Journal of Physiology 116:449–472

Horie M, Irisawa H, Noma A (1987) Voltage-dependent magnesium block of adenosine-triphosphate-sensitive potassium channel in guinea-pig ventricular cells. Journal of Physiology 387:251–272

Hoshi T, Zagotta WN, Aldrich RW (1990) Biophysical and molecular mechanisms of Shaker potassium channel inactivation. Science 250:533–538

Hoshi T, Zagotta WN, Aldrich RW (1991) Two types of inactivation in Shaker K^+ channels: effects of alterations in the carboxy-terminal region. Neuron 7:547–556

Inagaki N, Tsuura Y, Namba N, Masuda K, Gonoi T, Horie M et al. (1995) Cloning and functional characterization of a novel ATP-sensitive potassium channel ubiquitously expressed in rat tissues, including pancreatic islets, pituitary, skeletal muscle, and heart. Journal of Biological Chemistry 270:5691–5694

Inagaki N, Gonoi T, Clement JP, Namba N, Inazawa J, Gonzalez G et al. (1995) Reconstitution of I_{KATP}: an inward rectifier subunit plus the sulfonylurea receptor. Science 270:1166–1170

Inoue M, Nakajima S, Nakajima Y (1988) Somatostatin induces an inward rectification in rat locus coeruleus neurones through a pertussis toxin-sensitive mechanism. Journal of Physiology 407:177–198

Ishihara K, Hiraoka M (1994) Gating mechanisms of the cloned inward rectifier potassium channel from mouse heart. Journal of Membrane Biology 142:55–64

Ishihara K, Mitsuiye A, Noma A, Takano M (1989) The Mg^{2+} block and intrinsic gating underlying inward rectification of the K^+ current in guinea-pig cardiac myocytes. Journal of Physiology 419:297–320

Ishii K, Yamagashi T, Taira N (1994) Cloning and functional expression of a cardiac inward rectifier K^+ channel. FEBS Letters 338:107–111

Isomoto S, Kondo C, Takahashi N, Matsumoto S, Yamada M, Takumi T et al. (1996) A novel ubiquitously distributed isoform of GIRK2 (GIRK2B) enhances GIRK1 expression of the G-protein-gated K^+ current in *Xenopus* oocytes. Biochemical and Biophysical Research Communications 218:286–291

Kandel E, Tauc L (1966) Anomalous rectification in the metacerebral giant cells and its consequences for synaptic transmission. Journal of Physiology 183:287–304

Karschin C, Schreibmayer W, Dascal N, Lester H, Davidson N, Karschin A (1994) Distribution and localization of a G protein-coupled inwardly rectifying K^+ channel in the rat. FEBS Letters 348:139–144

Katz B (1949) Les constantes electriques de la membrane du muscle. Archives de Science et Physiologie 2:285–299

Kelly ME, Dixon SJ, Sims SM (1992) Inwardly rectifying potassium current in rabbit osteoclasts: a whole-cell and single-channel study. Journal of Membrane Biology 126:171–181

Ketchum KA, Joiner WJ, Sellers AJ, Kaczmarek LK, Goldstein SA (1995) A new family of outwardly rectifying potassium channel proteins with two pore domains in tandem. Nature 376:690–695

Kofuji P, Davidson N, Lester HA (1995) Evidence that neuronal G-protein-gated inwardly rectifying K^+ channels are activated by G beta gamma subunits and function as heteromultimers. Proceedings of the National Academy of Sciences of the United States of America 92:6542–6546

Koumi S, Sato R, Hayakawa H (1994) Modulation of voltage-dependent inactivation of the inwardly rectifying K^+ channel by chloramine-T. European Journal of Pharmacology 258(3):281–284

Krapivinsky G, Gordon EA, Wickman K, Velimirovic B, Krapivinsky L, Clapham DE (1995) The G-protein-gated atrial K^+ channel IKACh is a heteromultimer of two inwardly rectifying $K(^+)$–channel proteins. Nature 374:135–141

Kubo Y, Baldwin TJ, Jan YN, Jan LY (1993) Primary structure and functional expression of a mouse inward rectifier potassium channel. Nature 362:127–133

Kubo Y, Reuveny E, Slesinger PA, Jan YN, Jan LY (1993b) Primary structure and functional expression of a rat G- protein coupled muscarinic potassium channel. Nature 364:802–806

Kurachi Y (1985) Voltage-dependent activation of the inward rectifier potassium channel in the ventricular cell membrane of guinea-pig heart. Journal of Physiology 366:365–385

Lesage F, Duprat F, Fink M, Guillemare E, Coppola T, Lazdunski M, Hugnot JP (1994) Cloning provides evidence for a family of inward rectifier and G–protein coupled K^+ channels in the brain. FEBS Letters 353:37–42

Lesage F, Guillemare E, Fink M, Duprat F, Lazdunski M, Romey G, Barhanin J (1996) TWIK-1, a ubiquitous human weakly inward rectifying K^+ channel with a novel structure. EMBO Journal 15:1004–1011

Liman ER, Hess P, Weaver F, Koren G (1991) Voltage-sensing residues in the S4 region of a mammalian K channel. Nature 353:752–756

Liu Y, Holmgren M, Jurman ME, Yellen G (1997) Gated access to the pore of a voltage-dependent K^+ channel. Neuron 19:175–184

Lopatin AN, Makhina EN, Nichols CG (1994) Potassium channel block by cytoplasmic polyamines as the mechanism of intrinsic rectification. Nature 372:366–369

Lopatin AN, Makhina EN, Nichols CG (1995) The mechanism of inward rectification of potassium channels. Journal of General Physiology 106:923–955

Lopatin AN, Nichols CG (1994) Inward rectification of outward rectifying DRK1 (Kv21) potassium channels. Journal of General Physiology 103:203–216

Lopatin AN, Nichols CG (1996) $[K^+]$-dependence of polyamine induced rectification in inward rectifier potassium channels (IRK1, Kir2.1). Journal of General Physiology 108:105–113

Lu Z, Mackinnon R (1994) Electrostatic tuning of Mg^{2+} affinity in an inward rectifier K^+ channel Nature 371:243–246

Lu Z, Mackinnon R (1995) Probing a potassium channel pore with an engineered protonatable site. Biochemistry 34:13133–13138

MacKinnon R (1991) Determination of the subunit stoichiometry of a voltage–activated potassium channel. Nature 350:232–235

MacKinnon R, Yellen G (1990) Mutations affecting TEA blockade and ion permeation in voltage-activated K^+ channels. Science 250:276–279
Makhina EN, Kelly AJ, Lopatin AN, Mercer RW, Nichols CG (1994) Cloning and expression of a novel inward rectifier potassium channel from human brain. Journal of Biological Chemistry 269:20468–20474
Matsuda H (1988) Open-state substructure of inwardly rectifying potassium channels revealed by magnesium block in guinea-pig heart cells. Journal of Physiology 397:237–258
Matsuda H (1991) Magnesium gating of the inwardly rectifying K^+ channel. Annual Review of Physiology 53:289–298
Matsuda H, Matsuura H, Noma A (1989) Triple–barrel structure of inwardly rectifying K^+ channels revealed by Cs^+ and Rb^+ block in guinea–pig heart cells. Journal of Physiology 413:139–157
Matsuda H, Saigusa A, Irisawa H (1987) Ohmic conductance through the inwardly rectifying K^+ channel and blocking by internal Mg^{2+}. Nature 325:156–159
Miller C (1982) Bis–quaternary ammonium blockers as structural probes of the sarcoplasmic reticulum K^+ channel. J Gen Physiol 79:869–891
Nakajima Y, Nakajima S, Inoue M (1988) Pertussis toxin-insensitive G protein mediates substance P-induced inhibition of potassium channels in brain neurons. Proceedings of the National Academy of Sciences of the United States of America 85:3643–3647
Newman EA (1993) Inward-rectifying potassium channels in retinal glial (Muller) cells. Journal of Neuroscience 13:3333–3345
Nichols CG (1993) The "inner core" of inward rectifier potassium channels. Trends in Pharmacological Sciences 14:320–323
Nichols CG, Ho K, Hebert S (1994) Mg^{2+} dependent inward rectification of ROMK1 potassium channels expressed in *Xenopus* oocytes. Journal of Physiology 476:399–409
Nichols CG, Shyng S-L, Nestorowicz A, Glaser B, Clement JP IV, Gonzales G, Aguilar-Bryan L, Permutt AM, Bryan J (1996). Adenosine diphosphate as an intracellular regulator of insulin secretion. Science 272:1785–1787
Noble D (1965) Electrical properties of cardiac muscle attributable to inward going (anomalous) rectification. Journal of Cellular and Comparative Physiology 66:127–136
Noma A (1983) ATP–regulated K^+ channels in cardiac muscle. Nature 305:147–148
Oliva C, Cohen IS, Pennefather P (1990) The mechanism of rectification of I_{K1} in canine Purkinje myocytes. Journal of General Physiology 96:299–318
Papazian DM, Timpe LC, Jan YN, Jan L (1991) Alteration of voltage-dependence of Shaker potassium channel by mutations in the S4 sequence. Nature 349:305–310
Pearson WL, Nichols CG (1998) Block of Kir21 channels by alkylamine analogues of endogenous polyamines. J Gen Physiol 112:351–363
Perier F, Radeke CM, Vandenberg CA (1994) Primary structure and characterization of a small–conductance inwardly rectifying potassium channel from human hippocampus. PNAS 91:6240–6244
Pessia M, Bond CT, Kavanaugh MP, Adelman JP (1995) Contributions of the C–terminal domain to gating properties of inward rectifier potassium channels. Neuron 14:1039–1045
Pessia M, Tucker SJ, Lee K, Bond CT, Adelman JP (1996) Subunit positional effects revealed by novel heteromeric inwardly rectifying K^+ channels EMBO. Journal 15:2980–2987
Raab-Graham KF, Radeke CM, Vandenberg CA (1994) Molecular cloning and expression of a human heart inward rectifier potassium channel. Neuroreport 5:2501–2505
Rettig J, Wunder F, Stocker M, Lichtinghagen R, Mastiaux F, Beckh S et al. (1992) Characterization of a Shaw-related potassium channel family in rat brain. EMBO Journal 11:2473–2486

Romano C, Williams K, DePriest S, Seshadri R, Marshall GR, Israel M, Molinoff PB (1992) Effects of mono-, di-, and triamines on the *N*-methyl-D-aspartate receptor complex: a model of the polyamine recognition site. Mol Pharmacol 41:785–792

Ruppersberg JP, Fakler B, Brandle U, Zenner H-P, Schultz JH (1996) An N-terminal site controls blocker-release in Kir21 channels. Biophys J 70:A361

Ruppersberg JP, vanKitzing E, Schoepfer R (1994) The mechanism of magnesium block of NMDA receptors. Seminars in the Neurosciences 6:87–96

Sanguinetti CM, Jiang C, Curran ME, Keating MT (1995) A mechanistic link between an inherited and an acquired cardiac arrhythmia: HERG encodes the IKr potassium channel. Cell 81:299–307

Schreibmayer W, Dessauer CW, Vorobiov D, Gilman AG, Lester HA, Davidson N, Dascal N (1996) Inhibition of an inwardly rectifying K^+ channel by G-protein a subunits. Nature 380:624–627

Seiler N (1994) Formation, catabolism and properties of the natural polyamines. In: Carter C (ed) The neuropharmacology of polyamines. Academic Press, Harcourt Brace, London and New York, chap 1

Shuck ME, Bock JH, Benjamin CW, Tsai TD, Lee KS, Slightom JL, Bienkowski MJ (1994) Cloning and characterization of multiple forms of the human kidney ROM-K potassium channel. Journal of Biological Chemistry 269:24261–24270

Shyng SL, Ferrigni T, Nichols CG (1997) Control of rectification and gating of cloned K_{ATP} channels by the Kir62 subunit. Journal of General Physiology 110:141–153

Shyng SL, Nichols CG (1997) Octameric stoichiometry of the K_{ATP} channel complex. Journal of General Physiology 110:655–664

Shyng SL, Sha Q, Ferrigni T, Lopatin AN, Nichols CG (1996) Depletion of intracellular polyamines relieves inward rectification of potassium channels. Proc Natl Acad Sci USA 93:12014–12019

Silver MR, DeCoursey TE (1990) Intrinsic gating of inward rectifier in bovine pulmonary artery endothelial cells in the presence or absence of internal magnesium. Journal of General Physiology 96:109–133

Smith PL, Baukrowitz T, Yellen G (1996) The inward rectification mechanism of the HERG cardiac potassium channel. Nature 379:833–836

Spauschus A, Lentes KU, Wischmeyer E, Dissmann E, Karschin C, Karschin A (1996) A G-protein-activated inwardly rectifying K^+ channel (GIRK4) from human hippocampus associates with other GIRK channels. Journal of Neuroscience 16:930–938

Stanfield PR, Davies NW, Shelton PA, Sutcliffe MJ, Khan IA, Brammar WJ et al. (1994) A single aspartate residue is involved in both intrinsic gating and blockage by Mg^{2+} of the inward rectifier, IRK1. Journal of Physiology 478:1–6

Steglich C, Scheffler IE (1982) An ornithine decarboxylase-deficient mutant of Chinese hamster ovary cells. Journal of Biological Chemistry 257:4603–4609

Stocker M, Pongs O, Hoth M, Heinemann SH, Stuhmer W, Schroter KH, Ruppersberg JP (1991) Swapping of functional domains in voltage-gated K^+ channels. Proceedings of the Royal Society of London B245:101–107

Stuhmer W, Conti F, Suzuki H, Wang XD, Noda M, Yahagi N et al. (1989) Structural parts involved in activation and inactivation of the sodium channel. Nature 339:597–603

Tabor CW, Tabor H (1984) Polyamines. Annual Review of Biochemistry 53:749–790

Taglialatela M, Wible BA, Caporoso R, Brown AM (1994) Specification of the pore properties by the carboxyl terminus of inwardly rectifying K^+ channels. Science 264:844–847

Takumi T, Ishii T, Horio Y, Morishige K, Takahashi N, Yamada M et al. (1995) A novel ATP-dependent inward rectifier potassium channel expressed predominantly in glial cells. Journal of Biological Chemistry 270:16339–16346

Trudeau MC, Warmke JW, Ganetzky B, Robertson GA (1995) HERG, a human inward rectifier in the voltage-gated potassium channel family. Science 269:92–95

Tytgat J, Hess P (1992) Evidence for cooperative interactions in potassium channel gating. Nature 359:420–423

Tytgat J, Nakazawa K, Gross A, Hess P (1993) Pursuing the voltage sensor of a voltage-gated mammalian potassium channel. Journal of Biological Chemistry 268:23777–23779

Vandenberg CA (1987) Inward rectification of a potassium channel in cardiac ventricular cells depends on internal magnesium ions. Proceedings of the National Academy of Sciences 84:2560–2566

Vandenberg CA (1994) Cardiac inward rectifier potassium channel. In: Spooner PM, Brown AM (eds) Ion channels in the cardiovascular system. Futura Publishing, NY, chap 8

van Leeuwenhoek A (1678) Observationes D Anthonii Leeuwenhoek, de natis e semine genitali animalculis. Philosophical Transactions of the Royal Society 12:1040–1043

Wible BA, Taglialatela M, Ficker E, Brown AM (1994) Gating of inwardly rectifying K^+ channels localized to a single negatively charged residue. Nature 371:246–249

Williams JT, Colmers WF, Pan ZZ (1988) Voltage- and ligand-activated inwardly rectifying currents in dorsal raphe neurons in vitro. Journal of Neuroscience 8:3499–3506

Yang J, Jan YN, Jan LY (1995) Control of rectification and permeation by residues in two distinct domains in an inward rectifier K^+ channel. Neuron 14:1047–1054

Yang J, Jan YN, Jan LY (1995) Determination of the subunit stoichiometry of an inwardly potassium channel. Neuron 15:1441–1447

Yano H, Philipson LH, Kugler JL, Tokuyama Y, Davis EM, Le Beau MM et al. (1994) Alternative splicing of human inwardly rectifying K^+ channel ROMK1 mRNA. Molecular Pharmacology 45:854–860

Yellen G (1984) Relief of Na^+ block of Ca^{2+}-activated K^+ channels by external cations. Journal of General Physiology 84:187–199

Yool AJ, Schwartz TL (1991) Alteration of ionic selectivity of a K^+ channel by mutation in the H5 region. Nature 349:700–704

Zagotta WN, Hoshi T, Aldrich RW (1990) Restoration of inactivation in mutants of Shaker potassium channels by a peptide derived from ShB. Science 250:568–571

Zhou H, Tate SS, Palmer LG (1994) Primary structure and functional properties of an epithelial K channel. American Journal of Physiology 266:C809–C824

CHAPTER 10

ATP-Dependent Potassium Channels in the Kidney

G. GIEBISCH, W. WANG, and S.C. HEBERT

A. Introduction

The application of the patch-clamp technique (NEHER and SAKMANN 1976) to the kidney has led to the discovery of well-defined potassium (K) channels in the apical and basolateral membrane of tubule cells along the nephron. Such studies have permitted the biophysical characterization of renal K channels and defined several factors modulating their activity. A subfamily of these channels is distinguished by their sensitivity to alterations in metabolism including changes in cell pH, the level of hormones and cell messengers and ATP. These K channels play an important role in several transport processes in the proximal tubule, the thick ascending limb (TAL) of Henle's loop, and in principal cells of the cortical collecting duct (CCD). ATP-sensitive channels are inhibited by an increase in the concentration of ATP in the cytosol of tubule cells, and such channel block can be relieved by ADP. ATP-sensitive channels have also been detected in several extrarenal tissues including brain, smooth, skeletal and heart muscle and β-cells of pancreatic islands (WANG W and HEBERT 1999). The molecular structure of one ATP-sensitive K channel has been defined by expression cloning (Ho et al. 1993) and disturbances of its function shown to play a key role in the inherited electrolyte disorder of Bartters's syndrome (BARTTER et al. 1962).

B. The Function of ATP-Sensitive K Channels in the Proximal Tubule

Microelectrode studies on single amphibian and mammalian tubule cells have shown that the large cell-negative electrical potential depends on a large K conductance in the basolateral membrane and the steep transmembrane concentration gradient of K (GIEBISCH 1998; STANTON and GIEBISCH 1992). Figure 1 shows a model of a renal tubule cell and several transport processes that are relevant to a consideration of the role of ATP-sensitive K channels in tubule function. A model of a proximal tubule cell is shown in Fig. 2. It is well established that the activity of the Na,K-ATPase in the basolateral membrane is responsible for the high concentration of K in tubule cells, and that a large K conductance in the basolateral membrane generates the transmembrane

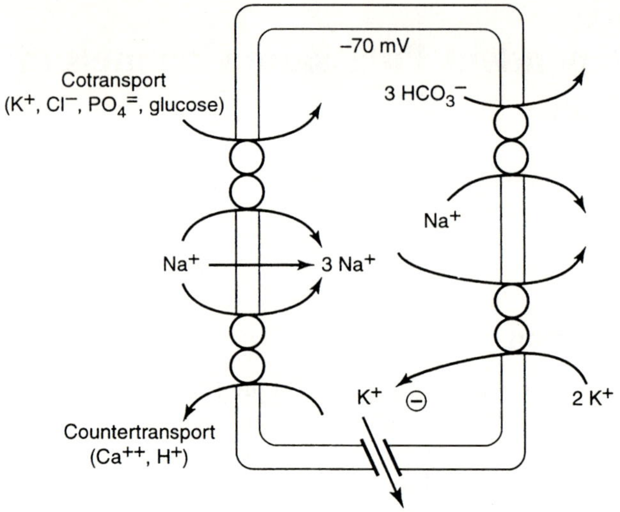

Fig. 1. Cell schema of renal tubule cell with basolateral Na-K pump (*right*) and potassium channel, and several apical and basolateral transporters including electrogenic and electroneutral mechanisms (modified from CLAUSEN 1996)

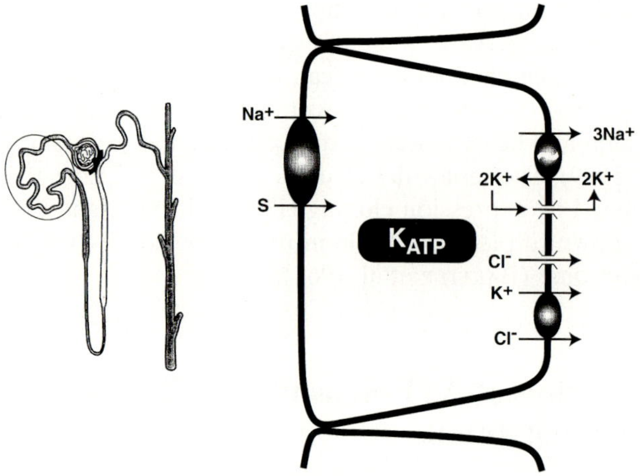

Fig. 2. Model for ion transport in the proximal tubule. A K_{ATP} channel is present in the basolateral membrane and has been shown to be involved in the coupling between Na^+,K^+-ATPase turnover and K conductance

potential difference. Several basolateral K channels have been identified, including one with significant sensitivity to ATP (WANG W and HEBERT 1999). Supporting evidence includes the inhibitory action of ATP (MAURER et al. 1998; KUBOKAWA et al. 1998; BECK et al. 1991, 1993; HURST AM et al. 1991), the reduction of the basolateral K conductance and cell potential by glibenclamide, an agent that inhibits ATP-sensitive K channels (NEHER and SAKMANN 1976; TSUCHIYA et al. 1992), and the activation by diazoxide, an opener of ATP-sensitive K channels (BECK et al. 1993). The ATP-sensitive K channel is also inhibited by lowering cell pH (OHNO-SHOSAKU et al. 1990).

The large cell-negative potential provides an important driving force for the transport of passively moving solutes and carriers across the apical and basolateral membrane of proximal tubule cells. Thus, the entry of positively charged ions such as sodium (Na) – especially in the distal nephron where Na channels are more abundant – and of several positively charged cotransporters, especially those involved in Na-dependent glucose, amino-acid, phosphate, and sulfate transport, is enhanced by the cell-negative potential (GIEBISCH 1995; CLAUSEN 1996). This potential also generates a significant driving force across the basolateral membrane for such electrogenic transporters as Na-Ca exchange and Na-HCO_3 cotransport. Cell depolarization leads to reduction of both transport processes and a corresponding increase in the concentration of Ca and HCO_3 in the cytoplasm. Taken together, it is apparent that ATP-sensitive K channels play an important role in ion transport in the proximal tubule.

ATP-dependent K channels are also involved in the tight coupling between Na,K-ATPase activity and basolateral K conductance (TSUCHIYA et al. 1992; BECK et al. 1991, 1993: HURST et al. 1991; WELLING 1995). Thus, inhibition of basolateral Na,K-ATPase activity by deletion of organic solutes from the tubule fluid or application of ouabain to the basolateral fluid decreases channel activity at a time when the concentration of ATP rises in the cytoplasm of proximal tubule cells (TSUCHIYA et al. 1992; BECK et al. 1993). In contrast, the addition of amino acids serving as substrate for cotransport with Na lowered the concentration of ATP and augmented the activity of basolateral K channels.

C. The Function of ATP-Sensitive K Channels in the Thick Ascending Limb (TAL) of Henle's Loop

Figure 3 shows a cell model of the main transport mechanisms of the TAL. Two K channels have been identified in the apical membrane and both demonstrated a sensitivity to inhibition by ATP (WANG W et al. 1990, 1997; HEBERT 1995a,b; BLEICH et al. 1990; WANG W 1994). Also shown are the electroneutral Na-2Cl-K cotransporter and several basolateral transporters including the electrogenic Na,K- ATPase, a K and chloride (Cl) channel and a KCl cotransporter (HEBERT and ANDREOLI 1984; GREGER 1985). It should also be noted that in addition to transcellular pathways for ion movement, several cations are reabsorbed by passive transport via the paracellular pathway. The main driving force for such movement is the lumen-positive potential.

The significant apical K conductance plays an important role in several transport processes. First, the low-conductance K channel shares most prop-

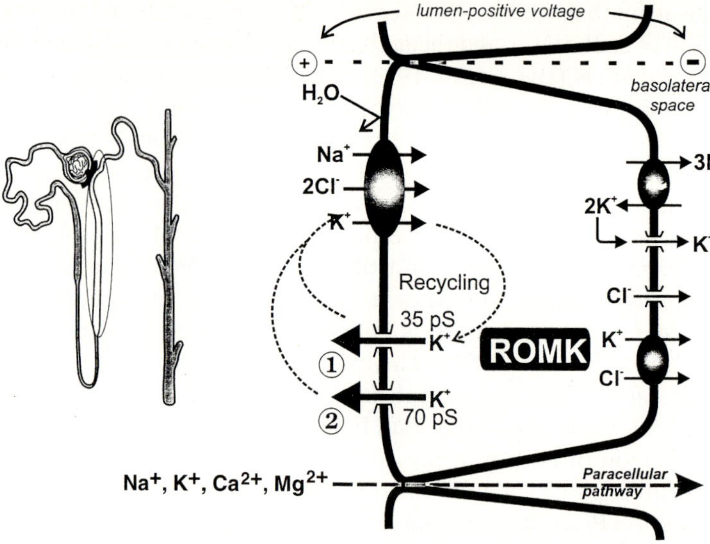

Fig. 3. Model for ion transport in the thick ascending limb. The two types of apical K^+ channels are shown; 35 pS (also called the small K^+ or SK channel) and 70 pS channels. ROMK and SK functional and regulatory characteristics are essentially identical. It has been proposed that the 70 pS channel is also formed by ROMK in association with another channel subunit, but this remains to be demonstrated (from WANG W and Hebert 1999)

erties with a similar ATP-sensitive K channel in the apical membrane of principal tubule cells, but it should be noted that the 70 pS K channel contributes more significantly to the apical K conductance (WANG W and HEBERT 1999; WANG W et al. 1997; WANG W 1994). Both channels permit recycling of K across the apical membrane to safeguard adequate supply to the Na-2Cl-K cotransporter (WANG W and HEBERT 1999; GIEBISCH 1995; WANG W et al. 1997; HEBERT 1995a,b). The importance of apical K recycling is demonstrated by observations that either deletion of K from the lumen or addition of K channel blockers leads to a significant decrease in net NaCl reabsorption (WANG T et al. 1995a,b). Accordingly, the state of activity of apical K channels has evolved as an important modulator of Na transport at this nephron site. A K channel that shares many biophysical properties with the low-conductance apical K channel in the TAL has also been discovered in the apical membrane of macula densa cells (HURST et al. 1994). Its functional importance is shown in experiments in which interference with apical K channel activity – either by decreasing K in the lumen or addition of potent K channel blockers – led to interference with the tubulo-glomerular feedback response: an increase in lumen flow rate in the perfused loop of Henle became ineffective in reducing single nephron glomerular filtration rate following inhibition of K channels (VALLON et al. 1997, 1998).

Finally, attention should be drawn to the important effect of apical K channel modulation on the transepithelial potential difference. Inhibition of the apical K channels has been shown to decrease the lumen positive potential (HEBERT and ANDREOLI 1984; GREGER 1985). As a consequence, passive transport of cationic solutes such as Na, Ca, K, and Mg falls and frequently leads to enhanced urinary loss.

D. The Function of ATP-Sensitive K Channels in the Cortical Collecting Duct (CCD)

There is general agreement that the low-conductance K channel in the apical membrane of principal tubule cells is a key element in K secretion. Many micropuncture and microperfusion studies have identified the initial and cortical collecting tubule as the main site of regulated K secretion in the kidney (WANG W and HEBERT 1999; GIEBISCH 1998; STANTON and GIEBISCH 1992; MALNIC et al. 1999). Figure 4 provides a cell model of a principal tubule cell including the main transporters that account for K secretion.

The basolateral membrane is the site of an Na-K-ATPase that is responsible for uptake of K from the peritubular fluid. The turnover of this ATPase is enhanced upon raising the plasma level of K and is further stimulated by mineralocorticoids (GIEBISCH 1998; STANTON and GIEBISCH 1992). At very high transport rates of active Na-K exchange, the magnitude of the membrane potential may exceed the K equilibrium potential so that the direction of K

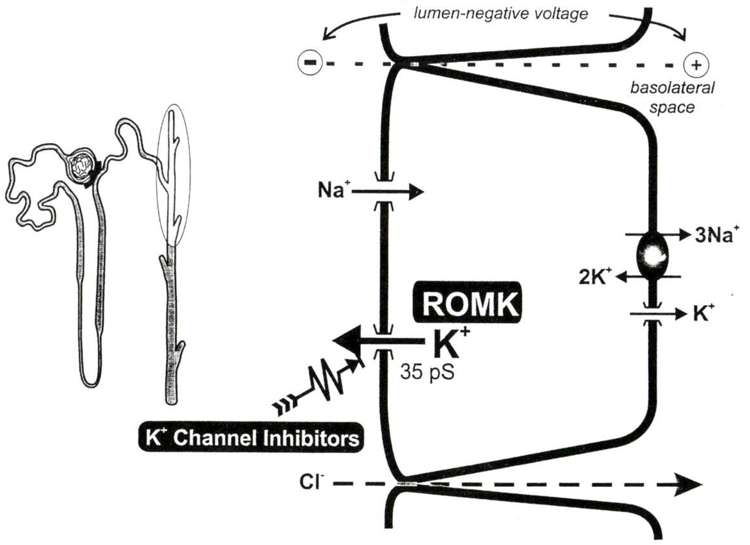

Fig. 4. Model for ion transport by the principal cell in the collecting duct. The apical K^+ secretory channel in this cell is ROMK

movement reverses and K uptake occurs both by pump-mediated transport and passive accumulation (GIEBISCH 1998; STANTON and GIEBISCH 1992; KOEPPEN and GIEBISCH 1985; SANSOM et al. 1989; STOKES 1993).

Two transport mechanisms, an ATP-sensitive K channel and a Na channel, determine the transfer of K from cell to lumen. A combination of microelectrode and patch clamp studies have shown that net K secretion depends on the electrochemical potential gradient of K ions across the apical membrane. While the electrochemical potential of K across the basolateral membrane is close to the K equilibrium potential, it is far removed from it across the apical membrane owing to the significant Na permeability. Accordingly, the activity of apical Na channels has a major effect on K movement across the apical membrane: a decrease in lumen Na concentration or a decline in apical Na channel opening reduces K secretion. In contrast, a high concentration of Na in the lumen or increased activity of Na channels stimulates K secretion (GIEBISCH 1998; STANTON and GIEBISCH 1992; MALNIC et al. 1999).

As pointed out above, the small-conductance K channel shares many properties with a similar K channel in the apical membrane of the cells of the TAL (WANG and HEBERT 1999; GIEBISCH 1995, 1998; STANTON and GIEBISCH 1992; WANG W et al. 1990a, 1997; HEBERT 1995a,b; FRINDT and PALMER 1989; WANG WH 1995). Its characteristics include high open probability, mild inward rectification, sensitivity to inhibition by low cell pH and ATP. Since the open probability of these channels is high, increase in apical K conductance is mediated largely by the recruitment of additional K channels and not by increasing open probability. The channel has a significant permeability to rubidium, which explains the observed ability of rubidium secretion in isolated collecting ducts. An unresolved problem concerns the mechanism of K channel recruitment: it is unknown whether the activation process involves stimulation of dormant channels already present in the apical membrane or, alternatively, the recruitment and insertion of "new" K channels from subapical pools of K channels.

E. The Regulation of ATP-Sensitive K Channels

I. Proximal Tubule

The basolateral 60 pS K channel is sensitive to inhibition by ATP (TSUCHIYA et al. 1992; MAURER et al. 1998; WANG W et al. 1997). The channel is also inhibited by glyburide, an agent known to block ATP-sensitive K channels in nonrenal tissues (WANG W and HEBERT 1999). Moreover, the channel is stimulated by diazoxide, an opener of ATP-sensitive channels (BECK et al. 1991). An important function of the basolateral channel is to link the activity of the basolateral Na,K-ATPase to the K conductance (TSUCHIYA et al. 1992; WELLING 1995). Inhibition of Na,K-ATPase decreases the channel's activity whereas luminal addition of amino acids or glucose, both serving as substrate for apical Na^+ entry, increase basolateral Na,K-ATPase activity. Such stimulation of transport decreases intracellular ATP and activates the basolateral channel. The tight coupling of Na,K-ATPase turnover to basolateral K channels prevents volume changes with alterations in net sodium and fluid transport (BECK et al. 1993; HURST et al. 1991; BECK et al. 1994).

II. Thick Ascending Limb of Henle's Loop

Figure 5 summarizes the most important factors shown to modulate the activity of apical K channels in native tubules. Of these, calcium, pH and ATP are the most important ones. An increase in Ca in the cytosol inhibits both K channels, an effect that may be mediated by PKC (WANG W and HEBERT 1999; WANG W et al. 1996, 1997). Both apical channels are sensitive to changes in cell pH (BLEICH et al. 1990), except in the rabbit (WANG W et al. 1990a), and acidosis decreases channel activity whereas alkalosis has the opposite effect (WANG W and HEBERT 1999). ATP-sensitive K channels in the apical membrane are inhibited by mM ATP, consistent with the notion that cell metabolism controls apical K conductance (WANG W and HEBERT 1999; WANG W et al. 1997). Arachidonic acid has also been shown to be an effective inhibitor of apical K channels (WANG W and LU 1995; MACICA et al. 1997a,b). Patch-clamp studies on both cell-attached patches and inside-out patches demonstrate a sharp decline in channel activity following arachidonic acid exposure, and this effect can be shown to be mediated by 20-hydroxytetraenoic acid (20 HETE),

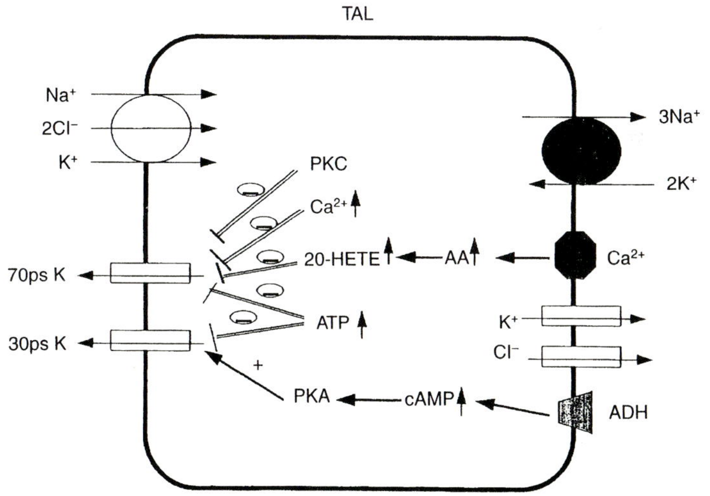

Fig. 5. Cell models of the thick ascending limb (TAL). Note two apical K channels and the effects of several cell messengers. Shown also are Ca and ADH receptors and the effects of cAMP and arachidonic acid (AA) on apical K^+ channels. Volume-sensitive, Ca-stimulated maxi-K channels have been described in cell cultures of TAL, but have not been observed in intact mammalian TAL (based on data from GIEBISCH G, WANG W (1996) Kidney Int 49:1624 – used with permission)

a major metabolite of P450 oxygenase. The activation of the cytochrome P450 pathway has recently been shown to be importantly involved in the transmission of the effect on apical K channel activity of raising the basolateral Ca concentration (WANG W et al. 1996). Following the demonstration of Ca-receptors in the basolateral membrane of TAL (Brown et al. 1993), their stimulation by receptor agonists mimicked the effect of increasing Ca, and such activation was abolished by inhibition of the P450-dependent metabolic pathway (WANG W and HEBERT 1999; WANG W et al. 1997). Additional studies have also shown that the inhibitory effect of angiotensin-II on the activity of the 70pS K channel is also mediated by P450 metabolites such as 20-HETE, because P450 monooxygenase inhibitors abolish the effect (LIU et al. 1999). The presence of arachidonic acid metabolites in TAL was directly demonstrated by gas chromatography/mass spectroscopy, and supports the thesis that they are involved in the regulation of apical K channels (LU M et al. 1996).

Channel activity is also inhibited by activation of PKC [Macica et al. 1997b). Apical K channels are inhibited by stimulation of phorbol esters and inhibition of PKC with calphostin C stimulated channel activity. Moreover, direct application of exogenous PKC inhibited apical K channels in inside-out patches. The important role of PKC in mediating the downregulation of apical K channels by elevation of Ca has already been mentioned.

Activation of apical K 70pS channels by cAMP is also of physiological significance. Vasopressin also stimulates the activity of the low-conductance K channel and this effect has been shown to be mediated by PKA (WANG W 1994; LIU et al. 1999). Increased channel activity by PKA-mediated stimulation of phosphorylation is likely to play an important role in the augmentation of NaCl reabsorption that has been observed after vasopressin (HEBERT and ANDREOLI 1984; GREGER 1985).

III. Cortical Collecting Tubules – Apical Membrane of Principal Cells

Figure 6 shows the factors that regulate the apical low-conductance K channel (WANG and HEBERT 1999; GIEBISCH 1995, 1998; STANTON and GIEBISCH 1992; WANG W et al. 1997). Those modulating basolateral K channels are included for comparison. The kidney responds to changes in K intake by varying the rate of distal tubule K secretion, and such adaptations are associated with either an increase or decrease of the apical K permeability. Patch-clamp studies indicate that a high K intake increases the channel density of apical low-conductance K channels, whereas channel density decreases in principal cells when a low K diet is administered (WANG WH et al. 1999; FRINDT et al. 1998; PALMER 1999). It is of interest that such functional adaptations are not solely mediated by K-related changes in aldosterone (PALMER et al. 1994). Rather, there appears to be a direct effect of K intake on channel density, independent of aldosterone and related to changes in protein tyrosinase activity (WANG WH et al. 1999). This conclusion is based on patch-clamp studies

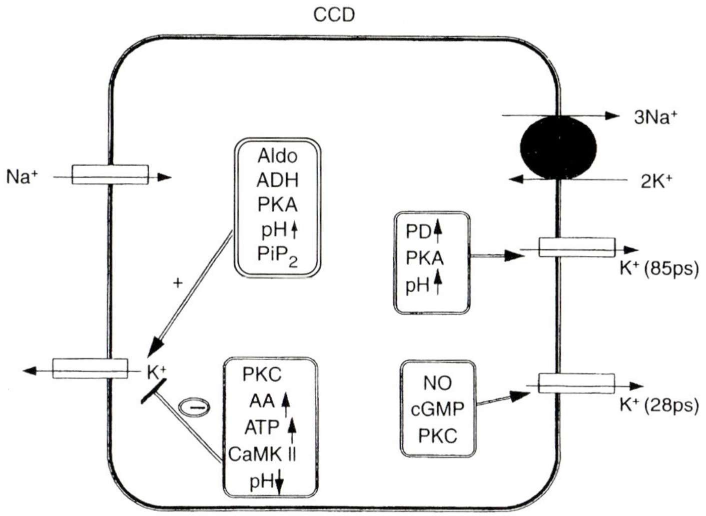

Fig. 6. Cell model of principal cell of cortical collecting duct. Four different K channels are shown; the Ca- and depolarization-activated maxi-K channel in the apical membrane is not included. The apical K channel is not voltage-sensitive whereas the 85 ps K channel in the basolateral membrane is activated by hyperpolarization so that basolateral K conductance increases with stimulation of electrogenic Na,K-ATPase activity. Regulation of renal ATP-sensitive K channels by membrane-bound protein phosphates has also been demonstrated. Whereas PKA-mediated phosphorylation induces channel opening, channel activity is inhibited by protein phosphatase PP-2 A and Mg^{2+}-dependent protein phosphatase PP-2 C, both of which dephosphorylate PKA-mediated phosphorylation sites (based on data from GIEBISCH G, WANG W (1996) Kidney Int 49:1624 – used with permission)

showing that the number of active channels in apical patches from animals on a low K diet could be effectively stimulated by inhibitors of protein tyrosine kinase, whereas this kinase inhibitor did not change the channel density in principal cells of animals on a high K diet. It thus appears safe to conclude that a fall in K intake – perhaps by lowering cell pH (ADLER and FRALEY 1997) – stimulates the activity of protein tyrosine kinase, thereby reducing channel activity. It is tempting to speculate that altered membrane trafficking may be the mechanism by which channel number is modulated during changes in external K balance.

The effects of aldosterone on K channel activity appear to be largely permissive and can best be demonstrated in conditions of chronic K excess or following adrenalectomy (PALMER 1999; PALMER et al. 1994; WALD et al. 1998). While an increase in plasma K alone is sufficient to stimulate K secretion and activation of apical K channels, the full effect of this response requires the elevation of aldosterone. An increase in aldosterone, independent of changes in K balance, is insufficient to alter channel density since neither the infusion of aldosterone nor a low Na diet – a maneuver increasing aldosterone release – changes apical K channel function (PALMER 1999; PALMER et al. 1994).

The apical K channel of principal cells is strongly dependent on cell pH. A fall in cell pH reversibly decreases both open probability and channel number, whereas channel activity increases when cell pH is raised from the normal value of 7.2 (WANG W et al. 1990; SCHLATTER et al. 1993). These findings are consistent with the decrease of K secretion observed in distal tubules during metabolic acidosis and with the well-established fact that metabolic alkalosis enhances K secretion.

Additional factors decreasing apical K channel activity – largely channel density – include PKC (WANG W and GIEBISCH 1991a), protein phosphatases (KUBOKAWA et al. 1995a), arachidonic acid [WANG W et al. 1992), calcium-calmodulin dependent kinase II (KUBOKAWA et al. 1995b) and elevated levels of cell ATP (WANG W and GIEBISCH 1991b). Patch clamp studies have shown that the inhibitory effect on apical K channels of increasing cell Ca is mediated by activation of PKC and calcium-calmodulin dependent kinases (KUBOKAWA et al. 1995b; WANG W et al. 1993). These inhibitory effects of high Ca are indirect since exposure of inside-out membrane patches to high Ca does not affect channel activity (WANG W et al. 1990a). Studies of the mechanisms of K channel inhibition after decreasing the activity of basolateral Na,K-ATPase (WANG W et al. 1993) or following exposure to cyclosporin [LING and EATON 1993) have shown that high Ca levels play an important role in mediating the observed reduction in apical K channels.

F. Properties of Cloned ATP-Sensitive K Channels (ROMK)

I. Channel Structure

A potassium channel, cloned from the medulla of the kidney (ROMK $K_{IR}1$), has many properties of the apical low-conductance K channel in the apical membrane of CCD. The cloned channel is a member of a large family of inwardly rectifying K channels which is distinguished by high K selectivity, inward rectification and structure characterized by two membrane-spanning segments, and a pore-forming region with high homology to the pore-forming H5-segment of voltage-gated channels (WANG W and HEBERT 1999; HO et al. 1993; GIEBISCH 1998; WANG W et al. 1997; HEBERT 1995a,b). Available evidence suggests that the N and C-terminals are extending into the cytosol and function as important regulatory domains; they interact with protons, nucleotides, kinases such as PKA and PKC, and phosphoinositides (WANG W and HEBERT 1999; HO et al. 1993; GIEBISCH 1998; WANG W et al. 1997; HEBERT 1995a,b). Figure 7 summarizes the present state of knowledge of the topology of the ROMK channel (WANG W and HEBERT 1999). Inspection of Fig. 7 highlights several important structural elements of the channel. These include an amino-acid sequence (GYG) typically conserved in H5-like regions of K channels

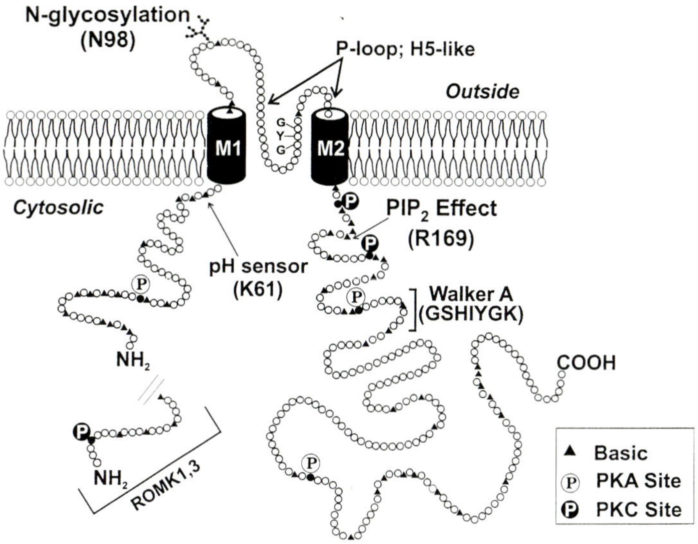

Fig. 7. Topology of ROMK ($K_{IR}1.1$) K^+ channel. M1 and M2 represent the two membrane-spanning domains characterizing the inward-rectifier family of potassium channel. Some important functional sites are indicated. A short amphipathic segment in the M1-M2 linking segment in ROMK is homologous to the pore-forming[P-loop] or H5 region of classic voltage-gated *Shaker* K^+ channels cloned from the fruit fly. See text for discussion. The canonical G-Y-G amino acid sequence found in all K^+ channels is shown in the H5 segment (from Wang W and Hebert 1999)

with high K selectivity (Jan and Jan 1997; Heginbotham et al. 1992), and an external segment that links the M1 and M2 transmembrane alpha helices and contains an N-linked glycosylation site (Ho et al. 1993). The proposed structure of a K channel from *Streptomyces lividans* supports the view that the selectivity barrier includes elements such as the H5 loop and the linking region between the first membrane-spanning helix (M1) and the H5 loop (MacKinnon et al. 1998). Mutations of single amino acids in this area significantly alter channel block (Wang W and Lu 1995; Sabirov et al. 1997; Zhou et al. 1996).

Figure 8 provides information on the proposed multimeric structure of ROMK channels. Although not yet directly proven, it is highly likely that, analogous to voltage-gated K channels, ROMK channels have a tetrameric structure. Such a model has been proposed for several K_{IR} channels (Clement et al. 1997; Yang et al. 1995; Fakler et al. 1996; Koster et al. 1998) and suggests that each subunit forms part of the channel pore and the selectivity barrier. It has also been shown that specific elements of the N and C termini as well as the M1, H5 and M2 core region participate in the assembly of the subunits to form functional ROMK channels.

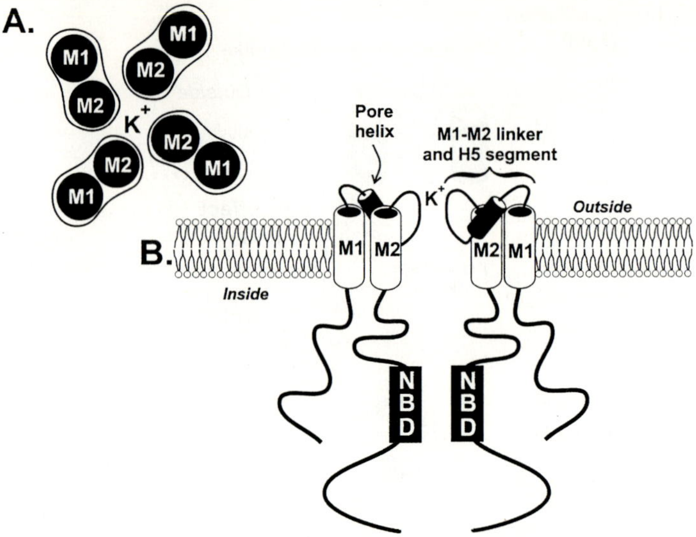

Fig. 8A,B. Multimeric structure of the K_{IR} family: **A** ROMK, like all K_{IR} channels, are formed from a tetrameric assembly of subunits. M2 segments line the channel pore and are surrounded by M1 segments that also participate in subunit-subunit interactions in the tetrameric channel complex; **B** two of the four subunits forming the tetrameric ROMK channel are depicted. The nucleotide binding domain on the channel C-terminus is shown [from WANG W and HEBERT 1999)

II. Channel Isoforms and Localization

Several isoforms of ROMK channels have been identified: ROMK1 (K_{IR}1a), ROMK2 (K_{IR}1b), ROMK3 (K_{IR}1c), and ROMK6 (K_{IR}1d). The difference between these isoforms is based on different properties of the N terminus (BOIM et al. 1995; ZHOU et al. 1994). ROMK2 (identical to rat ROMK6) has the shortest N terminus (see Figs. 8 and 9), and the addition of 19 or 26 amino acids leads to formation of ROMK1 or ROMK3, respectively (SHUCK 1994). Rat ROMK 1–3 are expressed in segments of the nephron ranging from the TAL of Henle's loop to the outer medullary collecting duct (BOIM et al. 1995; LEE and HEBERT 1995). Inspection of Fig. 9 indicates that the rat TAL and distal convoluted tubule express ROMK2 and ROMK3, whereas principal cells in the CCD express ROMK1 and 2 channel transcripts. Outer medullary collecting duct cells are distinguished by expression of ROMK1 only. No specific role for the functions of these three isoforms has been detected with the exception of the serine in position 4 in the extended N terminus of ROMK1. This site appears to be necessary for the sensitivity to arachidonic acid and PKC (MACICA et al. 1997a,b). An unresolved problem concerns the possibility that ROMK channels are formed by assembly of different subunits, and that such heterotetramers may display subtle differences in function. It should be noted (see Fig. 10) that the single channel conductance of the various isoforms of

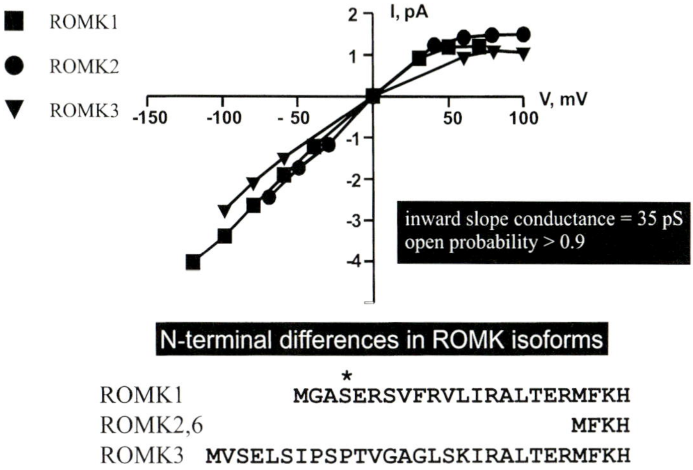

Fig. 9. The distribution of the ROMK 1, 2, and 3 isoforms along the rat nephron. The *shaded regions* indicate the localization of ROMK transcripts and protein. CCD, cortical collecting duct; CTAL, cortical thick ascending limb; DCT, distal convoluted tubule; MTAL, medullary thick ascending limb; OMCD, outer medullary collecting duct. In the CCD and OMCD, ROMK is expressed only in principal cells (from WANG W and HEBERT 1999)

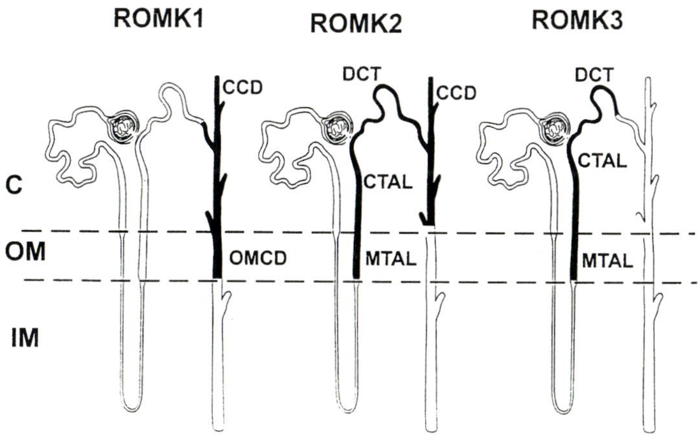

Fig. 10. I/V curve of ROMK channels (from WANG W and HEBERT 1999)

ROMK is identical (Wang and Hebert 1999). As expected from the expression of ROMK channels in nephron segments with low water permeability, water movement through the channel pore is minimal (SABIROV et al. 1998). There is general agreement that ROMK channels target to the apical membranes of TAL and principal cells in CCD (XU JZ et al. 1997; MENNIT et al. 1997; KOHDA et al. 1998).

III. Comparison of ROMK with the Native Secretory ATP-Sensitive K Channel

A large body of evidence, based on patch-clamp studies of apical K channels in cells of the TAL and the CCD, provides convincing evidence that the properties of ROMK channels are similar to those native K channels [WANG W and HEBERT 1999; HO et al. 1993; GIEBISCH 1998; MACICA et al. 1997a,b; BLEICH et al. 1990). The similarities include kinetic behavior of penetrating cations such as K, NH_4, and Tl (PALMER et al. 1997; CHEPILKO et al. 1995), lack of sensitivity to tetraethylammonium (HO et al. 1993; WANG W et al. 1997, 1990a; FRINDT and PALMER 1989), weak inward rectification (HO et al. 1993; PALMER et al. 1997), activation by PKA-dependent phosphorylation (XU Z-C et al. 1996; MCNICHOLAS et al. 1998), inhibition by mmol/l concentrations of ATP (MCNICHOLAS et al. 1994, 1998), block by low pH [MCNICHOLAS et al. 1998; CHOE et al. 1997; TSAI et al. 1995), by arachidonic acid and PKC (MACICA et al. 1997a,b). The inhibitory effect on K channel activity of high cell Ca (WANG WH et al. 1996), arachidonic acid (MACICA et al. 1997a,b), and prostaglandin E (LIU et al. 1999) appears to be mediated by PKC. Finally, the predominant localization of ROMK antibodies to the apical membranes of cells lining the TAL and CCD underscores the notion that ROMK is a major component of the pore-forming subunit of the secretory K channel in the distal nephron (XU JZ et al. 1997; MENNIT et al. 1997; KOHDA et al. 1998).

IV. The Channel Pore-Rectification

One of the distinguishing features of the kinetics of ROMK channels is their very high open probability (HO et al. 1993; WANG WH 1995; MALNIC et al. 1971; SCHLATTER et al. 1992). The predominant closed state is very short, and an infrequent second closed state of much longer duration has been shown to depend on the inhibitory action of divalent cations (SCHLATTER et al. 1992; CHOE et al. 1998). The closed state is sensitive to K ions inhibiting its own passage through the pore (CHOE et al. 1998). Owing to the high open probability, changes in apical conductance of K are predominantly achieved by the recruitment of additional channels. An unresolved problem concerns the mechanism by which such channel activation is initiated. It could involve activation of dormant channels already in the membrane or, alternatively, the targeting and insertion of new channels from a subapical pool.

Inward rectification is another important property of ROMK, shared with all K_{IR} channels (WANG W and HEBERT SC 1999; HO et al. 1993; WANG W et al. 1997; HEBERT 1995a,b; MALNIC et al. 1971; NICHOLS et al. 1994). ROMK channels are weak inward rectifiers and they continue to pass current in the outward direction, albeit less so than in the inward direction. The principal cell's ability to secrete large amounts of K despite weak inward rectification may be explained by the large number of K channels in the membrane. Moreover, varying rectification by changes in intracellular cations such as Mg

(NICHOLS et al. 1994) or polyamines such as spermine or spermidine (LOPATIN et al. 1994; FICKER et al. 1994) may serve as a mechanism modulating secretory K current. Finally, inward rectification may also allow apical depolarization to persist in the presence of an increase in apical K conductance. Studies on the kinetics of inward rectification by Mg and polyamines show that it depends on several factors including voltage, K concentration and dependence on the K equilibrium potential (CHOE et al. 1998; NICHOLS et al. 1994; SPASSOVA and LU Z 1998; OLIVER et al. 1998). An interesting result of kinetic studies was the observation that the M2 region of the channel pore as well as the extracellular loop linking M1 and M2 plays a key role in determining the extent to which K_{IR} displays inward rectification (CHOE et al. 1999). In addition to the role of the H5 region being responsible for the properties of ion selectivity, the M2 region as well as the C-terminus can be shown to determine inward rectification, not only in ROMK but also in other K_{IR} channels (MINOR et al. 1999; KUBO et al. 1993), and two amino acid residues have been identified to play a major role. Whereas strong rectification requires the presence of a negatively charged aspartic acid residue – in position D172 in $K_{IR}1$ in the M2 membrane segment – weak rectification is conferred on ROMK channels by replacement of aspartic acid by asparagine [LU Z and MACKINNON 1994; WIBLE et al. 1994). In addition, a glutamine residue (E 224 in IRK1) also contributes to inward rectification (YANG et al. 1995b), and its replacement by glycine in ROMK renders the latter weakly rectifying. It is noteworthy that this glycine residue is part of the Walker A site (see Fig. 7) which contributes to the C terminals' ability to interact with nucleotides (TAGLIALATELA et al. 1994). The importance of the C terminus in K channel rectification is fully supported by experiments in which the exchange of the ROMK C terminus with that of IRK1 leads to strong rectification when this mutant K channel is expressed in oocytes (TAGLIALATELA et al. 1994).

V. Regulation by Phosporylation: Protein Kinase A (PKA)

Similar to native channels in the TAL and CCD, ROMK channels are regulated by PKA-dependent phosphorylation processes (WANG W and HEBERT 1999; GIEBISCH 1995, 1998; WANG W et al. 1997; WANG W and GIEBISCH 1991a,b; WIBLE et al. 1994). Such channel modulation may occur either in response to receptor-mediated stimulation or by changes in second messenger such as cyclic AMP (cAMP), and channel activity in excised patches expressing ROMK channels has been shown to be affected by phosphorylation-dephosphorylation processes. Channel "run-down," frequently seen in excised patches, depends critically on the channel's state of phosphorylation: it can often be reversed by activation of PKA-dependent phosphorylation (WANG W and GIEBISCH 1991b; MCNICHOLAS et al. 1994). In contrast, channel activity declines whenever protein phosphatase-mediated dephosphorylation exceeds PKA-dependent phosphorylation (KUBOKAWA et al. 1995).

Three PKA-dependent serine phosphorylation sites on the channel protein have been identified (XU Z-C et al. 1996). As shown in Fig. 11, one residue is located on the N terminus (serine 25 in ROMK2) and two on the C terminus (serine 200 and 294 in ROMK2). Whereas K currents decline by about 40% following any single mutation of these PKA phosphorylation sites, mutation of two or more serine residues to alanine led to expression of non-functional channels (XU Z-C et al. 1996; MACGREGOR et al. 1998). Inspection of Fig. 11 shows that the effects on channel activity of mutations of specific phosphorylation sites differ. Although none of the mutations affects single channel conductances, mutations on thc C-terminus (residues 200 and 295 in ROMK2) reduce open probability whereas the replacement of serine with alanine on position 25 does not reduce the channel's open probability but decreases the number of active channels in the excised patch. The observed changes in K channel activity following manipulation of the PKA-dependent phosphorylation site on position 25 on the C terminus are consistent with similar changes in whole cell K currents in oocytes (XU Z-C et al. 1996).

Studies on the mechanism of channel activation by PKA have also demonstrated the importance of kinase anchoring proteins such as AKAP 79 (ALI et al. 1998). Experiments in oocytes which lack sufficient expression of this protein show that channel activation by cAMP is severely compromised, but can be restored provided the anchoring protein is co-expressed with ROMK2. These experiments indicate that an anchoring protein such as AKAP, known to bind PKA, is necessary to maintain the latter's specific binding site in the membrane. Experiments with agents interfering with the integrity of the cytoskeleton have demonstrated that its disruption severely compromises K channel function in the native tubule membrane (WANG W et al. 1995).

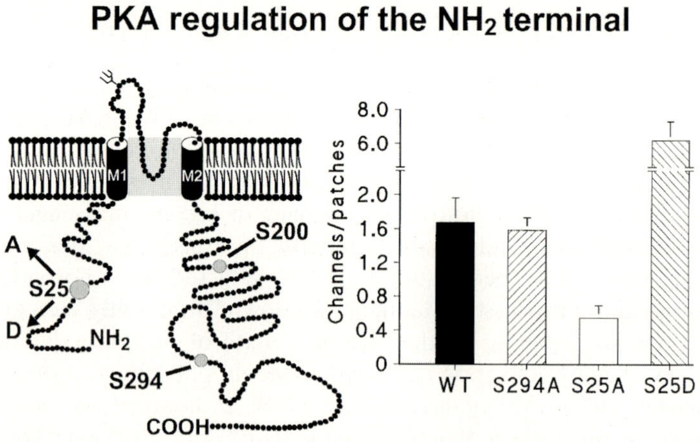

Fig. 11. Effect of PKA on ROMK channel activity (for details, see text)

VI. Regulation by Phosphorylation: Protein Kinase C (PKC)

PKC-dependent phosphorylation has been shown to inhibit the apical K channel and this effect is enhanced by Ca [WANG W and GIEBISCH 1991a). There are isoform-dependent differences in the number of potential PKC phosphorylation sites: ROMK1, expressed in the CCD, has three PKC-dependent phosphorylation sites: one of the serine residues is localized in the N terminus and two in the carboxy terminus (Fig. 7). Only two PKC-dependent phosphorylation sites are present in ROMK2 and ROMK3 (WANG W and HEBERT 1999; BOIM et al. 1995; ZHOU et al. 1994).

Although the apical ATP-sensitive K channel is insensitive to directly applied Ca, changes in channel activity can be induced by modulating cell Ca (WANG W et al. 1993; LING and EATON 1993). Thus, an elevation of cell Ca inhibits the apical ATP-sensitive K channel when basolateral Na,K-ATPase activity declines, and the effect can be completely abolished either by application of PKC inhibitors or by preventing the rise in cell Ca (WANG W et al. 1993). Indeed, one of the proposed mechanisms of coupling the apical K conductance with Na,K-ATPase involves changes in cell Ca. The measured elevation of cell Ca after pump inhibition is best explained by a rise in cell Na followed by decreased Ca extrusion by basolateral Ca-Na exchange. Ca-mediated changes in apical K channels may also mediate the rise in apical K conductance that occurs after stimulation of basolateral Na,K-ATPase by acutely raising the basolateral K concentration (MUTO et al. 1999). It is also of interest that the inhibitory action of arachidonic acid requires an intact serine phosphorylation site on the C terminus, and it is reasonable to conclude that these effects of arachidonic acid are mediated by activation of PKC in the apical membrane of principal cells (MACICA et al. 1997a,b).

VII. Regulation by Nucleotides

The secretory apical K channel in the TAL and CCD is inhibited by MgATP (WANG W and HEBERT 1999; WANG W et al. 1990a, 1997; WANG WH 1995; MCNICHOLAS et al. 1994, 1998). Depending on their concentration, nucleotides have been shown to both inhibit or stimulate the native secretory K channel and ROMK (WANG W and GIEBISCH G 1991b; MCNICHOLAS et al. 1994). Activation involves stimulation by PKA-mediated phosphorylation (see PKA above) or by modulating the formation of phosphatidylinositol 4,5 biphosphate (PIP_2). At higher mmol/l concentrations, ATP inhibits both native secretory K channels and ROMK, but renal K channels are less sensitive to inhibition than ATP-sensitive K channels in extrarenal tissues such as pancreatic beta cells, heart and skeletal muscle (WANG W and HEBERT 1999). Inhibition of the channel by MgATP involves binding to a segment of the carboxy terminus that contains a Walker site motif (MCNNICHOLAS et al. 1998); its mutation alters the sensitivity to inhibition by MgATP. The channel block after exposure to MgATP can be relieved by ADP and it is possible that changes in the cytosolic ATP/ADP ratio, brought about by fluctuations in Na,K-ATPase turnover, could provide a mechanism to link basolateral pump activity to apical K transport (WANG W and HEBERT 1999; GIEBISCH G 1995, 1998; TSUCHIYA 1992; MAURER et al. 1998; WELLING 1995; WANG W et al. 1997; BECK et al. 1994). External ATP can also inhibit the low-conductance K channel in the native cortical collecting duct. The effect of ATP can be mimicked by UTP and ADP, and is inhibited by suramin, a specific inhibitor of purinergic receptor, P2. Moreover, the effect of external ATP is abolished by okadaic acid, an inhibitor of phosphatases as well as by stimulation of cAMP productions is mediated by facilitating dephosphorylation or inhibiting PKA-induced phosphorylation (LU M et al. 1999).

VIII. Regulation by Interaction with Cystic Fibrosis Transmembrane Conductance Regulator (CFTR)

Several observations suggest that CFTR modifies the activity of ROMK channels. Patch clamp studies of apical membranes in the TAL and CCD have shown that glibenclamide inhibits apical ATP-sensitive K channels although significantly higher concentrations are necessary for channel block than in other ATP-sensitive channels (MCNICHOLAS et al. 1996, 1997; RUKNUDIN et al. 1998). Interactions of CFTR with ROMK channels are demonstrated by experiments in oocytes showing that channel inhibition by glibenclamide depends on the presence of CFTR. Further studies demonstrated that NBD1, a nucleotide binding site of CFTR, is required to confer upon ROMK sensitivity to glibenclamide (MCNICHOLAS et al. 1997). CFTR is present in the apical regions of both cortical TAL and CCD, suggestive of a role of CFTR in channel function (CRAWFORD et al. 1991; MORALES et al. 1996). The recent discovery of a renal isoform of a potent pancreatic sulphonylurea receptor (SUR2B) that is structurally related to CFTR suggests possible interaction with ROMK (TANEMOTO et al. 1999). The physiological role for either CFTR or SUR interaction with ROMK is incompletely understood but may involve modulation of the inhibitory action of ATP upon ROMK (RUKNUDIN et al. 1998).

IX. Regulation by pH

ATP-sensitive K channels in principal cells are highly sensitive to small changes in cell pH [WANG W, HEBERT SC (in press), WANG W et al. (1997), WANG W et al. (1990), SCHLATTER E et al. (1993)]. Figure 12 (left panel) shows the effects of progressive cytosolic acidification upon the activity of native secretory K channels [WANG W et al. (1990)]. It is apparent that a decline of pH from 7.4 to 7.0 virtually abolishes channel activity. Similar results have been found in ROMK expressed in oocytes [MCNICHOLAS CM et al. (1998), TSAI TD et al. (1995), CHOE H et al. (1997)]. Studies into the mechanisms of pH-related gating of ROMK have led to a complex picture. Mutation of Lysine 80 on the N terminus of ROMK1 (K61 on ROMK2) abolishes pH sensitivity

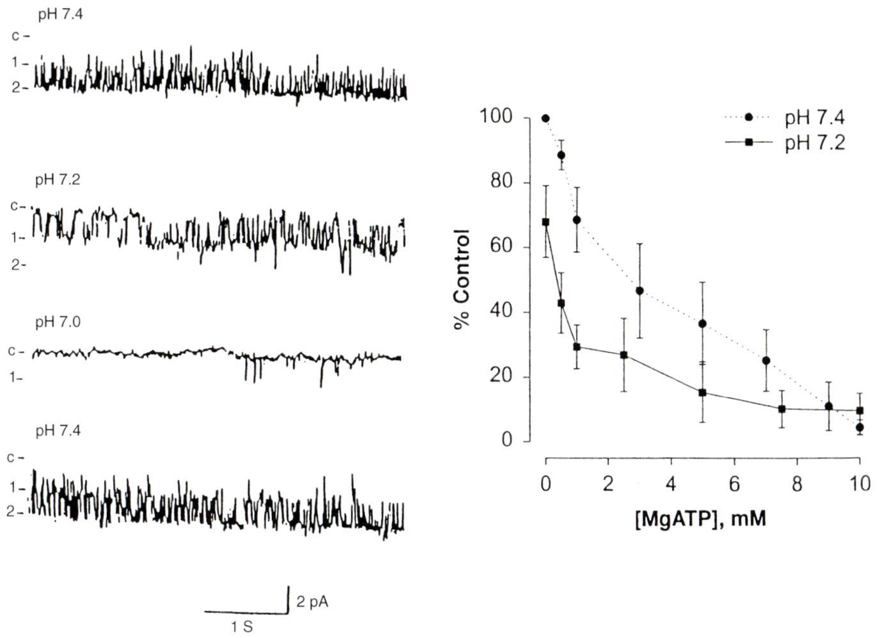

Fig. 12. *Left*, pH-sensitivity of native low-conductance K channel in the apical membrane of principal tubule cells: cortical collecting tubule. (WANG W et al. 1990a). *Right*, effect of pH in presence of Mg-ATP. Dose-response relationship for single channel activity expressed as percentage of control where control activity was measured in Mg^{2+}/ATP-free solutions at pH 7.4 and intracellular Mg-ATP concentration at pH 7.4 and pH 7.2. Some of these data obtained at pH 7.4 were obtained for a previous study (HEBERT 1995b). Note that at pH 7.2, 0.5 mm/ Mg-ATP significantly inhibits channel activity compared with the effect of the same concentration of ATP at pH 7.4 ($P < 0.005$). Data are means ± SE for 5 separate experiments at pH 7.4 and 11 experiments at pH 7.2. (MCNICHOLAS et al. 1998)

[FAKLER B et al. (1996a)], but additional amino residues also affect the cloned channels pH sensitivity [CHOE H et al. (1997)]. Evidence has also implicated conformational changes in both N and C termini [SCHULTE U et al. (1998)]. It was observed that prolonged exposure to acid media resulted in irreversible decline in channel activity, in contrast to short exposures that were fully reversible. It was suggested that such treatment leads to interactions of N and C termini that leads to formation of disulfide bridges that result in a stable closed confirmation. Such disulfide bridges can be broken by reducing agents like dithiothreitol (DTT) resulting in re-opening of the K channels [SCHULTE U et al. (1998)].

An additional mechanism by which changes in cytosolic pH affect channel activity involves altered sensitivity to inhibition by MgATP (MCNICHOLAS et al. 1998). A summary of experimental results is shown in Fig. 12 (right panel).

It is apparent that a modest decrease in pH on the cytosolic surface decreases the K 1/2 for MgATP inhibition, reflecting a large increase in affinity. This effect was independent of the lysine 80 residue and suggests an additional mechanism by which cell pH controls apical K channels.

X. Regulation by Phosphoinositides

Phospholipids such as PIP_2 (phosphatidylinositol 4,5 biphosphate) alter the activity of several ATP-sensitive K channels including ROMK (HUANG et al. 1998; HILGEMANN and BALL 1996; BAUKROWITZ et al. 1998). Patch-clamp experiments in oocytes expressing ROMK1 show that PIP_2 reduces the sensitivity to inhibition by ATP so that increasing PIP_2 activates K channels. Experiments involving site-directed mutagenesis have identified arginine 186 in ROMK1 as the main binding site for PIP_2. It was further suggested that the observed channel stimulation by ATP depends on its ability to generate PIP_2 through lipid kinases and that activation of channels by cAMP depends on PIP_2.

XI. Regulation of ROMK Density in CCD

Administration of a high K diet induces upregulation of K secretion in the distal nephron and patch-clamp studies have shown that such treatment leads to a significant increase in K channel density (WANG WH et al. 1999; FRINDT et al. 1998; PALMER 1999; FRINDT and PALMER 1998). In contrast, K depletion lowers the number of open channels (WANG WH et al. 1999; FRINDT and PALMER 1998).

Given that ROMK ($K_{IR}1.1$) likely forms the K secretory channel in principal cells of the collecting duct, ROMK mRNA expression in rat kidney has been studied following alterations in aldosterone, K adaptation and vasopressin. WALD et al. (1998) found that rats fed a K deficient diet had reduced ROMK mRNA expression in both cortex and medulla. Moreover, K loading increased ROMK transcript slightly only in medulla. The specific ROMK isoforms that changed with potassium were not assessed. In contrast, FRINDT et al. (1998) found that ROMK transcript abundance by in situ hybridization in the CCD was not affected by a high-K diet. Thus, the high-K diet induced increase in density of active K channels in principal cells in the CCD may not be due to increased abundance of ROMK mRNA. Accordingly, changes in ROMK protein abundance, channel activation, or ROMK channel translocation to the membrane are possible mechanisms to account for the high-K adaptation effect on K channels.

Mineralocorticoids also regulate ROMK abundance. Adrenalectomy decreased ROMK mRNA abundance in cortex but increased transcript abundance in the medulla (COLLINS et al. 1998). Inducing K deficiency in these adrenalectomized rats reduced ROMK mRNA to control levels suggesting that the hyperkalemia associated with adrenalectomy may have been the

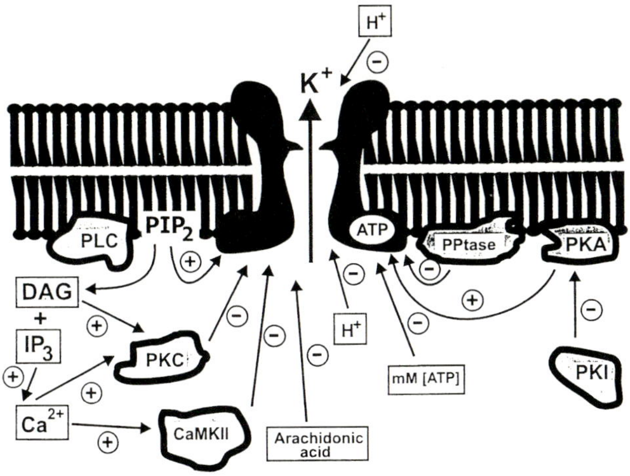

Fig. 13. Structural model of ROMK channel protein with factors modulating its activity. CaMKII: calcium-calmodulin activated kinase II; PPT: protein phosphatase; PKI: protein kinase inhibition (modified from WANG W et al. 1990a, used with permission)

cause for the increased ROMK message in the medulla. Consistent with a role for mineralocorticoids in ROMK regulation, administration of aldosterone by minipump to adrenal intact rats increased ROMK2, 3 and 6 transcripts in whole kidney (BEESLEY et al. 1998). A model of ROMK including the main regulatory mechanisms is shown in Fig. 13.

G. ROMK and Bartter's Syndrome

Antenatal Bartter's syndrome (BARTTER et al. 1962) comprises a set of autosomal recessive disorders characterized by hypokalemic metabolic alkalosis, salt wasting, hyperreninemia, and hyperaldosteronism, and elevated PGE_2 levels (KAROLYI et al. 1998; ASTERIA 1997; RODRIGUEZ-SORIANO 1998; GUAY-WOODFORD 1995). It is now established that antenatal Bartter's syndrome results from mutations in the genes encoding the ion transporters in TAL cells mediating NaCl absorption. These genes are: the *SLC12A1* gene encoding the apical Na-K-2Cl cotransporter (SIMON et al. 1996a; VARGUS-POUSSOU et al. 1998), the *CLCKB* basolateral Cl^- channel (SIMON et al. 1997), and the *KCNJ1* apical K^+ recycling channel, ROMK (SIMON et al. 1996b; KAROLYI et al. 1997; VOLLMER et al. 1998). The effect of mutations in ROMK on TAL function can be understood since apical K^+ recycling is crucial both to supplying K to the Na-K-2Cl cotransporter and to generation of the lumen positive transepithelial voltage (BLEICH et al. 1990; GREGER et al. 1990).

Mutant ROMK channels containing Bartter's mutations express either no or little function in *X. laevis* oocytes (DERST et al. 1997; SCHWALBE et al. 1998). While these studies establish the importance of ROMK in TAL function, several questions regarding the resulting Bartter's phenotype remain unanswered. First, why would loss of ROMK function produce a severe blunting of TAL NaCl transport if the 70pS K channel predominates in K recycling? While this question remains unanswered, one possible explanation is that the intermediate conductance channel requires ROMK for function (e.g., either as a subunit or as a regulator). ROMK gene deletion (knockout) mice currently being developed should help answer this puzzle. Second, ROMK comprises the apical K secretory channel in principal cells of the CCD. Yet, Bartter's individuals with *KCNJ1* mutations are hypokalemic. This suggests that these Bartter's individuals may have yet unknown adaptive mechanisms for K secretion in TAL or distal nephron segments like the distal convoluted tubule or collecting duct. Answers to these issues will likely uncover new adaptive mechanisms for renal K secretion and provide important insights to potassium handling by distal nephron segments including the collecting duct.

References

Adler S, Fraley DS (1997) Potassium and intracellular pH. Kidney Int 11:433–442

Ali S, Chen X, Lu M, Xu J-C, Lerea KM, Hebert SC, Wang W (1998) A kinase anchoring protein (AKAP) is required for mediating the effect of PKA on ROMK1. Proc Natl Acad Sci USA 95:10274–10278

Asteria C (1997) Molecular basis of Bartter's syndrome: new insights into correlation between genotype and phenotype. Eur J Endocrinol 137:613–615

Bartter FC, Pronove P, Gill JR Jr, MacCardle RC, Diller E (1962) Hyperplasia of the juxtaglomerular complex with hyperaldosteronism and hypokalemic alkalosis. Am J Med 33:811–828

Baukrowitz T, Schulte U, Oliver D, Herlitze S, Krauter T, Tucker SJ, Ruppersberg JP, Fakler B (1998) PIP_2 and PIP as determinants for ATP inhibition of K_{ATP} channels [see comments]. Science 282:1141–1144

Beck JS, Breton S, Mairbäurl H, Laprade R, Giebisch G (1991) The relationship between sodium transport and intracellular ATP in the isolated perfused rabbit proximal convoluted tubule. Am J Physiol 261:F634–F639

Beck JS, Hurst AM, Lapointe J-Y, Laprade R (1993) Regulation of basolateral K channels in proximal tubule studied during continuous microperfusion. Am J Physiol 264:F496–F501

Beck JS, Laprade R, Lapointe JY (1994) Coupling between transepithelial Na^+ transport and basolateral conductance in renal proximal tubule. Am J Physiol 266:F117–F127

Beesley AH, Hornby D, White SJ (1998) Regulation of distal nephron K^+ channels (ROMK) mRNA expression by aldosterone in rat kidney. J Physiol 509:629–634

Bleich M, Schlatter E, Greger R (1990) The luminal K^+ channel of the thick ascending limb of Henle's loop. Pflügers Arch 415:449–460

Bleich M, Schlatter E, Greger R (1990a) The luminal K^+ channel of the thick ascending limb of Henle's loop. Renal Physiol Biochem 13:37–50

Boim MA, Ho K, Shuck ME, Bienkowski MJ, Block JH, Slightom JL, Yang Y, Brenner BM, Hebert SC (1995) The ROMK inwardly rectifying ATP-sensitive K^+ channel. II. Cloning and intra-renal distribution of alternatively spliced forms. Am J Physiol 268:F1132–F1140

Brown EM, Gamba G, Riccardi D, Lombardi M, Butters R, et al. (1993) Cloning and characterization of an extracellular Ca-sensing receptor from bovine parathyroid. Nature 366:575–580

Chepilko S, Zhou H, Sackin H, Palmer LG (1995) Permeation and gating properties of a cloned renal K^+ channel. Am J Physiol 268:C389–C401
Choe H, Palmer LG, Sackin H (1999) Structural determinants of gating in inward-rectifier K^+ channels. Biophys J 76:1988–2003
Choe H, Sackin H, Palmer LG (1998) Permeation and gating of an inwardly rectifying potassium channel. Evidence for a variable energy well. J Gen Physiol 112:433–446
Choe H, Zhou H, Palmer LG, Sackin H (1997) A conserved cytoplasmic region of ROMK modulates pH sensitivity, conductance, and gating. Am J Physiol 273:F516–F529
Clausen T (1996) Long- and short-term regulation of the Na^+,K^+ pump in skeletal muscle. News Physiol Sci 11:24–30
Clement JP, Kunjilwar K, Gonzalez G, Schwanstecher M, Panten U, Aguilar-Bryan L, Bryan J (1997) Association and stoichiometry of K_{ATP} channel subunits. Neuron 18:827–838
Collins MT, Skarulis MC, Bilezikian JP, Silverberg SJ, Spiegel AM, Marx SJ (1998) Treatment of hypercalcemia secondary to parathyroid carcinoma with a novel calcimimetic agent. J Clin Endocrinol Metab 83:1083–1088
Crawford I, Maloney PC, Zeitlin PL, Guggino WB, Hyde SC, Turley H, Gatter KC, Harris A, Higgins CF (1991) Immunocytochemical localization of the cystic fibrosis gene product CFTR. Proc Natl Acad Sci USA 88:9262–9266
Derst C, Konrad M, Köckerling A, Karschin A, Daut J, Seyberth HW (1997) Mutations in the ROMK gene in antenatal Bartter syndrome are associated with impaired K^+ channel function. Biochem Biophys Res Comm 203:641–645
Fakler B, Bond CT, Adelman JP, Ruppersberg JP (1996) Heterooligomeric assembly of inward-rectifier K^+ channels from subunits of different subfamilies: Kir2.1 (IRK1) and Kir4.1(BIR10). Pflügers Arch 433:77–83
Fakler B, Schultz JH, Yang J, Schulte U, Brändle U, Zenner HP, Jan LY, Ruppersberg JP (1996a) Identification of a titratable lysine residue that determines sensitivity of kidney potassium channels (ROMK) to intracellular pH. The Embo J 15:4093–4099
Ficker E, Taglialatela M, Wible BA, Henley CM, Brown AM (1994) Spermine and spermidine as gating molecules for inward rectifier K^+ channels. Science 266:1068–1072
Frindt G, Palmer LG (1989) Low-conductance K channels in apical membrane of rat cortical collecting tubule. Am J Physiol 256:F143–F151
Frindt G, Palmer LG (1998) Short-term regulation of luminal K channels in the rat CCT by K intake. J Am Soc Nephrol 9:34 A
Frindt G, Zhou H, Sackin H, Palmer LG (1998) Dissociation of K channel density and ROMK mRNA in rat cortical collecting tubule during K adaptation. Am J Physiol 274:F525–F531
Giebisch G (1995) Renal potassium channels: an overview. Kidney Int 48:1004–1009
Giebisch G (1998) Renal potassium transport: mechanisms and regulation. Am J Physiol 274:F817
Greger R (1985) Ion transport mechanisms in thick ascending limb of Henle's loop of mammalian nephron. Physiol Rev 65:760–797
Greger R, Bleich M, Schlatter E (1990) Ion channels in the thick ascending limb of Henle's loop. Renal Physiol Biochem 13:37–50
Guay-Woodford LM (1995) Molecular insights into the pathogenesis of inherited renal tubular disorders. Curr Opinion in Nephrol and Hypertens 4:121–129
Hebert SC (1995a) Potassium secretory channels in the kidney, in Jameson JL (ed): Principles of Molecular Medicine. Totowa, NJ, Humana Press Inc., 1998, Ch. 67
Hebert SC, Andreoli TE (1984) Control of NaCl transport in the thick ascending limb. Am J Physiol 246:F745–F756
Hebert SC (1995b) An ATP-regulated, inwardly rectifying potassium channel from rat kidney (ROMK). Kidney Int 48:1010–1016
Heginbotham L, Abramson T, MacKinnon R (1992) A functional connection between the pores of distantly related ion channels as revealed by mutant K^+ channels. Science 258:1152–1155

Hilgemann DW, Ball R (1996) Regulation of cardiac Na^+, Ca^{2+} exchange and K_{ATP} potassium channels by PIP_s. Science 273:956–959

Ho K, Nichols CG, Lederer WJ, Lytton J, Vassilev PM, Kanazirska MV, Hebert SC (1993) Cloning and expression of an inwardly rectifying ATP-regulated potassium channel. Nature 362:31–38

Huang C-L, Feng S, Hilgemann DW (1998) Direct activation of inward rectifier potassium channels by PIP_2 and its stabilization by Gbg. Nature 391:803–806

Hurst AM, Beck JS, Laprade R, Lapointe JY (1993) Na pump inhibition downregulates an ATP-sensitive K channel in rabbit proximal convoluted tubule. Am J Physiol 264:F760–F764

Hurst AM, Lapointe J-Y, Laamarti A, Bell PD (1994) Basic properties and potential regulators of the apical K^+ channel in macula densa cells. J Gen Physiol 103:1055–1070

Jan LY, Jan YN (1997) Voltage-gated and inwardly rectifying potassium channels. J Physiol 505:267–282

Karolyi L, Koch MC, Grzeschik KH, Seyberth HW (1998) The molecular genetic approach to "Bartter's syndrome." J Molec Med 76:317–325

Karolyi L, Konrad M, Kockerling A, Ziegler A, Zimmermann DK, Roth B, Wieg C, Grzeschik K-H, Koch MC, Seyberth HW et al. (1997) Mutations in the gene encoding the inwardly-rectifying renal potassium channel, ROMK, cause the antenatal variant of Bartter syndrome: evidence for genetic heterogeneity. Human Molec Genetics 6:17–26

Koeppen B, Giebisch G (1985) Cellular electrophysiology of potassium transport in the mammalian cortical collecting tubule. Pflügers Arch. 405:S143–S146

Kohda Y, Ding W, Phan E, Housini I, Wang J, Star RA, Huang CL (1998) Localization of the ROMK potassium channel to the apical membrane of distal nephron in rat kidney. Kidney Int 54:1214–1223

Koster JC, Bentle KA, Nichols CG, Ho K (1998) Assembly of ROMK1 (Kir1.1a) inward rectifier K^+ channel subunits involves multiple interaction sites. Biophys J 74:1821–1829

Kubo Y, Baldwin TJ, Jan YN, Jan LY (1993) Primary structure and functional expression of a mouse inward rectifier potassium channel. Nature 362:127–133

Kubokawa M, McNicholas CM, Higgins MA, Wang W, Giebisch G (1995a) Regulation of ATP-sensitive K^+ channel by membrane-bound protein phosphatases in rat principal tubule cell. Am J Physiol 269:F355–F362

Kubokawa M, Mori Y, Fujimoto K, Kubota T (1998) Basolateral pH-sensitive K^+ channels mediate membrane potential of proximal tubule cells in bullfrog kidney. Jap J Physiol 48:1–8

Kubokawa M, Wang W, McNicholas CM, Giebisch G (1995b) Role of Ca^{2+}/CaMK II in Ca^{2+}-induced K^+ channel inhibition in rat CCD principal cell. Am J Physiol 268:F211-F219

Lee W-S, Hebert SC (1995) The ROMK inwardly rectifying ATP-sensitive K^+ channel. I. Expression in rat distal nephron segments. Am J Physiol 268:F1124–F1131

Ling BN, Eaton DC (1993) Cyclosporin A inhibits apical secretory K^+ channels in rabbit cortical collecting tubule principal cells. Kidney Int 44:974–984

Liu H, Fererri NR, Nasjletti A, Wang WH (1999) Vasopressin and PGE_2 regulate the activity of the apical 70 pS K^+ channel in the thick ascending limb of the rat kidney. Am J Physiol (in press)

Lopatin AN, Makhina EN, Nichols CG (1994) Potassium channel block by cytoplasmic polyamines as the mechanism of intrinsic rectification. Nature 372:366–369

Lu M, MacGregor GG, Wang W, Giebisch G (1999) Small-conductance K^+ channels in the apical membrane of cortical collecting duct from wild-type and CFTR knockout mice are regulated by extracellular ATP. J Gen Physiol, submitted

Lu M, Zhu Y, Balazy M, Falck JR, Wang WH (1996) Effect of angiotensin II on the apical K channels in the thick ascending limb of the rat kidney. J Gen Physiol 108:537–547

Lu Z, MacKinnon R (1994) Electrostatic tuning of Mg^{2+} affinity in an inward-rectifier K^+ channel. Nature 371:243–246

MacGregor GG, Xu J, McNicholas CM, Giebisch G, Hebert SC (1998) Partially active channels produced by PKA site mutation of the cloned renal K^+ channel ROMK2. Am J Physiol 275:F415–F422

Macica CM, Yang Y, Lerea K, Hebert SC, Wang W (1997a) Arachidonic acid inhibits the activity of the cloned renal K^+ channel, ROMK1. Am J Physiol 40:F588–F594

Macica CM, Yang Y, Lerea K, Hebert SC, Wang W (1997b) Role of the NH_2 terminus of the cloned renal K^+ channel, ROMK1, in arachidonic acid-mediated inhibition. Am J Physiol 274:F175–F181

MacKinnon R, Cohen SL, Kuo A, Lee A, Chait BT (1998) Structural conservation in prokaryotic and eukaryotic potassium channels. Science 280:106–109

Malnic G, Aires MM, Giebisch G (1971) Potassium transport across renal distal tubules during acid-base disturbances. Am J Physiol 221:1192–1208

Malnic G, Muto S, Giebisch G (1999) Regulation of Potassium Excretion. In: Seldin DS, Giebisch G (eds), The Kidney, Physiology and Pathophysiology 3rd edition, Lippincott-Raven, Philadelphia, PA (in press)

Maurer UR, Boulpaep EL, Segal AS (1998) Properties of an inwardly rectifying ATP-sensitive K^+ channel on the basolateral membrane of renal proximal tubule. J Gen Physiol 111:139–160

McNicholas CM, Guggino WB, Schwiebert EM, Hebert SC, Giebisch G, Egan ME (1996) Sensitivity of a renal K^+ channel (ROMK2) to the inhibitory sulfonylurea compound, glibenclamide, is enhanced by co-expression with the ATP-binding cassette transporter cystic fibrosis transmembrane regulator. Proc Natl Acad Sci USA 93:8083–8088

McNicholas CM, MacGregor GG, Islas LD, Yang Y, Hebert SC, Giebisch G (1998) pH-dependent modulation of the cloned renal K^+ channel, ROMK. Am J Physiol 275:F972-F981

McNicholas CM, Nason MW, Guggino WB, Schwiebert EM, Hebert SC, Giebisch G, Egan ME (1997) A functional CFTR-NBF1 is required for ROMK2-CFTR interaction. Am J Physiol 273:F843-F848

McNicholas CM, Wang W, Ho K, Hebert SC, Giebisch G (1994) Regulation of ROMK1 K^+ channel activity involves phosphorylation processes. Proc Natl Acad Sci USA 91:8077–8081

Mennit PA, Wade JB, Ecelbarger CA, Palmer LG, Frindt G (1997) Localization of ROMK channels in the rat kidney. J Am Soc Nephrol 8:1823–1830

Minor DL, Masseling SJ, Jan YN, Jan LY (1999) Transmembrane structure of an inwardly rectifying potassium channel. Cell 97:879–891

Morales MM, Carroll TP, Morita T, Schwiebert EM, Devuyst O, Wilson PD, Lopes AG, Stanton BA, Dietz HC, Cutting GR et al. (1996): Both the wild type and a functional isoform of CFTR are expressed in kidney. Am J Physiol 270:F1038–F1048

Muto S, Asano Y, Seldin D, Giebisch G (1999) Basolateral Na^+ pump modulates apical Na^+ and K^+ conductances in rabbit cortical collecting ducts. Am J Physiol 276:F143-F158

Neher E, Sakmann B (1976) Single channel currents recorded from membrane of denervated frog muscle fibers. Nature 260:779–802

Nichols CG, Ho K, Hebert SC (1994) Mg^{2+}-dependent inward rectification of ROMK1 potassium channels expressed in Xenopus oocytes. J Physiol 476:399–409

Ohno-Shosaku T, Kubota T, Yamaguchi J, Fujimoto M (1990) Regulation of inwardly rectifying K^+ channels by intracellular pH in opossum kidney cells. Pflügers Arch, 416:138–143

Oliver D, Hahn H, Antz C, Ruppersberg JP, Fakler B (1998) Interaction of permeant and blocking ions in cloned inward-rectifier K^+ channels. Biophys J 74:2318–2326

Palmer LG (1999) Potassium secretion and the regulation of distal nephron K channels. Am J Physiol 277:F821–F825

Palmer LG, Antonian L, Frindt G (1994) Regulation of apical K and Na channels and Na/K pumps in rat cortical collecting tubule by dietary K. J Gen Physiol 104:693–710

Palmer LG, Choe H, Frindt G (1997) Is the secretory K channel in the rat CCT ROMK? Am J Physiol 273:F404–F410

Rodriguez-Soriano J (1998) Bartter and related syndromes: the puzzle is almost solved. Pediatric Nephrol 12:315–327

Ruknudin A, Schulze DH, Sullivan SK, Lederer WJ, Welling PA (1998) Novel subunit composition of a renal epithelial K_{ATP} channel. J Biol Chem 273:14165–14171

Sabirov RZ, Morishima S, Okada Y (1998) Probing the water permeability of ROMK1 and amphotericin B channels using *Xenopus* oocytes. Biochim Biophys Acta 1368:19–26

Sabirov RZ, Tominaga T, Miwa A, Okada Y (1997) A conserved arginine residue in the pore region of an inward rectifier K channel (IRK1) as an external barrier for cationic blockers. J Gen Physiol 110:665–677

Sansom SC, Agulian S, Muto S, Illig V, Giebisch G (1989) K activity of CCD principal cells from normal and DOCA-treated rabbits. Am J Physiol 256: F136–F142

Schlatter E, Bleich M, Hirsch J, Greger R (1993) pH-sensitive K^+ channels in the distal nephron. Nephrol Dialys Transpl 8:488–490

Schlatter E, Lohrmann E, Greger R (1992) Properties of the potassium conductances of principal cells of rat cortical collecting ducts. Pflügers Arch 420:39–45

Schulte U, Hahn H, Wiesinger H, Ruppersberg JP, Fakler B (1998) pH-dependent gating of ROMK (Kir1.1) channels involves conformational changes in both N and C termini. J Biol Chem 273:34575–34579

Schwalbe RA, Bianchi L, Accili EA, Brown AM (1998) Functional consequences of ROMK mutants linked to antenatal Bartter's syndrome and implications for treatment. Human Molec Genetics 7:975–980

Shuck ME, Block JH, Benjamin CW, Tsai T-D, Lee KS, Slightom JL, Bienkowski MJ (1994) Cloning and characterization of multiple forms of the human kidney ROMK potassium channel. J Biol Chem 269:24261–24270

Simon DB, Bindra RS, Mansfield TA, Nelson-Williams C, Mendonÿa E, Stone R, Schurman S, Nayir A, Alpay H, Bakkaloglu A et al. (1997) Mutations in the chloride channel gene, CLCNKB, cause Bartter's syndrome type III. Nature Genetics 17:171–178

Simon DB, Karet FE, Hamdan JM, DiPietro A, Sanjad SA, Lifton RP (1996a) Bartter's syndrome, hypokalaemic alkalosis with hypercalciuria, is caused by mutations in the Na-K-2Cl cotransporter NKCC2. Nature Genetics 13:183–188

Simon DB, Karet FE, Rodriguez-Soriano J, Hamdan JH, DiPietro A, Trachtman H, Sanjad SA, Lifton RP (1996b) Genetic heterogeneity of Bartter's syndrome revealed by mutations in the K^+ channel, ROMK. Nature Genetics 14:152–156

Spassova M, Lu Z (1998) Coupled ion movement underlies rectification in an inward-rectifier K^+ channel. J Gen Physiol 112:211–221

Stanton BA, Giebisch GH (1992) Renal potassium transport. In Windhager E (ed): Handbook of Physiology, Sect 8, Renal Physiology. Oxford University Press, New York, pp 813–874

Stokes PB (1993) Ion transport in the collecting duct. Sem Nephrol 13:202–212

Taglialatela M, Wible BA, Caporaso R, Brown AM (1994) Specification of pore properties by the carboxyl terminus of inwardly rectifying K^+ channels. Science 264:844–847

Tanemoto M, Vanoye CG, Abe T, Welch R, Dong K, Hebert SC, Xu JZ (1999) A rat homolog of sulfonylurea receptor 2B (SUR2B) determines the glibenclamide-sensitivity of ROMK2 in *X. laevis* oocyte. J Biochem, submitted

Tsai TD, Shuck ME, Thompson DP, Bienkowski MJ, Lee KS (1995) Intracellular H^+ inhibits a cloned rat kidney outer medulla K^+ channel expressed in *Xenopus* oocytes. Am J Physiol 268:C1173–C1178

Tsuchiya K, Wang W, Giebisch G, Welling PA (1992) ATP is a coupling modulator of parallel Na,K-ATPase – K channel activity in the renal proximal tubule. Proc Natl Acad Sci USA 89:6418–6422
Vallon V, Albinus M, Balch D (1998) Effect of K_{ATP} channel blocker U37883 A on renal function in experimental diabetes mellitus in rats. J Pharmacol Exp Ther 286(3):1215–1221
Vallon V, Osswald H, Blantz RC, Thomson S (1997) Potential role of luminal potassium in tubuloglomerular feedback. J Am Soc Nephrol 8:1831–1837
Vargus-Poussou R, Feldmann D, Vollmer M, Konrad M, Kelly RP, van den Heuvell LPWJ, Tebourbi L, Brandis M, Karolyi L, Hebert SC et al. (1998) Novel molecular variants of the Na-K-2Cl cotransporter gene are responsible for antenatal Bartter syndrome. Am J Hum Genet 62:1332–1340
Vollmer M, Koehrer M, Topaloglu R, Strahm B, Omran H, Hildebrandt F (1998) Two novel mutations of the gene for Kir 1.1 (ROMK) in neonatal Bartter syndrome. Pediatr Nephrol 12:69–71
Wald H, Garty H, Palmer LG, Popovtzer MM (1998) Differential regulation of ROMK expression in kidney cortex and medulla by aldosterone and potassium. Am J Physiol 275:F239-F245
Wang T, Wang W, Klein-Robbenhaar G, Giebisch G (1995a) Effects of glyburide on renal tubule transport and potassium-channel activity. Renal Physiol Biochem 18:169–182
Wang T, Wang W, Klein-Robbenhaar G, Giebisch G (1995b) Effects of a novel K_{ATP} channel blocker on renal tubule function and K channel activity. J Pharm Exp Therap 273:1382–1389
Wang W (1994) Two types of K^+ channel in thick ascending limb of rat kidney. Am J Physiol 267:F599–F605
Wang W, Cassola AC, Giebisch G (1995) Involvement of the cytoskeleton in modulation of apical K channel activity in rat CCD. Am J Physiol 267:F591–F598
Wang W, Cassola AC, Giebisch G (1992) Arachidonic acid inhibits the secretory K channel of cortical collecting duct of rat kidney. Am J Physiol 264:F554–F559
Wang W, Geibel J, Giebisch G (1993) Mechanism of apical K channel modulation in principal renal tubule cells: effect of inhibition of basolateral Na-K-ATPase. J Gen Physiol 101:673–694
Wang W, Giebisch G (1991a) Dual modulation of renal ATP-sensitive K-channel by protein kinase A and C. Proc Natl Acad Sci USA 88:9722–9725
Wang W, Giebisch G (1991b) Dual effect of adenosine triphosphate on the apical small conductance K^+ channel of the rat cortical collecting duct. J Gen Physiol 98:35–61
Wang W, Hebert SC (1999) The Molecular Biology of Renal K Channels. In: Seldin DS, Giebisch G (eds), The Kidney, Physiology and Pathophysiology 3rd edition, Lippincott-Raven, Philadelphia, PA (in press)
Wang W, Hebert SC, Giebisch G (1997) Renal K^+ channels: Structure and function. Annu Rev Physiol 59:413–436
Wang W, Lu M (1995) Effect of arachidonic acid on activity of the apical K channel in the thick ascending limb of the rat kidney. J Gen Physiol 106:727–743
Wang W, Schwab A, Giebisch, G (1990a) The regulation of the small conductance K^+ channel in the apical membrane of rat cortical collecting tubule. Am J Physiol 259:F494–F502
Wang W, White S, Geibel J, Giebisch G (1990b) A potassium channel in the apical membrane of rabbit thick ascending limb of Henle's loop. Am J Physiol 258:F244–F253
Wang WH (1995) View of K^+ secretion through the apical K channel of cortical collecting duct. Kidney Int 48:1024–1030
Wang WH, Lerea KM, Chan M, Giebisch G (2000) Protein tyrosine kinase regulates the number of renal secretory K channels. Am J Physiol 278:F165–F171
Wang WH, Lu M, Hebert SC (1996) P450 metabolites mediate extracellular Ca^{2+}-induced inhibition of apical K^+ channels in the thick ascending limb of the rat kidney. Am J Physiol 270:C103–C111

Welling PA (1995) Cross-talk and the role of K_{ATP} channels in the proximal tubule. Kidney Int 48:1017–1023

Wible BA, Taglialatela M, Ficker E, Brown AM (1994) Gating of inwardly rectifying K^+ channels localized to a single negatively charged residue. Nature 371:246–249

Xu JZ, Hall AW, Peterson LN, Bienkowski MJ, Eessalu TE, Hebert SC (1997) Localization of the ROMK protein on apical membranes of rat kidney nephron segments. Am J Physiol 273:F739–F748

Xu Z-C, Yang Y, Hebert SC (1996) Phosphorylation of the ATP-sensitive, inwardly rectifying K^+ channel, ROMK, by cyclic AMP-dependent protein kinase. J Biol Chem 271:9313–9319

Yang J, Jan YN, Jan LY (1995a) Determination of the subunit stoichiometry of an inwardly rectifying potassium channel. Neuron 1441–1447

Yang J, Jan YN, Jan LY (1995b) Control of rectification and permeation by residues in two distinct domains in an inward rectifier K^+ channel. Neuron 14:1047–1054

Zhou H, Chepilko S, Schutt W, Chlor H, Palmer LG, Sackin H (1996) Mutations in the pore region of ROMK enhance Ba^{2+} block. Am J Physiol 271:C1949–C1956

Zhou H, Tate SS, Palmer LG (1994) Primary structure and functional properties of an epithelial K channel. Am J Physiol 266:C809–C824

CHAPTER 11

Structure and Function of ATP-Sensitive K^+ Channels

T. Gonoi and S. Seino

A. Introduction

ATP-sensitive K^+ channels (K_{ATP} channels) were first described by Noma (1983) in cardiac muscle using the patch clamp technique. K_{ATP} channels are characterized by channel inhibition with an increase in intracellular ATP concentration and stimulation with an increase in intracellular MgADP concentration (Dunne and Petersen 1986; Kakei et al. 1986; Misler et al. 1986). K_{ATP} channels are also found in many other tissues including pancreatic β-cells (Cook and Hales 1984; Ashcroft and Rorsman 1989), skeletal muscles (Davies et al. 1991), neurons, kidney, and various smooth muscles (Kuriyama et al. 1995; Quayle et al. 1997), and also in mitochondria (Inoue et al. 1991; Paucek et al. 1992). In several tissues, however, the presence of K_{ATP} channels has not been shown directly by electrophysiology, but by other physiological and pharmacological methods. For example, the presence of K_{ATP} channels in brain (Grigg and Anderson 1989; Politi and Rogawski 1991; Murphy and Greenfield 1991; Jiang et al. 1992; Tromba et al. 1992) and vascular tissues (Von Beckerath et al. 1991; Silberberg and Van Breemen 1992: Dart and Standen 1995; Katnik and Adams 1995) has been shown by the increase of K^+ conductance in energy depleting conditions and pharmacological modifications of the response. Similarly, an increase in K^+ conductance after the addition of K_{ATP} channel openers such as diazoxide or cromakalim (Fig. 1), or a decrease in conductance by the addition of K_{ATP} channel blockers, sulfonylureas such as glibenclamide or tolbutamide (Fig. 1) which are widely used in the treatment of non-insulin dependent diabetes mellitus (NIDDM), has also revealed the presence of K_{ATP} channels in brain and various smooth muscle cells.

K_{ATP} channels couple the cell metabolic state to membrane potential. They play roles in various cellular functions including hormone secretion, controlling excitability of muscles and neurons, K^+ recycling in renal epithelia, and cytoprotection under ischemic, hypoxic, or hypoglycemic conditions in brain, heart, and vascular cells. They are also the target of endogenous vasoreactive substances such as calcitonin gene-related peptide (CGRP) and vasoactive intestinal peptide (VIP) (Standen et al. 1989; Nelson and Quayle 1995).

In 1995, the structure of the pancreatic β-cell type K_{ATP} channel was clarified by a combined technique of molecular biology and electrophysiology

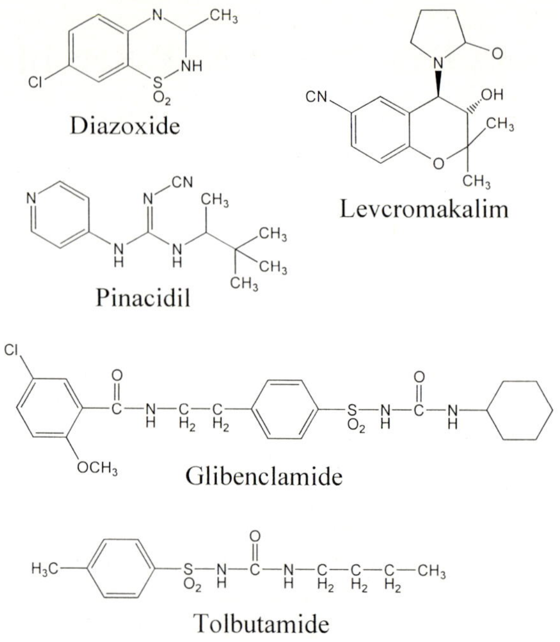

Fig. 1. Chemical structures of the drugs interacting with K_{ATP} channels

(INAGAKI et al. 1995b; SAKURA et al. 1995). The β-cell type K_{ATP} channel comprises two subunits: Kir6.2, a member of the inwardly rectifying K^+ channel subfamily Kir6.0, and SUR1 (formerly referred to as SUR), the receptor of the sulfonylureas, members of the ATP-binding cassette (ABC) protein superfamily (AGUILAR-BRYAN et al. 1995). Two Kir6.0 subfamily members (Kir6.1, Kir6.2) and two SUR family members (SUR1, SUR2) have so far been discovered, and reconstitution studies have shown that differing combinations of a Kir6.0 subfamily subunit and a SUR subunit constitute K_{ATP} channels with the distinct nucleotide sensitivities and pharmacological properties.

B. Properties of K_{ATP} Channels in Native Tissues

I. Heart

K_{ATP} channels in cardiac ventricular and atrial myocytes in mammals have a unitary conductance of approximately 80 pS with high $[K^+]$ on both sides of the membrane ($[150\,mmol/l]_o/[140\,mmol/l]_i$) and 35 pS at a physiological $[K^+]$ gradient (NOMA 1983; TRUBE and HESCHELER 1984; KAKEI et al. 1985). The channels show weak inward rectification in the presence of $[Mg^{2+}]_i$ (HORIE et al. 1987). The K_i of ATP for inhibition of cardiac K_{ATP} channel activity varies from 15 mmol/l to 110 mmol/l in the different cell preparations examined, a difference may be explained at least in part by G-protein modulation of K_{ATP} channel activity (ITO et al. 1994; TERZIC et al. 1994). K_{ATP} channels in cardiac cells are blocked by the sulfonylureas glibenclamide (Fig. 1) and tolbutamide (FINDLAY 1991; BELLES et al. 1987), and activated by application of pinacidil (Fig. 1) and cromakalim (ESCANDE et al. 1988; OSTERRIEDER 1988; SANGUINETTI et al. 1988), but are not activated by diazoxide, a potent activator of K_{ATP} channels in pancreatic β-cells.

II. Skeletal Muscles

K_{ATP} channels are present in mammalian and frog skeletal muscle fibers (SPRUCE et al. 1985; BURTON et al. 1988; DAVIES et al. 1991). K_{ATP} channels in skeletal muscle are thought to have a cytoprotective role in fatigue and to increase blood supply to the muscle by dilating vessels by increasing K^+ efflux from the muscle. The unitary conductance of the K_{ATP} channels in skeletal muscle ranges between 57 pS and 74 pS in symmetrical high $[K^+]$ (140–160 mmol/l) solutions (WEIK and NEUMCKE 1989; WOLL et al. 1989). The K_{ATP} channel activity of skeletal muscle is affected by cytosolic pH (DAVIES 1990; VIVAUDOU and FORESTIER 1995), $[Mg^{2+}]_i$, $[Ca^{2+}]_i$ (HEHL et al. 1994), and $[MgADP]_i$ (ALLARD and LAZDUNSKI 1992). The K_{ATP} channels in skeletal muscle are blocked by glibenclamide with a K_i significantly greater than that in mammalian pancreatic β-cells – 63–190 nmol/l (BARRETT-JOLLEY and DAVIES 1997; ALLARD and LAZDUNSKI 1993), 3–5 mmol/l (LIGHT and FRENCH 1994), and by tolbutamide with a K_i of approximately 60 μmol/l (WOLL et al. 1989). K_{ATP} channels in mammalian skeletal muscle are activated by cromakalim and pinacidil, but not by diazoxide (WEIK and NEUMCKE 1990).

III. Pancreatic β-Cells

The K_{ATP} channels in pancreatic β-cells have a unitary conductance ranging between 50 pS and 80 pS with high $[K^+]$ on both sides of the membrane and 20–30 pS in physiological $[K^+]$ gradients (COOK and HALES 1984; ASHCROFT et al. 1984; RORSMAN and TRUBE 1985; ASHCROFT and RORSMAN 1989). The single channel currents show weak inward rectification. An increase in cytosolic ATP inhibits channel activity with K_i of approximately 15 μmol/l and Hill coefficient between 1 and 2, while an increase in cytosolic MgADP activates the channels (KAKEI et al. 1986; DUNNE and PETERSEN 1986; MISLER et al. 1986). K_{ATP} channels in pancreatic β-cells are inhibited by glibenclamide with K_i of 4–27 nmol/l (ZÜNKLER et al. 1988; STURGESS et al. 1988). Diazoxide activates the channels (STURGESS et al. 1988), while cromakalim, which activates the K_{ATP} channels in cardiac cells, has little effect.

The physiological role of K_{ATP} channels has been best characterized in pancreatic β-cells (ASHCROFT and RORSMAN 1989). An increase in the intracellular ATP concentration (or an increase in the ATP/ADP ratio) closes the

K_{ATP} channels, which depolarizes the β-cell membrane and leads to the opening of the voltage-dependent calcium channels, allowing calcium influx. The rise in the intracellular calcium concentration in the β-cell triggers insulin granule exocytosis. Thus, the K_{ATP} channels in pancreatic β-cells, as ATP and ADP sensors, are thought to be key molecules in the regulation of glucose-induced insulin secretion (COOK and HALES 1984; ASHCROFT et al. 1984; ASHCROFT and RORSMAN 1989).

IV. Brain

Radio-labeled glibenclamide binding experiments in rat brain show that the highest level of binding is found in the substantia nigra (MOURRE et al. 1989; GEHLERT et al. 1990). High binding activity is also found in the globus and ventral pallidus, motor neocortex, molecular layer of the cerebellar cortex, limbic system, hippocampus, dentate gyrus, and caudate-putamen, while low binding activity is found in the hypothalamus (MOURRE et al. 1989; GEHLERT et al. 1990).

In the substantia nigra (SCHWANSTECHER and PANTEN 1993), cerebral cortex, hippocampus (OHNO-SHOSAKU and YAMAMOTO 1992), and caudate nucleus (SCHWANSTECHER and PANTEN 1994; SCHWANSTECHER and BASSEN 1997), K_{ATP} channels with electrophysiological and pharmacological properties similar to those in pancreatic β-cells are described. On the other hand, K_{ATP} channels with properties very different from those in pancreatic and cardiac K_{ATP} channels are also reported in brain. In rat substantia nigra, for example, the K_{ATP} channels, which have a unitary conductance of 226 pS in symmetrical 150 mmol/l $[K^+]$, are activated by membrane depolarization, and an increase in $[Ca^{2+}]_i$ is required for the K_{ATP} channel activation. These channels are inhibited by cytosolic ATP with K_i of 135 μmol/l. In the hippocampus K^+ channels activated under hypoglycemic and energy-depleted conditions and by cromakalim, and which are inhibited by glibenclamide (POLITI and ROGAWSKI 1991; TROMBA et al. 1992) have been found. The unitary conductance of the K_{ATP} channels in hippocampus ranges from 26 pS to 100 pS in physiological [K] gradients (POLITI and ROGAWSKI 1991; TROMBA et al. 1992). In hypothalamus the properties of the K_{ATP} channels are also different from those in pancreatic β-cells or cardiac cells. They are blocked by cytosolic ATP with a K_i value more than one order larger (a few mmol/l) than that for the K_{ATP} channels in β-cells or cardiac cells (ASHFORD et al. 1990). The unitary conductance, 146 pS in symmetrical high $[K^+]$ solutions of 140 mmol/l, is also larger than in the peripheral cells.

The roles of the K_{ATP} channels in the brain are not yet known, but may play a role in cytoprotection in hypoxic, hypoglycemic, and ischemic conditions. The K_{ATP} channels in the hypothalamus are activated by leptin, a protein encoded by the obese (ob) gene (SPANSWICK et al. 1997), and are thought to be involved in the control of satiety and energy expenditure.

V. Smooth Muscles

K_{ATP} channels are present in smooth muscle cells of arteries, veins, and capillaries. However, their electrophysiological and pharmacological properties vary significantly from one preparation to another (KURIYAMA et al. 1995; NELSON and QUAYLE 1995; QUAYLE et al. 1997). The K_{ATP} channels in rat and rabbit mesenteric arteries, for example, are activated by cromakalim, diazoxide, pinacidil, vasoactive intestinal polypeptide (VIP), and acetylcholine, and they are inhibited by glibenclamide (STANDEN et al. 1989; NELSON and QUAYLE 1995; QUAYLE et al. 1997). The channels have a unitary conductance of approximately 135 pS (60 mmol/l $[K^+]_o$/120 mmol/l $[K^+]_i$), significantly greater than those of the K_{ATP} channels in cardiac muscle and pancreatic β-cells.

In rabbit portal vein, K_{ATP} channels with unitary conductance of 15 pS in a physiological $[K^+]$ gradient are reported. These K_{ATP} channels are activated by bath application of pinacidil (KAJIOKA et al. 1991). In contrast to cardiac and pancreatic β-cell K_{ATP} channels, the channel activity disappears immediately after patch excision, but is recovered by application of GDP to the cytosolic side of the membrane. Similar channels have been described in rat portal vein (ZHANG and BOLTON 1996), and mesenteric artery (ZHANG and BOLTON 1995). Because these channels are activated by cytosolic nucleotide diphosphates they are called K_{NDP} channels (BEECH et al. 1993).

Unitary conductance of the K_{ATP} channels in vascular smooth muscle varies widely from 15 pS to 258 pS (KAJIOKA et al. 1991; LORENZ et al. 1992; KURIYAMA et al. 1995; QUAYLE et al. 1997), suggesting a molecular diversity of vascular K_{ATP} channels in different tissues.

In vascular smooth muscle cells, K_{ATP} channels are activated under hypoxic conditions (KAMOUCHI et al. 1994) or by endogenous vasodilators such as CGRP, VIP, adenosine, and prostacyclin, resulting in membrane hyperpolarization and vasodilation (QUAYLE et al. 1997).

Airway smooth muscles (trachea and trachealis) are relaxed in hypoxic conditions (LINDEMAN et al. 1994) and by cromakalim (ARCH et al. 1988; BLACK et al. 1990; RAEBURN et al. 1991), diazoxide (LONGMORE et al. 1991), and pinacidil (NIELSEN-KUDSK et al. 1990), and the relaxation is antagonized by glibenclamide, suggesting the presence of K_{ATP} channels in the tissue.

In gastrointenstinal tract, the K_{ATP} channel openers cromakalim and levcromakalim dilate smooth muscle of esophagus (HATAKEYAMA et al. 1995), stomach (ITO et al. 1992; KORTEZOVA et al. 1992; KATAYAMA et al. 1993), ileum (MCPHERSON et al. 1990), and colon (POST et al. 1991), and the dilation is blocked by glibenclamide.

VI. Kidney

ATP-regulated K^+ channels in kidney are described in Chap. 11.

VII. Mitochondria

K_{ATP} channels with pharmacological properties similar to those of the K_{ATP} channels in the plasma membrane are present in mitochondrial inner membrane of liver and heart (Inoue et al. 1991; Paucek et al. 1992; Szewczyk et al. 1997). The mitochondrial K_{ATP} channels are thought to be involved in energy metabolism by regulating mitochondrial volume (Paucek et al. 1992). The mitochondrial K_{ATP} channels are activated by diazoxide and cromakalim and are inhibited by ATP and glibenclamide (Paucek et al. 1996) and 5-hydroxydecanoic acid. The unitary conductance of the channel is approximately 10 pS in 100 mmol/l $[K^+]_{cytosol}$/33 mmol/l $[K^+]_{matrix}$ (Inoue et al. 1991). Whether the ATP-sensing site is facing the matrix side (Inoue et al. 1991) or is facing the cytosolic side (Yarov-Yarovoy et al. 1997) is controversial.

C. Structure and Functional Properties of Reconstituted K_{ATP} Channels

I. The Pancreatic β-Cell Type K_{ATP} Channel

1. The Inwardly Rectifying K^+ Channel Subfamily Kir6.0

Using GIRK1 cDNA as a probe, cDNA encoding a novel member of the Kir family, Kir6.1 (formerly referred to as uK_{ATP}-1), was isolated from a rat pancreatic islet cDNA library (Inagaki et al. 1995a). Rat Kir6.1 is a protein of 424 amino acids with two putative transmembrane segments (Fig. 2). Since Kir6.1 shares only 40%–50% identity with previously cloned inwardly rectifying K^+ channel members, it represents a new subfamily, Kir6.0. The glycine-tyrosine-glycine motif in the H5 region, which is critical for K^+ selectivity and is highly conserved among K^+ channels, is not conserved in Kir6.1. The motif in Kir6.1 is glycine-phenylalanine-glycine. Kir6.1 is ubiquitously expressed (Table 1), but is not expressed in the insulin-secreting cell lines HIT-T15 (hamster-derived), RINm5F (rat-derived), and MIN6 (mouse-derived), all of which are known to have K_{ATP} channels, indicating that Kir6.1 is not a component of the pancreatic β-cell type K_{ATP} channels which are responsible for insulin secretion. Immunohistological experiments show that Kir6.1 is present primarily in the inner membrane of mitochondria (Suzuki et al. 1998). The human Kir6.1 gene is located at chromosome 12p11.23.

An isoform of Kir6.1, BIR (the β-cell inward rectifier, currently referred to as Kir6.2), was subsequently cloned from a human genomic and the MIN6 cDNA libraries (Inagaki et al. 1995b). Kir6.2 is a protein of 390 amino acids and shares 71% amino acid identity with Kir6.1 (Fig. 2). As in Kir6.1, Kir6.2 has the glycine-phenylalanine-glycine motif in the H5 region. The strongly inwardly rectifying K^+ channel subunits such as Kir2.1 have aspartic acid in the second transmembrane segment (residue 172 of Kir2.1), a crucial determinant of the rectifying property (Lu and Mackinnon 1994; Stanfield et al.

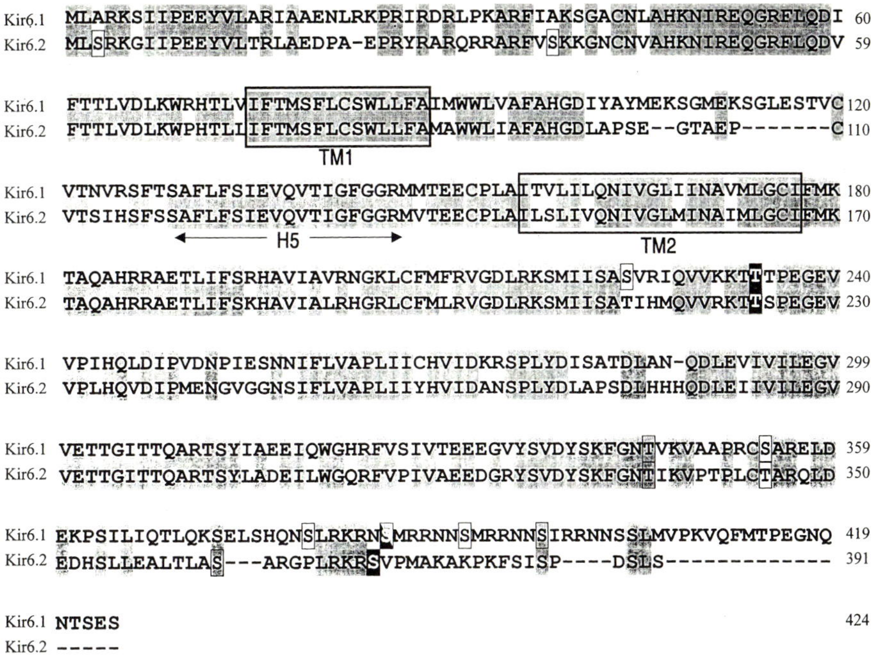

Fig. 2. Amino acids sequences of human Kir6.1 and human Kir6.2. The predicted transmembrane domains (TM1 and TM2) are *boxed*. The pore (H5) region is indicated by *arrows*. Amino acid residues for potential phosphorylation sites by cyclic AMP-dependent protein kinase and protein kinase C are indicated by *open and filled boxes*, respectively. Serine at residue 385 of Kir6.1 is a potential phosphorylation site for both kinases. References and DDBJ accession numbers, human Kir6.1, (Inagaki et al. 1995a), D50312; human Kir6.2, (Inagaki et al. 1995b), D50582

1994; Wile et al. 1994). In contrast, both Kir6.1 (residue 170) and Kir6.2 (residue 160) have asparagine at the corresponding position, as is found in the weakly inwardly rectifying K^+ channel subunit Kir1.1a (residue 171) (Ho et al. 1993). Kir6.2 is expressed at high levels in pancreatic islets and islet derived cell lines (MIN6, HIT-T15, aTC-6), and at low levels in heart, skeletal muscle, brain, and pituitary (Table 1). However, we have found that the expression of Kir6.2 protein alone either in mammalian cells or in *Xenopus* oocytes does not form functional K_{ATP} channels.

2. The Sulfonylurea Receptor SUR1

The sulfonylurea receptor, SUR1 (originally called SUR), was first cloned from rat and hamster insulinoma cDNA libraries by Aguilar-Bryan et al. (1995). Human SUR1 protein consists of 1581 amino acid residues, and has multiple transmembrane segments (Fig. 3). It also has two consensus

Table 1. Tissue distribution of K_{ATP} channel subunits

	Kir6.1[a]	Kir6.2[a]	SUR1[a]	SUR2[b]	SUR2A[c]	SUR2B[c]	SUR2C[d]
Heart	+++	++	+/–	+++	Ventricle + Atrium +	ventricle + atrium +	+++
Skeletal muscle	++	++	+/–	+++	+	+	hind limb +++ soleus ++
Pancreatic islets	++	+++	+++	++	Whole pancreas –	Whole pancreas +	
Brain	++	++	+	++	Forebrain – Cerebellum +	Forebrain + Cerebellum +	+
Eye					+	+	
Adrenal grand	+++			+			+
Pituitary	+			+/–			
Thyroid	++			+/–			
Thymus							+++
Lung	++			+	–	+	+++
Tongue				++			
Liver	+			–	–	+	+/–
Kidney	+			–	–	+	++
Spleen					–	+	++
Stomach	++			+/–	–	+	
Small intestine	+			–	–	+	
Colon	++			+/–	–	+	
Urinary bladder					+	+	++
Uterus					–	+	
Testis	++			+			
Ovary	+++			+++	–	+	
Aorta							+
Fat tissue					–	+	++
References	INAGAKI et al. (1995a)	INAGAKI et al. (1995b)	INAGAKI et al. (1995b)	INAGAKI et al. (1996)	IOSOMOTO et al. (1996)	IOSOMOTO et al. (1996)	CHUTKOW et al. (1996)

+++, High level expression; ++ modelate level; + low level; +/–, very low level; –, not detected; blank boxes, not determined.
[a] Rat tissues, by RNA blotting experiments.
[b] Rat tissues, by RNA blotting experiments. Note that the cDNA probe used cannot differentiate SUR2A, SUR2B, and SUR2C subunits.
[c] Mouse tissues by reverse transcription-polymerase chain reaction (RT-PCR) experiments.

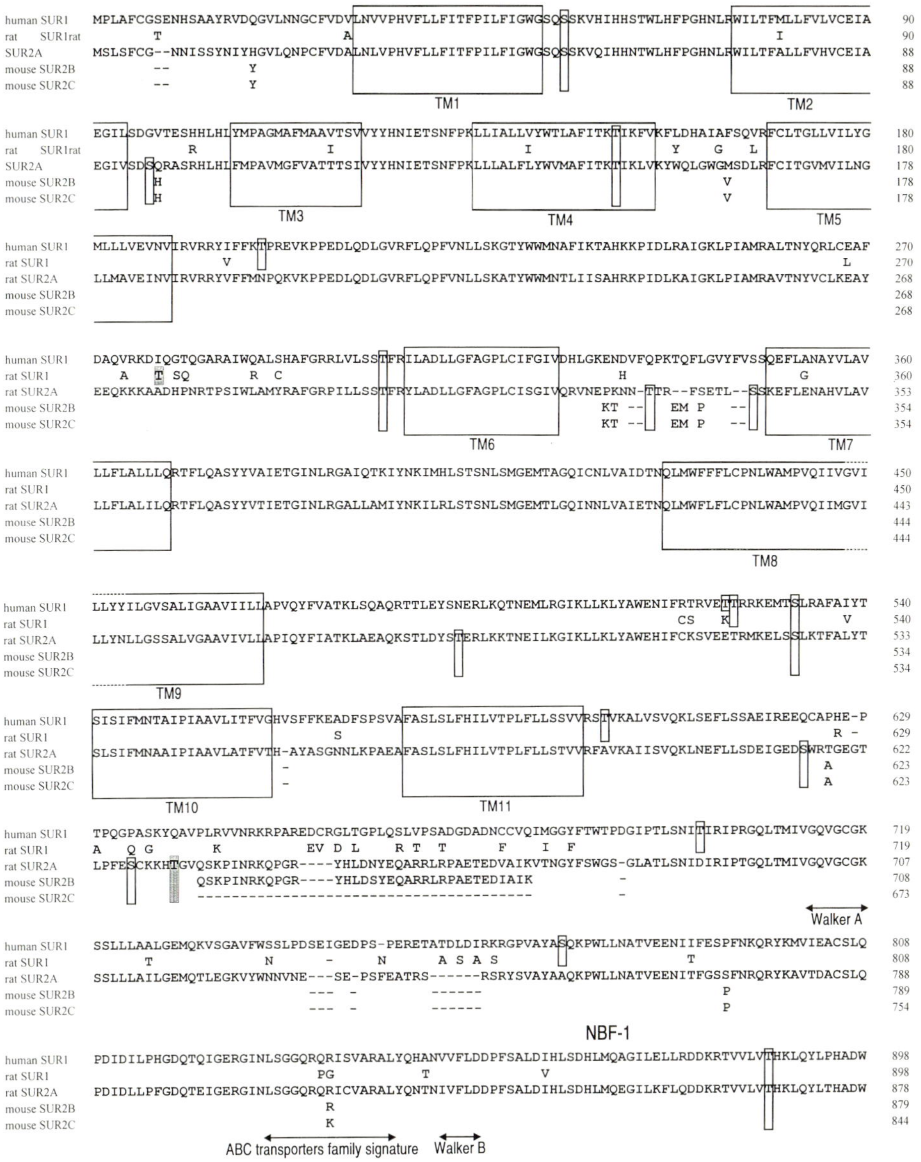

Fig. 3. Amino acids sequences of the human SUR1, rat SUR1, rat SUR2A, mouse SUR2B, and mouse SUR2C. The transmembrane segments (TM) are predicted according to the transfer free energy regions by the Goldman-Engleman-Steitz method (ENGLEMAN et al. 1986). The Walker A and Walker B motifs and the ABC transporters family signatures of the two nucleotide binding folds (NBFs) are indicated by *arrows*. Amino acid residues for potential phosphorylation sites for cyclic AMP-dependent protein kinase and protein kinase C are indicated by *shaded and filled boxes*, respectively. References and DDBJ accession numbers are: human SUR1 (THOMAS et al. 1996), U63455; rat SUR1 (AGUILAR-BRYAN et al. 1995), L40624; rat SUR2A (INAGAKI et al. 1996), D83598; mouse SUR2B (ISOMOTO et al. 1996), D86038; mouse SUR2C (CHUTKOW et al. 1996), and AF003531

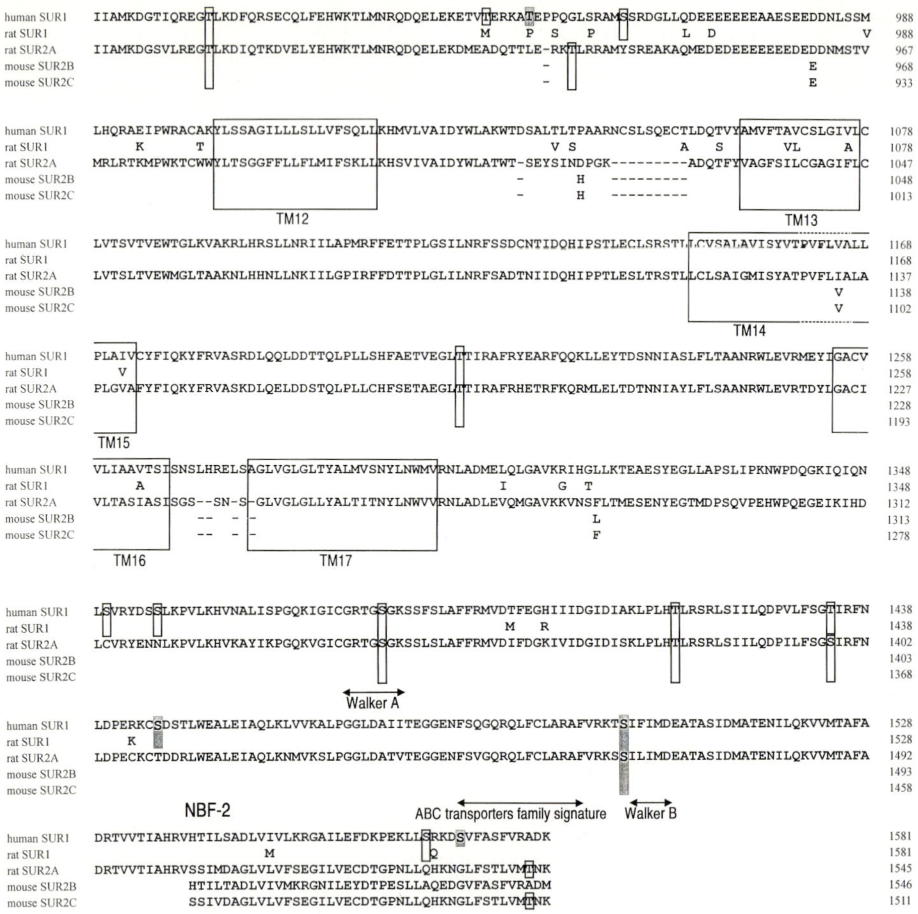

Fig. 3. *Continued*

sequences for nucleotide binding folds (NBFs), and therefore is a member of the ATP-binding cassette (ABC) superfamily, as are cystic fibrosis transmembrane conductance regulator (CFTR), P-glycoprotein (P-gp), and multi-drug resistance associated proteins (MRP) (HIGGINS 1992). Although SUR1 was originally proposed to have 13 transmembrane segments (AGUILAR-BRYAN et al. 1995), TUSNÁDY et al. (1997) recently proposed a 17 transmembrane segment model, based on sequence alignments of SUR1 and members of the MRP gene subfamily (Figs. 3 and 4). SUR1 mRNA is present at high levels in pancreatic islets and insulinoma cells, at low levels in brain, but is not present at detectable levels in heart, skeletal muscle, and other tissues (Table 1) (INAGAKI et al. 1995b). The human SUR1 gene has 39 exons and spans more than 100 kb at chromosome 11p15.1 (AGUILAR-BRYAN et al. 1998), with the

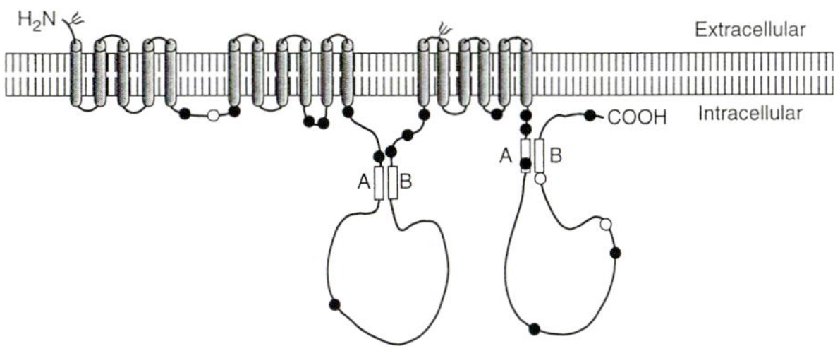

Fig. 4. Proposed membrane topology of SUR1. The membrane topology is based on TUSNÁDY et al. (1997). Location of potential N-linked glycosylation site is indicated by Ψ. Potential cyclic AMP-dependent protein kinase and protein kinase C phosphorylation sites are indicated by *open circles and filled circles*, respectively

Kir6.2 gene immediately downstream of the SUR1 gene (INAGAKI et al. 1995b). Expression of SUR1 cDNA alone in COSm6 cells shows glibenclamide binding activity (AGUILAR-BRYAN et al. 1995). However, neither the expression of the SUR1 protein alone nor the SUR1 protein in combination with ROMK1 (Kir1.1a), IRK1(Kir2.1), or CIR (Kir3.4) in *Xenopus* oocytes produces ATP-sensitive K^+ channel currents (AGUILAR-BRYAN et al. 1995; GRIBBLE et al. 1997a).

3. Reconstitution of the Pancreatic β-Cell Type K_{ATP} Channel

Coexpression of SUR1 and Kir6.2 mRNA at high levels in pancreatic islets and in various insulinoma cell lines suggested that SUR1 and Kir6.2 might couple functionally to form a novel K_{ATP} channel. Cotransfection of Kir6.2 and SUR1 into COS cells elicited weakly inwardly rectifying K^+ channel currents with a unitary conductance of 76 pS in symmetric 140 mmol/l $[K^+]$ (INAGAKI et al. 1995b; SAKURA et al. 1995). The activity of the reconstituted channel is inhibited by ATP applied to the cytosolic side of the membrane with K_i of 10 μmol/l. The reconstituted K^+ channel currents are also inhibited by adenyl-5'-yl imidodiphosphate (AMP-PNP), a non-hydrolysable ATP analog, and by the sulfonylureas glibenclamide and tolbutamide. This channel activity is stimulated by diazoxide (INAGAKI et al. 1995b), and pinacidil (GRIBBLE et al. 1997a) but not by cromakalim (Table 2) (GRIBBLE et al. 1997a; Gonoi and Seino, unpublished observation). Metabolic poisoning with oligomycin and 2-deoxy-glucose remarkably stimulates $^{86}Rb^+$ efflux from COS cells coexpressing Kir6.2 and SUR1. $^{86}Rb^+$ efflux through the reconstituted channels is inhibited by glibenclamide and stimulated by diazoxide. The properties of the K^+ channel currents reconstituted from Kir6.2 and SUR1 are identical to those of the K_{ATP} channels in native pancreatic β-cells, indicating that the pancreatic β-cell type K_{ATP} channel comprises the two subunits, Kir6.2, a member of the Kir subfamily 6.0, and SUR1, a member of the ABC superfamily.

Table 2. Electrophysiological and pharmacological properties of reconstituted channels

Combination of subunits	Unitary conductance (pS)[a]	K_I for ATP inhibition (μmol/l)	Sulfonylurea sensitivity	Openers	Other characters	Type of K_{ATP} channel	References
SUR1/Kir6.1	45[b]		Low	Diazoxide	Run down after patch excision		Ämmälä et al. (1996); Gribble et al. (1997a)
SUR1/Kir6.2	76	10 (+Mg^{2+})	High	Diazoxide Pinacidil		Pancreatic β-cell	Inagaki et al. (1995b); Gribble et al. (1997a)
SUR2A/Kir6.2	80	100–170 (Mg^{2+} have no significant effects)	Low	Cromakalim Pinacidil Nicorandil		Cardiac and skeletal muscle	Inagaki et al. (1996); Okuyama et al. (1998)
SUR2B/Kir6.1	33	>1000	Low	Pinacidil Nicorandil	Run down after patch excision, and activated by cytosolic NDP	Vascular smooth muscle, K_{NDP}	Yamada et al. (1997)
SUR2B/Kir6.2	80	68 (Mg^{2+} free) 300 (MgATP)		Diazoxide Pinacidil		Smooth muscle[c]	Isomoto et al. (1996)

[a] In symmetrical 140–145 mM [K^+].
[b] Gonoi and Seino, unpublished observation.
[c] It is unknown whether the K_{ATP} channels in smooth muscles comprise Kir6.2 subunit.

II. The Cardiac and Skeletal Muscle Type K_{ATP} Channel

1. The Sulfonylurea Receptor SUR2A

SUR2A (originally referred to as SUR2), an isoform of SUR1, was cloned from a rat brain cDNA library using SUR1 cDNA as a probe (INAGAKI et al. 1996). SUR2A protein consists of 1545 amino acid residues having 68% identity with SUR1 (Fig. 3). RNA blotting analysis reveals that SUR2A mRNA is expressed at high levels in heart, skeletal muscle, and ovary, at moderate levels in brain, tongue, and pancreatic islets, at low levels in lung, testis, and adrenal grand, and at very low levels in stomach, colon, thyroid, and pituitary (Table 1) (INAGAKI et al. 1996). The human SUR2 gene consists of 38 exons (AGUILAR-BRYAN et al. 1998) located at chromosome12p11.12 (CHUTKOW et al. 1996).

2. Reconstitution of the Cardiac and Skeletal Muscle Type K_{ATP} Channel

K^+ channel currents reconstituted from SUR2A and Kir6.2 (SUR2A/Kir6.2 channels) are inhibited by ATP with K_i of approximately 100 μmol/l, being much less sensitive to ATP than the pancreatic β-cell type K_{ATP} channel (Table 2) (INAGAKI et al. 1996). SUR2A/Kir6.2 channels are also much less sensitive to glibenclamide than β-cell type SUR1/Kir6.2 K_{ATP} channels (INAGAKI et al. 1996). SUR2A/Kir6.2 channels have a unitary conductance of 80 pS in symmetrical 140 mmol/l [K^+] and are activated by cromakalim and pinacidil, but not by diazoxide (Table 2) (INAGAKI et al. 1996), in contrast to the pharmacological properties of β-cell K_{ATP} channels. These observations together with the tissue specific expressions of SUR2A and Kir6.2 mRNAs suggest that SUR2A/Kir6.2 channels are cardiac and skeletal muscle type K_{ATP} channels.

III. The Smooth Muscle Type K_{ATP} Channel

1. The Sulfonylurea Receptor SUR2B

Two variants of SUR2A have been reported (ISOMOTO et al. 1996; CHUTKOW et al. 1996). One is identical to SUR2A except for the 42 (rat) amino acid residues in the C-terminals (Fig. 3) (ISOMOTO et al. 1996). The other has a deletion of 35 amino acids near NBF-1 of SUR2 (Fig. 3) (CHUTKOW et al. 1996). These variants are likely to be produced by alternative splicing of the SUR2 gene. The nomenclature for these variants is currently in flux, but SUR2B and SUR2C for the former and latter, respectively, are proposed (ASHCROFT and GRIBBLE 1998). Reverse transcription-polymerase chain reaction (RT-PCR) analyses show that SUR2B and SUR2C mRNAs are expressed in diverse tissues including brain, heart, liver, urinary bladder, and skeletal muscle (Table 1) (ISOMOTO et al. 1996; CHUTKOW et al. 1996).

2. Reconstitution of the Smooth Muscle Type K_{ATP} Channel

The unitary conductance and ATP-sensitivity of K^+ channels reconstituted from SUR2B and Kir6.2 are similar to those of SUR2A/Kir6.2 channels, but, unlike SUR2A/Kir6.2 channels, SUR2B/Kir6.2 channels are activated by diazoxide (Table 2) (ISOMOTO et al. 1996). Based on these pharmacological data and the RT-PCR analysis, it is proposed that SUR2B is the SUR subunit in smooth muscle type K_{ATP} channels (ISOMOTO et al. 1996). However, since the tissue distribution of SUR2B is different from that of Kir6.2, whether or not the K_{ATP} channels in native smooth muscle consist of the SUR2B subunit and the Kir6.2 subunit is not certain.

IV. The Vascular Smooth Muscle Type K_{ATP} Channel

1. Reconstitution of the Vascular Smooth Muscle Type K_{ATP} Channel

The channels reconstituted from SUR2B and Kir6.1 are activated by the K^+ channel openers, pinacidil and nicorandil, and are inhibited by glibenclamide in the cell-attached mode of patch-clamp recordings (Table 2) (YAMADA et al. 1997). These channels do not open spontaneously upon patch excision, but addition of nucleotide diphosphates or ATP at 10^{-6} to 10^{-4}mol/l ranges to the cytosolic side of the membrane activates the channels (YAMADA et al. 1997). The activity of SUR2B/Kir6.1 channels is inhibited by ATP only at high concentrations (>10^{-4}mol/l). SUR2B/Kir6.1 channels have a unitary conductance of 33pS in symmetrical 145mmol/l [K^+]. These properties of the SUR2B/Kir6.1 channel resemble those of the nucleotide diphosphate activated K^+ (K_{NDP}) channels described in some vascular smooth muscle cells (KAJIOKA et al. 1991; BEECH et al. 1993; NELSON and QUAYLE 1995; ZHANG and BOLTON 1996).

D. Physical Interaction and Stoichiometry of the Pancreatic β-Cell Type K_{ATP} Channel Subunits

I. Physical Interaction Between the SUR1 Subunit and the Kir6.2 Subunit

When SUR1 or histidine-tagged SUR1 is cotransfected with Kir6.2 into COS cells, they are copurified as a complex of glycosylated SUR1/Kir6.2 proteins by wheat germ agglutinin agarose chromatography, or by Ni^{2+}-agarose chromatography (CLEMENT et al. 1997). The molecular weight of the largest form of the complex is approximately 950kDa, a molecular weight close to that expected for a heterooctamer of four Kir6.2 subunits and four glycosylated SUR1 subunits (CLEMENT et al. 1997). SUR1 and Kir6.2 proteins are shown to be co-immunoprecipitated from a mixture of SUR1 and Kir6.2 subunits in in vitro-translated proteins and from COS cells transfected with both subunits (LORENZ et al. 1998). MAKHINA and NICHOLS (1998) suggested that Kir6.2 and SUR1 independently traffic to the plasma membrane, while LORENZ et al. (1998) suggested that interaction between Kir6.2 and SUR1 affects cellular distribution of the molecules.

II. Subunit Stoichiometry of the SUR1/Kir6.2 Channel

The subunit stoichiometry of the pancreatic β-cell type K_{ATP} channel was determined by constructing dimeric (SUR1-Kir6.2) and trimeric fusion (SUR1-Kir6.2-Kir6.2) proteins (INAGAKI et al. 1997; CLEMENT et al. 1997; SHYNG and NICHOLS 1997). Expression of the dimeric SUR1-Kir6.2 protein in COS cells produces K_{ATP} channels with physiological properties almost indistinguishable from those of the channels reconstituted from monomeric SUR1 and Kir6.2 subunits, including current density, ATP-sensitivity, and unitary conductance. The channels reconstituted from the trimeric fusion protein show smaller $^{86}Rb^+$ efflux and K_{ATP} channel currents with less sensitivity to ATP than the channels reconstituted from the dimeric fusion protein, but the properties of this triple fusion protein are restored by supplementation with monomeric wild-type SUR1 (INAGAKI et al. 1996). These and other supplementary experiments with mutant SUR1 and mutant Kir6.2 proteins suggest that the SUR1/Kir6.2 channels are optimally expressed as a heterooctamer of four Kir6.2 subunits and four SUR1 subunits, with a K^+ ion conducting pore formed by the Kir6.2 subunits (Fig. 5).

E. Domains Conferring Sensitivities to the Nucleotides and Pharmacological Agents

I. ATP-Sensitivity

Since the SUR subunit contains the Walker A and B motifs (WALKER et al. 1982) in each NBF and there are no obvious consensus nucleotide binding

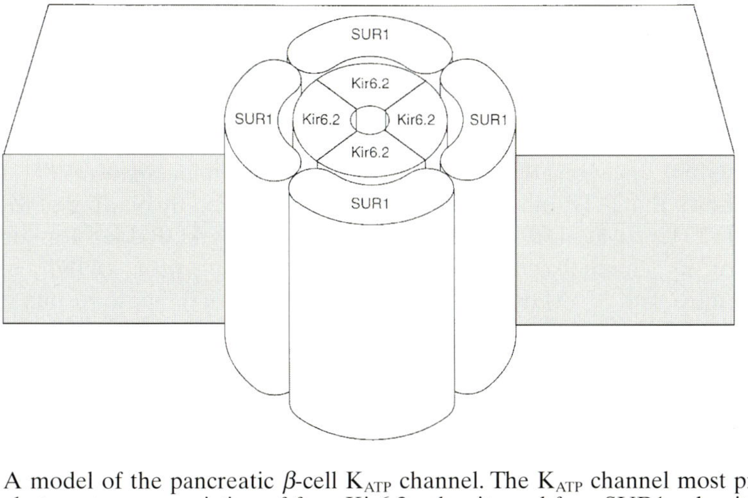

Fig. 5. A model of the pancreatic β-cell K_{ATP} channel. The K_{ATP} channel most probably is a hetrooctamer consisting of four Kir6.2 subunits and four SUR1 subunits, the K^+ ion permeable pore being formed by the tetramer of the Kir6.2 subunit. See text for details

sites in Kir6.1 or Kir6.2, it was at first thought that the SUR subunit and not the Kir6.2 subunit confers both the ATP and MgADP-sensitivities to the K_{ATP} channels. A biochemical study has shown that SUR1 is efficiently photolabeled with 8-azido-[α-^{32}P]ATP and 8-azido-[γ-^{32}P]ATP in the absence of Mg^{2+} (UEDA et al. 1997). In addition mutations of the lysine in the Walker A motif and a mutation of the aspartic acid in the Walker B motif in NBF-1 of SUR1 impair Mg^{2+}-independent high-affinity ATP binding. MgADP antagonizes ATP binding to NBF-1 ($IC_{50} < 10\,\mu$mol/l), and a mutation of the lysine in the Walker A motif of NBF-2 reduces the MgADP antagonism. Thus, ATP binds to NBF-1 of SUR1 with high affinity, and MgADP, through binding to NBF-2, antagonizes the ATP binding to NBF-1. However, mutations of the lysine residues in the Walker A motifs in NBF-1 and/or the equivalent mutation in NBF-2 of SUR1 do not prevent inhibition of channel activity by ATP (GRIBBLE et al. 1997b). In addition, a C-terminal truncation of Kir6.2 (Kir6.2ΔC26 or Kir6.2ΔC36) has been shown to generate K_{ATP} channel currents in the absence of the SUR subunit, although coexpression with SUR1 increases the ATP-sensitivity (TUCKER et al. 1997). Furthermore, when the lysine residue at position 185 in Kir6.2ΔC26 is mutated to glutamine, the ATP-sensitivity of the channel is substantially reduced (TUCKER et al. 1997). These results suggest that the Kir6.2 subunit confers the ATP-sensitivity of the K_{ATP} channel, but the precise domain is yet to be determined.

II. Nucleotide Diphosphate (NDP)-Sensitivity

Mutation of the lysine residue in the Walker A motif in NBF-1 of SUR1(K719 A) and/or the equivalent mutation in NBF-2 (K1384M) abolish activation of the SUR1/Kir6.2 channel (the β-cell K_{ATP} channel) by MgADP (GRIBBLE et al. 1997b). While the mutations in the linker region and the Walker B motif of NBF-2 mostly abolish channel activation by MgADP, the equivalent mutations in NBF-1 do not interfere with channel activation by MgADP, but alter the kinetic properties (NICHOLS et al. 1996; SHYNG et al. 1997b). Accordingly, both the Walker A motifs in NBF-1 and NBF-2 and the Walker B motif in NBF-2 are essential for MgADP activation. Binding experiments with ATP analogs have shown that ATP-binding to NBF-1 is antagonized by binding of MgADP to NBF-2 (UEDA et al. 1997). On the other hand, MgADP does not stimulate Kir6.2ΔC26 channel currents, and coexpression with SUR1 endows Kir6.2ΔC26 with MgADP-sensitivity (TRAPP et al. 1997). These results suggest that both NBFs of SUR1 are essential for the normal response of the channel by MgADP activation. Whether channel activation requires ATP hydrolysis at the NBF region(s) remains to be determined.

III. Diazoxide-Sensitivity

A mutation of the lysine residue in the Walker A motif of NBF-1 (K719A), but not the equivalent mutation in NBF-2 (K1384A), abolishes the channel activation by diazoxide, suggesting that the Walker A motif of NBF-1 is more important in diazoxide activation (GRIBBLE et al. 1997b). Mutations in the linker region and the Walker B motif of NBF-2 (G1479D, G1479R, G1485D, G1485R, Q1486H, and D1506A in hamster, corresponding to G1478, G1484, and G1485 in rat and human, see Fig. 3) abolish or diminish channel activation by diazoxide, and mutations in the linker region of NBF-1 (G827D, and G827R) alter the kinetics of diazoxide activation (SHYNG et al. 1997b). Thus, both NBFs are involved in the normal activation of the pancreatic β-cell type (SUR1/Kir6.2) K_{ATP} channel by diazoxide. On the other hand, it has been reported that, while SUR2A/Kir6.2 channels have no sensitivity to diazoxide, SUR2B/Kir6.2 channels are activated by diazoxide (ISOMOTO et al. 1996). Because the amino acid sequence of the C-terminus in SUR2B is similar to that of SUR1, this region might also confer diazoxide sensitivity to the channel.

IV. Sulfonylurea-Sensitivity

Expression of SUR1 protein alone in COS cells exhibits high-affinity glibenclamide-binding, indicating that the glibenclamide-binding site resides in SUR1 ($K_D = 2–10$ nmol/l) (AGUILAR-BRYAN et al. 1995). In contrast, expression of the SUR2A protein in COS cells shows a much lower binding-affinity for glibenclamide (KD ~1.2 μmol/l) than does SUR1 (INAGAKI et al. 1996). On the other hand, using a Kir6.2ΔC36 mutant, and K719A or K1384M SUR1 mutants, it is has been shown that tolbutamide interacts with SUR1 with high affinity (K_i = 2 μmol/l), and with Kir6.2 with low affinity (K_i = 1.8 mmol/l) (GRIBBLE et al. 1997c).

V. Mg^{2+}- and Spermin-Sensitivity

Mg^{2+} and spermine are known to induce inward rectification in Kir1.1 (ROMK1) channels acting from the cytosolic side of the membrane. The inward rectification of SUR1/Kir6.2 channels is also modulated by Mg^{2+} and spermine, and the aspargine residue at position 160 of Kir6.2 is a critical determinant of the inward rectification by Mg^{2+} and spermine (SHYNG et al. 1997a).

VI. Phentolamine-Sensitivity

Using Kir6.2ΔC36 and Kir6.2ΔC26, the imidazoline phentolamine, which is a potent stimulator of insulin secretion, has been shown to close β-cell K_{ATP} channels by interacting with the Kir6.2 subunit, at a site different from the ATP-inhibitory site (PROKS and ASHCROFT 1997).

VII. G-Protein Sensitivity

The activity of K_{ATP} channels in pancreatic β-cells and cardiac ventricular myocytes is modulated by the α-subunits of the G proteins G_i and G_o (RIBALET and EDDELSTONE 1995; ITO et al. 1994; TERZIC et al. 1994). It is suggested that G proteins activate both types of K_{ATP} channel by directly interacting with the SUR subunit (SÁNCHEZ et al. 1998).

F. Pathophysiology of the Pancreatic β-Cell K_{ATP} Channel

I. Persistent Hyperinsulinemic Hypoglycemia of Infancy

Familial persistent hyperinsulinemic hypoglycemia of infancy (PHHI) is an autosomal recessive disorder of childhood characterized by severe, recurrent, and fasting hypoglycemia associated with inappropriate hypersecretion of insulin in human (PERMUTT et al. 1996). It has been reported that mutations of the SUR1 gene or the Kir6.2 gene can cause PHHI. Mutations were found in the noncoding sequences of the SUR1 gene which are required for RNA processing (THOMAS et al. 1996a), and the NBF-1 (THOMAS et al. 1996a) and NBF-2 regions (THOMAS et al. 1995) of the SUR1 gene. A nonsense mutation (Y12X) (NESTOROWICZ et al. 1997), or a point mutation in the 2nd transmembrane (TM2) domain of the Kir6.2 gene (THOMAS et al. 1996b) also causes PHHI. Mutations of either the SUR1 or Kir6.2 gene, therefore, inactivate the K_{ATP} channels in pancreatic β-cells, causing constant membrane depolarization and continuous calcium influx, leading to unregulated insulin secretion and hypoglycemia (KANE et al. 1996).

II. Transgenic Mice

A substitution of the first residue of the glycine-phenylalanine-glycine motif in the H5 region of Kir6.2 with serine (Kir6.2G132S) blocks ion current, indicating that the motif is critical for K^+ permeation. Kir6.2G132S acts as a dominant-negative inhibitor of the K_{ATP} channels when coexpressed with SUR1 and wild-type Kir6.2. Transgenic mice expressing Kir6.2G132S specifically in the pancreatic β-cells develop hypoglycemia as neonates, similarly to that observed in PHHI in human, but develop hyperglycemia as adults (MIKI et al. 1997). Apoptotic pancreatic β-cells are frequently observed in the transgenic mice (MIKI et al. 1997). These observations suggest that the K_{ATP} channels in pancreatic β-cells are important for glucose-induced insulin secretion and also for survival of the β-cells during development.

G. Conclusions

K_{ATP} channels link the metabolic energy of the cell to the membrane K^+ conductance, thereby having many important roles in the various tissues in which they occur. In pancreatic β-cells K_{ATP} channels play a key role in glucose-induced insulin secretion. In cardiac muscles and brain the channels have a cyto-protective role in energy-depleted states. The channels are also the target of endogenous vasoactive substances and are involved in controlling energy supply through the vascular system.

Because K_{ATP} channels are also the target of many clinically used drugs such as the antidiabetic sulfonylureas, the K^+ channel openers, and the imidazoline drugs which are used as α-adrenoceptor blocking agents, detailed analyses of the structure and function relationships of K_{ATP} channels should be useful pharmacologically.

An approach by a combination of molecular biology and electrophysiology has clarified the molecular basis of K_{ATP} channels in some tissues. The K_{ATP} channel in pancreatic β-cells comprises the Kir6.2 subunit and the SUR1 subunit. The K_{ATP} channel in cardiac muscle (and probably in some skeletal muscles) comprises the Kir6.2 subunit and the SUR2A subunit. The K_{ATP} channel in some vascular smooth muscles comprises Kir6.1 and SUR2B. However, the constituents of K_{ATP} channels in many other tissues, including brain, other smooth muscles, and gastrointestinal tract remain to be determined. It appears that a Kir6.0 subfamily subunit forms a pore for K^+ permeation in K_{ATP} channels and that ATP might inhibit the K_{ATP} channels by acting primarily on the Kir6.2 subunit. What then is the role of the SUR subunit? The SUR subunit is a sensor for the cytosolic nucleotide-dihposphates concentration, and probably controls the ATP-sensitivity of the channel. It may also be a target molecule of G-proteins and protein kinases for modulating channel activity. Other roles of the SUR subunit remain to be determined. The SUR subunit may have a role in maintaining the channels in functional states and in trafficking or inserting the channel molecules into the plasma membrane. Whether, like other ABC protein family members, the SUR subunit hydrolyzes ATP for energy to perform the roles mentioned above should also be investigated.

Acknowledgments. This work is supported by scientific research grants from the Ministry of Education, Science, and Culture and from the Ministry of Health and Welfare, Japan.

References

Aguilar-Bryan L, Nichols CG, Wechsler SW, Clement IV JP, Boyd AE III, González G, Herrera-Sosa H, Nguy K, Bryan J, Nelson DA (1995) Cloning of the β cell high affinity sulfonylurea receptor: A regulator of insulin secretion. Science 268: 423–426

Aguilar-Bryan L, Clement IV JP, Gonzalez G, Kunjilwar K, Babenko A, Bryan J (1998) Toward understanding the assembly and structure of K_{ATP} channels. Physiol Rev 78:227–245

Allard B, Lazdunski M (1992) Nucleotide diphosphates activate the ATP-sensitive potassium channel in mouse skeletal muscle. Pflügers Arch 422:185–192

Allard B, Lazdunski M (1993) Pharmacological properties of ATP-sensitive K^+ channels in mammalian skeletal muscle cells. Eur J Pharmacol 236:419–426

Ämmälä C, Moorhouse A, Ashcroft FM (1996) The sulphonylurea receptor confers diazoxide sensitivity on the inwardly rectifying K^+ channel Kir6.1 expressed in human embryonic kidney cells. J Physiol (Lond) 494:709–714

Arch JRS, Buckle DR, Bumstead J, Clarke GD, Taylor JF, Taylor SG, (1988) Evaluation of the potassium channel activator cromakalim (BRL 34915) as a bronchodilator in the guinea-pig: comparison with nifedipine. Br J Pharmacol 95:763–770

Ashcroft FM, Harrison DE, Ashcroft SJH (1984) Glucose induces closure of single potassium channels in isolated rat pancreatic β-cells. Nature 312:446–448

Ashcroft FM, Rorsman P (1989) Electrophysiology of the pancreatic β-cell. Prog Biophys Molec Biol 54:87–143

Ashcroft FM, Gribble FM (1998) Correlating structure and function in ATP-sensitive potassium channels. Trends Neurosci (in press)

Ashford ML, Boden PR, Treherne JM (1990) Glucose-induced excitation of hypothalamic neurones is mediated by ATP-sensitive K^+ channels. Pflügers Arch 415:479–483

Barrett-Jolley R, Davies NW (1997) Kinetic analysis of the inhibitory effect of glibenclamide on K_{ATP} channels of mammalian skeletal muscle. J Membr Biol 155: 257–262

Beech DJ, Zhang H, Nakao K, Bolton TB (1993) Single channel and whole-cell K-currents evoked by levcromakalim in smooth muscle cells from the rabbit portal vein. Br J Pharmacol 110:583–590

Belles B, Hescheler J, Trube G (1987) Changes of membrane currents in cardiac cells induced by long whole-cell recordings and tolbutamide. Pflügers Arch 409:582–588

Black JL, Armour CL, Johnson RA, Alouan LA, Barnes PJ (1990) The action of a potassium channel activator, BRL 38227 (lemakalim), on human airway smooth muscle. Am Rev Respir Dis 142:1384–1389

Burton F, Dorstelmann U, Hutter OF (1988) Single-channel activity in salcolemmal vesicles from human and other mammalian muscles. Muscle Nerve 11:1029–1038

Chutkow WA, Simon MC, Le Beau MM, Burant CF (1996) Cloning, tissue expression, and chromosomal localization of SUR2, the putative drug-binding subunit of cardiac, skeletal muscle, and vascular K_{ATP} channels. Diabetes 45:1439–1445

Clement JP IV, Kunjilwar K, Gonzales G, Schwanstecher M, Panten U, Aguilar-Bryan L, Bryan J (1997) Association and stoichiometry of K_{ATP} channel subunits. Neuron 18:827–838

Cook DL, Hales N (1984) Intracellular ATP directly blocks K^+ channels in pancreatic B-cells. Nature 311:271–273

Dart C, Standen NB (1995) Activation of ATP-dependent K^+ channels by hypoxia in smooth muscle cells isolated from the pig coronary artery. J Physiol (Lond) 483:29–39

Davies NW (1990) Modulation of ATP-sensitive K^+ channels in skeletal muscle by intracellular protons. Nature 343:375–377

Davies NW, Standen NB, Stanfield PR (1991) ATP-dependent potassium channels of muscle cells: Their properties, regulation and possible functions. J Bioenerge Biomembr 23:509–535

Dunne MJ, Petersen MJ (1986) Intracellular ADP activates K^+ channels that are inhibited by ATP in an insulin-secreting cell line. FEBS Lett 208:59–62

Engleman DM, Steitz TA, Goldman A (1986) Identifying nonpolar transbilayer helices in amino acid sequences of membrane proteins. Ann Rev Biophys Biophy Chem 15:321–354

Escande D, Thuringer D, Leguern S, Cavero I (1988) The potassium channel opener cromakalim (BRL 34915) activates ATP-dependent K^+ channels in isolated cardiac myocytes. Biochem Biophys Res Commun 154:620–625

Findlay I (1992) Inhibition of ATP-sensitive K^+ channels in cardiac muscle by the sulphonylurea drug glibenclamide. J Pharmacol Exp Ther 261:540–545

Gehlert DR, Mais DE, Gackenheimer SL, Krushinski JH, Robertson DW (1990) Localization of ATP sensitive potassium channels in the rat brain using a novel radioligand, [^{125}I]iodoglibenclamide. Eur J Pharmacol 186:373–375

Gribble FM, Ashfield R, Ämmälä C, Ashcroft FM (1997a) Properties of cloned ATP-sensitive K^+ currents expressed in *Xenopus* oocytes. J Physiol (Lond) 498:87–98

Gribble FM, Tucker SJ, Ashcroft FM (1997b) The essential role of the Walker A motifs of SUR1 in K-ATP channel activation by Mg-ADP and diazoxide. EMBO J 16:1145–1152

Gribble FM, Tucker SJ, Ashcroft FM (1997c) The interaction of nucleotides with the tolbutamide block of cloned ATP-sensitive K^+ channel currents expressed in *Xenopus* oocytes: a reinterpretation. J Physiol (Lond) 504:35–45

Grigg JJ, Anderson EG (1989) Glucose and sulfonylureas modify different phases of the membrane potential change during hypoxia in rat hippocampal slices. Brain Res 489:302–310

Hatakeyama N, Wang Q, Goyal RK, Akbarali HI (1995) Muscarinic suppression of ATP-sensitive K^+ channel in rabbit esophageal smooth muscle. Am J Physiol 268:C877-C885

Hehl S, Moser C, Weik R, Neumcke B (1994) Internal Ca^{2+} ions inactivate and modify ATP-sensitive potassium channels in adult mouse skeletal muscle. Biochim Biophys Acta 1190:257–263

Higgins (1992) ABC transporters: from microorganisms to man. Ann Rev Cell Biol 8:67–113

Ho K, Nichols CG, Lederer WJ, Lytton J, Vassilev PM, Kanazirska MV, Hebert SC (1993) Cloning and expression of an inwardly rectifying ATP-regulated potassium channel. Nature 362:31–38

Horie M, Irisawa H, Noma A (1987) Voltage-dependent magnesium block of adenosine-triphosphate-sensitive potassium channel in guinea-pig ventricular cells. J Physiol (Lond) 387:251–272

Inagaki N, Tsuura Y, Namba N, Masuda K, Gonoi T, Seino Y, Seino S (1995a) Cloning and functional expression of a novel ATP-sensitive potassium channel ubiquitously expressed in rat tissues, including pancreatic islet, pituitary, skeletal muscle, and heart. J Biol Chem 270:5691–5694

Inagaki N, Gonoi T, Clement IV JP, Namba N, Inazawa J, Gonzalez G, Aguilar-Bryan L, Seino S, Bryan J (1995b) Reconstitution of I_{KATP}: An inward rectifier subunit plus the sulfonylurea receptor. Science 270:1166–1170

Inagaki N, Inazawa J, Seino S (1995c) cDNA sequence, gene structure, and chromosomal localization of human ATP-sensitive potassium channel, uK_{ATP}-1 gene (KCNJ8). Genomics 30:102–104

Inagaki N, Gonoi T, Clement IV JP, Wang CZ, Aguilar-Bryan L, Bryan J, Seino S (1996) A family of sulfonylurea receptors determines the pharmacological properties of ATP-sensitive K^+ channels. Neuron 16:1011–1017

Inagaki N, Gonoi T, Seino S (1997) Subunit stoichiometry of the pancreatic β-cell ATP-sensitive K^+ channel. FEBS Lett 409:232–236

Inoue I, Nagase H, Kishi K, Higuti T (1991) ATP-sensitive K^+ channel in the mitochondrial inner membrane. Nature 352:244–247

Isomoto S, Kondo C, Yamada M, Matsumoto S, Higashiguchi O, Horio Y, Matsuzawa Y, Kurachi Y (1996) A novel sulfonylurea receptor forms with BIR (Kir6.2) a smooth muscle type ATP-sensitive K^+ channel. J Biol Chem 271:24321–24324

Ito H, Vereecke J, Carmeliet E (1994) Mode of regulation by G protein of the ATP-sensitive K^+ channel in guinea-pig ventricular cell membrane. J Physiol (Lond) 478:101–107

Ito K, Kanno T, Suzuki K, Masuzawa-Ito K, Takewaki T, Ohashi H, Asano M, Suzuki H (1992) Effects of cromakalim on the contraction and the membrane potential of the circular smooth muscle of guinea-pig stomach. Br J Pharmacol 105:335–40

Jiang C, Xia Y, Haddad GG (1992) Role of ATP-sensitive K^+ channels during anoxia: major differences between rat (newborn and adult) and turtle neurons. J Physiol (Lond) 448:599–612

Kajioka S, Kitamura K, Kuriyama H (1991) Guanosine diphosphate activates an adenosine 5'-triphosphate-sensitive K^+ channel in the rabbit portal vein. J Physiol (Lond) 444:397–418

Kakei M, Kelly RP, Ashcroft AJH, Ashcroft FM (1986) The ATP-sensitivity of K^+ channels in rat pancreatic B-cells is modulated by ADP. FEBS Lett 208:63–66

Kakei M, Noma A, Shibasaki T (1985) Properties of adenosine-triphosphate-regulated potassium channels in guinea-pig ventricular cells. J Physiol (Lond) 363:441–462

Kamouchi M, Kitamura K (1994) Regulation of ATP-sensitive K^+ channels by ATP and nucleotide diphosphate in rabbit portal vein. Am J Physiol 266:H1687–H1698

Kane C, Shepherd RM, Squires PE, Johnson PRV, James RFL, Milla PJ, Aynsley-Green A, Lindley KJ, Dunne MJ (1996) Loss of functional K_{ATP} channels in pancreatic β-cells causes persistent hyperinsulinemic hypoglycemia of infancy. Nature Medicine 2:1344–1347

Katayama N, Huang SM, Tomita T, Brading AF (1993) Effects of cromakalim on the electrical slow wave in the circular muscle of guinea-pig gastric antrum. Br J Pharmacol 109:1097–1100

Katnik C, Adams DJ (1995) An ATP-sensitive potassium conductance in rabbit arterial endothelial cells. J Physiol (Lond) 485:595–606

Kortezova N, Bayguinov O, Boev K, Papasova M (1992) Effect of cromakalim on the smooth muscle of the cat gastric antrum. J Pharm Pharmacol 44:875–878

Kuriyama H, Kitamura K, Nabata H (1995) Pharmacological and physiological significance of ion channels and factors that modulate them in vascular tissues. Pharmacol Rev 47:387–573

Light PE, French RJ (1994) Glibenclamide selectively blocks ATP-sensitive K^+ channels reconstituted from skeletal muscle. Eur J Pharmacol. 259:219–222

Lindeman KS, Fernandes LB, Croxton TL, Hirshman CA (1994) Role of potassium channels in hypoxic relaxation of porcine bronchi in vitro. Am J Physiol 266: L232–L237.

Longmore J, Bray KM, Weston AH (1991) The contribution of Rb-permeable potassium channels to the relaxant and membrane hyperpolarizing actions of cromakalim, RP49356 and diazoxide in bovine tracheal smooth muscle. Br J Pharmacol 102:979–85

Lorenz E, Alekseev AE, Krapivinsky GB, Carrasco AJ, Clapham DE, Terzic A (1998) Evidence for direct physical association between a K^+ channel (Kir6.2) and an ATP-binding cassette protein (SUR1) which affects cellular distribution and kinetic behavior of an ATP-sensitive K^+ channel. Mol Cell Biol 18:1652–1659

Lorenz JN, Schnermann J, Brosius FC, Briggs JP, Furspan PB (1992) Intracellular ATP can regulate afferent arteriolar tone via ATP-sensitive K^+ channels in the rabbit. J Clin Invest 90:733–740

Lu Z, Mackinnon R (1994) Electrical tuning of Mg^{2+} affinity in an inward rectifier K^+ channel. Nature 371:243–246

Makhina EN, Nichols CG (1998) Independent trafficking of K_{ATP} channel subunits to the plasma membrane. J Biol Chem 273:3369–3374

McPherson GA, Angus JA (1990) Characterization of responses to cromakalim and pinacidil in smooth and cardiac muscle by use of selective antagonists. Br J Pharmacol 100:201–206

Miki T, Tashiro F, Iwanaga T, Nagashima K, Yoshitomi H, Aihara H, Nitta Y, Gonoi T, Inagaki N, Miyazaki J, Seino S (1997) Abnormalities of pancreatic islets by targeted expression of a dominant-negative K_{ATP} channel. Proc Natl Acad Sci USA 94:11969–11973

Misler S, Falke LC, Gillis K, McDaniel ML (1986) A metabolite-regulated potassium channel in rat pancreatic B cells. Proc Natl Acad Sci U S A 83:7119–7123

Mourre C, Ben-Ari Y, Bernardi H, Fosset M, Lazdunski M (1989) Antidiabetic sulfonylureas: localization of binding sites in the brain and effects on the hyperpolarization induced by anoxia in hippocampal slices. Brain Res 486:159–164

Murphy KPSJ, Greenfield SA (1991) ATP-sensitive potassium channels counteract anoxia in neurones of the substantia nigra. Exp Brain Res 84:355–358

Nelson MT, Quayle JM (1995) Physiological roles and properties of potassium channels in arterial smooth muscle. Am J Physiol 268:C799–C822

Nestorowicz A, Inagaki N, Gonoi T, Schoor KP, Wilson BA, Glaser B, Landau H, Stanley CA, Thornton PS, Seino S, Permutt MA (1997) A nonsense mutation in the inward rectifier potassium channel gene, Kir6.2, is associated with familial hyperinsulinism. Diabetes 46:1743–1748

Nichols CG, Shyng SL, Nestorowicz A, Glaser B, Clement JP IV, Gonzalez G, Aguilar-Bryan L, Permutt MA, Bryan J (1996) Adenosine diphosphate as an intracellular regulator of insulin secretion. Science 272:1785–1787

Nielsen-Kudsk JE, Bang L, Bronsgaard AM (1990) Glibenclamide blocks the relaxant action of pinacidil and cromakalim in airway smooth muscle. Eur J Pharmacol 180: 291–296

Noma A (1983) ATP-regulated K^+ channels in cardiac muscle. Nature 305:147–148

Ohno-Shosaku T, Yamamoto C (1992) Identification of an ATP-sensitive K^+ channel in rat cultured cortical neurons. Pflügers Arch 422:260–266

Okuyama Y, Yamada M, Kondo C, Satoh E, Isomoto S, Shindo T, Horio Y, Kitakaze M, Hori M, Kurachi Y (1998) The effects of nucleotides and potassium channel openers on the SUR2 A/Kir6.2 complex K^+ channel expressed in a mammalian cell line, HEK 293T cells. Pflügers Arch (in press)

Osterrieder W (1988) Modification of K^+ conductance of heart cell membrane by BRL 34915. Naunyn Schmiedebergs Arch Pharmacol 337:93–97

Paucek P, Mironova G, Mahdi F, Beavis AD, Woldegiorgis G, Garlid KG (1992) Reconstitution and partial purification of the glibenclamide-sensitive, ATP-dependent K^+ channel from rat liver and beef heart mitochondria. J Biol Physiol 267:26062–26069

Paucek P, Yarov-Yarovoy V, Sun X, Garlid K (1996) Inhibition of the mitochondrial K_{ATP} channel by long-chain acyl-CoA esters and activation by guanine nucleotides. J Biol Chem 271:32084–32088

Permutt MA, Nestorowicz A, Benjamin G (1996) Familial hyperinsulinism: an inherited disorder of spontaneous hypoglycemia in neonates and infants. Diabetes Reviews 4:347–355

Politi DM, Rogawski MA (1991) Glyburide-sensitive K^+ channels in cultured rat hippocampal neurons: activation by cromakalim and energy-depleting conditions. Mol Pharmacol 40:308–315

Post JM, Stevens RJ, Sanders KM, Hume JR (1991) Effect of cromakalim and lemakalim on slow waves and membrane currents in colonic smooth muscle. Am J Physiol 260:C375–382

Proks P, Ashcroft FM (1997) Phentolamine block of K_{ATP} channels is mediated by Kir6.2. Proc Natl Acad Sci USA 94:11716–11720

Quayle JM, Nelson MT, Standen NB (1997) ATP-sensitive and inwardly rectifying potassium channels in smooth muscle. Physiol Rev 77:1165–1232

Raeburn D, Brown TJ (1991) RP49356 and cromakalim relax airway smooth muscle in vitro by opening a sulphonylurea-sensitive K^+ channel: a comparison with nifedipine. J Pharmacol Exp Ther 256:480–485

Ribalet B, Eddelstone GT (1995) Characterization of the G protein coupling of a somatostatin receptor to the K_{ATP} channel in insulin-secreting mammalian HIT and RIN cell lines. J Physiol (Lond) 485:73–86

Rorsman P, Trube G (1985) Glucose dependent K^+-channels in pancreatic β-cells are regulated by intracellular ATP. Pflügers Arch 405:305–309

Sakura H, Ämmälä C, Smith PA, Gribble FM, Ashcroft FM (1995) Cloning and functional expression of the cDNA encoding a novel ATP-sensitive potassium channel subunit expressed in pancreatic beta-cells, brain, heart and skeletal muscle. FEBS Lett 377:338–344

Sanguinetti MC, Scott AL, Zingaro GJ, Siegl PK (1988) BRL 34915 (cromakalim) activates ATP-sensitive K^+ current in cardiac muscle. Proc Natl Acad Sci USA 85: 8360–8364

Schwanstecher C, Basse D (1997) K_{ATP}-channel on the somata of spiny neurones in rat caudate nucleus: regulation by drugs and nucleotides. Br J Pharmacol 121:193–198

Schwanstecher C, Panten U (1993) Tolbutamide- and diazoxide-sensitive K^+ channel in neurons of substantia nigra pars reticulata. Naunyn Schmiedebergs Arch Pharmacol 348:113–117

Schwanstecher C, Panten U (1994) Identification of an ATP-sensitive K^+ channel in spiny neurons of rat caudate nucleus. Pflügers Arch 427:187–189

Shyng SL, Ferrigni T, Nichols CG (1997a) Control of rectification and gating of cloned K_{ATP} channels by the Kir6.2 subunit. J Gen Physiol 110:141–153

Shyng SL, Ferrigni T, Nichols CG (1997b) Regulation of K_{ATP} channel activity by diazoxide and MgADP. Distinct functions of the two nucleotide binding folds of the sulfonylurea receptor. J Gen Physiol 110:643–654

Shyng SL, Nichols CG (1997) Octameric stoichiometry of the K_{ATP} channel complex. J Gen Physiol 110:655–664

Silberberg SD, van Breemen C (1992) A potassium current activated by lemakalim and metabolic inhibition in rabbit mesenteric artery. Pflügers Arch 420:118–120

Spanswick D, Smith MA, Groppi VE, Logan SD, Ashford ML (1997) Leptin inhibits hypothalamic neurons by activation of ATP-sensitive potassium channels. Nature 390:521–525

Spruce AE, Standen NB, Stanfield PR (1985) Voltage-dependent ATP-sensitive potassium channels of skeletal muscle membrane. Nature 316:736–738

Standen NB, Quayle JM, Davies NW, Brayden JE, Huang Y, Nelson MT (1989) Hyperpolarizing vasodilators activate ATP-sensitive K^+ channels in arterial smooth muscle. Science 245:177–180

Stanfield PR, Davies NW, Shelton PA, Sutcliffe MJ, Khan IA, Brammar WJ, Conley EC (1994) A single aspartate residue is involved in both intrinsic gating and blocking by Mg^{2+} of the inward rectifier, IRK1. J Physiol (Lond) 478:1–6

Sturgess NC, Kozlowski RZ, Carrington CA, Hales CN, Ashford ML (1988) Effects of sulphonylureas and diazoxide on insulin secretion and nucleotide-sensitive channels in an insulin-secreting cell line. Br J Pharmacol 95:83–94

Suzuki M, Kotake K, Fujikura K, Inagaki N, Suzuki T, Gonoi T, Seino S, Takata K (1998) Kir6.1: a possible subunit of ATP-sensitive K^+ channels in mitochondria. Biochemi Biophys Res Commun 241:693–697

Szewczyk A, Wojcik G, Lobanov NA, Nalecz MJ (1997) The mitochondrial sulfonylurea receptor: identification and characterization. Biochem Biophys Res Commun 230:611–615

Sánchez JA, Gonoi T, Inagaki N, Katada T, Seino S (1998) Modulation of reconstituted ATP-sensitive K^+ channels by GTP-binding proteins in a mammalian cell line. J Physiol (Lond) 507:315–324

Terzic A, Tung RT, Inanobe A, Katada T, Kurachi Y (1994) G proteins activate ATP-sensitive K^+ channels by antagonizing ATP-dependent gating. Neuron 12:885–893

Thomas PM, Cote GJ, Wohllk N, Haddad B, Mathew PM, Rabl W, Aguilar-Bryan L, Gagel RF, Bryan J (1995) Mutations in the sulfonylurea receptor gene in familial persistent hyperinsulinemic hypoglycemia of infancy. Science 268:426–429

Thomas PM, Wohllk N, Huang E, Kuhnle U, Rabl W, Gagel RF, Cote GJ (1996a) Inactivation of the first nucleotide-binding fold of the sulfonylurea receptor, and familial persistent hyperinsulinemic hypoglycemia of infancy. Am J Hum Genet 59:510–518

Thomas PM, Ye Y, Lightner E (1996b) Mutation of the pancreatic islet inward rectifier Kir6.2 also leads to familial persistent hyperinsulinemic hypoglycemia of infancy. Hum Mol Genet 5:1809–1812

Trapp S, Tucker SJ, Ashcroft FM (1997) Activation and inhibition of K-ATP currents by guanine nucleotides is mediated by different channel subunits. Proc Natl Acad Sci US 94:8872–8877

Tromba C, Salvaggio A, Racagni G, Volterra A (1992) Hypoglycemia-activated K^+ channels in hippocampal neurons. Neurosci Lett 143:185–189

Trube G, Hescheler J (1984) Inwardly-rectifying channels in isolated patches of the heart cell membrane: ATP-dependence and comparison with cell-attached patches. Pflügers Arch 401:178–184

Tucker SJ, Gribble FM, Zhao C, Trapp S, Ashcroft FM (1997) Truncation of Kir6.2 produces ATP-sensitive K^+ channels in the absence of the sulphonylurea receptor. Nature 387:179–183

Tusnády GE, Bakos E, Váradi A, Sarkadi B (1997) Membrane topology distinguishes a subfamily of the ATP-binding cassette (ABC) transporters. FEBS Lett 402:1–3

Ueda K, Inagaki N, Seino S (1997) MgADP antagonism to Mg^{2+}-independent ATP binding of the sulfonylurea receptor SUR1. J Biol Chem 272:22983–22986

Vivaudou M, Forestier C (1995) Modification by protons of frog skeletal muscle K_{ATP} channels: effects on ion conduction and nucleotide inhibition. J Physiol (Lond) 486:629–645

von Beckerath N, Cyrys S, Dischner A, Daut J (1991) Hypoxic vasodilatation in isolated, perfused guinea-pig heart: an analysis of the underlying mechanisms. J Physiol (Lond) 442:297–319

Walker JE, Saraste MJ, Runswick MJ, Gay NJ (1982) Distantly related sequences in the A- and B-subunits of ATP synthase, myoson, kinase and other ATP-requiring enzymes and a common nucleotide binding fold. EMBO J 1:945–951

Weik R, Neumcke B (1989) ATP-sensitive potassium channels in adult mouse skeletal muscle: characterization of the ATP-binding site. J Membr Biol 110:217–226

Weik R, Neumcke B (1990) Effects of potassium channel openers in mouse skeletal muscle. Naunyn Schmiedebergs Arch Pharmacol 342:258–263

Wile BA, Taglialatela M, Ficker E, Brawn AM (1994) Gating of inwardly rectifying K^+ channels localized to a single negatively charged residue. Nature 371:246–249

Woll KH, Lonnendonker U, Neumcke B (1989) ATP-sensitive potassium channels in adult mouse skeletal muscle: different modes of blockage by internal cations. ATP and tolbutamide. Pflügers Arch 414:622–628

Yamada M, Isomoto S, Matsumoto S, Kondo C, Shindo T, Horio Y, Kurachi Y (1997) Sulphonylurea receptor 2B and Kir6.1 form a sulphonylurea-sensitive but ATP-insensitive K^+ channel. J Physiol (Lond) 499:715–720

Yarov-Yarovoy V, Paucek P, Jaburek M, Garlid KD (1997) The nucleotide regulatory sites on the mitochondrial K_{ATP} channel face the cytosol. Biochim Biophys Acta 1321:128–136

Zhang H, Bolton TB (1995) Activation by intracellular GDP, metabolic inhibition and pinacidil of a glibenclamide-sensitive K-channel in smooth muscle cells of rat mesenteric artery. Br J Pharmacol 114:662–672

Zhang HL, Bolton TB (1996) Two types of ATP-sensitive potassium channels in rat portal vein smooth muscle cells. Br J Pharmacol 118:105–114

Zünkler BJ, Lenzen S, Manner K, Panten U, Trube G (1988) Concentration-dependent effects of tolbutamide, meglitinide, glipizide, glibenclamide and diazoxide on ATP-regulated K^+ currents in pancreatic B-cells. Naunyn Schmiedebergs Arch Pharmacol 337:225–230

CHAPTER 12

G Protein-Gated K^+ Channels

A. INANOBE and Y. KURACHI

A. Introduction

Upon stimulation of vagal nerves, "Vagusstoff," which was afterwards identified as acetylcholine (ACh), is released from the axonal terminals of the vagal nerve and decelerates the heart beat. This historical discovery by OTTO LOEWI in the 1920s, established the concept of synaptic chemical transmission (LOEWI 1921; LOEWI and NAVARATIL 1926). Since then, many physiologists have been trying to elucidate the mechanisms underlying ACh-induced bradycardia. DEL-CASTILLO and KATZ (1955) described hyperpolarization of the membrane induced by ACh in frog heart. HUTTER and TRAUTWEIN (1955) measured an increase of the K^+ efflux across the cardiac cell membrane under vagal stimulation. TRAUTWEIN and DUDEL (1958) showed an increase of K^+ conductance under the voltage clamp condition. TRAUTWEIN and his colleagues (NOMA and TRAUTWEIN 1978; OSTERRIEDER et al. 1981) further analyzed the relaxation kinetics of the ACh-induced K^+ current in the rabbit sinoatrial node and proposed that ACh induces activation of a specific population of K^+ channels, named muscarinic K^+ (K_{ACh}) channels, to decelerate the pacemaker activity. The single channel currents of the K_{ACh} channels were recorded for the first time by SAKMANN et al. (1983), who showed that the channel exhibits an inwardly rectifying property but gating kinetics different from that of the background inwardly rectifying K^+ (I_{K1}) channel in cardiac myocytes. In 1985–6, it was discovered that pertussis toxin (PTX)-sensitive heterotrimeric G proteins are involved in the activation of the K_{ACh} channel by m_2-muscarinic and A_1-purinergic receptors (PFAFFINGER et al. 1985; BREITWIESER and SZABO 1985; KURACHI et al. 1986a,b,c). Because the K_{ACh} channel could be activated by intracellular GTP (in the presence of agonists) and GTPγS (even in the absence of agonists) in cell-free inside-out patches (KURACHI et al. 1986a,b,c), the system is delimited to the cell membrane, leading to the proposal that the channel is directly activated by G protein subunits. The G protein responsible for activation of K_{ACh} channels was designated G_K according to its function (BREITWIESER and SZABO 1985).

It was quite a surprise that the $\beta\gamma$ subunit ($G_{\beta\gamma}$), but not the α subunit (G_α) of G proteins, were proposed to mediate the G_K-induced activation of K_{ACh} channels (LOGOTHETIS et al. 1987, 1988; KURACHI et al. 1989), because it was strongly believed in those days that regulation of various effectors by G proteins is mediated by only G_α, while $G_{\beta\gamma}$ merely binds to the GDP-form of G_α ($G_{\alpha\text{-GDP}}$) in order to anchor the trimeric G proteins to cell membrane (GILMAN 1987). Actually, Brown, Birnbaumer and their colleagues strongly proposed that $G_{K\alpha}$, but not the $G_{K\beta\gamma}$, is the physiological activator of K_{ACh} channels (YATANI et al. 1987, 1988; CODINA et al. 1987; for review see BROWN and BIRNBAUMER 1990). The dispute between the two proposals continued for nearly a decade (ITO et al. 1992; YAMADA et al. 1993, 1994; NANAVATI et al. 1990; KURACHI 1989, 1990, 1993, 1994, 1995; KURACHI et al. 1992; WICKMAN and CLAPHAM 1995). The functional interaction between the channel and $G_{\beta\gamma}$, but not G_α, was further confirmed at molecular level with the cloned G protein-gated K (K_G) channel and/or G protein subunits (KUBO et al. 1993b; DASCAL et al. 1993; WICKMAN et al. 1994; REUVENY et al. 1994; INANOBE et al. 1995b; KRAPIVINSKY et al. 1995a). Now it is well established that $G_{K\beta\gamma}$ is the physiological activator of K_G channels not only in cardiac myocytes, but also in neurons and endocrine cells (KURACHI 1995). Recently, it was indicated that G protein-inhibition of neuronal Ca^{2+} channel is also mediated by $G_{\beta\gamma}$ but not by G_α (IKEDA 1996; HERLITZE et al. 1996). The current efforts are now being made to elucidate the molecular mechanisms underlying $G_{\beta\gamma}$-control of K_G and N-type Ca^{2+} channels.

The importance of the G protein-activation of K_G channel system in receptor-mediated regulation of cell responses is now more appreciated than before, because a wide variety of membrane receptors, such as m_2-muscarinic, A_1-purinergic, α_2-adrenergic, D_2-dopamine, μ-, δ-, and κ-opioid, 5-HT_{1A}-serotonin, somatostatin, galanin, m-Glu, and $GABA_B$ receptors, have been shown to utilize this system in inhibiting cell excitation in the brain and various endocrine organs, in addition to the heart (NORTH et al. 1987; LACEY et al. 1988; HILLE 1992; GRUDT and WILLIAMS 1993). Recent rapid progresses in cloning and functional analyses of K_G channel molecules will further uncover the yet unknown functional roles of these molecules in various organs. In this chapter, we will first briefly summarize the acetylcholine-activation of cardiac K_{ACh} channels, the prototype of this system, and then recent progresses in molecular dissection of the K_G channel system.

B. Acetylcholine-Activation of Muscarinic K^+ Channels

Acetylcholine (ACh) added to the extracellular solution elicits an inwardly rectifying K^+ current in cardiac atrial myocytes (Fig. 1). The activation time-course is sigmoidal. It takes several hundreds of milliseconds before the current reaches a peak. With high concentrations of ACh (>0.3 μmol/l), the evoked current gradually decreases to a quasi-steady state level within 1 min after the peak. This is called "short-term" desensitization (KURACHI et al. 1987). After wash-out of the agonist from the bathing solution, the ACh-induced K^+ current quickly disappears within several seconds (deactivation). Many molecules are involved in these three phases of the ACh-response. The activation

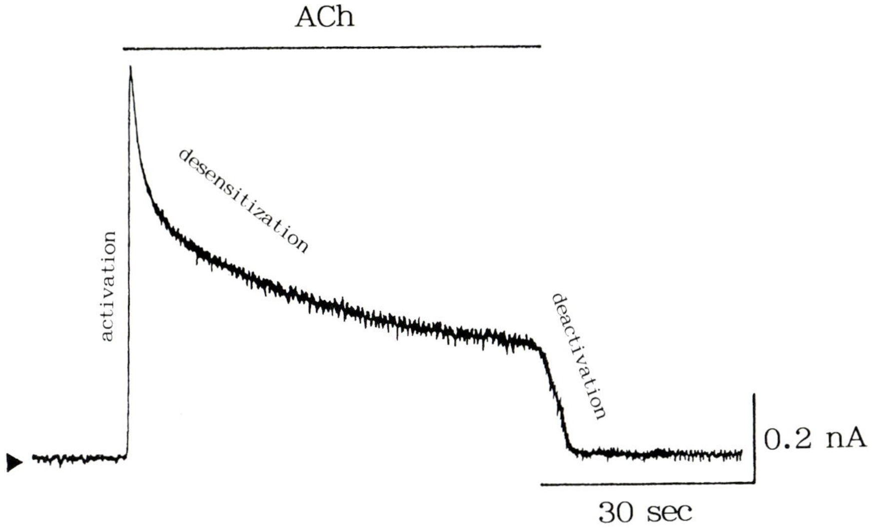

Fig. 1. Time-dependent response of the whole-cell muscarinic K^+ channel current to acetylcholine. By using the whole-cell clamp method of the patch-clamp technique, the response of the whole-cell current of a guinea-pig atrial myocyte to 11 μmol/l acetylcholine (ACh) was measured. In the presence of 5.4 mmol/l external K^+, the cell was held at –53 mV. ACh was applied to the bath for the duration indicated by a *horizontal bar* above the current trace. An *arrow head* indicates the zero current level

process includes an agonist (ACh), m_2-muscarinic receptor, a PTX-sensitive G protein, and K_{ACh} channel. The mechanism of the short-term desensitization is unclear but may include the G protein-mediated shift of m_2-muscarinic receptors from the high-affinity to the low-affinity binding states, phosphorylation of receptors, phosphorylation-related changes in gating behavior of K_{ACh} channels and alteration of G_K protein function. The deactivation may be largely influenced by the intrinsic GTPase activity of $G_{K\alpha}$ and its modulatory factors such as RGS (*R*egulator of *G* protein *S*ignaling) proteins (BERMAN et al. 1996; DRUEY et al. 1996; KOLLE et al. 1996; DOUPNIK et al. 1997; SAITOH et al. 1997) and intracellular anions (GILMAN 1987; NAKAJIMA et al. 1992).

Among three phases of K_{ACh} channel response, the main interests in this chapter will be the molecules of G protein-gated K^+ channels and their activation by G protein $\beta\gamma$ subunits. We will not deal with desensitization and deactivation mechanisms of ACh-response.

I. G Protein's Cyclic Reaction

The G protein-activation of K_{ACh} channel is mediated by $G_{K\beta\gamma}$. Therefore, the dissociation/association kinetics of G protein subunits must be taken into

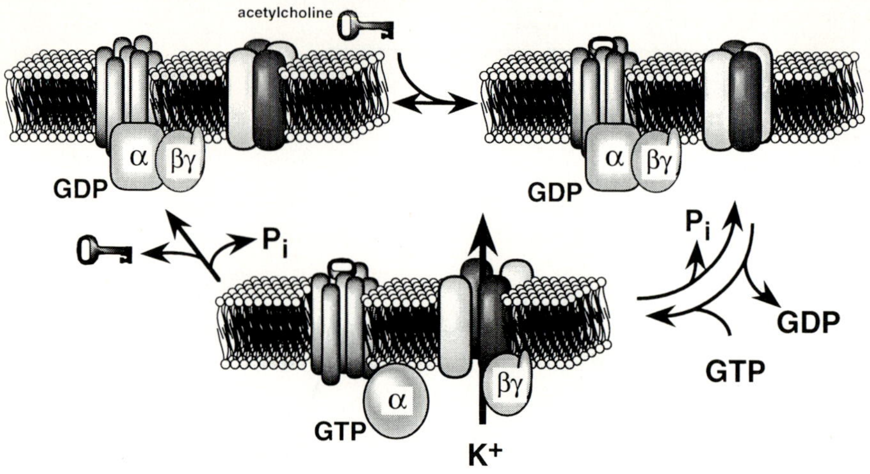

Fig. 2. G Protein cycle. Schematic representation of the G protein cycle. See text for details

consideration in the analysis of the receptor-mediated activation of K_{ACh} channels (Fig. 2) (GILMAN 1987). G proteins are membrane-bound proteins which transduce signals from receptors to various effectors, such as adenylyl cyclase, phospholipase C and ion channels. These proteins are heterotrimers composed of α, β, and γ subunits (G_α, G_β, and G_γ, respectively). Around 20 G_α, 5 G_β, and 7 G_γ have been identified (CLAPHAM and NEER 1993; SIMON et al. 1991). In the absence of an agonist, GDP is bound to G_α (Fig. 2). Upon binding of an agonist on a receptor, G_α releases GDP and instead binds GTP. This reaction in turn causes the dissociation of the GTP-bound form of G_α ($G_{\alpha\text{-GTP}}$) from $G_{\beta\gamma}$. $G_{\beta\gamma}$ is always a dimer under the physiological condition. $G_{\beta\gamma}$ activates K_{ACh} channels. In the other signaling systems, either or both of $G_{\alpha\text{-GTP}}$ and $G_{\beta\gamma}$ regulate effectors (Table 1) (CLAPHAM and NEER 1993; IÑIGUEZ-LLUHI et al. 1993; KURACHI 1995). G_α has intrinsic GTPase activity and, when GTP on G_α is hydrolyzed to GDP, the GDP-bound form of G_α ($G_{\alpha\text{-GDP}}$) reassociates with $G_{\beta\gamma}$ and forms the heterotrimeric form, resulting in the cessation of effector regulation.

One of the goals of the physiological studies of this system is to establish the functional model to explain the behavior of K_{ACh} channels activated by an agonist, ACh, via G proteins. For this purpose, we have analyzed this system by dividing it into two steps. One is to construct a functional model for the G protein subunit ($G_{K\beta\gamma}$) activation of K_{ACh} channel molecules (HOSOYA et al. 1996), and the second step is to incorporate the receptor-G protein interaction to this functional model, although the second step is still in the middle of progress (HOSOYA and KURACHI 1998).

Table 1. Effector subtype-specific effect of G protein subunit

Effector	Receptor	G Proteins	G Protein Subunits	
			G_α [a]	$G_{\beta\gamma}$ [b]
Ion channels				
Cardiac Kchannels				
K_{ACh} channel	Cardiac m_2, A_1	PTX-sensitive G_K	0 (= No effect)	Activation
K_{ATP} channel	Cardiac m_2, A_1	PTX-sensitive G (G_i ?)	Activation	0
Ca Channels				
N-type (neurons)	Various receptors	PTX-sensitive G (G_o)	0	Inhibition
L-type (endocine)	GH_3 cell m_4	G_{o1}	0	Inhibition
	GH_3 cell somatostatin	G_{o2}	0	Inhibition
Enzymes				
Retinal PLA_2		Transducin	0	Activation
Adenylyl cyclases				
Type I	(Brain)	G_s	Stimulation	Inhibition
Type II	(Brain, lung)		Stimulation	Further stimulation
Type III	(Olfactory)		Stimulation	0
Type IV	(Brain others)		Stimulation	Further stimulation
Type V	(Heart, brain, others)		Stimulation	0
Type VI	(Heart, brain, others)		Stimulation	0
Phospholipase C				
PLCβ1	G_q family	Stimulation	± Weak	
PLCβ2			Stimulation	Stimulation
PLCβ3			Stimulation	Stimulation
MAP kinase (ras-dependent pathway)		G_i, G_q family	0	Stimulation
βARK		?	0	Essential for translocation
cGMP PDE		Transducin	Stimualtion	0
Phosphoinoside 3 kinase (a novel subtype)			0	Activation
Unknown effectors				
Pheromon-induced mating (yeast)		?	0	Stimulation
Oocyte maturation (starfish)	?	0	Stimulation	

[a] Note: Actions of G_α and $G_{\beta\gamma}$ are independent.
[b] $G_{\beta\gamma}$ itself did not have any effect on AC. Each subtype of AC should be prestimulated by $G_{s\alpha\text{-}GTP\gamma S}$ before application of $G_{\beta\gamma}$.

II. Positive Cooperative Effect of GTP on the Muscarinic K^+ Channel Activity

The activation of K_{ACh} channels by intracellular GTP (GTP_i) can be reproduced in the inside-out patch of cardiac atrial cell membrane in the presence of ACh in the pipette (KURACHI et al. 1986a,b,c). Figure 3 shows the concentration-dependent effect of GTP_i in the presence of various concentrations of ACh in the extracellular solution (i.e., in the pipette) (ITO et al. 1991).

GTP_i activates the K_{ACh} channel in a highly positive cooperative manner: i.e., the Hill coefficient is around 3 (KURACHI et al. 1990; ITO et al. 1991; KARSHIN et al. 1991; YAMADA et al. 1993). When the concentration of extracellular ACh was increased; (a) the threshold concentration of GTP_i for the K_{ACh} channel activation decreased; (b) the half-maximum effective concentration of GTP_i decreased; (c) the maximum channel activity increased; and (d) the Hill coefficient remained constant around 3 (Fig. 3). These results may indicate that receptor stimulation by ACh increases both the efficacy and potency of GTP in activating the K_{ACh} channel without changing the cooperativity. This is probably due to the facilitation of the subunit dissociation of G_K by the agonist. The cooperativity may arise from an intrinsic nature of the interaction between $G_{K\beta\gamma}$ and the K_{ACh} channel. One may intuitively speculate that the positive cooperativity is caused by the binding of multiple number of $G_{K\beta\gamma}$ to each channel.

The functional interaction between the $G_{K\beta\gamma}$ and the cardiac K_{ACh} channel was first analyzed based on the concentration-dependent effect of GTP_i on channel activity in the presence of supermaximum concentrations of ACh (HOSOYA et al. 1996). Under these conditions, $G_{\beta\gamma}$ exogenously applied to the internal side of inside-out patch membranes did not further increase channel activity induced by more than 1 μmol/l of GTP_i. Therefore, the maximum channel activity under these conditions may be determined by the number of K_{ACh} channels but not by that of $G_{K\beta\gamma}$ available in the patch membranes. The channel activity and kinetics in the presence of each $[GTP]_i$ were analyzed with the spectral analysis because multiple K_{ACh} channels were usually included in a single I-O patch membrane of atrial myocytes. The power density spectra were well fitted with the sum of two Lorenzian functions at various $[GTP]_i$. Because the channel has one open state (SAKMANN et al. 1983; KURACHI et al. 1986a,b), the open-close transitions of the channel gate represented by the spectra can be described as $C2 \leftrightarrow C1 \leftrightarrow O$. When the channel activity was increased as $[GTP]_i$ was progressively raised, the powers of the two Lorenzian components increased, while the corner frequencies and the ratio of the powers at 0 Hz remained almost constant. Therefore, G protein-activation does not affect the gating of each channel but mainly increases the number of functionally active channels in a patch. Such regulation can be described with a slow transition of the two distinct channel states, U (unavailable) $\leftrightarrow$ A (available), which is practically independent of the gating. The equilibrium of this slow transition is shifted from U to A by GTP_i,

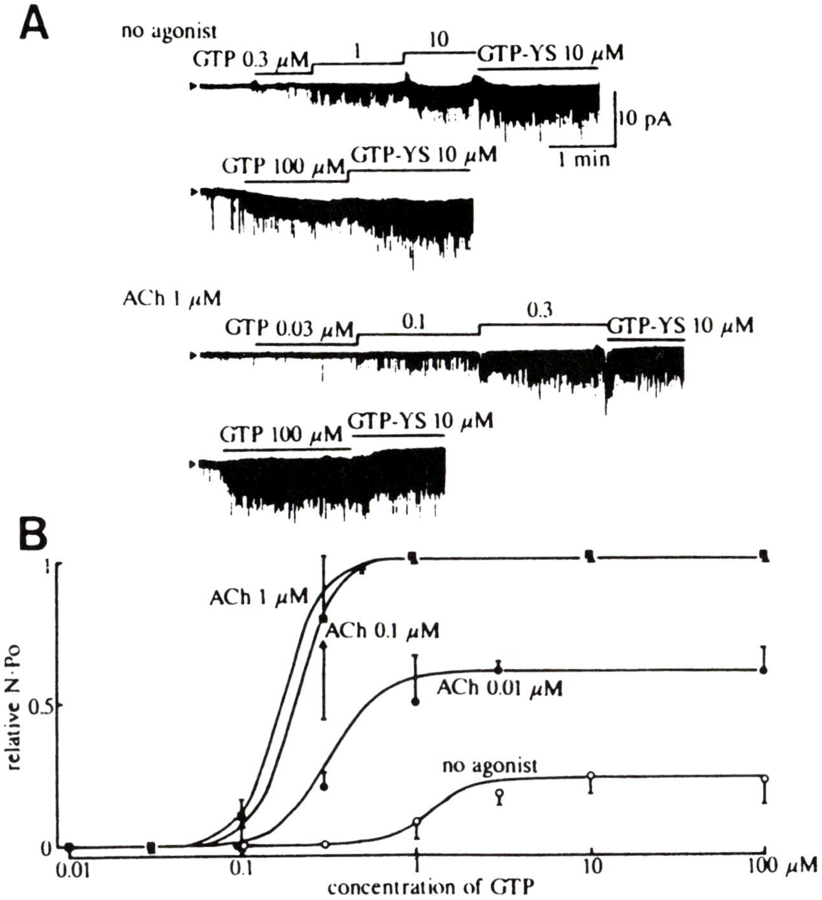

Fig. 3A,B. Concentration-dependent effect of intracellular GTP on the muscarinic channel in the absence and presence of acetylcholine. **A** examples of inside-out patch experiments obtained with guinea pig atrial myocytes. The concentration of acetylcholine (ACh) in the pipette was 0 μmol/l or 1 μmol/l as indicated. The *bars* above each trace indicates the protocol of perfusing various concentrations of GTP and 10 μmol/l GTPγS. The holding potential was –80 mV. Note that a 3–10-fold increase in GTP concentration resulted in a dramatic increase of $N{\bullet}P_o$ of K_{ACh} channels, indicating the existence of a highly cooperative process. **B** the relation between the concentration of GTP and the relative $N{\bullet}P_o$ of K_{ACh} channels with reference to the maximum $N{\bullet}P_o$ induced by 10 μmol/l GTPγS in each patch. Symbols and bars are mean±SD. 0 μmol/l ACh (*open circles*, $n = 7$), 0.01 μmol/l ACh (*closed circles*, $n = 6$), 0.1 μmol/l ACh (*closed triangles*, $n = 6$), 1 μmol/l ACh (*closed squares*, $n = 6$). The relationship between GTP and channel activity at each concentration of ACh (*continuous lines*) was fitted by the Hill equation with use of the least-squares method: $f = V_{max}/\{1+(K/[GTP]_n\}$ where f = the relative $N{\bullet}P_o$, V_{max} = the maximum N•Po, K = the half-maximum GTP concentration, and n = the Hill coefficient. The channel activity was expressed as $N{\bullet}P_o$, where N is the number of the channel in the patch and P_o is the open probability of each channel. Reproduced from Ito et al. (1991) with permission

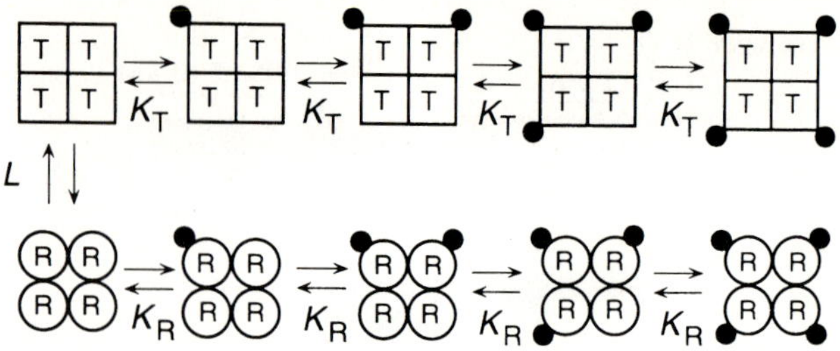

Fig. 4. Concerted allosteric model of Monod, Wyman and Changeux. Two different states of the protomers, tense (T) and relaxed (R) states, are represented by *squares* and *circles*, respectively (MONOD et al. 1965). The latter has a higher affinity with the activated G protein subunit (G_{K*}), which is represented by a *small solid circle*. In this illustration the K_{ACh} channel is supposed to be an oligomeric protein composed of four functionally identical protomers. Reproduced from HOSOYA et al. (1997) with permission

i.e., $G_{K\beta\gamma}$. Monod-Wyman-Changeux's (MWC) allosteric model (MONOD et al. 1965) (Fig. 4) for the channel state transition ($U \leftrightarrow A$) can well describe the positive cooperative increase in the channel availability by GTP_i, assuming that the concentration of $G_{K\beta\gamma}$ linearly increases in the membrane in the presence of physiological range of concentrations of GTP_i (HOSOYA et al. 1996). The model indicates that the cardiac K_{ACh} channel can be described as a multimer composed of four or more functionally identical subunits, to each of which one $G_{K\beta\gamma}$ might bind.

III. Incorporation of Receptor-G Protein Reaction to the Model of K_{ACh} Channel

The next step is to incorporate the concentration-dependent effect of ACh into the model (HOSOYA and KURACHI 1998). For the receptor-G protein cycle, two models have been proposed by THOMSEN et al. (1988) and MACKAY (1990a) (Fig. 5A). At each [ACh], the channel activity was calculated based on the assumption that (1) each K_{ACh} channel is activated by $G_{K\beta\gamma}$ as described by the MWC model, and (2) each $G_{K\beta\gamma}$ is supplied as described by either Thomsen or Mackay. The results are depicted in Fig. 5B. With both models; we could reproduce the major characteristics of the concentration-response effect of GTP on K_{ACh} channel in the presence of various [ACh]; i.e., as [ACh] is increased, the V_{max} of K_{ACh} channel activity increases and K_d value of [GTP] decreases (Fig. 3). However, there exist some differences between the models (HOSOYA and KURACHI 1998).

The concentration-dependent activation of K_{ACh} channels by GTP_i in the presence of 0.01, 0.1, or 1 μmol/l ACh were well fitted with Thomsen's model. Even in the absence of ACh, high concentrations of GTP_i can induce channel activity in the inside-out patches of cardiac atrial cell membrane in the

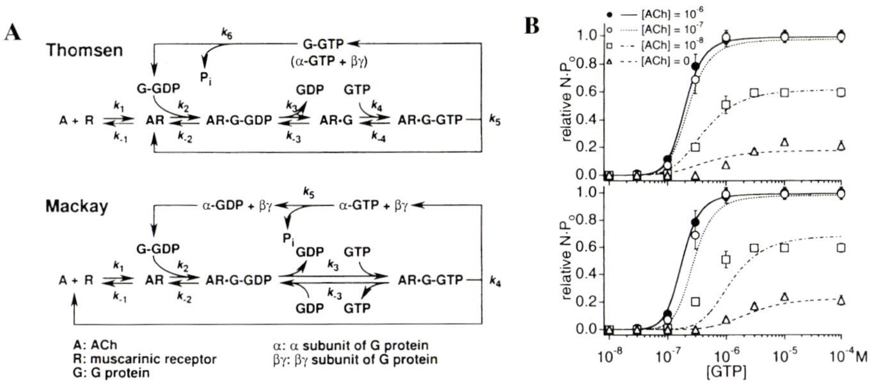

Fig. 5. A Thomsen's and Mackay's models for receptor-G protein reaction. **B** fitting used Mackay's and Thomsen's models. See text for details

internal solution containing chloride ions. This activity may be caused by the basal turn-on reaction of G_K stimulated by receptors even in the absence of agonist and the retarded turn-off reaction by intracellular chloride ions (Ito et al. 1991; Nakajima et al. 1992). The relation of agonist-independent background channel activity induced by GTP_i was well fitted by Mackay's but not by Thomsen's model (Fig. 5B).

By combining the MWC allosteric model for $G_{K\beta\gamma}$-activation of K_{ACh} channels and either of Thomsen's or Mackay's model for ACh-activation of G proteins, the concentration-response relationships between GTP_i and K_{ACh} channel activity in the presence of various concentrations of ACh could be reasonably simulated (Hosoya and Kurachi 1998). However, in this model we could not reconstitute either the rapid activation upon ACh-application or the quick deactivation after ACh-washout. Therefore, further improvement of the model is clearly needed to provide a functional basis to clarify the molecular mechanism underlying the receptor/G-protein/K_{ACh} channel interaction. The result, however, suggests a possibility that a multiple number (four or more) of m_2-receptors may be involved in activation of one functional K_{ACh} channel. Further studies using molecular biological techniques are needed to determine the stoichiometry.

C. Molecular Analyses of G Protein-Gated K⁺ Channels

I. Cloning of Inwardly Rectifying K⁺ Channels and Kir Subunits for G Protein-Gated K⁺ Channels

In 1993, the molecular structure of inwardly rectifying K⁺ channels (Kir) was disclosed. The cDNAs encoding ATP-dependent Kir channel, ROMK1/Kir1.1 (Ho et al. 1993), and a classical Kir channel, IRK1/Kir2.1 (Kubo et al. 1993a), were isolated by the expression cloning technique from the outer medulla of rat kidney and a mouse macrophage cell line, respectively. They have a common molecular motif in the primary structure: i.e., two putative membrane-spanning regions (M1 and M2) and one potential pore-forming region (H5). The primary structure of these Kir channel subunits resembles that of the S5, H5 and S6 segments of the voltage-gated K^+ (K_v) channels. Because the voltage-sensor of K_v channel subunit exists in its S4 segment which possesses repeated positively-charged amino acid residues, the Kir channel subunits lack the voltage-sensor region. This is consistent with the results obtained from electrophysiological studies that the kinetics of Kir channels apparently depends on the shift of the membrane potential from E_K but not on the membrane potential itself. After the cloning of ROMK1 and IRK1, the cDNAs encoding the main subunits of K_G and K_{ATP} channels have also been cloned (GIRK1/Kir3.1 and uK_{ATP}–1/Kir6.1) (KUBO et al. 1993b; DASCAL et al. 1993; INAGAKI et al. 1995b). All of these Kir channel subunits exhibit basically the same primary structure. So far, more than ten cDNAs encoding Kir channel subunits have been isolated.

The evolutionary tree of this family is depicted in Fig. 6. These cloned Kir subunit cDNAs encode proteins composed of 327–501 amino acids. The identity of the predicted amino acid sequence is 30~40% among the members of different Kir subfamilies and more than 60% among those in the same subfamilies. The highest level of sequence identity (50–60%) is found in the H5 region and the proximal part of the C-terminal cytosolic domain. These cloned Kir channel subunits can be classified into four groups (DOUPNIK 1995a):

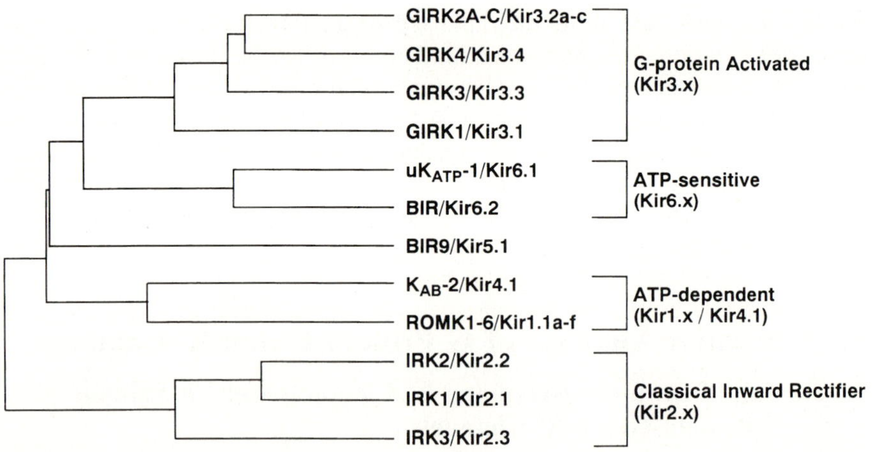

Fig. 6. Evolutionary tree of Kir subunits. The tree was made using the UPGMA (Unweighted Pair Group Method with Arithmetic Mean) Tree Window in Geneworks (IntelliGenetics, Inc., Mountain View, CA)

1. IRK (Kir2.0) subfamily, classical Kir channels: IRK1/Kir2.1 (KUBO et al. 1993a; MORISHIGE et al. 1993), IRK2/Kir2.2 (KOYAMA et al. 1994; TAKAHASHI et al. 1994) and IRK3/Kir2.3 (MORISHIGE et al. 1994; MAKHINA et al. 1994; PÉRIER et al. 1994).
2 GIRK (Kir3.0) subfamily, G protein-activated K^+ channels: GIRK1/Kir3.1 (KUBO et al. 1993b, DASCAL et al. 1993), GIRK2/Kir3.2 (LESAGE et al. 1994, 1995; ISOMOTO et al. 1996), GIRK3/Kir3.3 (LESAGE et al. 1994) and GIRK4/CIR/Kir3.4 (KRAPIVINSKY et al. 1995a,b), and GIRK5/XIR/Kir3.5 (HEDIN et al. 1996). GIRK5 was cloned from *Xenopus* oocytes, and no mammalian homologs of GIRK5 have been reported.
3. ROMK subfamily, ATP-dependent K^+ channels: ROMK1/Kir1.1 (HO et al. 1993) and KAB-2/BIR10/Kir4.1/Kir1.2 (BOND et al. 1994; TAKUMI et al. 1995).
4. K_{ATP} subfamily, ATP-sensitive K^+ channels: uK_{ATP}–1/Kir6.1 and BIR/Kir6.2 (INAGAKI et al. 1995a, b; SAKURA et al. 1995).

Recent progress in the molecular biology of Kir channels enables us to understand the structure-function relationship of biophysics, physiological regulation, and pharmacology of these channels at the molecular level.

II. GIRK Subfamily

GIRK1/Kir3.1, which encodes the main subunit of K_G channels, was first isolated from rat atrium (KUBO et al. 1993b; DASCAL et al. 1993). From a mouse brain cDNA library, two additional homologs of GIRK1 were further isolated and designated GIRK2/Kir3.2 and GIRK3/Kir3.3 (LESAGE et al. 1994). Furthermore, it has been shown that at least three different isoforms of mouse GIRK2 are generated by alternative splicing of transcripts from a single gene, and we designated them as GIRK2A/Kir3.2a, GIRK2B/Kir3.2b and GIRK2C/Kir3.2c in the order of identification (ISOMOTO et al. 1996) (Fig. 7). GIRK2A and GIRK2C correspond to GIRK2 and GIRK2A designated by LESAGE et al. (1994), respectively. These alternatively spliced transcripts share the N-terminal end and the central core, but differ at their C-terminal ends. GIRK2B was isolated from mouse brain cDNA library and shown to be ubiquitously expressed in various tissues (ISOMOTO et al. 1996). Its amino acid sequence is shorter than that of GIRK2A by 87 amino acids, and the 8 amino acid residues in the C-terminal end of GIRK2B are different from those of GIRK2A. GIRK2C has a C-terminus which is longer than that of GIRK2A by 11 amino acids. GIRK2C was isolated from cDNA libraries of insulinoma cells and brain (LESAGE et al. 1994, 1995; TSAUR et al. 1995; STOFFEL et al. 1995; BOND et al. 1995; FERRER et al. 1995).

The GIRK clones contain various known functional motifs in their amino acid sequences, which may be important for the physiological functions of the subunits in the K_G channels (Fig. 7). For examples, GIRK1 possesses an amino

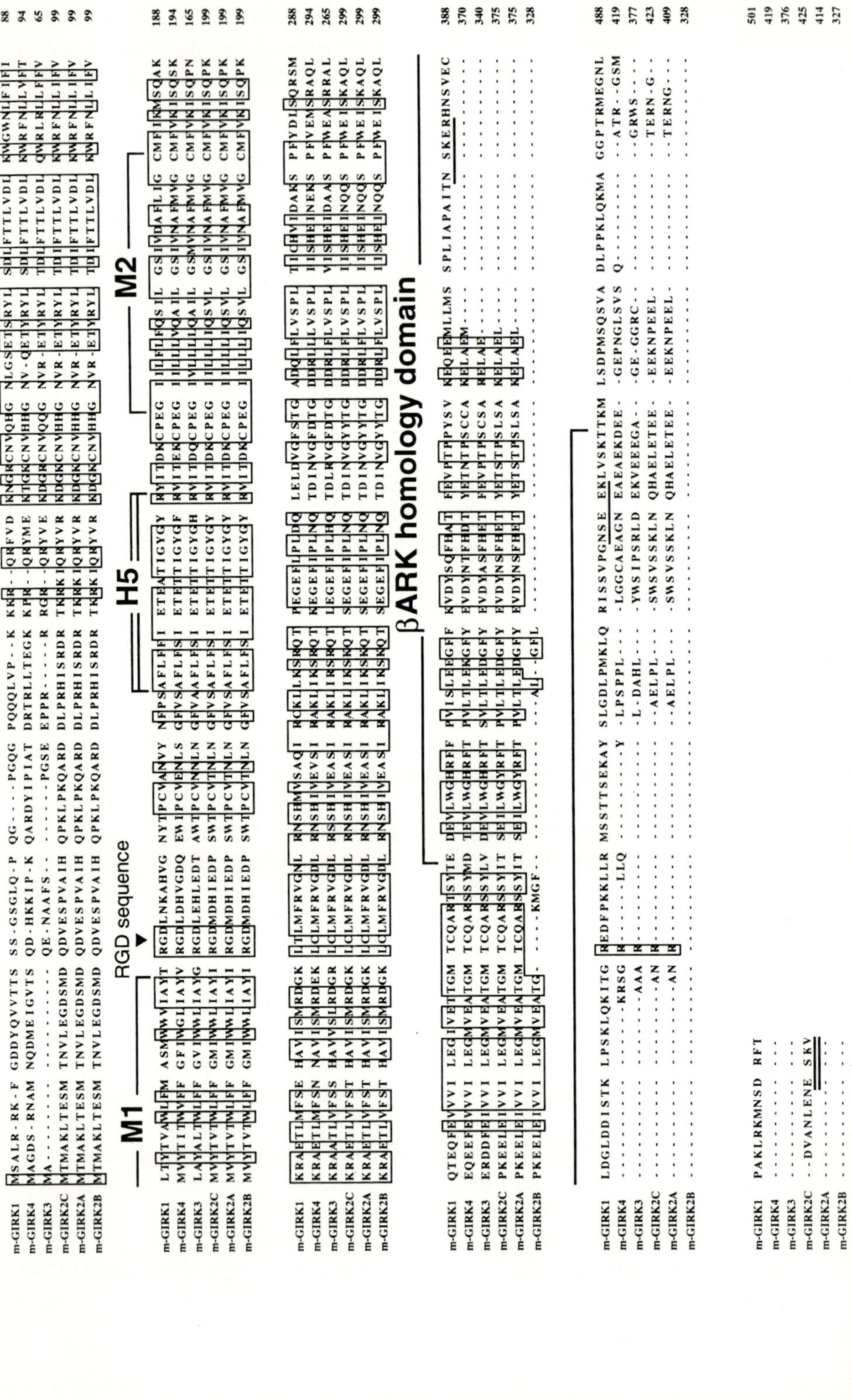
M1
M2
H5
RGD sequence
βARK homology domain
m-GIRK1
m-GIRK4
m-GIRK3
m-GIRK2C
m-GIRK2A
m-GIRK2B

acid sequence homologous to the $G_{\beta\gamma}$-binding domain of βARK1 in their C-terminus, which is the candidate for the site for $G_{\beta\gamma}$-binding (Reuveny et al. 1994). All of the GIRK clones have an arginine-glycine-aspartate (RGD) motif in their linker region between M1 and H5. This motif can be an integrin receptor-site (Hynes et al. 1992), whose role in K_G channels has not yet been examined. The characteristic feature of GIRK2C is the serine/threonine-X-valine (S/T-X-V) motif at its C-terminus end (Gomperts 1996). This motif has been shown to be important for channel's interaction with PSD-95/SAP90 family of anchoring proteins, not only for K_v or NMDA receptor channels (Kim et al. 1995; Kornau et al. 1995) but also for Kir channels such as IRK3 and K_{AB}–2 (Cohen et al. 1996; Horio et al. 1997).

III. Expression of GIRK Channels

When cRNAs for GIRK1 and m_2-muscarinic receptor are coinjected into *Xenopus* oocytes, Kir currents induced by ACh were observed (Kubo et al. 1993b; Dascal et al. 1993). This current well mimicked at least some of the characteristics of the K_{ACh} channel current. GIRK1 expressed in *Xenopus* oocytes has, therefore, been successfully used to investigate the structure-function relationship of K_G channels (Reuveny et al. 1994; Slesinger et al. 1995; Kofuji et al. 1996a). Either GIRK2A or GIRK4 could also form functional K_G channels when expressed alone in *Xenopus* oocytes (Lesage et al. 1994; Bond et al. 1995). Therefore, some K_G channels might be homomultimers of these GIRK subunits. A recent work proposed that the K_{ACh} channel in cardiac atria is a heteromultimer of GIRK1 and GIRK4/CIR rather than a homomultimer of GIRK1 (Krapivinsky et al. 1995a). When GIRK4 was coexpressed with GIRK1 in *Xenopus* oocytes or CHO cells, the K_G channel current was prominently enhanced compared with that observed when GIRK1 was expressed alone (Fig. 8A). When K_G channels in atrial membranes were immunoprecipitated with an antibody against GIRK1, GIRK4 was coim-

Fig. 7. Alignment of amino-acid sequences of GIRK1, GIRK2A, B, C, GIRK3, and GIRK4. Positions at which all six amino acid sequences are identical are *boxed*. The putative transmembrane segments (M1 and M2) and pore-forming region (H5) are indicated above the sequences. βARK homology domain indicates the amino acid sequence of GIRK1 which is homologous to the G protein $\beta\gamma$ subunit-binding site in β-adrenergic receptor kinase 1; RGD sequence: the sequence observed in the integrin-binding site of fibronectin, vitronectin, and a variety of other adhesive proteins. The *two underlined* sequences of GIRK1 are similar to that included in a region of adenylyl cyclase 2 which is critical for activation of the enzyme by G protein $\beta\gamma$ subunits (asparagine-X-X-glutamate-arginine). The *double-underlined* sequence of GIRK2C includes the consensus sequence (serine/threonine-X-valine/isoleucine) for interaction with PSD-95/SAP90 anchoring protein. The glutamate prior to the consensus sequence is also proposed to be important for the interaction of Shaker-type K+ channels with-these anchoring proteins

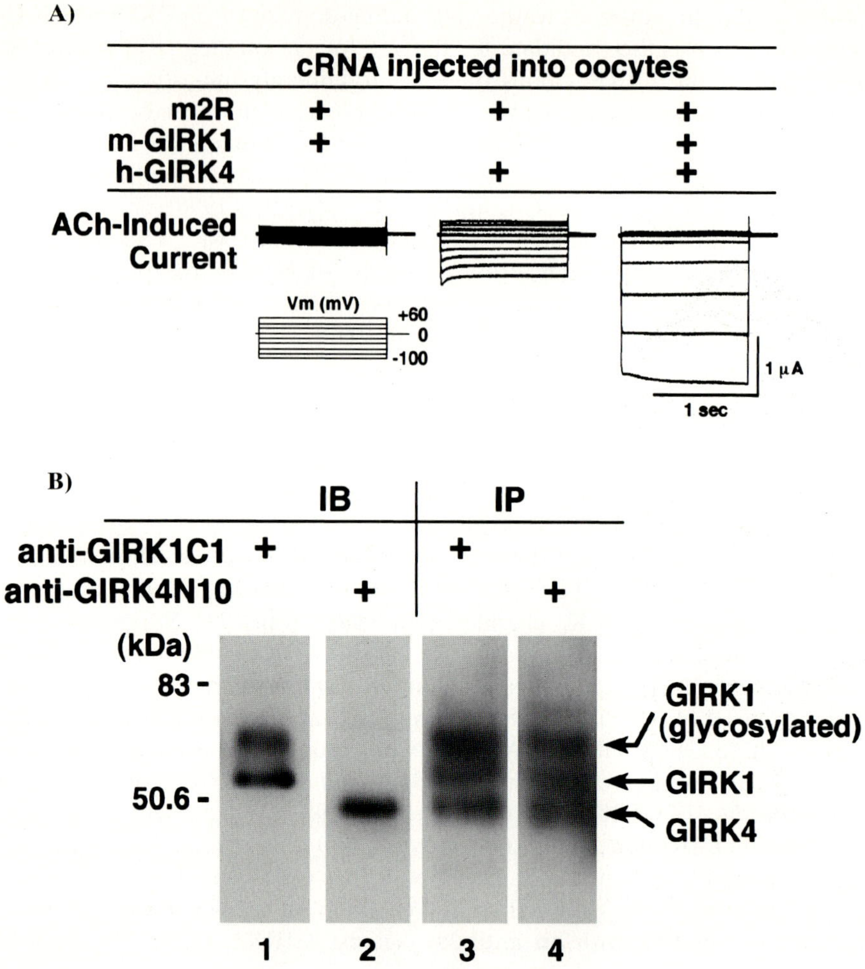

Fig. 8A,B. Heteromultimeric G protein-gated K^+ channel expressed in *Xenopus* oocytes and atrial myocytes. **A)** acetylcholine (ACh)-induced K^+ currents observed in *Xenopus* oocytes expressing m_2-muscarinic receptors (m_2R) plus mouse GIRK1 (m-GIRK1) and/or human GIRK4 (h-GIRK4). The cRNA of the m_2-muscarinic receptor was injected into oocytes with that of GIRK1 and/or GIRK4 as indicated in the table. ACh (1 μmol/l)-induced K^+ currents at different membrane potential were measured in the presence of 96 mmol/l external K^+ and are shown under the table. The voltage-clamp protocol is depicted at the *left lower corner*. **B)** immunological analyses for mouse atrial K_G channel. Atrial membrane proteins were immunoblotted with anti GIRK1C1 (*lane 1*) and anti-GIRK4N10 (*lane 2*) antibodies specific to GIRK1 and GIRK4 proteins, respectively. Some part of GIRK1 protein, but not GIRK4, appeared to be glycosylated in the atrium. Immunoprecipitants of both antibodies were compared. Both immunocomplexes of anti-GIRK1C1 (*lane 3*) and anti-GIRK4N10 (*lane 4*) seemed to be comprised of almost three proteins which had molecular weights identical to GIRK1 (glycosylated and non-glycosylated) and GIRK4 proteins on the gel munoprecipitated, and *vice versa* with the antibody for GIRK4 (Fig. 8B). It was further suggested that the efficient functional expression of GIRK1 in some *Xenopus* oocytes might be due to the endogenously expressed GIRK subunit in oocytes, XIR (GIRK5), whose amino acid sequence is 78% homologous to that of GIRK4 (HEDIN et al. 1996). Moreover, coinjection of GIRK1 plus either of GIRK2A-C, GIRK1 plus GIRK3, and GIRK2A plus GIRK4 into *Xenopus* oocytes also resulted in prominent enhancement of current expression, although coexpression of GIRK3 with GIRK2A suppressed the expression of the GIRK2A channel current (FERRER et al. 1995; LESAGE et al. 1995; KOFUJI et al. 1995; DUPRAT et al. 1995; VELIMIROVIC et al. 1996; ISOMOTO et al. 1996). These data indicate that K_G channels in various tissues can be either homo- or heteromultimer of GIRK subunits.

Differential distribution of the mRNAs for GIRK subunits has been shown. In northern blot analyses, GIRK1 is mainly expressed in heart and brain, GIRK2A and GIRK3 in brain, GIRK2C in brain and pancreas, and GIRK4 in heart (KUBO et al. 1993b; DASCAL et al. 1993; KRAPIVINSKY et al. 1995a; ASHFORD et al. 1994). *In situ* hybridization and the reverse transcription-polymerase chain reaction analyses have also revealed diversity in their distribution patterns, especially within the brain (KOBAYASHI et al. 1995; KARSCHIN et al. 1994, 1996; DIXON et al. 1995, SPAUSCHUS et al. 1996). Thus, the different combinations of GIRK subunits expressed in different tissues may generate many distinct types of K_G channels (see also Sect. D).

It has not been fully understood why many K_G channels need to be formed as heteromultimers of different GIRK subunits. Recent studies on GIRK1 have disclosed specific characteristics of the clone different from other members, which may provide some explanations for the heteromultimeric assembly of GIRK1 with GIRK2 or GIRK4. First KENNEDY et al. (1996) expressed epitope-tagged GIRK1 and GIRK4 in COS cell alone or in combination, and examined the localizations of the subunits with immunofluorescence labeling. GIRK1 alone appeared to be associated with the intracellular intermediate filament protein but not with the plasma membrane. It was detected on the plasma membrane when co-expressed with GIRK4. Therefore, GIRK4 may have a function to promote the translocation of GIRK1 to the cell membrane. Then CHAN et al. (1996) reported that a functional expression of a homomeric GIRK1 channel is prevented because of the phenylalanine (F) at a.a. 137 in the H5 region. This phenylalanine residue exists only in GIRK1, while GIRK2–4 possess a conserved serine (S) at the corresponding position. They showed that GIRK1 whose phenylalanine 137 was replaced with serine (F137S) could form a functional homomeric channel which exhibited fast relaxation. They did not show whether this was due to the facilitated translocation of the subunit by the mutation to the plasma membrane. They further showed that GIRK4 whose serine 143 was replaced with phenylalanine (GIRK4(S143F)) behaved as a GIRK1 analog; i.e., co-expression of GIRK4(S143F) with GIRK2A or GIRK4 resulted in an enhanced channel activity. At a single channel level, a homomeric GIRK4 channel showed very short open time, while the channels composed of GIRK4 plus GIRK1 and GIRK4 plus GIRK4(S143F) exhibited a longer open time of ~1–3 msec. KOFUJI et al. (1996a) also identified phenylalanine at a.a. 137 in the H5 region of GIRK1 to be responsible for the slow relaxation. These data indicate that the K_G channels exhibit slow relaxation kinetics only when the phenylalanine and serine residues coexist at the analogous position in their putative channel pore region. Finally HUANG et al. (1997) reported that both of the N- and C-terminal domains of GIRK1–4 possess $G_{\beta\gamma}$-binding activity, and that the C-terminal domain of GIRK1 and the N-terminal domain of GIRK4 can physically interact with each other, which thereby synergistically enhances the $G_{\beta\gamma}$-binding activity. In addition, the C-terminal domain of GIRK1 appears to have higher $G_{\beta\gamma}$-binding activity than that of the other types of GIRK subunits. Therefore, heteromultimeric K_G channels may possess different $G_{\beta\gamma}$-sensitivities depending on their subunit compositions.

IV. Tetrameric Structure of Kir Channels

YANG et al. (1995) examined the subunit stoichiometry of IRK1 (Kir2.1) by linking multiple cDNAs encoding the coding region of IRK1/Kir2.1 in tandem in a head-to-tail fashion, and showed that the IRK1 channel is composed of four IRK1 subunits. Biochemical measurement of the molecular weight of brain K_G channel proteins was consistent with the notion that the channels also have tetrameric structure composed of GIRK subunits (INANOBE et al. 1995a). TUCKER et al. (1996) assessed the stoichiometry and relative subunit positions within the heteromeric K_G channel composed of GIRK1 and GIRK4 by coexpressing with $G_{\beta1\gamma2}$ these subunits as a tandemly linked tetramers with different relative subunit positions in *Xenopus* oocytes. They found that the most efficient channel comprises two subunits of each type in an alternative array within the tetramer. Through a similar approach, SILVERMAN et al. (1996) also found that the functional K_G channel composed of GIRK1 and GIRK4 has a stoichiometry of (GIRK1)2(GIRK4)2 and that more than one kind of arrangement, such as G1G4G1G4 and G1G1G4G4, may be viable.

TINKER et al. (1996) studied the mechanism of homomeric assembly of IRK1. They concluded that among IRK1, IRK2, and IRK3, the proximal C-terminus and the M2 region contribute to polymerization. The proximal C-terminus plays a more significant role in prevention of heteromultimerization between more distantly related channel subunits, such as IRK1 and ROMK1. TUCKER et al. (1996) also found that, for the subunit assembly between GIRK1 and GIRK4 and potentiation of the current by coexpression of these subunits, the core region of GIRK subunit (i.e., M1-H5-M2), but neither the C- nor N-terminal domain, is most important. Thus, the mechanism of heteromultimerization of GIRK subunits might not be the same as that of homomeric assembly of IRK subunits. Further studies are clearly needed.

V. Molecular Mechanism Underlying Activation of the G Protein-Gated K^+ Channels by $\beta\gamma$ Subunits of G Protein

1. The G Protein $\beta\gamma$ Subunit-Binding Domains in GIRK Subunits

GIRK1 has a significantly longer C-terminal domain than the other clones of the constitutively active Kir channels like IRK1 (Fig. 7). Thus, the GIRK1 C-terminus (a.a. 180–501) may have $G_{\beta\gamma}$-binding site(s). It was first pointed out that the C-terminal domain of GIRK1 includes an amino acid sequence (a.a. 317–455) which shows a limited level of similarity (~26%) with that of the $G_{\beta\gamma}$-binding site of the βARK1 (REUVENY et al. 1994). Indeed, truncation of the C-terminal domain of GIRK1 at leucine 403 but not at proline 462 resulted in loss of functional expression of a K_G channel in *Xenopus* oocytes expressing $G_{\beta1\gamma2}$. To examine a possible direct interaction between $G_{\beta\gamma}$ and the C-terminal domain of GIRK1, INANOBE et al. (1995b) analyzed the ability of a glutathione *S*-transferase (GST) fusion proteins of the whole C terminus of GIRK1 (a.a. 180–501) to bind $G_{\beta\gamma}$ in vitro. They showed that the fusion protein actually bound $G_{\beta\gamma}$ when incubated with purified brain $G_{\beta\gamma}$ or with trimeric G_i in the presence of GTPγS but not GDP. When incubated with the fusion protein and $G_{\beta\gamma}$, $G_{\alpha\text{-GDP}}$ but not $G_{\alpha\text{-GTP}\gamma}$S prevented the binding of $G_{\beta\gamma}$ to the fusion protein. Therefore, the C-terminal domain of GIRK1 has an ability to bind directly to free $G_{\beta\gamma}$ but not to the trimeric G protein. HUANG et al. (1995) also found direct binding of $G_{\beta\gamma}$ to GST-fusion proteins of the C-terminal domain of GIRK1. They constructed various C-terminal deletion fusion proteins and compared their abilities to interact with $G_{\beta\gamma}$. They narrowed down the $G_{\beta\gamma}$-binding region of GIRK1 to a 190 amino acid stretch (a.a. 273–462) in the C-terminal domain. $G_{\beta\gamma}$ bound to the GST-fusion protein of the C-terminus through ~1:1 stoichiometry with calculated K_d of ~0.5 μmol/l. They found in the same study that the fusion protein of the N-terminus of GIRK1 was also capable of binding $G_{\beta\gamma}$ through 1:1 reaction, although its affinity was ~10 times lower than that of the C-terminus (HUANG et al. 1995). Thus, $G_{\beta\gamma}$ may directly interact with both N- and C-terminal domains of GIRK1.

The functional significance of the $G_{\beta\gamma}$-binding to the N- and C-terminal domains of GIRK1 was indicated mainly by the following two observations. First, HUANG et al. (1995) showed that the synthetic peptides corresponding to the $G_{\beta\gamma}$-binding domain of either N- or C-terminals of GIRK1 inhibited the K_G channel current in *Xenopus* oocytes coexpressing GIRK1 and $G_{\beta1\gamma2}$. Each synthetic peptide also antagonized the binding of $G_{\beta\gamma}$ to the fusion proteins containing the N- or C-terminal $G_{\beta\gamma}$-binding site, respectively. Thus, it is suggested that the direct interaction of $G_{\beta\gamma}$ with the N- and C-terminal domains of GIRK1 is indispensable for activation of the K_G channel. The second evidence was derived from the studies using chimeras of GIRK1 and constitutively active Kir channel clones such as IRK1 and IRK2. The chimeras of IRK1 and GIRK1 could be activated by $G_{\beta1\gamma2}$ in *Xenopus* oocytes, when they contained either of the N-(a.a. 31–85) or the C-terminal (a.a. 325–501) domain of GIRK1 (SLESINGER et al. 1995). A similar result was obtained with the chimeras of GIRK1 and IRK2 as well (KUNKEL and PERALTA 1995). Nevertheless, it was indicated that the 137 amino acid stretch in the GIRK1 C-terminus between histidine 325 and proline 462 primarily participates in the $G_{\beta\gamma}$-binding of GIRK1. This segment of GIRK1 largely overlaps a part of the C-terminal domain homologous to the $G_{\beta\gamma}$-binding domain of the βARK1 as previously predicted (REUVENY et al. 1994).

More precise localization of $G_{\beta\gamma}$-binding sites in the N- and C-terminal domains of GIRK1 were recently identified. HUANG et al. (1997) again constructed fusion proteins containing various truncated N- and C-terminal domains of GIRK1 and concluded that two separate segments in the C-terminal domain (a.a. 318–374 and 390–462) and a segment in the N-terminus (a.a. 24–86) contribute to the $G_{\beta\gamma}$-binding to GIRK1 subunit (Fig. 9). The segment a.a. 390–462 did not exhibit a significant $G_{\beta\gamma}$-binding activity by itself but enhanced the $G_{\beta\gamma}$-binding activity of the other site in the proximal C-terminal domain of GIRK1 (i.e., a.a. 318–374). As indicated by underlines in Fig. 7, GIRK1 possesses two sets of aminoacid sequences similar to the motif of asparagine-X-X-glutamate-arginine (N-X-X-E-R), which is supposed to be critical for regulation of adenylyl cyclase 2 by $G_{\beta\gamma}$ (CHEN et al. 1995; HUANG et al. 1995). However, the sequences are located between the two identified C-terminal $G_{\beta\gamma}$-binding domains, and thus may not be critical in the interaction between $G_{\beta\gamma}$ and GIRK1.

The binding site of $G_{\beta\gamma}$ on K_G channels was also examined by the yeast two-hybrid system by YAN and GAUTAM (1996). They showed that G_β can bind

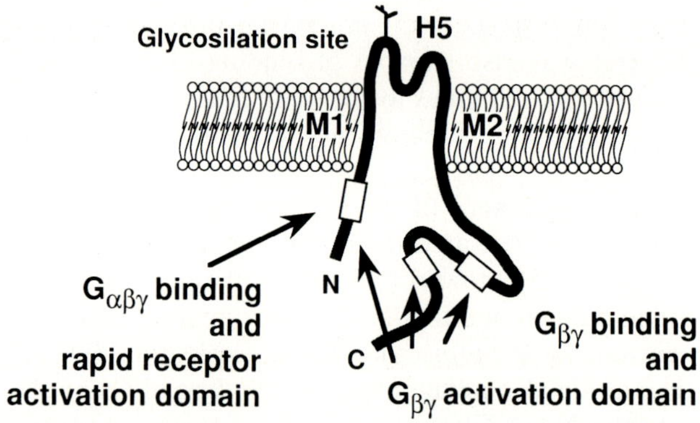

Fig. 9. Identified or putative functional domains of GIRK1. Schematic representation of approximate positions of identified or putative functional domains of GIRK1. N119 is the putative *N*-linked glycosylation site. The approximate positions of three identified G protein $\beta\gamma$-binding sites (one in the N- and two in the C-terminal domain) and one trimeric G protein-binding site in the N-terminal domain are depicted (HUANG et al. 1995, 1997)

with the N-terminus of GIRK1. Different G_β subunit types interact with the N-terminal domain of GIRK1 with different efficacies. Furthermore, an N-terminal fragment of 100 amino acids of G_β interacts with the N-terminal domain of GIRK1 as effectively as the whole G_β. This domain includes the region where the G_β subunit contacts the G_γ subunit in the crystal structure and may therefore explain the ability of the G_γ to shut off the activity of $G_{\beta\gamma}$. In this study, however, they could not detect the binding of G_β with the C-terminus of GIRK1, which was shown to possess the $G_{\beta\gamma}$-binding site.

Although both the N- and C-terminal domains of GIRK1 could independently interact with $G_{\beta\gamma}$ at least *in vitro* (HUANG et al. 1995), these domains might coordinately interact with $G_{\beta\gamma}$ in K_G channels. Actually, a study using the chimeras of GIRK1 and IRK1 demonstrated that the N- and the C-terminal domains of GIRK1 synergistically increased the ratio of the G protein-dependent to -independent current amplitude of the chimeric channel (SLESINGER et al. 1995). Also, it was pointed out that the apparent K_d for the $G_{\beta\gamma}$-binding of either the N- or C-terminal fusion protein (~μmol/l) was much higher than that estimated from the concentration-response relationship of $G_{\beta\gamma}$-induced activation of native K_{ACh} channels (~3 nmol/l) (ITO et al. 1992). Consistently, HUANG et al. (1997) demonstrated that the fusion proteins of the N- and C-terminal domains of GIRK1 actually bound with each other and thereby synergistically enhanced the $G_{\beta\gamma}$-binding activity. Therefore, a functional complex of the N- and C-terminal domains may form a binding site for $G_{\beta\gamma}$.

All of the data described above are on the interaction between $G_{\beta\gamma}$ and GIRK1. Only limited data are available for the other types of GIRKs. Nevertheless, homomeric channels composed of GIRK2 isoforms or GIRK4 have been reported to be activated by $G_{\beta\gamma}$ (KRAPIVINSKY et al. 1995a; VELIMIROVIC et al. 1996). In addition, it is likely that GIRK4 also mediates the $G_{\beta\gamma}$-activation of the heteromultimeric K_G channel composed of GIRK1 and GIRK4 (SLESINGER et al. 1995; TUCKER et al. 1996). GIRK2–4 possess the amino acid sequences highly homologous to that of GIRK1 but shorter C-terminal domains than GIRK1 (Fig. 7). GIRK2–4 have domains similar to the N- and the proximal C-terminal $G_{\beta\gamma}$-binding domains of GIRK1 (a.a. 24–86 and 318–374) but do not possess the sequence corresponding to the segment a.a. 390–462 of GIRK1 (HUANG et al. 1997). Consistent with these primary structures, the $G_{\beta\gamma}$-binding activity was similar among the N-terminal domains of GIRK1–4, while the $G_{\beta\gamma}$-binding activity of the C-terminal domains of GIRK2–4 was lower than that of the C-terminal domain of GIRK1 (HUANG et al. 1997). It was further shown that the fusion protein of C-terminal domain of GIRK1 interacts with that of the N-terminal domain of GIRK4. Therefore, in the K_{ACh} channel composed of GIRK1 and GIRK4, the N- and C-termini of the same subunit or those from adjacent subunits may interact with each other and form the high-affinity $G_{\beta\gamma}$-binding site (HUANG et al. 1997). The possibility, therefore, exists that various homo- or heteromultimeric K_G channels have distinct $G_{\beta\gamma}$-binding sites and thus different $G_{\beta\gamma}$-binding activities, depending on the subunit composition of each channel. Such a complexity of the $G_{\beta\gamma}$-binding sites and activities of K_G channels might be important in the diversity of the receptor-mediated activation of K_G channels in various tissues. At present, the whole aspects of the interaction between $G_{\beta\gamma}$ and K_G channel subunits have not yet been clarified.

2. Putative Mechanism Underlying the G Protein $\beta\gamma$ Subunit-Induced Activation of the G Protein-Gated K^+ Channels

It has not been clearly understood how the binding of $G_{\beta\gamma}$ to GIRK subunits leads to opening of K_G channels. From functional analyses of chimeras of GIRK1 and IRK1 expressed in *Xenopus* oocytes, SLESINGER et al. (1995) raised a possibility that the hydrophilic N-terminal domain of GIRK1 may have a function to suppress the $G_{\beta\gamma}$-independent basal current. On the other hand, DASCAL et al. (1995) proposed that the C-terminal domain of GIRK1 may block the K_G channel pore in a way similar to the 'Shaker ball' of the K_v channels because a myristoylated cytosolic C-terminal tail of GIRK1 decreased not only GIRK1 but also ROMK1 currents. Thus, the K_G channel might be intrinsically inhibited by either or both of the C- and N-terminal domains of GIRK1, and the $G_{\beta\gamma}$-binding to these domains might activate the channel by removing the inhibition. However, such "de-inhibition" might not be a sole mechanism by which $G_{\beta\gamma}$ activates the K_G channel. SLESINGER et al. (1995) showed that a chimera possessing a part of the C-terminal domain of GIRK1 (a.a. 325–501) and the other parts derived from IRK1 exhibited ~2 times larger current in *Xenopus* oocytes coexpressing $G_{\beta\gamma}$ than in the control oocytes. However, single channel recordings revealed that the open probability of the chimera channel was as high as ~0.8 in the control oocytes. Therefore, the twofold increase in the current in the presence of $G_{\beta\gamma}$ cannot be accounted for only by an increase of P_o of each channel but may require an increase in the number of functional channels. This is consistent with the notion that $G_{\beta\gamma}$ activates the native cardiac K_{ACh} channel by increasing the functional number of the channel (HOSOYA et al. 1996). Further studies are needed to elucidate the molecular mechanism responsible for the closure of K_G channel in the absence of $G_{\beta\gamma}$-stimulation and its opening in the presence of $G_{\beta\gamma}$.

3. PIP_2-Mediation of $G_{\beta\gamma}$-Activation of K_G Channels

Recently, it was shown that phosphatidylinositol 4,5-bisphospahte (PIP_2) regulates the activity of native and recombinant inwardly rectifying K^+ channels, the IP_3 receptor and transporters, such as sodium-calcium exchanger (SUI et al. 1998). Actually, PIP_2 was shown to bind directly to proteins as diverse as phospholipases, kinases, cytoskeletal, and channel proteins. In Kir channels, it was shown that intracellular ATP prevents run-down of the functional channel and restores the run-down channel in the inside-out patches. This phenomenon has been recognized for a long time especially in the ATP-sensitive K^+ channels but also in the G protein-gated K_{ACh} channels. PIP_2 could mimic these effects of intracellular ATP. The antibody for PIP_2 blocked the effects of ATP. Although it has not yet been clarified how PIP_2 affects the $G_{\beta\gamma}$-activation of K_G channels, it is suggested that the substance is necessary for the channel proteins to be functional in responding to $G_{\beta\gamma}$. Further studies are needed to clarify the functional role of PIP_2 in K_G channel signaling system.

VI. The Possible Role of G Protein α Subunits in the G Protein-Gated K^+ Channel Regulation

1. Possibility of Microdomain Composed of Receptor, G Protein and the G Protein-Gated K^+ Channel.

The activation of the K_{ACh} channel by $G_{\beta\gamma}$ but not by G_α does not necessarily indicate that the channel is unable to interact with G_α. HUANG et al. (1995) found that the GST fusion protein of the N-terminal domain of GIRK1 was also capable of binding to $G_{\alpha\text{-GDP}}$ and the heterotrimeric G protein *in vitro*. SLESINGER et al. (1995) demonstrated that the chimeras of IRK1 and GIRK1 responded to both application and washout of carbachol more promptly when they contained the N-terminal domain of GIRK1 than when they did not. Because deletion of the first 30 amino acids (a.a. 2–31) of GIRK1 did not impair the fast response of GIRK1 to the receptor stimulation, the remaining ~50 residues in the N-terminal domain may harbor the structural elements necessary for the fast activation and deactivation. These data may indicate that a part of the N-terminal domain of GIRK1 may have a function to keep the heterotrimeric G protein in the vicinity of itself and thereby facilitate the interaction between G protein and K_G channels.

SLESINGER et al. (1995) also raised an interesting possibility that GIRK1 may directly interact with muscarinic receptors through its hydrophobic core region (M1-H5-M2). They found that a chimera of IRK1 and GIRK1 containing the C-terminus but not the hydrophobic core of GIRK1 could be activated by $G_{\beta1\gamma2}$, but not by m_2-muscarinic receptor. The ability to respond to the receptor stimulation was endowed by transplantation of the hydrophobic core region of GIRK1 to the chimera. Other investigators, however, reported the data inconsistent with their proposal: KOFUJI et al. (1996a) showed that a chimera containing the N- and C-terminal domains of GIRK1 and the hydrophobic core of ROMK1, a clone of a G-protein-independent Kir channel, could respond to the m_2-muscarinic receptor stimulation. Thus, the hypothesis of the direct interaction between the GIRK1 hydrophobic core domain and the m_2-muscarinic receptor may need further verification.

2. Specificity of Signal Transduction Based on the Receptor/G Protein/G Protein-Gated K^+ Channel Interaction

The possibility that GIRK1 may physically interact with all of $G_{\beta\gamma}$, $G_{\alpha\text{-GDP}}$, heterotrimeric G proteins and receptors is very attractive, because it implies that the receptor, G protein, and GIRK1 can be compartmentalized into a certain microdomain of the cell membrane. It has been suggested that, to achieve the rapid response of the atrial K_{ACh} channel to acetylcholine, the m_2-muscarinic receptor, G_K, and the channels need to be located within 0.35 μm of each other (HILLE 1992). The intermolecular interaction among these signaling molecules may fulfill such a topological requirement. However, the activation of GIRK1 mediated by the m_2-muscarinic receptor in oocytes does not occur as fast as that of the K_{ACh} channel by acetylcholine or adenosine in cardiac atrial myocytes. The "short-term" desensitization of the atrial K_{ACh} channel following the application of these agonists has also not been well reconstituted in oocytes coexpressing the m_2-muscarinic receptor and GIRK1 with or without GIRK4. Thus, there might exist additional factor(s) to modify the coupling between these molecules. RGS protein may be one of the candidates (DOUPNIK et al. 1997; SAITOH et al. 1997).

In atrial myocytes, the K_{ACh} channel is activated by m_2-muscarinic and A_1-adenosine stimulation. However, β_1-adrenergic stimulation, which should also increase the free $G_{\beta\gamma}$ concentration in the membrane through activating G_s protein, never activates the K_{ACh} channel in cardiac atrial myocytes. It was shown that such a specific signal transduction cannot be explained in terms of the different affinities of distinct types of $G_{\beta\gamma}$ for the K_{ACh} channel (WICKMAN et al. 1994; YAMADA et al. 1994). The hypothetical compartmentalization of receptors, G proteins, and GIRK subunits may provide a plausible alternative explanation for the specificity of signal transduction. However, it is still difficult to have a realistic view on the microdomain composed of these signal transduction molecules. For example, it is not easy to explain the saturative effect of ACh and adenosine on the K_{ACh} channel activity (KURACHI et al. 1986b), when strict compartmentalization of those signaling molecules exists.

D. Localization of the G Protein-Gated K^+ Channel Systems in Various Organs

$G_{K\beta\gamma}$ appears to be the physiologically functional arm of G_K activating K_G channels not only in the heart but also in the brain and endocrine organs. However, the molecular mechanisms of G protein-regulation of ion channels have been shown to be more complicated than we had thought. In AtT20 cells which had been transfected with the α_{2A}-adrenergic receptor, adrenergic agonists can inhibit the Ca^{2+} current and adenylyl cyclase and activate a K^+ current. A point mutation of the receptor removes activation of the K^+ current, but not inhibition of Ca^{2+} current and adenylyl cyclase (SURPRENANT et al. 1992). This indicates that the G protein coupling to the K^+ channel is different from those to the Ca^{2+} channel and adenylyl cyclase although the receptor is the same. G proteins may thus be more specific to each receptor and to each signaling system than we are currently assuming. In *Xenopus* oocytes, however, when β_2-adrenergic receptors, G_s protein, and K_G channels

(GIRK1/KGA) are expressed together, β-adrenergic agonists could induce activation of K_G channel current (LIM et al. 1995). Accordingly, the affinity of particular G protein subunits for the K_G channel may not be sufficient to explain specific activation of K_{ACh} channel by G_K. Actually, various combinations of recombinant $G_{\beta\gamma}$ (except for $G_{\beta1\gamma1}$) have similar efficacy and potency in activating K_{ACh} channels (WICKMAN et al. 1994). However, receptor specificity in cardiac atrial myocytes is well documented by extensive studies (KURACHI 1995). It is often argued that receptor specificity could arise from compartmentalization of the appropriate receptors and channels, although little evidence exists for such compartmentalization. We do not know whether different mechanisms underlie receptor specificity in different organs. In other words, we have not yet fully answered the question how information specifically passes from a membrane receptor to the effector, the K_G channel, via G proteins.

Because the signal transduction mechanisms are not necessarily the same among heart, neurons, and endocrine cells, it is worthwhile at present to summarize the current observations on the localization of K_G channels in these organs. Apparently, GIRK1 and GIRK4 immunoreactivities diffusely distribute in the cell membrane of cardiac myocytes (Fig. 10A), while those of GIRK1 and/or GIRK2 are localized to specialized segments of neuronal membrane, such as presynaptic axonal termini and postsynaptic dendritic regions (Fig. 10C,D). Thus, we may tentatively classify the system based on the apparent distribution in the cell into two categories: (1) homogeneously distributed system and (2) localized system. In these systems, different mechanisms may potentially underlie the receptor specificity.

I. Cardiac Atrial Myocytes

The cardiac K_{ACh} channel is the prototype of K_G channels (KURACHI 1995). By forming the heterotetramer of GIRK1 and GIRK4, cardiac K_{ACh} channels are localized on the cardiac cell membrane (Fig. 9). Immunohistochemistry using the specific antibody showed that the GIRK1 are homogeneously localized on the cell membrane of atrial, but not of ventricular myocytes (Fig. 10A,B). This is consistent with the electrophysiological studies of cardiac cells. The electrophysiological experiments also suggested that some topological restriction may exist in cardiac atrial myocytes, because K_{ACh} channels in the cell-attached membrane patch are activated by ACh or adenosine when they are applied to the pipette solution, but not when they are added to the bathing solution (SOEJIMA and NOMA 1994). It was found that either G_i- or G_s-coupled receptors, when expressed together with GIRK channels in *Xenopus* oocytes, can activate the channels (LIM et al. 1995). This may indicate that under conditions where compartmentalization does not exist, as in *Xenopus* oocytes, the $\beta\gamma$ subunits released from G_s may be able to activate K_G channels. Because β-adrenergic agonists never activate these channels in cardiac myocytes, there

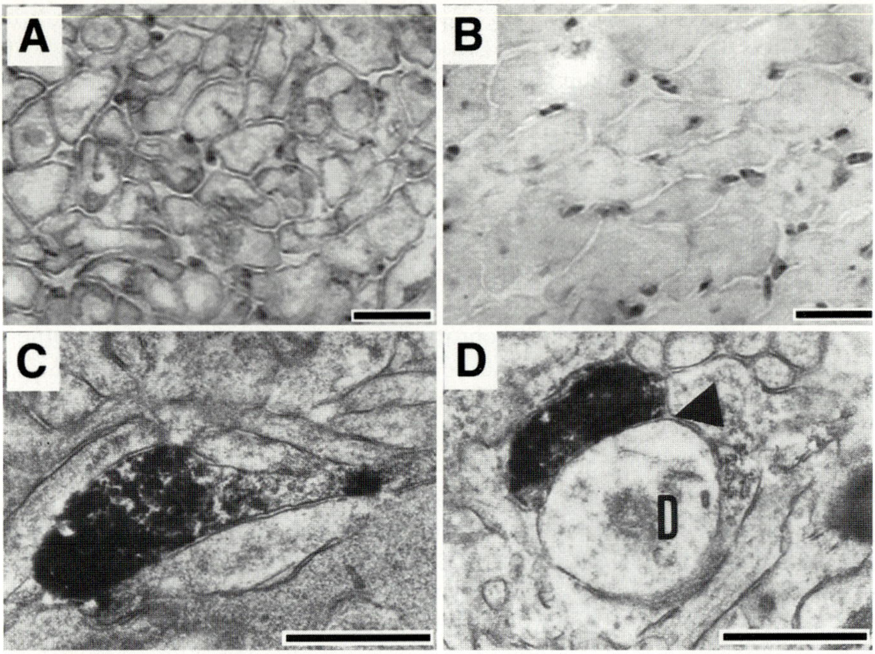

Fig. 10A–D. Different subcellular localization of GIRK1 proteins. **A,B** immunohistochemical analysis for the GIRK1 proteins in the rat atrium (A) and ventricle (B). Homogeneous immunoreactivity was found on the plasma membranes of the atrial, but not of the ventricular myocytes. **C,D** electron microscopic analysis for the GIRK1 immunoreactivity in the rat paraventricular nucleus of the hypothalamus. Obvious GIRK1 immunoreactivity was present at the axonal termini (probably on the vesicles) neighboring upon a dendrite (**D**). All GIRK1 immunoreactivities were developed with diaminobenzidine-horse radish peroxidase method. **C,D** reproduced with permission (MORISHIGE et al. 1996)

should be some mechanism to guarantee the specificity in the native G_K-K_{ACh} channel system in cardiac atrial myocytes.

One candidate for such a topological factor may be caveolae. Cardiac myocytes are rich in caveolae, which are microdomains of 50–80 nm diameter that are enriched in cholesterol, and express caveolin 1, caveolin 3, and low levels of caveolin 2. Caveolae contains proteins which can function as GAP or GDI for $G_{\alpha i2}$ and $G_{\alpha o}$ (SCHERER et al. 1996; TANG et al. 1996). Caveolae-association of G proteins, G protein-coupled receptors and their effectors such as IP_3-sensitive calcium channel and Ca^{2+}-ATPase has been shown (LI et al. 1995). Thus, caveolae are one of the candidates of the sites for microdomain of the system. However, no experiments to address this question has been reported.

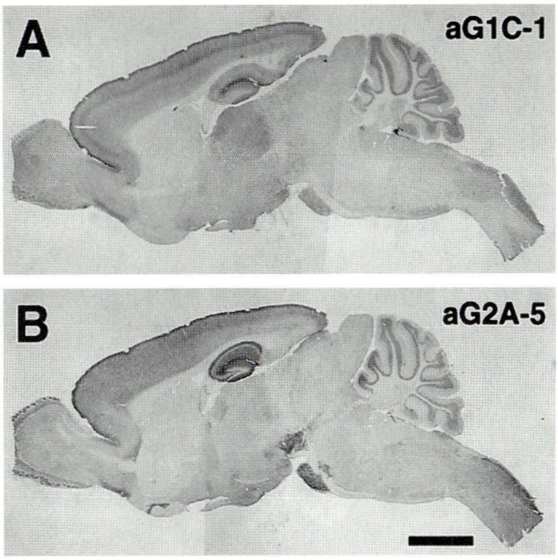

C

	Kir3.1	Kir3.2
Olfactory bulb	++	++
Neocortex	+++	+++
Hippocampus	+++	+++
Cerebellum	+++	+++
Thalamus	+	−
Substantia nigra	−	+++
Ventral tegmental area	±	+++

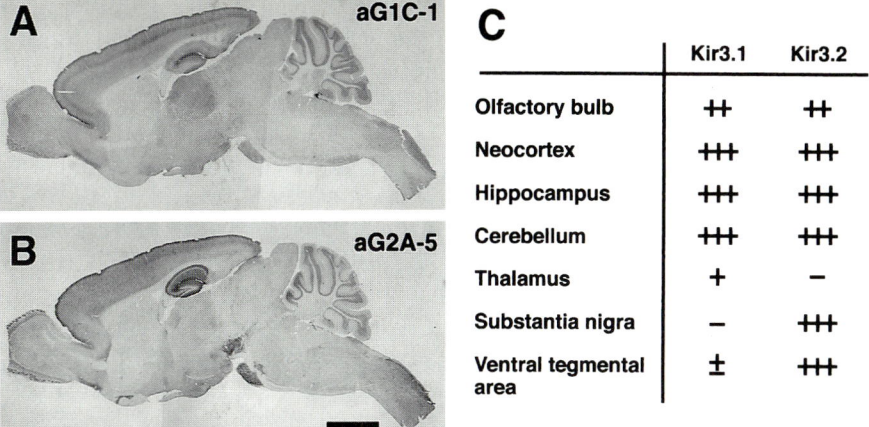

Fig. 11A–C. Immunoreactivities of GIRK1 and GIRK2 in mouse brain. The polyclonal antibodies specific to the C-termini of GIRK1 (aG1C-1) and GIRK2 (aG2A-5) were developed in rabbits (Inanobe et al. 1995a,b, 1999). aG2A-5 recognizes both GIRK2A and GIRK2C. **A** the sagittal section of the mouse brain was stained with aG1C-1. **B** the sagittal section of the mouse brain was stained with aG2A-5. **C** the distribution of immunoreactivities for GIRK1/Kir3.1 and GIRK2/Kir3.2 in various regions of the brain

II. Neurons

The mRNAs of GIRK1, 2 and 3 are coexpressed in many areas of brain (Kobayashi et al. 1995; Karschin et al. 1994, 1996), which suggested that diversity of K_G channels arise from the different combinations of these GIRK subunits in various brain areas.

Coexpression of GIRK1 and GIRK2A results in reconstitution of the K_G channels exhibiting neuronal K_G channels (Velimirovic et al. 1996). Thus, it was proposed that neuronal K_G channels of brain are mainly composed of GIRK1 and GIRK2A. Because knock-out of GIRK2 resulted not only in the disappearance of GIRK2 immunoreactivity but also in decrease of GIRK1 immunoreactivity in such areas as hippocampus and cerebral cortex (Signorini et al. 1997), it is indicated that GIRK1 and GIRK2A may form heterotetramers of K_G channels in neurons.

But the recent studies on the distributions of each GIRK subunit protein using specific antibodies are providing a slightly different view. Fig. 11 shows the distribution of immunoreactivities for GIRK1 and GIRK2 in mouse brain. Consistent with the notion, in many parts of the brain, the immunoreactivities of GIRK1 and GIRK2 overlapped in many parts of the brain, such as the neocortex and hippocampus, but not in some of the others: Dominant immunoreactivity of GIRK1 was detected in thalamus, where that of GIRK2 was not evident. In these parts, GIRK3 might be the partner, although so far no studies are available on the GIRK3 protein in the *in vivo* brain K_G channels. In substantia nigra and ventral tegmental area, there detected prominent GIRK2-immunoreactivity but marginal level of GIRK1, which is consistent with the *in situ* hybridization study (KARSCHIN et al. 1996). Our recent study strongly indicates that the K_G channels in these area may be composed of homomeric GIRK2 subunits (YOSHIMOTO et al. 1997; INANOBE et al. 1999).

The studies on GIRK1 protein further showed that GIRK1 proteins are localized not only in somata and dendrites (DRAKE et al. 1997), where GIRK channels may mediate postsynaptic inhibition, but also in axons and these terminals (PONCE et al. 1996; MORISHIGE et al. 1996). This suggests that K_G channels can also modulate presynaptic events, although the formation of slow inhibitory postsynaptic potential is the major task of the channels (HILLE 1992). Furthermore, it was found that the distribution of the protein to either somatodendritic or axonal-terminal regions of neurons varied in different brain regions. This may be related to the features of GIRK1 that it is by itself inactive, but it can associate with the other family members (GIRK2–GIRK4) to enhance their activity and alter their single-channel kinetics (KRAPIVINSKY et al. 1995a; DUPRAT et al. 1995; KOFUJI et al. 1995; FERRER et al. 1995; LESAGE et al. 1995; ISOMOTO et al. 1996; VELIMIROVIC et al. 1996; CHAN et al. 1996).

The distribution of GIRK4-immunoreactivity was also examined (IIZUKA et al. 1997), although the expression level of GIRK4 mRNA is low in the brain compared with those of GIRK1, 2, and 3. The GIRK4 protein and mRNA were detected in the cerebellar cortex, hippocampal formation, olfactory system, cerebral cortex, basal ganglia, several nuclei of the lower brain stem and the choroid plexus. In contrast to the mRNA, which was concentrated in the cell soma, the GIRK4 protein was found in a subset of nerve fibers and in axon terminals, but not on the somatodendritic regions. In the cerebellar cortex and hippocampus, the GIRK4 protein was concentrated in the axon terminals of basket cells which are GABAergic interneurons. However, the GIRK1 immunoreactivity was not detected in this region.

Because expression of mRNAs of each GIRK subunit differ among various neurons in the brain (KARSCHIN et al. 1994, 1996), combinations of GIRK subunits in a specific neuron should be an important factor to determine the distribution of K_G channel proteins in a neuron, i.e., whether on its somatodendritic segments or on axonal-terminal regions.

III. Endocrine Cells

Electrophysiological studies indicate that K_G channels activated by somatostatin and/or dopamine exist in endocrine cells of anterior pituitary lobe (PENNEFATHER et al. 1988; EINHORN and OXFORD 1993). This is a mechanism to inhibit secretion of such hormones as TSH, ACTH, and prolactin, from the endocrine cells, and thus may be important for negative feedback control of the hormome-secretion. The molecular properties of the K_G channels have not yet been fully clarified.

E. *Weaver* Mutant Mice and GIRK2 Gene

Weaver mice (*wv*) have been studied intensively over the past 25 years for insights into the normal processes of neuronal development and differentiation (HESS 1996). Homozygous animals suffer from severe ataxia, due to death of cerebellar granular cells. The animal was also used as a model of Parkinsonism because dopaminergic input to the striatum is lost during the first few weeks after birth due to the death of dopaminergic neurons in the substantia nigra. Male homozygous mice are sterile: spermatogenesis fails to proceed normally past the third postnatal week leading to a complete failure of sperm production. Recently, it was shown that *weaver* mutant mice have their neurologic abnormalities because of point mutation of guanine 953 to adenine in the GIRK2 gene (PATIL et al. 1995). This mutation causes a change of amino acid from glycine (G) at a.a. 156 to serine (S), which is in the selective filter of the potassium channel in H5 region. Thus, the fingerprint of K^+ channel sequence of glycine-tyrosine-glycine (GYG) is altered to serine-tyrosine-glycine (SYG) in *wv* mice.

This change of amino acid causes several functional alterations in K_G channels composed of the *weaver* allele of the GIRK2 subunit (GIRK2*wv*). In the GIRK2-homomultimeric K_G channel, the channel loses its selectivity to K^+ ions and allows other monovalent cations such as Na^+ and Cs^+ to pass though the channel. The channel does not respond to receptor or $G_{\beta\gamma}$-stimulation but is constitutively active without agonist. Similar loss of ion selectivity and G protein-independent activation was observed in GIRK1-GIRK2*wv* heteromultimeric K_G channels (LIAO et al. 1996; NAVARRO et al. 1996; SLESINGER et al. 1996). KOFUJI et al. (1996b) actually recorded such a constitutive Na^+ current probably flowing through the mutated K_G channels in cerebellar granule cells isolated from *weaver* mice. Cation channel blockers, QX-314 and MK-801, resulted in survival and differentiation of the *weaver* granule cells, supporting the notion that Na^+ influx through *weaver* K_G channels interferes the cell proliferation and differentiation. For this channel activation, intracellular Na^+-mediated G protein-independent activation of K_G channels might be involved (SUI et al. 1996).

F. Conclusions

The G protein-activation of inwardly rectifying K^+ channel system has been mainly studied in cardiac myocytes until 1993 with electrophysiological techniques. The recent rapid progresses in the molecular biology of K_G channels have disclosed the complexity of this channel system, which was not imagined before. Although many aspects of the regulation of the K_G channels have been elucidated by the efforts of many laboratories listed in this review, there have also emerged many unclarified but possibly important mechanisms which may underlie the physiological regulation of this system in various organs, including heart, brain, and endocrine organs. At present, we cannot yet explain even the molecular mechanisms responsible for the receptor-specific control of K_G channels in these organs. Because the mRNAs of GIRK clones are widely distributed in the brain, and because this K_G channel system can be utilized by many receptors in the brain, its functional role in various brain functions should be further studied. Various phenomena described in K_{ACh} channels in cardiac atrial myocytes, such as desensitization, deactivation, and cross-talk with the other signaling systems have not been examined at all in other tissues including brain. These phenomena may also be important for future studies.

References

Ashford MLJ, Bond CT, Blair TA, Adelman JP (1994) Cloning and functional expression of a rat heart K_{ATP} channel. Nature 370:456–459

Berman DM, Wilkie TM, Gilman AG (1996) GAIP and RGS4 are GTPase-activating proteins for the G_i subfamily of G protein α subunits. Cell 86:445–452

Bond CT, Ämmälä C, Ashfield R, Blair TA, Gribble F, Khan RN, Lee K, Proks P, Rowe IC, Sakura H, Ashford MLJ, Adelman JP, Ashcroft FM (1995) Cloning and functional expression of cDNA encoding an inwardly-rectifying potassium channel expressed in pancreatic β-cells and in the brain. FEBS Lett 367:61–66

Bond CT, Pessia M, Xia XM, Lagrutta A, Kavanaugh MP, Adelman JP (1994) Cloning and expression of a family of inward rectifier potassium channels. Receptors and Channels 2:183–191

Brown AM, Birnbaumer L (1990) Ionic channels and their regulation by G protein subunits. Annu Rev Physiol 52:197–213

Breitwieser GE, Szabo G (1985) Uncoupling of cardiac muscarinic and β-adrenergic receptors from ion channels by a guanine nucleotide analogue. Nature 317:538–540

Chan KM, Sui JL, Vivaudou M, Logothetis DE (1996) Control of channel activity through a unique amino acid residue of a G protein-gated inwardly rectifying K^+ channel subunit. Proc Natl Acad Sci USA 93:14193–14198

Chen J, DeVivo M, Dingus J, Harry A, Li J, Sui J, Carty DJ, Blank JL, Exton JH, Stoffel RH, Inglese J, Lefkowitz RJ, Logothethis DE, Hidebrandt JD, Iyengar R (1995) A region of adenylyl cyclase 2 critical for regulation by G protein $\beta\gamma$ subunits. Science 268:1166–1169

Clapham DE, Neer EJ (1993) New roles for G-protein $\beta\gamma$-dimers in transmembrane signalling. Nature 365:403–406

Codina J, Yatani A, Grenet D, Brown AM, Birnbaumer L (1987) The α subunit of the GTP binding protein G_K opens atrial potassium channels. Science 236:442–445

Cohen NA, Brenman JE, Snyder SH, Bredt DS (1996) Binding of the inward rectifier K^+ channel Kir2.3 to PSD-95 is regulated by protein kinase A phosphorylation. Neuron 17:759–767

Dascal N, Doupnik CA, Ivanina T, Bausch S, Wang W, Lin C, Garvey J, Chavkin C, Lester HA, Davidson N (1995) Inhibition of function in *Xenopus* oocytes of the inwardly rectifying G-protein-activated atrial K channel (GIRK1) by overexpression of a membrane-attached form of the C-terminal tail. Proc Natl Acad Sci USA 92:6758–6762

Dascal N, Schreibmayer W, Lim NF, Wang W, Chavkin C, DiMagno L, Labarca C, Kieffer BL, Gaveriauz-Ruff C, Trollinger D, Lester H, Davidson N (1993) Atrial

G protein-activated K^+ channel: Expression cloning and molecualr properties. Proc Natl Acad Sci USA 90:10235–10239
Del-Castillo J, Katz B (1955) Production of membrane potential changes in the frog's heart by inhibitory nerve impulses. Nature 175:1035
Dixon AK, Gubitz AK, Ashford MLJ, Richardson PJ, Freeman TC (1995) Distribution of mRNA encoding the inwardly rectifying K^+ channel, BIR1 in rat tissues. FEBS Lett 374:135–140
Doupnik CA, Davidson N, Lester HA (1995a) The inward rectifier potassium channel family. Curr Opin Neurobiol 5:268–277
Doupnik CA, Davidson N, Lester HA, Kofuji P (1997) RGS proteins reconstitute the rapid gating kinetics of $G_{\beta\gamma}$-activated inwardly rectifying K^+ channels. Proc Natl Acad Sci USA 94:10461–10466
Doupnik CA, Lim NF, Kofuji P, Davidson N, Lester HA (1995b) Intrinsic gating properties of a cloned G protein-activated inward rectifier K^+ channel. J Gen Physiol 106:1–23
Drake CT, Bausch SB, Milner TA, Chavkin C (1997) GIRK1 immunoreactivity is present predominantly in dendrites, dendritic spines, and somata in the CA1 region of the hippocampus. Proc Natl Acad Sci USA 94:1007–1012
Druey KM, Blumer KJ, Kang VH, Kehrl JH (1996) Inhibition of G-protein-mediated MAP kinase activation by a new mammalian gene family. Nature 379:742–746
Duprat F, Lesage F, Guillemare E, Fink M, Hugnot J-P, Bigay J, Lazdunski M, Romey G, Barhanin J (1995) Heterologous multimeric assembly is essential for K^+ channel activity of neuronal and cardiac G-protein-activated inward rectifiers. Biochem Biophys Res Commun 212:657–663
Einhorn LC, Oxford GS (1993) Guanine nucleotide binding proteins mediate D_2 dopamine receptor activation of a potassium channel in rat lactotrophs. J Physiol (Lond.) 462:563–578
Ferrer J, Nichols CG, Makhina EN, Salkoff L, Bernstein J, Gerhard D, Wasson J, Ramanadham S, Permutt A (1995) Pancreatic islet cells express a family of inwardly rectifying K^+ channel subunits which interact to form G-protein-activated channels. J Biol Chem 270:26086–26091
Gilman AG (1987) G proteins: transducers of receptor-generated signals. Annu Rev Biochem 56:615–649
Gomperts SN (1996) Clustering membrane proteins: it's all coming together with the PSD-95/SAP90 protein family. Cell 84:659–662
Grudt TJ, Williams JY (1993) κ-Opioid receptors also increase potassium conductance. Proc Natl Acad Sci USA 90:11429–11432
Hedin KE, Lim NF, Clapham DE (1996) Cloning of a Xenopus laevis inwardly rectifying K^+ channel subunit that permits GIRK1 expression of I_{KACh} currents in oocytes. Neuron 16:423–429
Herlitze S, Garcia DE, Mackie K, Hille B, Scheuer T, Catterall WA (1996) Modulation of Ca^{2+} channels by G-protein $\beta\gamma$ subunits. Nature 380:258–262
Hess EJ (1996) Identification of the weaver mouse mutation: the end of the beginning. Neuron 16:1073–1076
Hille B (1992) G protein-coupled mechanisms and nervous signaling. Neuron 9:187–195
Ho K, Nichols CG, Lederer WJ, Lytton J, Vassilev PM, Kanazirska MV, Hebert SC (1993) Cloning and expression of an inwardly rectifying ATP-regulated potassium channel. Nature 362:31–38
Horio Y, Hibino H, Inanobe A, Yamada M, Ishii M, Tada Y, Satoh E, Hata Y, Takai Y, Kurachi, Y (1997) Clustering and enhanced activity of an inwardly rectifying potassium channel, Kir4.1, by an anchoring protein, PSD-95/SAP90. J Biol Chem 272: 12885–12888
Hosoya Y, Kurachi Y (1998) Functional analyses of G protein activation of cardiac K_G channel. In Potassium Ion Channels; their structure, function and diseases, Academic Press, Edited by Kurachi Y, Jan LY, Lazdunski M (in press)

Hosoya Y, Yamada M, Ito H, Kurachi Y (1996) A functional model for G protein activation of the muscarinic K^+ channel in guinea pig atrial myocytes. J Gen Physiol 108:485–495

Huang C-L, Jan YN, Jan LY (1997) Binding of the G protein $\beta\gamma$ subunit to multiple regions of G protein-gated inward-rectifying K^+ channels. FEBS Lett 405:291–298

Huang C-L, Slesinger PA, Casey PJ, Jan YN, Jan LY (1995) Evidence that direct binding of $G_{\beta\gamma}$ to the GIRK1 G protein-gated inwardly rectifying K^+ channel is important for channel activation. Neuron 15:1133–1143

Hutter OF, Trautwein W (1955) Vagal and sympathetic effects on the pacemaker fibers in the sinus venosus of the heart. J Gen Physiol 39:715–733

Hynes RO (1992) Integrins: versatility, modulation, and signaling in cell adhesion. Cell 69:11–25

Iizuka M, Tsunenari I, Momota Y, Akiba I, Kono T (1997) Localization of a G-protein-coupled inwardly rectifying K^+ channel, CIR, in the rat brain. Neurosci 77:1–13

Ikeda SR (1996) Voltage-dependent modulation of N-type calcium channels by G-protein $\beta\gamma$ subunits. Nature 380:255–258

Inagaki N, Gonoi T, Clement IV JP, Namba N, Inazawa J, Gonzalez G, Aguilar-Bryan L, Seino S, Bryan J (1995a) Reconstitution of I_{KATP}: An inward rectifier subunit plus the sulfonylurea receptor. Science 270:1166–1170

Inagaki N, Tsuura Y, Namba N, Masuda K, Gonoi T, Horie M, Seino Y, Mizuta M, Seino S (1995b) Cloning and functional characterization of a novel ATP-sensitive potassium channel ubiquitously expressed in rat tissues, including pancreatic islets, pituitary, skeletal muscle and heart. J Biol Chem 270:5691–5694

Inanobe A, Ito H, Ito M, Hosoya Y, Kurachi Y (1995a) Immunological and physical characterization of the brain G protein-gated muscarinic potassium channel. Biochem Biophys. Res. Commun. 217:1238–1244

Inanobe A, Morishige K, Takahashi N, Ito H, Yamada M, Takumi T, Nishina H, Takahashi K, Kanaho Y, Katada T, Kurachi Y (1995b) $G_{\beta\gamma}$ directly binds to the carboxyl terminus of the G protein-gated muscarinic K^+ channel, GIRK1. Biochem Biophys Res Commun 212:1022–1028

Inanobe A, Yoshimoto Y, Horio Y, Morishige K, Hibino H, Matsumoto S, Tokunaga Y, Maeda T, Hata Y, Takai Y, Kurachi Y (1999) Characterization of G-protein-gated K^+ channels composed of Kir3.2 subunits in dopaminergic neurons of the substantia nigra. J Neurosci 19:1006–1017

Iñiguez-Lluhi J, Kleuss C, Gilman AG (1993) The importance of G-protein $\beta\gamma$ subunits. Trends Cell Biol 3:230–236

Isomoto S, Kondo C, Takahashi N, Matsumoto S, Yamada M, Takumi T, Horio Y, Kurachi Y (1996) A novel ubiquitously distributed isoform of GIRK2 (GIRK2B) enhances GIRK1 expression of the G-protein-gated K^+ current in *Xenopus* oocytes. Biochem Biophys Res Commun 218:286–291

Ito H, Sugimoto T, Kobayashi I, Takahashi K, Katada T, Ui M, Kurachi Y (1991) On the mechanism of basal and agonist-induced activation of the G protein-gated muscarinic K^+ channel in atrial myocytes of guinea pig heart. J Gen Physiol 98: 517–533

Ito H, Tung RT, Sugimoto T, Kobayashi I, Takahashi K, Katada T, Ui M, Kurachi Y (1992) On the mechanism of G protein $\beta\gamma$ subunit activation of the muscarinic K^+ channel in guinea pig atrial cell membrane: comparison with the ATP-sensitive K^+ channel. J Gen Physiol 99:961–983

Karschin C, Dißmann E, Stühmer W, Karschin A (1996) IRK(1–3) and GIRK(1–4) inwardly rectifying K^+ channel mRNAs are differentially expressed in the adult rat brain. J Neurosci 16:3559–3570

Karschin A, Ho BY, Labarca C, Elroy-Stein O, Moss B, Davidson N, Lester HA (1991) Heterologously expressed serotonin 1 A receptors couple to muscarinic K^+ channels in heart. Proc Natl Acad Sci USA 88:5694–5698

Karschin C, Schreibmayer W, Dascal N, Lester HA, Davidson N, Karschin A (1994) Distribution and localization of a G protein-coupled inwardly rectifying K^+ channel in the rat. FEBS Lett 348:139–144

Kennedy ME, Nemec J, Clapham DE (1996) Localization and interaction of epitope-tagged GIRK1 and CIR inward rectifier K^+ channel subunits. Neuropharmacol. 35:831–839

Kim E, Niethammer M, Rothschild A, Jan YN, Sheng M (1995) Clustering of Shaker-type K^+ channels by interaction with a family of membrane-associated guanylate kinases. Nature 378:85–88

Kobayashi T, Ikeda K, Ichikawa T, Abe S, Togashi S, Kumanishi T (1995) Molecular cloning of a mouse G-protein-activated K^+ channel (mGIRK1) and distinct distributions of three GIRK (GIRK1, 2 and 3) mRNAs in mouse brain. Biochem. Biophys. Res Commun 208:1166–1173

Kofuji P, Davidson N, Lester HA (1995) Evidence that neuronal G-protein-gated inwardly rectifying K^+ channels are activated by $G_{\beta\gamma}$ subunits and function as heteromultimers. Proc Natl Acad Sci USA 92:6542–6546

Kofuji P, Doupnik CA, Davidson N, Lester HA (1996a) A unique P-region residue is required for slow voltage-dependent gating of a G protein-activated inward rectifier K^+ channel expressed in *Xenopus* oocytes. J Physiol (Lond) 490:633–645

Kofuji P, Hofer M, Millen KJ, Millonig JH, Davidson N, Lester HA, Hatten ME (1996b) Functional analysis of the weaver mutant GIRK2 K^+ channel and rescue of *weaver* granule cells. Neuron 16:941–952

Koelle MR, Horvitz HR (1996) EGL-10 regulates G protein signaling in the C. elegans nervous system and shares a conserved domain with many mammalian proteins. Cell 84: 115–125

Kornau HC, Schenker LT, Kennedy MB, Seeberg PH (1995) Domain interaction between NMDA receptor subunits and the postsynaptic density protein PSD-95. Science 269:1737–1740

Koyama H, Morishige K, Takahashi N, Zanelli JS, Fass DN, Kurachi Y (1994) Molecular cloning, functional expression and localization of a novel inward rectifier potassium channel in the rat brain. FEBS Lett 341:303–307

Krapivinsky G, Gordon EA, Wickman K, Velimirovic B, Krapivinsky L, Clapham DE (1995a) The G-protein-gated atrial K^+ channel I_{KACh} is a heteromultimer of two inwardly rectifying K^+-channel protein. Nature 374:135–141

Krapivinsky G, Krapivinsky L, Velimirovic B, Wickman K, Navarro B, Clapham DE (1995b) The cardiac inward rectifier K^+ channel subunit, CIR, does not comprise the ATP-sensitive K^+ channel, I_{KATP}. J Biol Chem 270:28777–28779

Kubo Y, Baldwin TJ, Jan YN, Jan LY (1993a) Primary structure and functional expression of a mouse inward rectifier potassium channel. Nature 362:127–133

Kubo Y, Reuveny E, Slesinger PA, Jan YN, Jan LY (1993b) Primary structure and functional expression of a rat G-protein-coupled muscarinic potassium channel. Nature 364:802–806

Kunkel MT, Peralta EG (1995) Identification of domains conferring G protein regulation on inward rectifier potassium channels. Cell 83:443–449

Kurachi Y (1989) Regulation of G protein-gated K^+ channels. News Physiol. Sci. 4:158–161

Kurachi Y (1990) Muscarinic acetylcholine-gated K^+ channels in mammalian heart. In Regulation of potassium transport across biological membranes, Edited by Reuss L, Russell JM, Szabo G, The University of Texas Press, Texas, pp. 403–428

Kurachi Y (1993) G-protein regulation of cardiac K^+ channels. In *Handbook of Experimental Pharmacology, GTPases in Biology II*, edited by Dickey BF, Birnbaumer L, vol. 108/II, pp. 500–526, Springer-Verlag, Berlin

Kurachi Y (1994) G-protein control of cardiac potassium channels. Trends Cardiovasc Med 4:64–69

Kurachi Y (1995) G protein regulation of cardiac muscarinic potassium channel. Am J Physiol 269:C821–C830

Kurachi Y, Ito H, Sugimoto T (1990): Positive cooperativity in activation of the cardiac muscarinic K^+ channel by intracellular GTP. Pflügers Arch 416:216–218
Kurachi Y, Ito H, Sugimoto T, Katada T, Ui M (1989) Activation of atrial muscarinic K^+ channels by low concentrations of $\beta\gamma$ subunits of rat brain G protein. Pflügers Arch 413:325–327
Kurachi Y, Nakajima T, Sugimoto T (1986a) Acetylcholine activation of K^+ channels in cell-free membrane of atrial cells. Am J Physiol 251:H681–H684
Kurachi Y, Nakajima T, Sugimoto T (1986b) On the mechanism of activation of muscarinic K^+ channels by adenosine in isolated atrial cells: involvement of GTP-binding proteins. Pflügers Arch 407:264–274
Kurachi Y, Nakajima T, Sugimoto T (1986c) Role of intracellular Mg^{2+} in the activation of muscarinic K^+ channel in cardiac atrial cell membrane. Pflügers Arch 407: 572–574
Kurachi Y, Nakajima T, Sugimoto T (1987) Short-term desensitization of muscarinic K^+ channel current in isolated atrial myocytes and possible role of GTP-binding proteins. Pflügers Arch 410:227–233
Kurachi Y, Tung RT, Ito H, Nakajima T (1992) G protein activation of cardiac muscarinic K^+ channels. Prog Neurobiol 39:229–246
Lacey MG, Mercuri NB, North RA (1988) On the potassium conductance increase activated by $GABA_B$ and dopamine D_2 receptor in rat substantia nigra neurons. J Physiol (Lond.) 401:437–453
Lesage F, Duprat F, Fink M, Guillemare E, Coppola T, Lazdunski M, Hugnot J-P (1994) Cloning provides evidence for a family of inward rectifier and G-protein coupled K^+ channels in the brain. FEBS Lett 353:37–42
Lesage F, Guillemare E, Fink M, Duprat F, Heurteaux C, Fosset M, Romey G, Barhanin J, Lazdunski M (1995) Molecular properties of neuronal G-protein-activated inwardly rectifying K^+ channels. J Biol Chem 270:28660–28667
Li S, Okamoto T, Chun M, Sargiacomo M, Casanova JE, Hansen SH, Nishimoto I, Lisanti MP (1995) Evidence for a regulated interaction between heterotrimeric G proteins and caveolin. J Biol Chem 270:15693–15701
Liao YJ, Jan YN, Jan LY (1996) Heteromultimerization of G-protein-gated inwardly rectifying K^+ channel proteins GIRK1 and GIRK2 and their altered expression in *weaver* brain. J Neurosci 16:7137–7150
Lim NF, Dascal N, Labarca C, Davidson N, Lester HA (1995) A G protein-gated K channel is activated via β_2-adrenergic receptors and $G_{\beta\gamma}$ subunits in *Xenopus* oocytes. J Gen Physiol 105:421–439
Loewi O (1921) Über humorale Übertragbarkeit der Herznervenwirkung. Pflügers Arch 189:239–242
Loewi O, Navaratil E (1926) Übertragbarkeit der Herznevenwirking, X. Mittteilung. Über das Schicksal des Vagusstoffs. Pflügers Arch 214:678–688
Logothetis DE, Kim DH, Northup JK, Neer EJ, Clapham DE (1988) Specificity of action of guanine nucleotide-binding regulatory protein subunits on the cardiac muscarinic K^+ channel. Proc Natl Acad Sci USA 85:5814–5818
Logothetis DE, Kurachi Y, Galper J, Neer EJ, Clapham DE (1987) The $\beta\gamma$ subunits of GTP-binding proteins activate the muscarinic K^+ channel in heart. Nature 325: 321–326
Mackay D (1990a) Interpretation of relative potencies, relative efficacies and apparent affinity constants of agonist drugs estimated from concentration-response curves. J Theor Biol 142:415–427
Mackay D (1990b) Agonist potency and apparent affinity: interpretation using classical and steady-state ternary-complex models. Trend Pharmacol Sci 11:17–22
Makhina EN, Kelly AJ, Lopatin AN, Mercer RW, Nichols CG (1994) Cloning and expression of a novel human brain inward rectifier potassium channel. J Biol Chem 269:20468–20474
Monod J, Wyman J, Changeux JP (1965) On the nature of allosteric transitions: A plausible model. J Mol Biol 12:88–118

Morishige K, Inanobe A, Takahashi N, Yoshimoto Y, Kurachi H, Miyake A, Tokunaga Y, Maeda T, Kurachi Y (1996) G protein-gated K^+ channel (GIRK1) protein is expressed presynaptically in the paraventricular nucleus of the hypothalamus. Biochem Biophys Res Commun 220:300–305
Morishige K, Takahashi N, Findlay I, Koyama H, Zanelli JS, Peterson C, Jenkins NA, Copeland NG, Mori N, Kurachi Y (1993) Molecular cloning, functional expression and localization of an inward rectifier potassium channel in the mouse brain. FEBS Lett 336:375–380
Morishige K, Takahashi N, Jahangir A, Yamada M, Koyama H, Zanelli JS, Kurachi Y (1994) Molecular cloning and functional expression of a novel brain-specific inward rectifier potassium channel. FEBS Lett 346:251–256
Nakajima T, Sugimoto T, Kurachi Y (1992) Effects of anions on the G protein-mediated activation of the muscarinic K^+ channel in the cardiac atrial cell membrane; intracllular chloride inhibition of the GTPase activity of G_K. J Gen Physiol 99:665–682
Nanavati C, Clapham DE, Ito H, Kurachi Y (1990) A comparison of the roles of purified G protein subunits in the activation of the cardiac muscarinic K^+ channel. In: G Protein and Signal Transduction, pp. 29-pp. 41, Rockefeller University Press, New York
Navarro B, Kennedy ME, Velimirovic B, Bhat D, Peterson AS, Clapham DE (1996) Nonselective and $G_{\beta\gamma}$-insensitive *weaver* K^+ channels. Science 272:1950–1953
Noma A, Trautwein W (1978) Relaxation of the ACh-induced potassium current in the rabbit sinoatrial node cell. Pflügers Arch 377:193–200
North RA, Williams JT, Surprenant A, Christie MJ (1987) μ and δ receptors belong to a family of receptors that are coupled to potassium channels. Proc Natl Acad Sci USA 84:5487–5491
Osterrieder W, Yang QF, Trautwein W (1981) The time course of the muscarinic response to inophoretic acetylcholine application to the S-A node of the rabbit heart. Pflügers Arch. 389:283–291
Patil N, Cox DR, Bhat D, Faham M, Myers RM, Peterson AS (1995) A potassium channel mutation in weaver mice implicates membrane excitability in granule cell differentiation. Nature Genetics 11:126–129
Pennefather PS, Heisler S, MacDonald JF (1988) A potassium conductance contributes to the action of somatostatin-14 to suppress ACTH secretion. Brain Res 444: 346–350
Périer F, Radeke CM, Vandenberg CA (1994) Primary structure and characterization of a small-conductance inwardly rectifying potassium channel from human hippocampus. Proc Natl Acad Sci USA 91:6240–6244
Pfaffinger PJ, Martin JM, Hunter DD, Nathanson NM, Hille B (1985) GTP-binding proteins couple cardiac muscarinic receptors to a K channel. Nature 317:536–538
Ponce A, Bueno E, Kentros C, Vega-Saenz de Miera E, Chow A, Hillman D, Chen S, Zhu L, Wu MB, Wu X, Rudy B, Thornhill B (1996) G-protein-gated inward rectifier K^+ channel proteins (GIRK1) are present in the soma and dendrites as well as in nerve terminals of specific neurons in the brain. J Neurosci 16:1990–2001
Reuveny E, Slesinger PA, Inglese J, Morales JM, Iñiguez-Lluhi JA, Lefkowitz RJ, Bourne HR, Jan YN, Jan LY (1994) Activation of the cloned muscarinic potassium channel by G protein $\beta\gamma$ subunits. Nature 370:143–146
Saitoh O, Kubo Y, Miyatani Y, Asano T, Nakata H (1997) RGS8 accelerates G-protein-mediated medullation of K^+ currents. Nature 390:525–529
Sakmann B, Noma A, Trautwein W (1983) Acetylcholine activation of single muscarinic K^+ channels in isolated pacemaker cells of the mammalian heart. Nature 303: 250–253
Sakura, H, Smith PA, Gribble FM, Ashcroft FM (1995) Cloning and functional expression of the cDNA encoding a novel ATP-sensitive potassium channel subunit expressed in pancreatic β-cells, brain, heart and skeletal muscle. FEBS Lett 377:338–344

Scherer PE, Okamoto T, Chun M, Nishimoto I, Lodish HF, Lisanti MP (1996) Identification, sequence, and expression of caveolin-2 defines a caveolin gene family. Proc Natl Acad Sci USA 93:131–135

Signorini S, Liao YJ, Duncan SA, Jan LY, Stoffel M (1997) Normal cerebellar development but susceptibility to seizures in mice lacking G protein-coupled, inwardly rectifying K^+ channel GIRK2. Proc. Natl Acad Sci USA 94:923–937

Silverman SK, Lester HA, Dougherty DA (1996) Subunit stoichiometry of a heteromultimeric G protein-coupled inward-rectifier K^+ channel. J Biol Chem 271: 30524–30528

Simon MI, Strathmann MP, Gautam N (1991) Diversity of G proteins in signal transduction. Science 252:802–808

Slesinger PA, Patil N, Liao YJ, Jan YN, Jan LY, Cox DR (1996) Functional effects of the mouse *weaver* mutation on G protein-gated inwardly rectifying K^+ channels. Neuron 16:321–331

Slesinger PA, Reuveny E, Jan YN, Jan LY (1995) Identification of Structural elements involved in G protein gating of the GIRK1 potassium channel. Neuron 15:1145–1156

Soejima M, Noma A (1984) Mode of regulation of the ACh-sensitive K-channel by the muscarinic receptor in rabbit atrial cells. Pflügers Arch 400:424–431

Spauschus A, Lentes K-U, Wischmeyer E, Di*β*mann E, Karschin C, Karschin A (1996) A G-protein-activated inwardly rectifying K^+ channel (GIRK4) from human hippocampus associates with other GIRK channels. J Neurosci 16:930–938

Stoffel M, Tokuyama Y, Trabb JB, German MS, Tsaar M-L, Jan LY, Polonsky KS, Bell GI (1995) Cloning of rat K_{ATP}-2 channel and decreased expression in pancreatic islets of male zucker diabetic fatty rats. Biochem Biophys Res Commun 212: 894–899

Sui JL, Chan KW, Logothetis DE (1996) Na^+ activation of the muscarinic K^+ channel by a G-protein-independent mechanism. J Gen Physiol 108:381–391

Sui JL, Petit-Jacques J, Logothetis DE (1998) Phosphatidylinositol phosphates exert a permissive influence on the gating of G protein-activated K^+ channels. In Potassium Ion Channels; their structure, function and diseases, Academic Press, Edited by Kurachi Y, Jan LY, Lazdunski M, in press

Surprenant A, Horstman DA, Akbarali H, Limbird LE (1992) A point mutation of the α_2-adrenoceptor that blocks coupling to potassium but not calcium currents. Science 257:977–980

Takahashi N, Morishige K, Jahangir A, Yamada M, Findlay I, Koyama H, Kurachi Y (1994) Molecular cloning and functional expression of cDNA encoding a second class of inward rectifier potassium channels in the mouse brain. J Biol Chem 269:23274–23279

Takumi T, Ishii T, Horio Y, Morishige K, Takahashi N, Yamada M, Yamashita T, Kiyama H, Sohmiya K, Nakanishi S, Kurachi Y (1995) A novel ATP-dependent inward rectifier potassium channel expressed predominantly in glial cells. J Biol Chem 270:16339–16346

Tang Z, Scherer PE, Okamoto T, Song K, Chu C, Kohtz DS, Nishimoto I, Lodish HF, Lisanti MP (1996) Molecular cloning of caveolin-3, a novel member of the caveolin gene family expressed predominantly in muscle. J Biol Chem 271:2255–2261

Thomsen WJ, Jacquez JA, Neubig RR (1988) Inhibition of adenylate cyclase is mediated by the high affinity conformation of the α_2-adrenergic receptor. Mol Pharmacol 34:814–822

Tinker A, Jan YN, Jan LY (1996) Regions responsible for the assembly of inwardly rectifying potassium channels. Cell 87:857–868

Trautwein W, Dudel J (1958) Zum Mechanismus der Membranwirkung des Acetylcholines as der Herzmusklefaser. Pflügers Arch 398:283–291

Tsaur M-V, Menzel S, Lai F-P, Espinosa III R, Concannon P, Spielman RS, Hanis CL, Cox NJ, Le Beau MM, German MS, Jan LY, Bell GI, Stoffel M (1995) Isolation of a cDNA clone encoding a K_{ATP} channel-like protein expressed in insulin-secreting cells, localization of the human gene to chromosome band 21q22.1, and linkage studies with NIDDM. Diabetes 44:592–596

Tucker SJ, Pessia M, Adelman JP (1996) Muscarine-gated K^+ channel: subunit stoichiometry and structural domains essential for G protein stimulation. Am J Physiol 271:H379–H385

Velimirovic BM, Gordon EA, Lim NF, Navarro B, Clapham DE (1996) The K^+ channel inward rectifier subunits form a channel similar to neuronal G protein-gated K^+ channel. FEBS Lett 379:31–37

Wickman KD, Clapham DE (1995) Ion channel regulation by G proteins. Physiol Rev 75:865–885

Wickman KD, I–iguez-Lluhl JA, Davenport PA, Taussig R, Krapivinsky GB, Linder ME, Gilman AG, Clapham DE (1994) Recombinant G-protein $\beta\gamma$-subunits activate the muscarinic-gated atrial potassium channel. Nature 368:255–257

Yamada M, Ho YK, Lee RH, Kontani K, Takahashi K, Katada T, Kurachi Y (1994) Muscarinic K^+ channels are activated by $\beta\gamma$ subunits and inhibited by the GDP-bound form of a subunit of transducin. Biochem Biophys Res Commun 200:1484–1490

Yamada M, Jahangir A, Hosoya Y, Inanobe A, Katada T, Kurachi Y (1993) G_{K*} and brain $G_{\beta\gamma}$ activate muscarinic K^+ channel through the same mechanism. J Biol Chem 268:24551–24554

Yan K, Gautam N (1996) A domain on the G protein β subunit interacts with both adenylyl cyclase 2 and the muscarinic atrial potassium channel. J Biol Chem 271: 17597–17600

Yang J, Jan YN, Jan LY (1995) Determination of the subunit stoichiometry of an inwardly rectifying potassium channel. Neuron 15:1441–1447

Yatani A, Codina J, Brown AM, Birnbaumer L (1987) Direct activation of mammalian atrial muscarinic potassium channels by GTP regulatory protein G_K. Science 235: 207–211

Yatani A, Mattera R, Codina J, Graf R, Okabe K, Padrell E, Iyengar R, Brown AM, Birnbaumer L (1988) The G protein-gated atrial K^+ channel is simulated by three distinct Giα-subunits. Nature 336:680–682

Yoshimoto Y, Morishige K-I, Inanobe A, Tokunaga Y, Maeda T, Kurachi Y (1997) Localization of G protein-gated inwardly rectifying K^+ channel in rat forebrain. Jpn J Pharmacol 73 (Supple 1) (abstract):80P

CHAPTER 13
Potassium Channels with Two Pore Domains

F. LESAGE and M. LAZDUNSKI

A. K⁺ Channels with One Pore Domain

Among ion channels, the K^+-selective channels form the largest family and probably the most puzzling. Electrophysiological studies have revealed a wide variety of K^+ currents which differ by their gating properties, their unitary conductance, their pharmacology and their regulations (RUDY 1988; HILLE 1992). A probable explanation for this diversity is to offer to each excitable cell the repertoire of K^+ currents that is the most suitable for its function. Molecular characterization of K^+ channels is recent. From the original cloning of the *Shaker* gene from the *Drosophila* in 1987, a deluge of data concerning the structure of K^+ channels has occurred. Today, more than 100 pore-forming K^+ channel subunits as well as a variety of auxiliary subunits have been cloned. Heterologous expression of these proteins has allowed to reconstitute voltage-gated (Kv), Ca^{2+}-activated (KCa), inwardly rectifying (IRK), G-protein-coupled (GIRK) and ATP-sensitive (KATP) K^+ channels and to determine their biophysical, pharmacological, and regulation properties (for reviews see CHANDY et al. 1995; ROEPER et al. 1996; ISOMOTO et al. 1997; JAN et al. 1997; NICHOLS et al. 1997). On the other hand, the association of site-directed mutagenesis and electrophysiology techniques has allowed us to define structural features that are associated with particular K^+ functional properties.

Despite the number of functional classes and the number of different channels in each class, all these cloned proteins fall into only two structural types of pore-forming K^+channel subunits. The *Shaker* -type comprises the subunits forming the pore of the Kv and the KCa channels. They have a common hydrophobic core containing six transmembrane segments (TMS) and a particular domain called the P domain (P for pore). This domain has been shown to be involved in the formation of the K^+-selective filter of the pore (for review see MACKINNON 1995). The IRK-type comprises the pore-forming subunits of the IRK, GIRK, and KATP channels. These proteins have two TMS and one P domain. The existence of only two different structural classes evoked a question concerning the molecular nature of yet unidentified K^+ channels that belong to other functional classes: is it possible that a part or all these channels have another structure and if so, how to isolate such structures?

B. K⁺ Channels with Two Pore Domains

I. TWIK, the Archetype of a Novel Structural Class of K^+ Channel

1. Cloning and Gene Organization

We recently isolated such a novel structure (Fig. 1). The advent of the systematic sequencing projects, and the easy computational analysis of the data released in public DNA databases allowed us to identify a partial sequence from human cDNA (LESAGE 1998). This anonymous sequence encoded a sequence similar to the P domain of *Shaker* and IRK channels. The full cDNA was isolated by screening a human library and sequenced (LESAGE et al.

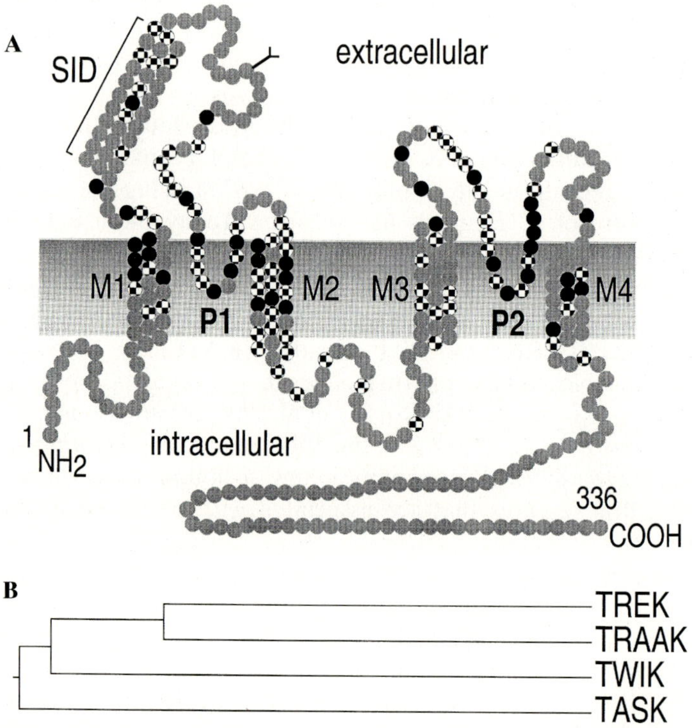

Fig. 1. A Schematic representation of the human TWIK K^+ channel subunit. The individual residues are denoted by *rounds*. The four potential transmembrane segments are noted as M1–M4 and the two P domains P1 and P2. The absence of signal peptide and the demonstration of the extracellular localization of the linker M1P1 loop suggest this membrane topology where both NH2- and COOH-termini are cytoplasmic. The potential hydrophilic α-helix that is involved in the formation of disulfide-bridged homodimers is noted SID for Self-Interacting Domain. The residues which are identical or conserved in TWIK, TREK, TASK, and TRAAK are shown in *black or hatched*, respectively. **B** Dendrogramm derived from the sequence alignment of TWIK, TREK, TASK, and TRAAK

1996b). It encodes a 336 amino acid polypeptide containing two P domains (P1 and P2). The hydrophobicity analysis predicts the presence of four TMS (M1 to M4), two TMS flanking each P domain as expected for a K^+ channel. An unusual, large loop of 60 amino acids is present between M1 and P1 that extends the length of M1-P1 linker (Fig. 1). This structure (4TMS/2P) is very different of that of *Shaker* (6TMS/1P) and IRK (2TMS/1P). The striking feature is the presence of two P domains instead of only one. This novel class of K^+ channel has been called the two P domain class to emphasize this point. Beside the two P domains, no sequence homology was found between this protein and the *Shaker*- and IRK-type channels. A TWIK cDNA was also cloned from mouse brain (LESAGE et al. 1997). It encodes a protein 94% identical to its human counterpart. At the gene level, the TWIK coding sequence is contained within three exons on a region larger than 40 kbp (ARRIGHI et al. 1998). The mouse gene has been mapped to chromosome 8 (ARRIGHI et al. 1998), consistent with its localization to 1q42–43 in human (LESAGE et al. 1996c).

2. Functional Expression

This TWIK K^+ channel directs the expression of K^+-selective currents which are instantaneous and sustained, i.e., that do not have kinetics of activation, inactivation or deactivation (LESAGE et al. 1996b, 1997) (Fig. 2). The channels are open at all membrane potentials. However, a saturation of outward currents is observed for high depolarizations suggesting a weak inward rectification. This channel was called TWIK–1 for *T*andem of P domains in a *W*eak *I*nward rectifier *K*$^+$ channel. As expected from its biophysical properties, TWIK activity controls the resting membrane potential (E_m). *Xenopus* oocytes expressing TWIK are more polarized than control oocytes, their resting potential reaching a value close to the K^+ equilibrium potential (E_K). TWIK channels are flickering and their unitary conductance is 19 pS at +80 mV and 34 pS at –80 mV. The inward rectification is abolished in the absence of internal Mg^{2+}. TWIK currents are blocked by Ba^{2+}, quinine, and quinidine (50 μmol/l > IC_{50} > 100 μmol/l) and slightly blocked by tetraethylammonium (TEA) (30% inhibition at 10 mmol/l). They are insensitive to 4-aminopyridine (4-AP) and to toxins that block Kv or KCa K^+ channels (charybdotoxin, apamin, and dendrotoxin). TWIK activity is regulated in opposite ways by the activation of the protein kinase C (PKC) and by acidification of the intracellular medium. Activation of the PKC increases TWIK. This effect seems to be indirect since the mutation of the unique consensus site for phosphorylation by PKC did not modify the sensitivity of these channels to agents that activate PKC. The inhibition of TWIK by internal acidification is also indirect. This effect is not seen in the inside-out patch configuration when the internal face of the channel is faced to the acidic medium. The functional properties of TWIK are novel. Another weak inward rectifier of the IRK-type called ROMK1 has been previously cloned from rat kidney, but it is not expressed in brain and, except for

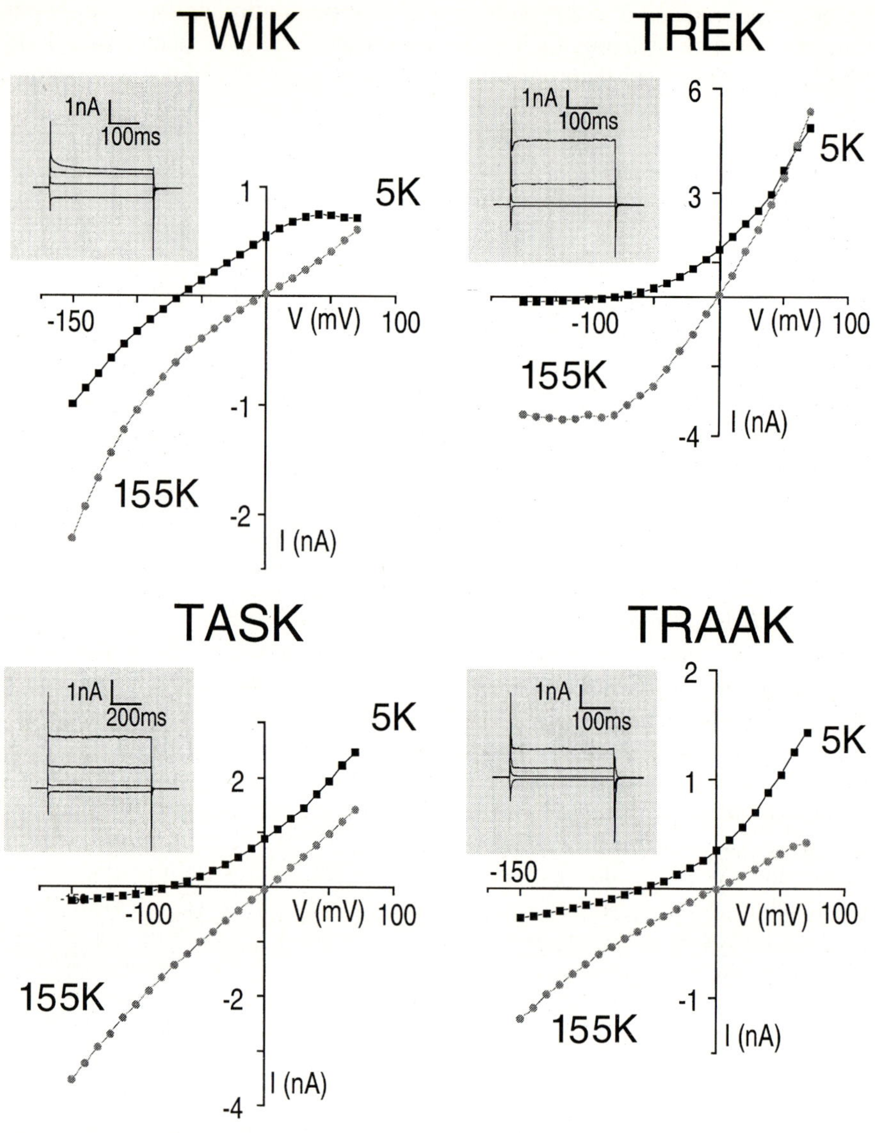

Fig. 2. Functional properties of two P domain K^+ channels expressed in COS cells. Current-voltage relationships recorded at the end of 500 ms voltage pulses from –150 to 50 mV, in 10 mV steps, in low (5 mmol/l) or high (155 mmol/l) external K^+ solutions. *insets*: currents recorded at –120, –60, 0, 60 mV from a holding potential of –80 mV

its weak rectification, the other biophysical, pharmacological, and regulation properties are very different from those displayed by TWIK (Ho et al. 1993).

3. Structure of the Channel

TWIK mRNA is widely distributed in human tissues, and is particularly abundant in heart and brain (Lesage et al. 1996b). In situ hybridization from mouse brain showed that TWIK expression is restricted to a few regions, with the highest levels in cerebellar granule and Purkinje cells, brainstem, hippocampus and cerebral cortex (Lesage et al. 1997). The same results were obtained by immunolocalization using anti-TWIK affinity-purified antibodies (unpublished results). The apparent molecular weight of TWIK in mouse brain is 81 kDa (Lesage et al. 1997). A 40 kDa form is revealed after treatment with a reducing agent strongly suggesting that TWIK dimerizes via a disulfide bridge. We showed that TWIK, expressed in insect cells by using a baculovirus recombinant, self-associates to give dimers containing an interchain disulfide bond (Lesage et al. 1996d). This assembly involves an extracellular domain of 44 amino acids forming a potential hydrophilic α-helix (Fig. 1). Cysteine 69 which is part of this interacting domain is implicated in the formation of the disulfide bridge. Replacing this cysteine with a serine residue results in the loss of functional K^+ channel expression (Lesage et al. 1996d). What are the implications of these first results concerning the structure-function relationships of TWIK? It has been demonstrated that subunits belonging to the *Shaker* and IRK structural classes form non-covalent tetramers (Mackinnon 1991; Yang et al. 1995). This probably means that four P domains are necessary for the formation of the K^+-selective pore in these classes of K^+ channels. TWIK forms homodimers stabilized by an inter-subunit disulfide bond which is necessary for the function and each constitutive subunit contains 2 P domains. This probably means that, in active TWIK covalent homodimers, both P domains in each subunits are fully functional.

II. Related K^+ Channels in Mammals

Shaker- and IRK-type channels form large gene families. To test the possibility that TWIK could be the founding member of a novel family, we tried to clone homologous proteins by using two different approaches. The first one was to carry out degenerate PCR experiments by taking advantage of moderate sequence conservations between TWIK and several potential related proteins from *C. Elegans* (see Sect. III). This approach led to the cloning of TREK (Fink et al. 1996). The second strategy was to search the public DNA databases to find sequences encoding protein fragments homologous to TWIK. This approach resulted in the cloning of TASK (Duprat et al. 1997) and TRAAK (Fink et al. 1998).

1. TREK is an Unusual Outward Rectifier K$^+$ Channel

TREK for *T*WIK-*RE*lated *K*$^{\pm}$ channel is a 370 amino acid polypeptide which has the same overall structure as TWIK, i.e., 4 TMS and two P domains (FINK et al. 1996). Despite this similar topology, the amino acid identity between TREK and TWIK is very low (around 26%). TREK is larger that TWIK, 370 vs 336 aa, because its amino and carboxy termini are more extended. Like TWIK, it contains an extended M1P1 loop and a cysteine residue at a position equivalent to the cysteine 69 of TWIK. This suggests that the observed covalent homodimerization of TREK (unpublished results) could occur via the same mechanism as TWIK. TREK is expressed in most mouse tissues, and is particularly abundant in lung and brain (FINK et al. 1996) (Fig. 3) . In situ hybridization in the latter tissue showed that the TREK expression is high in the olfactory bulb, hippocampus, and cerebellum. The gene encoding TREK has been mapped to human chromosome 1q41 (LESAGE et al. 1998).

Currents generated by TREK have been studied in oocytes and in transfected COS cells (FINK et al. 1996). They are K$^+$-selective and instantaneous. The current-potential (IV) curve of these currents is quite different of the IV curve of TWIK (Fig. 2). TREK channels pass preferentially outward currents. A similar outward rectification is seen with the *Shaker*-type Kv channels. For the Kv channels, this apparent rectification is due to their intrinsic voltage-dependence. They are open only from a fixed threshold potential which is determined by their voltage-sensor. In the case of TREK, the mechanism is completely different because the inversion potential of TREK currents is not fixed and closely follows the K$^+$ equilibrium potential. Because of this prop-

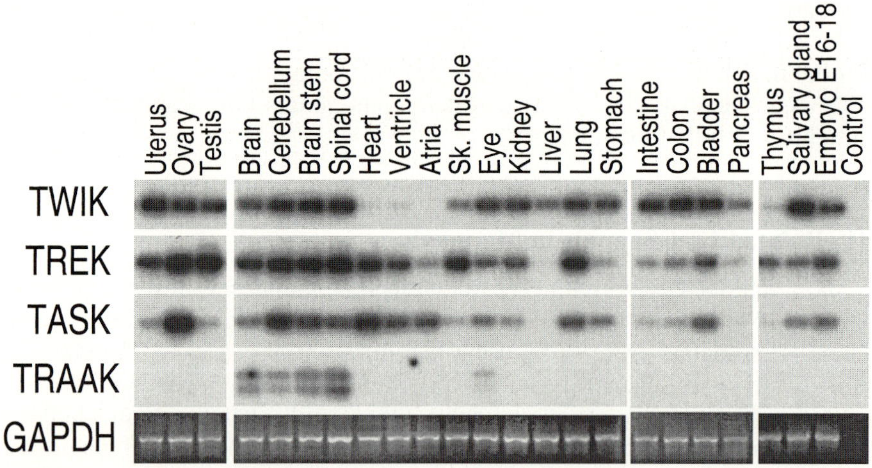

Fig. 3. Distribution of TWIK, TREK, TASK, and TRAAK in mouse adult tissues. DNAs were amplified by PCR using specific primers and analyzed by Southern blot using internal ^{32}P–labeled oligonucleotides. Two bands were obtained for TRAAK that correspond to splice variants

erty, TREK, like TWIK, is able to drive E_m close to E_K. TREK currents are inhibited by agents activating PKC and protein kinase A (PKA).

2. TASK is an Open Rectifier Channel Highly Sensitive to External pH

The third two P domain K^+ channel that we have cloned and expressed, is TASK, for *T*WIK-related *A*cid-*S*ensitive *K*$^+$ channel (DUPRAT et al. 1997). This human 395 aa long polypeptide has the same structural topology as TWIK and TREK and a low amino-acid sequence identity (around 25%). TASK mRNA is widely expressed in human and is particularly abundant in pancreas, placenta, and brain. TASK has also been cloned in rat (rTASK, r for rat) (LEONOUDAKIS et al. 1998) and mouse (cTBAK-1 for *c*ardiac *T*wo-pore *BA*ckground *K*$^{\pm}$ channel) (KIM et al. 1998). In rodents, TASK is highly expressed in heart (DUPRAT et al. 1997; KIM et al. 1998; LEONOUDAKIS et al. 1998) and particularly in atria (DUPRAT et al. 1997). In mouse brain, it was detected throughout the cell layers of the cerebral cortex, in the CA1-CA4 pyramidal cell layers, in the granule cells of the dentate gyrus, in the habenula, in the paraventricular thalamic nuclei, in the amyloid nuclei, in the substantia nigra and in the Purkinje and granular cells of the cerebellum (DUPRAT et al. 1997). The human TASK gene has been mapped to chromosome 2p23 (LESAGE et al. 1998).

TASK channels produce K^+-selective, instantaneous, and sustained currents. When external (K^+) is low (2 mmol/l), TASK current are outwardly rectifying like TREK currents. However, this rectification is not observed for high K^+ concentrations (155 mmol/l) (Fig. 2). In fact, this rectification can be approximated by the Goldman-Hodgkin-Katz equation that predicts a curvature of the current-voltage plot in asymmetric K^+ conditions. This strongly suggests that TASK currents show no rectification other than that predicted from the constant-field assumptions for an open channel and that TASK lacks intrinsic voltage-sensitivity. Moreover, the open probability of single TASK currents is independent from the patch potential (KIM et al. 1998; LEONOUDAKIS et al. 1998). For these reasons, TASK can be described as an open rectifier. These properties, absence of activation and inactivation, as well as voltage-independence, are characteristic of conductances that are referred to as background or leak conductances. As expected from these properties, TASK, as TWIK and TREK, is able to drive the resting membrane potential of expressing oocytes very close to E_K. The unitary conductance of the channel is 14–16 pS in symmetrical K^+ concentrations. TASK is slightly sensitive to Ba^{2+} (less than 20% of inhibition at 0.1 mmol/l) and insensitive to TEA and 4-AP. The rat TASK was shown to be blocked by Zn^+ ($IC_{50} = 175\,\mu$mol/l) and the local anesthetic bupivicaine ($IC_{50} = 70\,\mu$mol/l) (LEONOUDAKIS et al. 1998). An essential property of TASK is that it is very sensitive to variations of extracellular pH in narrow physiological range (DUPRAT et al. 1997; LEONOUDAKIS et al. 1998). As much as 90% of the maximum current is recorded at pH 7.7 and only 10% at pH 6.7. On the other hand, activation of PKA produces inhibition of TASK whereas activation of PKC has no effect (LEONOUDAKIS et al. 1998).

3. TRAAK Forms K^+ Channels Activated by Unsaturated Fatty Acids

TRAAK, for *T*WIK-*R*elated *A*rachidonic acid-*A*ctivated *K*$^+$ channel, was cloned from a mouse brain cDNA library (FINK et al. 1998). This 398 amino acid polypeptide is more related to TREK (38% of aa identity) than to other two P domain K^+ channels (25% of aa identity) (Fig. 1B). A shorter form of TRAAK that probably results from alternative splicing has also been identified. Deletion of 126 bp nucleotides in the coding sequence leads to the formation of a premature stop codon. The resulting open reading frame encodes a protein of 67 aa containing only the amino terminus, the M1 domain, and a short part of the M1P1 loop of TRAAK. This truncated form is not functional. TRAAK is unique among the K^+ channels with two pore domains because it is exclusively expressed in neuronal cells in brain, cerebellum, and spinal cord as well as in the retina (FINK et al. 1998) (Fig. 3).

In transfected cells, TRAAK produces K^+-currents very similar to TASK, i.e., open rectifying and time-independent currents (Fig. 2). The unitary conductance of TRAAK is 45 pS. TRAAK currents are only partially inhibited by Ba^{2+} at high concentrations and are insensitive to the other classical K^+ channels blockers TEA, 4–AP, and Cs^+. The particularly salient feature of TRAAK is to be activated by arachidonic acid (AA). This activation is completely reversible and concentration-dependent. It is not prevented when the AA perfusion is supplemented with a mixture of inhibitors of the AA metabolism pathway. This demonstrates that AA effect on TRAAK is direct and does not require the production of another active eicosanoid. Moreover, the reversible effect of AA on TRAAK is observed in both outside-out and inside-out patches as expected for a direct effect of AA. This effect is specific for unsaturated FAs. Oleate, linoleate, linoleate, arachidonate, eicosapentaenoate, and docosahexaenoate (DOHA) all strongly activate TRAAK, while saturated FAs such as palmitate, stearate, and arachidate are ineffective. Derivatives of active FAs (AA and DOHA), where the carboxylic function was substituted with an alcohol or a methyl ester function, are also inactive on TRAAK. In both oocytes and transiently-transfected COS cells, a basal TRAAK current can be recorded in the absence of any AA application. In stably-transfected COS cells, this basal current is almost undetectable (unpublished results). This result suggests that TRAAK channels are normally closed in physiological conditions in the absence of free AA. The observed basal currents in oocytes and transiently-transfected cells are probably due to the presence of low endogenous levels of free polyunsaturated FAs.

III. Related Channels in Worm, Fly, Yeast, and Plant

K^+ channels with two pore domains are not found only in mammals but seem to be largely expressed in the animal kingdom, from worms to human throughout fly. Data issued from the systematic sequencing of the genome of *Caenorhabditis elegans* provide information concerning the diversity of K^+

channels found in a single multicellular organism. From the part of the genome already sequenced (around 80%), 39 genes potentially encode proteins structurally similar to TWIK (25–28% of amino acid identity) as compared to only 14 genes for *Shaker* and IRK subunits (SALKOFF et al. 1997). In mammals, more than 40 genes encoding *Shaker* and IRK subunit have already been identified. On one hand, we do not know whether these *C. elegans* proteins are all functional. On the other hand, it cannot be excluded that this diversity of two P domain K^+ channels corresponds to a special adaptation of *C. elegans*. However, it is certainly possible that a high number of 2P domains-containing channels is also present in mammals. In such a case, they will certainly be characterized rapidly with the help of data coming from systematic gene and cDNA sequencing programs.

A TWIK-related channel with 4TMS and 2P domain has also been cloned from the fly *Drosophila* for its capacity to complement a K^+-transport deficient strain of yeast (GOLDSTEIN et al. 1996). In oocyte, this channel directs the expression of K^+ currents with electrophysiological properties very similar to TASK and TRAAK currents. As TASK and TRAAK, this channel called DORK for *Drosophila Open Rectifier K^+* channel, behaves as a background K^+ channel and lacks intrinsic voltage-dependence. In the fly, DORK mRNA is found in the nervous system. No information concerning the regulation and a possible sensitivity of this channel to extracellular pH variations or unsaturated fatty acids has yet been reported.

To close this review on the two P domain K^+ channel distribution among living species, it should be noted that two other related channels have been isolated from the yeast *Saccharomyces cerevisiae* and the plant *Arabidopsis thaliana*. The yeast channel, called TOK/DUK/YKC/YORK, has eight potential TMS (KETCHUM et al. 1995; ZHOU et al. 1995; LESAGE et al. 1996a; REID et al. 1996) and the plant channel, called KCO1, four TMS (CZEMPINSKI et al. 1997). Both have two pore domains but they cannot really be classified in the TWIK family since no significant sequence similarity is found outside the P domains. On the other hand, neither the yeast nor the plant channel present the extended M1P1 linker loop characteristic of TWIK-related channels. TOK and KCO1 channels produce outward rectifying currents whose inversion potentials closely follow E_K as observed for TREK currents. KCO1 has a steep Ca^{2+} dependence, and its activation is strongly dependent on the presence of nanomolar concentration of cytosolic free Ca^{2+}.

C. Concluding Remarks

The literature on background or leak K^+ channels is not abundant compared to other types of K^+ channels. This probably originates from the fact that these channels are difficult to study: they are voltage- and time-independent, and they have no specific pharmacology. Some native currents that seem to be voltage- and time-independent have been previously reported in the literature. They have been observed in invertebrates, in *Aplysia* sensory neurons (Siegelbaum et al. 1982) and in lobster stretch receptor neurons (Theander et al. 1996), as well as in vertebrates, in bullfrog sympathetic ganglia (Koyano et al. 1992) and smooth muscle cells (Ordway et al. 1991), in *Xenopus* myelinated nerve (Koh et al. 1992) and demyelinated axons (Wu et al. 1993), in guinea pig submucosal neurons (Shen et al. 1992), in rat carotid bodies (Buckler 1997), ventricular myocytes (Yue et al. 1988; Backx et al. 1993), and hippocampal (Premkumar et al. 1990a,b; Kim et al. 1995) and premotor respiratory neurons (Wagner et al. 1997).

TWIK, TREK, TASK, and TRAAK are time-independent. Only TASK and TRAAK are strictly independent from the potential. However, for TWIK and TREK, the voltage-dependence is weak. These K^+ channels with two P domains are not gated by the voltage and are active at the resting potential. Moreover, they have a poor pharmacology. For these reasons, they are referred to as background K^+ channels. They probably play a major role in the control of the resting membrane potential and, in turn, in the modulation of electrical activity of cells. A salient feature of these channels is the modulation of their activity by a variety of intracellular and extracellular messages. TWIK is activated by stimulating PKC, and inhibited by the intracellular acidification, TREK is inhibited by activating both PKC and PKA, TASK activity is very sensitive to extracellular pH changes, and TRAAK is sensitive to unsaturated fatty acids. This suggests that these channels could control the membrane potential in response to a variety of factors. For example, the modulation of TASK by external protons probably has important implications for its physiological function. Stimulus-elicited pH shifts have been characterized in a wide variety of neural tissues by using extracellular pH-sensitive electrodes. Electrical stimulation of Schaeffer collateral fibers in the hippocampal slice, or light stimulation of the retina, or parallel fibers in cerebellum, produce pH-shifts corresponding to bursts of H^+ or OH^- creating small pH variations from the external physiological pH value of 7.4 (up to 0.3 pH unit in the alkaline or acidic directions) and are rapid, in the second to the 30 s range. They might actually be larger or much larger in range or shorter in time course in the vicinity of the synaptic cleft. The strong modulation of TASK at external pH values favors the idea that H^+ can be a natural modulator of neuronal activity. Large acidic pH variations can be observed in physiopathological situations such as epileptiform activity and spreading depression. They can also be observed in the course of brain and cardiac ischemia.

The regulation of TRAAK by AA as well as other unsaturated fatty acids also has important implications for its physiological functions. Native fatty acids (FA-activated K^+ currents that have been described in neurons from rat brain) are very similar to TRAAK in terms of both electrophysiological behavior and pharmacological properties (Premkumar et al. 1990b; Kim et al. 1995). Such FA-activated currents have also been described in the heart (Kim et al. 1989) and in smooth muscle cells (Ordway et al. 1991). It is probable that TRAAK is an essential component of these FA-activated K^+ channels in the neuronal cells. All situations associated with variations of AA levels (pathological like ischemia or physiological like receptor-dependent hydrolysis of AA-containing phospholipids) are expected to lead to a modification of the activity of this peculiar class of K^+ channels.

In the past years, the cloning of the *Shaker* gene and of the first inward rectifiers has rapidly led to the identification of a lot of related channels. Molecular characterization of all these channels has shed considerable light on the structures and functions of Kv, KCa, Kir, and KATP channels as well as associated pathologies. By analogy, the identification of TWIK has provided access to a new structural and functional class of background K^+ channels that comprises H^+-gated (TASK) and FA-activated K^+ (TRAAK) channels. There still remain many open questions: How many different types of TWIK-related K^+ channels are there? Where and when are they expressed exactly? What is the exact relationship between the regulation observed in recombinant systems and the in vivo regulation of these channels? Are there genetic diseases associated with these channels, particularly in the nervous and the cardio-vascular systems as well as related to kidney, lung, and even immune system dysfunction? Can we find a specific pharmacology for these different classes of 2P domain-K^+ channels?

Acknowledgements. Thanks are due to all the investigators of the laboratory that have taken part to the study of K^+channels with two Pdomains. We also thank V. Briet for secretarial assistance.

References

Arrighi I, Lesage F, Scimeca JC, Carle GF, Barhanin J (1998) Structure, chromosome localization, and tissue distribution of the mouse TWIK K^+ channel gene. FEBS Lett 425:310–316

Backx PH, Marban E (1993) Background potassium current active during the plateau of the action potential in Guinea-pig ventricular myocytes. Circulation Research 72:890–900

Buckler KJ (1997) A novel oxigen-sensitive potassium current in rat carotid body type I cells. J Physiol (London) 498:649–662

Chandy KG, Gutman GA (1995) Voltage-gated potassium channel genes. In: (ed) Ligand and voltage-gated ion channels, FL: CRC, Boca Raton, 1–71

Czempinski K, Zimmermann S, Ehrhardt T, MullerRober B (1997) New structure and function in plant K^+ channels: KCO1, an outward rectifier with a steep $Ca2^+$ dependency. EMBO J 16:2565–2575

Duprat F, Lesage F, Fink M, Reyes R, Heurteaux C, Lazdunski M (1997) TASK, a human background K^+ channel to sense external pH variations near physiological pH. EMBO J 16:5464–5471

Fink M, Duprat F, Lesage F, Reyes R, Romey G, Heurteaux C, Lazdunski M (1996) Cloning, functional expression and brain localization of a novel unconventional outward rectifier K^+ channel. EMBO J 15:6854–6862

Fink M, Lesage F, Duprat F, Heurteaux C, Reyes R, Fosset M, Lazdunski M (1998) A neuronal two P domain K^+ channel stimulated by arachidonic acid and polyunsaturated fatty acids. EMBO J (in press)

Goldstein SAN, Price LA, N. RD, Pausch MH (1996) ORK1, a potassium-selective leak channel with two pore domains cloned from *Drosophila melanogaster* by expression in *Saccharomyces cerevisiae*. Proc Natl Acad Sci USA 93:13256–13261

Hille B (1992) Ionic channels of excitable membranes (second edition). Sinauer, Sunderland, Massachusetts

Ho K, Nichols CG, Lederer WJ, Lytton J, Vassilev PM, Kanazirska MV, Hebert SC (1993) Cloning and expression of an inwardly rectifying ATP-regulated potassium channel. Nature 362:31–38

Isomoto S, Kondo C, Kurachi Y (1997) Inwardly rectifying potassium channels: their molecular heterogeneity and function. Jpn J Physiol 47:11–39

Jan LY, Jan YN (1997) Voltage-gated and inwardly rectifying potassium channels. J Physiol 505:267–282

Ketchum KA, Joiner WJ, Sellers AJ, Kaczmarek LK, Goldstein SAN (1995) A new family of outwardly rectifying potassium channel proteins with two pore domains in tandem. Nature 376:690–695

Kim D, Clapham DE (1989) Potassium channels in cardiac cells activated by arachidonic acid and phospholipid. Science 244: 1174–1176

Kim D, Fujita A, Horio Y, Kurachi Y (1998) Cloning and functional expression of a novel cardiac two-pore background K^+ channel (cTBAK-1). Circ Res 82:513–518

Kim DH, Sladek CD, Aguadovelasco C, Mathiasen JR (1995) Arachidonic acid activation of a new family of K^+ channels in cultured rat neuronal cells. J Physiol-London 484:643–660

Koh DS, Jonas P, Bräu ME, Vogel W (1992) A TEA-insensitive flickering potassium channel active around the resting potential in myelinated nerve. J Membrane Biol 130:149–162

Koyano K, Tanaka K, Kuba K (1992) A patch-clamp study on the muscarine-sensitive potassium channel in bullfrog sympathetic ganglion cells. J Physiol (London) 454:231–246

Leonoudakis D, Gray AT, Winegar BD, Kindler CH, Harada M, Taylor DM, Chavez RA, Forsayeth JR, Yost CS (1998) An open rectifier potassium channel with two pore domains in tandem cloned from rat cerebellum. J Neurosci 18:868–877

Lesage F (1998) An internet-based computational strategy to identify and clone K^+ channels with new structures. Methods Mol Biol (in press)

Lesage F, Guillemare E, Fink M, Duprat F, Lazdunski M, Romey G, Barhanin J (1996a) A pH-sensitive yeast outward rectifier K^+ channel with two pore domains and novel gating properties. J Biol Chem 271:4183–4187

Lesage F, Guillemare E, Fink M, Duprat F, Lazdunski M, Romey G, Barhanin J (1996b) TWIK-1, a ubiquitous human weakly inward rectifying k^+ channel with a novel structure. EMBO J 15:1004–1011

Lesage F, Lauritzen I, Duprat F, Reyes R, Fink M, Heurteaux C, Lazdunski M (1997) The structure, function and distribution of the mouse TWIK-1 K^+ channel. FEBS Lett 402:28–32

Lesage F, Lazdunski M (1998) Mapping of human potassium channel genes TREK-1 (*KCNK2*) and TASK (*KCNK3*) to chromosomes 1q41 and 2p23. Genomics (in press)

Lesage F, Mattei MG, Fink M, Barhanin J, Lazdunski M (1996c) Assignment of the human weak inward rectifier K^+ channel TWIK-1 gene to chromosome 1q42-q43. Genomics 34:153–155

Lesage F, Reyes R, Fink M, Duprat F, Guillemare E, Lazdunski M (1996d) Dimerization of TWIK-1 K^+ channel subunits via a disulfide bridge. EMBO J 15:6400–6407

Mackinnon R (1991) Determination of the subunit stoichiometry of a voltage-activated potassium channel. Nature 350:232–235

Mackinnon R (1995) Pore loops: an emerging theme in ion channel structure. Neuron 14:889–892

Nichols CG, Lopatin AN (1997) Inward rectifier potassium channels. Annu Rev Physiol 59:171–191

Ordway RW, Singer JJ, Walsh Jr JV (1991) Direct regulation of ion channels by fatty acids. TINS 14:96–100

Premkumar LS, Chung SH, Gage PW (1990a) GABA-induced potassium channels in cultured neurons. Proc R Soc Lond B 241:153–158

Premkumar LS, Gage PW, Chung SH (1990b) Coupled potassium channels induced by arachidonic acid in cultured neurons. Proc R Soc Lond B 242:17–22

Reid JD, Lukas W, Shafaatian R, Bertl A, Scheurmannkettner C, Guy HR, North RA (1996) The S. cerevisiae outwardly-rectifying potassium channel (DUK1) identifies a new family of channels with duplicated pore domains. Recept Channel 4:51–62

Roeper J, Pongs O (1996) Presynaptic potassium channels. Curr Opin Neurobiol 6:338–341

Rudy B (1988) Diversity and ubiquity of K^+ channels. Neurosciences 25:729–749

Salkoff L, Butler A, Nonet M, Wei A (1997) The impact of the C. elegans genome-sequencing project on K^+ channel biology. Pflugers Arch Eur J Physiol R79

Shen KZ, North RA, Surprenant A (1992) Potassium channels opened by noradrenaline and other transmitters in excised membrane patches of guinea-pig submucosal neurones. J Physiol (London) 445:581–599

Siegelbaum SA, Camardo JS, Kandel E (1982) Serotonin and cyclic AMP close single K^+ channels in Aplysia sensory neurones. Nature 229:413–417

Theander S, Fahraeus C, Grampp W (1996) Analysis of leak current properties in the lobster stretch receptor neurone. Acta Physiol Scand 157:493–509

Wagner PG, Dekin MS (1997) cAMP modulates an S-type K^+ channel coupled to GABA(B) receptors in mammalian respiratory neurons. Neuroreport 8:1667–1670

Wu JV, Rubinstein T, Shrager P (1993) Single channel characterization of multiple types of potassium channels in demyelinated Xenopus axons. J Neurosci 13:5153–5163

Yang J, Jan YN, Jan LY (1995) Determination of the subunit stoichiometry of an inwardly rectifying potassium channel. Neuron 15:1441–1447

Yue TY, Marban E (1988) A novel cardiac potassium channel that is active and conductive at depolarized potentials. Pflügers Arch Eur J Physiol 413:127–133

Zhou XL, Vaillant B, Loukin SH, Kung C, Saimi Y (1995) YKC1 encodes the depolarization-activated k^+ channel in the plasma membrane of yeast. FEBS Lett 373:170–176

CHAPTER 14

Cardiac K^+ Channels and Inherited Long QT Syndrome

M.-D. DRICI and J. BARHANIN

A. Long QT Syndromes

The hallmark of all long QT syndromes (LQTS) is an abnormal ventricular repolarization characterized by a prolonged QT interval on the electrocardiogram. LQTS have a drastically different prognosis whether or not they are congenital. Congenital Long QT syndrome is a rare cardiac disorder associating the occurrence of syncopes often triggered in adrenergic setting, like strenuous exercise or emotional stress (RODEN et al. 1996). Most of the times, syncopes result from polymorphic ventricular tachycardia, called *torsades de pointes*, that were described first by Dessertene who characterized their pause-dependency and their distinctive time-dependent change in electrical axis (DESSERTENNE 1966). They may degenerate into entricular fibrillation, possibly causing sudden death, and are remarkably prevented by β-adrenergic antagonists. In fact, the clinical diagnosis of LQTS can be fairly difficult in the absence of typical rhythmic problems or if the QT interval remains within the normal limits (QTc<0.46s in women, or 0.45s in men). In that case, the diagnosis relies upon an array of elements including other electrocardiographic abnormalities like the presence of a U wave, a T wave alternans phenomenon, a notched T wave in three leads and a clinical history including seizures or unexplained cardiac death among immediate family members (WANG et al. 1997).

According to their mode of transmission, two inherited forms of LQTS can be characterized: the autosomal dominant Romano-Ward (RW) syndrome (WARD 1964; ROMANO 1965) that bears most of the cardiac-related criteria of diagnosis, and the autosomal recessive Jervell and Lange-Nielsen (JLN) syndrome (JERVELL et al. 1957) that is a rare particular syndrome associating a profound bilateral congenital deafness to the cardiac symptomatology. Typically, if patients with either affection are predisposed to *torsades de pointes*, relatively fewer cases of JLN syndrome are currently recognized, due to its recessive mode of transmission (RODEN et al. 1996; WANG et al. 1997).

Congenital LQTS is genetically heterogeneous, implicating at least five chromosomal *loci*, LQT1 to LQT5, three of which correspond to mutations concerning the coding of K^+ channel proteins (RODEN et al. 1996; SANGUINETTI et al. 1997b; WANG et al. 1997). One of them, the chromosome 11-linked LQT1 concerns the K^+ channel protein KvLQT1 (WANG et al. 1996), that associates with IsK, a small transmembrane protein encoded by a gene mapped on chromosome 21, to generate the slowly activating K^+ channel I_{Ks} (BARHANIN et al. 1996; SANGUINETTI et al. 1996b). *ISK (KCNE1)* has been identified as the LQT5 responsible gene (SCHULZE-BAHR et al. 1997; SPLAWSKI et al. 1997b). The third potassium gene implicated in LQT syndrome is *HERG* that encodes the I_{Kr} current and is responsible for the chromosome 7-linked LQT2 (CURRAN et al. 1995; SANGUINETTI et al. 1995). The LQT3 *locus* corresponds to the *SCN5 A* Na^+channel gene (WANG et al. 1995), and the LQT4 gene is not yet known (SCHOTT et al. 1995). Among the congenital LQTS, LQT1 is undoubtedly the most frequent.

Clinical symptomatology remains largely heterogeneous, even within a group bearing the same dysfunctional locus, indicating that environmental factors and/or modifier genes can modify their phenotypic manifestations.

In vitro experiments determined that mutations responsible for LQT1, LQT2, or LQT5 were all inducing a loss of function and/or a modification of the kinetics of the delayed rectifier I_K which is responsible for the repolarization of the cardiac action potential during the phase 3 (SANGUINETTI et al. 1990). The delayed rectifier is mainly composed of two components, a rapidly activated current named I_{Kr}, which function is altered in LQT2 and a slowly activated, time-dependent component, I_{Ks}, affected in either LQT1 or LQT5 (Fig. 1).

B. *HERG* and LQT2

I. The *HERG* Gene

The role of I_{Kr} in the process of cardiac repolarization and the genesis of LQTS had long been suspected since drugs targeting I_K had the potentiality to prolong the action potential duration and to induce *torsades de pointes* in animal models (SALATA et al. 1997a). In 1995, the key link between a mutated *HERG* gene encoding I_{Kr} channel and LQT has been established.

The HERG potassium channel is a human homologue of the *Drosophila ether-a-go-go* (*eag*) gene, mapped to chromosome 7q35–36 (CURRAN et al. 1995). The HERG protein has a typical six-membrane-spanning-segment *Shaker* related structure which assembles in a tetrameric fashion to form the channel. Originally cloned from a human hippocampal cDNA library (WARMKE et al. 1994), HERG expression is prominent in the heart, and is present in the brain, retina, adrenal glands, lungs, and thymus (Table 1); its physiological role in those tissues remains unclear.

II. I_{Kr} Current and LQT2

I_{Kr} is characterized by certain electrophysiological parameters (SANGUINETTI et al. 1995), including:

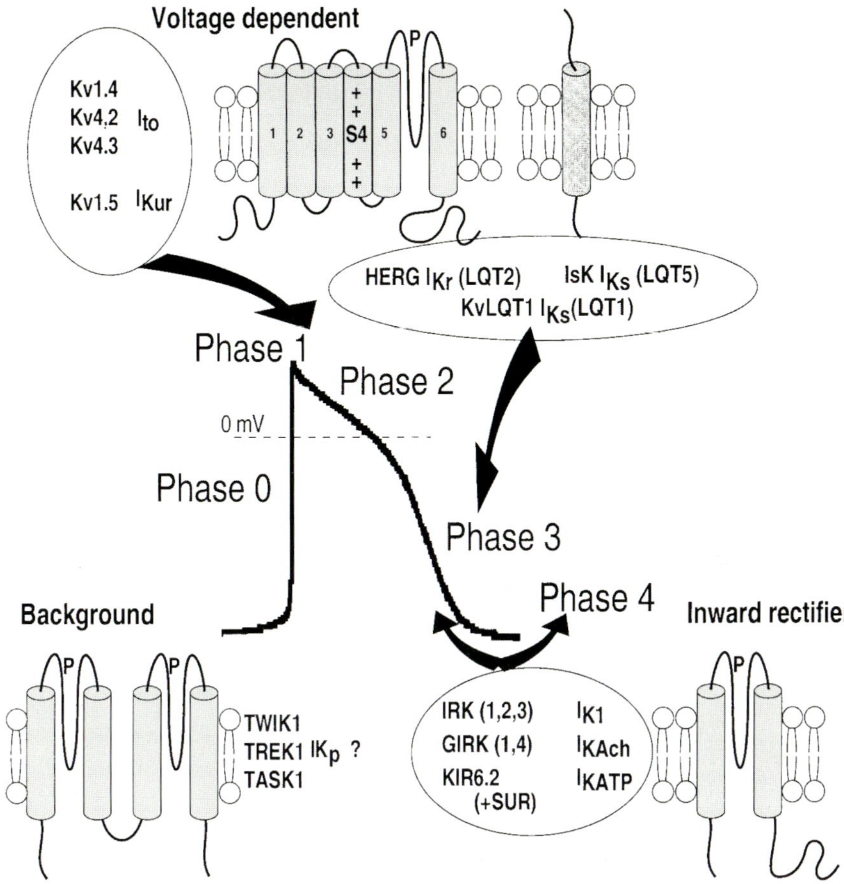

Fig. 1. The cardiac action potential and the main K^+ channels. The fast inward Na^+ current produces the ascending phase 0, while the plateau phase 2 is mainly due to the inward L-type Ca^{2+} current. Outward K^+ currents are major determinant of the resting potential in phase 4 and of the repolarizing phases 1 and 3. These currents are due to the activity of three main K^+ channel structural and functional families. The first family comprises the voltage-dependent *Shaker* type K^+ channels (6 transmembrane domains, S1 to S6, one P domain). The transient (I_{to}) and sustained (I_{Kur}) outward rectifying currents produce the rapid and partial repolarization in phase 1, while the delayed rectifiers currents I_{Kr} and I_{Ks} are mainly active in phase 3 and in the begining of phase 4. The genes corresponding to these K^+ channels are indicated. The single transmembrane domain protein IsK is an auxillary subunit that associates with KvLQT1 to form the I_{Ks} channel. The second family includes channels of the inward-rectifier type (I_{K1}), G-protein-coupled (I_{Kach}) and ATP-sensitive (I_{KATP}) K^+ channels with two hydrophobic segments (M1 and M2) and one P-domain. These currents are mainly responsible for the stabilization of the resting potential. The third group represents a novel K^+ channel architecture that exhibits the unique feature of having in tandem two pore motifs. Their functional role during the action potential is not yet known, although a current designed I_{Kp} (Backx et al. 1993) that is involved in the control of the duration of the plateau phase 2 could belong to this channel class (Duprat et al. 1997)

Table 1. Distribution of HERG, KvLQT1. and IsK mRNAs

Organ	HERG	KvLQT1	IsK
Heart	+++	+++	++
Brain	++	–	–
Retina	++	ND	+
Kydney	–	++	++
Lung	+	++	+
Pancreas	ND	+++	–
Colon	ND	+	+
Small intestine	–	++	–
Spleen	ND	+	–
Thymus	+	+	+
Leucocytes	ND	++	+
Adrenal glands	++	++++	+
Thyroid gland	ND	++++	ND
Submandibular gland	ND	++	++
Uterus	ND	++	++
Placenta	ND	++	+
Prostate	ND	++	–
Testis	ND	+	++
Ovaries	ND	+	++

Summary of current available data obtained by in situ hybridization and Northern blot analyses. The + represents a relative index and not a quantitative estimate. ND means not determined.

1. An important inward rectification: with increasing depolarizations the activating current decreases progressively (Fig. 2). This results from a rapid and voltage dependent C-type inactivation, that is more rapid than the activation. Therefore, at very positive potentials, inactivation predominates, and the current appears to rectify. The inactivation being almost instantaneously removed upon repolarization, study of native I_{Kr} in cardiomyocytes usually relies upon the measurement of its tail currents (Fig. 2A,B).
2. The fact that, conversely to I_{Ks}, the amplitude of I_{Kr} decreases when external concentrations of K^+ decreases. This particularity has been used in the treatment of LQT2 with potassium intake that has shown to correct some electrocardiographic abnormalities (CHOY et al. 1997).

Because of its proximity to the LQT2 locus, *HERG* was the most relevant candidate gene for the LQT2 linked LQT syndrome. As a fact, mutations in *HERG* induced an autosomal dominant RW syndrome in six families affected with congenital LQTS (CURRAN et al. 1995). Many other mutations have since been found (WANG et al. 1997). As a general rule, all mutations induce a loss of function that causes a decrease of the I_{Kr} current (SANGUINETTI et al. 1996a). This diminution often exceeds the expected value of 50% (for a simple haploinsufficiency), due to a dominant negative effect of the mutated channel

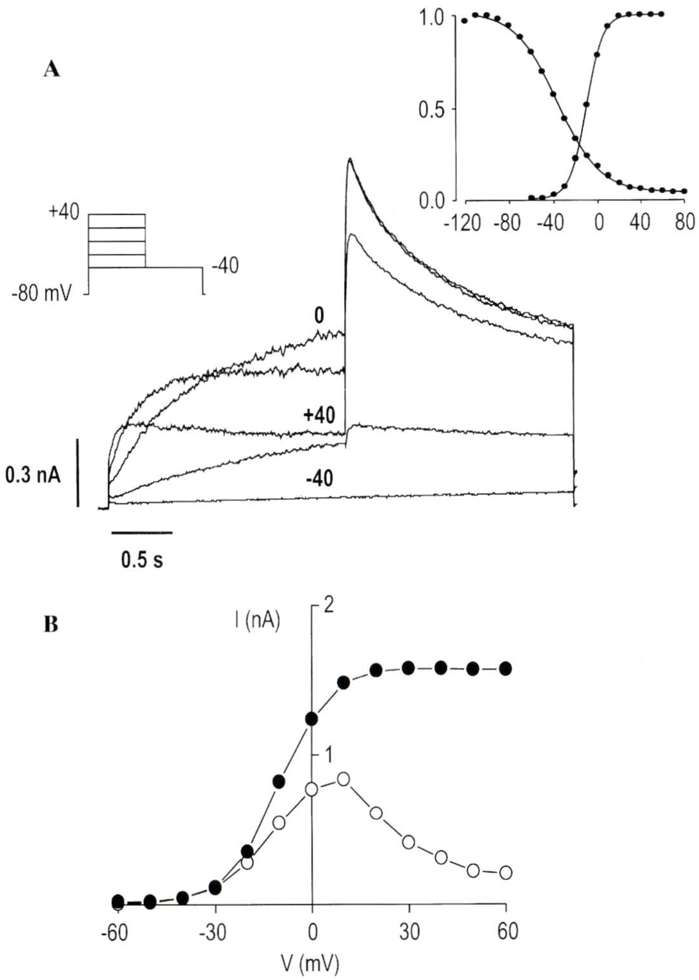

Fig. 2A,B. HERG Current (HERG). **A** representative HERG current evoked by increasing depolarizations from a holding potential of –80 mV in a HERG transfected COS cell. Currents develop and rapidly inactivate, due to a voltage dependent C type inactivation. However, when the cell is rapidly repolarized to –40 mV, inactivated channels recover to open state, accounting for the initial rise in outward current (tail currents) and slowly de-activate. Steady state activation and inactivation curves (*insert*) yield values for half activation and inactivation of the current of –10 mV for $V_{50\ act}$ (slope: 8 mV) and –36 mV for $V_{50\ inact}$ (slope : 18 mV), respectively. **B** With pulses to less depolarized potentials, I_{Kr} inactivation is less prominent, resulting in greater outward current; therefore, current voltage relationships from the same cell display an apparent strong inward rectification subunit, i.e., the inhibition of the activity of the unaffected subunits after co-assembly into heterotetramers. The severity of the mutations appears to be variable, from no dominant negative effect – the mutated subunit does not interfere with the wild type (WT), resulting in the loss of 50% of the channels – to a full dominant negative effect in which only one mutated subunit is sufficient to impair the heterotetrameric channel, resulting in the loss of function of 95% of the channels (SANGUINETTI et al. 1997a). This is probably accounting for the large clinical disparity of such mutations when present in patients affected by LQT syndrome.

C. KvLQT1/IsK, LQT1, and LQT5

I. *KVLQT1* and *ISK* Genes

The I_{Ks} channel has two components – KvLQT1 and IsK – which are very different in structure and function (BARHANIN et al. 1998). Unlike other K^+ channel genes that were identified by sequence homologies or functional expression, *KVLQT1* was identified by positional cloning on chromosome 11p15.5 as the gene responsible for the LQT1 syndrome. KvLQT1 (676 amino acids) has the classical structure of K^+ channel proteins with 6 transmembrane regions including a voltage sensor S4 segment and one P domain which confers the K^+ selectivity (HEGINBOTHAM et al. 1994; DOYLE et al. 1998). KvLQT1 is prominently expressed in the human heart as well as in the kidney, adrenal and thyroid gland, pancreas, placenta, lungs, and in the *stria vascularis* of the inner ear (Table 1).

Unlike KvLQT1, IsK is a small protein (130 amino-acids in mice, 129 in humans), with a single transmembrane domain and no P domain as found in other voltage-sensitive channels (TAKUMI et al. 1991; BUSCH et al. 1997b; KACZMAREK et al. 1997; BARHANIN et al. 1998). The IsK transcripts are present in the heart, kidney, thymus, eye, ear, and term uterus, where it seems extremely sensitive to 17-β estradiol. IsK has unique properties of interaction with KvLQT1 and serves as its essential modulator (ATTALI et al. 1993; BARHANIN et al. 1996; SANGUINETTI et al. 1996b). Therefore, it confers to the I_{Ks} channel functional features that probably have important physio-pathological implications. However, if a direct interaction between IsK and KvLQT1 has been clearly demonstrated in vitro and definitively confirmed in vivo by recent genetic findings concerning human and mouse LQT mutations, an interaction of IsK with HERG, which increases the amplitude of I_{Kr} has also been reported both in AT1 and in transfected cells (YANG et al. 1995; MCDONALD et al. 1997).

II. I_{Ks} Current, LQT1, and LQT5

The outward rectifying I_{Ks} current was first described in sheep Purkinje fibers as one of the multiple components responsible for the delayed rectifier I_K

current (NOBLE et al. 1969). Later on, the properties of this current could be discriminated in isolated guinea pig cardiac myocytes and its unique slow activation kinetics accounted for its name I_{Ks} (SANGUINETTI et al. 1990). The importance of this current during the plateau phase of the action potential is strongly dependent upon the animal species considered. A major component in guinea pig heart (SANGUINETTI et al. 1990) it is also present in cat (WOOSLEY et al. 1993), rabbit (SALATA et al. 1996) and mouse ventricular myocytes (HONORÉ et al. 1991; DAVIES et al. 1996) and is almost nonexistent in rats (APKON et al. 1991). So far, the physiological role of such a slow developing current in human ventricular cells remains debatable, but special experimental conditions are required to elicit this current in cardiomyocytes, that render it difficult to record except in certain species like guinea-pig (LI et al. 1996).

In mammalian cells, transfection of KvLQT1 alone produces a rapidly activating and slowly deactivating outward K^+ current that has not been characterized in heart cells so far (Fig. 3A,B). No current can be recorded when expressing IsK alone in mammalian cell lines. Co-transfection of those two subunits evoke upon long depolarization a slow time dependent outward K^+ current with very slow activation and deactivation kinetics, a small single channel conductance (ROMEY et al. 1997), and a regulation by protein kinase C and intracellular Ca^{2+} (BARHANIN et al. 1996). All these properties correspond to those that characterize the I_{Ks} current (Fig. 3C,D). Therefore, with the cloning and functional expression of KvLQT1 and IsK, not only the molecular nature of the I_{Ks} channel was elucidated, but also its role in cardiac repolarization and the lack of protection against arrhythmias featuring its loss in LQTS.

KVLQT1 is a very large gene (more than 300kb) and at least 33 distinct mutations have already been described. They are mostly missense mutations located in the S-S3, S3-S4 loops, the P domain, the S6 segment, and in a conserved sequence of the C-terminal tail (CHOUABE et al. 1997; DONGER et al. 1997; TYSON et al. 1997). Only few mutations are found in the small *ISK* gene (the coding sequence is only ~400bp long) that characterize the LQT5 (TYSON et al. 1997). The most common, D76N, occurs in the cytoplasmic C-terminal segment proximal to the transmembrane domain and suppresses the function of I_{Ks} with a strong dominant negative effect (SPLAWSKI et al. 1997b).

Due to the ubiquity of both KvLQT1 and IsK, mutations may affect not only the cardiac function but also that of other organs. The best described example to date is the sensorineural deafness displayed by patients with JLN syndrome, who not only exhibit a long QT interval, but also a profound deafness from birth. This results from the expression of the corresponding genes *KvLQT1* and *IsK* in the inner ear, where they control the endolymph homeostasis (VETTER et al. 1996; NEYROUD et al. 1997b). Interestingly, an insertional mutation leading to a truncated protein was identified in members of a family that was affected by both RW and JLN syndromes (NEYROUD et al. 1997b). One consanguineous child, affected by JLNS was homozygote for the muta-

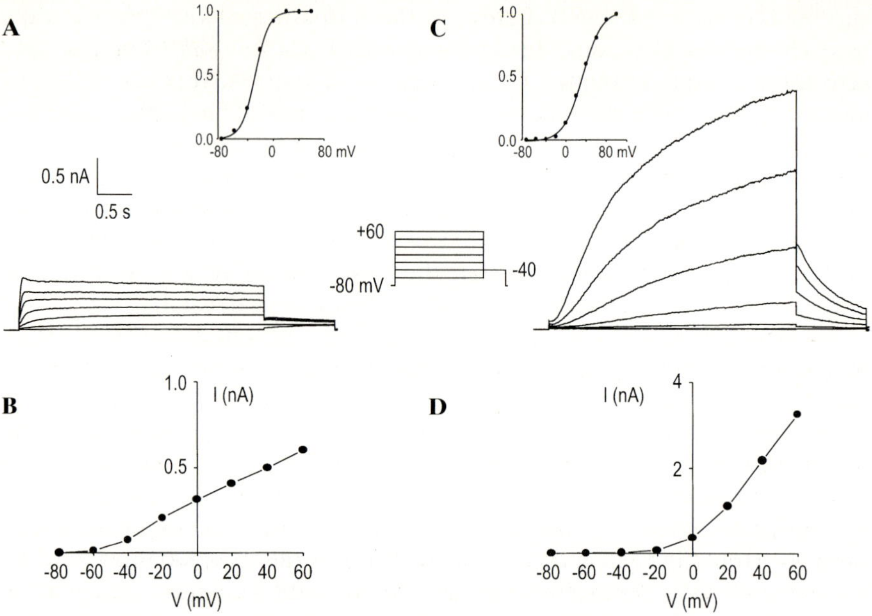

Fig. 3A–D. I_{Ks} Current (KvLQT1 and IsK). **A** Typical tracings of KvLQT1, transiently transfected in COS cells. Currents are elicited by 20 mV incremental depolarizations up to 60 mV, from a holding potential of –80 mV. KvLQT1 currents increase in amplitude according to the depolarization with fast kinetics of activation. Upon repolarization to –40 mV, KvLQT1 de-activation is characterized by slow tail currents. *Insert* is the steady state activation curve of the corresponding current, with a $V_{50\ act}$ of –28 mV (slope : 10 mV). **B** I-V Curve of the corresponding cell currents. No current of this type has been discriminated in isolated cardiomyocytes yet. **C** A typical trace of KvLQT1+IsK, transiently transfected in COS cells. As for KvLQT1, currents were elicited by incremental 20 mV depolarizations up to 60 mV, from a holding potential of –80 mV. KvLQT1+IsK currents increase in amplitude according to the depolarization with slow kinetics of activation and a time-dependent increase in the current amplitude, that is similar to the I_{Ks} component of the delayed rectifier I_K. Upon repolarization to –40 mV, KvLQT1+ISK de-activation is characterized by characteristic slow tail currents. The steady state activation curve of the corresponding cell current yields a $V_{50\ act}$ of 33 mV (slope : 19 mV). **D** I-V Curve of the corresponding cell currents. I_{Ks} can easily be identified in guinea-pig cardiomyocytes. It requires special settings (temperature at 37 °C, low external potassium, beta adrenergic agonists) to be elicited and/or discriminated from I_{Kr}, in rabbit, cat, mouse, dog, and human cardiomyocytes

tion whereas the mutation was found at a heterologous state in RW family members. More recent evidence that *KvLQT1*- and *IsK*-linked gene alterations could lead to RW and JLN symptomatology within the same family comforts the hypothesis that JLN syndrome results from a complete loss of KvLQT1 or IsK that is only partly affected in RW syndrome (Neyroud et al. 1997a; Schulze-Bahr et al. 1997; Splawski et al. 1997a,b; Duggal et al. 1998).

Again, mutations seem to display a broad range in the severity of their clinical manifestations that can render their clinical diagnosis difficult. The functional location of the mutation, its type (missense mutation, deletion, or insertion) and above all its dominant negative effect are accounting for this diversity.

Expression studies determined that, in most cases, mutant proteins fail to produce functional channels. On less frequent occasions, the mutations encode functional channels with altered gating properties, likely to correspond to a milder clinical impairment (Chouabe et al. 1997; Shalaby et al. 1997; Wollnik et al. 1997).

When coexpressed with IsK and WT KvLQT1 subunits, KvLQT1 mutants have been shown to decrease strongly the resulting currents in all RW mutations while the dominant negative effect was very mild for the JLN recessive mutations (Chouabe et al. 1997; Wollnik et al. 1997). However, these JLN mutations resulted in a complete loss of function when expressed alone, as the case is in homozygous carriers. Since deafness only affects JLN patients, it can be concluded that a total disparition of I_{Ks} current is required to produce deafness whereas a partial loss is sufficient to induce repolarization abnormalities. It is unsure as yet, whether parent carriers of "mild" mutations in KvLQT1 display subtle defects in hearing. Hence, mice carrying a null mutation on the IsK gene display a profound inner ear dysfunction associated with drastically altered K^+secretion into the endolymph of the inner ear leading to hair cell degeneration (Vetter et al. 1996). Thus, KvLQT1/IsK assembly forms a K^+ channel that has a key electrogenic role in ventricular repolarization and a key secretory role in the control of endolymph homeostasis associated with normal hearing.

III. Physiological Role of I_{Ks} in Cardiac Repolarization

IsK is abundant in sinoatrial node (in about one third of the cells), but less abundantly expressed in ventricular myocytes (10–15%) (Davies et al. 1996; Drici et al. 1998). This is consistent with a recent report that, at least in the mouse, IsK expression appears largely restricted to the conducting system (Kupershmidt et al. 1996).

Experimental data suggest a role of I_{Ks} in shortening the action potential duration at fast heart rates. Due to the slow kinetics of deactivation, the summation of tail currents effectively permits an increase of the outward K^+ conductance, on a beat to beat basis, when the stimulation frequency increases (Hauswirth et al. 1972; Jurkiewicz et al. 1993; Romey et al. 1997). No convincing data based on a more physiologically relevant model have yet confirmed these data.

However, comparative murine experiments involving knockout mice (Drici et al. 1998) have led to an interesting feature, that has been previously decribed in human (Hirao et al. 1996; Krahn et al. 1997) without drawing much attention. Since I_{Ks} has the unique particularity to develop slowly with time, its most important role might be during action potentials of long durations, i.e., during bradycardia, in order to limit the extent of the action potential prolongation. In fact, a mouse JLN model presenting typical inner ear defects has been created by knocking out the *isk* gene (*isk–/–*) (VETTER et al. 1996). In *isk–/–* mice isolated cardiac cells, the I_{Ks} current was abolished, leading to a longer QT interval at slow heart rates and an overall exacerbated QT-heart rate adaptation, compared to wild type (WT). An increase of 300 ms in the heart cycle length induced a highly significant increase, by 300 %, in the QT duration of the WT mice vs 500 % in *isk–/–* mice. It was concluded that the *isk* gene product and/or I_{Ks}, when present, had a preponderant role in blunting the QT adaptation to heart rate variations (DRICI et al. 1998). Steeper QT-RR relationships in mice could reflect the greater susceptibility to arrhythmias that patients with LQT1 or LQT5 are prone to. This is relevant in view of what happens during *torsades de pointes*. The onset of these particular arrhythmias is constantly preceded by a long pause with an abnormally prolonged QT interval. In case of a functional impairment of I_{Ks}, it is likely that the following action potential lengthens more rapidly and extensively up to a deleterious level, facilitating such arrhythmia in LQTS patients.

D. Pharmacological Considerations in the Acquired LQTS

I. Determinants of Cardiac Repolarization

EINTHOVEN (1913) showed that the pattern of the cardiac repolarization could be easily monitored by a surface electrocardiogram, the T wave and the duration of the QT interval being relevant surrogate markers of the ventricular repolarization itself. The prolonged QT interval on a surface ECG can be explained rather logically through the characteristic of the cardiac action potential. The repolarization of cardiac ventricular myocytes is a complex and finely tuned electrophysiological phenomenon. Myocardial cells express multiple types of channels. Among them, voltage-dependent K^+ channels probably form the most diverse family (DEAL et al. 1996). Furthermore, differences in the types and/or densities of K^+ channels expressed contribute to determine the variability in action potential waveforms recorded in different regions of the heart (ANTZELEVITCH et al. 1991, 1996).

During the plateau phase of the action potential, the membrane resistance is increased and its voltage is clamped by the inward calcium conductance, allowing I_{Kr}, activated by the depolarization, to be in an inactivated state, and I_{Ks} to develop slowly with time (ZENG et al. 1995b). In case of a fully efficient I_K, the rapid repolarization of the membrane during phase 3 removes I_{Kr} inactivation which enhances the outward K^+ conductances. I_{Ks} and I_{Kr} slow deactivation kinetics resulting in high conductance/low resistance of the membrane which repolarizes very rapidly, dumping any post potential that could occur during this period. It is clear that a blockade of I_{Kr} sharply reduces the steep phase 3 slope of the action potential, whereas an enhanced sodium conductance, such as *SCN5A* defect for example, prolongs the plateau phase with little effect on the repolarizing slope (SANGUINETTI et al. 1997a).

The proposed underlying mechanism of *torsades de pointes* in the setting of a LQTS is the triggering of oscillations known as "early after depolarizations" (EADs) secondary to the reactivation of inward currents. They interrupt the normal repolarizing time-course of the action potential, especially at slow heart rates (ZENG et al. 1995a). Therefore, the blockade of I_{Kr}, I_{Ks}, or both may facilitate the occurrence of EADs in two ways: first, by delaying the repolarization phase and lengthening the action potential, enabling inward currents to reactivate and, second, by opposing much weakened outward conductances upon the emergence of such depolarizations. Furthermore, experimental evidence obtained from *isk* knockout mice attribute to I_{Ks} an important role in the adaptation of the action potential duration to the heart rate.

II. Pharmacological Modulation of Cardiac Repolarization and Acquired Long QT Syndromes

The blockade of the delayed rectifier I_K is an important feature of the treatment of ventricular arrhythmia (RODEN 1994; NAIR et al. 1997); it increases the refractory period of cardiac myocytes through a prolongation of the repolarization phase of the action potential (SATOH et al. 1996). But extensively prolonging the cardiac repolarization has its drawbacks. Not only specific class 3 antiarrhythmics fail to protect patients from mortality compared to placebo in long term prescription studies, but they dose dependently induce *torsades de pointes* in a non-negligible proportion of the population (possibly corresponding to the population bearing asymptomatic genetic defects?) (CAST INVESTIGATORS 1989; RODEN 1994, 1998).

Far more common than patients suffering from congenital LQT syndrome are patients with acquired LQTS (aLQTS) that are mostly drug-induced. Patients usually have normal or subnormal QTc values when untreated and show an excessive QT prolongation when exposed to numerous drugs which all block the I_{Kr} current. The prognosis of aLQTS is burdened by the same predisposition to develop *torsades de pointes* as its congenital counterpart (NAPOLITANO et al. 1994). These cardiac arrhythmias, based on an abnormally prolonged repolarization, are also associated with a prolonged QT interval on an electrocardiogram during drug treatment. Nevertheless, a genetic predisposition for the aLQTS has been speculated upon, and premature results indicate the pertinence of such a hypothesis. A much larger that previously thought proportion of the population is affected by "asymptomatic" mutations of the LQT genes (VINCENT et al. 1992; PRIORI et al. 1997). Due to the peculiar mode of interaction of I_{Kr} and I_{Ks}, it is reasonable to think that, if the QT of these patients is normal without any drug, they can easily be affected by a blockade of a current that sufficiently compensates an ailing repolarization in normal conditions. Identification of patients whose genetic defect in LQTS genes have been unmasked by drug prescription has already been reported (NAPOLITANO et al. 1997).

Several settings interfering with cardiac repolarization – besides congenital LQTS – and including hypokaliemia, hypomagnesemia, bradycardia, and feminine gender, have shown to facilitate the occurrence of *torsades de pointes* (NAPOLITANO et al. 1994) This probably results from a direct or indirect modification of the regulation of the I_{Kr} and/or I_{Ks} functions. This has been well documented for the effect of hypokaliemia on the I_{Kr} blockade exerted by drugs (see above, Sect. II.). Interestingly, 17ß-estradiol experimentally blocks I_{Ks}, through a non-genomic effect, though at concentrations >1 μmol/l (BUSCH et al. 1997a). I_{Kr} current has also been shown to be modulated by genomic effects of sex steroid hormones at concentrations that are more relevant to physiology (DRICI et al. 1996). This could help to explain the propensity for cardiac arrhythmias in women (MAKKAR et al. 1993).

As a general rule, like for erythromycin or terfenadine, the occurrence of *torsades de pointes* results from a blockade of I_{Kr} (WOOSLEY et al. 1993; ANTZELEVITCH et al. 1996). Fewer drugs, like indapamide, block I_{Ks} at relevant therapeutic concentrations (TURGEON et al. 1994; FISET et al. 1997) Several of them, like amiodarone, azimilide, or bepridil, block both components of I_K (SALATA et al. 1997b). No drug so far has shown to block selectively KvLQT1 current without also blocking I_{Ks}, even though the two components may share different affinities to the same drug. However, one must keep in mind that *torsades de pointes* may occur in case of impairment of either I_{Kr} or I_{Ks} in congenital LQTsyndrome (CURRAN et al. 1995; NEYROUD et al. 1997b).

E. Conclusion

The extent to which potentially lethal arrhythmia like *torsades de pointes* occur in patients depends of the degree of abnormality of the channel function in the particular setting of a disease, and/or its inbalance under physiological stress or pharmacological modifications. Molecular cloning of the genes is of a tremendous help in this area, by providing models on which to study the effects of modulating factors on the corresponding currents expressed. This enables a better comprehension of arrhythmogenesis and could permit one to identify new targets for a better treatment of arrhythmias, that has not yet been achieved.

Acknowledgements. We are indebted to C. Chouabe, for the help provided in the discussion of the manuscript. We thank Ms Y. Benhamou and V. Briet for skillfull secretarial help.

References

Antzelevitch C, Sicouri S, Litovsky SH, Lukas A, Krishnan SC, Di DJ, Gintant GA, Liu DW (1991) Heterogeneity within the ventricular wall. Electrophysiology and pharmacology of epicardial, endocardial, and M cells. Circ Res 69:1427–1449

Antzelevitch C, Zhuo-Quian S, Zi-Qing Z, Gan-Xin Y (1996) Cellular and ionic mechanisms underlying erythromycin-induced long QT intervals and Torsade de Pointe. J Am Coll Cardiol 28:1836–1848

Apkon M, Nerbonne JM (1991) Characterization of two distinct depolarization-activated K^+ currents in isolated adult rat ventricular myocytes. J Gen Physiol 97:973–1011

Attali B, Guillemare E, Lesage F, Honore E, Romey G, Lazdunski M, Barhanin J (1993) The protein IsK is a dual activator of K^+ and Cl^- channels. Nature 365:850–852

Backx PH, Marban E (1993) Background potassium current active during the plateau of the action potential in Guinea-pig ventricular myocytes. Circ Res 72:890–900

Barhanin J, Attali B, Lazdunski M (1998) I_{Ks}, a very slow and very intriguing cardiac K^+ channel and its associated long QT diseases. Trends Cardiovasc Med 8:207–214

Barhanin J, Lesage F, Guillemare E, Fink M, Lazdunski M, Romey G (1996) K(v)LQT1 and IsK (minK) proteins associate to form the I_{Ks} cardiac potassium current. Nature 384:78–80

Busch AE, Busch GL, Ford E, Suessbrich H, Lang HJ, Greger R, Kunzelmann K, Attali B, Stuhmer W (1997a) The role of the IsK protein in the specific pharmacological properties of the I_{Ks} channel complex. Br J Pharmacol 122:187–189

Busch AE, Suessbrich H (1997b) Role of the ISK protein in the IminK channel complex. Trends Pharmacol Sci 18:26–29

Chouabe C, Neyroud N, Guicheney P, Lazdunski M, Romey G, Barhanin J (1997) Properties of KvLQT1 K^+ channel mutations in Romano-Ward and Jervell and Lange-Nielsen inherited cardiac arrhythmias. EMBO J 16:5472–5479

Choy AM, Lang CC, Chomsky DM, Rayos GH, Wilson JR, Roden DM (1997) Normalization of acquired QT prolongation in humans by intravenous potassium. Circulation 96:2149–2154

Curran ME, Splawski I, Timothy KW, Vincent GM, Green ED, Keating MT (1995) A molecular basis for cardiac arrhythmia: HERG mutations cause long QT syndrome. Cell 80:795–803

Davies MP, Doevendans P, An RH, Kubalak S, Chien KR, Kass RS (1996) Developmental changes in ionic channel activity in the embryonic murine heart. Circ Res 78:15–25

Deal KK, England SK, Tamkun MM (1996) Molecular physiology of cardiac potassium channels. Physiol Rev 76:49–67

Dessertenne F (1966) La tachycardie ventriculaire à deux foyers opposés variables. Arch Malcoeur 59:263–272

Donger C, Denjoy I, Berthet M, Neyroud N, Cruaud C, Bennaceur M, Chivoret G, Schwartz K, Coumel P, Guicheney P (1997) KVLQT1 C-terminal missense mutation causes a forme fruste long-QT syndrome. Circulation 96:2778–2781

Doyle DA, Cabral JM, Pfuetzner RA, Kuo AL, Gulbis JM, Cohen SL, Chait BT, MacKinnon R (1998) The structure of the potassium channel: Molecular basis of K^+ conduction and selectivity. Science 280:69–77

Drici MD, Arrighi I, Chouabe C, Mann JR, Lazdunski M, Romey G, Barhanin J (1998) Involvment of IsK associated K^+ channel in heart rate control of repolarization in a murine engineered model of Jervell and Lange-Nielsen syndrome. Circ Res 83:95–102

Drici MD, Burklow TR, Haridasse V, Glazer RI, Woosley RL (1996) Sex hormones prolong the QT interval and downregulate potassium channel expression in the rabbit heart. Circulation 94:1471–1474

Duggal P, Vesely MR, Wattanasirichaigoon D, Villafane J, Kaushik V, Beggs AH (1998) Mutation of the gene for IsK associated with both Jervell and Lange-Nielsen and Romano-Ward forms of Long-QT syndrome. Circulation 97:142–146

Duprat F, Lesage F, Fink M, Reyes R, Heurteaux C, Lazdunski M (1997) TASK, a human background K^+ channel to sense external pH variations near physiological pH. EMBO J 16:5464–5471

Fiset C, Drolet B, Hamelin BA, Turgeon J (1997) Block of I_{Ks} by the diuretic agent indapamide modulates cardiac electrophysiological effects of the class III antiarrhythmic drug dl-sotalol. J Pharmacol Exp Ther 283:148–156

Hauswirth O, Noble D, Tsien RW (1972) Separation of the pace-maker and plateau components of delayed rectification in cardiac Purkinje fibres. J Physiol 225:211–235

Heginbotham L, Lu Z, Abramson T, Mackinnon R (1994) Mutations in the K^+ channel signature sequence. Biophys J 66:1061–1067

Hirao H, Shimizu W, Kurita T, Suyama K, Aihara N, Kamakura S, Shimomura K (1996) Frequency-dependent electrophysiologic properties of ventricular repolarization in patients with congenital long QT syndrome. J Am Coll Cardiol 28:1269–1277

Honoré E, Attali B, Romey G, Heurteaux C, Ricard P, Lesage F, Lazdunski M, Barhanin J (1991) Cloning, expression, pharmacology and regulation of a delayed rectifier K^+ channel in mouse heart. EMBO J 10:2805–2811

Cast Investigators (1989) Preliminary report: effect of encainide and flecainide on mortality in a randomized trial arrhythmia suppression after myocardial infarction. N Engl J Med 321:406–412

Jervell A, Lange-Nielsen F (1957) Congenital deaf-mutism, functional heart disease with prolongation of Q-T interval and sudden death. Am Heart J 54:59–68

Jurkiewicz NK, Sanguinetti MC (1993) Rate-dependent prolongation of cardiac action potentials by a methanesulfonanilide class-III antiarrhythmic agent – specific block of rapidly activating delayed rectifier K^+- current by dofetilide. Circ Res 72:75–83

Kaczmarek LK, Blumenthal EM (1997) Properties and regulation of the minK potassium channel protein. Physiol Rev 77:627–641

Krahn AD, Klein GJ, Yee R (1997) Hysteresis of the RT interval with exercise: a new marker for the long-QT syndrome? Circulation 96:1551–1556

Kupershmidt S, Sutherland M, King D, Magnuson MA, Roden DM (1996) Replacement by homologous recombination of the minK gene with LacZ reveals cell-specific minK expression (Abstract). Biophys J 72:A226

Li GR, Feng JL, Yue LX, Carrier M, Nattel S (1996) Evidence for two components of delayed rectifier K^+ current in human ventricular myocytes. Circ Res 78:689–696

Makkar RR, Fromm BS, Steinman RT, Meissner MD, Lehmann MH (1993) Female gender as a risk factor for *torsades de pointes* associated with cardiovascular drugs. JAMA 270:2590–2597

McDonald TV, Yu ZH, Ming Z, Palma E, Meyers MB, Wang KW, Goldstein SAN, Fishman GI (1997) A minK-HERG complex regulates the cardiac potassium current I-Kr. Nature 388:289–292

Nair LA, Grant AO (1997) Emerging class III antiarrhythmic agents: Mechanism of action and proarrhythmic potential. Cardiovasc Drug Therapy 11:149–167

Napolitano C, Priori S, Schwartz P, Cantù F, Paganini V, De Fusco M, Pinnavia A, Aquaro G, Casari G (1997) Identification of a long QT syndrome molecular defect in drug-induced torsade de pointes. Circulation 96:211 (Abstract)

Napolitano C, Priori SG, Schwartz PJ (1994) *Torsade de pointes*. Mechanisms and management. Drugs 47:51–65

Neyroud N, Denjoy I, Donger C, Villain E, Leenhardt A, Gary F, Coumel P, Schwartz K, Guicheney P (1998) A heterozygous mutation in the pore of the potassium channel gene KvLQT1 causes a benign phenotype in the long QT syndrome. Europ J Hum Genet 19(1):158–165

Neyroud N, Tesson F, Denjoy I, Leibovici M, Donger C, Barhanin J, Faure S, Gary F, Coumel P, Petit C, Schwartz K, Guicheney P (1997) A novel mutation in the potassium channel gene KVLQT1 causes the Jervell and Lange-Nielsen cardioauditory syndrome. Nature Genet 15:186–189

Noble D, Tsien RW (1969) Outward membrane currents activated in the plateau range of potential in cardiac Purkinje fibers. J Physiol 200:205–231

Priori SG, Napolitano C, Paganini V, Cantu F, Schwartz PJ (1997) Molecular biology of the long QT syndrome: Impact on management. Pac Clin Electrophys 20:2052–2057

Roden D (1994) Risks and benefits of antiarrhythmic drug therapy. N Engl J Med 331:785–791

Roden DM, Lazzara R, Rosen M, Schwartz PJ, Towbin J, Vincent GM (1996) Multiple mechanisms in the long-QT syndrome. Current knowledge, gaps, and future directions. The SADS Foundation Task Force on LQTS. Circulation 94:1996–2012

Romano C (1965) Congenital cardiac arrhythmia. (Letter) Lancet 1:658–659

Romey G, Attali B, Chouabe C, Abitbol I, Guillemare E, Barhanin J, Lazdunski M (1997) Molecular mechanism and functional significance of the MinK control of the KvLQT1 channel activity. J Biol Chem 272:16713–16716

Rosen MR (1998) Antiarrhythmic drugs: Rethinking targets, development strategies, and evaluation tools. Am J Cardiol 81:D21–D23

Salata JJ, Brooks RR (1997a) Pharmacology of azimilide dihydrochloride (NE–10064), a class III antiarrhthmic agent. Cardiov Drug Rev 15:137–156

Salata JJ, Jurkiewicz NK, Jow B, Folander K, Guinosso PJ, Raynor B, Swanson R, Fermini B (1996) I_K Of rabbit ventricle is composed of two currents: Evidence for I-Ks. Amer J Physiol 40:H2477–H2489

Sanguinetti M, Keating M (1997a) Role of delayed rectifier potassium channels in cardiac repolarization and arrhythmias. News Physiol Sci 12:152–157

Sanguinetti MC, Spector PS (1997b) Potassium channelopathies. Neuropharmacology 36:755–762

Sanguinetti MC, Curran ME, Spector PS, Keating MT (1996a) Spectrum of HERG K^+-channel dysfunction in an inherited cardiac arrhythmia. Proc. Natl. Acad Sci USA 93:2208–2212

Sanguinetti MC, Curran ME, Zou A, Shen J, Spector PS, Atkinson DL, Keating MT (1996b) Coassembly of K(v)LQT1 and MinK (IsK) proteins to form cardiac I_{Ks} potassium channel. Nature 384:80–83

Sanguinetti MC, Jiang CG, Curran ME, Keating MT (1995) A mechanistic link between an inherited and an acquired cardiac arrhythmia: HERG encodes the I_{Kr} potassium channel. Cell 81:299–307

Sanguinetti MC, Jurkiewicz NK (1990) Two components of cardiac delayed rectifier K^+ current – Differential sensitivity to block by class-III antiarrhythmic agents. J Gen Physiol 96:195–215

Satoh T, Zipes DP (1996) Rapid rates during bradycardia prolong ventricular refractoriness and facilitate ventricular tachycardia induction with cesium in dogs. Circulation 94:217–227

Schott JJ, Charpentier F, Peltier S, Foley P, Drouin E, Bouhour JB, Donnelly P, Vergnaud G, Bachner L, Moisan JP (1995) Mapping of a gene for long QT syndrome to chromosome 4q25–27. Am J Hum Genet 57:1114–1122

Schulze-Bahr E, Wang Q, Wedekind H, Haverkamp W, Chen Q, Sun Y (1997) KCNE1 mutations cause jervell and Lange-Nielsen syndrome. Nature Genet 17:267–268

Shalaby FY, Levesque PC, Yang WP, Little WA, Conder ML, JenkinsWest T, Blanar MA (1997) Dominant-negative K_vLQT1 mutations underlie the LQT1 form of long QT syndrome. Circulation 96:1733–1736

Splawski I, Timothy KW, Vincent GM, Atkinson DL, Keating MT (1997a) Molecular basis of the long-QT syndrome associated with deafness. N Engl J Med 336:1562–1567

Splawski I, Tristani-Firouzi M, Lehmannn MH, Sanguinetti MC, Keating MT (1997b) Mutations in hminK gene cause long-QT syndrome and suppress I_{Ks} function. Nature Genet 17:338–340

Takumi T, Moriyoshi K, Aramori I, Ishii T, Oiki S, Okada Y, Ohkubo H, Nakanishi S (1991) Alteration of Channel Activities and Gating by Mutations of Slow-IsK Potassium Channel. J Biol Chem 266:22192–22198

Turgeon J, Daleau P, Bennett PB, Wiggins SS, Selby L, Roden DM (1994) Block of I_{Ks}, the slow component of the delayed rectifier K^+ current, by the diuretic agent indapamide in guinea pig myocytes. Circ Res 75:879–886

Tyson J, Tranebjærg L, Bellman S, Wren C, Taylor J, Bathen J, Aslaksen B, Sørland SJ, Lund O, Malcolm S, Pembrey M, Bhattacharya S, Bitner-Glindzicz M (1997) IsK and KvLQT1: mutation in either of the two subunits of the slow component of the delayed rectifier potassium channel can cause Jervell and Lange-Nielsen syndrome. Hum Mol Genet 6:2179–2185

Vetter DE, Mann JR, Wangemann P, Liu JZ, McLaughlin KJ, Lesage F, Marcus DC, Lazdunski M, Heinemann SF, Barhanin J (1996) Inner ear defects induced by null mutation of the IsK gene. Neuron 17:1251–1264

Vincent G, Timothy K, Leppert M, Keating M (1992) The spectrum of symptoms and QT intervals in carriers of the gene for long QT syndrome. N Engl J Med 327:846–852

Wang Q, Chen Q, Li H, Towbin JA (1997) Molecular genetics of long QT syndrome from genes to patients. Curr Opin Cardiol 12:310–320

Wang Q, Curran ME, Splawski I, Burn TC, Millholland JM, Vanraay TJ, Shen J, Timothy KW, Vincent GM, Dejager T, Schwartz PJ, Towbin JA, Moss AJ, Atkinson DL, Landes GM, Connors TD, Keating MT (1996) Positional cloning of a novel potassium channel gene: KVLQT1 mutations cause cardiac arrhythmias. Nature Genet 12:17–23

Wang Q, Shen JX, Splawski I, Atkinson D, Li ZZ, Robinson JL, Moss AJ, Towbin JA, Keating MT (1995) SCN5 A mutations associated with an inherited cardiac arrhythmia, long QT syndrome. Cell 80:805–811

Ward OC (1964) A new familial cardiac syndrome in children. J Irish Med Assoc 54:103–106

Warmke JW, Ganetzky B (1994) A family of potassium channel genes related to eag in drosophila and mammals. Proc Natl Acad Sci USA 91:3438–3442

Wollnik B, Schroeder BC, Kubisch C, Esperer HD, Wieacker P, Jentsch TJ (1997) Pathophysiological mechanisms of dominant and recessive KVLQT1 K^+ channel mutations found in inherited cardiac arrhythmias. Hum Mol Genet 6:1943–1949

Woosley RL, Chen YW, Freiman JP, Gillis RA (1993) Mechanism of the cardiotoxic actions of terfenadine. JAMA 269:1532–1536

Yang T, Kupershmidt S, Roden DM (1995) Anti-minK antisense decreases the amplitude of the rapidly activating cardiac delayed rectifier K^+ current. Circ Res 77:1246–1253

Zeng J, Rudy Y (1995a) Early afterdepolarizations in cardiac myocytes: mechanism and rate dependence. Biophys J 68:949–964

Zeng JL, Laurita KR, Rosenbaum DS, Rudy Y (1995b) Two components of the delayed rectifier K^+ current in ventricular myocytes of the guinea pig type – theoretical formation and their role in repolarization. Circ Res 77:140–152

Section II
Ligand Operated Ion Channels

CHAPTER 15

Gating of Ion Channels by Transmitters: The Range of Structures of the Transmitter-Gated Channels

E.A. BARNARD

A. Introduction: The Scope of the Transmitter-Gated Channel Class

Although a great variety of types of ion channels in membranes has become known in recent years, they can still in all cases be separated logically into two classes. These comprise (A) those which open spontaneously or in response to changes in the membrane potential, and (B) those which require a native transmitter to open the channel (or in rare cases to close it). Class A is dealt with in Sect. I of this volume (and also in parts of Sect. III). The present section considers class B, the transmitter-gated channels – in this chapter, comparing the whole range thereof, and in subsequent chapters detailing some individual cases where a great deal of information is available. A few cases are borderline between classes A and B, e.g., the "SK" Ca^{2+}-dependent K^+ channels (voltage-independent), where the Ca^{2+} ion might strictly be considered to be a transmitter (HIRSCHBERG et al. 1998); since these channels belong structurally to the family of voltage-gated K^+ channels, they are conveniently treated in the discussion of Class A channels (see Chap. 9). The subunits of the G-protein-gated K^+ channels (see Chap. 13) are entirely in the Kir family, and while the channel opening frequency is greatly increased upon binding G-protein subunits it is not dependent on them (KRAPIVINSKY et al. 1995), so these are definitely in Class A. The K_{ATP} channel type (see Chap. 11) is complex but should be placed in the transmitter-gated context, as discussed below. The proton-gated cation channels (BASSILANA et al. 1997) are structurally in the same family as the amiloride-sensitive Na^+ channels (the latter being in Class A and reviewed in Chap. 28) but undergo no channel opening without excess protons; they are, therefore, treated here as in Class B.

In the definition of the Class A channels given above, the terms "native transmitter" and "require" must be strictly applied if a logical distinction from other channels is to be maintained. Thus, there are many ligands which can control voltage-gated ion channels without being native transmitters, e.g., a variety of exogenous toxins (including some that open such channels or maintain their open state). It is not useful to speak of "the tetrodotoxin-gated Na^+ channel," although the commonly-used term "ligand-gated channel" would include such cases. That term is now replaced by "transmitter-gated channel," where a transmitter is a messenger molecule acting in vivo to relay specifically a transducing signal into a cell or between compartments within a cell (BARNARD 1996, 1997a). "Transmitter" used thus has a wider meaning than "neurotransmitter" and embraces the latter plus all other natural receptor-activating ligands – hormones, trophins, growth factors, cytokines and other immunomediators, morphogens, sensory stimulants and chemoattractants, and intracellular messengers acting as agonists. The transmitter-gated channels (TGCs) are therefore not confined to the cell membrane – although the vast majority of them occur only there – but include members specific to organelle membranes, e.g. the IP_3 receptor of the endoplasmic reticular membrane (Chap. 26). The TGCs are, of course, receptors for the transmitter involved.

B. Structural Elements of the Membrane Domains of the Transmitter-Gated Channels

I. Transmembrane Domains

It is obvious that any transmitter-gated channel must contain a minimum of three basic elements: (i) a predominantly hydrophobic domain transversing the membrane and enclosing the hydrophilic ion permeation pathway; (ii) a domain in or exposed to the aqueous medium on that side of the membrane where the transmitter is incident and specialized for binding the transmitter; and (iii) the structural linkage that relays the conformational change evoked by the transmitter binding into the conformational change from the closed to the open channel state(s). Thus far, the techniques available to us have provided significant but incomplete information on component (i) for a wide range of transmitter-gated channels, on component (ii) only fragmentary information on some contributions to the binding site domain in a few of those channels (exemplified in Chaps. 18, 20, and 21 here), and on component (iii) virtually no definite information. This situation will, of course, change radically when a three-dimensional structure at atomic resolution for any transmitter-gated channel in a membranous or membrane-like environment is obtained; in the light of such a determination for a bacterial K^+ channel (DOYLE et al. 1998), that development can be predicted to arrive before too long. It is now instructive to review the information already acquired on component (i), the channel domain, over the full range of the transmitter-gated channels.

All of the protein sequences in the TGCs contain discrete, strongly hydrophobic segments of a length suitable to span the membrane (as discussed below). For the usual interpretation of these as transmembrane domains (TMs) to be confirmed, adequate experimental evidence on the protein/membrane topology is required; in a few cases in the TGCs such investigation has shown that interpretation to be incorrect, e.g., one of the originally-deduced four TMs of the glutamate-gated cation channels is now known not to span the membrane, as detailed below. In some of the larger polypeptides the hydropathy plot is not clear enough for the number of TMs present to be generally agreed, e.g., for the "ryanodine receptor" (Table 1). Even in some of the shorter channel subunits some segments generally assumed to be TMs by analogy with related receptors or by glycosylation or accessibility evidence on the topology are unimpressive in the hydropathy plots, e.g., some cases within the $GABA_A$ receptor or cyclic nucleotide receptor subunits.

There is no uniformity between the TGCs in the number or the spacing of the TMs in a single polypeptide chain and that number can vary from two to twelve (Table 1). Even larger numbers have occasionally been deduced, as in a subunit of the K_{ATP} channel (INAGAKI et al. 1996; TUCKER and ASHCROFT 1998).

Excepting, possibly, some of the transporter-related TGCs (subclasses 1.6 and 1.7 in Table 1) which contain of the order of twelve TMs, a single subunit is clearly insufficient structurally to enclose an ion channel and the TGCs are generally multi-subunit, with selected TMs from different subunits contributing to the channel structure. A minority of the TGCs are, on present information, homomeric – among others, perhaps some transporters and most, but not all, of the P2X ATP-gated channels (subclass 1.5.2 in Table 1). The great majority of TGCs are, however, heteromers, and the number and relatedness of the different subunits which then assemble to form one channel can vary between the receptor types. There can be two structurally-related subunit types combined, e.g., in a P2X receptor (LEWIS et al. 1995), in some of the neuronal nicotinic ACh receptors (Chap. 17) or in the glycine receptors. The number of related subunit types in one TGC molecule can also be four, as in the muscle/electric organ nicotinic receptors, or three or four, as in some neuronal nicotinic receptors (LINDSTROM 1997; FORSAYETH and KOBRIN 1997) and in most of the $GABA_A$ receptors (BARNARD et al. 1998). On the other hand, two completely unrelated subunit types may be required, as in the K_{ATP} receptor (CLEMENT et al. 1997; GRIBBLE et al. 1997).

Three structural motifs have been discerned in the channel architecture of the TGCs, based upon several analytical approaches. These are: (i) the α-helix; (ii) the β-barrel and related non-helical TM structures; and (iii) the pore (P) domain.

II. The α-Helix in Channel Transmembrane Domains

The hydrophobic segments presumed to be TMs in the TGCs range, in different subunits, from 17 to 27 amino acids in length. These lengths could form a membrane-spanning α-helix; with the longer values that helix must be presumed to be tilted from the perpendicular to the membrane, or kinked, curved, or otherwise distorted, or to project beyond the membrane. α-Helical structure of that type has long been deduced for those hydrophobic segments, this being supported by the three-dimensional structures which it has been possible to determine by diffraction studies (X-ray, electron or optical) for a few non-TGC membrane proteins. Thus membrane-spanning α-helices, either straight or kinked, have been demonstrated to correspond to all or most of

Table 1. Structural sub-classes of the transmitter-gated channels

Superfamily[a]	Receptor types	Code[b]	Membrane topology	Transmitter origin
Cys-loop superfamily				
1.1	Anion channels	1.1 GABA	4 TM	EC
1.1	Anion channels	1.1 GLY		
1.1	Anion channels	1.1 GLU		
1.1	Cation channels	1.1 ACH		
1.1	Cation channels	1.1 5HT		
Glutamate-gated cation channels				
1.2	Non-NMDA receptors	1.2 GLU	3 TM + P	EC
1.2	NMDA receptors	1.2 GLU		
Related to voltage-gated cation channels				
1.3.1	Cyclic nucleotide receptors	1.3 CNUCT	6 TM + P	IC
1.3.2	IP_3 receptors	1.3 IP3	6 TM + P	IC
1.3.2	Ca^{2+} release channels ("ryanodine receptors")	[c]1.3 ryan	4 TM + P	IC
[d]1.3.3	Vanilloid receptor	1.3 VAN	6 TM + P	EC
Related to epithelial Na^+ channels; non-peptide agonists				
1.4.1	ATP-gated cation channels (P_{2X})	1.4 NUCT	2 TM	EC
1.4.2	H^+-gated Na^+ channels	1.4 H H.ASIC.01 H.ASIC.02 H.ASIC.01.ASIC.02.M H.ASIC.03	2 TM	EC
Related to epithelial Na^+ channels; peptide agonists				
1.4.2	FMRFamide-gated Na^+ channel	1.4 FMRF	2 TM	EC

Related to inward rectifier K+ channels				
1.5.1	ATP-antagonised K^+ channel (K_{ATP})	[e] 1.6 NUCT.n.M	2TM + P	IC
1.5.1	Nucleotide-dependent K^+ channel (K_{NDP})	[e] 1.6 NUCT.n.M	2TM + P	IC
1.5.2	OH^--activated K^+ channel	1.5 OH.TASK	4 TM + 2P	IC
Related to ATPase-linked transporters				
1.6	CFTR (ATP-activated anion channel)	1.6 NUCT.CFTR	12 TM	EC
Related to neurotransmitter transporters				
1.7	glutamate-activated Cl^- channel/transporter	1.7.GLU	12 TM	EC
1.7	GABA-activated channel/transporter	.GABA		
1.7	5HT-activated channel/transporter	.5HT		
1.7	dopamine-activated channel/transporter	.DA		

EC, extracellular; IC, intracellular. This Table is modified and up-dated from Barnard, 1997a.

[a] Where a structural sub-class (e.g., 1.3) contains more than one known superfamily, these are numbered as a third sub-division (e.g., 1.3.1, 1.3.2). References to all of the receptor types listed are given in the text.

[b] The coding shown is based on the IUPHAR Receptor Code of Humphrey and Barnard (1998), where details of the receptor coding system are given. M, multiple subunit types occur in the receptor, of known composition. S, a component subunit.

[c] The lower case denotes that ryanodine is not a native agonist and this receptor code name is temporary, pending agreement on natural agonists for this receptor. On the assigment of 4TMs in this receptor, see Chap. 23.

[d] This superfamily also includes the putative "store-operated" Ca^{2+} channels, since so far as is known they are homologous (Caterina et al. 1997). The (intracellular) native ligand for the latter, and whether they are all of one receptor type, are currently debated questions. Hence a further entry under 1.3.3 for this type is not yet made but may soon become available.

[e] 1.6.KIR.61.S + 1.6.SUR.01.S or etc., for the Kir 6.1/SUR1 or similar combinations. In the K_{NDP} type these are Kir 6.1/SUR2B, etc. "n" denotes sequential numbers in the code for these combinations. The topology given is for the Kir component.

such hydrophobic domains in the transducing membrane proteins which have been crystallized (Deisenhofer et al. 1984; Kühlbrandt and Wang 1991; Doyle et al. 1998) and (using two-dimensional crystals) to all of the previously-deduced TMs in bacteriorhodopsin (Henderson and Schertler 1990; Grigorieff et al. 1996) and in vertebrate rhodopsin (Schertler et al. 1993; Unger et al. 1997). In fact, direct evidence for the presence of a membrane-spanning α-helix in TGCs has been obtained in one case so far: in the nicotinic acetylcholine receptor of the electric organ, it has been established by three-dimensional electron image analysis of the receptor crystallised in *Torpedo* membranes, at 9 Å resolution, that all of the subunits span the membrane and that the second TM along the chain, the M2 segment, is essentially α-helical (Unwin 1993, 1995). Mapping of chemical accessibilities and of the effects of mutations have provided independent evidence that M2 lines the channel (see Chap. 17) and that it is α-helical (with a central 3-residue stretch which is non-helical) in the closed state of the channel (Revah et al. 1990; Akabas et al. 1994). That central stretch corresponds to a bend seen by Unwin (1995) in the α-helix in the resting channel state. The extreme N-terminal (intracellular) end of M2 may also not be part of the helix.

In theory, types of polypeptide helix other than the α-helix could be present in TMs in other channels, e.g., the 3_{10} helix (which does occur in some transmembrane structures formed by certain peptides) or a π-helix, but those types are rare in the very large number of crystal structures of proteins in general that are now known, whereas the α-helix is found universally. This, together with the diffraction evidence for α-helices in the integral membrane proteins just noted, means that at present when any TM in the TGCs is interpreted as a helix this will be as an α-helix, although possible future exceptions cannot be excluded.

For the nicotinic ACh receptor, the question immediately arises whether M2 contains the sole α-helical structure of the four TMs, M1–M4, present in each subunit. Unwin (1993, 1995) has found, from the electron image analysis, no evidence for helices in the other three TMs. Consistent with this, Akabas et al. (1994) and Karlin and Akabas (1995) deduced from chemical accessibility studies that a part of M1 is non-helical and that a part of that sub-domain is exposed in the channel. Also, lipid-phase photo-reactivity labeling (Verrall and Hall 1992; Blanton and Cohen 1994) has placed M4 outside the channel lining, at the lipid/protein interface. M4 shows a low degree of conservation between the receptor subunits and between species and has the highest hydrophobicity. Consistent with this, the Numa group made the interesting finding on M4 that an unrelated viral protein hydrophobic segment could replace it without loss of function (Tobimatsu et al. 1987). This evidence suggests an indirect, supporting role for M4. Despite continuing debate on the structures present in the membrane outside the ring of M2 helices, it should be emphasized that the status of M2 as a channel-lining kinked α-helix has not been challenged, and this serves as a prototype for such TMs in the ion permeation pathway of other TGCs in the same superfamily.

III. Supporting Transmembrane Structures

The muscle/electric organ ACh receptor has been established by direct structural analysis to be pentameric (UNWIN 1993) and hence contains 5 M2 helices enclosing the central channel plus 15 other TMs. The diffraction analyses noted above have led to the interpretation that these 15 form a β-stranded barrel surrounding the 5 M2 α-helices (UNWIN 1995). Such circular β-structures are known in some bacterial pore-forming proteins whose three-dimensional structure has been determined, e.g., the porins (WEISS and SCHULZ 1992; COWAN et al. 1992; SCHIRMER et al. 1995). Exactly the device deduced for this model of the nicotinic receptor channel, i.e., an inner ring of 5 α-helices supported by an enclosing barrel of β-structure, is indeed known to be stable in the three-dimensional structures established for certain bacterial toxin proteins, e.g., enterotoxin (SIXMA et al. 1991) or verotoxin-1 (STEIN et al. 1992), which, although not pore-forming, associate with vertebrate cell membranes.

Infra-red spectroscopic techniques applied to the protease-resistant membrane-embedded domains of the *Torpedo* nicotinic receptor have confirmed the presence of considerable β-structure there, and that this is not oriented along a single axis (GÖRNE-TSCHELNOKOW et al. 1994). However, the numerical apportionment by such methods of the TM region between α-helix, β-structure and other non-helical (e.g., extended) structure remains an open question.

IV. Pore Loops (P-Domains)

In some TGCs outside the superfamily containing the nicotinic ACh receptors (Table 1), the original assignment of the TMs could not be reconciled with fuller investigations. This is exemplified in the glutamate-gated cation channels, i.e., the NMDA and the non-NMDA glutamate receptors. Although their hydropathy plots had initially led to the assignment of 4TMs per subunit, more recent mapping of the locations of extracellular and intracellular segments within the sequence in these glutamate receptors has led to a major re-interpretation of that transmembrane topology, as described in Sect. C.III below (HOLLMANN et al. 1994; WO and OSWALD 1994; BENNETT and DINGLEDINE 1995). The new evidence establishes that only three domains can be transmembrane, and hence the previously-assigned M2 is a re-entrant loop, of the type of the P (pore)-domain known in the ion conduction pathway (MACKINNON 1995) of the voltage-gated cation channels (Fig. 1A). Furthermore, there are some points of sequence homology between the P-domain of K^+ channels and the M2 region of the glutamate receptors (WO and OSWALD 1994; KUNER et al. 1995). In this region of the voltage-gated and related cation channels, both ends of the loop face the extracellular side, but in the glutamate receptors both ends face the cytoplasm. As in the voltage-gated and related cation channels, so in the glutamate-gated channels this segment has been shown in mutagenesis studies to be the major determinant

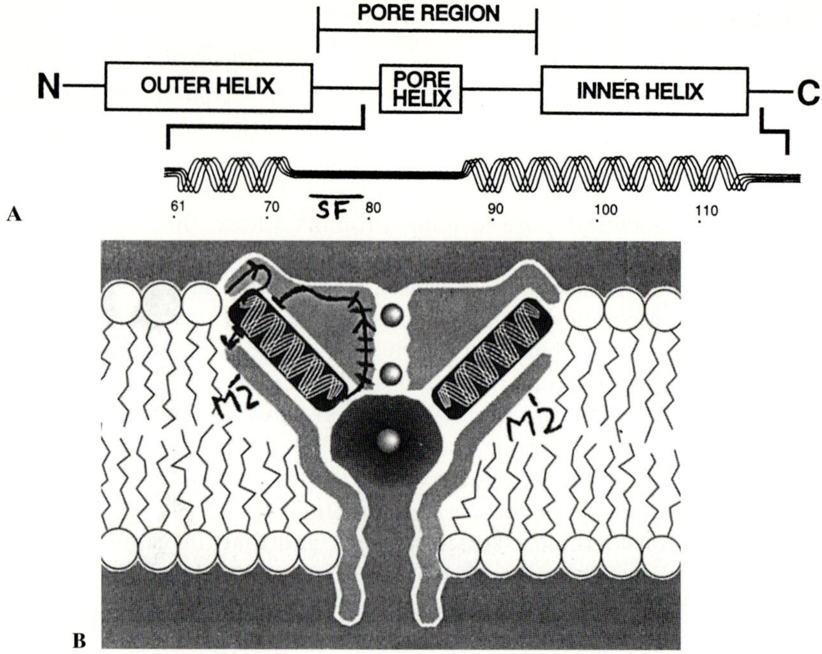

Fig. 1A,B. Diagrams illustrating how the P-domain is used in constructing a pore through the membrane, and its selectivity filter, in an ion channel protein. The structure is that obtained (DOYLE et al. 1998) for a member of the 2-TM K^+ channel family: one TM is an α-helix facing the pore ("inner helix") and the other is an α-helix facing the lipid membrane ("outer helix"). The region joining these two contributes, from the four copies in each molecule: (i) the outer vestibule of the pore; (ii) a "pore α-helix" sequence; and (iii) an extended chain sequence which forms the selectivity filter and loops back to the top of the inner helix at the extracellular surface. **A** the orientations of these sections are shown in a ribbon representation of the chains, with the sequence T/S. V/I. G.Y.G, a K^+ channel signature, forming the selectivity filter (SF). The inner helix continues the pore lining below this down to the intracellular surface. **B** How the pore is formed (parts of only 2 of the 4 subunits are shown and the outer helices are omitted). The pore helices point towards a central aqueous cavity, giving a helix dipole stabilization of the hydrated K^+ ion in that cavity. The contiguous SF segment (*hatched line*) lines a narrow pore which contains 2 dehydrated K^+ ions, co-ordinated to backbone carbonyl groups (which is not possible with Na^+). These elements form the "P loop". Part of the inner helix (M2) forms the rest of the pore, down to the intracellular surface. The four inner helices are tilted about 25% from the perpendicular, packing together (not shown) in forming the pore on the intracellular side to form an "inverted tepee". Based upon the crystallographic structure of a K^+ channel of DOYLE et al. (1998)

of the ionic selectivity (DINGLEDINE et al. 1992; SEEBURG 1993, HOLLMANN et al. 1994).

A model for the P-domain in the structure of certain TGCs is given by the bacterial porins noted above. In their known three-dimensional structures the channels are each formed by β-barrels of 16 or 18 transmembrane segments, but this channel is very wide and is constricted by re-entrant loops which, as MacKinnon (1995) and Sun et al. (1996) have pointed out, act structurally and functionally as P-domains. Although several known porins have a modest cation selectivity, there is no significant homology between these loops and the P-loops of ion channels of animals.

Definite structural information on the P-domains of cation-selective channels has more recently become available, however, since the crystallographic structure at 3.2 Å resolution of a bacterial K^+ channel has established the reality of P-domains (four in the tetrameric molecule) in its ion conduction pathway (Doyle et al. 1998). Each subunit has two helical TMs with a P-loop inserted in the chain between them: the P-loop sequence has a central section which is α-helical while the rest of it is in an extended chain structure (Fig. 1).

In summary, a number of TGCs, while being of diverse types, have been deduced to contain P-domains as well as TMs. The membrane-embedded regions of a TGC may be built from TMs, or from P-domains plus TMs (but P-domains alone are not known in them). The term "membrane-inserted segment" (MIS) is needed (Barnard 1996) to cover both of those elements; it will also cover glycolipid anchors in the membrane where they occur in other receptor classes and it allows for the ambiguity present where the transmembrane status of a segment is still uncertain. If the convention is adopted that the membrane-inserted segments are numbered along the sequence as M1, M2, etc. (avoiding the term "TM" in proteins containing P-domains), this allows the previous assignment as M2, of the hydrophobic domain in glutamate receptors which is now known to be a P-domain, to continue in use without confusion.

The involvement of P-domains in certain types of TGC will be specified under their respective headings below.

C. The Subclasses of the Transmitter-gated Channels

I. The TGCs are in Completely Diverse Superfamilies

On a functional basis the TGCs constitute one receptor Class. Despite their common transduction system (as specified at the start of Sect. B.I above) they are, nevertheless, greatly divergent. They exist in at least twelve superfamilies (Table 1), i.e., they show some sequence homology within each superfamily (subclass), but no significant sequence similarities between any pair of those subclasses. Their designation by codes is given by Humphrey and Barnard (1998). There is no relationship between the subclass and the charge or valence of the permeating ions, nor with the homomeric or heteromeric type of subunit composition. Nevertheless, the elements which form or control the channel show basic similarities in secondary (but not primary) structure throughout the Class. The walls of the pore are, as far as has been analyzed, transmembrane α-helices, plus in many cases (but far from all) P-domains projecting into the channel (Fig. 1). These inserted P-domains may function as the channel ion selectivity control, as they do (see below) in voltage-gated channels (MACKINNON 1995; DOYLE et al. 1998).

II. The Cys-Loop Receptors

The well-studied nicotinic acetylcholine receptors, both muscle and neuronal types, were found to be part of a superfamily of receptors, initially with the GABA and the glycine receptors (BARNARD et al. 1987). Later, the existence of other receptors of this type was unveiled, also having low but significant homology to the nicotinic receptors, to give a present total of five families in this superfamily (Table 1). They show a common pattern of secondary structure and membrane topology, with four TMs at equivalent positions in the C-terminal half of the ~50000-dalton subunit, and no P-domain (Fig. 2A). In all cases where it has been investigated, the M2 region is the segment primarily involved in the channel lining (GALZI and CHANGEUX 1994).

A highly characteristic structural feature of this subclass is the presence (starting at about one-quarter of the way along each subunit) of the "Cys-loop", formed by a pair of cysteines separated by 13 other residues. Where analyzed, those two cysteine residues are disulphide-bonded and were shown to be on the extracellular side of the receptor (KAO and KARLIN 1986). The residues in this loop are strongly conserved in all of the superfamily (Fig. 3). This motif is a defining feature of all these receptors (BARNARD et al. 1987), which have therefore been termed the Cys-loop superfamily (COCKCROFT et al. 1990; KARLIN and AKABAS 1995). From molecular-modeling studies on the various subunit types of these receptors, a β-folded amphipathic structure for this loop, very similar in all of them, is predicted (COCKCROFT et al. 1990). This motif does not appear in other types of TGCs, even those (discussed below) with apparent slight homologies with the Cys-loop superfamily. Another signature element in this superfamily is a conserved pattern in the center of the M1 domain, around an invariant proline (BARNARD 1992). Indeed, these signature structural features found in the subunits of the Cys-loop superfamily (BARNARD et al. 1987; BARNARD 1992) are maintained in all of the five families now recognized in it (Table 1) and in all animal species so far investigated for

Fig. 2A–G. The wide range of structures which form the ion channels gated by transmitters. (See Table 1 for the coding noted). **A** *Left*, the Cys-loop (C-C) subunit structure (1.1), *right*, the 3TM+P structure of the glutamate-gated cation channels (1.2). **B** The 6TM+P structure of the cGMP and IP_3 receptors (1.3.1 and 1.3.2). **C** *Left*, the P2X receptors (1.4.1), *right*, the proton-gated and FMRFamide-gated Na^+ channels (1.4.2). **D** The K_{ATP} and K_{NDP} channels (1.5.1). **E** The 4TM+2P structure of an OH-activated K^+ channel (1.5.2). **F** The CFTR Cl^- channel (1.6). *NBD*, nucleotide-binding domain. **G** Subunit of a neurotransmitter-activated channel/transporter (1.7) as deduced from hydropathy plots. The precise topology is, however, controversial (see text). Alternative versions are given by GRUNEWALD et al. (1998) and YU et al. (1998)

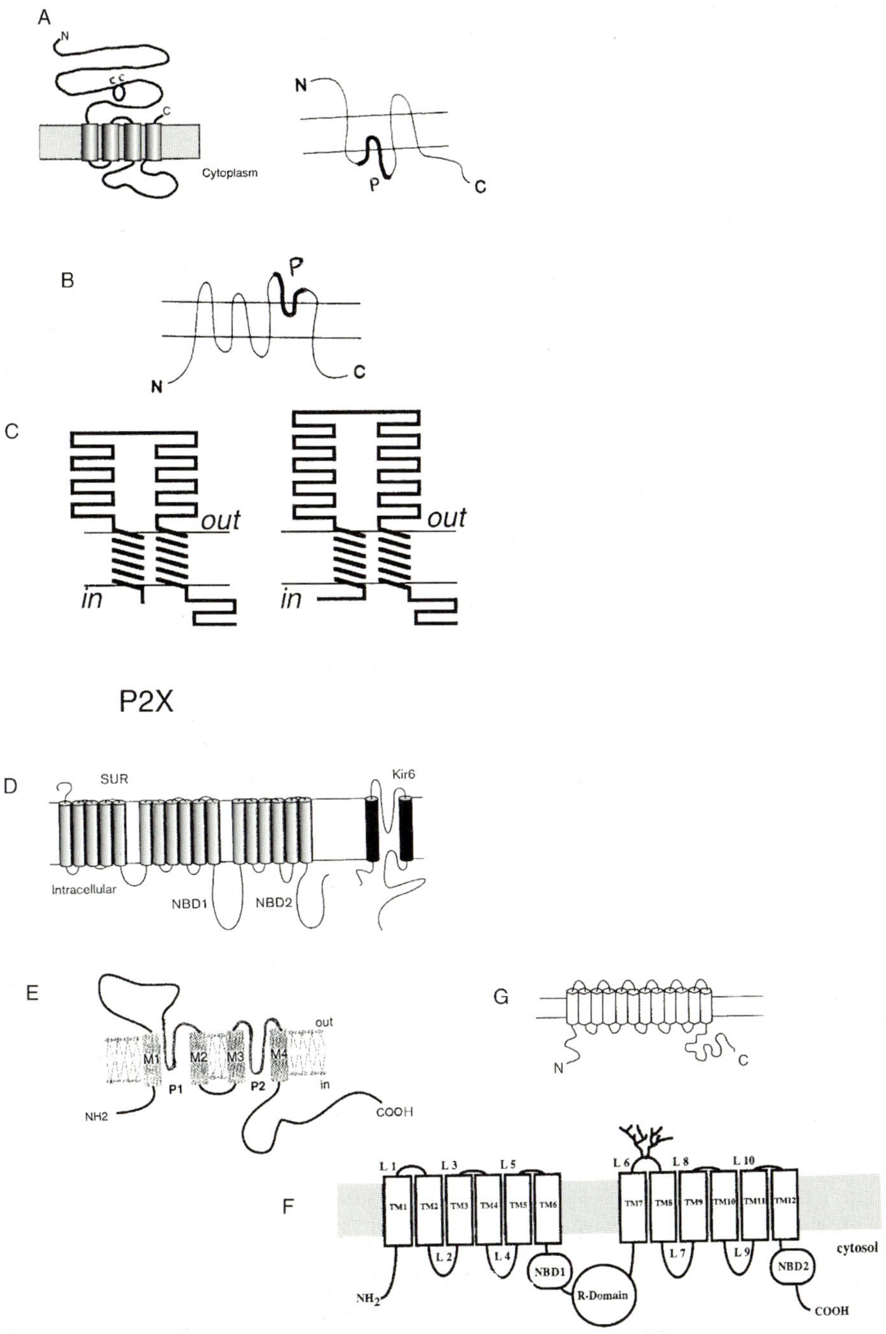
A
N
C C
C
Cytoplasm
N
P
C
B
P
N
C
C
out
in
out
in
P2X
D
SUR
Kir6
Intracellular
NBD1
NBD2
E
out
M1
M2
M3
M4
P1
P2
in
NH2
COOH
G
N
C
F
L 1
L 3
L 5
L 6
L 8
L 10
TM1
TM2
TM3
TM4
TM5
TM6
TM7
TM8
TM9
TM10
TM11
TM12
L 2
L 4
L 7
L 9
NBD1
R-Domain
NBD2
cytosol
NH_2
COOH

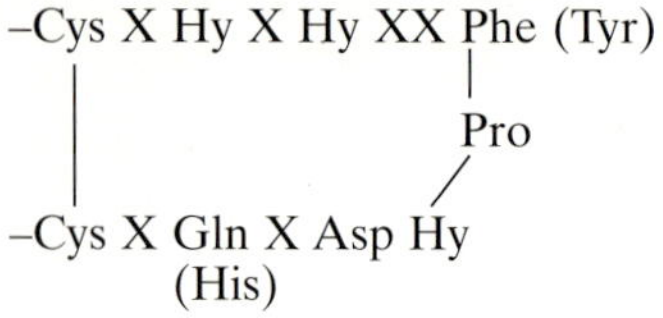

Fig. 3. The constant "Cys-loop" of superfamily 1.1. Hy = one of a hydrophobic group of amino acids, restricted to Ile, Leu, Met, Phe, Trp, Tyr, Val. X = any amino acid. This consensus sequence, across a disulphide-bridged (Kao and Karlin, 1986) pair of cysteines which are always at the same (15-residue) spacing, occurs in the N-terminal domain in all of the subunit types of the diverse receptors in this group. That is, in all of the muscle nicotinic acetylcholine, neuronal nicotinic acetylcholine, $GABA_A$, glycine, 5-HT_3 and glutamate (anion channel) receptors. This is true for all subtypes and species (both vertebrate and invertebrate) investigated. The position in the sequence of this loop relative to the M1 domain, as exemplified in Fig. 2A, is also constant for all cases. It is interesting that this motif is independent of the ion selectivity or kinetics of the channel and of whether the subunit is agonist-binding or not. This (and mutagenesis evidence on it) suggests that this loop is important in acting as a core for the correct folding to form the external vestibule structure and the connection of the agonist binding sites to the channel gate.

The sequences screened for conformity with this concensus were all those in the receptor types noted above, as given by N. Le Novère and J.-P. Changeux (1998) at http//:www.pasteur.fr/units/neubiomol/, plus additional *C. elegans* receptor sequences (to a total of 17 subunits from that nematode) of Fleming et al. (1997) and Mongan et al. (1998). The latter group, in fact, shows a greater diversity of the α subunits of the acetylcholine receptor than is known even in mammals; it is unclear at present whether this represents more specialisations in the nematode use of this receptor or the incompleteness of the mammalian genome sequencing.

Hy, one of a hydrophobic group of amino acids, restricted to Ile, Leu, Met, Phe, Trp, Tyr, Val. *X*, any amino acid. This consensus sequence, across a disulphide-bridged (KAO and KARLIN 1986) pair of cysteines which are always at the same (15-residue) spacing, occurs in the N-terminal domain in all of the subunit types of the diverse receptors in this group. That is, in all of the muscle nicotinic acetylcholine, neuronal nicotinic acetylcholine, $GABA_A$, glycine, 5-HT_3 and glutamate (anion channel) receptors. This is true for all subtypes and species (both vertebrate and invertebrate) investigated. The position in the sequence of this loop relative to the M1 domain, as exemplified in Fig. 2A, is also constant for all cases. It is interesting that this motif is independent of the ion selectivity or kinetics of the channel and of whether the subunit is agonist-binding or not. This (and mutagenesis evidence of it) suggests that this loop is important in acting as a core for the correct folding to form the external vestibule structure and the connection of the agonist binding sites to the channel gate. The sequences screened for conformity with this concensus were all those in the receptor types noted above, as given by N. Le Novère and J.-P. Changeux (1998) at http//:www.pasteur.fr/units/neubiomol/, plus additional *C. elegans* receptor sequences (to a total of 17 subunits from that nematode) of FLEMING et al. (1997) and MONGAN et al. (1998). The latter group, in fact, shows a greater diversity of the α subunits of the acetylcholine receptor than is known even in mammals; it is unclear at present whether this represents more specializations in the nematode use of this receptor or the incompleteness of the mammalian genome sequencing these from man down to the nematode (CULLY et al. 1994; SQUIRE et al. 1995; TREININ and CHALFIE 1995; FLEMING et al. 1997).

It is interesting that an anion channel in the Cys-loop superfamily is gated by glutamate (evidenced, to date, only from invertebrates (CULLY et al. 1994, 1996)). This receptor is heteromeric, several subunit types which co-assemble now being known (VASSILATIS et al. 1997). The strong homology of these subunits to the vertebrate glycine receptor subunits (especially high in the M2 region), their complete dissimilarity in sequence to the subunits of the glutamate-gated cation channels, and their structural features characteristic of the Cys-loop set, show that the anion channel design within the Cys-loop superfamily has been adapted to several different transmitter-binding sites. Glutamate is, therefore, the only known example of a transmitter that can gate alternative anion and cation native channels.

A convenient collection from the databases of the protein sequences of the Cys-loop receptor class of subunits subunits (at present ~180 sequences, from all species investigated) is maintained by N. Le Novère and J.-P. Changeux at the website www.pasteur.fr/units/neubiomol/ and includes the original references, accession numbers, and a tree showing the relatedness of the sequences.

In the Cys-loop superfamily, heteromeric assembly of different subunits is the usual case. The serotonin-gated channel form (the $5HT_3$ receptor) is no longer an exception to this (DAVIES et al. 1999). A few homomeric subtypes have been proposed for certain neuronal nicotinic ($\alpha 7$) and $GABA_A$ receptors (ρ subunits), but in each of those receptor types heteromers greatly predominate. The glutamate-gated anion channels are as yet not well explored, but evidence for heteromeric assembly of two subunit types has been reported (ETTER et al. 1996). The other receptor types in this superfamily also exist as heteromeric assemblies.

The muscle/electric organ ACh receptor forms a distinct sub-set of the large family of nicotinic ACh receptors, because in any given organism its subunits are drawn from a different set of genes than the neuronal type. It is known in only two subtypes ($\alpha 1_2 \beta 1 \gamma \delta$, and in adult muscle $\alpha 1_2 \beta 1 \varepsilon \delta$) in all vertebrate species and skeletal muscle types studied. The neuronal ACh receptors are highly multiple, being formed from a pool of subunit types (11 so far known) in a number of specific combinations. The glycine receptors exhibit lower multiplicity, but the $GABA_A$ receptors exist in an extreme of multiplicity. These latter three cases are described in detail elsewhere in this volume.

It is a hallmark of the Cys-loop set of ion channels that each molecule is constructed from five subunits, arranged radially so as to enclose a central ion channel. In the muscle/electric organ ACh receptor this structure, proposed originally from a variety of biochemical studies, has been proven by the electron image analysis discussed above (UNWIN 1993). Further, electron microscope image analysis has established five subunits in native $GABA_A$ receptors (NAYEEM et al. 1994) and in $5HT_3$ receptors (BOESS et al. 1995). Biochemical and biophysical studies have also given evidence for five subunits per molecule in neuronal nicotinic receptors and in glycine receptors (Anand et al. 1991; Cooper et al. 1991; Langosch et al. 1988).

In the muscle/electric organ ACh receptor a unique cyclic order of the subunits ($\alpha.\gamma.\alpha.\beta.\delta$) has been deduced (Karlin et al. 1983). This specificity of interaction of the subunit interfaces in assembling a unique oligomer is also known as a universal occurrence in the quaternary structures of enzymes having a given heteromeric subunit composition, wherever determined by crystallography. It is an important principle in relation to the structures of all heteromeric TGCs, in that a given subunit composition for one of their subtypes will lead to only one of the various possible cyclic orders in that oligomer.

III. Glutamate-Gated Cation Channels

These comprise the well-known glutamate receptors of the NMDA and the non-NMDA types. The sequence homology between the NMDA and the non-NMDA glutamate receptor subunits varies from low (~30% maximum, in the NR2 series) to almost non-existent, except for some degree of common homology in the "M2" segment. Both types differ from the TGC family considered above in that neither possesses a Cys-loop and they have a different topology: as noted above, when this is carefully mapped their subunits show a common pattern of three TMs plus one P-domain, located as shown in Fig. 2A. A variety of methods for locating intracellular and extracellular segments of the chain has concurred in establishing that the M2 segment forms a P-domain in several types of non-NMDA receptor subunits investigated (Hollmann et al. 1994; Wo and Oswald 1995; Bennett and Dingledine 1995) and likewise in NMDA receptor subunits (Hirai et al. 1996; Kuner et al. 1996). Further, at ~15 of the positions in the P-domain sequence of K^+ channels an identical or very similar amino acid is found in the GluR/kainate subunit series and in the NR1 subunits and at ~10 in the NR2s. In summary, despite similarities with the Cys-loop superfamily at occasional points in the sequence, they form a single, although highly divergent, superfamily of their own (Barnard 1992, 1996).

The non-NMDA receptors are constituted from a set of subunits of which 11 are known so far (GluR1–7, KA1–2, $\delta 1$ and $\delta 2$), a set which is expanded by an alternatively-spliced exon for some of the subunits and by mRNA editing (Seeburg 1993). Combinations between certain of these subunits have been shown to be functional (Seeburg 1993; Hollmann and Heinemann 1994). The NMDA receptors, on present evidence, contain one of eight possible isoforms of the NR1 subunit, plus one or more of the NR2 A, B, C, or D, or NMDAR-L (Ciabarra et al. 1995; Sucher et al. 1995) subunit types (the latter set being products of five different genes). There exists, therefore, a considerable repertoire of subunits to build those two sets of receptors (see M Mishina, this volume). Furthermore, the observed properties of native glutamate receptors at some locations differ from those known from the expression or co-expression of any of the recombinant subunits, indicating that not all the subunits or combinations are as yet known (Sucher et al. 1996; Lerma et al. 1997). Another indication of this occurs with the delta subunits, which in vitro fail to assemble with any of the known glutamate receptor subunits but which occur at certain brain synapses (Kashiwabuchi et al. 1995; Mayat et al. 1995); these are putative glutamate receptor subunits, which can be part of a functional receptor, as is shown by the identification of a point mutation in the $\delta 2$ gene as the cause of the lurcher mouse mutant, producing an aberrant channel (Zuo et al. 1997). Likewise, at least in amphibia a combination of an NR1 subunit with a new type of non-NMDA subunit has been observed both in vitro and in situ, and is a functional receptor (Soloviev et al. 1996; Barnard 1997b). This suggests that yet further possibilities for glutamate receptor subtypes may exist, with unknown subunits, in the mammals.

What is the total number of these subunits which assemble to form the receptor in each case? In both the non-NMDA and the NMDA receptors there is strong evidence, from both biophysical and biochemical approaches, that this number is four (Wu et al. 1996; Laube et al. 1997; Mano and Teichberg 1998; Rosenmund et al. 1998), although some other studies have proposed five (Blackstone et al. 1992; Ferrer-Montiel and Montal 1996; Premkumar and Auerbach 1997). It was shown, for a recombinant homomeric GluR3 receptor, that the agonist molecules (after the first one) binding sequentially produce incremental steps in the channel conductance, up to a total of four agonists bound (Rosenmund et al. 1998). For the NMDA receptors, the evidence has suggested a parallel situation in which two Glu and two Gly molecules together must bind (to NR2 and NR1 units respectively) for full channel opening (Clements and Westbrook 1991; Laube et al. 1997). A tetrameric channel would further emphasise the difference in construction between glutamate receptors and the Cys-loop receptors which, as noted above, are pentameric.

IV. Channels Structurally Related to Voltage-Gated Channels

1. Cyclic Nucleotide-Gated Channels

These are best known from retinal photoreceptor cells and from olfactory neurones, although also known elsewhere, being cation channels gated by intracellular cGMP and cAMP (for details see Chap. 22). They constitute a separate superfamily but are nevertheless built upon the same plan as the voltage-gated K^+ channels, with intracellular N-terminal and C-terminal domains and six TMs (TM1–6), plus a P-domain again situated between TM5 and TM6 (Goulding et al. 1993). The P-domain sequence is clearly homologous in those two types for a 12-residue stretch around the known (Doyle et al. 1998) selectivity filter region for the K^+ channels, and mutagenesis evidence shows that it contains that filter also in the cyclic nucleotide receptors (Heginbotham et al. 1992; Goulding et al. 1993; Sun et al. 1996). However the main determinant sequence for selectivity in K^+ channels, TVGYG (in the one-letter amino acid code) is partly changed in the cyclic nucleotide-gated channels. The backbone carbonyls of this stretch in the voltage-gated K^+ channels co-ordinate dehydrated K^+ ions (DOYLE et al. 1998), and the modified structure in the cyclic nucleotide channel must be related to its wider cation selectivity ($Ca^{2+} > K^+ = Na^+$). This is supported by the ability of corresponding mutations to change the K^+ channel to the latter specificity (HEGINBOTHAM et al. 1992).

2. Inositol Trisphosphate (IP_3) Receptors

These receptors have six deduced TMs plus an apparent P-domain between TM5 and TM6 (YAMAMOTO-HINO et al. 1994). This is the same transmembrane protein topology as in the voltage-gated channels, but this putative P-domain has negligible sequence homology with the P-domains of the latter, whether for K^+ or Ca^{2+}. However, such sequence dissimilarity is likewise found for the established P-domains of Ca^{2+} vs K^+ voltage-gated channels.

The IP_3 receptors are members of a further protein superfamily within this sub-class 1.3 (Table 1). They are located primarily in the membrane of the endoplasmic reticulum or equivalent organelle, being Ca^{2+} channels gated by IP_3, for the mobilization of Ca^{2+} stores. Detailed information on these receptors is given in Chap. 24.

3. Ryanodine Receptors

These are another type of organellar Ca^{2+}-release channels. It is important to note, for the overall view taken here, that they are in the same superfamily as the IP_3 receptors, and have been proposed to contain a P-domain located similarly (YAMAMOTO-HINO et al. 1994). That domain has homology between the ryanodine and IP_3 receptors. These are very large proteins, the subunit containing ~5000 amino acids. Several TMs are present, all in the C-terminal one-tenth of the sequence; the structure, subtypes, agonists, and oligomeric composition are described in Chap. 23. Their channel is selective for Ca^{2+}, with the ratio of permeabilities for Ca^{2+}/K^+ (for RyR1, the heterologously expressed subtype 1), being 6.8 (CHEN et al. 1997).

4. Vanilloid Receptors and Store-Operated Channels

A receptor present on sensory neurones and activated by capsaicin, the pain-producing agent of chilli peppers, etc., has been cloned by CATERINA et al. (1997) and shown to be a Ca^{2+} preferring cation channel. It is apparently of the 6TM+P structural class (see Table 1), but is in a separate superfamily. This new superfamily also contains several putative "store-operated Ca^{2+} channels", the transmitter for which is still uncertain (PAREKH and PENNER 1997).

The capsaicin receptor, termed VR1 as a vanilloid receptor (CATERINA et al. 1997), has been presumed to have an as yet unidentified native ligand as its agonist (although capsaicin itself and some vanilloids are natural exogenous plant-defensive ligands for this receptor, on tongue sensory detectors). However, protons alone (even at pH 6.4) act as agonists at this receptor at 37 °C or lower temperatures and this may provide the natural activation at many nociceptive neurones (TOMINAGA et al. 1998).

This is only one of two channel types at present known to use the proton as a gating transmitter. The second is in another sub-class of entirely different topology (1.4) as detailed below.

V. Channels Topologically Related to Epithelial Na^+ Channels

1. P2X Channels

Extracellular ATP gates a cation channel (for Na^+, K^+, and Ca^{2+}) in the P2X series of nucleotide receptors. Two subtypes were first cloned in 1994 and at least five other subtypes, to give $P2X_1$–$P2X_7$, are now known (reviewed by NORTH and BARNARD 1997). These all differ in pharmacology and desensitisation behavior. They occur across a wide range of cell types. The P2X receptors are described in detail in Chap. 20.

It is of great interest that these TGCs have a structural pattern different from any of the series considered above, in that they have 2 TMs (TM1 and TM2) connected by a very long extracellular loop (rich in cysteines), their N-terminal and C-terminal ends being intracellular, and they have no obvious P-domain sequence. This is the same pattern as in the long-known epithelial Na^+ (ENaC) channels, which, in general are not ligand-gated. The relationship with P2X channels is purely in the membrane topology of the polypeptide, there being no sequence homology with any ENaC series. However, while the cysteine-scanning accessibility analysis for the P2X receptor of RASSENDREN et al. (1997b) showed no evidence for a P-domain but major involvement of some residues of TM2 in the channel, EGAN et al. (1997) using similar methods on the same receptor deduced that an N-terminal sequence of TM2 does form a loop in the membrane, which together with some residues in the rest of TM2 lines the channel. The question of the structural type of the P2X receptor subunit is therefore still open at present.

An additional channel type is presented within the P2X family, namely the large pore which can be formed specifically in the $P2X_7$ subtype (RASSENDREN et al. 1997a). This subunit forms, in heterologous expression, homomeric receptors with a cation-selective channel equivalent to that in the $P2X_1$–$P2X_6$ series, but upon repeated application of agonist a much larger non-selective pore opens. The latter will pass all types of solutes up to 900 daltons in size. The molecular basis of the $P2X_7$ large pore is as yet unknown. This intriguing phenomenon is reported in detail in Chap. 20.

2. Proton-Gated Channels

The simplest possible ligand, the proton, is able to gate a specific class of channels, mostly selective for Na^+ ions. These are a sub-class of the epithelial Na^+ channels, related both in membrane topology (Fig. 2C) and in sequence homology to the non-ligand-gated majority (ENaC family) of those channels. The H^+-gated channels have been cloned and characterized by M. Lazdunski and co-workers (WALDMANN et al. 1997a,b; BASSILANA et al. 1997; LINGUEGLIA et al. 1997). Their subunits so far comprise the ASIC1, ASIC2a (or MDEG1), ASIC2b (or MEDG2), and ASIC3 (or DRASIC) subtypes (WALDMANN and LAZDUNSKI 1998), acting in various homomeric or heteromeric assemblies. The ASIC subunits are mammalian relatives of the degenerin series, members of the ENaC superfamily known previously from *C.elegans*, and proposed to be components there of stretch-activated cation channels. The ASIC2 subunit is known in two splice variants (2a and 2b), with the first 236 residues differing between these (LINGUEGLIA et al. 1997). The sequence relationships between all of these channels are shown in Fig. 4.

These channels are truly proton-gated, rather than being merely affected by pH change after opening: activation of ASIC1, for example, is not detectable above pH 6·9, and below it a fast-rising current is elicited with steep dependence on the H^+ concentration, half-maximal activation being at pH 6·2 (BASSILANA et al. 1997). The H^+ ion is the only agent known so far to activate these channels. The sensitivity to slight acidification, and other properties of

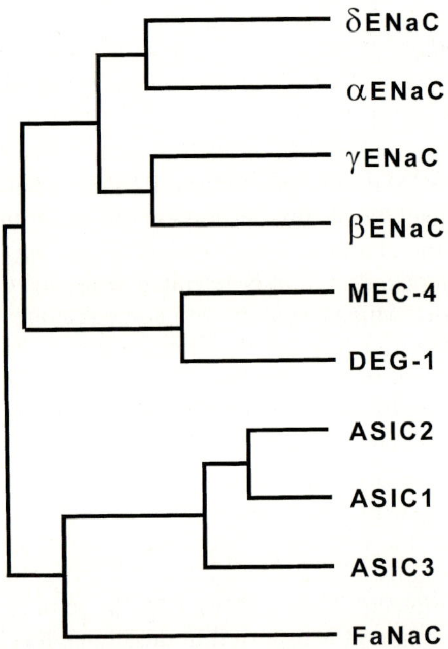

Fig. 4. The ENaC/DEG family of cation channels. Phylogenetic tree showing the relatedness between the H^+-gated cation channel subunits (ASIC1, ASIC2, ASIC3), two of the related degenerins, the peptide-gated Na^+ channel FaNaC from the snail *Helix aspersa*, and the amiloride-sensitive Na^+ channel subunits (αENaC, βENaC, γENaC, δENaC). Tree provided by R. Waldmann and M. Lazdunski (CNRS, Nice)

the subtypes formed by these subunits correspond to features of various native channels found in acid-sensitive sensory neurones and pathways where, indeed, the mRNAs of these subunits occur (LINGUEGLIA et al. 1997; WALDMANN and LAZDUNSKI 1998).

Interestingly, although based on the structure of the Na^+-selective ENaC channel, the channels discussed here are not all as Na^+-selective as the latter. Thus, unlike ASIC2a, the ASIC1 homomer has a permeability ratio for Na^+:Ca^{2+} of only 2·5, while the sustained current generated by the ASIC2b/ASIC3 heteromer is equi-permeable to Na^+ and K^+ (WALDMANN et al. 1997b; LINGUEGLIA et al. 1997).

3. Peptide-Gated Channels

An amiloride-sensitive channel protein of the ENaC superfamily which is gated by the excitatory peptide FMRFamide in the snail *Helix* was the first TGC in this series to be cloned (LINGUEGLIA et al. 1995). It has been shown now that it is functional as a homomer of four subunits (COSCOY et al. 1998).

This is the only TGC so far which is known to have a peptide as its natural agonist. This is a significant development, because of the enormous potential range of structures offered by peptides as a ligand class. Therefore, although it is in the ENaC superfamily 1.4.2 (Table 1) it is listed here in a separate subdivision, allowing for others such to join it in future without interrupting the series of non-peptide TGCs under the main 1.4.2 heading. It will be of great importance to see if equivalent channels occur in mammals, and if other peptide-gated channels fall into the 1.4.2 class only: if not, such sub-divisions of other classes may be needed.

VI. Channels Related to Inward Rectifier K^+ Channels

The inward rectifier sub-class of K^+ channels (Kir family) is known to possess two TMs and a P-domain, as discussed in Chap. 9. In contrast to the ENaC superfamily, there is no large extracellular domain linking the two TMs. A few Kir channels have evolved to a TGC fuction, as follows.

1. Nucleotide-Sensitive K^+ Channels

A few Kir channels are specialized to form part of a TGC assembly, principally in the K_{ATP} channel. In this, ATP acts as an intracellular transmitter to close a spontaneously-open K^+-selective channel. The structure, properties and the high biological significance of the K_{ATP} channel are described in Chap. 11.

The K_{ATP} channel, as so far known, is constructed from the 6·1 or the 6·2 subtype of Kir subunits, plus a protein with a transporter-type structure, the "sulphonylurea receptor" (SUR), being the site of binding of the sulphonylurea modulators of this channel. The SUR subunits are members of the ATP-binding cassette (ABC) transporter superfamily (INAGAKI et al. 1996). They have been deduced to have 12TM domains, as occurs in other ABC transporters (either in one chain or by dimerization), although larger numbers of TMs have also been proposed in SUR (reviewed by TUCKER and ASHCROFT 1998). Evidence has been presented to propose that one K_{ATP} molecule contains a ring of four Kir subunits, forming the channel, with an outer ring of four SUR subunits (Fig. 2D) (CLEMENT et al. 1997; see also the review by TUCKER and ASHCROFT 1998). TUCKER et al. (1997) and GRIBBLE et al. (1997) showed that it is the Kir subunit assembly which has the ATP-antagonist site and the SUR subunits which interact with MgADP (which enhances the channel openings). The MgADP binding is at two nucleotide binding consensus sites on the SUR subunit.

2. Nucleotide-Dependent K^+ Channels

A few K^+ channels require an intracellular nucleotide for their opening. This topic is reviewed in Chap. 10. In some cases where channels were described thus, the activation by ATP is due to a requirement for phosphorylation to give an active form and these are only apparently transmitter-gated and must be excluded from the present series of TGCs.

There is evidence for a true activation by intact nucleotide in two well-characterized cases so far. One is the K_{NDP} channel of smooth muscles, where both adenosine and uridine diphosphates can gate a channel of the Kir type. The K_{NDP} channel is strongly related structurally to the K_{ATP} channel, with Kir 6.1/SUR2B or Kir6.1/SUR2A forming such a channel (YAMADA et al. 1997; KONDO et al. 1998). This channel is ATP-insensitive and differs also in other properties from the K_{ATP} channel. However, its overall structure is surmised to be similar to the latter, but with different isoforms of the two subunit types being required for it.

The second case is that of a few members of the G-protein $\beta\gamma$-stimulated inward rectifier family, notably GIRK1,-2 or GIRK2,-4. LESAGE et al.(1995) showed that, for these heteromeric K^+ channels, ATP or a non-hydrolyzable analogue gates this channel. This was independent of protein kinase A or C, or protein phosphatase treatments.

3. Channels Containing Bi-Functional Kir Subunits

A new type of K^+ channel subunit which is equivalent to a pair of Kir subunits joined covalently, to give four transmembrane domains and two P-domains (Fig. 2E), was described from mammalian tissues by M. Lazdunski and co-workers (LESAGE et al. 1996; DUPRAT et al. 1997). The pore-forming function is therefore present twice in one subunit. These are described in full in Chap. 13. At least one of these acts as a transmitter-gated channel, the TASK K^+ channel. Here, interestingly, the ligand is the hydroxyl ion. The channel activation begins near pH 6.7 and reaches 90% of the maximum current at pH 7.7. This steep dependence upon the extracellular OH^- concentration can endow this channel with a specific communication role in the nervous system (DUPRAT et al. 1997).

VII. Channels Related to ATP-Binding Transporters

The only such TGC known so far is the CFTR (cystic fibrosis transmembrane regulator) Cl^- channel. This has a transporter structure with 12 apparent TMs, separated in two loops of six in the sequence. Between those two groups is a large region, largely or entirely intracellular, which contains two nucleotide-binding domains (NBDs) and a highly polar regulatory (R) domain (Fig. 2F). Hydrolysis of ATP at one of the NBDs has a role in the Cl^- channel gating cycle, but there is also evidence that a non-hydrolyzed ATP, bound at the second NBD domain, is required for the open state of the channel (SHEPPARD et al. 1994; BAUKROWITZ et al. 1994; MATHEWS et al.1998). ATP is required for a phosphorylation by protein kinase A of the R domain; however, this is not essential for CFTR activation, but increases it (WINTER and WELSH 1997).The evidence indicates that this is due to an allosteric increase in the binding of ATP at the second NBD.Gating by ATP continues after a removal of Mg^{2+} or inhibition of ATPases (REDDY and QUINTON 1996; SCHULTZ et al. 1996). A non-hydrolyzable analogue of ATP can sustain gating of CFTR, after pre-activation by phosphorylation (REDDY and QUINTON 1996).

The CFTR gating cycle is complex and completely exceptional in its energetics, among both TGCs and passive transporters.The mechanism is currently much debated, but a recent conclusion is: "The energy of the CFTR-ATP-Mg interaction in the transition state is responsible for the CFTR ion channel opening rather than the energy of ATP hydrolysis" (ALEKSANDROV and RIORDAN 1998). The CFTR is considered here, therefore, in the present state of knowledge, as an ATP-gated channel.

VIII. Channels Related to Neurotransmitter Transporters

A surprising finding in recent studies of transporters for neurotransmitters is that they can be linked to a specific ionic conductance activated by the transported substrate. The transporters in question belong to a large superfamily of sodium-dependent or sodium-and-chloride-dependent transporters (AMARA and KUHAR 1993), with 12 apparent TMs (Fig. 2G). That membrane topology has been supported by accessibility studies on several of these transporters (see BENNETT and KANNER (1997) and references therein) for a large part of it but not for all of it. Parts of the topology remain at present controversial. Thus, to accommodate the accessibility patterns which are observed, the hydropathy plot has been re-interpreted to propose that a P-domain is present within a set of eight transmembrane helices in the glutamate transporters (GRUNEWALD et al. 1998), and a somewhat similar deviation has been found for the GABA transporter (YU et al. 1998).

A chloride channel activated by glutamate in an excitatory amino acid transporter (EAAT) was described by FAIRMAN et al. (1995). This has been confirmed for the series EAAT 1–5 in several studies since, as reviewed by SONDERS and AMARA (1997). The EAAT can thus function as a combined transporter and inhibitory glutamate receptor (DEHNES et al. 1998).

The range of this phenomenon has been widened to show cation channel activation by the substrate in the cases of the serotonin transporter (GALLI et al. 1997), the noradrenaline transporter (GALLI et al. 1996), the GABA transporter (CAMMACK and SCHWARTZ 1996), and the dopamine transporter (SONDERS et al. 1997) both in recombinant expression and (PICAUD et al. 1995) in situ. These transporters can operate alternatively in transport (T-mode) or in channel opening (C-mode), and, e.g., ~500 serotonin molecules are translocated for 10,000 ions passing through the channel (GALLI et al. 1966, 1967). How such a channel is formed in the transporter protein, and whether the same permeation pathway serves for both the transport of substrate, its co-substrate ions and this larger ion flux, are at present obscure; as noted above, even the structure present which could form any such pathway is still uncertain.

D. Conclusion

Any case where the opening or the closing of a channel is totally dependent upon the binding of a transmitter is a TGC. Although a wide range of activities is then encompassed in this Class, as is seen above, it is logically not possible to restrict the TGC category to exclude any of these. A remarkably wide range of structures (Fig. 2) is, therefore, found in the designs employed for the gating of a channel by a transmitter.

References

Akabas MH, Karlin A (1995) Identification of acetylcholine receptor channel-lining residues in the M1 segment of the α subunit. Biochemistry 34:12496–12500

Akabas MH, Kaufmann C, Archdeacon P, Karlin A (1994) Identification of acetylcholine receptor channel-lining residues in the entire M2 segment of the α subunit. Neuron 13:919–927

Aleksandrov AA, Riordan JR (1998) Regulation of CFTR ion channel gating by MgATP. FEBS Lett 431:97–101

Amara SG, Kuhar MJ (1993) Neurotransmitter transporters: recent progress. Annu Rev Neurosci 16:73–93

Anand R, Conroy WG, Schoepfer R, Whiting P, Lindstrom J (1991) Neuronal nicotinic acetylcholine receptors expressed in *Xenopus* oocytes have a pentameric quaternary structure. J Biol Chem 266:11192–11198

Arriza JL, Eliasof S, Kavanaugh MP, Amara SG (1997) Excitatory amino acid transporter 5, a retinal glutamate transporter coupled to a chloride conductance. Proc Natl Acad Sci USA 94:4155–4160

Barnard EA, Darlison MG, Seeburg PH (1987) Molecular biology of the $GABA_A$ receptor: the receptor/channel superfamily. Trends Neurosci 10:502–509

Barnard EA (1992) Receptor classes and the transmitter-gated ion channels. Trends Biochem Sci 17:368–374

Barnard EA (1996) The transmitter-gated channels: a range of receptor types and structures. Trends Pharmacol Sci 17:305–308

Barnard EA (1997a) Protein structures in receptor classification. Ann NY Acad Sci 812:14–28

Barnard EA (1997b) Ionotropic glutamate receptors: new types and new concepts. Trends Pharmacol Sci 18:141–148

Barnard EA, Skolnick P, Olsen RW, Möhler H, Sieghart W, Biggio G, Braestrup C, Bateson AN, Langer SZ (1998) Subtypes of $GABA_A$ receptors: classification on the basis of subunit structure and receptor function. Pharm Revs 50: 291–314.

Bassilana F, Champigny G, Waldmann R, de Weille JR, Heurteaux C, Lazdunski M (1997) The acid-sensitive ionic channel subunit ASIC and the mammalian degenerin MDEG form a heteromultimeric H^+-gated Na^+ channel with novel properties. J Biol Chem 272:28819–28822

Baukrowvitz T, Hwang TC, Nairn AC, Gadsby DC (1994) Coupling of CFTR Cl^- channel gating to an ATP hydrolysis cycle. Neuron 12:473–482

Bennett JA, Dingledine R (1995) Topology profile for a glutamate receptor: three transmembrane domains and a channel-lining reentrant membrane loop. Neuron 14:373–84

Bennett ER, Kanner BI (1997) The membrane topology of GAT-1, a (Na^+ + Cl^-)-coupled γ-aminobutyric acid transporter from rat brain. J Biol Chem 272:1203–1210

Blackstone CD, Moss SJ, Martin LJ, Levey AI, Price DL, Huganir RL (1992) Biochemical characterization and localization of a non-N-methyl-D-aspartate glutamate receptor in rat brain. J Neurochem 58:1118–1126

Blanton MP, Cohen JB (1994) Identifying the lipid-protein interface of the *Torpedo* nicotinic acetylcholine receptor: Secondary structure implications. Biochemistry 33:2859–2872

Boess FG, Beroukhim R, Martin IL (1995) Ultrastructure of the 5-hydroxytryptamine$_3$ receptor. J Neurochem 64:1401–1405

Cammack JN, Schwartz EA (1996) Channel behavior in a γ-aminobutyrate transporter. Proc Natl Acad Sci USA 93:723–727

Caterina MJ, Schumacher MA, Tominaga M, Rosen TA, Levine JD, Julius D (1997) The capsaicin receptor: a heat-activated ion channel in the pain pathway. Nature 389:816–24

Chen SR, Leong P, Imredy JP, Bartlett C, Zhang L, MacLennan DH (1997) Single-channel properties of the recombinant skeletal muscle Ca^{2+} release channel (ryanodine receptor). Biophys J 73:1904–1912

Ciabarra AM, Sullivan JM, Gahn LG, Pecht G, Heinemann S, Sevarino KA (1995) Cloning and characterization of χ-1: a developmentally regulated member of a novel class of the ionotropic glutamate receptor family. J Neurosci 15:6498–6508

Clement JP, Kunjilwar K, Gonzalez G, Schwanstecher M, Panten U, Aguilar-Bryan L, Bryan J (1997) Association and stoichiometry of K_{ATP} channel subunits. Neuron 18:827–838

Clements JD, Westbrook GL (1991) Activation kinetics reveal the number of glutamate and glycine binding sites on the N-methyl-D-aspartate receptor. Neuron 7:605–613

Cockroft VB, Ostedgaard DJ, Barnard EA, Lunt GG (1990) Modelling of agonist binding to the ligand-gated ion channel superfamily of receptors. Proteins 8:386–397

Conroy W, Berg D (1995) Neurons can maintain multiple classes of nicotinic acetylcholine receptors distinguished by different subunit compositions. J Biol Chem 270:4424–4431

Cooper E, Couturier S, Ballivet M (1991) Pentameric structure and subunit stoichiometry of a neuronal nicotinic acetylcholine receptor. Nature 350:235–238
Coscoy S, Lingueglia E, Lazdunski M, Barbry P (1998) The FMRFamide activated sodium channel is a tetramer. J Biol Chem 273:8317–8322
Cowan SW, Schirmer T, Rummel G, Steiert M, Ghosh R, Pauptit RA, Jansonius JN, Rosenbusch JP (1992) Crystal structures explain functional properties of two *E. coli* porins. Nature 358:727–733
Cully DF, Vassilatis DK, Liu KK, Paress PS, Van der Ploeg LH, Schaeffer JM, Arena JP (1994) Cloning of an avermectin-sensitive glutamate-gated chloride channel from *Caenorhabditis elegans*. Nature 371:707–711
Cully DF, Paress PS, Liu KK, Schaeffer JM, Arena JP (1996) Identification of a *Drosophila melanogaster* glutamate-gated chloride channel sensitive to the antiparasitic agent avermectin. J Biol Chem 271:20187–20191
Davies PA, Pistis M, Hanna MC, Peters JA, Lambert JJ, Hales TG, Kirkness EF (1999) The 5-HT_{3B} subunit is a major determinant of serotonin-receptor function. Nature 397:359–363
Dehnes Y, Chaudhry FA, Ullensvang K, Lehre KP, Storm–Mathisen J, Danbolt NC (1998) The glutamate transporter EAAT4 in rat cerebellar Purkinje cells: A glutamate-gated chloride channel concentrated near the synapse in parts of the dendritic membrane facing astroglia. J Neurosci 18:3606–3619
Deisenhofer J, Epp O, Miki K, Huber R, Michel H (1985) Structure of the protein subunits in the photosynthetic center of *Rhodopseudomonas viridis* at 3 Å resolution. Nature 318:618–624
Dingledine R, Hume RI, Heinemann SF (1992) Stuctural determinants of barium permeation and rectification in non-NMDA glutamate receptor channels. J Neurosci 12:4080–4087
Doyle DA, Morais Cabral J, Pfuetzner RA, Kuo A, Gulbis JM, Cohen SL, Chait BT, MacKinnon R (1998) The structure of the potassium channel: molecular basis of K^+ conduction and selectivity. Science 280, 69–77
Duprat F, Lesage F, Fink M, Reyes R, Heurteaux C, Lazdunski M (1997) TASK, a human background K^+ channel to sense external pH variations near physiological pH. EMBO J 16:5464–5471
Egan TM, Haines WR, Voigt MM (1998) A domain contributing to the ion channel of ATP-gated $P2X_2$ receptors identified by the substituted cysteine accessibility method. J Neurosci 18:2350–2359
Etter A, Cully DF, Schaeffer JM, Liu KK, Arena JP (1996) An amino acid substitution in the pore region of a glutamate-gated chloride channel enables the coupling of ligand binding to channel gating. J Biol Chem 271:16035–16039
Fairman WA, Vandenberg RJ, Arriza JL, Kavanaugh MP, Amara SG (1995) An excitatory amino-acid transporter with properties of a ligand-gated chloride channel. Nature 375:599–603
Ferrer-Montiel AV, Montal M (1996) Pentameric subunit stoichiometry of a neuronal glutamate receptor. Proc Natl Acad Sci USA 93:2741–2744
Fleming JT, Squire MD, Barnes TM, Tornoe C, Matsuda K, Ahnn J, Fire A, Sulston JE, Barnard EA, Sattelle DB, Lewis JA (1997) *Caenorhabditis elegans* levamisole-resistance genes lev-1, unc-29, and unc-38 encode functional nicotinic acetylcholine receptor subunits.. J Neurosci 17:5843–5857
Forsayeth JR, Kobrin E (1997) Formation of oligomers contain the $\beta3$ and $\beta4$ subunits of the rat nicotinic receptor. J Neurosci 17:1531–1538
Galli A, Blakely R, DeFelice LJ (1996) Norepinephrine transporters have channel modes of conduction. Proc Natl Acad Sci USA 93:8671–8676
Galli A, Petersen CI, de Blaquiere M, Blakely RD, DeFelice LJ (1997) *Drosophila* serotonin transporters have voltage-dependent uptake coupled to a serotonin-gated ion channel. J Neurosci 17:3401–3411
Galzi J, Changeux J-P (1994) Neurotransmitter-gated ion channels as unconventional allosteric proteins. Curr Opin Struct Biol 4:554–565

Görne-Tschelnokow U, Strecker A, Kaduk C, Naumann D, Hucho F (1994) The transmembrane domains of the nicotinic acetylcholine receptor contain α-helical and β-structures. EMBO J 13:338–341

Goulding EH, Tibbs GR, Liu DT, Siegelbaum SA (1993) Role of H5 domain in determining pore diameter and ion permeation through cyclic nucleotide-gated channels. Nature 364:61–64

Gribble FM, Tucker SJ, Ashcroft, FM (1997) The essential role of the Walker A motifs of SUR1 in K-ATP activation of Mg-ADP and diazoxide. EMBO J 16:1145–1152

Grigorieff N, Ceska TA, Downing KH, Baldwin JM, Henderson R (1996) Electron-crystallographic refinement of the structure of bacteriorhodopsin. J Mol Biol 259:393–421

Grunewald M, Bendahan A, Kanner I (1998) Biotinylation of single cysteine mutants of the glutamate transporter GLT-1 from rat brain reveals its unusual topology. Neuron 21:623–632

Gunderson KL, Kopito RR (1995). Conformational states of CFTR associated with channel gating: the role of ATP binding and hydrolysis. Cell 82:231–239

Heginbotham L, Abramson T, MacKinnon R (1992) A functional connection between the pores of distantly related ion channels as revealed by mutant K^+ channels. Science 258:1152–1155

Henderson R, Schertler GF (1990) The structure of bacteriorhodopsin and its relevance to the visual opsins and other seven-helix G-protein coupled receptors. Phil Trans Roy Soc B 326:379–389

Hirai H, Kirsch J, Laube B, Betz H, Kuhse J (1996) The glycine binding site of the N-methyl-D-aspartate receptor subunit NR1: Identification of novel determinants of co-agonist potentiation in the extracellular M3-M4 loop region. Proc Natl Acad Sci USA 93:6031–6036

Hirschberg B, Maylie J, Adelman JP, Marrion NV (1998) Gating of recombinant small–conductance Ca–activated K^+ channels by calcium. J Gen Physiol 111:565–581

Hollmann M, Heinemann S (1994) Cloned glutamate receptors. Annu Rev Neurosci 17: 31–108

Hollmann M, Maron C, Heinemann S (1994) N-glycosylation site tagging suggests a three transmembrane domain topology for the glutamate receptor GluR1. Neuron 113:1331–1343

Humphrey PPA, Barnard EA (1998) The IUPHAR Receptor Code: A proposal for an alphanumeric classification system. Pharmacol Rev 50:271–277

Inagaki N, Gonoi T, Clement JP, Wang CZ, Aguilar-Bryan L, Bryan J, Seino S (1996) A family of sulfonylurea receptors determines the pharmacological properties of ATP-sensitive K^+ channels. Neuron 16:1011–1017

Kao P, Karlin A (1986) Acetylcholine receptor binding site contains a disulfide crosslink between adjacent half-cystinal residues. J Biol Chem 261:8085–8088

Karlin A, Holtzman E, Yodh N, Lobel P, Wall J, Hainfield J (1983) The arrangement of the subunits of the acetylcholine receptor of *Torpedo californica*. J Biol Chem 258:6678–6681

Karlin A, Akabas MH (1995) Toward a structural basis for the function of nicotinic acetylcholine receptors and their cousins. Neuron 15:1231–1244

Kashiwabuchi N, Ikeda K, Araki K, Hirano T, Shibuki K, Takayama C, Inoue Y, Kutsuwada T, Yagi T, KangY, Aizawa S, Mishima M (1995) Impairment of motor coordination, Purkinje cell synapse formation, and cerebellar long-term depression of GluR $\delta 2$ mutant mice. Cell 81:245–252

Kondo C, Repunte VP, Satoh E, Yamada M, Horio Y, Matsuzawa Y, Pott L, Kurachi Y (1998) Chimeras of Kir6.1 and Kir6.2 in spontaneous opening reveal structural elements involved and unitary conductance of the ATP–sensitive K^+ channels. Recept Chann 6:129–140

Krapivinsky G, Gordon EA, Wickman B, Velmirovic, Krapivinsky L, Clapham DE (1995) The G-protein-gated atrial K^+ channel I_{KACh} is a heteromultimer of two inwardly rectifying K^+-channel proteins. Nature 374:135–141

Kühlbrandt W, Wang DN (1991) Three-dimensional structure of plant light-harvesting complex determined by electron crystallography. Nature 350:130–134

Kuner T, Wollmuth LP, Karlin A, Seeburg PH, Sakmann B (1996) Structure of the NMDA receptor channel M2 segment inferred from the accessibility of substituted cysteines. Neuron 17:343–352

Langosch D, Thomas L, Betz H (1988) Conserved quaternary structure of ligand-gated ion channels: the postsynaptic glycine receptor is a pentamer. Proc Natl Acad Sci USA 85:7394–7398

Laube B, Hirai H, Sturgess M, Betz H, Kuhse J (1997) Molecular determinants of agonist discrimination by NMDA receptor subunits: analysis of the glutamate binding site on the NR2B subunuit. Neuron 18:493–503

Lerma J, Morales M, Vincente MA, Herreras O (1997) Glutamate receptors of the kainate type and synaptic transmission. Trends Neurosci 20:9–12

Lesage F, Guillemare E, Fink M, Duprat F, Heurteaux C, Fosset M, Romey G, Barhanin J, Lazdunski M (1995) Molecular properties of neuronal G-protein-activated inwardly rectifying K^+ channels. J Biol Chem 270:28660–28667

Lesage F, Reyes R, Fink M, Duprat F, Guillemar E, Lazdunski M (1996) Dimerization of TWIK-1 K^+ channel subunits via a disulfide bridge. EMBO J 15:6400–6407

Lewis C, Neidhart S, Holy C, North RA, Buell G, Surprenant A (1995) Co-expression of P_{2X2} and P_{2X3} receptor subunits can account for ATP-gated currents in sensory neurones. Nature 377:432–434

Lindstrom J (1997) Nicotinic acetylcholine receptors in health and disease. Molec Neurobiol 15:193–222

Lingueglia E, Champigny G, Lazdunski M, Barbry P (1995) Cloning of the amiloride-sensitive FMRFamide peptide-gated sodium channel. Nature 378:730–733

Lingueglia E, de Weille JR, Bassilana F, Heurteaux C, Sakai H, Waldmann R, Lazdunski M (1997): A modulatory subunit of acid sensing ion channels in brain and dorsal root ganglion cells. J Biol Chem 272:29778–29783.

Lomeli H, Sprengel R, Laurie DJ, Köhr G, Herb A, Seeburg PH, Wisden W (1993) The rat delta-1 and delta-2 subunits extend the excitatory amino acid receptor family. FEBS Lett. 15:318–322

MacKinnon R (1995) Pore loops: An emerging theme in ion channel structure. Neuron 14:889–892

Mano I, Teichberg VI (1998) A tetrameric subunit stoichiometry for a glutamate receptor-channel complex. NeuroReport 9:327–331

Mathews CJ, Tabcharani JA, Hanrahan JW (1998) The CFTR chloride channel: nucleotide interactions and temperature-dependent gating.J Membr Biol 163: 55–66

Mayat E, Petralia RS, Wang Y, Wenthold RJ (1995) Immunoprecipitation, immunoblotting, and immunocytochemistry studies suggest that glutamate receptor δ subunits form novel postsynaptic receptor complexes. J Neurosci 15:2533–2546

Mongan NP, Baylis HA, Adcock C, Smith GR, Sansom MSP, Sattelle DB (1998) An extensive and diverse gene family of nicotinic acetylcholine receptor α subunits in *Caenorhabditis elegans*. Recept Chann 6:213–228

Nayeem N, Green TP, Martin IL, Barnard E A (1994) Quaternary structure of the native $GABA_A$ receptor determined by electron microscopic image analysis. J Neurochem 62:815–818

North RA, Barnard EA (1997) Nucleotide receptors. Curr Opin Neurobiol 7:346–357

Parekh AB, Penner R (1997) Store depletion and calcium influx. Physiol Rev 77:901–30

Picaud SA, Larsson HP, Grant GB, Lecar H, Werblin FS (1995) Glutamate-gated chloride channel with glutamate-transporter-like properties in core photoreceptors of the tiger salamander. J Neurophysiol 74:1760–1771

Premkumar LS, Auerbach A (1997) Stoichiometry of recombinant N-methyl-D-aspartate receptor channels inferred from single-channel current patterns. J Gen Physiol 110:485–502

Price MP, Snyder PM, Welsh MJ (1996) Cloning and expression of a novel human brain Na^+ channel. J Biol Chem 271:7879–7882

Rassendren F, Buell GN, Virginio C, Collo G, North RA (1997a) The permeabilizing ATP receptor, $P2X_7$. Cloning and expression of a human cDNA. J Biol Chem 272:5482–5486

Rassendren F, Buell G, Newbolt A, North RA, Surprenant (1997b) Identification of amino acid residues contributing to the pore of a P2X receptor. EMBO J 16:3446–3454

Reddy MM, Quinton PM (1996). Hydrolytic and nonhydrolytic interactions in the ATP regulation of CFTR Cl^- conductance. Amer J Physiol 271:C35–42

Revah F, Glazi J-L, Giraudat J, Haumont PY, Lederer F, Changeux J-P (1990) The noncompetitive blocker [^{3}H]chlorpromazine labels three amino acids of the acetylcholine receptor γ subunit: implications for the α-helical organization of region MII and for the structure of the ion channel. Proc Natl Acad Sci USA 87: 4675–4679

Rosenmund C, Stern-Bach Y, Stevens CF (1998) The tetrameric structure of a glutamate receptor channel. Science 280:1596–1599

Schertler GF, Villa C, Henderson R (1993) Projection structure of rhodopsin. Nature 362:770–772

Schirmer T, Keller TA, Wang YF, Rosenbusch JP (1995) Structural basis for sugar translocation through maltoporin channels at 3.1 Å resolution. Science 267:512–514

Schultz BD, Bridges RJ, Frizzell RA (1996) Lack of conventional ATPase properties in CFTR chloride channel gating. J Membr Biol 105:63–75

Seeburg PH (1993) The molecular biology of mammalian glutamate receptor channels. Trends Pharmacol Sci 14:297–303

Sheppard DN, Ostedgaard LS, Rich DP, Welsh MJ (1994) The amino –terminal portion of CFTR forms a regulated Cl^- channel. Cell 76:1091–1098

Sixma TK, Pronk SE, Kalk KH, Wartna ES, van Zanten BAM, Witholt B, Hol WGJ (1991) Crystal structure of a cholera toxin-related heat-labile enterotoxin from *E. coli*. Nature 351:371–377

Soloviev MM, Abutidze K, Mellor I, Streit P, Grishin EV, Usherwood PN, Barnard EA (1998) Plasticity of agonist binding sites in hetero-oligomers of the unitary glutamate receptor subunit XenU1. J Neurochem 71:991–1001

Sonders MS, Amara SG (1996) Channels in transporters. Curr Opin Neurobiol 6:294–302

Sonders MS, Zhu SJ, Zahniser NR, Kavanaugh MP, Amara SG (1997) Multiple ionic conductances of the human dopamine transporter: the actions of dopamine and psychostimulants. J Neurosci 17:960–974.

Squire MD, Tornoe C, Baylis GA, Fleming JT, Barnard EA, Sattelle DB (1995) Expression of a recombinant *Caenorhabditis elegans* nicotinic acetylcholine receptor subunit (acr-2). Recept Chann 3:101–109

Stein PE, Boodhoo A, Tyrrell GJ, Brunton JL, Read RJ (1992) Crystal structure of the cell-binding B oligomer of verotoxin-1 from *E. coli*. Nature 355:748–750

Sucher NJ, Akbarian S, Chi CL, Leclerc CL, Awobuluyi M, Deitcher DL, Wu MK, Yuan JP, Jones EG, Lipton SA (1995) Developmental and regional expression pattern of a novel NMDA receptor-like subunit (NMDAR-L) in the rodent brain. J Neurosci 15:6509–6520

Sucher NJ, Awobuluyi M, Choi YB, Lipton SA (1996) NMDA receptors: From genes to channels. Trends Pharmacol Sci 17:348–355

Sun Z-P, Akabas MH, Goulding EH, Karlin A, Siegelbaum SA (1996) Exposure of residues in the cyclic nucleotide-gated channel pore: P region structure and function in gating. Neuron 16:141–149

Tobimatsu T, Fujita Y, Fukuda K, Tanaka K, Mori Y, Konno T, Mishina M, Numa S (1987) Effects of substitution of putative transmembrane segments on nicotinic acetylcholine receptor function. FEBS Lett 222:56–62

Tominaga M, Caterina MJ, Malmberg AB, Rosen TA, Gilbert H, Skinner K, Raumann BE, Basbaum AI, Julius D (1998) The cloned capsaicin receptor integrates multiple pain producing stimuli. Neuron 21:531–543

Treinin M, Chalfie M (1995) A mutated acetylcholine receptor subunit causes neuronal degeneration in *C. elegans*. Neuron 14:871–877

Tucker SJ, Gribble FM, Zhao C, Trapp S, Ashcroft FM (1997) Truncation of Kir 6.2 produces ATP-sensitive K^+ channels in the absence of the sulphonylurea receptor. Nature 387:179–183

Tucker SJ, Ashcroft FM (1998) A touching case of channel regulation: the ATP-sensitive K^+ channel. Curr Opin Neurobiol 8:316–320

Unger VM, Hargrave PA, Baldwin JM, Schertler GFX (1997) Arrangement of rhodopsin transmembrane α-helices. Nature 389:203–206

Unwin N (1993) Nicotinic acetylcholine receptor at 9 Å resolution.J Mol Biol 229: 1101–1124

Unwin N (1995) Acetylcholine receptor channel imaged in the open state. Nature 373:37–43

Vassilatis DK, Arena JP, Plasterk RH, Wilkinson HA, Schaeffer JM, Cully DF, Van der Ploeg LH (1997) Genetic and biochemical evidence for a novel avermectin–sensitive chloride channel in *C. elegans*. J Biol Chem 272:33167–33174

Verrall S, Hall ZW (1992) The N-terminal domains of acetylcholine receptor subunits contain recognition signals for the initial steps of receptor assembly. Cell 68:23–31

Waldmann R, Bassilana F, de Weille J, Champigny G, Heurteaux C, Lazdunski M (1997a) Molecular cloning of a non-inactivating proton-gated Na^+ channel specific for sensory neurons. J Biol Chem 272:20975–20978

Waldmann R, Champigny G, Bassilana F, Heurteaux C, Lazdunski M (1997b) A proton-gated cation channel involved in acid-sensing. Nature 836:173–177

Waldmann R, Lazdunski M (1998) Proton-gated cation channels – neuronal acid sensors in the amiloride–sensitive Na^+ channel/degenerin family of ion channels. Curr Opin Neurobiol 8:418–424

Weiss MS, Schulz GE (1992) Structure of porin refined at 1.8 Å resolution. J Mol Biol 227:493–509

Winter MC, Welsh MJ (1997) Stimulation of CFTR activity by its phosphorylated R domain. Nature 389:294–296

Wo ZG, Oswald RE (1994) Transmembrane topology of two kainate receptor subunits revealed by N-glycosylation. Proc Natl Acad Sci USA 91:7154–7158

Wu T-Y, Liu C-I, Chang Y-C (1996) A study of the oligomeric state of the alpha-amino-3-hydroxy-5-methyl-4-isoxazolepropionic acid-preferring glutamate receptors in the synaptic junctions of porcine brain. Biochem J 319:731–739.

Xie J, Drumm ML, Ma J, Davis PB (1995) Intracellular loop between transmembrane segments IV and V of CFTR is involved in recognition of Cl^- conductance state. J Biol Chem 270:28084–28091

Yamada YM, Isomoto S, Matsumoto S, Kondo C, Shindo T, Horio Y, Kurachi Y (1997) Sulphonylurea receptor SUR2B and Kir 6. 1 form a sulphonylurea–sensitive but ATP–insensitive K^+ channel. J Physiol 499:715–720.

Yamamoto-Hino M, Sugiyama T, Hikichi K, Mattei MG, Hasegawa K, Sekine S, Sakurada K, Miyawaki A, Furuichi T, Hasegawa M, Mikoshiba K (1994) Cloning and characterization of human type 2 and type 3 inositol 1, 4, 5-trisphosphate receptors. Recept Chann 2:9–22

Yu N, Cao Y, Mager S, Lester HA (1998) Topological localization of cysteine 74 in the GABA transporter, GAT1, and its importance in ion binding and permeation. FEBS Lett 426:174–178

Zuo J, De Jager PL, Takahashi KA, Jlang W, Linden DJ, Heintz N (1997) Neurodegeneration in Lurcher mice caused by mutation in $\delta 2$ glutamate receptor gene. Nature 388:769–773

CHAPTER 16

Molecular Diversity, Structure, and Function of Glutamate Receptor Channels

M. Mishina

A. Introduction

In 1954, Hayashi noted the excitatory action of L-glutamate in the motor cortex. Extensive studies by Watkins and colleagues revealed structure and function relationships of excitatory amino acids and their derivatives (Watkins and Olverman 1981). Since then, cumulative evidence indicates that glutamate receptor (GluR) channels mediate most fast excitatory synaptic transmission in the vertebrate central nervous system. The development of selective agonists and antagonists led to the classification of GluR channels into *N*-methyl-D-aspartate (NMDA) and non-NMDA subtypes. Subsequently, the non-NMDA subtype was further subdivided into the α-amino-3-hydroxy-5-methyl-4-isoxazole propionic acid (AMPA) and kainate subtypes. It is becoming clear that some of the most important functions of the nervous system, such as synaptic plasticity and synapse formation, critically depend on GluR channels and that neurological damage caused by a variety of pathological states can result from exaggerated activation of GluR channels. In 1989, Hollmann et al. cloned the first member of GluR channel subunit genes. Successful cloning and targeting of GluR channel subunit genes have made it possible to study the molecular and functional diversity of GluR channel families and their physiological roles in brain function.

B. Structure and Molecular Diversity of the GluR Channel

I. Subunit Families and Subtypes

There are at least 17 subunit genes belonging to mammalian GluR channel families (Table 1). These subunits can be classified into 7 subfamilies according to the amino-acid sequence identity (Hollmann and Heinemann 1994; Mishina et al. 1993; Seeburg 1993). Five of seven subfamilies correspond well to pharmacological classification of AMPA, kainate, and NMDA subtypes, while pharmacological characterization of two subfamilies remains to be established. Members of the first (GluR1–4 or GluRα) subfamily form GluR channels with high affinity for AMPA. The second (GluR5–7 or GluRβ) subfamily and the third (KA or GluRγ) subfamily correspond to the subunits of

Table 1. Molecular diversity of glutamate receptor channels

Subtype	Subfamily	Subunit	Putative mature protein	
			Amino acids	kDa
AMPA receptor	GluR1–4 (GluRα)	GluR1 (GluRA, α1)	889	100
		GluR2 (GluRB, α2)	862	96
		GluR3 (GluRC)	866	98
		GluR4 (GluRD)	881	101
Kinate receptor	GluR5–7 (GluRβ)	GluR5	875	103
		GluR6 (GluRβ2)	877	94
		GluR7	888	100
	KA (GluRγ)	KA1	936	105
		KA2 (GluRγ2)	965	109
GluRδ	GluRδ	GluRδ1	994	110
		GluRδ2	991	113
NMDA receptor	GluRε (NR2)	GluRε1 (NR2A)	1445	163
		GluRε2 (NR2B)	1456	163
		GluRε3 (NR2C)	1218	134
		GluRε4 (NR2D)	1329	141
	NR1 (GluRζ)	NMDAR1 (GluRζ1, NR1)	920	104
	GluRχ	GluRχ1 (NR3A)	1082	121

the kainate-selective GluR channel. Combination of the fifth (GluRε or NR2) and sixth (NR1 or GluRζ) subfamilies constitutes NMDA-type GluR channels. Heteromeric assembly of GluR channel subunits within and between subfamilies yields GluR channel heterogeneity in the CNS. Splice variants and editing contribute further genetic diversity of GluR channels.

II. Primary Structure and Transmembrane Topology Model

The GluR channel subunits have a putative signal peptide at the amino-terminus and four hydrophobic segments (M1–M4) in the middle of the molecules (Fig. 1). The amino-terminal domain preceding segment M1 contains numerous consensus sites for N-glycosylation is assigned to be extracellular. Segment M2 contains important determinants of GluR channel properties and forms narrow channel constriction (see below). The carboxyl-terminal region of the NMDA-type GluRζ1 subunit and AMPA-type GluR1 subunit are phosphorylated (Tingley et al. 1993; Roche et al. 1996). Furthermore, the NMDA-type GluRε2 and AMPA-type subunits interact with PDZ domain-containing post-synaptic proteins through their very end of the carboxyl-terminus (Kornau et al. 1995; Dong et al. 1997). Thus, it is likely that the carboxyl-terminal region resides on the cytoplasmic side. Mutational analyses of the

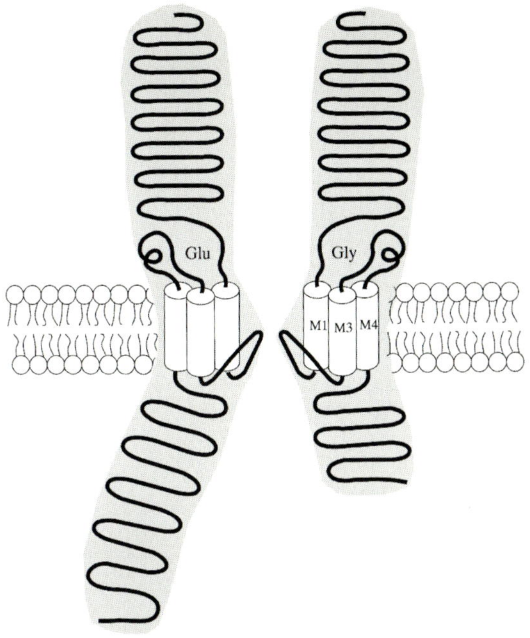

Fig. 1. A transmembrane topology model of the NMDA receptor channel subunits

glycine-binding and redox modulation sites of the GluRζ1 subunit suggest the possible extracellular localization of the region between segments M3 and M4 (KURYATOV et al. 1994; SULLIVAN et al. 1994). Additional evidence is provided by the fact that a consensus glycosylation site present in this domain of the GluR6 subunit is glycosylated in vivo (ROCHE et al. 1994; TAVERNA et al. 1994). N-glycosylation between segment M3 and M4 is also demonstrated for the structurally related goldfish kainate-binding protein GFKARα (WO and OSWALD 1994) as well as for the GluR1 subunit protein expressed in *Xenopus* oocytes (HOLLMANN et al. 1994). A three transmembrane segment model is proposed for the GluR channel subunits, in which putative channel-lining segment M2 loops into the membrane without traversing it (WO and OSWALD 1994; HOLLMANN et al. 1994).

C. AMPA Subtype

I. AMPA-Type Subunits

There are four members of the first GluR1–4 (GluRα) subfamily constituting the AMPA subtype of the GluR channel (Table 1). The GluR1 to GluR4

(GluRA to GluRD) subunits form homomeric or heteromeric GluR channels responsive to both AMPA and kainate (HOLLMANN et al. 1989; BOULTER et al. 1990; KEINÄNEN et al. 1990; NAKANISHI et al. 1990; SAKIMURA et al. 1990). Although kainate elicits larger responses, AMPA shows much higher agonist potency. The recombinant channels have high affinity for AMPA but low affinity for kainate. The four subunits are expressed widely in the brain with some regional and developmental differences.

Alternative splicing of a region immediately preceding segment M4 produces two forms of each AMPA-type subunit, termed flip and flop, that differ in kinetic properties (SOMMER et al. 1990). Alternative splicing at the carboxyl terminus is known for the GluR4 subunit (GALLO et al. 1992).

II. GluR2 Subunit and Ca^{2+} Permeability

Combination of the GluR2 subunit with either the GluR1, GluR3, or GluR4 subunit produces GluR channels with very low Ca^{2+} permeability and linear or outwardly rectifying current-voltage relationships, similar to most native AMPA-type GluR channels (MAYER and WESTBROOK 1987). On the other hand, subunit combinations without the GluR2 subunit exhibit significant Ca^{2+} permeability and an inwardly rectifying current-voltage relationship (HOLLMANN et al. 1991). Thus, the GluR2 subunit dominates the channel properties of AMPA-type GluR channels. Some inhibitory interneurons in the hippocampus and neocortex and Bergman glial cells have AMPA-type GluR channels with a high permeability to Ca^{2+} and an inwardly rectifying current-voltage relationship and these cells lack the GluR2 subunit or express this subunit in relatively low abundance (IINO et al. 1990; BURNASHEV et al. 1992; MÜLLER et al. 1992; BOCHET et al. 1994; JONAS et al. 1994).

Mutant mice lacking the GluR2 subunit exhibit increased motality, and those surviving show reduced exploration and impaired motor coordination (JIA et al. 1996). LTP in the CA1 region of hippocampal slices is markedly enhanced and nonsaturating.

III. Q/R Site as a Determinant of Channel Properties

Arginine residue in segment M2 of the GluR2 subunit and glutamine residue of the GluR1, GluR3, and GluR4 subunits in the homologous position (Q/R site) determine the Ca^{2+} permeability and rectification property of the channel (HUME et al. 1991; MISHINA et al. 1991; VERDOORN et al. 1991). The critical arginine codon of the GluR2 subunit is introduced at the precursor messenger RNA (pre-mRNA) stage by site-selective adenosine editing of a glutamine codon (SOMMER et al. 1991). Heterozygous mice harboring an editing-incompetent GluR2 allele synthesize unedited GluR2 subunit and express AMPA receptors with increased calcium permeability (BRUSA et al. 1995). These mice develop seizures and die by three weeks of age. Assembly of different subunit combinations, relative abundance of subunit specific mRNAs, and editing of mRNA are major mechanisms which control this wide range of Ca^{2+} inflow through different versions of GluR channels under physiological conditions (BURNASHEV et al. 1995). The single-channel conductance of GluR2/4 channels is dependent on the Q/R site editing state of the subunits comprising the channel (SWANSON et al. 1997). Unedited channels have resolvable single-channel events with main conductance states of 7–8 pS, whereas fully edited channels show very low conductances of approximately 0.3 pS estimated from noise analysis.

IV. Phosphorylation

GluR1 is phosphorylated on multiple sites that are all located on the C-terminus of the protein (ROCHE et al. 1996). Cyclic AMP-dependent protein kinase and protein kinase C specifically phosphorylates Ser845 and Ser831 of GluR1, respectively. The modulation of GluR1 by PKA suggests that phosphorylation of this residue may underlie the PKA-induced potentiation of AMPA receptors in neurons. Induction of LTP increased the P^{32} labeling of AMPA-type glutamate receptors (BARRIA et al. 1997). This AMPA-R phosphorylation appeared to be catalyzed by Ca^{2+}- and calmodulin-dependent protein kinase II.

V. Autoimmune Disease

Rasmussen's encephalitis is a progressive childhood disease characterized by severe epilepsy, hemiplegia, and inflammation of the brain. Rabbits immunized with GluR3 protein developed symptoms mimicking Rasmussen's encephalitis (ROGERS et al. 1994). There was a correlation between the presence of the disease and serum antibodies to GluR3 protein, suggesting that Rasmussen's encephalitis is mediated by autoantibodies against GluR3 protein.

VI. GRIP, an Associated Protein

A synaptic PDZ domain-containing protein GRIP (glutamate receptor interacting protein) specifically interacts with the carboxyl termini of AMPA receptors (DONG et al. 1997). GRIP has seven PDZ domains and appears to serve as an adapter protein that links AMPA receptors to other proteins and may be crucial for the clustering of AMPA receptors at excitatory synapses in the brain.

D. Kainate Subtype

The subunits of the second (GluR5–7 or GluRβ) and third (KA or GluRγ) subfamilies constitute the kainate subtype of the GluR channel (Table 1). The GluR5, GluR6, and GluR7 subunits form functional homomeric channels that desensitize in the presence of glutamate and kainate (Bettler 1990, 1992; Egebjerg et al. 1991; Schiffer et al. 1997). On the other hand, the KA1 and KA2 subunits show GluR channel activities only when expressed together with the GluR5 or GluR6 subunits (Werner et al. 1991; Herb et al. 1992; Sakimura et al. 1992). The binding affinity for kainate of the GluR5, GluR6, and GluR7 subunits is similar to the value of the low-affinity binding site found in the brain (K_D = ~50 nmol/l), whereas that of the KA1 and KA2 subunits is close to the value of the high-affinity binding site (K_D = ~5 nmol/l).

The GluR5 and GluR6 subunits are found in two forms with arginine or glutamine at the Q/R site in segment M2, specified by RNA editing (Sommer et al. 1992). In contrast to the GluR2 subunit of the AMPA subtype, both edited and unedited forms are present. Furthermore, two residues in M1 of GluR6 can be altered by RNA editing, with functional consequences for glutamate-activated Ca^{2+} permeability. Indeed, different homomeric and heteromeric GluR6 channels with respect to the unedited and edited M1/M2 positions display a considerable range of Ca^{2+} permeability (Burnashev et al. 1995). There are several splice variants of the GluR5 subunit with the differences in the amino-terminal and carboxyl-terminal regions. The homomeric unedited GluR5 channel functionally resembles the native kainate receptor found on dorsal root ganglia (DRG) cells. In contrast to the GluR6 subunit, GluR5 channels can be gated by AMPA, a feature also described for the DRG receptors. When KA subunits are co-expressed with GluR5 or GluR6 subunits, the channels display a different desensitization profile to kainate (Werner et al. 1991; Herb et al. 1992; Sakimura et al. 1992). Furthermore, AMPA can elicit a non-desensitizing current component on KA2/GluR6 combinations, although homomeric GluR6 receptors are not sensitive to AMPA.

The hippocampal neurons in the CA3 region of mutant mice lacking the GluR6 subunit fail to show post-synaptic kainate currents evoked by a train of stimulation of the mossy fiber system, indicating kainate receptors containing the GluR6 subunit are important in synaptic transmission (Mulle et al. 1998). GluR6-deficient mice are less susceptible to systemic administration of kainate, as judged by onset of seizures and by the activation of immediate early genes in the hippocampus.

The GluR6 subunit expressed in mammalian cells is directly phosphorylated by PKA (Raymond et al. 1993; Wang et al. 1993). Serine 684 in the region between segments M3 and M4 is a major phosphorylation site responsible for the potentiation of the glutamate response. It is to be noted that this assignment apparently contradicts the current transmembrane topology model of the GluR channel subunits.

E. NMDA Subtype

I. Heteromeric Nature of NMDA Receptor Channels

NMDA receptor channels are heteromeric in nature and composed of the GluRε (NR2) and GluRζ (NR1) subunits (Mori and Mishina 1995) (Table 1).

There are four GluRε subunit genes (IKEDA et al. 1992; KUTSUWADA et al. 1992; MEGURO et al. 1992; MONYER et al. 1992; NAGASAWA et al. 1996), while GluRζ subunit variants are derived from a single gene (MORIYOSHI et al. 1991; YAMAZAKI et al. 1992; HOLLMANN et al. 1993). Expression of the GluRε and GluRζ subunits together produces highly active NMDA receptor channels (IKEDA et al. 1992; ISHII et al. 1993; KUTSUWADA et al. 1992; MEGURO et al. 1992; MONYER et al. 1992). Most brain regions express both the GluRε and GluRζ subunits (see below). There are no detectable NMDA receptor channel activities in maturated cerebellar Purkinje cells which express the GluRζ1 subunit but none of the GluRε subunits (BROSE et al. 1993; MONYER et al. 1994; PERKEL et al. 1990; QUINLAN and DAVIES, 1985; WATANABE et al. 1992, 1994a) and in the hippocampal slices of the mutant mice that are lacking the GluRε2 subunit but expresses GluRζ1 subunit (KUTSUWADA et al. 1996). Heteromeric assembly of the NMDA receptor channel subunits is also suggested from the biochemical and immunological studies (SHENG et al. 1994; TINGLEY et al. 1993). Small NMDA responses observed in *Xenopus* oocytes injected with the GluRζ1 mRNA may be due to low activities of homomeric GluRζ1 channels or of heteromeric channels formed by the combination of the exogenous GluRζ1 and endogenous *Xenopus* subunits.

II. Dynamic Variations of the Distribution of the Subunits

Distributions of the NMDA receptor channel subunit mRNAs in the rodent brains are highly variable (Fig. 2). Four GluRε subunit mRNAs show characteristic distributions in the brain, while the GluRζ1 subunit mRNA is distributed ubiquitously (WATANABE et al. 1992, 1993, 1994a,b; AKAZAWA et al. 1994; MONYER et al. 1994). The GluRε1 subunit mRNA is distributed widely in the brain and the level of expression is higher in the cerebral cortex, the hippocampal formation, and cerebellar granule cells. In contrast, the GluRε2 subunit mRNA is expressed selectively in the forebrain. High levels of expression are observed in the cerebral cortex, the hippocampal formation, the septum, the caudate-putamen, the olfactory bulb, and the thalamus. The GluRε3 subunit mRNA is found predominantly in the cerebellum. Strong expression is observed in the granule cell layer of the cerebellum, while weak expression is detected in the olfactory bulb and the thalamus. Low levels of the GluRε4 subunit mRNA are found in the thalamus, the brainstem, and the olfactory bulb.

The expression of the respective GluRε subunit mRNAs are differentially regulated during development, while the GluRζ1 subunit mRNA is ubiquitously expressed in the brain throughout the developmental stages (WATANABE et al. 1992, 1993, 1994a,b; AKAZAWA et al. 1994; MONYER et al. 1994). Among four GluRε subunits, only the GluRε2 and GluRε4 subunit mRNAs are expressed in the embryonic brain. In contrast to the wide distribution of the GluRε2 subunit mRNA, the GluRε4 subunit mRNA is found exclusively in the diencephalon and the brainstem. During the first two weeks after birth, the expression patterns of the GluRε subunit mRNAs change drastically.

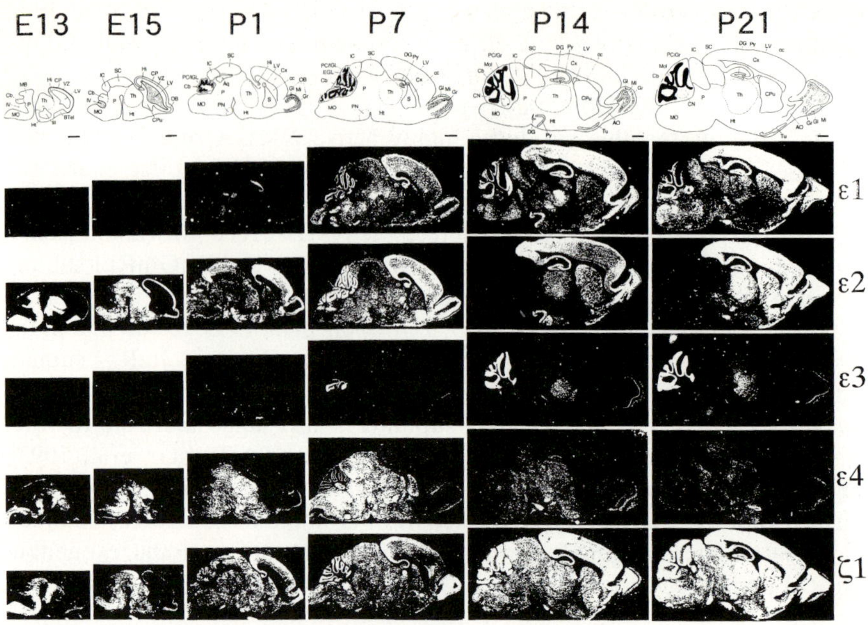

Fig. 2. Expression of the NMDA receptor channel subunits in the mouse brain during development. Parasagittal sections of the mouse brain at embryonic day 13 (*E13*), E15, P1, P7, P14 and P21 were hybridized with radiolabeled oligonucleotide probes specific for the GluRε1, GluRε2, GluRε3, GluRε4, or GluRζ1 subunit mRNA. Sections are shown schematically in the *top row*. Abbreviations: *AO*, anterior olfactory nucleus; *Aq*, cerebral aqueduct; *BTel*, basal telencephalon; *Cb*, cerebellum; *cc*, corpus callosum; *CN*, cerebellar nuclei; *CP*, cortical plate; *CPu*, caudate-putamen; *Cx*, cerebral cortex; *DG*, dentate gyrus; *EGL*, external granular layer; *Gl*, olfactory glomerular layer; *Gr*, olfactory granular layer; *Hi*, hippocampus; *Ht*, hypothalamus; *IC*, inferior colliculus; *IZ*, intermediate zone; *LV*, lateral ventricle; *MB*, midbrain; *Mi*, olfactory mitral cell layer; *Mol*, cerebellar molecular layer; *MO*, medulla oblongata; *OB*, olfactory bulb; *P*, pons; *PC/Gr*, Purkinje cell layer/granule cell layer; *PC/IGL*, Purkinje cell layer/internal granular layer; *PN*, pontine nuclei; *Py*, hippocampal pyramidal cell layer; *S*, septum; *SC*, superior colliculus; *Th*, thalamus; *Tu*, olfactory tubercle; *VZ*, ventricular zone; *III*, the third ventricle; *IV*, the fourth ventricle. Scale bars, 1 mm. (From WATANABE et al. 1992)

The GluRε1 subunit mRNA appears in the entire brain and the GluRε3 subunit mRNA in the cerebellum. In contrast, the expression of the GluRε2 subunit mRNA becomes restricted in the forebrain and that of the GluRε4 subunit mRNA is decreased. Thus, the molecular compositions of NMDA receptor channels varies depending on brain regions and developmental stages.

III. Splice Variants

There are eight alternative splice variants of the GluRζ1 subunit and two variants of GluRε4 subunit (NAKANISHI et al. 1992; SUGIHARA et al. 1992; YAMAZAKI et al. 1992; HOLLMANN et al. 1993; ISHII et al. 1993). Differential distributions of the GluRζ1 subunit splice variants mRNAs were suggested in the rat basal ganglia (LAURIE and SEEBURG 1994; STANDAERT et al. 1994).

IV. Channel Pore and Gating

All of the NMDA receptor channel subunits possess asparagine in segment M2 at the position corresponding to glutamine or arginine that determine the Ca^{2+} permeability of the AMPA-selective GluR channel. Replacement by glutamine of the asparagine in segment M2 of the GluRε2 and GluRζ1 subunits strongly reduces the sensitivity to Mg^{2+} block of the heteromeric NMDA receptor channel (BURNASHEV et al. 1992; MORI et al. 1992; SAKURADA et al. 1993). Since there is strong evidence that Mg^{2+} produces a voltage-dependent block of the channel by binding a site deep within the ionophore (ASCHER and NOWAK 1988), these results are consistent with the view that segment M2 constitutes the ion channel pore of the NMDA receptor channel. The mutation in the GluRζ1 subunit decreases the Ca^{2+} permeability of the heteromeric channels whereas that in the GluRε1 or GluRε3 subunit exerts little effect (BURNASHEV et al. 1992). The patterns of individually mutated residues of M2 segment affecting the sensitivity to external and internal Mg^{2+} block or the accessibility to charged sulfhydryl-specific reagents provide evidence that segment M2 loops into the plasma membrane from the intracellular side without actually traversing it (KUNER et al. 1996; KUPPER et al. 1996).

The relative permeability of different sized organic cations suggests that the narrow constriction of NMDA receptor channels is 0.55 nm and the asparagine residue in GluRζ M2 segment and the carboxyl-terminal residue of two adjacent asparagine residues of GluRε M2 segment constitute the narrow constriction of the NMDA receptor channel (WOLLMUTH et al. 1996). The two adjacent GluRε asparagine residues form a critical blocking site for extra cellular Mg^{2+}, while the GluRζ asparagine residue is a dominant blocking site for intracellular Mg^{2+} (KUPPER et al. 1998; WOLLMUTH et al. 1998a,b). Thus, the asparagine residues in M2 segments of both the GluRε and GluRζ subunits, positioning asymmetrically in the horizontal plane of the channel, form the narrow constriction (selective filter) of the channel. The site of non-competitive antagonists such as (+)-MK-801, phencyclidine (PCP), ketamine and *N*-allynormetazocine (SKF-10,047 overlaps the Mg^{2+} site (MORI et al. 1992; YAMAKURA et al. 1993).

The GluRε1/ζ1 and GluRε2/ζ1 channels were more sensitive to Mg^{2+} blocking than the GluRε3/ζ1 and GluRε4/ζ1 channel (IKEDA et al. 1992; ISHII et al. 1993; KUTSUWADA et al. 1992; MONYER et al. 1992). These subtypes are also more sensitive to Zn^{2+}, and (+)-MK-801 (IKEDA et al. 1992; KUTSUWADA et al. 1992; MISHINA et al. 1993), whereas the sensitivities to PCP, ketamine and SKF-10,047 are only slightly variable among four subtypes (YAMAKURA et al. 1993).

Both the GluRε1/ζ1 and GluRε2/ζ1 channels have two conductance levels, 50 pS openings and brief 40 pS sublevels, with similar mean life times and frequencies (Stern et al. 1992, 1994; Tsuzuki et al. 1994). The conductance levels and amplitude histograms for these subtypes are similar to those reported for CA1 hippocampal pyramidal cells (Gibb and Colquhoun 1991, 1992). The GluRε3/ζ1 channel shows 36 pS and 19 pS conductances of similar duration. NMDA receptor channels are characterized by slow gating. The offset decay time constant of the GluRε1/ζ1 channel is ~120 ms and that of the GluRε2/ζ1 and GluRε3/ζ1 channels is ~400 ms (Monyer et al. 1994). On the other hand, the GluRε4/ζ1 channel shows a very long offset decay time constant (~5000 ms). The duration of NMDA receptor channel-mediated excitatory post-synaptic currents (EPSCs) in neurons of the visual cortex layer IV and superior colliculas is longer at early developmental stages and becomes progressively shorter (Hestrin 1992; Carmignoto and Vicini 1992). Neurons of post-natal neocortex expressing GluRε1 subunit have faster EPSCs and the proportion of cells expressing this subunit increases developmentally (Flint et al. 1997).

In the cerebellar granule cells, the GluRε2 subunit is expressed transiently at early post-natal stage, whereas the GluRε1 and GluRε3 subunits are expressed during later stages. Pre-migratory and migrating granule cells express the NMDA receptor channel with 50 pS and 40 pS, whereas mature post-migratory cells express 33 pS and 20 pS channels in addition to 50 pS and 40 pS channels (Farrant et al. 1994). At late developmental stages, the high-conductance channels disappeared in GluRε1 mutant mice, and the low-conductance channels in GluRε3 mutant mice (Ebralidze et al. 1996; Takahashi et al. 1996). The openings of the 33 pS and 20 pS channels are briefer than those of the 50 pS and 40 pS channels, which are similar to those reported for the GluRε3/ζ1 channels (Stern et al. 1992). The decay time-course of NMDA receptor channel currents in cerebellar granule cells is slower in the GluRε1 mutant mice and faster in GluRε3 mutant mice than in the wild-type mice, suggesting that the GluRε1 subunit determines the fast kinetics, and the GluRε3 subunit the slow ones (Ebralidze et al. 1996; Takahashi et al. 1996). In GluRε3 mutant mice, the voltage-dependent Mg^{2+} block of synaptic currents is decreased (Takahashi et al. 1996).

V. Agonist Binding

Full activation of NMDA receptor channels requires both L-glutamate and glycine (Johnson and Ascher 1987; Kleckner and Dingledine 1988; Meguro et al. 1992). The EC_{50} values for L-glutamate were 1.7 mmol/l, 0.8 mmol/l, 0.7 mmol/l, and 0.4 mmol/l for the GluRε1/ζ1, GluRε2/ζ1, GluRε3/ζ1, and GluRε4/ζ1 channels, respectively, whereas those for glycine were 2.1 mmol/l, 0.3 mmol/l, 0.2 mmol/l, and 0.09 mmol/l, respectively (Ikeda et al. 1992; Kutsuwada et al. 1992). The sensitivity to APV is in the order of the GluRε1/ζ1 > GluRε2/ζ1 > GluRε3/ζ1 > GluRε4/ζ1 channels, whereas that to

7-chlorokynurenic acid is in the order of the GluRε3/ζ1 > GluRε2/ζ1 > GluRε1/ζ1 ≥ GluRε4/ζ1 channels (IKEDA et al. 1992; KUTSUWADA et al. 1992; MISHINA et al. 1993). Ligand binding studies in situ show pharmacological heterogeneity of NMDA receptor channels (MONAGHAN et al. 1988). There is a parallelism between the distributions in the cerebrum of the GluRε1 and GluRε2 subunits and those of the antagonist-preferring and agonist-preferring NMDA receptors, respectively (BULLER et al. 1994; WATANABE et al. 1993).

Amino acid sequence homology is noted between the region preceding segment M1 and the loop region between segments M3 and M4 of the NMDA receptor channel subunits and bacterial amino acid binding proteins (NAKANISHI et al. 1990; O'HARA et al. 1993; KURYATOV et al. 1994). Mutations of the amino acid residues in the two regions of the GluRζ1 subunit affect the EC_{50} values for glycine with little effect on the glutamate efficacy (KURYATOV et al. 1994; HIRAI et al. 1996). Mutations of residues within the homologous regions of the GluRε2 subunit significantly reduced the efficacy of glutamate, but not glycine, in channel gating (LAUBE et al. 1997). The GluRζ1 subunit expressed alone shows a high affinity binding for glycine but neither NMDA nor a competitive glutamate antagonist CGP-39653 binds to the polypeptide (LAURIE and SEEBURG 1994; LYNCH et al. 1994), while the GluRε1 subunit expressed alone has shows high affinity binding of glutamate and NMDA (KENDRICK et al. 1996). Thus, the glutamate binding site of NMDA receptor channels resides on the GluRε subunit, and the glycine binding site on the GluRζ subunit. Molecular modeling of the respective domains by LAUBE et al. (1997) will be useful for rationalizing the design of novel NMDA receptor ligands.

VI. Phosphorylation

The carboxyl-terminal region of the GluRζ1 subunit is phosphorylated and most of these sites are contained within a single alternatively spliced exon (TINGLEY et al. 1993). Treatment with TPA potentiates the GluRε1/ζ1 and GluRε2/ζ1 channels expressed in *Xenopus* oocytes, but not the GluRε3/ζ1 and GluRε4/ζ1 channels (KUTSUWADA et al. 1992; MORI et al. 1993). The carboxyl-terminal region of the GluRε2 subunit is responsible for the activation of the GluRε2/ζ1 channel by the TPA treatment. The GluRε2 subunit is a prominent tyrosine-phosphorylated protein in the post-synaptic density (MOON et al. 1994). Incubation with src and fyn kinases potentiates the GluRε1/ζ1 channel but not the GluRε2/ζ1, GluRε3/ζ1, and GluRε4/ζ1 and carboxyl-terminally truncated GluRε1/ζ1 channels (KÖHR and SEEBURG 1996). In membrane patches excised from mammalian central neurons, the endogenous tyrosine kinase Src was shown to regulate the activity of NMDA channels (YU et al. 1997).

VII. Modulation

The NMDA receptor channels are modulated by various endogenous compounds, such as sulfhydryl (redox) reagents with strong oxidizing or reducing potentials, ethanol, spermine, nitric oxide, and proton. The insertion at the near amino terminus of some GluRζ1 splice variants and the region between segments M3 and M4 are involved in the modulation by dithiothreitol, spermine, and proton (Köhr et al. 1994; Sullivan et al. 1994; Traynelis et al. 1995). Ethanol, in concentrations associated with intoxication in humans, inhibits the GluRε1/ζ1 and GluRε2/ζ1 channels expressed in *Xenopus* oocytes, but not the GluRε3/ζ1 channel (Masood et al. 1993).

VIII. Synaptic Plasticity, Learning, and Neural Development

The NMDA receptor channel acts as the associative switch for the induction of LTP, turning on only when post-synaptic depolarization is paired temporally with the synaptic release of glutamate (Bliss and Collingridge 1993). Chronic intraventricular infusion of APV impaired both hippocampal LTP and spatial learning in rat (Davis et al. 1992; Morris et al. 1986). Disruption of the GluRε1 gene results in reduction of hippocampal LTP and impairment of Morris water maze learning (Sakimura et al. 1995). The ablation of the GluRε2 subunit also impaired synaptic plasticity in the hippocampus (Kutsuwada et al. 1996; Ito et al. 1997). Selective elimination of the GluRζ1 subunit in the hippocampal CA1 region impairs LTP and spatial learning (Tsien et al. 1996). However, NMDA receptor channel-dependent hippocampal LTP may not be essential for spatial memory itself, though required for some component of water maze learning, since pretraining eliminates the APV inhibition (Bannerman et al. 1995; Saucier and Cain 1995). In GluRε1 mutant mice, thresholds for both hippocampal LTP and contextual learning increase (Kiyama et al. 1998). These observations suggest that NMDA-receptor channel-dependent synaptic plasticity is the cellular basis of certain forms of learning.

Chronic infusion of APV suggests the involvement of the NMDA receptor channel in experience-dependent synaptic plasticity during development (Cline et al. 1987; Kleinschmidt et al. 1987). Of five subunits of the NMDA receptor channel, the GluRε2, GluRε4, and GluRζ1 subunits are expressed in the embryonic brain (Watanabe et al. 1992). Mice lacking the GluRζ1 subunit fail to form whisker-related neural pattern in the brainstem trigeminal complex and die after birth (Forrest et al. 1994; Li et al. 1994). GluRε2 mutant mice die shortly after birth and fail to form the whisker-related neural pattern (barrelettes) in the brainstem trigeminal complex (Kutsuwada et al. 1996). In contrast, the barrelette formation is normal in GluRε4 mutant mice (Ikeda et al. 1995). These results show the involvement of the GluRε2 subunit in the refinement of the synapse formation of periphery-related neural patterns in the mammalian brain.

IX. Associated Post-Synaptic Proteins

GluRζ1 variants containing the first carboxyl-terminal exon cassette expressed in fibroblasts are located in discrete, receptor-rich domains associated with the plasma membrane, while those lacking this exon are distributed throughout the cell and large amounts of the protein are present in the cell interior (EHLERS et al. 1995). Furthermore, protein kinase C phosphorylation of specific serines within this exon disrupted the receptor-rich domains (TINGLY and HUGANIR 1994). Thus, the alternative splicing and phosphorylation of the GluRζ1 subunit may regulate the subcellular distribution of NMDA receptor channels.

The carboxyl-terminal tail of the GluRε2 subunit interacts directly with PSD-95 family of post-synaptic density (PSD) proteins, including PSD-95/SAP90, PSD-93/chapsyn-110, and SAP102 (KORNAU et al. 1995; KIM et al. 1996; MÜLLER et al. 1996). The PSD-95 family proteins have three tandem PDZ domains, an src homology 3 (SH3) domain, and an inactive guanylate kinase (GK) domain. The terminal T/SXV motif (where X is any amino acid) common to GluRε subunits and certain GluRζ1 splice forms is essential for binding to the PDZ domains. PSD-95 family proteins can form homomeric or heteromeric mulitimers through the head to head interaction or PDZ domains (KIM et al. 1996; HSUEH et al. 1997). Furthermore, neuronal nitric oxide synthase (nNOS), a neural cell adhesion molecule, neroligin, and a novel ras-GTPase activating protein, synGAP, bind to the PDZ domains of PSD-95 family proteins (BRENMAN et al. 1996; IRIE et al. 1997; KIM et al. 1998). In addition, GK-associated protein (GKAP) and SAP90/PSD-95 associated proteins (SAPAPs) interact with the GK domain of PSD-95 family proteins (KIM et al. 1997; TAKEUCHI et al. 1997). The carboxyl-terminal region of the GluRζ1 subunit binds to calmodulin and filamentous proteins (EHLERS et al. 1996, 1997; LIN et al. 1998). Calmodulin binding to the GluRζ1 subunit is Ca^{2+}-dependent and causes a reduction in channel open probability, suggesting a possible mechanism for activity-dependent feedback inhibition and Ca^{2+}-dependent inactivation of NMDA receptors (EHLERS et al. 1996). α-Actinin-2, a member of the spectrin/dystrophyn family of actin-binding proteins, binds to the cytoplasmic tail of both GluRε2 and GluRζ1 subunits and GluRζ1-α-actinin binding is directly antagonized by Ca^{2+}/calmodulin (WYSZYNSKI et al. 1997). The interaction with PSD-95 family proteins and α-actinin may mediate the clustering and synaptic targeting of NMDA receptor channels. In addition, the binding ability of PSD-95 family proteins to post-synaptic proteins such as nNOS, synGAP, neuroligins, and GKAP may form a large complex of signal transduction molecules.

F. Additional Members of the GluR Channel Family

I. GluRδ Subfamily

The GluRδ subfamily positions in between the NMDA and non-NMDA receptor channel subunits with respect to the amino acid sequence identity (YAMAZAKI et al. 1992; ARAKI et al. 1993; LOMELI et al. 1993). Thus far, no GluR channel functions have been detected after expression in *Xenopus* ooctes and mammalian cells. The GluRδ1 subunit distributes widely in the brain, while the GluRδ2 subunit is selectively expressed in cerebellar Purkinje cells (Araki et al. 1993; Lomeli et al. 1993; Takayama et al. 1996). GluRδ2 proteins are localized in parallel fiber-Purkinje cell dendritic spine synapses, but not in climbing fiber synapses (Landsend et al. 1997).

GluRδ2 mutant mice exhibited severe disturbance of motor coordination and impaired motor learning (Kashiwabuchi et al. 1995; Funabiki et al. 1995). The gene disruption abolished LTD of synaptic transmission between parallel fibers and Purkinje cells (Kashiwabuchi et al. 1995). Furthermore, the number of parallel fiber-Purkinje cell synapses was decreased and Purkinje cells remained to be multiply innervated by climbing fibers in the mutant mice (Kashiwabuchi et al. 1995; Kurihara et al. 1997). Thus, GluRδ2 subunit plays a central role in formation and plasticity of cerebellar Purkinje cell synapses and in motor coordination and motor learning.

Lurcher is a spontaneous, semidominant neurodegeneration mutation in mouse (Phillips 1960). The mutation causes a selective, cell-autonomous, and apoptic death of cerebellar Purkibnje cells and ataxia in heterozygous mice and a massive loss of mid- and hindbrain neurons during late embryogenesis and neonatal death in homozygous mice (Caddy and Biscoe 1979; Cheng and Heinz 1997). Positional cloning shows that Lucher is a gain-of-function mutation of the GluRδ2 gene that change highly conserved alanine in M3 region to threonine (Zuo et al. 1997). The Lurcher GluRδ2 forms constitutive active channels permeable to Na^+ and K^+.

II. GluRχ Subfamily

The last member of GluR channel subfamily is GluRχ subfamily. The GluRχ1 (NR3 A) subunit is expressed mainly in late embryonic and early neonatal stages but sharply declines thereafter (Ciabarra et al. 1995; Sucher et al. 1995). Coexpression of GluRχ1 subunit with NMDA receptor channel subunits in *Xenopus* oocytes reduces the current responses. Disruption of the NR3 A gene in mice results in enhanced NMDA responses and increased dendritic spines in early post-natal cerebrocortical neurons (Das et al. 1998). The NR3 A subunit may be involved in the modulation of NMDA receptor channels.

Acknowledgments. I thank Ms. M. Senbonmatsu for help in the preparation of the manuscript. This work was supported by research grants from Core Research for Evolutional Science and Technology of Japan Science and Technology Corporation, the Ministry of Education, Science, Sports and Culture of Japan, and the Asahi Glass Foundation.

References

Akazawa C, Shigemoto R, Bessho Y, Nakanishi S, Mizuno N (1994) Differential expression of five *N*-methyl-D-aspartate receptor subunit mRNAs in the cerebellum of developing and adult rats. J Comp Neurol 347:150–160

Araki K, Meguro H, Kushiya E, Takayama C, Inoue Y, Mishina M (1993) Selective expression of the glutamate receptor channel $\delta 2$ subunit in cerebellar Purkinje cells. Biochem Biophys Res Commun 197:1267–1276

Ascher P, Nowak L (1988) The role of divalent cations in the *N*-methyl-D-aspartate responses of mouse central neurones in culture. J Physiol 399:247–266

Bannerman DM, Good MA, Butcher SP, Ramsay M, Morris RG (1995) Distinct components of spatial learning revealed by prior training and NMDA receptor blockade. Nature 378:182–186

Barria A, Muller D, Derkach V, Griffith LC, Soderling TR (1997) Regulatory phosphorylation of AMPA-type glutamate receptors by CaM-KII during long-term potentiation. Science 276(5321):2042–2045

Bettler B, Boulter J, Hermans-Borgmeyer I, O'Shea-Greenfield A, Deneris ES, Moll C, Borgmeyer U, Hollmann M, Heinemann S (1990) Cloning of a novel glutamate receptor subunit. GluR5: expression in the nervous system during development. Neuron 5:583–595

Bettler B, Egebjerg J, Sharma G, Pecht G, Hermans-Borgmeyer I, Moll C, Stevens CF, Heinemann S (1992) Cloning of a putative glutamate receptor: a low affinity kainate-binding subunit. Neuron 8:257–265

Bliss TVP, Collingridge GL (1993) A synaptic model of memory: long-term potentiation in the hippocampus. Nature 361:31–39

Bochet P, Audinat E, Lambolez B, Crépel F, Rossier J, Iino M, Tsuzuki K, Ozawa S (1994) Subunit composition at the single-cell level explains functional properties of a glutamate-gated channel. Neuron 12:383–388

Boulter J, Hollmann M, O'Shea-Greenfield A, Hartley M, Deneris ES, Heinemann S (1990) Molecular cloning and functional expression of glutamate receptor subunit genes. Science 249:1033–1037

Brenman JE, Chao DS, Gee SH, McGee AW, Craven SE, Santillano DR, Wu Z, Huang F, Xia H, Peters MF, Froehner SC, Bredt DS (1996) Interaction of nitric oxide synthase with the postsynaptic density protein PSD-95 and alpha1-syntrophin mediated by PDZ domains. Cell 84:757–767

Brose N, Gasic GP, Vetter DE, Sullivan JM, Heinemann SF (1993) Protein chemical characterization and immunocytochemical localization of the NMDA receptor subunit NMDA R1. J Biol Chem 268:22663–22671

Brusa R, Zimmermann F, Koh DS, Feldmeyer D, Gass P, Seeburg PH, Sprengel R (1995) Early-onset epilepsy and postnatal lethality associated with an editing-deficient *GluR-B* allele in mice. Science 270:1677–80

Buller AL, Larson HC, Schneider BE, Beaton JA, Morrisett RA, Monaghan DT (1994) The molecular basis of NMDA receptor subtypes: native receptor diversity is predicted by subunit composition. J. Neurosci. 14:5471–5484

Burnashev N, Schoepfer R, Monyer H, Ruppersberg JP, Günther W, Seeburg PH, Sakmann B (1992) Control by asparagine residues of calcium permeability and magnesium blockade in the NMDA receptor. Science 257:1415–1419

Burnashev N, Zhou Z, Neher E, Sakmann B (1995) Fractional calcium currents through recombinant GluR channels of the NMDA, AMPA and kainate receptor subtypes. Journal of Physiology 485:403–18

Carmignoto G, Vicini S (1992) Activity-dependent decrease in NMDA receptor responses during development of the visual cortex. Science 258:1007–1011

Caddy KW, Biscoe TJ (1979) Structural and quantitative studies on the normal C3H and Lurcher mutant mouse. Phil Trans R Soc Lond B 287:167–201

Cheng SS, Heinz N (1997) Massive loss of mid- and hindbrain neurons during embryonic development of homozygous Lurcher mice. J Neurosci 17:2400–2407

Ciabarra AM, Sullivan JM, Gahn, LG, Pecht G, Heinemann S, Sevarino KA (1995) Cloning and characterization of χ–1: a developmentally regulated member of a novel class of the ionotropic glutamate receptor family. J Neurosci 15:6498–6508

Cline HT, Debski EA, Constantine-Paton M (1987) *N*-methyl-D-aspartate receptor antagonist desegregates eye-specific stripes. Proc Natl Acad Sci USA 84:4342–4345

Das S, Sasaki YF, Rothe T, Premkumar LS, Takasu M, Crandall JE, Dikkes P, Conner DA, Rayudu PV, Cheung Wing, Chen H-SV, Lipton SA, Nakanishi N (1998) Increased NMDA current and spine density in mice lacking the NMDA receptor subunit NR3A. Nature 393:377–381.

Davis S, Butcher SP, Morris RGM (1992) The NMDA receptor antagonist D-2-amino-5-phosphonopentanoate (D-AP5) impairs spatial learning and LTP *in vivo* at intracerebral concentrations comparable to those that block LTP *in vitro*. J Neurosci 12:21–34

Dong H, O'Brien RJ, Fung ET, Lanahan AA, Worley PF, Huganir RL (1997) GRIP: a synaptic PDZ domain-containing protein that interacts with AMPA receptors. Nature 386:279–284

Ebralidze AK, Rossi DJ, Tonegawa S, Slater NT (1996) Modification of NMDA receptor channels and synaptic transmission by targeted disruption of the NR2C gene. J Neurosci 16:5014–5025

Egebjerg J, Bettler B, Hermans-Borgmeyer I, Heinemann S (1991) Cloning of a cDNA for a glutamate receptor subunit activated by kainate but not AMPA. Nature 351:745–748

Ehlers MD, Tingley WG, Huganir RL (1995) Regulated subcellular distribution of the NR1 subunit of the NMDA receptor. Science 269:1734–1737

Ehlers MD, Zhang S, Bernhadt JP, Huganir RL (1996) Inactivation of NMDA receptors by direct interaction of calmodulin with the NR1 subunit. Cell 84:745–755

Farrant M, Feldmeyer D, Takahashi T, Cull-Candy SG (1994) NMDA-receptor channel diversity in the developing cerebellum. Nature 368:335–339

Flint AC, Maisch US, Weishaupt JH, Kriegstein AR, Monyer H (1997) NR2A Subunit Expression Shortens NMDA Receptor Synaptic Currents in Developing Neocortex. J Neurosci 17:2469–2476.

Forrest D, Yuzaki M, Soares HD, Ng L, Luk DC, Sheng M, Stewart CL, Morgan JI, Connor JA, Curran T (1994) Targeted disruption of NMDA receptor 1 gene abolishes NMDA response and results in neonatal death. Neuron 13:325–338

Funabiki K, Mishina M, Hirano T (1995) Retarded vestibular compensation in the mutant mice deficient of $\delta 2$ glutamate receptor subunit. NeuroReport 7:189–192

Gallo V, Upso LM, Hayes WP, Vyklicky L, Winters CA, Buonanno A (1992) Molecular cloning and developmental analysis of a new glutamate receptor isoform in cerebellum. J Neurosci 12:1010–1023

Gibb AJ, Colquhoun D (1991) Glutamate activation of a single NMDA receptor-channel produces a cluster of channel openings. Proc R Soc B 243:39–45

Gibb AJ, Colquhoun D (1992) Activation of *N*-methyl-D-aspartate receptors by L-glutamate in cells dissociated from adult rat hippocampus. J Physiol 456:143–179

Herb A, Burnashev N, Werner P, Sakmann B, Wisden W, Seeburg PH (1992) The KA-2 subunit of excitatory amino acid receptors shows widespread expression in brain and forms ion channels with distantly related subunits. Neuron 8:775–785

Hestrin S (1992) Developmental regulation of NMDA receptor-mediated synaptic currents at a central synapse. Nature 357:686–689

Hirai H, Kirsch J, Laube B, Betz H, Kuhse J (1996) The glycine binding site of the *N*-methyl-D-aspartate receptor subunit NR1: identification of novel determinants of co-agonist potentiation in the extracellular M3-M4 loop region. Proc Natl Acad Sci USA 93(12):6031–6

Hollmann M, O'Shea-Greenfield A, Rogers SW, Heinemann S (1989) Cloning by functional expression of a member of the glutamate receptor family. Nature 342:643–648

Hollmann M, Hartley M, Heinemann S (1991) Ca^{2+} permeability of KA-AMPA-gated glutamate receptor channels depends on subunit composition. Science 252:851–853

Hollmann M, Boulter J, Maron C, Beasley L, Sullivan J, Pecht G, Heinemann S (1993) Zinc potentiates agonist-induced currents at certain splice variants of the NMDA receptor. Neuron 10:943–954

Hollmann M, Heinemann S (1994) Cloned glutamate receptors. Annu Rev Neurosci 17:31–108

Hollmann M, Maron C, Heinemann S (1994) *N*-glycosylation site tagging suggests a three transmembrane domain topology for the glutamate receptor GluR1. Neuron 13:1331–1343

Hsueh YP, Kim E, Sheng M (1997) Disulfide-linked head-to-head multimerization in the mechanism of ion channel clustering by PSD-95. Neuron 18:803–814

Hume RI, Dingledine R, Heinemann SF (1991) Identification of a site in glutamate receptor subunits that controls calcium permeability. Science 253:1028–1031

Iino M, Ozawa S, Tsuzuki K (1990) Permeation of calcium through excitatory amino acid receptor channels in cultured rat hippocampal neurons. J Physiol 424:151–165

Ikeda K, Nagasawa M, Mori H, Araki K, Sakimura K, Watanabe M, Inoue Y, Mishina M (1992) Cloning and expression of the $\varepsilon 4$ subunit of the NMDA receptor channel. FEBS Lett 313:34–38

Ikeda K, Araki K, Takayama C, Inoue Y, Yagi T, Aizawa S, Mishina M (1995) Reduced spontaneous activity of mice defective in the $\varepsilon 4$ subunit of the NMDA receptor channel. Mol Brain Res 33:61–71

Irie M, Hata Y, Takeuchi M, Ichtchenko K, Toyoda A, Hirao K, Takai Y, Rosahl TW, Sudhof TC (1997) Binding of neuroligins to PSD-95. Science 277:1511–1515

Ishii T, Moriyoshi K, Sugihara H, Sakurada K, Kadotani H, Yokoi M, Akazawa C, Shigemoto R, Mizuno N, Masu M, Nakanishi S (1993) Molecular characterization of the family of the *N*-methyl-D-aspartate receptor subunits. J Biol Chem 268: 2836–2843

Ito I, Futai K, Katagiri H, Watanabe M, Sakimura K, Mishina M, Sugiyama H (1997) Synapse-selective impairment of NMDA receptor functions in mice lacking NMDA receptor $\varepsilon 1$ or $\varepsilon 2$ subunit. J Physiol 500:401–408

Jia Z, Agopyan N, Miu P, Xiong Z, Henderson J, Gerlai R, Taverna FA, Velumian A, MacDonald J, Carlen P, Abramow-Newerly W, Roder J (1996) Enhanced LTP in mice deficient in the AMPA receptor GluR2. Neuron 17:945–956

Johnson JW, Ascher P (1987) Glycine potentiates the NMDA response in cultured mouse brain neurons. Nature 325:529–531

Jonas P, Racca C, Sakmann B, Seeburg PH, Monyer H (1994) Differences in calcium permeability of AMPA-type glutamate receptor channels in neocortical neurons caused by differential GluR-B subunit expression. Neuron 12:1281–1289

Kashiwabuchi N, Ikeda K, Araki K, Hirano T, Shibuki K, Takayama C, Inoue Y, Kutsuwada T, Yagi T, Kang Y, Aizawa S, Mishina M (1995) Impairment of motor coordination, Purkinje cell synapse formation and cerebellar long-term depression in GluR$\delta 2$ mutant mice. Cell 81:245–252

Keinänen K, Wisden W, Sommer B, Werner P, Herb A, Verdoorn TA, Sakmann B Seeburg PH (1990) A family of AMPA-selective glutamate receptors. Science 249:556–560

Kendrick SJ, Lynch DR, Pritchett DB (1996) Characterization of glutamate binding sites in receptors assembled from transfected NMDA receptor subunits. J Neurochem 67:608–616

Kim E, Cho K-O, Rothschild A, Sheng M (1996) Heteromultimerization and NMDA receptor-clustering activity of chapsyn-110, a member of the PSD-95 family of proteins. Neuron 17:103–113

Kim E, Naisbitt S, Hsueh YP, Rao A, Rothschild A, Craig AM, Sheng M (1997) GKAP, a novel synaptic protein that interacts with the guanylate kinase-like domain of the PSD-95/SAP90 family of channel clustering molecules. J Cell Biol 136:669–678

Kim J-H, Liao D, Lau L-F, Huganir RL (1998) SynGAP: a synaptic RasGAP that associates with the PSD-95/SAP90 protein family. Neuron 20:683–691

Kiyama Y, Manabe T, Sakimura K, Kawakami F, Mori H, Mishina M (1998) Increased thresholds for LTP and contextual learning in mice lacking the GluR$\varepsilon 1$ subunit of the NMDA receptor channel. J Neurosci 18:6704–6712

Kleckner NW, Dingledine R (1988) Requirement for glycine in activation of NMDA-receptors expressed in *Xenopus* oocytes. Science 241:835–837

Kleinschmidt A, Bear MF, Singer W (1987) Blockade of "NMDA" receptors disrupts experience-dependent plasticity of kitten striate cortex. Science 238:355–358

Köhr G, Eckardt S, Lüddens H, Monyer H, Seeburg PH (1994) NMDA receptor channels: subunit-specific potentiation by reducing agents. Neuron 12:1031–1040

Köhr G, Seeburg PH (1996) Subtype-specific regulation of recombinant NMDA receptor-channels by protein tyrosine kinases of the *src* family. J Physiol 492:445–52

Kornau HC, Schenker LT, Kennedy MB, Seeburg PH (1995) Domain interaction between NMDA receptor subunits and the postsynaptic density protein PSD-95. Science 269:1737–40

Kuner T, Wollmuth LP, Karlin A, Seeburg PH, Sakmann B (1996) Structure of the NMDA receptor channel M2 segment inferred from the accessibility of substituted cysteines. Neuron 17:343–352

Kurihara H, Hashimoto K, Kano M, Takayama C, Sakimura K, Mishina M, Inoue Y, Watanabe M (1997) Impaired parallel fiber-Purkinje cell synapse stabilization during cerebellar development of mutant mice lacking the glutamate receptor $\delta 2$ subunit. J Neurosci 15:9613–9623

Kupper J, Ascher P, Neyton J (1996) Probing the pore region of recombinant *N*-methyl-D-aspartate channels using external and internal magnesium block. Proc Natl Acad Sci USA 93:8648–8653

Kupper J, Ascher P, Neyton J (1998) Internal Mg^{2+} block of recombinant NMDA channels mutated within the selectivity filter and expressed in *Xenopus* oocytes. J Physiol 507:1–12

Kuryatov A, Laube B, Betz H, Kuhse J (1994) Mutational analysis of the glycine-binding site of the NMDA receptor: structural similarity with bacterial amino acid-binding proteins. Neuron 12:1291–1300

Kutsuwada T, Kashiwabuchi N, Mori H, Sakimura K, Kushiya E, Araki K, Meguro H, Masaki H, Kumanishi T, Arakawa M, Mishina M (1992) Molecular diversity of the NMDA receptor channel. Nature 358:36–41

Kutsuwada T, Sakimura K, Manabe T, Takayama C, Katakura N, Kushiya E, Natsume R, Watanabe M, Inoue Y, Yagi T, Aizawa S, Arakawa M, Takahashi T, Nakamura Y, Mori H, Mishina M (1996) Impairment of suckling response, trigeminal neuronal pattern formation and hippocampal LTD in NMDA receptor $\varepsilon 2$ subunit mutant mice. Neuron 16:333–344

Landsend AS, Amiry-Moghaddam M, Matsubara A, Bergersen L, Usami S, Wenthold RJ, Ottersen OP (1997) Differential localization of δ glutamate receptors in the rat cerebellum: coexpression with AMPA receptors in parallel fiber-spine synapses and absence from climbing fiber-spine synapses. J Neurosci 17:834–842

Laube B, Hirai H, Sturgess M, Betz H, Kuhse J (1997) Molecular determinants of agonist discrimination by NMDA receptor subunits: analysis of the glutamate binding site on the NR2B subunit. Neuron 18:493–503

Laurie DJ, Seeburg PH (1994) Regional and developmental heterogeneity in splicing of the rat brain NMDAR1 mRNA. J Neurosci 14:3180–3194

Li Y, Erzurumlu RS, Chen C, Jhaveri S, Tonegawa S (1994) Whisker-related neuronal patterns fail to develop in the trigeminal brainstem nuclei of NMDAR1 knockout mice. Cell 76:427–437

Lin JW, Wyszynski M, Madhavan R, Sealock R, Kim JU, Sheng M (1998) Yotiao: a novel protein of neuromuscular junction and brain that interacts with specific splice variants of NMDA receptor subunit NR1. J Neurosci 18:2017–2027

Lomeli H, Sprengel R, Laurie DJ, Köhr G, Herb A, Seeburg PH, Wisden W (1993) The rat delta-1 and delta-2 subunits extend the excitatory amino acid receptor family. FEBS Lett 315:318–322

Lynch DR, Anegawa NJ, Verdoorn T, Pritchett DB (1994) *N*-methyl-D-aspartate receptors: different subunit requirements for binding of glutamate antagonists, glycine antagonists, and channel-blocking agents. Mol Pharmacol 45:540–545

Masood K, Wu C, Brauneis U, Weight FF (1993) Differential ethanol sensitivity of recombinant *N*-methyl-D-aspartate receptor subunits. Mol Pharmacol 45:324–329

Mayer ML, Westbrook GL (1987) Permeation and block of *N*-methyl-D-aspartic acid receptor channels by divalent cations in mouse cultured central neurons, J Physiol 394:501

Meguro H, Mori H, Araki K, Kushiya E, Kutsuwada T, Yamazaki M, Kumanishi T, Arakawa M, Sakimura K, Mishina M (1992) Functional characterization of a heteromeric NMDA receptor channel expressed from cloned cDNAs. Nature 357:70–74

Mishina M, Sakimura K, Mori H, Kushiya E, Harabayashi M, Uchino S, Nagahari K (1991) A single amino acid residue determines the Ca^{2+} permeability of AMPA-selective glutamate receptor channels. Biochem Biophys Res Commun 180:813–821

Mishina M, Mori H, Araki K, Kushiya E, Meguro H, Kutsuwada T, Kashiwabuchi N, Ikeda K, Nagasawa M, Yamazaki M, Masaki M, Yamakura T, Morita T, Sakimura K (1993) Molecular and functional diversity of the NMDA receptor channel. Ann NY Acad Sci 707:136–152

Monaghan DT, Olverman HJ, Nguyen L, Watkins JC, Cotman CW (1988) Two classes of *N*-methyl-D-aspartate recognition sites: differential distribution and differential regulation by glycine. Proc Natl Acad Sci USA 85:9836–9840

Monyer H, Sprengel R, Schoepfer R, Herb A, Higuchi M, Lomeli H, Burnashev N, Sakmann B, Seeburg PH (1992) Heteromeric NMDA receptors: molecular and functional distinction of subtypes. Science 256:1217–1221

Monyer H, Burnashev N, Laurie DJ, Sakmann B, Seeburg PH (1994) Developmental and regional expression in the rat brain and functional properties of four NMDA receptors. Neuron 12:529–540

Moon IS, Apperson ML, Kennedy MB (1994) The major tyrosine-phosphorylated protein in the postsynaptic density fraction is *N*-methyl-D-aspartate receptor subunit 2B. Proc Natl Acad Sci USA 91:3954–3958

Mori H, Masaki H, Yamakura T, Mishina M (1992) Identification by mutagenesis of a Mg^{2+}-block site of the NMDA receptor channel. Nature 358:673–675

Mori H, Yamakura T, Masaki H, Mishina M (1993) Involvement of the carboxyl-terminal region in modulation by TPA of the NMDA receptor channel. NeuroReport 4:519–522

Mori H, Mishina M (1995) Structure and function of the NMDA receptor channel. Neuropharmacology 34:1219–1237

Moriyoshi K, Masu M, Ishii T, Shigemoto R, Mizuno N, Nakanishi S (1991) Molecular cloning and characterization of the rat NMDA receptor. Nature 354:31–37

Morris RGM, Anderson E, Lynch GS, Baudry M (1986) Selective impairment of learning and blockade of long-term potentiation by an *N*-methyl-D-aspartate receptor antagonist, AP5. Nature 319:774–776

Mulle C, Sailer A, Pérez-Otaño I, Dickinson-Anson H, Castillo PE, Bureau I, Maron C, Gage FH, Mann JR, Bettler B, Heinemann SF (1998) Altered synaptic physiology and reduced susceptibility to kainate-induced seizures in GluR6-deficient mice. Nature 392:601–605

Müller T, Möller T, Berger T, Schnitzer J, Kettenmann H (1992) Calcium entry through kainate receptors and resulting potassium-channel blockade in Bergmann glial cells. Science 256:1563–1566

Müller BM, Kistner U, Kindler S, Chung WJ, Kuhlendahl S, Fenster SD, Lau L-F, Veh RW, Huganir RL, Gundelfinger ED, Garner CC (1996) SAP102, a novel postsynaptic protein that interacts with NMDA receptor complexes in vivo. Neuron 17:255–265

Nagasawa M, Sakimura K, Mori KJ, Bedell MA, Copeland NG, Jenkins NA, Mishina M (1996) Gene structure and chromosomal localization of the mouse NMDA receptor channel subunits. Mol Brain Res 36:1–11

Nakanishi N, Shneider NA, Axel R (1990) A family of glutamate receptor genes: evidence for the formation of heteromultimeric receptors with distinct channel properties. Neuron 5:569–581

Nakanishi N, Axel R, Shneider NA (1992) Alternative splicing generates functionally distinct *N*-methyl-D-aspartate receptors. Proc Natl Acad Sci USA 89:8552–8556

O'Hara PJ, Sheppard PO, Thøgersen H, Venezia D, Haldemann BA, McGrane V, Houamed KM, Thomsen C, Gilbert TL, Mulvihill ER (1993) The ligand-binding domain in metabotropic glutamate receptors is related to bacterial periplasmic binding proteins. Neuron 11:41–52

Perkel DJ, Hestrin S, Sah P, Nicoll RA (1990) Excitatory synaptic currents in Purkinje cells. Proc R Soc Lond B 241:116–121

Phillips RJS (1960) "Lurcher", new gene in linkage group XI of the house mouse. J Genet 57:35–42

Quinlan JE, Davies J (1985) Excitatory and inhibitory responses of Purkinje cells, in the rat cerebellum in vivo, induced by excitatory amino acids. Neurosci Lett 60:39–46

Raymond LA, Blackstone CD, Huganir RL (1993) Phosphorylation and modulation of recombinant GluR6 glutamate receptors by cAMP-dependent protein kinase. Nature 361:637–41

Roche KW, Raymond LA, Blackstone C, Huganir RL (1994) Transmembrane topology of the glutamate receptor subunit GluR6. J Biol Chem 269:11679–11682

Roche KW, O'Brien RJ, Mammen AL, Bernhardt J, Huganir RL (1996) Characterization of multiple phosphorylation sites on the AMPA receptor GluR1 subunit. Neuron 16:1179–1188

Rogers SW, Andrews PI, Gahring LC, Whisenand T, Cauley K, Crain B, Hughes TE, Heinemann SF, McNamara JO (1994) Autoantibodies to glutamate receptor GluR3 in Rasmussen's encephalitis. Science 265:648–651

Sakimura K, Bujo H, Kushiya E, Araki K, Yamazaki M, Yamazaki M, Meguro H, Warashina A, Numa S, Mishina M (1990) Functional expression from cloned cDNAs of glutamate receptor species responsive to kainate and quisqualate. FEBS Lett 272:73–80

Sakimura K, Morita T, Kushiya E, Mishina M (1992) Primary structure and expression of the $\gamma 2$ subunit of the glutamate receptor channel selective for kainate. Neuron 8:267–274

Sakimura K, Kutsuwada T, Ito I, Manabe T, Takayama C, Kushiya E, Yagi T, Aizawa S, Inoue Y, Sugiyama H, Mishina M (1995) Reduced hippocampal LTP and spatial learning in mice lacking NMDA receptor $\varepsilon 1$ subunit. Nature 373:151–155

Sakurada K, Masu M, Nakanishi S (1993) Alteration of Ca^{2+} permeability and sensitivity to Mg^{2+} and channel blockers by a single amino acid substitution in the *N*-methyl-D-aspartate receptor. J Biol Chem 268:410–415

Saucier D, Cain DP (1995) Spatial learning without NMDA receptor-dependent long-term potentiation. Nature 378:186–189

Schiffer HH, Swanson GT, Heinemann SF (1997) Rat GluR7 and a carboxy-terminal splice variant, GluR7b, are functional kainate receptor subunits with a low sensitivity to glutamate. Neuron 19:1141–1146

Seeburg PH (1993) The molecular biology of mammalian glutamate receptor channels. Trends Neurosci 16:359–365

Sheng M, Cummings J, Roldan LA, Jan YN, Jan LY (1994) Changing subunit composition of heteromeric NMDA receptors during development of rat cortex. Nature 368:144–147

Sommer B, Keinänen K, Verdoorn TA, Wisden W, Burnashev N, Herb A, Köhler M, Takagi T, Sakmann B, Seeburg PH (1990) Flip and flop: a cell-specific functional switch in glutamate-operated channels of the CNS. Science 249:1580–1585

Sommer B, Köhler M, Sprengel R, Seeburg PH (1991) RNA editing in the brain controls a determinant of ion flow in glutamate-gated channels. Cell 67:11–19

Sommer B, Burnashev N, Verdoorn TA, Keinänen K, Sakmann B, Seeburg PH (1992) A glutamate receptor channel with high affinity for domoate and kainate. EMBO J 11:1651–1656

Standaert DG, Testa CM, Young AB, Penny JB Jr (1994) Organization of *N*-methyl-D-aspartate glutamate receptor gene expression in the basal ganglia of the rat. J Comp Neurol 343:1–16

Stern P, Béhé P, Schoepfer R, Colquhoun D (1992) Single-channel conductances of NMDA receptors expressed from cloned cDNAs: comparison with native receptors. Proc R Soc Lond B 250:271–277

Sucher NJ, Akbarian S, Chi CL, Leclerc CL, Awobuluyi M, Deitcher DL, Wu MK, Yuan JP, Jones EG, Lipton SA (1995) Developmental and regional expression pattern of a novel NMDA receptor-like subunit (NMDAR-L) in the rodent brain. J Neurosci 15:6509–6520

Sugihara H, Moriyoshi K, Ishii T, Masu M, Nakanishi S (1992) Structures and properties of seven isoforms of the NMDA receptor generated by alternative splicing. Biochem . Biophys Res Commun 185:826–832

Sullivan JM, Traynelis SF, Chen H-SV, Escobar W, Heinemann SF, Lipton SA (1994) Identification of two cysteine residues that are required for redox modulation of the NMDA subtype of glutamate receptor. Neuron 13:929–936

Swanson GT, Kamboj SK, Cull-Candy SG (1997) Single-channel properties of recombinant AMPA receptors depend on RNA editing, splice variation, and subunit composition. J Neurosci 17:58–69

Takahashi T, Feldmeyer D, Suzuki N, Onodera K, Cull-Candy SG, Sakimura K, Mishina M (1996) Functional correlation of NMDA receptor ε subunits expression with the properties of single-channel and synaptic currents in the developing cerebellum. J Neurosci 16:4376–4382

Takayama C, Nakagawa S, Watanabe M, Mishina M, Inoue Y (1996) Developmental changes in expression and distribution of the glutamate receptor channel $\delta 2$ subunit according to the Purkinje cell maturation. Dev Brain Res 92:147–155

Takeuchi M, Hata Y, Hirao K, Toyoda A, Irie M, Takai Y (1997) SAPAPs. A family of PSD-95/SAP90-associated proteins localized at postsynaptic density. J Biol Chem 272:11943–11951

Taverna FA, Wang L-Y, MacDonald JF, Hampson DR (1994) A transmembrane model for an ionotropic glutamate receptor predicted on the basis of the location of asparagine-linked oligosaccharides. J Biol Chem 269:14159–14164

Tingley WG, Roche KW, Thompson AK, Huganir RL (1993) Regulation of NMDA receptor phosphorylation by alternative splicing of the C-terminal domain. Nature 364:70–73

Tingley WG, Huganir RL (1994) Generation of antibodies specific for phosphorylated NMDA receptors. Soc Neurosci Abstr 20:1466

Traynelis SF, Hartley M, Heinemann SF (1995) Control of proton sensitivity of the NMDA receptor by RNA splicing and polyamines. Science 268:873–876

Tsien JZ, Huerta PT, Tonegawa S (1996) The essential role of hippocampal CA1 NMDA receptor-dependent synaptic plasticity in spatial memory. Cell 87:1327–1338

Tsuzuki K, Mochizuki S, Iino M, Mori H, Mishina M, Ozawa S (1994) Ion permeation properties of the cloned mouse $\varepsilon 2/\zeta 1$ NMDA receptor channel. Mol Brain Res 26:37–46

Verdoorn TA, Burnashev N, Monyer H, Seeburg PH, Sakmann B (1991) Structural determinants of ion flow through recombinant glutamate receptor channels. Science 252:1715–1718

Wang LY, Taverna FA, Huang XP, MacDonald JF, Hampson DR (1993) Phosphorylation and modulation of a kainate receptor (GluR6) by cAMP-dependent protein kinase. Science 259(5098):1173–1175

Watanabe M, Inoue Y, Sakimura K, Mishina M (1992) Developmental changes in distribution of NMDA receptor channel subunit mRNAs. NeuroReport 3:1138–1140

Watanabe M, Inoue Y, Sakimura K, Mishina M (1993) Distinct distributions of five *N*-methyl-D-aspartate receptor channel subunit mRNAs in the forebrain. J Comp Neurol 338:377–390
Watanabe M, Mishina M, Inoue Y (1994a) Distinct spatiotemporal expressions of five NMDA receptor channel subunit mRNAs in the cerebellum. J Comp Neurol 343:513–519
Watanabe M, Mishina M, Inoue Y (1994b) Distinct distributions of five NMDA receptor channel subunit mRNAs in the brainstem. J Comp Neurol 343:520–531
Watkins JC, Olverman HJ (1981) Agonists and antagonists for excitatory amino acid receptors. Trends Neurosci 10:265–272
Werner P, Voight M, Keinänen K, Wisden W, Seeburg PH (1991) Cloning of a putative high-affinity kainate receptor expressed predominantly in hippocampal CA3 cells. Nature 351:742–744
Wo ZG, Oswald RE (1994) Transmembrane topology of two kainate receptor subunits revealed by *N*-glycosylation. Proc Natl Acad Sci USA 91:7154–7158
Wollmuth LP, Kuner T, Seeburg PH, Sakmann B (1996) Differential contribution of the NR1- and NR2A-subunits to the selectivity filter of recombinant NMDA receptor channels. J Physiol 491:779–797
Wollmuth LP, Kuner T, Sakmann G (1998a) Adjacent asparagines in the NR2-subunit of the NMDA receptor channel control the voltage-dependent block by extracellular Mg^{2+}. J Physiol 506:13–32
Wollmuth LP, Kuner T, Sakmann B (1998b) Intracellular Mg^{2+} interacts with structural determinants of the narrow constriction contributed by the NR1-subunit in the NMDA receptor channel. J Physiol 506:33–52
Wyszynski M, Lin J, Rao A, Nigh E, Beggs AH, Craig AM, Sheng M (1997) Competitive binding of α-actinin and calmodulin to the NMDA receptor. Nature 385:439–442
Yamakura T, Mori H, Masaki H, Shimoji K, Mishina M (1993) Different sensitivities of NMDA receptor channel subtypes to non-competitive antagonists. NeuroReport 4:687–690
Yamazaki M, Mori H, Araki K, Mori KJ, Mishina M (1992) Cloning, expression and modulation of a mouse NMDA receptor subunit. FEBS Lett 300:39–45
Yu XM, Askalan R, Keil GJ II, Salter MW (1997) NMDA channel regulation by channel-associated protein tyrosine kinase Src. Science 275:674–8
Zuo J, De Jager PL, Takahashi KA, Jiang W, Linden DJ, Heintz N (1997) Neurodegeneration in Lurcher mice caused by mutation in $\delta 2$ glutamate receptor gene. Nature 388:769–773

CHAPTER 17

Glutamate Receptor Ion Channels: Activators and Inhibitors

D.E. JANE, H.-W. TSE, D.A. SKIFTER, J.M. CHRISTIE and D.T. MONAGHAN

A. Introduction

I. Receptor Classification

The classification of excitatory amino acid (EAA) receptors into *N*-methyl-D-aspartic acid (NMDA) and non-NMDA receptors was based mainly on the discovery of selective agonists and antagonists (for a review see WATKINS and EVANS 1981). Thus the selective NMDA receptor antagonist, D-α-aminoadipate blocked responses due to NMDA but not those due to quisqualate or kainate in the frog and rat spinal cord. Evidence for more than one type of non-NMDA receptor came from the observation that responses due to quisqualate but not those due to kainate are depressed by L-glutamic acid diethyl ester (GDEE) (MCLENNAN and LODGE 1979; DAVIES and WATKINS 1979) while responses to kainate can be selectively antagonized by γ-D-glutamylglycine (γDGG) (DAVIES and WATKINS 1981). In addition, receptors present on dorsal root fibers were shown to be sensitive to kainate but not quisqualate (DAVIES et al. 1979). These observations resulted in the classification of EAA receptors into NMDA, quisqualate and kainate receptors. Such receptors are linked to ionic fluxes and are now known as ionotropic glutamate receptors (iGluRs) as distinct from the more recently discovered metabotropic glutamate receptors (mGluRs) linked to G-protein coupled metabolic changes (CONN and PIN 1997). The isoxazole analogue, (*S*)-2-amino-3-(3-hydroxy-5-methyl-4-isoxazolyl)propionic acid (AMPA) was reported to be more selective for the quisqualate type of iGluR than quisqualate itself (KROGSGAARD-LARSEN et al. 1980, 1982) and this led to this ionotropic receptor being renamed the AMPA receptor to avoid confusion with quisqualate-activated mGluRs (MONAGHAN et al. 1989; WATKINS et al. 1990). Molecular biology has identified a large number of subtypes of both iGluRs and mGluRs.

Progress in defining the role of EAA receptors in CNS function has also followed the development of receptor-selective antagonists. With the availability of NMDA receptor antagonists, it was possible to demonstrate a role of NMDA receptors in polysynaptic responses in the vertebrate spinal cord (DAVIES and WATKINS 1978; EVANS et al. 1979) and in hippocampal long term potentiation (COLLINGRIDGE et al. 1983; HARRIS et al. 1984). For several years, it was presumed that many central synapses used AMPA and/or kainate receptors for fast excitatory neurotransmission (for review see Watkins and Evans 1981), but it was not until the advent of antagonists selective for non-NMDA receptors (Davies and Watkins 1985; Honoré et al. 1988) that their role in neurotransmission became established (Watkins et al. 1990). In particular, it has become clear that AMPA receptors mediate fast excitatory neurotransmission in many pathways of the CNS. With the relatively recent development of kainate receptor antagonists, it has finally become possible to demonstrate that kainate receptors also mediate synaptic transmission (Clarke et al. 1997; Yamamoto et al. 1998).

II. Molecular Biology of AMPA, Kainate, and NMDA Receptors

The isolation of genes encoding for four AMPA receptor subunits, termed iGluR1–4 and five kainate receptor subunits, iGluR5–7 and KA1 and KA2 has lead to a considerable increase in the understanding of their structure and function (for reviews see Bettler and Mulle 1995; Bigge et al. 1996; Fletcher and Lodge 1996). Homomeric or heterooligomeric expression of iGluR1–4 in host cells produces complexes which have high affinity for AMPA and a lower affinity for kainate. Responses evoked by activation of such receptors by AMPA are strongly desensitizing while those due to kainate are relatively non-desensitizing, a characteristic which serves to distinguish AMPA from kainate receptors. It is known that AMPA receptors expressing the iGluR2 subunit are Ca^{2+}-impermeable while those without iGluR2 can gate Ca^{2+} (Fletcher and Lodge 1996). The Ca^{2+} impermeability is determined by the presence of a positively charged arginine (R) residue, which resides within the second transmembrane section of the ionic channel of iGluR2. For iGluR1, 3, and 4 this arginine residue is replaced with a neutral glutamine residue (Q). This switch from Q to R for GluR2 occurs by RNA editing and in rat brain this process is highly efficient, most iGluR2 being in the iGluR2(R) form. Flip and flop forms of AMPA receptors, resulting from alternative splicing, have different desensitization properties (Sommer et al. 1990), the flip form desensitizing less rapidly than the flop form. When expressed homomerically, iGluR5 and iGluR6 desensitize rapidly on application of kainate but not the far less potent AMPA, while KA1 and KA2 bind kainate with high affinity (Fletcher and Lodge 1996) but do not form functional ion-channels when expressed in host cells. iGluR7 when expressed homomerically forms functional ion-channels with a unique pharmacology, having low sensitivity to kainate and glutamate (Schiffer et al. 1997).

As described in Chap. 16, two, and perhaps three, families of NMDA receptor subunits have been cloned. Of the NR1 family (Moriyoshi et al. 1991; Yamazaki et al. 1992) there are eight alternative splice forms of one gene product (Sugihara et al. 1992; Yamazaki et al. 1992; Hollmann et al. 1993) while of the NR2s there are four distinct gene products (Ikeda et al. 1992; Meguro et al. 1992; Monyer et al. 1992, 1994; Ishii et al. 1993). Additionally, recent studies have indicated that an additional subunit may be present in NMDA receptors which can modify channel activity (CIABARRA et al. 1995; SUCHER et al. 1995; DAS et al. 1998). Studies generally agree that native NMDA receptors are composed of at least four subunits including both NR1 and NR2 subunits. While several studies have examined the properties of "homomeric" NR1 receptors expressed in *Xenopus* oocytes, it is possible that a recently identified *Xenopus* glutamate receptor subunit may be substituting for an NR2 subunit in these studies (SOLOVIEV and BARNARD 1997).

B. Pharmacology of AMPA Receptors

I. AMPA Receptor Agonists

AMPA (Fig. 1) and a range of analogues have been reported to be potent and selective agonists for AMPA receptors (KROGSGAARD-LARSEN et al. 1980, 1982; Krogsgaard-Larsen et al. 1993). A new AMPA analogue, (*RS*)-2-amino-3-[3-hydroxy-5-(2-methyl-2*H*-tetrazol-5-yl)isoxazol-4-yl]propionic acid (compound I, Fig. 1) was reported to be more potent than (RS)-AMPA at inducing depolarizations in the cortical slice preparation (EC_{50} values 0.92 μmol/l and 5.4 μmol/l respectively) (BANG-ANDERSEN et al. 1997). From an in-depth QSAR study it was concluded that a hydrogen bond between the protonated amino group and an ortho-positioned heteroatom of the ring substituent at

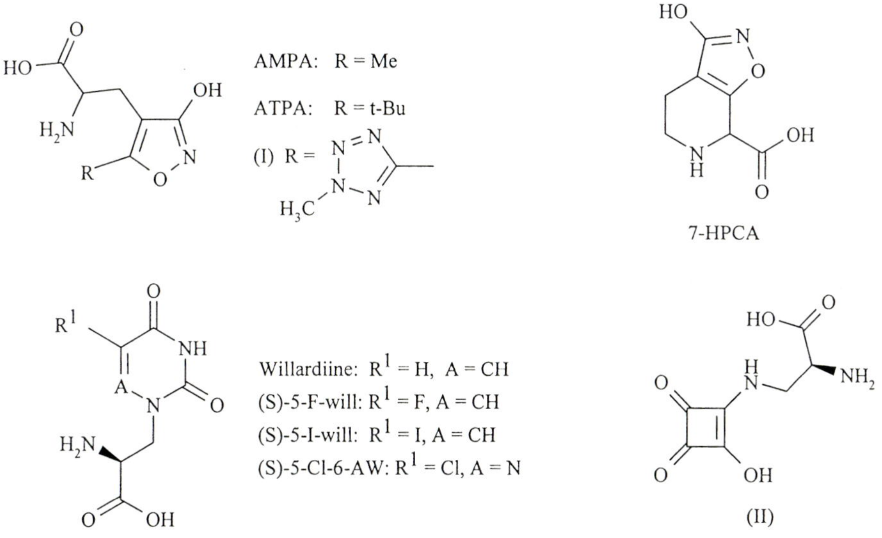

Fig. 1. AMPA receptor agonists

position 5 of the isoxazole ring stabilizes the active conformation of the molecule. It has been suggested that the receptor-active conformation of AMPA would be similar to the extended conformation of the potent and conformationally restricted AMPA receptor agonist (RS)-3-hydroxy-4,5,6,7-tetrhydroisoxazole[5,4-*c*]pyridine-7-carboxylic acid (7-HPCA, see Fig. 1) (CHAMBERLIN and BRIDGES 1993). Chain-extension of AMPA to give (*S*)-Homo-AMPA removed all activity at AMPA receptors, the compound being a selective agonist of the $mGlu_6$ subtype of metabotropic glutamate receptors (AHMADIAN et al. 1997). Data from human embryonic kidney (HEK) cells expressing human iGluRs shows that AMPA binds to homomeric iGluR1, 2, 3, or 4 with approximately equal affinity (K_i value ~100 nmol/l for displacement of [^{3}H]AMPA) but binds only weakly to homomeric iGluR5 (JANE et al. 1997; VARNEY et al. 1998).

The natural product willardiine (Fig. 1) was shown to be a potent quisqualate-like depolarising agent (EVANS et al. 1980). More recently, the 5-fluoro analogue, (*S*)-1-(2-amino-2-carboxyethyl)-5-fluoropyrimidine-2,4-dione ((*S*)-5-F-will, Fig. 1) has been reported to be more potent and selective than AMPA as a depolarizing agent in neonatal rat spinal motoneurones (JANE et al. 1991). Willardiine analogues have been shown to be strongly desensitising agonists in cultured hippocampal neurones (PATNEAU et al. 1992; WONG et al. 1994). A recent study has compared the affinity of a range of willardiine analogues (including 6-azawillardiines) for human homomeric iGluR1, 2, 4, or 5 expressed in HEK293 cells (JANE et al. 1997). (*S*)-5-F-will had high affinity for both iGluR1 and iGluR2 (K_i values 14.7 nmol/l and 25.1 nmol/l respectively for displacement of [^{3}H]-AMPA), being more potent and selective than AMPA, and showed a 20- and >100-fold selectivity respectively for iGluR1 over iGluR4 and iGluR5. VARNEY et al. 1998 have reported the affinity of (*S*)-5-F-will for human homomeric iGluR3 (K_i value 179 nmol/l for displacement of [^{3}H]-AMPA) expressed in HEK69–8 cells. Thus it would appear that (*S*)-5-F-will shows selectivity for the iGluR1 and iGluR2 subunits. The radiolabel, [^{3}H]5-F-will, binds to AMPA receptors with high affinity even in the absence of the chaotropic agent potassium thiocyanate (HAWKINS et al. 1995). A 6-azawillardiine analogue, (*S*)-2-(2-amino-2-carboxyethyl)-6-chloro-1,2,4-triazine-3,5-dione ((*S*)-5-Cl-6-AW, Fig. 1) showed high affinity for iGluR4 (K_i value 3.6 nmol/l) but little selectivity with respect to iGluR1 and iGluR2 (2- and 5-fold respectively). From both the electrophysiological and binding studies (JANE et al. 1997; WONG et al. 1994) it has been deduced that, for optimal activity at AMPA receptors, small electron-withdrawing substituents are required, while the 6-aza modification is favored for activity at iGluR4.

A structurally novel glutamate analogue, (*S*)-N^b-(2-hydroxy-3,4-dioxo-1-cyclobutenyl)-2,3-diaminopropionic acid (II, Fig. 1) showed high affinity for AMPA receptors in rat brain (IC_{50} value 190 nmol/l for displacement of [^{3}H]-AMPA) and showed comparable potency to AMPA for depolarizing pyramidal neurones (CHAN et al. 1995).

II. Competitive AMPA Receptor Antagonists

1. Quinoxalinediones and Related Compounds

The discovery of two quinoxalinediones, 6,7-dinitroquinoxaline-2,3-dione (DNQX) and 6-cyano-7-nitroquinoxaline-2,3-dione (CNQX) (for structures see Fig. 2) was an important milestone in the development of AMPA/kainate receptor antagonists (for a review see WATKINS et al. 1990). These compounds, however, suffered from a number of disadvantages such as significant antagonist activity at the glycine site of the NMDA receptor complex, poor selectivity between AMPA and kainate receptors, and low water solubility, which likely explains their low bioavailability. Starting from these two lead compounds, a number of analogues have been developed which have overcome at least some of these limitations. One such compound, 6-nitro-7-sulphamoyl[*f*]quinoxaline-2,3-dione (NBQX) displays a 30-fold selectivity in $[^{3}\mathrm{H}]$AMPA vs $[^{3}\mathrm{H}]$kainate radioligand binding studies while showing no affinity for NMDA receptors (SHEARDOWN et al. 1990). Since these initial com-

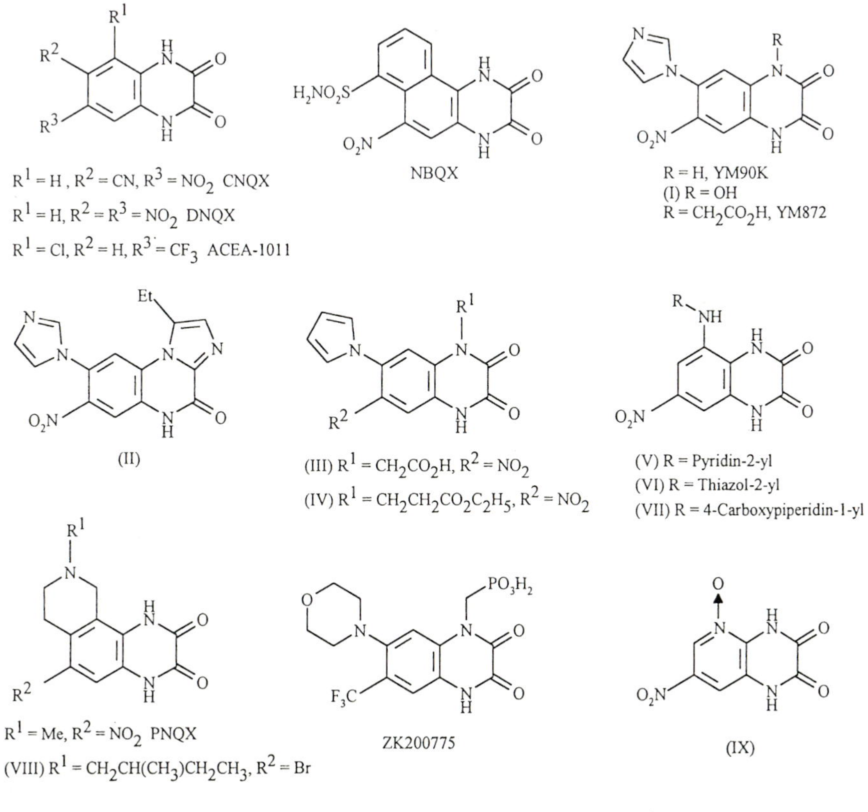

Fig. 2. Quinoxaline-2,3-diones, potent AMPA receptor antagonists

pounds were discovered there has been an explosion of interest in quinoxalinediones and this has lead to reports of analogues with similarly high affinity for AMPA receptors. These include 6-(1*H*-imidazol-1-yl)-7-nitro-2,3-(1*H*,4*H*)-quinoxalinedione (YM90K, Fig. 2) which has been reported to be a potent and selective AMPA receptor antagonist (see Table 1) with a 30-fold weaker affinity for the glycine binding-site of the NMDA receptor complex (OHMORI et al. 1994; SHIMIZU-SASAMATA et al. 1996). OKADA et al. (1996) have reported that both YM90K and NBQX are potent inhibitors of AMPA receptor-mediated currents in rat cortical mRNA-injected *Xenopus* oocytes (pA_2 values 6.83 ± 0.01 and 7.24 ± 0.01 respectively). In the same study, cyclothiazide was found to reduce the potency of YM90K for inhibition of AMPA-induced currents. This suggests that, as well as potentiating the binding of agonists via interaction with an allosteric site, cyclothiazide can modify binding of competitive antagonists to the AMPA receptor.

Recent structure-activity studies based on YM90K have led to the development of more potent and selective compounds such as 1-hydroxy-7-(1*H*-imidazol-1-yl)-6-nitro-2,3(1*H*,4*H*)-quinoxalinedione (compound I, Fig. 2) reported to have high affinity for the AMPA receptor (Table 1) and over 100-fold selectivity for this receptor than for the glycine binding site on the NMDA receptor (OHMORI et al. 1996). In a follow up study OHMORI et al. (1997) introduced a new analogue of YM90K, 8-(1*H*-imidazol-1-yl)-7-nitro-4(5*H*)-imidazo[1,2-*a*]quinoxalinone (compound II, Fig. 2) again with high affinity for the AMPA receptor (K_i value 0.057 μmol/l) and over 5000-fold selectivity for AMPA receptors than for either the NMDA or glycine binding-sites of the NMDA receptor. A pharmacophore model to account for the action of YM90 K and its analogues on AMPA receptors has been proposed based on an in-depth structure-activity study (Ohmori et al. 1994, 1996, 1997). A water soluble analogue of YM90K has been synthesized [2,3-dioxo-7-(1*H*-imidazol-1-yl)-6-nitro-1,2,3,4-tetrahydro-1-quinoxalinyl]-acetic acid (YM872, Fig. 2) which has high potency and selectivity for AMPA receptors (Table 1) (TAKAHASHI et al. 1998). TURSKI et al. (1996) have also reported a water-soluble quinoxalinedione analogue 7-(morpholin-4-yl)-1-phosphonomethyl-6-trifluoromethyl-2,3-(4*H*)-quinoxalinedione (ZK200775, Fig. 2) with high affinity for AMPA receptors (Table 1).

LUBISCH et al. (1996) have reported a series of pyrrolylquinoxalinediones as potent and selective AMPA receptor antagonists. Within this series, the pyrrol analogue of YM90K (compound III, Fig. 2) was found to be equipotent with NBQX at AMPA and to have low affinity for the glycine binding-site of the NMDA receptor and for kainate receptors (Table 1). Interestingly the propionic ester (compound IV, Fig. 2) was almost as potent as the corresponding acid on AMPA receptors and was also more selective for AMPA receptors than compound III (Table 1). AUBERSON et al. (1998a,b) have reported a structure-activity study based on a series of 5-aminomethylquinoxaline-2,3-diones. Compounds V, VI, and VII (Fig. 2) displayed high affinity for AMPA receptors but little affinity for the glycine binding-site of the NMDA receptor

Table 1. Binding data for a range of quinoxalinediones and related structures

Compound	Ki (μmol/l) for displacement of:			
	[^{3}H]Kainate	[^{3}H]AMPA	[^{3}H]CGS19755	[^{3}H]Glycine
CNQX[a]	1.5 ± 0.3	0.30 ± 0.15	25[b]	14
DNQX[a]	2.0 ± 0.1	0.5 ± 0.1	40[b]	9.5
NBQX[a]	4.80 ± 0.47	0.15 ± 0.01	>90[b]	>100
YM90K[c]	2.2	0.084	>100[d]	37
NBQX[c]	4.1	0.060	>100[d]	>100
CNQX[c]	1.8	0.27	25[d]	5.6
YM872[e]	2.2	0.096	100	>100
ZK200775[f]	ND	0.105	ND	ND
PNQX[g]	0.368 ± 0.050	0.063 ± 0.012	ND	0.37
NBQX[g]	0.079	0.052	ND	>100
NS257[h]	13 ± 2	0.70 ± 0.08	44 ± 6.4[d]	>100
S 17625[i]	ND	0.9	ND	111
NBQX[i]	ND	0.06	ND	>500
I[c]	ND	0.021	ND	ND
III[j]	0.41	0.07	ND	>25
IV[j]	8.5	0.26	ND	>30
V[k]	ND	0.50 ± 0.21	ND	0% (1 μmol/l)[l]
VI[k]	ND	0.19 ± 0.03	ND	41% (1 μmol/l)[l]
VII[m]	ND	0.07 ± 0.02	ND	3.9[n]
IX[o]	2.4 ± 0.3	1.3 ± 0.8		0.69 ± 0.06[n]
II (Fig 3)[p]	37% (10 μmol/l)[q]	1.8	17% (10 μmol/l)[q]	ND
DNQX[p]	ND	0.28	ND	ND

ND, not determined.
[a] IC_{50} values taken from Sheardown et al. (1990).
[b] [^{3}H]CPP used as radioligand.
[c] NBQX and CNQX are reference substances for YM90 K; Ohmori et al. (1994).
[d] NMDA-sensitive [^{3}H]glutamate binding.
[e] Taken from Takahashi et al. (1998).
[f] Taken from Turski et al. (1996).
[g] NBQX is reference substance for PNQX; IC_{50} values taken from Bigge et al. (1995).
[h] IC_{50} value taken from Wätjen et al. (1994).
[i] NBQX is reference substance for S 17625; IC_{50} values taken from Desos et al. (1996).
[j] Taken from Lubisch et al. (1996).
[k] IC_{50} values taken from Auberson et al. (1998a).
[l] Percent inhibition of [^{3}H]-DCKA binding at 1 μmol/l.
[m] IC_{50} values taken from Auberson et al. (1998b).
[n] IC_{50} value for inhibition of [^{3}H]-DCKA binding.
[o] IC_{50} values taken from Cai et al. (1997).
[p] DNQX is reference substance; IC_{50} values taken from Subramanyam et al. (1995).
[q] Percent inhibition at 10 μmol/l.

(Table 1). The hydrobromide salt of compound VII (Fig. 2) has the advantage of good aqueous solubility (1.68 g/l) at physiological pH.

A series of novel quinoxalinediones have been developed by combining the fused cyclic amine structure of NS257 and the quinoxalinedione ring of NBQX (Bigge et al. 1995). One of these, 1,4,7,8,9,10-hexahydro-9-methyl-6-

nitropyrido[3,4-*f*]-quinoxaline-2,3-dione (PNQX, Fig. 2) had high affinity for AMPA (similar to that determined for NBQX), kainate and glycine binding sites (Table 1). Both PNQX and NBQX had similar potencies for antagonism of AMPA-induced depolarizations in the cortical wedge preparation. In the same study, pharmacophore models for both the AMPA and glycine binding-sites were proposed based on a wide range of novel and previously reported compounds. These models highlighted differences in the structural requirements between the AMPA receptor binding-site and the glycine binding-site within the NMDA receptor, the most important difference being the presence of a site of steric intolerance in the glycine binding-site. This information may be useful in increasing the selectivity of compounds for AMPA receptors. In addition, QSAR analysis established that, for optimal activity at AMPA receptors, the 6-position substituent should be small, electron withdrawing, and lipophilic. Due to these limitations, it was concluded that 6-nitro substitution was optimal.

WÄTJEN et al. (1994) have reported that 1,2,3,6,7,8-hexahydro-3-(hydroxyimino)-*N*,*N*,7-trimethyl-2-oxobenzo[2,1-*b*:3,4-*c'*]dipyrrole-5-sulfonamide (NS 257) (Fig. 3) is a potent systemically active AMPA receptor antagonist. Although NS 257 does not possess the 2,3-quinoxalinedione nucleus, it is likely that it binds to the receptor in a similar manner. In binding studies, NS 257 proved to have selectively high affinity for AMPA over, NMDA (or glycine binding sites therein) (Table 1). The radiolabeled form, $[^3H]$NS 257 (NIELSEN et al. 1995) in the presence of thiocyanate (100 mmol/l) binds to a single population of binding sites (K_i value 225 ± 8 nmol/l). Autoradiographic studies showed that the distribution of $[^3H]$NS 257 binding sites was similar to that of $[^3H]$AMPA.

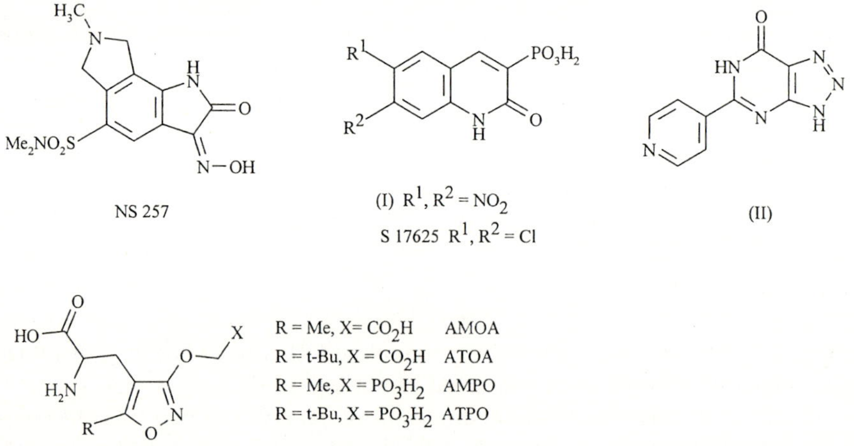

Fig. 3. Selective AMPA receptor antagonists

A number of AMPA receptor antagonists, NBQX, DNQX, YM90K, CNQX, and NS 257 competitively antagonized glutamate-evoked increases in intracellular Ca^{2+} in HEK cells expressing human iGluR3 flip with IC_{50} values of 0.38 μmol/l, 2.09 μmol/l, 2.45 μmol/l, 2.64 μmol/l, and 10.5 μmol/l respectively (VARNEY et al. 1998), these results being in agreement with the binding and functional assays for the compounds described above.

A series of 2(1*H*)-quinolones bearing different acidic groups at the 3-position have been reported to be potent and selective AMPA receptor antagonists (DESOS et al. 1996). A structure-activity study revealed that a 3-phosphono substituent conferred optimal potency and selectivity for AMPA receptors. Indeed, the 6,7-dinitro analogue (compound I, Fig. 3) and NBQX had similar potency (IC_{50} values for inhibition of AMPA receptor-mediated currents 0.15 μmol/l and 0.09 μmol/l respectively) and selectivity for AMPA receptors. The 6,7-dichloro analogue (6,7-dichloro-2(1*H*)-oxoquinoline-3-phosphonic acid, S 17625, Fig. 3) had lower affinity than NBQX for displacement of [^{3}H]AMPA (Table 1) but was equipotent in vivo as an anticonvulsant (DESOS et al. 1996).

A structurally novel series of AMPA receptor antagonists based around the 1,2,3-triazolo[4,5-d]pyrimidin-4(5*H*)-one nucleus (compound II, Fig. 3) has been described (SUBRAMANYAM et al. 1995). A limited structure-activity study revealed that the carbonyl group at the 4-position and the 6-(4-pyridyl) substituent were necessary for optimal activity, whereas the proton at N1 was not, as the *N*-cyclopentyl analogue retained high affinity for AMPA receptors. The pyrimidin-4(5*H*)-one (compound II, Fig. 3) was selective for AMPA receptors over NMDA and kainate receptors (Table 1). Schild analysis of the data for the antagonism of AMPA-induced responses in frog oocytes by compound II (Fig. 3) revealed a pA2 value of 6.08.

2. Decahydroisoquinolines

ORNSTEIN et al. (1996a,b) have reported the design and synthesis of a range of decahydroisoquinolines as selective AMPA receptor antagonists. These compounds having evolved from decahydroisoquinoline analogues initially synthesized as potential NMDA receptor antagonists. From the structure-activity analysis of a wide range of analogues conclusions have been reached for obtaining optimal affinity and selectivity for AMPA receptors.

LY293558 ((3S,4aR,6R,8aR)-6-(2-(1*H*-tetrazol-5-yl)ethyl)-1,2,3,4,4*a*,5,6,7,8,8*a*-decahydroisoquinoline-3-carboxylic acid, Fig. 4), with the optimal length of the linker between the tetrazole group and the isoquinoline ring (carbon chain of two atoms), had high affinity for AMPA receptors (Tables 2 and 3). Shorter or longer linkers resulted in a reduction in affinity and selectivity for AMPA receptors. Substitution of the C-atom in the linker adjacent to the tetrazole ring with an S atom but not an NH group resulted in increased AMPA receptor antagonist potency. A methyl substituent on the 2-carbon linker either on the atom adjacent to the tetrazole or the isoquinoline ring did

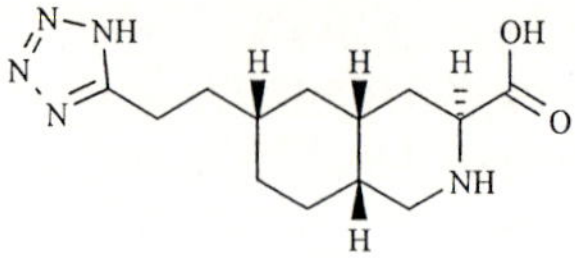

LY293558

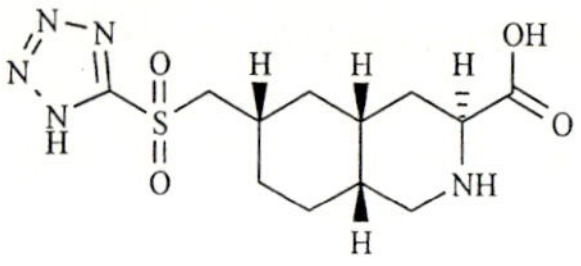

LY302679

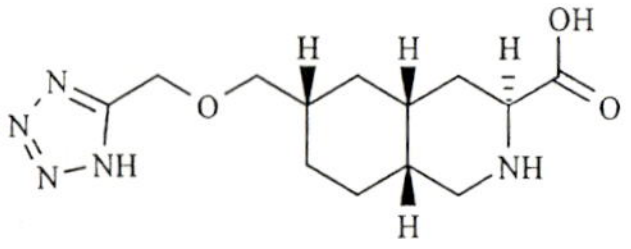

LY294486

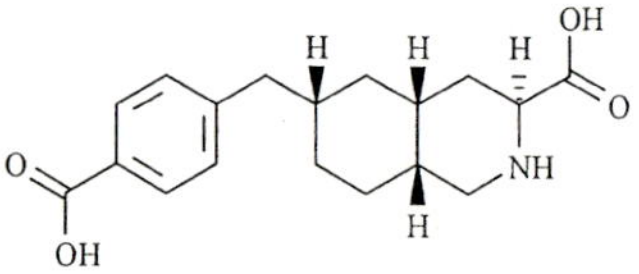

LY382884

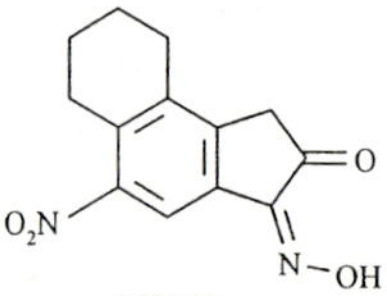

NS102

Fig. 4. Competitive kainate receptor antagonists

Table 2. Data from binding and electrophysiological assays for a range of isoxazole and decahydroisoquinoline analogues

Compound	IC_{50} (μmol/l) for displacement of:			IC_{50} (μmol/l) antagonism of depol induced by AMPA[a]
	[^{3}H]CNQX	[^{3}H]CPP	[^{3}H]KA	
AMOA[b]	8 ± 0.7	>100	>100	320 ± 25
ATOA[b]	12 ± 5	>100	>100	150 ± 14
AMPO[b]	6.9 ± 2.6	>100	>100	60 ± 7
ATPO[b]	5.7 ± 3.2	>100	>100	28 ± 3
CNQX[b]	0.038 ± 0.004	25	1.5	0.6
LY293558[c]	14 ± 0.1[d]	12.1 ± 2.0[e]	28.1 ± 1.7	1.8 ± 0.2

[a] Rat cortical slice preparation.
[b] Taken from MADSEN et al. (1996), CNQX is reference substance.
[c] Taken from Ornstein et al. (1996b).
[d] Ligand used [^{3}H]AMPA.
[e] Ligand used [^{3}H]CGS19755.

not significantly alter affinity for AMPA receptors but reduced affinity for NMDA receptors and increased affinity for kainate receptors. A phenyl substituent on the linker adjacent to the isoquinoline ring was tolerated but was less selective being more potent than the parent compound at kainate receptors. Thus it would appear that kainate but not AMPA receptors tolerate bulky substituents on the linker.

Replacement of the tetrazole with a sulphonyltriazole moiety resulted in an enhancement of potency and selectivity for AMPA receptors ($IC_{50} = 0.16 \pm 0.79\,\mu mol/l$ for antagonism of AMPA-mediated depolarizations in the cortical wedge assay) (ORNSTEIN et al. 1996b). The activity was found to reside in the (–)-(3S,4*a*R,6S,8*aR*)-isomer (IC_{50} 0.6 μmol/l for displacement of [^{3}H]AMPA binding). Interestingly, replacement of the tetrazole with a phosphono group resulted in complete loss of affinity for ionotropic glutamate receptors.

Evidence for the influence of conformation on AMPA receptor antagonist potency stems from the observation that several analogues of LY293558 with less conformational restriction but the same chain-length (including the open chain analogue, 2-amino-8-tetrazolyloctanoic acid) were inactive (ORNSTEIN et al. 1996a). Thus it would appear that the decahydroisoquinoline ring imposes a unique conformation on the molecule. It was observed that for both C-6 epimers it is the compound with the S absolute stereochemistry at C-3 which has the AMPA receptor antagonist activity (ORNSTEIN et al. 1996b). This contrasts with observation that for the decahydroisoquinoline series of NMDA receptor antagonists the activity was found to reside in the isomer with R absolute stereochemistry at C-3. Thus the pharmacophore for NMDA receptor antagonists would appear to be much different from that for AMPA receptors.

3. Isoxazoles

MADSEN et al. (1996) have reported a series of AMPA analogues with carboxymethyl or phosphonomethyl substituents on the 3-hydroxy group attached to the isoxazole ring (for structures see Fig. 3). These compounds were highly selective AMPA receptor antagonists, showing little affinity for either kainate or NMDA receptors (Table 2). The 5-*t*-butyl substituted analogue, (*RS*)-2-amino-3-[5-*tert*-butyl-3-(phosphonomethoxy)-4-isoxazolyl]propionic acid (ATPO) was the most potent antagonist and generally phosphono substituted analogues were more potent than those with carboxy substituents. In contrast to (*RS*)-2-amino-3-[3-(carboxymethoxy)-5-methyl-4-isoxazolyl] propionic acid (AMOA), neither (*RS*)-2-amino-3-[5-*tert*-butyl-3-(carboxymethoxy)-4-isoxazolyl]propionic acid (ATOA) nor ATPO (1 mmol/l) antagonized NMDA-induced depolarizations in the cortical wedge preparation. Although the *t*-butyl group at the 5-position of the isoxazole ring of ATOA and (*RS*)-2-amino-3-(5-*tert*-butyl-3-hydroxy-4-isoxazolyl)propionic acid (ATPA) is well accommodated by the receptor this is not the case for the corresponding agonists, the *t*-butyl substituted analogue of AMPA, ATPA being 14 times less potent than AMPA. One possible explanation for this is that agonists and antagonists are binding to different residues within the AMPA receptor-binding site or to different conformations of the receptor (MADSEN et al. 1996).

WAHL et al. (1998) have recently reported that ATPO antagonizes responses due to kainate in *Xenopus* oocytes expressing homomeric iGluR1,

3, or 4 (IC_{50} values 5.3 ± 0.8, 12 ± 0.38, 34 ± 2.8 μmol/l respectively) as well as cells expressing iGluR1/2 (IC_{50} value 6.6 ± 0.21 μmol/l). ATPO is a partial antagonist at either iGluR5 (EC_{50} value 24 ± 3.3 μmol/l) or iGluR5/KA2 and is inactive on kainate responses at homomeric iGluR6 or heterooligomeric iGluR6/KA2 expressed in HEK cells.

4. Phenylglycine and Phenylalanine Analogues

Substituted phosphonoethylphenylalanines have been reported to be selective AMPA receptor antagonists (HAMILTON et al. 1992, 1994). A structure-activity study has demonstrated that a 5-substituent on the phenyl ring is critical for activity at AMPA receptors as compounds without this substituent were either inactive or were NMDA receptor antagonists. Bulky substituents at the 5-position were not well tolerated but 3,5-dimethyl substitution did not result in loss of activity. Lipophilic, electron-withdrawing substituents such as iodo- or trifluoromethyl at the 5-position were required for optimal antagonist activity. Reduction of the aromatic ring completely abolished activity. The most potent compound, 5-iodo-2-(2-phosphonethyl)phenylalanine (I, Fig. 5) antagonised AMPA receptor mediated currents in rat brain mRNA-injected *Xenopus* oocytes (K_i value 3.6 μmol/l).

Two novel phenylglycine analogues, (*RS*)-3,5- and (*R*)-3,4-dicarboxyphenylglycine (DCPG, see Fig. 5) have recently been reported to selectively antagonise AMPA- (K_D values 167 μmol/l and 77 μmol/l respectively) over kainate- (K_D value for (*R*)-3,4-DCPG >3 mmol/l) induced depolarizations in the neonatal rat spinal cord preparation (THOMAS et al.1997). Indeed 3,5-DCPG (1 mmol/l) potentiated responses due to kainate in this preparation by an as yet undetermined mechanism. However, both (*RS*)-3,4- and (*RS*)-3,5-DCPG weakly antagonise NMDA-induced depolarizations (K_D values 472 μmol/l and 346 μmol/l respectively) so caution must be exercised when using these compounds as selective AMPA receptor antagonists.

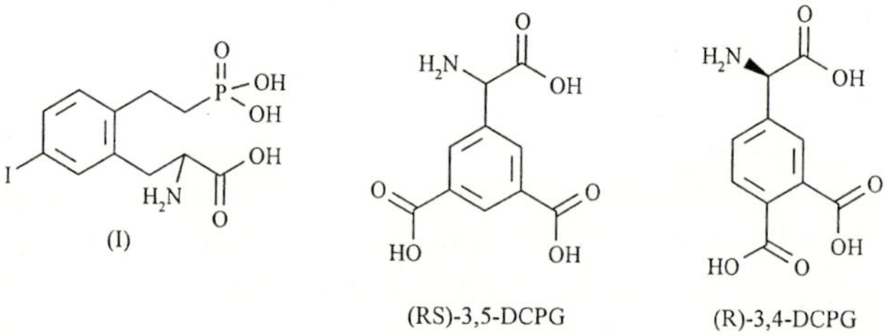

Fig. 5. Phenylalanine and phenylglycine analogues with AMPA receptor antagonist action

III. Benzodiazepine Analogues as Non-Competitive AMPA Receptor Antagonists

A range of 2,3-benzodiazepine analogues have been shown to be selective non-competitive AMPA receptor antagonists (Fig. 6) (DONEVAN and ROGAWSKI 1993). The first 2,3-benzodiazepine which potently and selectively antagonized AMPA-induced responses was (±)-1-(4-aminophenyl)-4-methyl-7,8-(methylenedioxy)-5*H*-2,3-benzodiazepine (GYKI 52466). Many groups have investigated the structural requirements for optimizing the selectivity and potency of GYKI 52466, by synthesizing 2,3-benzodiazepine analogues (Fig. 6). It is difficult to make evaluations on structure-activity relationships for 2,3-benzodiazepine analogues as in many cases these compounds have been assessed on different test systems. Hence, caution must be exercised when comparing potencies and activities.

Substitution at N-3 of GYKI 52466 with a methylcarbamyl group gave (±)-1-(4-aminophenyl)-3-methylcarbamyl-4-methyl-3,4-dihydro-7,8-(methylene-dioxy)-5*H*-2,3-benzodiazepine (GYKI 53655, Fig. 6) which had increased antagonist potency at recombinant human iGluR1 and 4 receptors expressed in HEK293 cells (IC_{50} values 6 μmol/l and 5 μmol/l respectively) compared to GYKI 52466 (IC_{50} values 18 μmol/l and 22 μmol/l respectively) (BLEAKMAN et al. 1996). A further increase in potency was observed, when the active (–)-isomer of GYKI 53655 (LY303070) was tested for antagonist action on recombinant human GluR4 receptors (IC_{50} value 0.7 μmol/l). GYKI 53655 was selective for AMPA receptors having only weak antagonist activity on DRG neurones (a source of iGluR5-containing kainate receptors). PELLETIER et al. (1996) investigated N-3 substitution of 2,3-benzodiazepine compounds further,

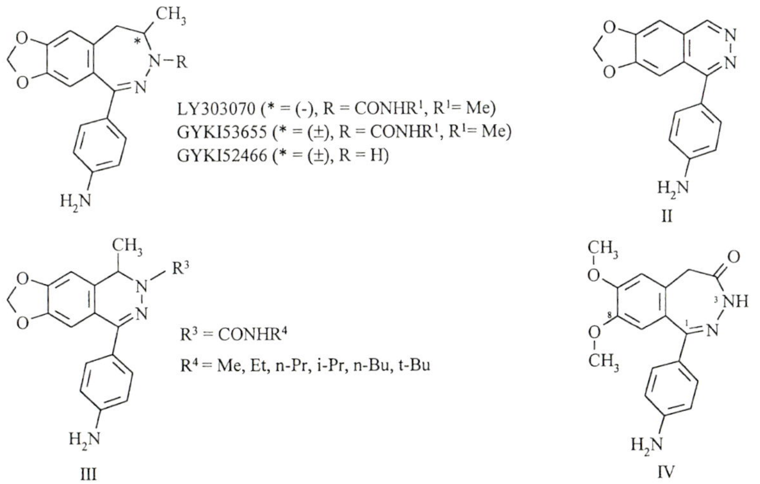

Fig. 6. Non-competitive AMPA receptor antagonists

to assess the influence of different alkyl carbamoyl groups on antagonist potency. It was observed that a reduction in ring size from a seven membered ring (GYKI 53655) to a six membered ring (III, R^3 = $CONHCH_3$), led to a decrease in potency for the inhibition of kainate-induced AMPA currents in rat cortical cells (IC_{50} values 1 μmol/l and 23 μmol/l respectively). This possibly explains the weak inhibition by compound II (Fig. 6) of kainate-induced AMPA-receptor mediated currents in rat cortical neurones. Increasing the carbon chain length of the alkyl group R^4 in compound III (R^3 = $CONHR^4$, Fig. 6) resulted in an increase in AMPA receptor antagonist potency (R^4 = Me, Et, *n*-Pr, *n*-Bu; IC_{50} values 23, 7.2, 2.8, and 1.8 μmol/l respectively). These compounds had only weak antagonist action on homomeric iGluR6 expressed in HEK cells (10% inhibition of kainate induced current at 100 μmol/l). It has been noted that a bulky R^4 group, such as *t*-Bu and *i*-Pr, attenuates the antagonist potency with respect to their straight chain equivalent (*n*-Bu and *n*-Pr respectively).

IV. Positive Allosteric Modulators

Aniracetam (Fig. 7) was reported to potentiate quisqualate-induced responses (ITO et al. 1990) by interacting with an allosteric site on the receptor. Diazoxide (Fig. 7) was subsequently reported to have inhibitory actions on rapid glutamate-induced desensitisation (YAMADA and ROTHMAN 1992). A range of benzothiadiazides has been shown to inhibit effectively AMPA-induced desensitisation (YAMADA and TANG 1993). In contrast, the lectin Concanavalin A (Con A) has little effect on AMPA-induced desensitization, but strongly attenuates desensitization at kainate receptors (PARTIN et al. 1993). Cyclothiazide, one of the most potent benzothiadiazide analogues tested, enhanced

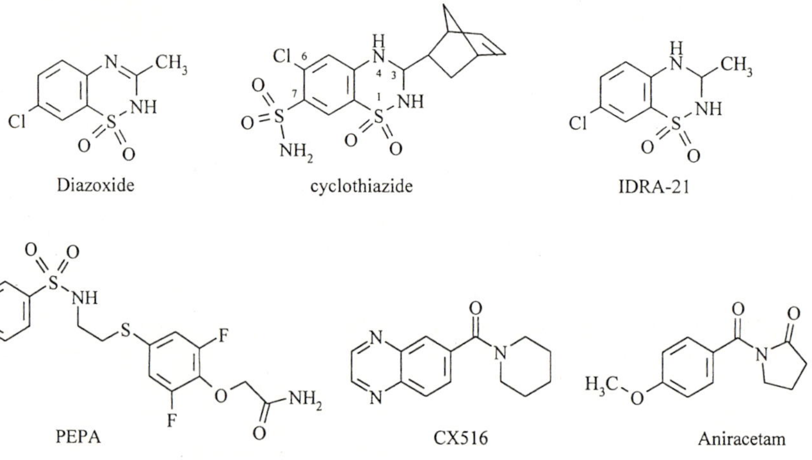

Fig. 7. Positive allosteric modulators of AMPA receptors

the peak and steady-state currents induced by quisqualate (1 mmol/l) on voltage-clamped hippocampal neurones by 4- and 400-fold respectively (EC_{50} values of 12 μmol/l and 14 μmol/l respectively) (YAMADA and TANG 1993; PATNEAU et al. 1993). It has been suggested that cyclothiazide acts at a site distinct from both the glutamate recognition site and the non-competitive 2,3-benzodiazepine antagonists site (DESAI et al. 1995). YAMADA and TANG (1993) have undertaken a structure-activity relationship study for the benzothiadiazide sensitive non-NMDA receptor. The sulphonamide link within the heterocyclic ring was suggested to be necessary for activity; however the type and position of the halogen on the phenyl ring was thought unimportant. Benzothiadiazide compounds with or without sulphonamide substituents at C-7 or substituents at C-3 inhibit glutamate-induced desensitization. One compound, 7-chloro-3-methyl-3,4-dihydro-2*H*-1,2,4-benzothiadiazine-*S,S*-dioxide (IDRA-21, Fig. 7) (10 μmol/l) potentiated the glutamate-evoked non-NMDA current in rat hippocampal neurones (ZIVKOIC et al. 1995). It was noted that unsaturation at position 3–4 of IDRA-21 attenuated the effect in reducing desensitization.

4-[2-(Phenylsulphonylamino)ethylthio]-2,6-difluoro-phenoxyacetamide (PEPA, Fig. 7) potentiated glutamate-induced currents in *Xenopus* oocytes expressing AMPA receptors but not those due to kainate or NMDA receptors (SEKIGUCHI et al. 1997). Similar to aniracetam, PEPA shows preferential selectivity for the flop isoform but is at least 100 times more potent. PEPA displayed similar selectivity when tested on homomeric AMPA receptors (rank order of potency: iGluR3≤iGluR4>iGluR1) expressed in HEK 293 cells, where the rate of onset of desensitization evoked by glutamate was abolished or reduced considerably for the flop but not the flip splice-variant. PEPA has been used to investigate the heterogeneity of AMPA receptors expressed in hippocampal cultures (SEKIGUCHI et al. 1998).

V. Channel Blockers

Open channel blockers, such as certain spider and wasp toxins, are known to block recombinant AMPA receptors lacking the edited iGluR2 subunit (BLASCHKE et al. 1993; HERLITZE et al. 1993). A range of polyamines have been investigated (WASHBURN and DINGLEDINE 1996) for their effectiveness as channel blockers of recombinant AMPA receptors. The polyamines, spermine, spermidine, *N*-(4-hydroxyphenylpropanoyl)spermine (HPP-SP), *N*-(4-hydroxyphenylacetyl)spermine (HPA-SP), and Ageltoxin-489 (Agel-489, Fig. 8), have been shown to block selectively iGluR3 over iGluR1 or 4 receptors expressed in oocytes (Agel-489≤HPP-SP>HPA-SP>>spermine>spermidine; IC_{50} values for blockade of iGluR3, 0.060 μmol/l, 0.08 μmol/l, 0.580 μmol/l, 120 μmol/l, and 820 μmol/l respectively). These results indicate that monoacylation at the terminal position of polyamine toxin, to give an amide with an aromatic end group leads to a dramatic increase in the potency of the channel block. A number of factors could explain this increase in potency, such as an

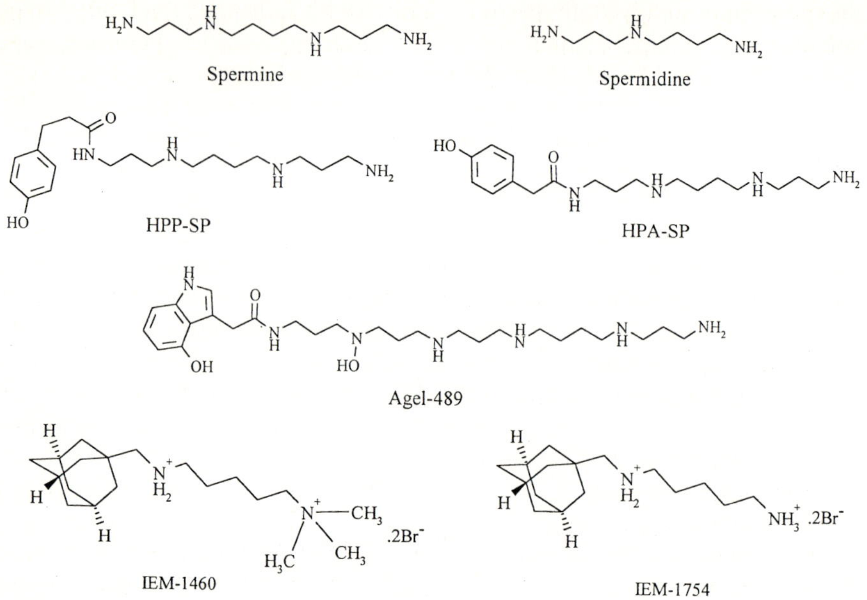

Fig. 8. Channel blockers acting at AMPA receptors

alteration in hydrophobicity, which is influenced by the presence of the aromatic group. The number of amino groups available for protonation has also been proposed to influence the channel blocking activity.

Recently, MAGAZANIK et al. (1997) investigated adamantane compounds 1-trimethylammonio-5-(1-adamantanemethylammoniopentane dibromide) (IEM-1460) and 1-ammonio-5-(1-adamantanemethylammoniopentane dibromide) (IEM-1754) (Fig. 8) for their antagonist action on kainate-induced currents in *Xenopus* oocytes expressing recombinant AMPA receptors and isolated neurones from rat hippocampal slices. The kainate induced-currents recorded from cells expressing homomeric iGluR1 and 3 receptors are similar to those from heteromeric receptors containing the iGluR2 subunit; however, their sensitivity to IEM-1460 and IEM-1754 proved to be different. IEM-1460 and IEM-1754 potently inhibited kainate-induced currents in cells expressing homomeric iGluR1 and 3 receptors (IC_{50} values 1.6 μmol/l and 6.0 μmol/l respectively). In contrast, both compounds exhibited poor inhibition of kainate-induced responses on heteromeric AMPA receptors comprising iGluR3 and edited iGluR2 subunits (100 μmol/l IEM-1460 inhibited the kainate response by 7.8 ± 2.4%). Although the adamantane polyamines are not as potent or as selective as the monoacylated polyamines, such as HPP-SP (Fig. 8), they are considerably more potent than spermidine, which is of a similar chain length. The high potency observed for IEM-1460 and 1754, may be partly due to the bulky adamantane terminal group and the doubly charged amino groups at physiological pH. The permanently charged quaternary amino group in IEM-1460 may also contribute to the observed higher potency.

C. Kainate Receptor Pharmacology

I. Kainate Receptor Agonists

The conformationally restricted glutamate analogue, kainic acid, isolated from the seaweed *Digenea simplex* (Takemoto 1978) is the prototypic agonist for kainate receptors. Other natural products based on the kainoid structure such as domoic acid (Biscoe et al. 1976), acromelic acid A and B (Konno et al. 1983) activate kainate receptors more potently than kainic acid itself (for structures see Fig. 9). A novel photoaffinity label for the kainate receptor, (2'*S*,3'*S*,4'*R*)-2'-carboxy-4'-(2-diazo-1-oxo-3,3,3-trifluoropropyl)-3'-pyrrolidinyl acetate (Fig. 9, DZKA) has recently been reported (Willis et al. 1997).

The activity of kainoid analogues at AMPA receptors has limited their usefulness as tools to distinguish AMPA and kainate receptors. Considerable effort has gone into designing more selective agonists for kainate receptors. Amongst recent examples, (2*S*,4*R*)-4-methylglutamate ((2*S*,4*R*)-4MG, Fig. 9) (Jones et al. 1997) was shown to produce desensitizing responses in HEK293 cells expressing iGluR6 with a potency similar to kainate (EC_{50} values 1.0 μmol/l and 1.8 μmol/l respectively). However, (2*S*,4*R*)-4MG also completely desensitized responses in dorsal root ganglion (DRG) cells (which have iGluR5-containing kainate receptors) with an IC_{50} value of 11 nmol/l (compared to 3.4 μmol/l for glutamate). Shimamoto and Ohfune (1996) have synthesized a range of 3-methoxymethyl-substituted cyclopropylglycines. One such analogue, (2*S*,1'*R*,2'*R*,3'*R*)-2-[2-carboxy-3-(methoxymethyl)cyclo-

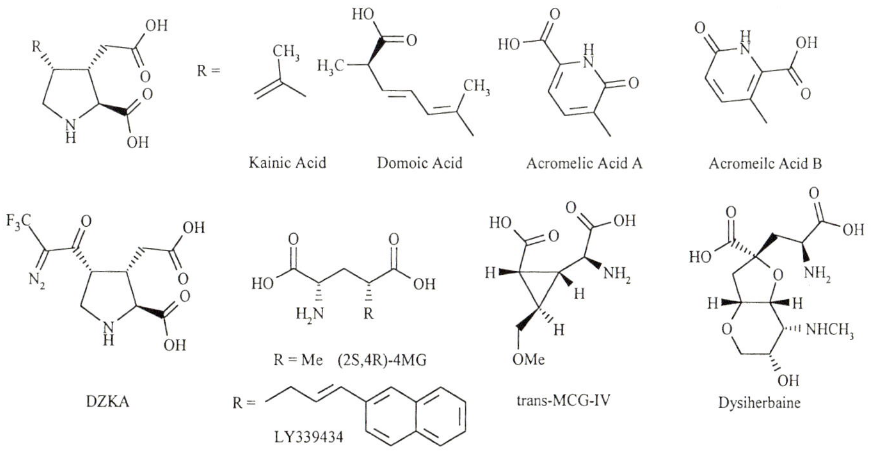

Fig. 9. Potent kainate receptor agonists

propyl]glycine (*trans*-MCG-IV) was identified as a potent depolarizing agent of dorsal root C-fibers (known to be a source of iGluR5 containing receptors; Partin et al. 1993). This data suggests that a folded conformation of glutamate is required for activating iGluR5-containing kainate receptors.

A number of recent publications have highlighted compounds with selectivity for iGluR5 containing kainate receptors. (*RS*)-5-Bromowillardiine was one of the first compounds reported to depolarize isolated immature rat dorsal roots (Agrawal and Evans 1986). A recent study of the binding affinities of a series of willardiine analogues for human homomeric iGluR5 expressed in HEK293 cells revealed a rank order of potency of 5-I-will >5-Br-will >5-Cl-will >5-F-will which is the reverse of that determined for homomeric iGluR1, 2, or 4 (Jane et al. 1997). Indeed (*S*)-1-(2-amino-2-carboxyethyl)-5-iodopyrimidine-2,4-dione (5-I-will, Fig. 1) has the highest affinity for iGluR5 yet reported (K_i value 0.24 nmol/l for displacement of [^{3}H]kainate binding) with greater than 700-fold selectivity for iGluR5 over iGluR1, 2 or 4. (*S*)-5-I-will also shows >400,000-fold selectivity between iGluR5 and iGluR6 as it does not displace [^{3}H]kainate binding to iGluR6 at a concentration of 100 μmol/l. The rank order of potency for willardiine analogues on iGluR5 is in excellent agreement with earlier electrophysiological studies on the immature isolated dorsal root (Blake et al. 1991) and dorsal root ganglion cells (Wong et al. 1994). A QSAR study revealed a strong correlation between activity on DRG cells and size, electronegativity, and lipophilicity of the substituent at the 5-position of the uracil ring (Wong et al. 1994). The corresponding 6-aza analogue of 5-I-will was less potent and selective on iGluR5 (Jane et al. 1997).

The AMPA analogue ATPA (Fig. 1), previously reported to be a selective AMPA receptor agonist (Sløk et al. 1997), binds to homomeric iGluR5 expressed in HEK293 cells with high affinity (K_i value 4.3 ± 1.1 nmol/l) (Clarke et al. 1997). ATPA displayed only weak affinity for AMPA receptors (Ki values 6–14 μmol/l) and had no activity at iGluR6 (>1 mmol/l). Thus ATPA and 5-I-will have similarly high affinity and selectivity for iGluR5. However, ATPA is approximately 10-fold weaker than (*S*)-5-I-will at depolarizing immature dorsal roots (EC_{50} values 1.3 ± 0.3 μmol/l and 0.127 ± 0.01 μmol/l respectively (Thomas et al. 1998)). In agreement with earlier work on DRG cells (Wong et al. 1994), (*S*)-5-trifluoromethylwillardiine (EC_{50} value 0.108 ± 0.02 μmol/l) was found to be more potent than (*S*)-5-I-will at depolarizing dorsal roots (Thomas et al. 1998).

An analogue of (2*S*,4*R*)-4MG, LY339434 ((2*S*,4*R*)-4-[3-(2-naphthyl)-2(*E*)-propenyl]glutamic acid, Fig. 9), also has high affinity for iGluR5 expressed in HEK293 cells (K_i value 15 nmol/l), but only weak affinity for iGluR1, 2, 4, and 6 (K_i values >10 μmol/l) (Small et al. 1997). Preliminary pharmacological data on a naturally occurring neurotoxic 4-substituted glutamate analogue, dysiherbaine (Fig. 9), isolated from the marine sponge *Dysidea herbacea*, provides evidence of high affinity for both AMPA (IC_{50} value 224 ± 22 nmol/l for displacement of [^{3}H]AMPA binding to rat brain membranes) and kainate (IC_{50} value 59 ± 7.8 nmol/l for displacement of [^{3}H]kainate binding to rat brain membranes) but not NMDA (IC_{50} value >10,000 nmol/l for displacement of [^{3}H]CGS19755 binding to rat brain membranes) receptors (SAKAI et al. 1997).

II. Competitive Kainate Receptor Antagonists

Although much progress has been made in the development of selective AMPA receptor antagonists it is only recently that selective kainate receptor antagonists have begun to emerge. Progress has been made in the design of both quinoxalinediones and closely related analogues (BIGGE et al. 1995; VERDOON et al. 1994; WILDING and HUETTNER 1996) and notably decahydroisoquinolines as kainate receptor antagonists (BLEISCH et al. 1997; CLARKE et al. 1997; SIMMONS et al. 1998).

1. Quinoxalinediones and Related Compounds

The quinoxalinediones have provided a rich source for the design of selective antagonists for both AMPA receptors and the glycine binding-site of the NMDA receptor complex. A few quinoxalinediones have been shown to be selective antagonists for kainate receptors. These include 5-chloro-7-trifluoromethyl-2,3-quinoxalinedione (ACEA-1011, Fig. 2), which displays a 12-fold selectivity for kainate receptors present on dorsal root ganglion cells (K_B value 1 μmol/l for antagonism of kainate receptor mediated currents) over the AMPA-preferring subtype (K_B value 12 μmol/l for antagonism of AMPA receptor mediated currents) expressed in neurones in the cerebral cortex. In the same study 5-nitro-6,7-tetrahydrobenzo[*g*]indole-2,3-dione-3-oxime (NS-102, Fig. 4), previously reported to be selective for iGluR6-containing kainate receptors over AMPA receptors (VERDOON et al. 1994), was shown to block kainate receptor mediated currents (K_B value 6 μmol/l) selectively over those mediated by AMPA receptors (K_B value 114 μmol/l) (WILDING and HEUTTNER 1996). Two other quinoxalinediones, NBQX and CNQX (Fig. 2), displayed very little selectivity between AMPA and kainate receptors. It was noted in the PNQX (Fig. 2) series of compounds that a bromo substituent in the 6-position and a bulky N-alkyl substituent on the piperidine moiety (see compound VIII, Fig. 2) leads to selectivity for kainate receptors suggesting that hydrophobic interactions are important for kainate receptor binding (BIGGE et al. 1995).

2. Decahydroisoquinolines

A novel, selective antagonist of kainate receptors containing the iGluR5 subunit (see Table 3), (3*SR*,4a*RS*,6*SR*,8a*RS*)-6-((1*H*-tetrazol-5-yl)methyloxymethyl)-1,2,3,4,4*a*,5,6,7,8,8*a*-decahydroisoqiuinoline-3-carboxylic acid (LY294486, Fig. 4), has been used to show that iGluR5 containing kainate receptors regulate inhibitory synaptic transmission in the hippocampus (CLARKE et al. 1997). LY294486 was also shown to inhibit potently both kainate- and ATPA-evoked responses from DRG neurones (IC_{50} values 0.62

Table 3. Binding affinities for a series of decahydroisoquinolines on cloned human iGluR subtypes (K_i μmol/l)[a]

Compound	iGluR1	iGluR2	iGluR3	iGluR4	iGluR5	iGluR6	iGluR7 KA2
LY293558	9.21	3.25	32	50.52	4.80	>100	>100
LY302679	7.9	0.6	ND	14.8	4.7	>100	>100
LY294486	>30	>30	>30	>30	3.9	>100	ND
LY382884	>100	>100	>100	>100	6.8	>100	>100
NBQX	0.56	0.11	0.9	0.34	19.76	15.79	ND

[a] NBQX is reference substance, values taken from SIMMONS et al. (1998).

± 0.14 μmol/l and 1.3 ± 0.2 μmol/l respectively) and to block iGluR5 containing kainate receptors in area CA3 of the rat hippocampus (VIGNES et al. 1997). Recently, the binding affinity of a range of decahydroisoquinolines (see Fig. 4) at cloned human iGluRs (Table 3) has been reported (SIMMONS et al. 1998). One such compound, (3*S*,4a*R*,6*S*,8a*R*)-6-((4-carboxyphenyl)methyl-1,2,3,4,4*a*,5,6,7,8,8*a*-decahydroisoquinoline-3-carboxylic acid (LY382884, Fig. 4) (BLEISCH et al. 1997), is highly selective for the iGluR5 subunit (Table 3).

3. Positive Allosteric Modulators Acting on Kainate Receptors

Prolonged application of kainate and domoate on DRG cells induces a kainate-current, which is followed by desensitization of the peak response by 90% to a steady state current. Desensitization of the kainate-current can be blocked by brief pre-incubation with Concanavalin A (Con A) (HUETTNER 1990). However, it has recently been demonstrated that only modest enhancements of kainate-induced depolarisations on iGluR7 were obtained with Con A in comparison to those obtained with iGluR5 and iGluR6 (SCHIFFER et al. 1997). In a recent study by EVERTS et al. (1997) the actions of Con A were investigated on a range of functional iGluR subunits. It was observed that Con A potentiated effects on recombinant kainate receptors but had a much lesser effect on AMPA (no action on iGluR2) and NMDA receptor subtypes. It was also shown that the action of Con A was due to direct binding to the carbohydrate side chains of the receptor protein.

D. Therapeutic Potential of AMPA and Kainate Receptor Ligands

Excitatory amino acid neurotransmitters are required for the normal function of the CNS and dysfunction of this system is involved, in a direct or indirect way, in a number of neurological disorders, such as epilepsy and ischemic neuronal damage occurring in heart failure, head trauma injuries, and stroke. It has been observed that neuronal cell death can be caused by excessive neuronal excitation by EAA neurotransmitters (CHOI 1990; SZATKOWSKI and

ATTWELL 1994). The condition of ischemia can be classified as either focal ischemia (human stroke) or global ischemia (cardiac arrest and brain trauma). In both cases, the cause of neuronal cell death is due to a large increase in the release of glutamate and aspartate. High concentrations of glutamate can over excite glutamate receptors, leading to an influx of calcium ions into the cells, which eventually results in cell death.

The development of clinically useful EAA antagonists as drugs for the treatment of CNS disorders initially focused upon NMDA receptor antagonists (for more discussion see below), but more recently the focus has been on AMPA and kainate receptor antagonists. Investigation into the therapeutic potential of AMPA and kainate receptor ligands has intensified in recent years, mainly due to the discovery of more selective non-NMDA receptor ligands and the ability to clone, express, and localize EAA receptors using molecular biological techniques. Non-NMDA receptor antagonists, such as NBQX (Fig. 2), have been shown to protect against global ischemia, even when administrated 2h after the ischemic challenge (SHEARDOWN et al. 1990). As NBQX is highly selective for non-NMDA receptors, it was speculated that delayed neuronal cell death after a period of global ischemia is mediated not only by the NMDA receptor, but also by a mechanism involving AMPA and kainate receptors. The significance of Ca^{2+}-permeability for AMPA receptors becomes apparent in global ischemia, where pyramidal cells in the CA1 region in the hippocampus are particularly sensitive to post-ischemic damage. It has been reported that iGluR2 expression is dramatically reduced in the CA1 region of post-ischemic rat brain, and this may subsequently contribute to the delayed CA1 pyramidal cell death (PELLEGRINI-GIAMPIETRO et al. 1997; GORTER et al. 1997). The administration of NBQX during or after ischemic insult had little effect on the reduction of iGluR2 expression, implying that antagonist actions of NBQX may not be due to the interference of iGluR2 expression, but more likely to interference in conformational changes of the receptor.

NBQX was withdrawn from clinical trials due to poor water solubility and nephrotoxicity problems (NORDHOLM et al. 1997). A range of quinoxalinediones displayed similar or improved anticonvulsant activity compared to NBQX (OHMORI et al. 1994, 1997; WÄTJEN et al. 1994; BIGGE et al. 1995; DESOS et al. 1996; LUBISCH et al. 1996). No impairment of motor function was observed at anticonvulsant doses of NS257 (Fig. 3) (WÄTJEN et al. 1994). A number of quinoxalinedione analogues had improved neuroprotective properties over NBQX (OHMORI et al. 1994; WÄTJEN et al. 1994; BIGGE et al. 1995; DESOS et al. 1996; TAKAHASHI et al. 1998). These improvements in the in vivo activity are likely to be due to the higher water solubility of some of the quinoxalinedione analogues tested (WÄTJEN et al. 1994; DESOS et al. 1996; TAKAHASHI et al. 1998). Importantly, unlike NBQX, S 17625 (Fig. 3) is active when administered orally. At present S 17625 is undergoing trials as a potential therapeutic agent for the treatment of stroke (DESOS et al. 1996). Although both PNQX (Fig. 2) and NBQX were equipotent as antagonists of AMPA-induced excitotoxicity in cultured cortical neurones, only PNQX blocked glutamate-induced cell death (Bigge et al. 1995). The greater efficacy of PNQX in the latter test was thought to be due to the higher affinity of PNQX for the glycine binding-site of the NMDA receptor complex. It was therefore proposed that to achieve significant in vivo potency in animal models of stroke a more balanced affinity for AMPA, kainate and NMDA receptor glycine binding-sites was necessary. In agreement with the proposal that broad spectrum antagonists are likely to have potent in vivo activity, 7-nitro-5-(*N*-oxyaza)-1,4-dihydroquinoxaline-2,3-dione (compound IX, Fig. 2), reported to have high affinity for AMPA, kainate, and glycine binding-sites (Table 1), also displayed similar antinociceptive activity to NBQX (Cai et al. 1997).

Recently it was reported that, unlike NBQX, the iGluR5 selective decahydroisoquinoline (LY382884; Fig. 4) (Simmons et al. 1998) exhibited antinociceptive activity without ataxia. This result suggests that iGluR5-containing receptors may play a major role in the processing of nociceptive information.

The 2,3-benzodiazepine analogue GYKI 52466 (Fig. 6) displays both anticonvulsant and neuroprotective properties (Chapman et al. 1991; Smith and Meldrum 1992). The N-3 methylcarbamyl substituted analogue, GYKI 53655 (Fig. 6), was not only a more potent inhibitor of AMPA- and kainate-induced currents in cultured rat hippocampus neurones, but also more effective against kainate-induced seizures (Donevan et al. 1994). However, at doses that gave seizure protection GYKI 53655 also caused motor impairment. At present, compound IV (Fig. 6) is one of the most potent 2,3-benzodiazepine analogues reported with anticonvulsant activity against audiogenic seizures in DBA/2 mice (Chimirri et al. 1997; De Sarro et al. 1998).

Benzothiadiazine compounds, such as IDRA-21 (Fig. 7), which attenuate the rapid desensitization of AMPA-selective receptors, may inflict further neurological damage, rather than protect against seizures. However, Zivkovic et al. (1995) have reported that IDRA-21-treated rats showed improved cognition in the water maze test. Furthermore, IDRA-21 was administrated orally, an indication of effective blood-brain penetration. Memory improvements in rats over a variety of experimental paradigms, and, in some aspects of memory, in humans, by a benzoylpiperidine analogue, CX516 (Fig. 7), has also been reported (Davis et al. 1997; Hampson et al. 1998a,b; Ingvar et al. 1997). These results suggest that potentiation of AMPA-activated currents may play an important role in the enhancement of learning and memory.

E. Pharmacology of NMDA Receptors

I. Therapeutic Considerations

NMDA receptors are ligand-gated ion channels which are activated by the combined binding of glutamate and glycine (or D-serine). NMDA receptor channel currents are long lasting, high conductance currents carried by Na^+, K^+, and Ca^{++} ions. These properties, combined with the widespread distribution of NMDA receptors in the vertebrate CNS, account for the significant effects that NMDA receptors have on several aspects of CNS function. They participate in a number of neuronal processes such as long term potentiation (LTP), long term depression (LTD) (COLLINGRIDGE and BLISS 1995), experience-dependent formation of synaptic connections in development (SINGER 1990; BEAR 1996), neuronal differentiation/migration (KOMURO and RAKIC 1993), pain modulation (DICKENSON et al. 1997; BARANAUSKAS and NISTRI 1998; WIESENFELD-HALLIN 1998; MELLER and GEBHART 1993), locomotion (HOCHMAN et al. 1994; GRILLNER et al. 1995), baroreceptor (SAPRU 1996) and respiratory (BONHAM 1995) reflexes, peristalsis in the colon (COSENTINO et al. 1995), and other functions in various neuronal systems. While NMDA receptor activation plays a key role in neuronal plasticity and other normal functions, their ability to increase quickly the intracellular concentration of calcium ions also appears to account for the NMDA receptors involvement in a variety of neuropathological phenomena. In preclinical studies, the blockade of NMDA receptors has been shown to reduce significantly seizure activity and neuronal loss following focal ischemia, head trauma, and spinal cord injury (MELDRUM and GARTHWAITE 1990). Furthermore, the potent excitotoxic effect of NMDA receptor overactivation has made this receptor a prime suspect in various neurodegenerative diseases such as Alzheimer's, Parkinson's, and AIDS dementia (LIPTON and ROSENBERG 1994; MITCHELL and CARROLL 1997) and psychiatric disorders such as schizophrenia, depression, and alcoholism (COYLE 1996; HERESCO-LEVY and JAVITT 1998; TAMMINGA 1998). However, many agents being developed for stroke and epilepsy have been found to have unacceptable side effects (ROGAWSKI 1993; GASIOR et al. 1997; LEES 1997; LOSCHER et al. 1998; YENARI et al. 1998).

Consequently, the past two decades has seen intensive efforts to identify NMDA receptor antagonists that can have the therapeutic benefits of blocking excess NMDA receptor activity without the adverse side-effects of blocking normal NMDA receptor activity. Two general approaches have been taken to develop agents with a higher therapeutic index – using compounds that act at different regulatory domains on the NMDA receptor or identifying compounds that act at different subtypes of NMDA receptors. NMDA receptors display a rich diversity of sites at which pharmacological agents can modify activity. In addition to the glutamate and glycine agonist binding sites, there are sites for channel blockers, polyamines, redox reagents, ifenprodil, protons, steroids, Zn^{++}, Mg^{++}, and histamine (for reviews see MCBAIN and MAYER 1994; MORI and MISHINA 1995; WILLIAMS 1997). It is hoped that inhibition of NMDA receptor activity via one of these domains may be associated with fewer adverse effects. This may be feasible since the nature of the blockade is different at these various sites and thus the specific set of NMDA receptors blocked in vivo need not be the same for the different types of blockers. For example, blockade at the glutamate binding site would be expected to be reversed at those receptors exposed to a steady, high concentration of extracellular glutamate. Channel blockers, on the other hand, would be expected to become more effective due to their use-dependency. In contrast to both of these sites, glycine site antagonism would be relatively unaltered by extracellular glutamate levels, but would be altered by regional variations in glycine or D-serine concentrations. Further receptor selectivity is possible with channel blockers wherein low affinity channel blockers more rapidly block, and reverse from block, than high affinity channel blockers (ROGAWSKI 1993). The low affinity channel blockers have thus been proposed to have fewer effects upon normal synaptic activation of NMDA receptors, an observation that is consistent with clinical data. For example the low affinity channel blockers ketamine and dextromethorphan are better tolerated than the high affinity blockers phencyclidine (PCP) and MK-801.

The other approach to develop therapeutically useful NMDA receptor antagonists is to find agents that work selectively on discrete subtypes of NMDA receptors. At the moment, however, it is not known how many different types of NMDA receptors are found in vivo. As described above, NMDA receptors are made by the coassembly of subunits from at least two different families, NR1 and NR2 into a tetrameric (or possibly pentameric) structure. With eight alternative splice forms of the NR1 subunit (MORIYOSHI et al. 1991; SUGIHARA et al. 1992; YAMAZAKI et al. 1992; HOLLMANN et al. 1993) and four distinct NR2 gene products (IKEDA et al. 1992; MEGURO et al. 1992; MONYER et al. 1992. 1994; ISHII et al. 1993), there are many potential NMDA receptor subtypes having different subunit compositions. In spite of this large number, there appear to be only a small number of pharmacologically-distinct NMDA receptor subtypes. By targeting drugs at these subtypes, it should be possible to generate NMDA receptor antagonists with varied therapeutic and adverse effects.

II. The NMDA Receptor Glutamate Recognition Site

1. Glutamate Recognition Site Radioligands

With the development of radioligand binding procedures to determine agonist and antagonist affinities at NMDA receptors, there has been a rapid growth in the identification of NMDA receptor active compounds and in the understanding of the structural requirements for antagonist binding at the receptor. Initial studies used L-[^{3}H]glutamate as a ligand (MONAGHAN et al. 1983, 1985; MONAGHAN and COTMAN 1986; FOSTER and FAGG 1987; MONAHAN and MICHEL 1987) to characterize NMDA receptors. D-[^{3}H]AP5 was the first radiolabeled antagonist to be used to study NMDA receptors (OLVERMAN et al. 1984, 1988) and this has been replaced by the higher affinity antagonists [^{3}H]CPP (OLVERMAN et al. 1986; MURPHY et al. 1987) and [^{3}H]CGS19755 (MURPHY et al. 1988), and, of highest affinity, [^{3}H]CGP39653 (SILLS et al. 1991). Despite its low affinity, the most detailed structure-activity studies have been performed using D-[^{3}H]AP5 (OLVERMAN et al. 1984). Other studies have generated photoaffinity ligands that enable the molecular characterization of these binding sites

Table 4. Potencies of compounds at the NMDA receptor glutamate recognition site (μmol/l)

	Native NMDA receptors	NR2A[a]	NR2B	NR2C	NR2D
L-Glutamate[b]	0.3				
NMDA[b]	2				
L-CCG-IV[c]	0.02				
Homoquinolinate[d]	7	16	26	56	75
D-α-aminoadipate[b]	10				
R-AP5	2	0.3	0.5	1.6	3.7
R-CPPene[f,g]	0.1	0.11	0.14	1.5	1.8
PBPD[e,f]	20	16	5	9	4
CGS19755[d,g]	1.7	1.4	4	43	31
CGP 39653[g]	.3	.6	11	4	
EAB 515[e]	0.1	0.04	0.02	0.04	0.03

[a] NR2 subunits were coexpressed with NR1 subunits.
[b] Monaghan and Cotman (1986), L-[^{3}H]glutamate binding.
[c] Kawai et al. (1992), [^{3}H] binding.
[d] Monaghan and Beaton (1992).
[e] Buller and Monaghan (1997), activation and blockade of recombinant receptors expressed in *Xenopus* oocytes.
[f] Andaloro et al (1995), L-[^{3}H]glutamate binding.
[g] Laurie and Seeburg (1994), recombinant receptor L-[^{3}H]glutamate binding.

(Benke et al. 1993; Heckendorn et al. 1993; Marti et al. 1993). Of current radioligands, only L-[^{3}H]glutamate labels all known populations of NMDA receptors, the current radiolabeled antagonists generally label NR2A-, and to varying degrees NR2B-containing receptors (Monaghan et al. 1998). Recently, we have found that [^{3}H]homoquinolinate labels predominately NR2B-containing NMDA receptors in rat brain (Brown et al. 1998).

2. Glutamate Binding Site Agonists

In early electrophysiological studies, it was established that the optimal structure for activating NMDA receptors (and for activating EAA receptors in general) is represented by L-aspartate and L-glutamate (for review see Watkins and Evans 1981). For optimal agonist action, the two negative charge groups (preferably both carboxys) should be separated by three or four carbon-carbon bond lengths (aspartate and glutamate, respectively), the α-carbon should be in the S- (or L)-configuration, and the ω-charge group should be a carboxy. The ω-acid group can also be a sulphate, or a tetrazole group. In the latter case, the carbon chain should be shorter (as in the very potent NMDA receptor agonist tetrazol-5-ylglycine, Lunn et al. 1992).

Several rigid glutamate analogs have been constructed which are potent NMDA receptor agonists that provide insight into the optimal configuration of charges to obtain agonist activity (Fig. 10). These compounds include homoquinolinate, (2*S*,1′*R*,2′*S*)-2-(carboxycyclopropyl)glycine (L-CCG-IV)

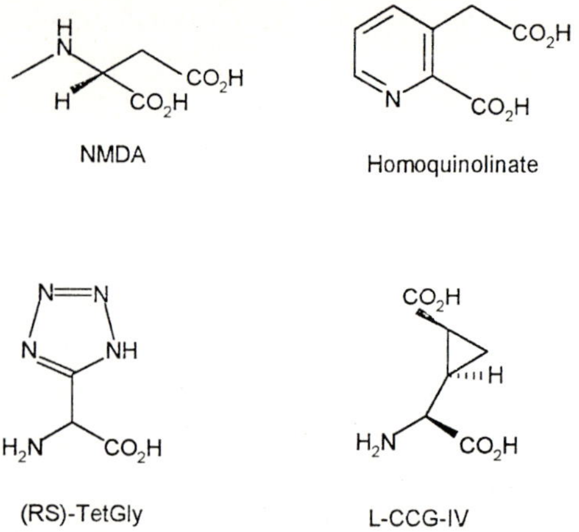

Fig. 10. NMDA receptor agonists (glutamate site)

(SHINOZAKI et al. 1989), (1*R*,3*R*) 1-aminocyclopentane-1,3-dicarboxylic acid (ACPD) and 1-aminocyclobutane-1,3-dicarboxylic acid (ACBD). The high potency of these structures suggests that L-glutamate is active in a folded conformation (O'CALLAGHAN et al. 1992).

3. Glutamate Recognition Site Competitive Antagonists

The initial discovery of NMDA receptors was made possible by the development of antagonists such as D-α-aminoadipate (D-α AA) which inhibited NMDA-evoked depolarizations while having little effect upon kainate- or quisqualate-evoked responses (BISCOE et al. 1977, 1978; EVANS and WATKINS 1978; WATKINS and EVANS 1981). In D-α AA, antagonism is found in the D-isomer and by extending the carbon chain of glutamate by one methylene group (Fig. 11). Soon after, yet greater antagonism potency was found by replacing the ω-carboxy group of D-α-AA with a phosphate group, resulting in D-2-amino-5-phosphonopentanoate (DAVIES et al. 1981; DAVIES and WATKINS 1982; EVANS et al. 1982) (D-AP5), also known as D-2-amino-5-phosphonovalerate (D-APV). For both D-αAA and D-AP5, extending the chain length by adding a methylene group diminished affinity, yet adding two methylene groups (D-α-aminosuberate and D-2-amino-7-phosphonoheptanoate, respectively) restored potency.

Structure-activity studies indicate several features that are important for antagonist action at the glutamate recognitions site of the NMDA receptor complex (for a detailed review see JANE et al. 1994). Antagonist binding requires at least two negative charge centers (generally provided by a carboxylic acid α to an amino group and by a distal phosphorate group) and a positive charge center (provided by a primary or a secondary amine). The distal phosphorate group may be providing two charge-charge interactions

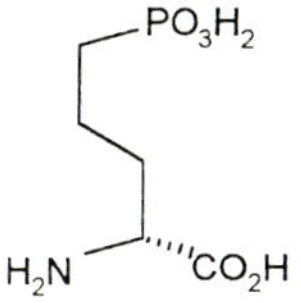
(R)-AP5

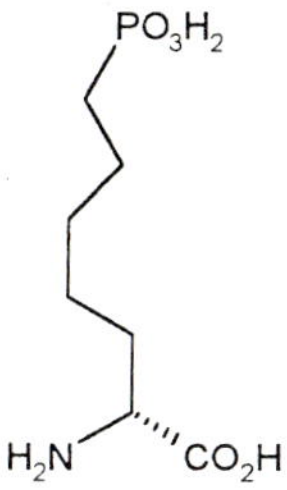
(R)-AP7

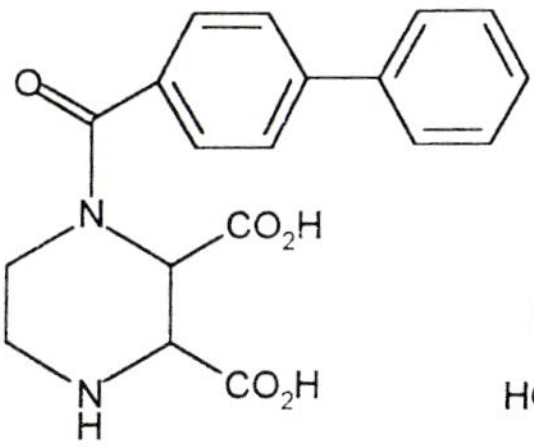
PBPD

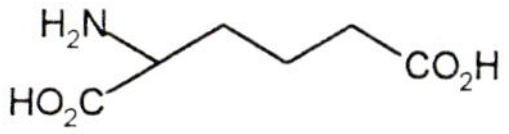
D-alpha-Aminoadipate

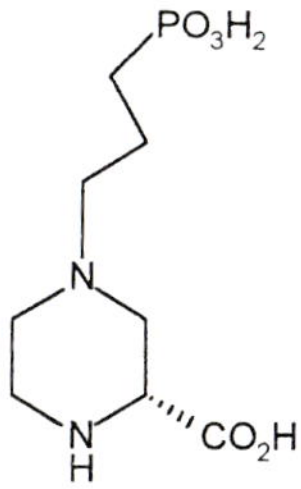
(R)-CPP

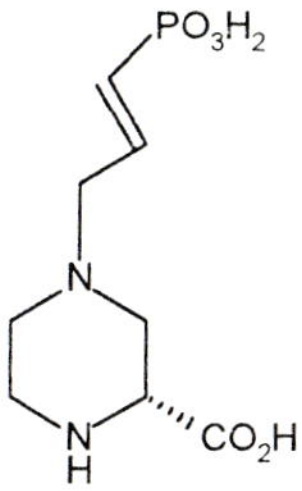
(R)-CPP-ene

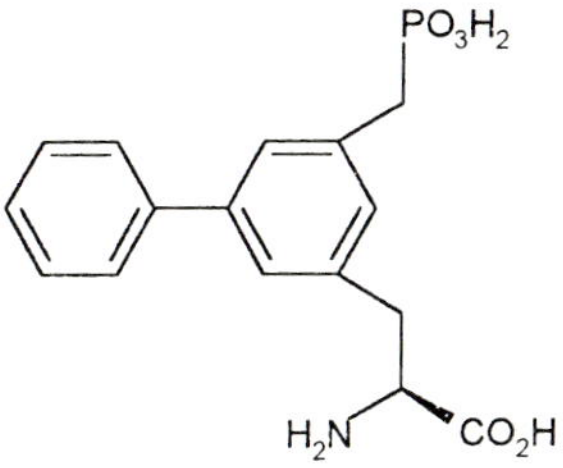
EAB 515

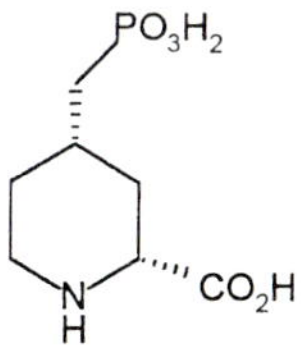
CGS 19755

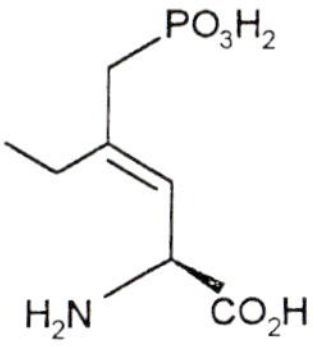
CGP 39653

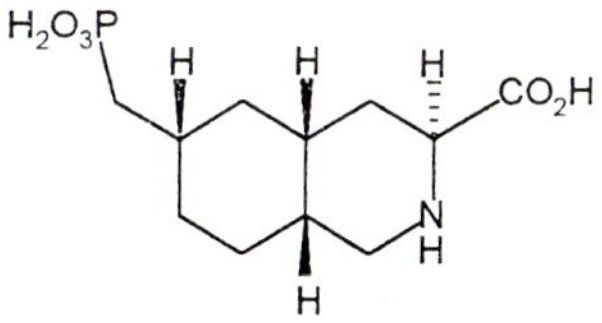
LY 235959

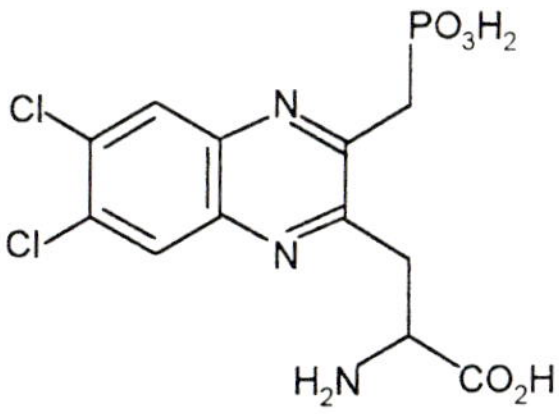
(RS)-alpha-amino-6,7-dichloro-3-(phosphonomethyl)-2-quinoxalinepropanoic acid

Fig. 11. NMDA receptor antagonists (glutamate site)

with the receptor since phosphorates provide significantly greater affinity than the corresponding carboxylate or sulphate (OLVERMAN et al. 1988). The ω-phosphorate group of NMDA receptor antagonists frequently can be replaced by a tetrazole (ORNSTEIN et al. 1991), but this modification reduces potency. The chiral carbon attached to both the carboxy and amino groups should be in the R configuration. Antagonist action is optimal with five or seven bond lengths between the negative charge groups.

Further increases in antagonist potency have been provided by constraining the AP5/AP7 chain in various ring structures and by adding specific groups (e.g., bulky hydrophilic groups, methyl groups, or double bonds) to this backbone. Several potent and selective NMDA receptor antagonists are generated by incorporating the AP5 or AP7 backbone into a piperidine and piperazine ring. Hence, 4-phosphonomethyl-2-piperidine carboxylic acid (CGS19755) (LEHMANN et al. 1988) is a potent AP5 analogue where the amino group is part of a piperidine ring, and 4-(3-phosphonopropyl)piperazine-2-carboxylic acid (CPP) (DAVIES et al. 1986; HARRIS et al. 1986) is a potent AP7 analogue incorporated into a piperazine ring (Fig. 11). A further increase in potency results when a double bond is introduced into the carbon chain of D-CPP to make D-CPPene ((*R,E*)-4-(3-phosphonoprop-2-enyl) piperazine-2-carboxylic acid) (LOWE et al. 1994).

A variety of other ring structures and additional groups have also been shown to increase the antagonist potency of the basic AP5/AP7 structure. The addition of a cyclohexane ring (NPC 17742), a biphenyl group (EAB 515; URWYLER et al. 1996), a methyl group plus a double bond (CGP 37849; FAGG et al. 1990), a quinoxaline ring (BAUDY et al. 1993) all yield compounds of increased affinity for NMDA receptors. NMDA receptor antagonists with a benzene ring include a variety of phenylglycine and phenylalanine derivatives that have a wide range of potencies (JANE et al. 1994). The incorporation of the unsaturated bicyclic decahydroisoquinoline ring or the partially unsaturated tetrahydroisoquinoline ring into the AP7 backbone, result in a wide variety of NMDA receptor antagonists of varying activities (ORNSTEIN et al. 1992). Of these, the phosphono derivative LY 274614 is the most potent. Interestingly, some of these compounds display distinctive NMDA receptor subtype selectivities (BEATON et al. 1992; BULLER and MONAGHAN 1997).

There are a few exceptions to the general rules listed above for NMDA receptor antagonist activity. For example, there are cases in which a six bond length between the acidic groups is preferred for optimal activity. The insertion of a chlorinated quinoxaline ring (BAUDY et al. 1993) into the D-AP6 structure results in α-amino-6,7-dichloro-3-(phosphonomethyl)-2-quinoxalinepropanoic acid which is a highly potent NMDA receptor antagonist. Likewise, the addition of a cyclobutane ring into D-AP6 yields two 1-aminocyclobutanecarboxylic acid derivatives which are antagonists (GAONI et al. 1994).

For most potent NMDA receptor antagonists, the R- configuration at the alpha carbon has greater activity than the corresponding S- isomer. However, for the EAB515-like antagonists in which a biphenyl (or triphenyl) group is incorporated into the AP7 chain, it is the S- (or L-) isomer which displays higher affinity (MULLER et al. 1992). Furthermore, the R- isomer decahydroisoquinoline antagonist LY 235959 also has higher activity than the S-isomer (ORNSTEIN et al. 1992). Another antagonist that does not fit the general antagonist structure described above is 4-(4-phenylbenzoyl) piperazine-2,3-dicarboxylic acid (PBPD; Fig. 11). In this structure there are two carboxylic acids separated by only three carbon-carbon bonds and an additional carbonyl group 4 bond lengths away from the amino carbon.

In the past several years various studies have used molecular modeling techniques to describe the probable optimal conformations for agonist and antagonist activity (DORVILLE et al. 1992; ORTWINE et al. 1992; WHITTEN et al. 1992). In general these studies are in reasonable agreement about the geometry of the glutamate binding site pharmacophore. The reader is referred to JANE et al. (1994) and BIGGE (1993) for a discussion of these results. Recently, a model for the ligand binding pocket of the glutamate binding site on the NMDA receptor has been proposed. Starting with the X-ray crystallography of homologous bacterial proteins, and using site-directed mutagenesis to identify critical residues for glutamate activation of NMDA receptors, Laube and colleagues have generated a three dimensional model of the glutamate recognition site (LAUBE et al. 1997).

4. Antagonist Specificity for Subtypes of Glutamate Recognition Sites

As described above, NMDA receptors are hetero-oligomeric structures generated from eight NR1 subunits, four NR2 subunits, and, potentially, various NR3 subunits. An individual NMDA receptor complex consists of probably at least four subunits (LAUBE et al. 1998), and contains two glutamate binding sites and two glycine binding sites (CLEMENTS and WESTBROOK 1991, 1994; BENVENISTE and MAYER 1991). This is consistent with observations that functional NMDA receptors appear to consist of two NR1 subunits and two NR2 subunits (BEHE et al. 1995; LAUBE et al. 1998) and that the NR1 subunit contains the glycine binding domain (KURYATOV et al. 1994; HIRAI et al. 1996) while the NR2 contains the glutamate binding domain (LAUBE et al. 1997; ANSON et al. 1998). Co-immunoprecipitation studies indicate that multiple types of NR1 subunits, as well as multiple types of NR2 subunits, can be coassembled into the same receptor complex (SHENG et al. 1994; CHAZOT and STEPHENSON 1997; LUO et al. 1997; DUNAH et al. 1998) however, the precise stoichiometry of specific NR1 and NR2 subunits has not been established for any given population of NMDA receptors. Thus, NMDA receptor complexes could be composed of many differing NMDA subunit combinations and their relationship to pharmacologically-distinct NMDA receptors has not been fully described. Nevertheless, radioligand binding studies and electrophysiological studies indicate a fairly straightforward correspondence between individual radioligand binding sites and the presence of specific NMDA receptor subunits.

Since the NR2 subunit has a glutamate binding site, the four different NR2 gene products might be expected to each contain a pharmacologically-distinct glutamate binding site. Indeed, recent studies have confirmed that four distinct pharmacological profiles can be seen for native and recombinant NMDA receptors containing the different NR2 subunits. Studies of native NMDA receptors expressed in rat brain have identified four pharmacologically-distinct populations of glutamate recognition sites (MONAGHAN et al. 1988; MONAGHAN and BEATON 1991; BEATON et al. 1992; CHRISTIE et al. 2000). The population of L-[^{3}H]glutamate binding sites in regions enriched in NR2B subunits display a higher affinity for D-CPPene and homoquinolinate than L-[^{3}H]glutamate binding sites in regions containing NR2C and NR2D subunits (BEATON et al. 1992). Similarly, recombinant NR2B-containing NMDA receptors display a higher affinity for D-CPPene and homoquinolinate than NR2C- and NR2D- containing NMDA receptors (BULLER et al. 1994; BULLER and MONAGHAN 1997). In contrast, the biphenyl compounds EAB515 and PBPD discriminate poorly between native NMDA receptor containing NR2B and NR2D receptors (ANDALORO et al. 1996) and these antagonists do not show a significantly higher affinity for NR2B subunits compared to NRC and NR2D subunit-containing receptors (BULLER and MONAGHAN 1997).

In recent studies we find that LY233536 displays an approximately tenfold greater selectivity for NR2B over NR2A-containing receptors at both recombinant (BULLER and MONAGHAN 1997) and native NMDA receptors (CHRISTIE et al. 2000). In contrast D-CPPene and D-AP5 display a higher affinity for NR2A-containing receptors than NR2B containing receptors at both recombinant (BULLER et al. 1994) and native receptors (CHRISTIE et al. 2000).

III. NMDA Receptor Channel Blockers

1. Channel Blocker Pharmacology

Subsequent to the finding by Lodge and colleagues that ketamine and phencyclidine can block NMDA receptor mediated responses (ANIS et al. 1983), many compounds have been identified that block NMDA receptor action in an uncompetitive manner by binding to a site(s) within the open ion channel. NMDA receptor channel blockers are typified by the high affinity compounds MK-801 (dizocilpine maleate), PCP (phencyclidine), and TCP (1-[1-(2-thienyl)-cyclohexyl] piperidine (Fig. 12). Each of these compounds display use-dependent and voltage-dependent blockade of the receptor complex. In both electrophysiological (HUETTNER and BEAN 1988) and radioligand binding (KLOOG et al. 1988) studies, channel blockade (or radiolabeled channel blocker binding) is dependent upon the activation of the receptor complex by agonist binding at both the glutamate and glycine binding sites. Furthermore, upon channel closure, the slowly dissociating channel blockers can become trapped in the channel until future channel activation allows blocker dissociation.

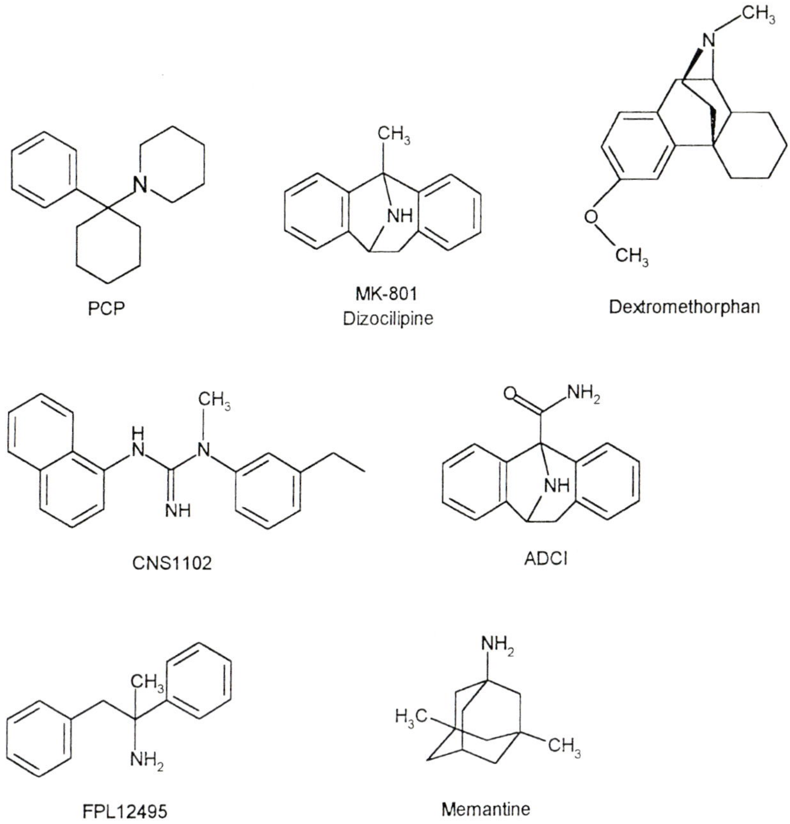

Fig. 12. NMDA channel blockers

The availability of high affinity, selective channel blocker radioligands has greatly aided the identification and development of NMDA receptor channel blockers. The most extensively used ligand has been [^{3}H]MK-801 (Reynolds and Miller 1988; Bakker et al. 1991) which has even higher affinity as the iodinated ligand (Jacobson and Cottrell 1993). A photoaffinity form of this ligand has also been developed (Sonders et al. 1990). Earlier studies have used [^{3}H] TCP which is also a useful ligand (Largent et al. 1986; Ogita et al. 1990).

Therapeutically, the high affinity channel blockers have been disappointing because of their association with various adverse side effects (motor impairment, learning impairment, psychotomimetic effects (Gasior et al. 1997; Lees 1997; Loscher et al. 1998; Yenari et al. 1998), and the appearance of vacuoles in the cingulate cortex (Olney et al. 1991; Olney 1994). Most recent efforts at developing NMDA receptor channel blockers have focused on low affinity antagonists which appear to have a better therapeutic index (ROGAWSKI 1993). Specifically, ketamine, dextromethorphan, memantine, remacemide, the remacemide analog FPL12495, and ADCI (5-aminocarbonyl-10,11-dihydro-5-*H*-dibenzo-(*a,d*)cyclohepten,5,10-imine) are low affinity channel blockers with generally more acceptable side effects (PALMER et al. 1995). The low affinity antagonists are associated with faster on kinetics (due to the higher concentrations necessary for binding) and faster off kinetics. As such, the low affinity blockers are thought to show less blockade of channel under normal activation conditions and greater blockade under pathological conditions (seizure or ischemia) (ROGAWSKI 1993).

The apparent structure-activity requirements for channel blockers are a relatively large T-shaped lipophilic domain surrounding a positive charge center provided by an amine. Frequently an aromatic group is present (LEESON et al. 1990; BIGGE 1993). NMDA receptor site-directed mutagenesis studies have identified residues along the M2 as well as M3 region of the receptor that are important for MK-801 binding to the channel (FERRER-MONTIEL et al. 1995).

2. Channel Blocker Receptor Subtype Selectivity

It has been reported that NR1 splice variants can distinguish between different NMDA receptor channel blockers (RODRIGUEZ PAZ et al. 1995). In other studies, however, a variety of channel blockers displayed no NR1 selectivity when examined under steady-state response conditions (MONAGHAN and LARSON 1997). In contrast, several compounds were able to distinguish between NMDA receptors containing different NR2 subunits. Overall, these findings were largely consistent with the differential blockade displayed by

Table 5. Potencies of NMDA receptor channel blockers (μmol/l)

	Native NMDA receptors	NR2A	NR2B	NR2C	NR2D
PCP[a]	0.06				
TCP[b,c]	0.1	0.4		0.3	
MK-801[b,c/d]	0.03	0.01/0.005	0.01/0.006	0.01/0.2	/0.15
dextromethorphan[b,c]	7	3		1	
FPL 12495[a]	0.5	6		3	
CNS 1102[a]	0.04				
ADCI[a]	9				
Memantine[a]	0.5				

[a] ROGAWSKI (1993).
[b] [^{3}H]MK-801 binding to rat brain, BEATON et al. (1992).
[c] Blockade of recombinant receptors expressed in *Xenopus* oocytes, MONAGHAN and LARSON (1997).
[d] [^{3}H]MK-801 binding to recombinant receptors, LAURIE and SEEBURG (1994).

these antagonists at native NMDA receptors (Ebert et al. 1991; Beaton et al. 1992; Porter and Greenamyre 1995). For example, NMDA receptors of the cerebellum (NR2C-containing) displayed a significantly higher affinity for dextromethorphan than did NR2B-containing receptors of the forebrain (Beaton et al. 1992). Likewise, NR2C-containing recombinant receptors display a higher affinity than NR2A- and NR2B-containing receptors for dextromethorphan (Monaghan and Larson 1997). For the high affinity channel blockers TCP and MK-801 the rate of channel block and channel unblock was significantly slower at NR2C-containing receptors than at NR2A- or NR2B-containing receptors. This may reflect the differing channel gating kinetics of these receptors (Monaghan and Larson 1997).

IV. The NMDA Receptor Glycine Recognition Site

The discovery that glycine enhances NMDA responses (Johnson and Ascher 1987) generated enormous interest in the role of glycine in NMDA function. In recent years considerable research has been conducted to describe its action in NMDA receptor function, radioligand binding properties, molecular biology, and its pharmacology.

1. Radioligand Binding and Functional Characteristics of the Glycine Receptor

The glycine recognition site on the NMDA receptor complex can be radiolabeled by a variety of ligands including the agonists [^{3}H]glycine and [^{3}H]D-serine and by the antagonists [^{3}H]5,7-dichlorokyurenic acid, [^{3}H]L-689,560 and [^{3}H]MDL 105,519 (Cotman et al. 1987; Kessler 1989; McDonald et al. 1990; Danysz et al. 1990; Baron et al. 1991, 1996; Hurt and Baron 1991; Grimwood et al. 1992). Glycine binding to the inhibitory glycine receptor, localized in the lower brain stem and spinal cord (Frostholm and Rotter 1985; Zarbin et al. 1981), can be distinguished from glycine binding to the NMDA receptor by using the inhibitory glycine receptor antagonist, strychnine.

Table 6. Potencies of compounds at the NMDA receptor glycine recognition site (μmol/l)

Glycine[a]	0.1
D-serine	0.4
ACPC	0.1
ACBC	25
Kynurenic acid	20
7-Chloro-5-iodokynurenate	20
R-(+)-HA966	10
L-689,560	5
MDL-105,519[b]	4

[a] Values obtained from Grimwood et al. (1992).
[b] Value obtained from Baron et al. (1996).

As mentioned above, the glycine-binding site has been localized to the NR1 subunit. Mutational analyses have identified two important extracellular domains which are critical for glycine binding, one between the transmembrane segments M3 and M4 and the second comprised of a segment of the N-terminus, 260 amino acids preceding the first transmembrane region (UCHINO et al. 1997; WILLIAMS et al. 1996; KUSHE et al. 1996). Important amino acid residues on the NR1 subunit involved in glycine site binding have begun to be elucidated. For example, the mutation of aspartate 732, an amino acid in the extracellular M3-M4 loop, to glycine, asparagine, or alanine reduced potency of glycine by 4000-fold (WILLIAMS et al. 1996). Important to note, while the mutation had a significant effect on the affinity of glycine for the NMDA receptor, glutamate affinity remained unchanged.

The requirement of glycine, working as a co-transmitter, to activate NMDA receptors (KLECKNER and DINGLEDINE 1988) is unique among the ionotropic glutamate receptor family and has led researchers to speculate on the functional significance of glycine modulation of NMDA receptors. Currently, glycine binding is thought regulate current flow through NMDA receptors by reducing desensitization (MAYER et al. 1989; BENVENISTE et al. 1990). In the presence of low concentrations of glycine, neuronal NMDA receptors, and recombinant NMDA receptors expressed in *Xenopus* oocytes, display a partially desensitizing inward current. When increasing concentrations of glycine are added, NMDA receptor desensitization is reduced (PARSONS et al. 1993; LERMA et al. 1990; LESTER et al. 1993).

Several studies have demonstrated an allosteric interaction between the glutamate and glycine binding sites. In radioligand binding experiments, agonist binding at either the glutamate or glycine site has been shown to increase the affinity of agonist binding at the other's binding site while decreasing antagonist affinity (FADDA et al. 1988; MONAGHAN et al. 1988). Conversely, antagonists decrease the affinity of agonists and increase the affinity of antagonists at the other site. More recently these findings have been expanded by the observation that glutamate site antagonists with the five carbon spacing between negative charge centers are differentially regulated by occupancy at the glycine binding domain (MONAHAN et al. 1990; GRIMWOOD et al. 1993, 1995). Since the glutamate antagonist binding site is on the NR2 subunit while the glycine binding site is on the NR1 subunit, it would appear that five and seven carbon spaced antagonists impose a different conformational change in the NR2 subunit, causing an altered allosteric interaction. Electrophysiological studies have reported a negative interaction between the glutamate and glycine agonist binding sites (LESTER et al. 1993). Since an electrophysiological response appears to require the binding of two glutamate and two glycine molecules, these results may be difficult to compare to radioligand binding results.

Although glycine appears to bind specifically to the NR1 subunit (UCHINO et al. 1997; WILLIAMS et al. 1996; KUSHE et al. 1996), the NR2 subunits confer subtype-specific pharmacological properties to the glycine binding site in a heteromeric receptor complex. Potencies for the agonists glycine, D-serine, D-alanine, and 1-amino-carboxycyclobutane are significantly lower at NR1/NR2A receptors than receptors composed of NR1/NR2B, NR1/NR2C and NR1/NR2D (ranked in order of increasing potency; PRIESTLEY et al. 1995, LAURIE and SEEBURG 1994; KUTSUWADA et al. 1992; BULLER et al. 1995; MATSUI et al. 1995; HESS et al. 1996). Recently a glycine-site antagonist ([^{3}H]CGP 61594) has been shown to display a high affinity selectively for NR2B-containing receptors (HONER et al. 1998).

2. NMDA Receptor Glycine Site Agonists

Since the discovery that glycine acts as a co-agonist at the NMDA receptor, a number of other glycine site agonists have been reported; the majority of these agonists are simple amino acids. D-Serine and D-alanine have the highest affinities, 0.3 μmol/l and 1.0 μmol/l, when measured with [^{3}H]glycine (McDONALD et al. 1990). The affinity of these compounds is close to that of glycine (0.2 μmol/l). The amino acid agonists with L stereochemistry are considerably less potent (REYNOLDS et al. 1987). Although the apparent requirement for amino and carboxyl groups have limited the development of compounds with selectivities greater than glycine, altering the ring structure of the cyclic homologue of glycine, 1-amino-1-carboxycyclopropane (ACPC; Fig. 13) reveals some structure activity rules for glycine-site specific ligands.

The structure of ACPC is similar to that of the amino acid agonists, while being incorporated into a cyclopropyl ring. ACPC is considered a selective agonist of the glycine binding site with an intrinsic activity of 92% (MARVIZON

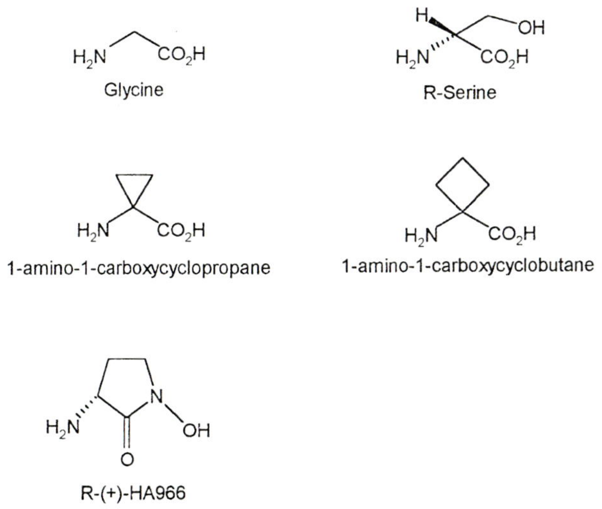

Fig. 13. NMDA receptor agonists and partial agonists (glycine site)

et al. 1989; Karcz-Kubicha et al. 1997). Expanding the cyclopropyl ring of ACPC to a cyclobutyl ring results in 1-aminocarboxycyclobutane (ACBC), a partial agonist with low efficacy (Hood et al. 1989). Increasing the size of the ring structure of ACBC to cyclopentane results in the amino acid derivative cycloleucine, a full antagonist of the NMDA glycine-binding site with weak potency (Hershkowitz and Rogawski 1989). Thus, increasing ring size results in a transformation from a ligand with agonist activity to a ligand with agonist/antagonist properties to a ligand with full antagonist activity (Watson and Lanthorn 1990).

Other partial agonists have also been described. HA-966 (Fig. 13) was one of the first compounds used to inhibit, through an unknown action, the actions of NMDA receptors. Subsequent studies revealed that it acts largely as a glycine site antagonist (Foster and Kemp 1989) and more specifically the *R*-(+)-enantiomer of HA-966 is a partial glycine agonist with low efficacy (13%) (Preistley and Kemp 1994; Karcz-Kubicha 1997). D-Cycloserine, structurally similar to HA-966, has partial agonist activity at the glycine binding site with an intrinsic activity of 40–50% (Watson et al. 1990; Hood et al. 1989).

A number of factors have led to the suggestion that D-serine may be the endogenous, strychnine-insensitive, glycine receptor ligand. D-Serine is found in abundance in the mammalian central nervous system (Hashimoto and Oka 1997; Wood et al. 1996) and has an overlapping distribution with the NMDA receptor in the central nervous system (Hashimoto et al. 1993; Schell et al. 1997). In addition, it is likely that the normal levels of endogenous glycine are not fully saturating at all, or a portion, of the native NMDA receptor population (Wood 1995; Fedele et al. 1997).

3. NMDA Receptor Glycine Site Antagonists

The development of potent antagonists at the glycine-binding site on the NMDA receptor was accelerated by the discovery that kynurenic acid blocked the stimulatory effects of glycine (Kessler et al. 1989). Derivatives of kynurenic acid, a weak non-selective excitatory amino acid antagonist, have been developed with selective NMDA receptor antagonist-activity at the glycine site. The first class of derivatives have chlorine and iodine group substitution, among these antagonists, 7-chlorokynurenic acid, 5,7-dichlorokynurenic acid, and 7-chloro-5-iodokynurenic acid (L-683,344) display high affinities, 0.56 μmol/l, 0.079 μmol/l, and 0.032 μmol/l, respectively (Kemp et al. 1988; Baron at al. 1990; Leeson et al. 1991).

Substitutions to the bicyclic ring of kynurenic have led to the development of a number of glycine antagonists, including the 2-carboxy-indoles, 4-hydroxy-2-quinolones and 2-carboxytetrahydroquinolines. The 2-carboxy-indoles include the high affinity antagonist, (*E*)-3-(2-phenyl-2-carboxyethenyl)-4,6-dichloro-1*H*-indole-2-carboxylic acid (MDL 105,519; Fig. 14); with a 10 nmol/l affinity when measured with [^{3}H]glycine (Baron et al. 1997). Among the 4-

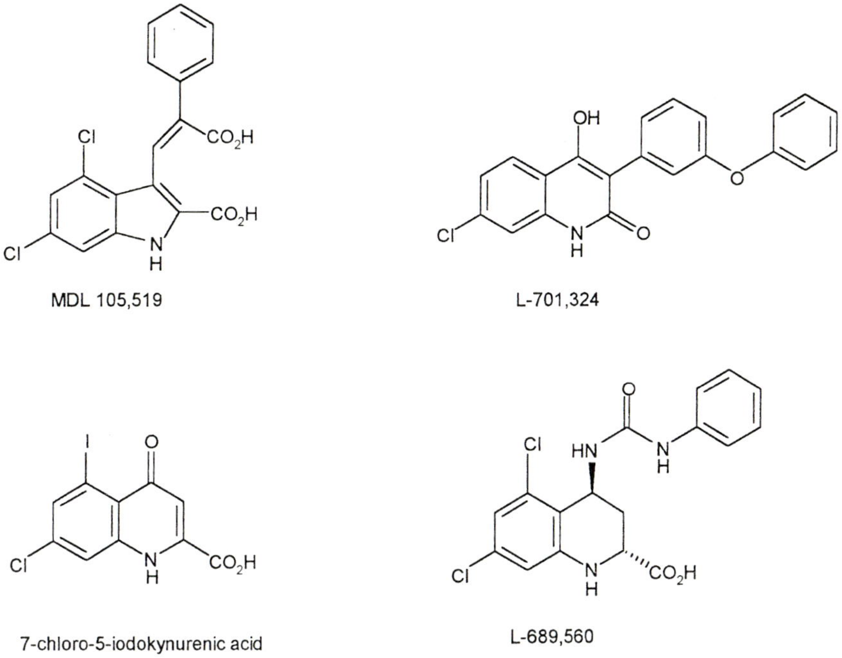

Fig. 14. NMDA receptor antagonists (glycine site)

hydroxy-2-quinolones, the ligand 7-chloro-4-hydroxy-3(3-phenoxy)phenyl-2(*H*)quinolone (L-701,324) is the most potent (Priestley et al. 1996). The tetrahydroquinoline (+/–)-4-(*trans*)-2-carboxy-5,7-dichloro-4-phenylaminocarbonylamino-1,2,3,4-tetrahydroquinoline (L-689,560; Fig. 14) is one of the most potent glycine site antagonists (Foster et al. 1992).

A class of glycine antagonists structurally related to kynurenic acid are the quinoxaline-2,3-diones: 6,7-dichloroquinoxaline-2,3-dione, 5,7-dinitro-1,4-dihydro-2,3-quinoxalinedione and 6-cyano-7-nitroquinoxaline-2,3-dione (DCQX, MNQX, and CNQX). Most of the quinoxaline-2,3-diones block AMPA and kainate receptor responses (Honore et al. 1989). CNQX and MNQX, two of the most potent of the quinoxaline-2,3-diones, display glycine-site antagonist activity (Watkins et al. 1990).

The pharmacological data generated from the kynurenic acid derivatives and quinoxaline-2,3-diones antagonists have been used to generate a theoretical antagonist pharmacophore (Leeson et al. 1991, 1992). The constituents of this pharmacophore include an electrostatic interaction with the 2-position carboxylate group, hydrogen-bonding by the proton on the 1-position nitrogen, a hydrophobic binding pocket for the aromatic ring bearing the chloro substituents, and a hydrogen bond donor from the receptor interacting with a 4-position carbonyl.

V. Allosteric Modulatory Sites on the NMDA Receptor

1. Polyamines

Polyamines, including putrescine, spermidine, and spermine, are found widely throughout the brain (SHAW and PATEMAN 1973; RUSSELL and GFELLER 1974; SEILER and SCHMIDT-GLENEWINKEL 1975) and some have been shown to bind and modulate a variety of ion channels including glutamate receptors. Polyamines are released extracellularly following neuronal depolarization (HARMAN and SHAW 1981; FAGE et al. 1992) where they may modulate endogenous NMDA receptor activity in the synapse. High affinity uptake of spermine has also been reported and may serve as an endpoint in polyamine neurotransmission (HARMAN and SHAW 1981).

Studies on both native and recombinant NMDA receptors have revealed three effects of polyamines on NMDA receptor activity. These include glycine-dependent stimulation characterized by an increase in glycine affinity for its binding site, glycine-independent stimulation characterized by an increase in the maximal amplitude of NMDA receptor responses at saturating concentrations of glycine, and voltage-dependent inhibition. In the absence of glutamate and glycine, polyamines have no effect on NMDA receptor activity. However, it has been shown that polyamines increase glycine affinity (SACAAN and JOHNSON 1989; MCGURK et al. 1990; RANSOM and DESCHENES 1990; BENVENISTE and MAYER 1993; REYNOLDS and ROTHERMUND 1995) and thus increase NMDA receptor responses at subsaturating glycine concentrations by increasing glycine association (LERMA 1992; ROCK and MACDONALD 1992; BENVENISTE and MAYER 1993; WILLIAMS 1994). Under saturating glycine conditions, polyamines still potentiate NMDA receptor responses, hence "glycine-independent" potentiation (BENVENISTE and MAYER 1993). In addition, at negative potentials, polyamines reduce channel conductance by partial channel blockade (ROCK and MACDONALD 1992; ARANEDA et al. 1993; BENVENISTE and MAYER 1993; WILLIAMS et al. 1994; IGARASHI and WILLIAMS 1995; KASHIWAGI et al. 1996, 1997; CHAO et al. 1997). Consistent with early studies (RANSOM and STEC 1988), these polyamine effects are noncompetitive with glutamate, glycine, and channel blockers suggesting distinct binding sites for polyamines. For reviews on polyamines see MCBAIN and MAYER (1994); WILLIAMS (1997).

Polyamine responses are dependent upon specific NR1 and NR2 subunits. Glycine-independent stimulation by spermine in recombinant receptors expressed in *Xenopus* oocytes is inhibited by the N-terminal insert of the NR1 subunit coded by exon 5 ($NR1_{1XX}$ isoforms) (DURAND et al. 1992, 1993; ZAPPIA et al. 1994; ZHANG et al. 1994; TRAYNELIS et al. 1995). In addition, the acidic amino acid, E342, in the amino terminus of the NR1 subunit, is necessary for glycine-independent spermine stimulation (WILLIAMS et al. 1995) but has no effect upon polyamine glycine-dependent potentiation or voltage-dependent channel block. Mutations at equivalent positions in NR2A and NR2B subunits had no effect on spermine stimulation.

The extracellular loop region between TM3 and TM4 of the NR1 subunit also participates in glycine-independent spermine stimulation as well as voltage-dependent channel block. Mutations in this region reduce glycine-independent polyamine potentiation and mutations of specific negatively charged amino acids in this same region on both NR1a and NR2B subunits reduced the voltage-dependent block by spermine (KASHIWAGI et al. 1996). However, when these negatively charged amino acids were mutated to other negatively charged amino acids there was no reduction in voltage-dependent block. Thus spermine may block NMDA receptor activity in a voltage-dependent manner by screening negative charges at amino acids in the first part of the extracellular loop region on both NR1 and NR2 subunits (ROCK and MACDONALD 1992; KASHIWAGI et al. 1996). Additionally, amino acids in a portion of the transmembrane spanning regions of the NR1 subunit (TM1,2,3) are involved in spermine stimulation and block by N^1-dansyl-spermine probably through allosteric effects or changes in gating processes (CHAO et al. 1997; KASHIWAGI et al. 1997).

In addition to the NR1 subunit, the NR2 subunit also contributes to both the stimulatory and inhibitory effects of polyamines at NMDA receptors (WILLIAMS 1994; WILLIAMS et al. 1994, 1995). Polyamines cause glycine-independent stimulation and a decrease in the affinity for glutamate-site agonists at NR1a/NR2B receptors but not at NR1a/NR2A, NR1a/NR2C, or NR1a/NR2D receptors (WILLIAMS 1994, 1995; WILLIAMS et al. 1994; ZHANG et al. 1994). However, glycine-dependent stimulation (WILLIAMS et al. 1994) and voltage-dependent inhibition (IGARASHI and WILLIAMS 1995) were seen at both NR1a/NR2A and NR1a/NR2B receptors. Taken together these data suggest that there are at least three distinct polyamine binding sites on NMDA receptors.

2. Spider and Wasp Toxins

A variety of spider and wasp polyamine toxins inhibit NMDA receptors by directly blocking the ion channel (JACKSON and USHERWOOD 1988; JACKSON and PARKS 1989). Argiotoxin$_{636}$ (Fig. 15), from the orb-web spider venom, selectively blocks receptors containing NR2A and NR2B subunits in a voltage-dependent manner while having a much lower affinity at receptors containing the NR2C subunit (PRIESTLEY et al. 1989; RADITSCH et al. 1993; WILLIAMS 1993). Additionally, argiotoxin$_{636}$ shows a 30-fold selectivity to NMDA receptors over non-NMDA types of glutamate receptors (PRIESTLEY et al. 1989). Other toxins, including philanthotoxin and the alpha agatoxins, have also been shown to inhibit both native and recombinant NMDA receptor activity (RAGSDALE et al. 1989; BRACKLEY et al. 1990, 1993; PARKS et al. 1991; WILLIAMS 1993; DONEVAN and ROGAWSKI 1996).

3. Ifenprodil and Other NR2B Selective Compounds

A variety of other pharmacological agents bind and modulate NMDA receptor activity with a selectivity similar to the polyamines. Ifenprodil, an NMDA

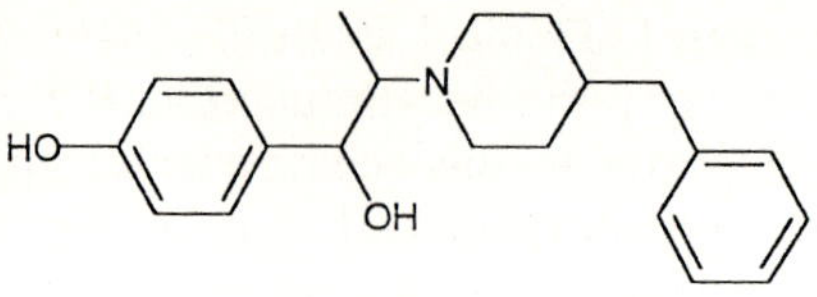

Ifenprodil

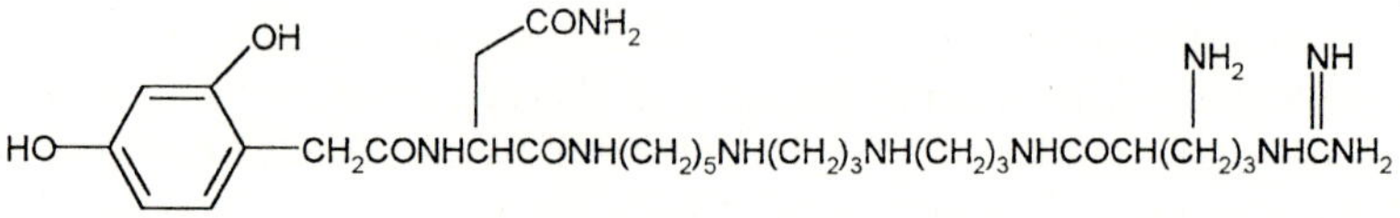

$Argiotoxin_{636}$

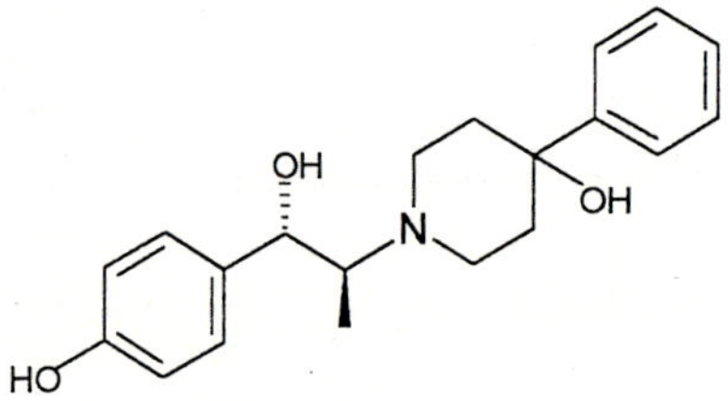

CP-101,606

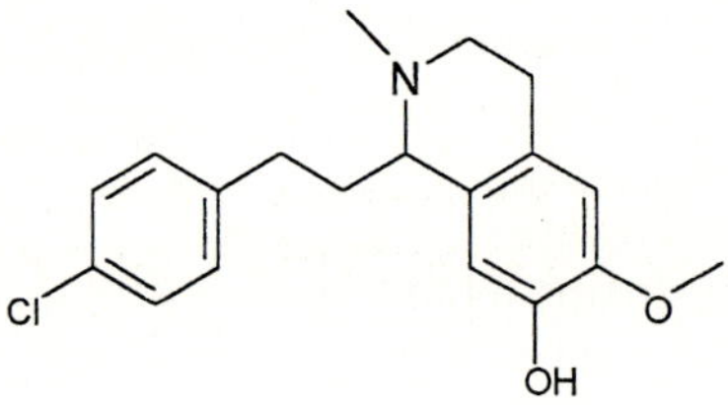

Ro 04-5595

Fig. 15. NR2B selective NMDA receptor modulators

receptor antagonist (CARTER et al. 1989) at sites separate from that of glutamate and glycine, is a phenylethanolamine (Fig. 15) displaying distinct high and low affinities at native NMDA receptors (REYNOLDS and MILLER 1989; LEGENDRE and WESTBROOK 1991). Ifenprodil exhibits greater than a 100-fold selectivity for NR2B over NR2A containing receptors (WILLIAMS 1993; GALLAGHER et al. 1996) and very low affinity at NR2C- and NR2D-containing receptors (WILLIAMS 1995). The precise binding sites for ifenprodil and ifenprodil-like compounds are not clear, but evidence (CARTER et al. 1990;

SCHOEMAKER et al. 1990; BEART et al. 1991; MERCER et al. 1993; TAMURA et al. 1993; WILLIAMS et al. 1995; GALLAGHER et al. 1996; KASHIWAGI et al. 1996) suggests that the binding site(s) may overlap with at least one polyamine binding site on the amino terminus and extracellular loop region of the NR2B subunit. Additionally, the NR1 insert (exon 5) effect of polyamine modulation of NMDA receptors has no effect on ifenprodil inhibition of NMDA receptor activity suggesting at least that the glycine-independent polyamine binding site on NMDA receptors is separate from that of the ifenprodil binding site (GALLAGHER et al. 1996).

A variety of other compounds show NR2B selectivity (BUTLER et al. 1998; GALLAGHER et al. 1998; KEW et al. 1998; MUTEL et al. 1998; STOCCA and VICINI 1998); these include haloperidol, CP-101,606, Ro 8–4304, Ro 25–6981. Site directed mutagenesis studies show that spermidine, haloperidol, and ifenprodil all have overlapping binding sites but the specific molecular determinants required for high affinity binding differ between each of these compounds (GALLAGHER et al. 1996, 1998). At the moment, these compounds display the highest degree of subtype selectivity among the different classes of NMDA receptor antagonists. As such these compounds should be useful for defining the actions of NR2B-containing receptors in brain.

4. Proton Inhibition

At low pH, NMDA receptor responses are inhibited (TANG et al. 1990). Increased external protons suppress NMDA receptor currents by decreasing channel open probability. The proton site appears independent of agonist binding sites since proton blockade was non competitive with NMDA and glycine. Proton inhibition may represent an intrinsic mechanism to protect neurons from NMDA receptor excitotoxicity during pathological acidosis. The absence of the N-terminal insert of the NR1 subunit is required, like that of glycine-independent stimulation by spermine, for proton inhibition. Thus the presence of exon 5, and more specifically K211 in exon 5, potentiates NMDA receptor function through relief of the tonic proton inhibition that is present at physiological pH (TRAYNELIS et al. 1995). Additionally, polyamine stimulation may be linked to the relief of tonic inhibition by protons suggesting that polyamines and protons share common molecular binding determinants (GALLAGHER et al. 1997), particularly within NR2B containing receptors for which both are most selective.

5. Zinc

Zinc displays subunit-specific actions at recombinant NMDA receptors. At low concentrations, zinc (1 μmol/l) enhances homomeric $NR1_{OXX}$ (NR1 lacking the N-terminal insert) receptor responses while having no effect on homomeric receptors containing $NR1_{1XX}$ subunits (HOLLMANN et al. 1993; ZHENG et al. 1994). At higher concentrations zinc inhibits both NR1 subunits with and without the N-terminal insert. Both of these phenomena occur without a change in the affinity for glutamate or glycine. The NR2 subunits also contribute to zinc's actions on NMDA receptors. Zinc displays a voltage-dependent inhibition of NMDA receptor responses in heteromeric NR1/NR2A and NR1/NR2B receptors and, at lower zinc concentrations, a voltage-independent inhibition of NR1/NR2A receptors (Williams 1996; Chen et al. 1997; Paoletti et al. 1997). Additional studies (Chen et al. 1997) have shown that the addition of heavy metal chelators to buffer solutions significantly potentiates NR1a/NR2A but not NR1a/NR2B receptor responses and this response is probably due to chelation of contaminant traces of heavy metals in solutions which tonically inhibit NR1a/NR2A NMDA receptor responses. Two effects of zinc were also seen in cultured murine cortical neurons (Christine and Choi 1990). At low concentrations (3 μmol/l) zinc produced a voltage-independent reduction in channel open probability and at higher concentrations (10–100 μmol/l) zinc produced a voltage-dependent reduction in single channel amplitude associated with an increase in channel noise suggesting a fast channel block. The consistent effects of zinc on both native and recombinant NMDA receptors suggest a dual interaction of zinc with NMDA receptors and may be a physiologically relevant response if zinc is co-released with glutamate from presynaptic terminals (Assaf and Chung 1984; Anikszteјn et al. 1987).

F. Conclusions

The advent of NMDA receptor antagonists in the late 1970s began a new era in the study of brain mechanisms. With potent and selective NMDA receptor antagonists it was soon shown that NMDA receptors play a pervasive role in neuronal plasticity and pathology. A decade later, the development of AMPA receptor antagonists likewise resulted in breakthroughs in the understanding of the role played by AMPA receptors in CNS function and pathology. The recent development of kainate receptor antagonists has enabled the demonstration of the kainate receptor's role in synaptic transmission and will no doubt help to define kainate receptor function.

With the plethora of glutamate receptor subtypes that have recently been cloned, the next series of discoveries would seem to be the development of subtype-specific antagonists which should yield many different classes of compounds with very different actions on CNS function and with very different therapeutic/adverse effect profiles.

Acknowledgment. The authors wish to thank Dr. Jeff Watkins for helpful comments on the manuscript.

List of Abbreviations

(2S,4R)-4MG	(2S,4R)-4-methylglutamate
γDGG	γ-glutamylglycine

3,4-DCPG	(*R*)-3,4-dicarboxyphenylglycine
3,5-DCPG	(*RS*)-3,5-dicarboxyphenylglycine
5-Cl-6-AW	(*S*)-2-(2-amino-2-carboxyethyl)-6-chloro-1,2,4-triazine-3,5-dione
5-F-will	(*S*)-1-(2-amino-2-carboxyethyl)-5-fluoropyrimidine-2,4-dione
5-I-will	(*S*)-1-(2-amino-2-carboxyethyl)-5-iodopyrimidine-2,4-dione
7-HPCA	(*RS*)-3-hydroxy-4,5,6,7-tetrhydroisoxazole[5,4-c]pyridine-7-carboxylic acid
ACEA-1011	5-chloro-7-trifluoromethyl-2,3-quinoxalinedione
ACBC	1-aminocarboxycyclobutane
ACBD	1-aminocyclobutane-1,3-dicarboxylic acid
ACPC	1-amino-1-carboxycyclopropane
ACPD	1-aminocyclopentane-1,3-dicarboxylic acid
ADCI	5-aminocarbonyl-10,11-dihydro-5-*H*-dibenzo-(*a,d*) cyclohepten-5,10-imine
Agel-489	Ageltoxin-489
AMOA	(*RS*)-2-amino-3-[3-(carboxymethoxy)-5-methyl-4-isoxazolyl]propionic acid
AMPA	(*S*)-2-amino-3-(3-hydroxy-5-methyl-4-isoxazolyl)propionic acid
AMPO	(*RS*)-2-amino-3-[5-methyl-3-(phosphonomethoxy)-4-isoxazolyl]propionic acid
ATOA	(RS)-2-amino-3-[5-*tert*-butyl-3-(carboxymethoxy)-4-isoxazolyl]propionic acid
ATPA	(*RS*)-2-amino-3-(5-*tert*-butyl-3-hydroxy-4-isoxazolyl) propionic acid
ATPO	(*RS*)-2-amino-3-[5-*tert*-butyl-3-(phosphonomethoxy)-4-isoxazolyl]propionic acid
CCG-IV	(2*S*,1′*R*,2′*S*) 2-(carboxycyclopropyl)glycine
CGP37849	(*RS*)-(*E*)-2-amino-4-methyl-5-phosphono-3-pentenoic acid
CGP39653	(*RS*)-(*E*)-2-amino-4-propyl-5-phosphono-3-pentenoic acid
CGP61594	(±)-*trans*-4-[2-(4-azidophenyl)acetylamino]-5,7-dichloro-1,2,3,4-tetrahydroquinoline-2-carboxylic acid
CGS19755	4-phosphonomethyl-2-piperidine carboxylic acid
CNQX	6-cyano-7-nitroquinoxaline-2,3-dione
CP-101,606	(1*S*,2*S*)-1-(4-hydroxyphenyl)-2-(4-hydroxy-4-phenylpiperidino)-1-propanol
CPP	4-(3-phosphonopropyl)piperazine-2-carboxylic acid
CPPene	(*R,E*)-4-(3-phosphonoprop-2-enyl) piperazine-2-carboxylic acid
DNQX	6,7-dinitroquinoxaline-2,3-dione
D-αAA	D-α-aminoadipate
D-AP5	D-2-amino-5-phosphonopentanoate

D-AP7	D-2-amino-7-phoshponoheptanoate
D-APV	D-2-amino-5-phosphonovalerate (D-AP5)
DZKA	(2'*S*,3'*S*,4'*R*) - 2' - carboxy - 4' - (2 - diazo - 1 - oxo - 3,3,3 - trifluoropropyl)-3'-pyrrolidinyl acetate
EAB515	alpha - amino - 5 - (phosphonomethyl)[1,1'biphenyl] - 3 - propanoic acid
FPL 12495	1, 2-diphenyl-2-proplamine monohdyrochloride
GDEE	L-glutamic acid diethyl ester
GYKI52466	(±)-1-(4-aminophenyl)-4-methyl-7,8-(methylenedioxy)-5*H*-2,3-benzodiazepine
GYKI53655	(±)-1-(4-aminophenyl)-3-methylcarbamyl-4-methyl-3,4-dihydro-7,8-(methylenedioxy)-5*H*-2,3-benzodiazepine
HA-966	3-amino-1-hydroxypyrrolid-2-one
Homo-AMPA	2-amino-4-(3-hydroxy-5-methylisoxazol-4-yl)butyric acid
HPA-SP	*N*-(4-hydroxyphenylacetyl)spermine
HPP-SP	*N*-(4-hydroxyphenylpropanoyl)spermine
IDRA-21	7-chloro-3-methyl-3,4-dihydro-2*H*-1,2,4-benzothiadiazine-S,S-dioxide
IEM-1460	1 - trimethylammonio - 5 - (1 - adamantanemethylammoniopentane dibromide)
IEM-1754	1-ammonio-5-(1-adamantanemethylammoniopentane dibromide)
L-689,560	(+/–)-4-(*trans*)-2-carboxy-5,7-dichloro-4-phenylaminocarbonylamino-1,2,3,4-tetrahydroquinoline
L-701,324	7-chloro-4-hydroxy-3(3-phenoxy)phenyl-2(*H*)quinolone
LTD	long term depression
LTP	long term potentiation
LY233536	(*RS*)-6-(1*H*-tetrazol-5-ylmethyl)decahydraisoquinoline-3-carboxylic acid
LY274614	(*RS*) - 6 - (phosphomethyl)decahydraisoquinoline - 3 - carboxylic acid
LY293558	(3*S*,4a*R*,6*R*,8a*R*)-6-(2-(1*H*-tetrazol-5-yl)ethyl)-1,2,3,4,4*a*,5,6,7,8,8*a*-decahydroisoquinoline-3-carboxylic acid
LY294486	(3S*R*,4a*RS*,6*SR*,8a*RS*) - 6 - ((1*H* - tetrazol - 5 - yl)methyloxymethyl)-1,2,3,4,4*a*,5,6,7,8,8*a*-decahydroisoqiuinoline-3-carboxylic acid
LY302679	(3*S*,4a*R*,6*S*,8a*R*) - 6 - (([1*H*]1,2,4 - triazol - 5 - ylsulphonyl)methyl)-1,2,3,4,4*a*,5,6,7,8,8*a*-decahydroisoquinoline-3-carboxylic acid
LY303070	(–)-1-(4-aminophenyl)-3-methylcarbamyl-4-methyl-3,4-dihydro-7,8-(methylenedioxy)-5*H*-2,3-benzodiazepine
LY339434	(2*S*,4*R*)-4-[3-(2-naphthyl)-2(*E*)-propenyl]glutamic acid
LY382884	(3*S*,4a*R*,6*S*,8a*R*) - 6 - (4 - carboxyphenyl)methyl-1,2,3,4,4*a*,5,6,7,8,8*a*-decahydroisoquinoline-3-carboxylic acid

MDL-105,519	(*E*)-3-(2-phenyl-2-carboxyethenyl)-4,6-dichloro-1*H*-indole-2-carboxylic acid
MK-801	dizocilpine maleate
MNQX	5, 7-dinitro-1,4-dihydro-2,3-quinoxalinedione
NBQX	6-nitro-7-sulphamoyl[*f*]quinoxaline-2,3-dione
NS 257	1,2,3,6,7,8-hexahydro-3-(hydroxyimino)-*N*,*N*,7-trimethyl-2-oxobenzo[2,1-*b*:3,4-*c*']dipyrrole-5-sulfonamide
NS-102	5-nitro-6,7-tetrahydrobenzo[*g*]indole-2,3-dione-3-oxime
PBPD	4-(4-phenylbenzoyl) piperazine-2,3-dicarboxylic acid
PCP	phencyclidine
PEPA	4 - [2 - (phenylsulphonylamino)ethylthio] - 2,6 - difluorophenoxyacetamide
PNQX	1,4,7,8,9,10 - hexahydro - 9 - methyl - 6 - nitropyrido[3,4 - *f*] - quinoxaline-2,3-dione
Ro 8–4304	4–3-[4-(4-fluoro-phenyl)-3,6-dihydro-2*H*-pyridin-1-yl]-2-hydroxy-propoxy -benzamide
Ro 25–6981	((*R*-(*R**,*S**))-α-(4-hydroxyphenyl)-beta-methyl-4-(phenylmethyl)-1- piperidinepropanol)
S 17625	6,7-dichloro-2(1*H*)-oxoquinoline-3-phosphonic acid
TCP	1-[1-(2-thienyl)-cyclohexyl] piperidine
trans-MCG-IV	(2*S*,1'*R*,2'*R*,3'*R*) - 2 - [2 - carboxy - 3 - (methoxymethyl)cyclopropyl]glycine
YM872	[2,3-dioxo-7-(1*H*-imidazol-1-yl)-6-nitro-1,2,3,4-tetrahydro-1-quinoxalinyl]acetic acid
YM90K	6-(1*H*-imidazol-1-yl)-7-nitro-2,3-(1*H*,4*H*)-quinoxalinedione
ZK200775	7-(morpholin-4-yl)-1-phosphonomethyl-6-trifluoromethyl-2,3-(4*H*)-quinoxalinedione

References

Agrawal SG, Evans RH (1986) The primary afferent depolarizing action of kainate in the rat. Br J Pharmacol 87:345–355

Ahmadian H, Nielsen B, Bräuner-Osborne H, Johansen TN, Stensbøl TB, Sløk FA, Sekiyama N, Nakanishi S, Krogsgaard-Larsen P, Madsen U (1997) (S)-Homo-AMPA, a specific agonist at the mGlu6 subtype of metabotropic glutamic acid receptors. J Med Chem 40:3700–3705

Andaloro VJ, Jane DJ, Tse HW, Watkins JC, Monaghan DT (1996) Pharmacology of NMDA receptor subtypes. Society for Neuroscience Abstracts 604

Aniksztejn L, Charton G, Ben-Ari Y (1987) Selective release of endogenous zinc from the hippocampal mossy fibers in situ. Brain Res 404:58–64

Anis NA, Berry SC, Burton NR, Lodge D (1983) The dissociative anaesthetics, ketamine and phencyclidine, selectively reduce excitation of central mammalian neurones by N-methyl-aspartate. Br J Pharmacol 79:565–575

Anson LC, Chen PE, Wyllie DJA, Colquhoun D, Schoepfer R (1998) Identification of amino acid residues of the NR2A subunit that control glutamate potency in recombinant NR1/NR2A NMDA receptors. J Neurosci 18:581–589

Araneda RC, Zukin RS, Bennett MV (1993) Effects of polyamines on NMDA-induced currents in rat hippocampal neurons: a whole-cell and single-channel study. Neurosci Lett 152:107–112

Assaf SY, Chung SH (1984) Release of endogenous Zn2+ from brain tissue during activity. Nature 308:734–736
Auberson YP, Acklin P, Allgeier H, Biollaz M, Bischoff S, Ofner S, Veenstra SJ (1998a) 5-Aminomethylquinoxaline-2,3-diones. Part II: N-aryl derivatives as novel NMDA/glycine and AMPA antagonists. Bioorg & Med Chem Letts 8:71–74
Auberson YP, Bischoff S, Moretti R, Schmutz M, Veenstra SJ (1998b) 5-Aminomethylquinoxaline-2,3-diones. Part I: a novel class of AMPA receptor antagonists. Bioorg & Med Chem Letts 8:65–70
Bakker MH, McKernan RM, Wong EH, Foster AC (1991) [3H]MK-801 binding to N-methyl-D-aspartate receptors solubilized from rat brain: effects of glycine site ligands, polyamines, ifenprodil, and desipramine. J Neurochem 57:39–45
Bang-Andersen B, Lenz SM, Skjærbæk N, Søby KK, Hansen HO, Ebert B, Bøgesø KP, Krogsgaard-Larsen P (1997) Heteroaryl analogues of AMPA. Synthesis and quantitative structure-activity relationships. J Med Chem 40:2831–2842
Baranauskas G, Nistri A (1998) Sensitization of pain pathways in the spinal cord: cellular mechanisms. Prog Neurobiol 54:349–365
Baron BM, Harrison BL, Kehne JH, Schmidt CJ, van Giersbergen PL, White HS, Siegel BW, Senyah Y, McCloskey TC, Fadayel GM, Taylor VL, Murawsky MK, Nyce P, Salituro FG (1997) Pharmacological characterization of MDL 105,519, an NMDA receptor glycine site antagonist. Eur J Pharmacol 323:181–192
Baron BM, Harrison BL, Miller FP, McDonald IA, Salituro FG, Schmidt CJ, Sorensen SM, White HS, Palfreyman MG (1990) Activity of 5,7-dichlorokynurenic acid, a potent antagonist at the N-methyl-D-aspartate receptor-associated glycine binding site. Mol Pharmacol 38:554–561
Baron BM, Siegel BW, Harrison BL, Gross RS, Hawes C, Towers P (1996) [3H]MDL 105,519, a high-affinity radioligand for the N-methyl-D-aspartate receptor-associated glycine recognition site. J Pharmacol Exp Ther 279:62–68
Baron BM, Siegel BW, Slone AL, Harrison BL, Palfreyman MG, Hurt SD (1991) [3H]5,7-dichlorokynurenic acid, a novel radioligand labels NMDA receptor-associated glycine binding sites. Eur J Pharmacol 206:149–154
Baudy RB, Greenblatt LP, Jirkovsky IL, Conklin M, Russo RJ, Bramlett DR, Emrey TA, et al. (1993) Potent quinoxaline-spaced phosphono alpha-amino acids of the AP-6 type as competitive NMDA antagonists: synthesis and biological evaluation. J Med Chem 36:331–342
Bear MF (1996) NMDA-receptor-dependent synaptic plasticity in the visual cortex. Prog Brain Res 108:205–18
Beart PM, Mercer LD, Jarrott B (1991) [125I]Ifenprodil: a convenient radioligand for binding and autoradiographic studies of the polyamine-sensitive site of the NMDA receptor. Neurosci Lett 124:187–189
Beaton JA, Stemsrud K, Monaghan DT (1992) Identification of a novel N-methyl-D-aspartate receptor population in the rat medial thalamus. J Neurochem 59: 754–757
Behe P, Stern P, Wyllie DJ, Nassar M, Schoepfer R, Colquhoun D (1995) Determination of NMDA NR1 subunit copy number in recombinant NMDA receptors. Proc R Soc Lond B Biol Sci 262:205–213
Benke D, Marti T, Heckendorn R, Rehm H, Kunzi R, Allgeier H, Angst C, et al (1993) Photoaffinity labeling of the NMDA receptor. Eur J Pharmacol 246:179–180
Benveniste M, Mayer ML (1991) Kinetic analysis of antagonist action at N-methyl-D-aspartic acid receptors. Two binding sites each for glutamate and glycine. Biophys J 59:560–573
Benveniste M, Clements J, Vyklicky L Jr, Mayer ML (1990) A kinetic analysis of the modulation of N-methyl-D-aspartic acid receptors by glycine in mouse cultured hippocampal neurones. J Physiol (Lond) 428:333–357
Benveniste M, Mayer ML (1993) Multiple effects of spermine on N-methyl-D-aspartic acid receptor responses of rat cultured hippocampal neurones. J Physiol Lond 464:131–163

Bettler B, Mulle C (1995) Neurotransmitter Receptors. 2. AMPA and kainate receptors. Neuropharmacology 34:123–139
Bigge CF (1993) Structural requirements for the development of potent N-methyl-D-aspartic acid (NMDA) receptor antagonists. Biochem Pharmacol 45:1547–1561
Bigge CF, Boxer PA, Ortwine DF (1996) AMPA/kainate receptors. Current Pharmaceutical Design 2:397–412
Bigge CF, Malone TC, Boxer PA, Nelson CB, Ortwine DF, Schelkun RM, Retz DM, Lescosky LJ, Borosky SA, Vartanian MG, Schwarz RD, Campbell GW, Robichaud LJ, Wätjen F (1995) Synthesis of 1,4,7,8,9,10-hexahydro-9-methyl-6-nitropyrido[3,4-*f*]-quinoxaline-2,3-dione and related quinoxalinediones: characterisation of α-amino-3-hydroxy-5-methyl-4-isoxazolepropionic acid (and N-methyl-D-aspartate) receptor and anticonvulsant activity. J Med Chem 38:3720–3740
Biscoe TJ, Davies J, Dray A, Evans RH, Martin MR, Watkins JC (1978) D-alpha-aminoadipate, alpha, epsilon-diominopimelic acid and HA-966 as antagonists of amino acid-induced and synpatic excitation of mammalian spinal neurones in vivo. Brain Res 148:543–548
Biscoe TJ, Evans RH, Francis AA, Martin MR, Watkins JC, Davies J, Dray A (1977) D-alpha-Aminoadipate as a selective antagonist of amino acid-induced and synaptic excitation of mammalian spinal neurones. Nature 270:743–745
Biscoe TJ, Evans RH, Headley PM, Martin MR, Watkins JC (1976) Structure-reactivity relations of excitatory amino acids on frog and spinal neurons. J Pharmacol 58:373–382
Blake JF, Jane DE, Watkins JC (1991) Action of willardiine analogues on immature rat dorsal roots. Br J Pharmacol 104:334P
Blaschke M, Keller BU, Rivosecchi R, Hollmann M, Heinemann S, Konnerth A (1993) A single amino acid determines the subunit-specific spider toxin block of α-amino-3-hydroxy-5-methylisoxazole-4-propionate/kainate receptorchannels. Proc Natl Acad Sci 90:6528–6532
Bleakman D, Ballyk BA, Schoepp DD, PalmerAJ, Bath CP, Sharpe EF, Woolley ML, Bufton HR, Kamboj RK, Tarnawa I, Lodge D (1996) Activity of 2,3-benzodiazepines at native rat and recombinant human glutamate receptors in vitro: stereospecificity and selectivity profiles. Neuropharmacol 35:1689–1702
Bleisch TJ, Ornstein PL, Allen NK, Wright RA, Lodge D, Schoepp DD (1997) Structure-activity studies of aryl-spaced decahydroisoquinoline-3-carboxylic acid AMPA receptor antagonists. Bioorg & Med Chem Letts 7:1161–1166
Bonham AC (1995) Neurotransmitters in the CNS control of breathing. Respir Physiol 101:219–230
Brackley P, Goodnow R Jr, Nakanishi K, Sudan HL, Usherwood PN (1990) Spermine and philanthotoxin potentiate excitatory amino acid responses of Xenopus oocytes injected with rat and chick brain RNA. Neurosci Lett 114:51–56
Brackley PT, Bell DR, Choi SK, Nakanishi K, Usherwood PN (1993) Selective antagonism of native and cloned kainate and NMDA receptors by polyamine-containing toxins. J Pharmacol Exp Ther 266:1573–580
Brown JC, Tse HW, Skifter DA, Christie JM, Andaloro VJ, Kemp MC, Watkins JC, Jane DE, and Monaghan DT (1998) [3H]Homoquinolinate binds to a subpopulation of NMDA receptors and to a novel binding site. J Neurochem (in press)
Buller AL, Larson HC, Morrisett RA, Monaghan DT (1995) Glycine modulates ethanol inhibition of heteromeric N-methyl-D-aspartate receptors expressed in Xenopus oocytes. Mol Pharmacol 48:717–723
Buller AL, Larson HC, Schneider BE, Beaton JA, Morrisett RA, Monaghan DT (1994) The molecular basis of NMDA receptor subtypes: native receptor diversity is predicted by subunit composition. J Neurosci 14:5471–484
Buller AL, Monaghan DT (1997) Pharmacological heterogeneity of NMDA receptors: characterization of NR1a/NR2D heteromers expressed in Xenopus oocytes. Eur J Pharmacol 320:87–94

Butler TW, Blake JF, Bordner J, Butler P, Chenard BL, Collins MA, DeCosta D, et al (1998) (3R,4S)-3-[4-(4-fluorophenyl)-4-hydroxypiperidin-1-yl]chroman-4,7-diol: a conformationally restricted analogue of the NR2B subtype-selective NMDA antagonist (1S,2S)-1-(4-hydroxyphenyl)-2-(4-hydroxy-4- phenylpiperidino)- 1-propanol. J Med Chem 41:1172–1184

Cai SX, Huang J-C, Espitia SA, Tran M, Ilyin VI, Hawkinson JE, Woodward RM, Weber E, Keana JFW (1997) 5-(*N*-Oxyaza)-7-substituted-1,4-dihydroquinoxaline-2,3-diones: novel, systemically active and broad spectrum antagonists for NMDA/glycine, AMPA and kainate receptors. J Med Chem 40:3679–3686

Carter C, Rivy JP, Scatton B (1989) Ifenprodil and SL 82.0715 are antagonists at the polyamine site of the N-methyl-D-aspartate (NMDA) receptor. Eur J Pharmacol 164:611–612

Carter CJ, Lloyd KG, Zivkovic B, Scatton B (1990) Ifenprodil and SL 82.0715 as cerebral antiischemic agents. III. Evidence for antagonistic effects at the polyamine modulatory site within the N-methyl-D-aspartate receptor complex. J Pharmacol Exp Ther 253:475–482

Chamberlin R, Bridges R (1993) Conformationally constrained acidic amino acids as probes of glutamate receptors and transporters. In: Drug design for neuroscience Ed. A. P. Kozikowski, Raven Press, New York, USA, pp. 231–259

Chan PCM, Roon RJ, Koerner JF, Taylor NJ, Honek JF (1995) A 3-amino-4-hydroxy-3-cyclobutene-1,2-dione-containing glutamate analogue exhibiting high affinity to excitatory amino acid receptors. J Med Chem 38:4433–4438

Chao J, Seiler N, Renault J, Kashiwagi K, Masuko T, Igarashi K, Williams K (1997) N1-dansyl-spermine and N1-(n-octanesulfonyl)-spermine, novel glutamate receptor antagonists: block and permeation of N-methyl-D-aspartate receptors. Mol Pharmacol 51:861–871

Chapman AG, Smith SE, Meldrum BS (1991) The anticonvulsant effect of the non-NMDA antagonists, NBQX and GYKI 52466, in mice. Epilepsy Res 9:92–96

Chazot PL, Stephenson FA (1997) Molecular dissection of native mammalian forebrain NMDA receptors containing the NR1 C2 exon: directo demonstration of NMDA receptors comprising NR1, NR2A, and NR2B subunits within the same complex. J Neurochem In Press

Chen N, Moshaver A, Raymond LA (1997) Differential sensitivity of recombinant N-methyl-D-aspartate receptor subtypes to zinc inhibition. Mol Pharmacol 51: 1015–1023

Christine CW, Choi DW (1990) Effect of zinc on NMDA receptor-mediated channel currents in cortical neurons. J Neurosci 10:108–116

Chimirri A, De Sarro G, De Sarro A, Gitto R, Grasso S, Quartarone S, Zappala M, Giusti P, Libri V, Constanti A, Chapman AG (1997) 1-Aryl-3,5-dihydro-4H-2,3-benzodiazepin-4-ones: Novel AMPA receptor antagonists. J Med Chem 40: 1258–1269

Choi DW (1990) The role of glutamate neurotoxicity in hypoxic-ischemic neuronal death. Ann Rev Neurosci 13:171–182

Ciabarra AM, Sullivan JM, Gahn LG, Pecht G, Heinemann S, Sevarino KA (1995) Cloning and characterization of chi-1: a developmentally regulated member of a novel class of the ionotropic glutamate receptor family. J Neurosci 15:6498–6508

Clarke VR, Ballyk BA, Hoo KH, Mandelzys A, Pellizzari A, Bath CP, Thomas J, et al (1997) A hippocampal GluR5 kainate receptor regulating inhibitory synaptic transmission [see comments]. Nature 389:599–603

Clements JD, Westbrook GL (1991) Activation kinetics reveal the number of glutamate and glycine binding sites on the N-methyl-D-aspartate receptor. Neuron 7:605–613

Clements JD, Westbrook GL (1994) Kinetics of AP5 dissociation from NMDA receptors: evidence for two identical cooperative binding sites. J Neurophysiol 71:2566–2569

Collingridge GL, Bliss TV (1995) Memories of NMDA receptors and LTP. Trends Neurosci 18:54–56
Collingridge GL, Kehl SJ, McLennan H (1983) Excitatory amino acids in synaptic transmission in the Schaffer collateral-commissural pathway of the rat hippocampus. J Physiol Lond 334:33–46
Conn PJ, Pin J-P. (1997) Pharmacology and functions of metabotropic glutamate receptors. Annu Rev Pharmacol Toxicol 37:205–237
Cosentino M, De Ponti F, Marino F, Giaroni C, Leoni O, Lecchini S, Frigo G (1995) N-methyl-D-aspartate receptors modulate neurotransmitter release and peristalsis in the guinea pig isolated colon. Neurosci Lett 183:139–142
Cotman CW, Monaghan DT, Ottersen OP, Storm-Mathisen J (1987) Anatomical organization of excitatory amino acid receptors and their pathways. Trends in Neuroscience 10:273–280[Monahan, 1990 #5467]
Coyle JT (1996) The glutamatergic dysfunction hypothesis for schizophrenia. Harv Rev Psychiatry 3:241–253
Christie J, Jane D, Monaghan D (2000) Native NMDA receptors containing NR2A NR2B subunits have pharmacologically-distinct competitive antagonist binding sites. J Pharmacol Exp Ther 292:1169–1174
Czuczwar SJ, Meldrum B (1982) Protection against chemically induced seizures by 2-amino-7-phosphonoheptanoic acid. Eur J Pharmacol 83:335–338
Danysz W, Fadda E, Wroblewski JT, Costa E (1990) [3H]D-serine labels strychnine-insensitive glycine recognition sites of rat central nervous system. Life Sci 46:155–164
Das S, Sasaki YF, Rothe T, Premkumar LS, Takasu M, Crandall JE, Dikkes P, et al (1998) Increased NMDA current and spine density in mice lacking the NMDA receptor subunit NR3 A. Nature 393:377–381
Davies J, Evans RH, Francis AA, Watkins JC (1979) Excitatory amino acid receptors and synaptic excitation in the mammalian central nervous system. J Physiol (Paris) 75:641–645
Davies J, Evans RH, Herrling PL, Jones AW, Olverman HJ, Pook P, Watkins JC (1986) CPP, a new potent and selective NMDA antagonist. Depression of central neuron responses, affinity for [3H]D-AP5 binding sites on brain membranes and anticonvulsant activity. Brain Res 382:169–173
Davies J, Francis AA, Jones AW, Watkins JC (1980) 2-Amino-5-phosphonovalerate (2APV), a highly potent and selective antagonist at spinal NMDA receptors. Br J Pharmac 70:52–53
Davies J, Francis AA, Jones AW, Watkins JC (1981) 2-Amino-5-phosphonovalerate (2APV), a potent and selective antagonist of amino acid-induced and synaptic excitation. Neurosci Lett 21:77–81
Davies J, Watkins JC (1978) Specific antagonism of amino acid-induced and dorsal root evoked synaptic excitation of Renshaw cells [proceedings]. Br J Pharmacol 62: 433P–434P
Davies J, Watkins JC (1979) Selective antagonism of amino acid-induced and synaptic excitation in the cat spinal cord. J Physiol (Lond) 297:621–635
Davies J, Watkins JC (1981) Differentiation of kainate and quisqualate receptors in the cat spinal cord by selective antagonism with γ-D (and L)-glutamylglycine. Brain Res 206:172–177
Davies J, Watkins JC (1982) Actions of D and L forms of 2-amino-5-phosphonovalerate and 2-amino-4-phosphonobutyrate in the cat spinal cord. Brain Res 235:378–386
Davies J, Watkins JC (1985) Depressant actions of gamma-D-glutamylaminomethyl sulfonate (GAMS) on amino acid-induced and synaptic excitation in the cat spinal cord. Brain Res 327:113–120
Davis CM, Moskovitz B, Nguyen MA, Tran BB, Arai A, Lynch G, Granger R (1997) A profile of the behavioural changes produced by facilitation of AMPA-type glutamate receptor. Psychopharmacol 133:161–167
Desai MA, Burnett JP, Ornstain PL, Schoepp DD (1995) Cyclothiazide acts at a site on the α-amino-3-hydroxy-5-methyl-4-isoxazole propanoic acid receptor complex

that does not recognise competitive or noncompetitive AMPA receptor antagonist. J Pharmacol Exp Ther 272:38–43

De Sarro A, De Sarro G, Gitto R, Grasso S, Micale N, Quartarone S, Zappala M (1998) 7,8-Methylenedioxy-4*H*-2,3-benzodiazepin-4-ones as novel AMPA receptor antagonists. Bioorg & Med Chem Letts 8:971–976

Desos P, Lepagnol JM, Morain P, Lestage P, Cordi AA (1996) Structure-activity relationships in a series of 2(1*H*)-quinolones bearing different acidic function in the 3-position: 6,7-dichloro-2(1*H*)-oxoquinoline-3-phosphonic acid, a new potent and selective AMPA/kainate antagonist with neuroprotective properties. J Med Chem 39:197–206

Dickenson AH, Chapman V, Green GM (1997) The pharmacology of excitatory and inhibitory amino acid-mediated events in the transmission and modulation of pain in the spinal cord. Gen Pharmacol 28:633–638

Donevan SD, Rogawski MA (1993) A 2,3-benzodiazepine, is a highly selective, noncompetitive antagonist of AMPA/ kainate receptor responses. Neuron 10:51–59

Donevan SD, Rogawski MA (1996) Multiple actions of arylalkylamine arthropod toxins on the N-methyl-D- aspartate receptor. Neuroscience 70:361–375

Donevan SD, Yamaguchi SI, Rogawski MA (1994) Non-N-methyl-D-aspartate receptor antagonism by 3-N-substituted 2,3-benzodiazepines: relationship to anticonvulsant activity. J Pharmacol Exp Ther 271:25–29

Dorville A, McCort Tranchepain I, Vichard D, Sather W, Maroun R, Ascher P, Roques BP (1992) Preferred antagonist binding state of the NMDA receptor: synthesis, pharmacology, and computer modeling of (phosphonomethyl)phenylalanine derivatives. J Med Chem 35:2551–2562

Dunah AW, Luo J, Wang YH, Yasuda RP, Wolfe BB (1998) Subunit composition of N-methyl-D-aspartate receptors in the central nervous system that contain the NR2D subunit. Mol Pharmacol 53:429–437

Durand GM, Bennett MV, Zukin RS (1993) Splice variants of the N-methyl-D-aspartate receptor NR1 identify domains involved in regulation by polyamines and protein kinase C. [published erratum appears in Proc Natl Acad Sci USA 1993 Oct 15;90(20):9739]. Proc Natl Acad Sci U S A 90:6731–6735

Durand GM, Gregor P, Zheng X, Bennett MV, Uhl GR, Zukin RS (1992) Cloning of an apparent splice variant of the rat N-methyl-D-aspartate receptor NMDAR1 with altered sensitivity to polyamines and activators of protein kinase C. Proc Natl Acad Sci U S A 89:9359–9363

Ebert B, Wong EH, Krogsgaard Larsen P (1991) Identification of a novel NMDA receptor in rat cerebellum. Eur J Pharmacol 208:49–52

Evans RH, Francis AA, Hunt K, Oakes DJ, Watkins JC (1979) Antagonism of excitatory amino acid-induced responses and of synaptic excitation in the isolated spinal cord of the frog. Br J Pharmacol 67:591–603

Evans RH, Francis AA, Jones AW, Smith DA, Watkins JC (1982) The effects of a series of omega-phosphonic alpha-carboxylic amino acids on electrically evoked and excitant amino acid-induced responses in isolated spinal cord preparations. Br J Pharmacol 75:65–75

Evans RH, Jones AW, Watkins JC (1980) Willardiine: A potent quisqualate-like excitant. J Physiol 308:71–72P

Evans RH, Watkins JC (1978) Specific antagonism of excitant amino acids in the isolated spinal cord of the neonatal rat. Eur J Pharmacol 50:123–129

Everts I, Villmann C, Hollmann M (1997) N-Glycosylation is not a prerequisite for glutamate receptor function but is essential for lectin modulation. Mol Pharmacol 52:861–873

Fadda E, Danysz W, Wroblewski JT, Costa E (1988) Glycine and D-serine increase the affinity of N-methyl-D-aspartate sensitive glutamate binding sites in rat brain synaptic membranes. Neuropharmacology 27:1183–1185

Fage D, Voltz C, Scatton B, Carter C (1992) Selective release of spermine and spermidine from the rat striatum by N- methyl-D-aspartate receptor activation in vivo. J Neurochem 58:2170–2175

Fagg GE, Olpe HR, Pozza MF, Baud J, Steinmann M, Schmutz M, Portet C, et al (1990) CGP 37849 and CGP 39551: novel and potent competitive N-methyl-D- aspartate receptor antagonists with oral activity. Br J Pharmacol 99:791–797
Fedele E, Bisaglia M, Raiteri M (1997) D-serine modulates the NMDA receptor/nitric oxide/cGMP pathway in the rat cerebellum during in vivo microdialysis. Naunyn Schmiedebergs Arch Pharmacol 355:43–47
Ferrer-Montiel AV, Sun W, Montal M (1995) Molecular design of the N-methyl-D-aspartate receptor binding site for phencyclidine and dizolcipine. Proc Natl Acad Sci U S A 92:8021–8025
Fletcher EJ, Lodge D (1996) New developments in the molecular pharmacology of α-amino-3-hydroxy-5-methyl-4-isoxazole propionate and kainate receptors. Pharmacology and Therapeutics 70:65–89
Foster AC, Fagg GE (1987) Comparison of L-[3H]glutamate, D-[3H]aspartate, DL-[3H]AP5 and [3H]NMDA as ligands for NMDA receptors in crude postsynaptic densities from rat brain. Eur J Pharmacol 133:291–300
Foster AC, Kemp JA (1989) HA-966 antagonizes N-methyl-D-aspartate receptors through a selective interaction with the glycine modulatory site. J Neurosci 9: 2191–2196
Foster AC, Kemp JA, Leeson PD, Grimwood S, Donald AE, Marshall GR, Priestley T, Smith JD, Carling RW (1992) Kynurenic acid analogues with improved affinity and selectivity for the glycine site on the N-methyl-D-aspartate receptor from rat brain. Mol Pharmacol 41:914–922
Frostholm A, Rotter A (1985) Glycine receptor distribution in mouse CNS: autoradiographic localization of [3H]strychnine binding sites. Brain Res Bull 15:473–486
Gallagher MJ, Huang H, Grant ER, Lynch DR (1997) The NR2B-specific interactions of polyamines and protons with the N- methyl-D-aspartate receptor. J Biol Chem 272:24971–24979
Gallagher MJ, Huang H, Lynch DR (1998) Modulation of the N-methyl-D-aspartate receptor by haloperidol: NR2B- specific interactions. J Neurochem 70:2120–2128
Gallagher MJ, Huang H, Pritchett DB, Lynch DR (1996) Interactions between ifenprodil and the NR2B subunit of the N-methyl-D- aspartate receptor. J Biol Chem 271:9603–9611
Gallagher MJ, Huang H, Pritchett DB, Lynch DR (1996) Interactions between ifenprodil and the NR2B subunit of the N-methyl-D-aspartate receptor. J Biol Chem 271:9603–9611
Gaoni Y, Chapman AG, Parvez N, Pook PC, Jane DE, Watkins JC (1994) Synthesis, NMDA receptor antagonist activity, and anticonvulsant action of 1-aminocyclobutanecarboxylic acid derivatives. J Med Chem 37:4288–4296
Gasior M, Borowicz K, Kleinrok Z, Starownik R, Czuczwar SJ (1997) Anticonvulsant and adverse effects of MK-801, LY 235959, and GYKI 52466 in combination with Ca2+ channel inhibitors in mice. Pharmacol Biochem Behav 56:629–635
Gorter JA, Petrozzino JJ, Aronica EM, Rosenbaum DM, Opitz T, Bennett MVL, Connor JA, Zukin RS (1997) Global ischemia induceds downregulation of GluR2 mRNA and increases AMPA receptor-mediated Ca^{2+} influx in hippocampal CA1 neurons of gerbil. J Neurosci 17:6179–6188
Grillner S, Deliagina T, Ekeberg O, el Manira A, Hill RH, Lansner A, Orlovsky GN, et al. (1995) Neural networks that co-ordinate locomotion and body orientation in lamprey. Trends Neurosci 18:270–279
Grimwood S, Moseley AM, Carling RW, Leeson PD, Foster AC (1992) Characterization of the binding of [3H]L-689,560, an antagonist for the glycine site on the NMDA receptor, to rat brain membranes. Mol. Pharmacol. 41:923–930
Grimwood S, Kulagowski JJ, Mawer IM, Rowley M, Leeson PD, Foster AC (1995) Allosteric modulation of the glutamate site on the NMDA receptor by four novel glycine site antagonists. Eur J Pharmacol 290:221–226
Grimwood S, Wilde GJ, Foster AC (1993) Interactions between the glutamate and glycine recognition sites of the N-methyl-D-aspartate receptor from rat brain, as revealed from radioligand binding studies. J Neurochem 60:1729–1738

Harman RJ, Shaw GG (1981) High-affinity uptake of spermine by slices of rat cerebral cortex. J Neurochem 36:1609–1615

Hamilton GS, Bednar D, Borosky SA, Huang Z, Zubrowski R, Ferkany JW, Karbon EW (1994) Synthesis and glutamate antagonist activity of 4-phosphonoalkylquinoline derivatives: a novel class of non-NMDA antagonist. Bioorg & Med Chem Letts 4:2035–2040

Hamilton GS, Huang Z, Patch RJ, Guzewska ME, Elliott RL, Borosky SA, Bednar DL, Ferkany JW, Karbon EW (1992) Phosphonoethylphenylalanine derivatives as novel antagonists of non-NMDA ionotropic glutamate receptors. Bioorg & Med Chem Letts 2:1269–1274

Hampson RE, Roger G, Lynch G, Deadwyler SA (1998a) Facilitative effects of the ampakine CX516 on short-term memory in rats: enhancement of delayed-non-match-to-sample performance. J Neurosci 18:2740–2747

Hampson RE, Roger G, Lynch G, Deadwyler SA (1998b) Facilitative effects of the ampakine CX516 on short-term memory in rats: correlations with hippocampal neuronal activity. J Neurosci 18:2748–2763

Harman RJ, Shaw GG (1981) High-affinity uptake of spermine by slices of rat cerebral cortex. J Neurochem 36:1609–615

Harman RJ, Shaw GG (1981) The spontaneous and evoked release of spermine from rat brain in vitro. Br J Pharmacol 73:165–174

Harris EW, Ganong AH, Cotman CW (1984) Long-term potentiation in the hippocampus involves activation of N- methyl-D-aspartate receptors. Brain Res 323:132–137

Harris EW, Ganong AH, Monaghan DT, Watkins JC, Cotman CW (1986) Action of 3-((+/-)-2-carboxypiperazin-4-yl)-propyl-1-phosphonic acid (CPP): a new and highly potent antagonist of N-methyl-D-aspartate receptors in the hippocampus. Brain Res 382:174–177

Hashimoto A, Nishikawa T, Oka T, Takahashi K (1993) Endogenous D-serine in rat brain: N-methyl-D-aspartate receptor-related distribution and aging. J Neurochem 60:783–786

Hashimoto A, Oka T (1997) Free D-aspartate and D-serine in the mammalian brain and periphery. Prog Neurobiol 52:325–353

Hawkins LM, Beaver KM, Jane DE, Taylor PM, Sunter DC, Roberts PJ (1995) Characterisation of the pharmacology and regional distribution of (S)-[^{3}H]-5-fluorowillardiine binding in rat brain. Br J Pharmacol 116:2033–2039

Heckendorn R, Allgeier H, Baud J, Gunzenhauser W, Angst C (1993) Synthesis and binding properties of 2-amino-5-phosphono-3-pentenoic acid photoaffinity ligands as probes for the glutamate recognition site of the NMDA receptor. J Med Chem 36:3721–3726

Heresco-Levy U, Javitt DC (1998) The role of N-methyl-D-aspartate (NMDA) receptor-mediated neurotransmission in the pathophysiology and therapeutics of psychiatric syndromes [In Process Citation]. Eur Neuropsychopharmacol 8:141–152

Herlitze S, Raditsch M, Ruppersberg JP, Jahn W, Monyer H, Schoepfer R, Witzemann V (1993) Argiotoxin detects molecular differences in AMPA receptor channels. Neuron 10:1131–1140

Herrling PL, Emre M, Watkins JC (1997) D-CPPene (SDZ EAA-494)-A competitive NMDA antagonist: Pharmacology and results in humans. In: Herrling P (ed) Excitatory amino acids clinical results with antagonists. Academic press, London, p 7

Hershkowitz N, Rogawski MA (1989) Cycloleucine blocks NMDA responses in cultured hippocampal neurones under voltage clamp: antagonism at the strychnine-insensitive glycine receptor. Br J Pharmacol 98:1005–1013

Hess SD, Daggett LP, Crona J, Deal C, Lu CC, Urrutia A, Chavez-Noriega L, Ellis SB, Johnson EC, Velicelebi G (1996) Cloning and functional characterization of human heteromeric N-methyl-D-aspartate receptors. J Pharmacol Exp Ther 278:808–816

Hirai H, Kirsch J, Laube B, Betz H, Kuhse J (1996) The glycine binding site of the N-methyl-D-aspartate receptor subunit NR1: identification of novel determinants

of co-agonist potentiation in the extracellular M3-M4 loop region. Proc Natl Acad Sci USA 93:6031–6036
Hochman S, Jordan LM, MacDonald JF (1994) N-methyl-D-aspartate receptor-mediated voltage oscillations in neurons surrounding the central canal in slices of rat spinal cord. J Neurophysiol 72:565–577
Hollmann M, Boulter J, Maron C, Beasley L, Sullivan J, Pecht G, Heinemann S (1993) Zinc potentiates agonist-induced currents at certain splice variants of the NMDA receptor. Neuron 10:943–954
Honer M, Benke D, Laube B, Kuhse J, Heckendorn R, Allgeier H, Angst C, Monyer H, Seeburg PH, Betz H, Mohler H (1998) Differentiation of glycine antagonist sites of N-methyl-D-aspartate receptor subtypes. Preferential interaction of CGP 61594 with NR1/2B receptors. J Biol Chem 273:11158–11163
Honore T, Davies SN, Drejer J, Fletcher EJ, Jacobsen P, Lodge D, Nielsen FE (1988) Quinoxalinediones: potent competitive non-NMDA glutamate receptor antagonists. Science 241:701–703
Honore T, Drejer J, Nielsen EO, Nielsen M (1989) Non-NMDA glutamate receptor antagonist 3H-CNOX binds with equal affinity to two agonist states of quisqualate receptors. Biochem Pharmacol 38:3207–3212
Hood WF, Compton RP, Monahan JB (1989) D-cycloserine: a ligand for the N-methyl-D-aspartate coupled glycine receptor has partial agonist characteristics. Neurosci Lett 98:91–95
Hood WF, Sun ET, Compton RP, Monahan JB (1989) 1-Aminocyclobutane-1-carboxylate (ACBC): a specific antagonist of the N-methyl-D-aspartate receptor coupled glycine receptor. Eur J Pharmacol 161:281–282
Huettner JE (1990) Glutamate receptor channels in rat DRG neurons: Activation by kainate and quisqualate and blockade of desensitization by Con A. Neuron 5:255–266
Huettner JE, Bean BP (1988) Block of N-methyl-D-aspartate-activated current by the anticonvulsant MK-801: selective binding to open channels. Proc Natl Acad Sci U S A 85:1307–1311
Hurt SD, Baron BM (1991) [3H] 5,7-dichlorokynurenic acid, a high affinity ligand for the NMDA receptor glycine regulatory site. Recept Res 11:215–220
Igarashi K, Williams K (1995) Antagonist properties of polyamines and bis(ethyl)polyamines at N-methyl-D-aspartate receptors. J Pharmacol Exp Ther 272:1101–1109
Ikeda K, Nagasawa M, Mori H, Araki K, Sakimura K, Watanabe M, Inoue Y, et al (1992) Cloning and expression of the epsilon 4 subunit of the NMDA receptor channel. FEBS Lett 313:34–38
Ingvar M, Ambros-ingerson J, Davis M, Granger R, Kessler M, Roger GA, Schehr RS, Lynch G (1997) Enhancement by an ampakine of memory encoding in humans. Exp Neurol 146:553–559
Ikeda K, Nagasawa M, Mori H, Araki K, Sakimura K, Watanabe M, Inoue Y, et al (1992) Cloning and expression of the epsilon 4 subunit of the NMDA receptor channel. FEBS Lett 313:34–38
Ishii T, Moriyoshi K, Sugihara H, Sakurada K, Kadotani H, Yokoi M, Akazawa C, et al (1993) Molecular characterization of the family of the N-methyl-D-aspartate receptor subunits. J Biol Chem 268:2836–2843
Ito I, Tanabe S, Kohda A, Sugiyama H (1990) Allosteric potentiation of quisqualate receptors by a nootropic drug aniracetam. J Physiol 424:533–543
Jackson H, Parks TN (1989) Spider toxins: recent applications in neurobiology. Annu Rev Neurosci 12:405–414
Jackson H, Usherwood PN (1988) Spider toxins as tools for dissecting elements of excitatory amino acid transmission. Trends Neurosci 11:278–83
Jacobson W, Cottrell GA (1993) Rapid visualization of NMDA receptors in the brain: characterization of (+)-3-[125I]-iodo-MK-801 binding to thin sections of rat brain. J Neurosci Methods 46:17–25

Jane DE, Hoo K, Kamboj R, Deverill M, Bleakman D, Mandelzys A (1997) Synthesis of willardiine and 6-azawillardiine analogs: pharmacological characterization on cloned homomeric human AMPA and kainate receptor subtypes. J Med Chem 40:3645–3650

Jane DE, Olverman HJ, Watkins JC (1994). Agonists and competitive antagonists: structure-activity and molecular modelling studies. In: Collingridge G, Watkins J (eds) The NMDA receptor. Oxford University Press, Oxford, pp 31–104

Jane DE, Pook PC-K, Sunter DC, Udvarhelyi PM, Watkins JC (1991) New willardiine analogues with potent, stereoselective actions on mammalian spinal neurones. Br J Pharamacol 104:333P

Jones KA, Wilding TJ, Huettner JE, Costa AM (1997) Desensitization of kainate receptors by kainate, glutamate and diastereoisomers of 4-methylglutamate. Neuropharmacology 36:853–863

Johnson JW, Ascher P (1987) Glycine potentiates the NMDA response in cultured mouse brain neurons. Nature 325:529–531

Karcz-Kubicha M, Jessa M, Nazar M, Plaznik A, Hartmann S, Parsons CG, Danysz W (1997) Anxiolytic activity of glycine-B antagonists and partial agonists– no relation to intrinsic activity in the patch clamp. Neuropharmacology 36:1355–1367

Kashiwagi K, Fukuchi J, Chao J, Igarashi K, Williams K (1996) An aspartate residue in the extracellular loop of the N-methyl-D-aspartate receptor controls sensitivity to spermine and protons. Mol Pharmacol 49:1131–1341

Kashiwagi K, Pahk AJ, Masuko T, Igarashi K, Williams K (1997) Block and modulation of N-methyl-D-aspartate receptors by polyamines and protons: role of amino acid residues in the transmembrane and pore- forming regions of NR1 and NR2 subunits. Mol Pharmacol 52:701–713

Kawai M, Horikawa Y, Ishihara T, Shimamoto K, Ohfune Y (1992) 2-(Carboxycyclopropyl)glycines: binding, neurotoxicity and induction of intracellular free Ca2+ increase. Eur J Pharmacol 211:195–202

Kemp JA, Foster AC, Leeson PD, Priestley T, Tridgett R, Iversen LL, Woodruff GN (1988) 7-Chlorokynurenic acid is a selective antagonist at the glycine modulatory site of the N-methyl-D-aspartate receptor complex. Proc Natl Acad Sci USA 85:6547–6550

Kessler M, Terramani T, Lynch G, Baudry M (1989) A glycine site associated with N-methyl-D-aspartic acid receptors: characterization and identification of a new class of antagonists. J Neurochem 52:1319–1328

Kew JN, Richards JG, Mutel V, Kemp JA (1998) Developmental changes in NMDA receptor glycine affinity and ifenprodil sensitivity reveal three distinct populations of NMDA receptors in individual rat cortical neurons. J Neurosci 18:1935–1943

Kew JN, Trube G, Kemp JA (1998) State-dependent NMDA receptor antagonism by Ro 8–4304, a novel NR2B selective, non-competitive, voltage-independent antagonist. Br J Pharmacol 123:463–472

Kleckner NW, Dingledine R (1988) Requirement for glycine in activation of NMDA-receptors expressed in Xenopus oocytes. Science 241:835–837

Kloog Y, Haring R, Sokolovsky M (1988) Kinetic characterization of the phencyclidine-N-methyl-D-aspartate receptor interaction: evidence for a steric blockade of the channel. Biochemistry 27:843–848

Komuro H, Rakic P (1993) Modulation of neuronal migration by NMDA receptors. Science 260:95–97

Konno K, Shirahama H, Matsumoto T (1983) Isolation and structure of acromelic acid A and B. New kainoids of Clitocybe acromelaga. Tetrahedron Lett 24:939–942

Krogsgaard-Larsen P (1993) Neurotransmitter receptors as pharmacological targets in Alzheimer's disease: design of drugs on a rational basis. In: Drug design for neuroscience Ed. A. P. Kozikowski, Raven Press, New York, USA pp 1–31

Krogsgaard-Larsen P, Hansen JJ, Lauridsen J, Peet MJ, Leah JD, Curtis DR (1982) Glutamic acid agonists, stereochemical and conformational studies of DL-α-amino-3-

hydroxy-5-methyl-4-isoxazolepropionic acid (AMPA) and related compounds. Neurosci Lett 31:313–317
Krogsgaard-Larsen P, Honoré T, Hansen JJ, Curtis DR, Lodge D (1980) New class of glutamate agonist structurally related to ibotenic acid. Nature (London) 284: 64–66
Kuryatov A, Laube B, Betz H, Kuhse J (1994) Mutational analysis of the glycine-binding site of the NMDA receptor: structural similarity with bacterial amino acid-binding proteins. Neuron 12:1291–300
Kutsuwada T, Kashiwabuchi N, Mori H, Sakimura K, Kushiya E, Araki K, Meguro H, Masaki H, Kumanishi T, Arakawa M (1992) Molecular diversity of the NMDA receptor channel. Nature 358:36–41
Largent BL, Gundlach AL, Snyder SH (1986) Pharmacological and autoradiographic discrimination of sigma and phencyclidine receptor binding sites in brain with (+)-[3H]SKF 10,047, (+)-[3H]-3-[3-hydroxyphenyl]-N-(1-propyl)piperidine and [3H]-1-[1-(2- thienyl)cyclohexyl]piperidine. J Pharmacol Exp Ther 238:739–748
Laube B, Hirai H, Sturgess M, Betz H, Kuhse J (1997) Molecular determinants of agonist discrimination by NMDA receptor subunits: analysis of the glutamate binding site on the NR2B subunit. Neuron 18:493–503
Laube B, Kuhse J, Betz H (1998) Evidence for a tetrameric structure of recombinant NMDA receptors. J Neurosci 18:2954–2961
Lees KR (1997) Cerestat and other NMDA antagonists in ischemic stroke. Neurology 49:S66–S69
Leeson PD, Baker R, Carling RW, Curtis NR, Moore KW, Williams BJ, Foster AC, Donald AE, Kemp JA, Marshall GR (1991) Kynurenic acid derivatives. Structure-activity relationships for excitatory amino acid antagonism and identification of potent and selective antagonists at the glycine site on the N-methyl-D-aspartate receptor. J Med Chem 34:1243–1252
Leeson PD, Carling RW, Moore KW, Moseley AM, Smith JD, Stevenson G, Chan T, Baker R, Foster AC, Grimwood S (1992) 4-Amido-2-carboxytetrahydroquinolines. Structure-activity relationshipsfor antagonism at the glycine site of the NMDA receptor. J Med Chem 35:1954–1968
Leeson PD, James K, Carling RW, Wong EH, Baker R (1990) MK-801 analogues: molecular properties required for recognition by the N-methyl-D-aspartate receptor. Prog Clin Biol Res 361:513–518
Legendre P, Westbrook GL (1991) Ifenprodil blocks N-methyl-D-aspartate receptors by a two-component mechanism. Mol Pharmacol 40:289–298
Lehmann J, Hutchison AJ, McPherson SE, Mondadori C, Schmutz M, Sinton CM, Tsai C, et al (1988) CGS 19755, a selective and competitive N-methyl-D-aspartate-type excitatory amino acid receptor antagonist. J Pharmacol Exp Ther 246:65–75
Lerma J (1992) Spermine regulates N-methyl-D-aspartate receptor desensitization. Neuron 8:343–52
Lerma J, Zukin RS, Bennett MV (1990) Glycine decreases desensitization of N-methyl-D-aspartate (NMDA) receptors expressed in Xenopus oocytes and is required for NMDA responses. Proc Natl Acad Sci USA 87:2354–2358
Lester RA, Tong G, Jahr CE (1993) Interactions between the glycine and glutamate binding sites of the NMDA receptor. J Neurosci 13:1088–1096
Lipton SA, Rosenberg PA (1994) Excitatory amino acids as a final common pathway for neurologic disorders [see comments]. N Engl J Med 330:613–622
Loscher W, Wlaz P, Szabo L (1998) Focal ischemia enhances the adverse effect potential of N-methyl-D- aspartate receptor antagonists in rats. Neurosci Lett 240: 33–6
Lowe DA, Emre M, Frey P, Kelly PH, Malanowski J, McAllister KH, Neijt HC, et al (1994) The pharmacology of SDZ EAA 494, a competitive NMDA antagonist. Neurochem Int 25:583–600
Lubisch W, Behl B, Hofmann HP (1996) Pyrrolylquinoxalinediones: a new class of AMPA receptor antagonists. Bioorg & Med Chem Letts 6:2887–2892

Luo J, Wang Y, Yasuda RP, Dunah AW, Wolfe BB (1997) The majority of N-methyl-D-aspartate receptor complexes in adult rat cerebral cortex contain at least three different subunits (NR1/NR2 A/NR2B). Mol Pharmacol 51:79–86

Madsen U, Bang-Andersen B, Brehm L, Christensen IT, Ebert B, Kristoffersen ITS, Lang Y, Krogsgaard-Larsen P (1996) Synthesis and pharmacology of highly selective carboxy and phosphono isoxazole amino acid AMPA receptor antagonists. J Med Chem 39:1682–1691

Magazanik LG, Buldakova SL, Samoilova MV, Gmiro VE, Mellor IR, Usherwood PNR (1997) Block of open channels of recombinant AMPA receptors and native AMPA/ kainate receptors by adamantane derivatives. J Physiol 505:655–663

Marti T, Benke D, Mertens S, Heckendorn R, Pozza M, Allgeier H, Angst C, et al (1993) Molecular distinction of three N-methyl-D-aspartate-receptor subtypes in situ and developmental receptor maturation demonstrated with the photoaffinity ligand 125I-labeled CGP 55802 A. Proc Natl Acad Sci USA 90:8434–8438

Marvizon JC, Lewin AH, Skolnick P (1989) 1-Aminocyclopropane carboxylic acid: a potent and selective ligand for the glycine modulatory site of the N-methyl-D-aspartate receptor complex. J Neurochem 52:992–994

Matsui T, Sekiguchi M, Hashimoto A, Tomita U, Nishikawa T, Wada K (1995) Functional comparison of D-serine and glycine in rodents: the effect on cloned NMDA receptors and the extracellular concentration. Neurochem 65:454–458

Mayer ML, Vyklicky L Jr, Clements J (1989) Regulation of NMDA receptor desensitization in mouse hippocampal neurons by glycine. Nature 338:425–427

McBain CJ, Mayer ML (1994) N-methyl-D-aspartic acid receptor structure and function. Physiol Rev 74:723–760

McDonald JW, Penney JB, Johnston MV, Young AB (1990) Characterization and regional distribution of strychnine-insensitive [3H]glycine binding sites in rat brain by quantitative receptor autoradiography. Neuroscience 35:653–668

McGurk JF, Bennett MV, Zukin RS (1990) Polyamines potentiate responses of N-methyl-D-aspartate receptors expressed in xenopus oocytes. Proc Natl Acad Sci USA 87:9971–9974

McLennan H, Lodge D (1979) The antagonism of amino acid-induced excitation of spinal neurons in the cat. Brain Res 169:83–90

Meguro H, Mori H, Araki K, Kushiya E, Kutsuwada T, Yamazaki M, Kumanishi T, et al (1992) Functional characterization of a heteromeric NMDA receptor channel expressed from cloned cDNAs. Nature 357:70–74

Meldrum B, Garthwaite J (1990) Excitatory amino acid neurotoxicity and neurodegenerative disease. Trends Pharmacol Sci 11:379–87

Meldrum B, Kerwin RW (1987) Glutamate receptors and schizophrenia. J Phychopharmacol 1:217–221

Mercer LD, Jarrott B, Beart PM (1993) 125I-ifenprodil: synthesis and characterization of binding to a polyamine-sensitive site in cerebral cortical membranes. J Neurochem 61:120–126

Mitchell IJ, Carroll CB (1997) Reversal of parkinsonian symptoms in primates by antagonism of excitatory amino acid transmission: potential mechanisms of action. Neurosci Biobehav Rev 21:469–75

Monaghan DT, Andaloro VJ, Skifter DA (1998) Molecular determinants of NMDA receptor pharmacological diversity. Prog Brain Research 116:158–177

Monaghan DT, Beaton JA (1991) Quinolinate differentiates between forebrain and cerebellar NMDA receptors. Eur J Pharmacol 194:123–125

Monaghan DT, Beaton JA (1992) Pharmacologically-distinct NMDA receptor populations of the cerebellum, medial thalamic nuclei and forebrain. Mol Neuropharm 2:71–72

Monaghan DT, Bridges RJ, Cotman CW (1989) The excitatory amino acid receptors: their classes, pharmacology, and distinct properties in the function of the central nervous system. Annu Rev Pharmacol Toxicol 29:365–402

Monaghan DT, Cotman CW (1986) Identification and properties of N-methyl-D-aspartate receptors in rat brain synaptic plasma membranes. Proc Natl Acad Sci USA 83:7532–7536

Monaghan DT, Holets VR, Toy DW, Cotman CW (1983) Anatomical distributions of four pharmacologically distinct 3H-L-glutamate binding sites. Nature 306:176–179

Monaghan DT, Larson H (1997) NR1 and NR2 subunit contributions to N-methyl-D-aspartate receptor channel blocker pharmacology. J Pharmacol Exp Ther 280: 614–620

Monaghan DT, Olverman HJ, Nguyen L, Watkins JC, Cotman CW (1988) Two classes of N-methyl-D-aspartate recognition sites: differential distribution and differential regulation by glycine. Proc Natl Acad Sci U S A 85:9836–9840

Monaghan DT, Yao D, Cotman CW (1985) L-[3H]Glutamate binds to kainate-, NMDA- and AMPA-sensitive binding sites: an autoradiographic analysis. Brain Res 340:378–383

Monahan JB, Biesterfeldt JP, Hood WF, Compton RP, Cordi AA, Vazquez MI, Lanthorn TH, et al. (1990) Differential modulation of the associated glycine recognition site by competitive N-methyl-D-aspartate receptor antagonists. Mol Pharmacol 37:780–784

Monahan JB, Michel J (1987) Identification and characterization of an N-methyl-D-aspartate-specific L-[3H]glutamate recognition site in synaptic plasma membranes. J Neurochem 48:1699–1708

Monyer H, Burnashev N, Laurie DJ, Sakmann B, Seeburg PH (1994) Developmental and regional expression in the rat brain and functional properties of four NMDA receptors. Neuron 12:529–540

Monyer H, Sprengel R, Schoepfer R, Herb A, Higuchi M, Lomeli H, Burnashev N, et al. (1992) Heteromeric NMDA receptors: molecular and functional distinction of subtypes. Science 256:1217–1221

Mori H, Mishina M (1995) Structure and function of the NMDA receptor channel. Neuropharmacology 34:1219–1237

Moriyoshi K, Masu M, Ishii T, Shigemoto R, Mizuno N, Nakanishi S (1991) Molecular cloning and characterization of the rat NMDA receptor [see comments]. Nature 354:31–37

Morris RGM, Anderson E, Lynch GS, Baudry M (1986) Selective impairment of learning and blockade of long-term potentiation by an N-methyl-D-aspartate receptor antagonist, AP5. Nature 319:774–776

Müller W, Lowe DA, Neijt H, Urwyler S, Herrling PL, Blaser D, Seebach D (1992) Synthesis and N-methyl-D-aspartate (NMDA) antagonist properties of the enantiomers of a-amino-5-(phosphonomethyl)[1,1'-biphenyl]-3-propanoic acid. Use of a new chiral glycine derivative. Helvetica Chimica Acta 75:855–864

Murphy DE, Hutchison AJ, Hurt SD, Williams M, Sills MA (1988) Characterization of the binding of [3H]-CGS 19755: a novel N-methyl-D-aspartate antagonist with nanomolar affinity in rat brain. Br J Pharmacol 95:932–938

Murphy DE, Schneider J, Boehm C, Lehmann J, Williams M (1987) Binding of [3H]3-(2-carboxypiperazin-4-yl)propyl-1-phosphonic acid to rat brain membranes: a selective, high-affinity ligand for N-methyl-D-aspartate receptors. J Pharmacol Exp Ther 240:778–784

Mutel V, Buchy D, Klingelschmidt A, Messer J, Bleuel Z, Kemp JA, Richards JG (1998) In vitro binding properties in rat brain of [3H]Ro 25–6981, a potent and selective antagonist of NMDA receptors containing NR2B subunits. J Neurochem 70:2147–2155

Nielsen EØ, Johansen TH, Wätjen F, Jørgen D (1995) Characterization of the binding of [^{3}H]NS 257, a novel competitive AMPA receptor antagonist, to rat brain membranes and brain sections. J Neurochem 65:1264–1273

Nordholm L, Sheardown M, Honore T (1997) The NBQX story. In: Herrling P (ed) Excitatory amino acids clinical results with antagonists. Academic press, London, p 89

O'Callaghan D, Wong MG, Beart PM (1992) Molecular modelling of N-methyl-D-aspartate receptor agonists. Mol Neuropharmacol 2:89–92

Ogita K, Suzuki T, Enomoto R, Ohgaki T, Katagawa J, Uchida S, Meguri H, et al (1990) Profiles of [3H]N-[1-(2-thienyl)cyclohexyl]piperidine binding in brain synaptic membranes treated with Triton X-100. Neurosci Res (N Y) 9:35–47

Ohmori J, Sakamoto S, Kubota H, Shimizu-Sasamata M, Okada M, Kawasaki S, Hidaka K, Togami J, Furuya T, Murase K (1994) 6-(1*H*-Imidazol-1-yl)-7-nitro-2,3-(1*H*,4*H*)-quinoxalinedione hydrochloride (YM90K) and related compounds: structure-activity relationships for the AMPA-type non-NMDA receptor. J Med Chem 37:467–475

Ohmori J, Shimizu-Sasamata M, Okada M, Sakamoto S (1996) Novel AMPA receptor antagonists: synthesis and structure-activity relationships of 1-hydroxy-7-(1*H*-imidazol-1-yl)-6-nitro-2,3(1*H*,4*H*)-quinoxalinedione and related compounds. J Med Chem 39:3971–3979

Ohmori J, Shimizu-Sasamata M, Okada M, Sakamoto S (1997) 8-(1*H*-imidazol-1-yl)-7-nitro-4(5*H*)-imidazo[1,2-a]quinoxalinone and related compounds: synthesis and structure-activity relationships for the AMPA-type non-NMDA receptor. J Med Chem 40:2053–2063

Okada M, Kohara A, Yamaguchi T (1996) Characterization of YM90K, a selective and potent antagonist of AMPA receptors, in rat cortical mRNA-injected *Xenopus* oocytes. Eur J Pharmacol 309:299–306

Olney JW (1994) Neurotoxicity of NMDA receptor antagonists: an overview. Psychopharmacol Bull 30:533–540

Olney JW, Labruyere J, Wang G, Wozniak DF, Price MT, Sesma MA (1991) NMDA antagonist neurotoxicity: mechanism and prevention. Science 254:1515–1518

Olverman HJ, Jones AW, Mewett KN, Watkins JC (1988) Structure/activity relations of N-methyl-D-aspartate receptor ligands as studied by their inhibition of [3H]D-2-amino-5-phosphonopentanoic acid binding in rat brain membranes. Neuroscience 26:17–31

Olverman HJ, Jones AW, Watkins JC (1984) L-glutamate has higher affinity than other amino acids for [3H]-D-AP5 binding sites in rat brain membranes. Nature 307:460–462

Olverman HJ, Monaghan DT, Cotman CW, Watkins JC (1986) [3H]CPP, a new competitive ligand for NMDA receptors. Eur J Pharmacol 131:161–162

Ornstein PL, Arnold MB, Allen NK, Bleisch T, Borromeo PS, Lugar CW, Leander JD, Lodge D, Schoepp DD (1996a) Structure-activity studies of 6-(tetrazolylalkyl)-substituted decahydroisoquinoline-3-carboxylic acid AMPA receptor antagonists. 1. Effects of stereochemistry, chain length, and chain substitution. J Med Chem 39:2219–2231

Ornstein PL, Arnold MB, Allen NK, Bleisch T, Borromeo PS, Lugar CW, Leander JD, Lodge D, Schoepp DD (1996b) Structure-activity studies of 6-substituted decahydroisoquinoline-3-carboxylic acid AMPA receptor antagonists. 2. Effects of distal acid bioisosteric substitution, absolute stereochemical preferences and in Vivo activity. J Med Chem 39:2232–2244

Ornstein PL, Schoepp DD, Arnold MB, Augenstein NK, Lodge D, Millar JD, Chambers J, et al (1992) 6-substituted decahydroisoquinoline-3-carboxylic acids as potent and selective conformationally constrained NMDA receptor antagonists. J Med Chem 35:3547–6350

Ornstein PL, Schoepp DD, Arnold MB, Leander JD, Lodge D, Paschal JW, Elzey T (1991) 4-(Tetrazolylalkyl)piperidine-2-carboxylic acids. Potent and selective N-methyl-D-aspartic acid receptor antagonists with a short duration of action. J Med Chem 34:90–97

Ortwine DF, Malone TC, Bigge CF, Drummond JT, Humblet C, Johnson G, Pinter GW (1992) Generation of N-methyl-D-aspartate agonist and competitive antagonist pharmacophore models. Design and synthesis of phosphonoalkyl-substituted tetrahydroisoquinolines as novel antagonists. J Med Chem 35:1345–1370

Palmer GC, Cregan EF, Borrelli AR, Willett F (1995) Neuroprotective properties of the uncompetitive NMDA receptor antagonist remacemide hydrochloride. Ann N Y Acad Sci 765:236–47; discussion 248

Paoletti P, Ascher P, Neyton J (1997) High-affinity zinc inhibition of NMDA NR1-NR2 A receptors [published erratum appears in J Neurosci 1997 Oct 15;17(20):following table of contents]. J Neurosci 17:5711–5725

Parks TN, Mueller AL, Artman LD, Albensi BC, Nemeth EF, Jackson H, Jasys VJ, et al (1991) Arylamine toxins from funnel-web spider (Agelenopsis aperta) venom antagonize N-methyl-D-aspartate receptor function in mammalian brain. J Biol Chem 266:21523–21529

Parsons CG, Zong X, Lux HD (1993) Whole cell and single channel analysis of the kinetics of glycine-sensitive N-methyl-D-aspartate receptor desensitization. Br J Pharmacol 109:213–221

Paschen W (1992) Polyamine metabolism in reversible cerebral ischemia. Cerebrovasc Brain Metab Rev 4:59–88

Partin KM, Patneau DK, Winters CA, Mayer ML, Buonanno A (1993) Selective modulation of desensitisation at AMPA versus kainate receptors by cyclothiazide and concanavlin A Neuron 11:1069–1082

Patneau DK, Mayer ML, Jane DE, Watkins JC (1992) Activation and desensitization of AMPA/kainate receptors by novel derivatives of willardiine. J Neurosci 12:595–606

Patneau DK, Vyklicky L, Mayer ML (1993) Hippocampal neurons exhibit cyclot iazide-sensitive rapidly desensitising responses to kainate. J Neurosci 13:3496–3509

Pellegrini-Giampietro DE, Gorter JA, Bennett MVL, Zukin RS (1997) The GluR2 (GluR-B) hypothesis: Ca^{2+}-permeable AMPA receptors in neurological disorder. Trends Neurosci 20:464–470

Pelletier JC, Hesson DP, Jones KA, Costa AM (1996) Substituted 1,2-dihydrophthalazines: potent, selective, and non-competitive inhibitors of the AMPA receptor. J Med Chem 39:343–346

Porter RH, Greenamyre JT (1995) Regional variations in the pharmacology of NMDA receptor channel blockers: implications for therapeutic potential. J Neurochem 64:614–23

Priestley T, Kemp JA (1994) Kinetic study of the interactions between the glutamate and glycine recognition sites on the N-methyl-D-aspartic acid receptor complex. Mol Pharmacol 46:1191–1196

Priestley T, Laughton P, Macaulay AJ, Hill RG, Kemp JA (1996) Electrophysiological characterisation of the antagonist properties of two novel NMDA receptor glycine site antagonists, L-695,902 and L-701,324. Neuropharmacology 35: 1573–1581

Priestley T, Laughton P, Myers J, Le Bourdelles B, Kerby J, Whiting PJ (1995) Pharmacological properties of recombinant human N-methyl-D-aspartate receptors comprising NR1a/NR2A and NR1a/NR2B subunit assemblies expressed in permanently transfected mouse fibroblast cells. Mol Pharmacol 48:841–848

Priestley T, Woodruff GN, Kemp JA (1989) Antagonism of responses to excitatory amino acids on rat cortical neurones by the spider toxin, argiotoxin636. Br J Pharmacol 97:1315–1323

Raditsch M, Ruppersberg JP, Kuner T, Gunther W, Schoepfer R, Seeburg PH, Jahn W, et al. (1993) Subunit-specific block of cloned NMDA receptors by argiotoxin636. FEBS Lett 324:63–66

Ragsdale D, Gant DB, Anis NA, Eldefrawi AT, Eldefrawi ME, Konno K, Miledi R (1989) Inhibition of rat brain glutamate receptors by philanthotoxin. J Pharmacol Exp Ther 251:156–163

Ransom RW, Deschenes NL (1990) Polyamines regulate glycine interaction with the N-methyl-D-aspartate receptor. Synapse 5:294–298

Ransom RW, Stec NL (1988) Cooperative modulation of [3H]MK-801 binding to the N-methyl-D- aspartate receptor-ion channel complex by L-glutamate, glycine, and polyamines. J Neurochem 51:830–836

Reynolds IJ, Miller RJ (1988) [3H]MK801 binding to the N-methyl-D-aspartate receptor reveals drug interactions with the zinc and magnesium binding sites. J Pharmacol Exp Ther 247:1025–1031

Reynolds IJ, Miller RJ (1988) Multiple sites for the regulation of the N-methyl-D-aspartate receptor. Mol Pharmacol 33:581–584

Reynolds IJ, Miller RJ (1989) Ifenprodil is a novel type of N-methyl-D-aspartate receptor antagonist: interaction with polyamines. Mol Pharmacol 36:758–765

Reynolds IJ, Murphy SN, Miller RJ (1987) 3H-labeled MK-801 binding to the excitatory amino acid receptor complex from rat brain is enhanced by glycine. Proc Natl Acad Sci USA 84:7744–7748

Reynolds IJ, Rothermund KD (1995) Characterization of the effects of polyamines on the modulation of the N-methyl-D-aspartate receptor by glycine. Neuropharmacology 34:1147–1157

Rock DM, Macdonald RL (1992) The polyamine spermine has multiple actions on N-methyl-D-aspartate receptor single-channel currents in cultured cortical neurons. Mol Pharmacol 41:83–88

Rock DM, MacDonald RL (1992) Spermine and related polyamines produce a voltage-dependent reduction of N-methyl-D-aspartate receptor single-channel conductance. Mol Pharmacol 42:157–164

Rodriguez Paz JM, Anantharam V, Treistman SN (1995) Block of the N-methyl-D-aspartate receptor by phencyclidine-like drugs is influenced by alternative splicing. Neurosci Lett 190:147–150

Rogawski MA (1993) Therapeutic potential of excitatory amino acid antagonists: channel blockers and 2,3-benzodiazepines. Trends Pharmacol Sci 14:325–331

Russell DH, Gfeller E (1974) Distribution of putrescine, spermidine, and spermine in rhesus monkey brain: decrease in spermidine and spermine concentrations in motor cortex after electrical stimulation. J Neurobiol 5:349–354

Sacaan AI, Johnson KM (1989) Spermine enhances binding to the glycine site associated with the N- methyl-D-aspartate receptor complex. Mol Pharmacol 36: 836–839

Sakai R, Kamiya H, Murata M, Shimamoto K (1997) Dysiherbaine: A new neurotoxic amino acid from the Micronesian marine sponge *Dysidea herbacea* J Am Chem Soc 119:4112–4116

Sapru HN (1996) Carotid chemoreflex. Neural pathways and transmitters. Adv Exp Med Biol 410:357–364

Schell MJ, Brady RO Jr, Molliver ME, Snyder SH (1997) D-serine as a neuromodulator: regional and developmental localizations in rat brain glia resemble NMDA receptors. Neurosci 17:1604–1615

Schiffer HH, Swanson GT, Heinemann SF (1997) Rat GluR7 and a carboxy-terminal splice variant, GluR7b, are functional kainate receptor subunits with a low sensitivity to glutamate. Neuron 19:1141–1146

Schoemaker H, Allen J, Langer SZ (1990) Binding of [3H]ifenprodil, a novel NMDA antagonist, to a polyamine- sensitive site in the rat cerebral cortex. Eur J Pharmacol 176:249–250

Seiler N, Schmidt-Glenewinkel T (1975) Regional distribution of putrescine, spermidine and spermine in relation to the distribution of RNA and DNA in the rat nervous system. J Neurochem 24:791–795

Sekiguchi M, Fleck MW, Mayer ML, Takeo J, Chiba Y, Yamashita S, Wada K (1997) A novel allosteric potentiator of AMPA receptor: 4-[2-(phenylsulphonylamino)ethylthio]-2,6-difluoro-phenoxyacetamide. J Neurosci 17:5760–5771

Sekiguchi M, Takeo J, Harada T, Morimoto T, Kudo Y, Yamashita S, Kohsaka S, Wada K (1998) Pharmacological detection of AMPA receptor heterogeneity by use of two allosteric potentiators in rat hippocampal cultures. Br J Pharmacol 123: 1294–1303

Shaw GG, Pateman AJ (1973) The regional distribution of the polyamines spermidine and spermine in brain. J Neurochem 20:1225–1230

Sheardown MJ, Nielsen Eø, Hansen AJ, Jacobsen P, Honore T (1990) 2,3-Dihydroxy-6-nitro-7-sulphamoyl-benzo(F)quinoxaline: A neuroprotectant for cerebral ischemia. Science 247:571–574

Sheng M, Cummings J, Roldan LA, Jan YN, Jan LY (1994) Changing subunit composition of heteromeric NMDA receptors during development of rat cortex. Nature 368:144–147

Shimamoto H, Ohfune Y (1996) Syntheses and conformational analyses of glutamate analogs: 2-(2-Carboxy-3-substituted-cyclopropyl)glycines as useful probes for excitatory amino acid receptors. J Med Chem 39:407–423
Shimizu-Sasamata M, Kawasaki-Yatsugi S, Okada M, Sakamoto S, Yatsugi S, Togami J, Hatanaka K, Ohmori J, Koshiya K, Usuda S, Murase K (1996) YM90K: Pharmacological characterisation as a selective and potent α-amino-3-hydroxy-5-methylisoxazole-4-propionate (AMPA)/kainate receptor antagonist. JPET 276: 84–92
Shinozaki H, Ishida M, Shimamoto K, Ohfune Y (1989) A conformationally restricted analogue of L-glutamate, the (2S,3R,4S) isomer of L-alpha-(carboxycyclopropyl)glycine, activates the NMDA-type receptor more markedly than NMDA in the isolated rat spinal cord. Brain Res 480:355–359
Sills MA, Fagg G, Pozza M, Angst C, Brundish DE, Hurt SD, Wilusz EJ, et al (1991) [3H]CGP 39653: a new N-methyl-D-aspartate antagonist radioligand with low nanomolar affinity in rat brain. Eur J Pharmacol 192:19–24
Simmons RMA, Li DL, Hoo KH, Deverill M, Ornstein PL, Iyengar S (1998) Kainate GluR5 receptor mediates the nociceptive response to formalin in the rat. Neuropharmacology 37:25–36
Singer W (1990) The formation of cooperative cell assemblies in the visual cortex. J Exp Biol 153:177–197
Sløk FA, Ebert B, Lang Y, Krogsgaard-Larsen P, Lenz SM, Madsen U (1997) Excitatory amino-acid receptor agonists. Synthesis and pharmacology of analogues of 2-amino-3-(3-hydroxy-5-methylisoxazol-4-yl)propionic acid. Eur J Med Chem 32:329–338
Small, BG, Baker SR, Rubio A, Ballyk BA, Hoo KE, Mandelzys A, Kamboj R, Lodge D, Bleakman D (1997) LY339434, a GluR5 selective kainate receptor agonist. Br J Pharmacol 122:P59
Smith SE, Meldrum BS (1992) Cerebroprotective effect of a non-N-methyl-D-aspartate antagonists, GYKI 52466, after focal ischemia in the rat. Stroke 23: 861–864
Soloviev MM, Barnard EA (1997) Xenopus oocytes express a unitary glutamate receptor endogenously. J Mol Biol 273:14–18
Sommer B, Keinanen K, Verdoon TA, Wisden W, Burnashev N, Herb A, Kohler M, Takagi T, Sakamann B, Seeburg PH (1990) Flip and flop: a cell specific functional switch in glutamate-operated channels of the CNS. Science 249:1580–1585
Sonders MS, Barmettler P, Lee JA, Kitahara Y, Keana JF, Weber E (1990) A novel photoaffinity ligand for the phencyclidine site of the N-methyl- D-aspartate receptor labels a Mr 120,000 polypeptide. J Biol Chem 265:6776–6781
Stocca G, Vicini S (1998) Increased contribution of NR2A subunit to synaptic NMDA receptors in developing rat cortical neurons. J Physiol (Lond) 507:13–24
Subramanyam C, Ault B, Sawutz D, Bacon ER, Singh B, Pennock PO, Kelly MD, Kraynak M, Krafte D, Treasurywala A (1995) 6-(4-Pyridinyl)-1H-1,2,3-triazolo[4,5-d]-pyrimidin-4(5H)-one: a structurally novel competitive AMAP receptor antagonist. J Med Chem 38:587–589
Sucher NJ, Akbarian S, Chi CL, Leclerc CL, Awobuluyi M, Deitcher DL, Wu MK, et al. (1995) Developmental and regional expression pattern of a novel NMDA receptor-like subunit (NMDAR-L) in the rodent brain. J Neurosci 15:6509–6520
Sugihara H, Moriyoshi K, Ishii T, Masu M, Nakanishi S (1992) Structures and properties of seven isoforms of the NMDA receptor generated by alternative splicing. Biochem Biophys Res Commun 185:826–832
Szatkowski M, Attwell D (1994) Triggering and execution of neuronal death in brain ischemia – 2 phases of glutamate release by different mechanisms. Trends Neurosci 17:359–365
Takahashi M, Ni JW, Kawasaki-Yatsugi S, Toya T, Yatsugi S-I, Shimizu-Sasamata M, Koshiya K, Shishikura J-I, Sakamoto S, Yamaguchi T (1998) YM872, a novel selec-

tive α-amino-3-hydroxy-5-methylisoxazole-4-propionic acid receptor antagonist, reduces brain damage after permanent focal cerebral ischemia in cats. JPET 284:467–473

Takemoto T (1978) Isolation and structural identification of naturally occurring excitatory amino acids. In McGreer EG, Olney JW, McGreer Pl eds. Kainic acid as a tool in neurobiology. New York: Raven; 1–15

Tamminga CA (1998) Schizophrenia and glutamatergic transmission. Crit Rev Neurobiol 12:21–36

Tamura Y, Sato Y, Yokota T, Akaike A, Sasa M, Takaori S (1993) Ifenprodil prevents glutamate cytotoxicity via polyamine modulatory sites of N-methyl-D-aspartate receptors in cultured cortical neurons. J Pharmacol Exp Ther 265: 1017–1025

Tang CM, Dichter M, Morad M (1990) Modulation of the N-methyl-D-aspartate channel by extracellular H+. Proc Natl Acad Sci USA 87:6445–6449

Thomas NK, Clayton P, Jane DE (1997) Dicarboxyphenylglycines antagonize AMPA- but not kainate-induced depolarizations in neonatal rat motoneurones. Eur J Pharm 338:111–116

Thomas NK, Hawkins LM, Troop HM, Miller JC, Roberts PJ, Jane DE (1998) Pharmacological differentiation of kainate receptors on neonatal rat spinal motoneurones and dorsal roots. Neuropharmacology (In Press)

Traynelis SF, Hartley M, Heinemann SF (1995) Control of proton sensitivity of the NMDA receptor by RNA splicing and polyamines. Science 268:873–8736

Turski L, Huth A, Sheardown MJ, Jacobsen P, Ottow E (1996) Pharmacology of ZK200775 and ZK202000, competitive non-NMDA glutamate receptor antagonists. Soc. Neurosci Abstr 22:1529

Uchino S, Nakajima-Iijima S, Okuda K, Mishina M, Kawamoto S (1997) Analysis of the glycine binding domain of the NMDA receptor channel zeta 1 subunit. Neuroreport 8:445–449

Urwyler S, Laurie D, Lowe DA, Meier CL, Muller W (1996) Biphenyl-derivatives of 2-amino-7-phosphonoheptanoic acid, a novel class of potent competitive N-methyl-D-aspartate receptor antagonist–I. Pharmacological characterization in vitro. Neuropharmacology 35:643–654

Varney MA, Rao SP, Jachec C, Deal C, Hess SD, Daggett LP, Lin FF, Johnson EC, Veliÿelebi G (1998) Pharmacological characterization of the human ionotropic glutamate receptor subtype GluR3 stably expressed in mammalian cells. JPET 285:358–370

Verdoon TA, Johansen TH, Drejer J, Nielsen EØ (1994) Selective block of recombinant GluR6 receptors by NS-102, a novel non-NMDA receptor antagonist. Eur J Pharmacol 269:43–49

Vignes M, Bleakman D, Lodge D, Collingridge GL (1997) The synaptic activation of the GluR5 subtype of kainate receptor in area CA3 of the rat hippocampus. Neuropharmacology 36:1477–1481

Wahl P, Anker C, Traynelis SF, Egebjerg J, Rasmussen JS, Krogsgaard-Larsen P, Madsen U (1998) Antagonist properties of a phosphono isoxazole amino acid at glutamate R1–4 (*R,S*)-2-amino-3-(3-hydroxy-5-methyl-4-isoxazolyl)propionic acid receptor subtypes. Molecular Pharmacol 53:590–596

Washburn MS, Dingledine R (1996) Block of α-amino-3-hydroxy-5-methyl-4-isoxazolepropanoic acid (AMPA) receptors by polyamines and polyamine toxins. J Pharmacol Exp Ther 278:669–678

Wätjen F, Bigge CF, Jensen LH, Boxer PA, Lescosky LJ, Nielsen Eø, Malone TC, Campbell GW, Coughenour LL, Rock DM, Drejer J, Marcoux FW (1994) NS 257 (1,2,3,6,7,8-Hexahydro-3-(hydroxyimino)-*N,N*,7-trimethyl-2-oxobenzo[2,1-b:3,4-c']dipyrrole-5-sulfonamide) is a potent systemically active AMPA receptor antagonist. Bioorg & Med Chem Letts 4:371–376

Watkins JC, Evans RH (1981) Excitatory amino acid transmitters. Annu Rev Pharmacol Toxicol 21:165–204

Watkins JC, Krogsgaard Larsen P, Honore T (1990) Structure-activity relationships in the development of excitatory amino acid receptor agonists and competitive antagonists. Trends Pharmacol Sci 11:25–33

Watkins JC, Pook PC, Sunter DC, Davies J, Honore T (1990) Experiments with kainate and quisqualate agonists and antagonists in relation to the sub-classification of 'non-NMDA' receptors. Adv Exp Med Biol 268:49–55

Watson GB, Bolanowski MA, Baganoff MP, Deppeler CL, Lanthorn TH (1990) D-cycloserine acts as a partial agonist at the glycine modulatory site of the NMDA receptor expressed in Xenopus oocytes. Brain Res 510:158–160

Watson GB, Lanthorn TH (1990) Pharmacological characteristics of cyclic homologues of glycine at the N-methyl-D-aspartate receptor-associated glycine site. Neuropharmacology 29:727–730

Whitten JP, Harrison BL, Weintraub HJ, McDonald IA (1992) Modeling of competitive phosphono amino acid NMDA receptor antagonists. J Med Chem 35:1509–1514

Wiesenfeld-Hallin Z (1998) Combined opioid-NMDA antagonist therapies. What advantages do they offer for the control of pain syndromes? Drugs 55:1–4

Wilding TJ, Heuttner JE (1996) Antagonist pharmacology of kainate- and α-amino-3-hydroxy-5-methyl-4-isoxazolepropionic acid-preferring receptors. Mol Pharmacol 49:540–546

Williams K (1993) Effects of Agelenopsis aperta toxins on the N-methyl-D-aspartate receptor: polyamine-like and high-affinity antagonist actions. J Pharmacol Exp Ther 266:231–236

Williams K (1993) Ifenprodil discriminates subtypes of the N-methyl-D-aspartate receptor: selectivity and mechanisms at recombinant heteromeric receptors. Mol Pharmacol 44:851–859

Williams K (1994) Mechanisms influencing stimulatory effects of spermine at recombinant N-methyl-D-aspartate receptors. Mol Pharmacol 46:161–168

Williams K (1995) Pharmacological properties of recombinant N-methyl-D-aspartate (NMDA) receptors containing the epsilon 4 (NR2D) subunit. Neurosci Lett 184:181–184

Williams K (1996) Separating dual effects of zinc at recombinant N-methyl-D-aspartate receptors. Neurosci Lett 215:9–12

Williams K (1997) Interactions of polyamines with ion channels [published erratum appears in Biochem J 1997 Sep 15;326(Pt 3):943]. Biochem J 325:289–297

Williams K (1997) Modulation and block of ion channels: a new biology of polyamines. Cell Signal 9:1–13

Williams K, Chao J, Kashiwagi K, Masuko T, Igarashi K (1996) Activation of N-methyl-D-aspartate receptors by glycine: role of an aspartate residue in the M3-M4 loop of the NR1 subunit. Mol Pharmacol 50:701–708

Williams K, Kashiwagi K, Fukuchi J, Igarashi K (1995) An acidic amino acid in the N-methyl-D-aspartate receptor that is important for spermine stimulation. Mol Pharmacol 48:1087–1098

Williams K, Russell SL, Shen YM, Molinoff PB (1993) Developmental switch in the expression of NMDA receptors occurs in vivo and in vitro. Neuron 10:267–78

Williams K, Zappia AM, Pritchett DB, Shen YM, Molinoff PB (1994) Sensitivity of the N-methyl-D-aspartate receptor to polyamines is controlled by NR2 subunits. Mol Pharmacol 45:803–809

Willis CL, Wacker DA, Bartlett RD, Bleakman D, Lodge D, Chamberlin AR, Bridges RJ (1997) Irreversible inhibition of high-affinity [3H]kainate binding by a novel photoactivatable analogue: (2'S,3'S,4'R)-2'-carboxy-4'-(2-diazo-1-oxo-3,3,3-trifluoropropyl)-3'-pyrrolidinyl acetate. J Neurochem 68:1503–1510

Wong LA, Mayer ML, Jane DE, Watkins JC (1994) Willardiines differentiate agonist binding sites for kainate-versus AMPA-preferring glutamate receptors in DRG and hippocampal neurones. J Neurosci 14:3881–3897

Wood PL (1995) The co-agonist concept: is the NMDA-associated glycine receptor saturated in vivo? Life Sci 57:301–310

Wood PL, Hawkinson JE, Goodnough DB (1996) Formation of D-serine from L-phosphoserine in brain synaptosomes. J Neurochem 67:1485–1490
Yamada KA, Rothman SM (1992) Diazoxide reversibly blocks glutamate desensitisation and prolongs excitatory postsynaptic currents in rat hippocampal neurons. J Physiol 458:385–407
Yamada KA, Tang CM (1993) Benzothiadiazides inhibit rapid glutamate receptor desensitisation and enhance glutamatergic synaptic currents. J Neurosci 13: 3904–3915
Yamamoto C, Sawada S, Ohno-Shosaku T (1998) Distribution and properties of kainate receptors distinct in the CA3 region of the hippocampus of the guinea pig. Brain Res 783:227–235
Yamazaki M, Mori H, Araki K, Mori KJ, Mishina M (1992) Cloning, expression and modulation of a mouse NMDA receptor subunit. FEBS Lett 300:39–45
Yenari MA, Bell TE, Kotake AN, Powell M, Steinberg GK (1998) Dose escalation safety and tolerance study of the competitive NMDA antagonist selfotel (CGS 19755) in neurosurgery patients. Clin Neuropharmacol 21:28–34
Zarbin MA, Wamsley JK, Kuhar MJ (1981) Glycine receptor: light microscopic autoradiographic localization with [3H]strychnine. J Neurosci 1:532–547
Zivkovic I, Thompson DM, Bertolino M, Uzunov D, Dibella M, Costa E, Guidotti A (1995) 7-Chloro-3-methyl-3,4-dihydro-2H-1,2,4-benzothiadiazine S,S-dioxide (IDRA-21): A benzothiadiazine derivative that enhances cognition by attenuating DL-α-amino-2,3-dihydro-5-methyl-3-oxo-4-isoxazolepropanoic acid (AMPA) receptor desensitisation. J Pharmacol Exp Ther 272:300–309
Zhang L, Zheng X, Paupard MC, Wang AP, Santchi L, Friedman LK, Zukin RS, et al (1994) Spermine potentiation of recombinant N-methyl-D-aspartate receptors is affected by subunit composition. Proc Natl Acad Sci USA 91:10883–10887
Zheng X, Zhang L, Durand GM, Bennett MV, Zukin RS (1994) Mutagenesis rescues spermine and Zn2+ potentiation of recombinant NMDA receptors. Neuron 12:811–818

CHAPTER 18

Structure, Diversity, Pharmacology, and Pathology of Glycine Receptor Chloride Channels

R.J. HARVEY and H. BETZ

A. Introduction

I. The Neurotransmitter Glycine

Glycine, the simplest of all amino acids, is highly enriched in spinal cord and brain stem compared with other regions of the central nervous system. Classical physiological analysis has revealed that glycine serves as a major inhibitory neurotransmitter in the control of motor and sensory pathways (APRISON 1990). In the nerve terminals of glycinergic interneurons in spinal cord and brain stem, cytosolic glycine is concentrated in small clear synaptic vesicles by an H^+-dependent vesicular transporter. Excitation of the interneurons causes Ca^{2+}-triggered fusion of these synaptic vesicles with the presynaptic plasma membrane, thus initiating glycine release into the synaptic cleft. This results in the activation of postsynaptic glycine receptors (GlyRs) which mediate an increase in chloride conductance by opening an integral anion channel in response to agonist binding. As the chloride equilibrium of mature neurons is close to their resting potential, glycine-mediated Cl^- influx normally antagonizes depolarization by Na^+ influx and thus inhibits the propagation of action potentials. However, glycine can also serve as an excitatory neurotransmitter. Immature neurons in the developing central nervous system often contain very high intracellular chloride concentrations (WANG et al. 1994). In these neurons, glycine-induced increases in chloride conductance cause Cl^- efflux, resulting in membrane depolarization and neurotransmitter release (REICHLING et al. 1994; BOEHM et al. 1997). Excitatory GlyRs may be especially relevant to synaptogenesis, since glycine-triggered rises in intracellular Ca^{2+} have recently been shown to be crucial for the correct formation of postsynaptic glycinergic membrane specializations (KIRSCH and BETZ 1998). Thus, the developmental regulation of intracellular Cl^- concentration critically controls the nature of the postsynaptic response to glycine.

B. Structure and Diversity of Glycine Receptor Channels

I. GlyRs are Ligand-Gated Ion Channels of the nAChR Superfamily

The GlyR was initially purified from adult rat spinal cord by affinity chromatography, utilizing aminostrychnine-agarose columns (PFEIFFER et al. 1982).

GlyRs purified in this manner contain two glycosylated integral membrane proteins of 48 kDa (α) and 58 kDa (β) and an associated peripheral membrane protein of 93 kDa, named gephyrin. The primary structures of GlyR α and β subunits have been deduced by molecular cloning methods (GRENNINGLOH et al. 1987, 1990a) and show significant sequence and structural similarity to nicotinic acetylcholine receptor (nAChR), γ-aminobutyric acid type A ($GABA_A$) receptor and serotonin type 3 ($5HT_3$) receptor subunits (BETZ 1990). This sequence conservation is particularly evident in a conserved cysteine motif in the large N-terminal extracellular domain and in the four hydrophobic membrane-spanning domains (M1–M4; see Fig. 1). The quaternary structure of GlyRs has been analyzed using cross-linking reagents and subunit-specific monoclonal antibodies in combination with electrophoretic analysis and sedimentation techniques (LANGOSCH et al. 1988). GlyRs purified from adult rat spinal cord are pentameric in structure and contain three α and two β subunits. This stoichiometry strongly resembles that of nAChR and $GABA_A$ receptors, which are also thought to contain five membrane-spanning subunits (see NAYEEM et al. 1994). Given the known sequence relatedness and topological similarity of the components of these receptors, a pentameric arrangement of membrane-spanning subunits around a central ion pore is generally believed to represent the common quaternary structure of receptors of this ligand-gated ion channel superfamily.

II. Glycine Receptor Heterogeneity

Heterogeneity of the GlyR was first suggested by the discovery of a neonatal GlyR in rat spinal cord, whose α subunit differs in strychnine-binding affinity, molecular weight (49 kDa) and immunological properties from the adult 48 kDa polypeptide (BECKER et al. 1988). Such diversity was confirmed when molecular cloning methods were applied to GlyR analysis. Initially, peptide sequences derived from affinity-purified adult rat spinal cord GlyRs were used to isolate cDNAs for the 48 kDa (α1) and 58 kDa (β) subunits (GRENNINGLOH et al. 1987, 1990a). Subsequently, cDNA clones corresponding to two novel GlyR α subunits (α2 and α3) were cloned by homology screening (GRENNINGLOH et al. 1990b; KUHSE et al. 1990b, 1991; AKAGI et al. 1991). A partial mouse genomic sequence encoding part of a fourth α subunit (MATZENBACH et al. 1994) recently allowed the isolation of a full-length α4 subunit cDNA (HARVEY et al. in preparation). In situ hybridization studies have revealed that the different α subunit genes exhibit unique spatial and temporal patterns of expression in spinal cord, brain stem, and some higher brain regions (KUHSE et al. 1991; FUJITA et al. 1991; MALOSIO et al. 1991a,b; SATO et al. 1991, 1992; WATANABE et al. 1995). GlyR α2 subunit transcripts predominate in the embryonic and neonatal brain and spinal cord, and are replaced postnatally by the α1 or α3 subunit mRNAs. Interestingly, transcripts for the GlyR β subunit are very widely expressed, and are even found in some adult brain regions that lack α1, α2, and α3 subunit transcripts (FUJITA et al. 1991;

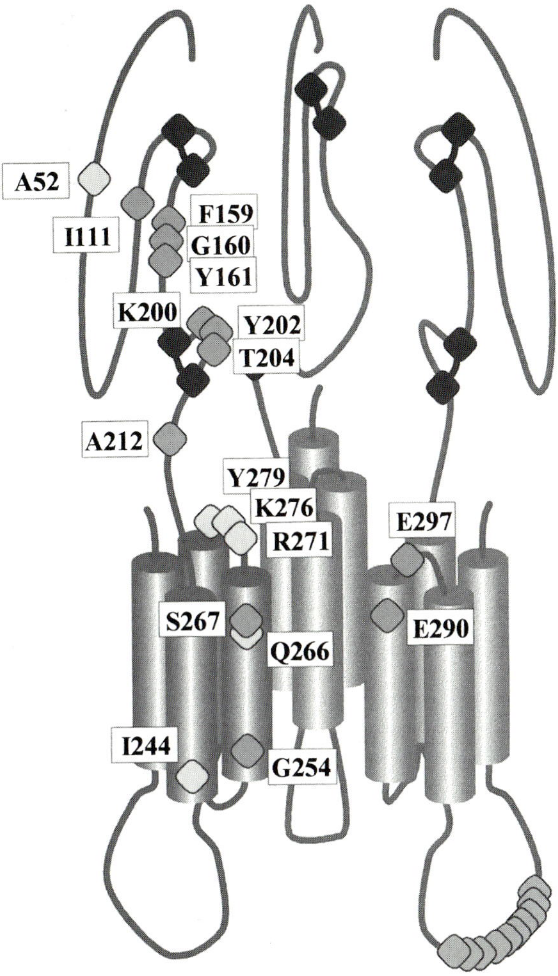

Fig. 1. A schematic representation of the membrane-spanning topology and location of functionally-important amino-acid residues in the GlyR $\alpha1$ (*grey*) and β (*dark-grey*) subunits. For simplicity, three out of the five subunits are shown. *Cylinders* represent the four membrane-spanning domains (M1–M4), and conserved cysteine residues thought to form disulphide bridges are indicated by *black diamonds*. *Natural GlyR mutants* (*white diamonds*): mutation A52S is found in the GlyR $\alpha1$ subunit gene in *spasmodic* mice; mutations I244N, Q266H, R271Q/L, K276E, and Y279C are found in the GlyR $\alpha1$ subunit gene in different hyperekplexia families. *Binding site determinants* (*grey diamonds*): in the GlyR $\alpha1$ subunit residues G160, K200, and Y202 are involved in strychnine binding, the efficacy of taurine is determined by residues I111 and A212, while F159, Y161, and T204 are determinants of agonist affinity and specificity. S267 is a target for alcohol and volatile anesthetics. *Channel function*: G254 in the $\alpha1$ subunit is a determinant of main-state conductances; E290 and E297 in the β subunit are involved in resistance to picrotoxin blockade. *Intracellular interactions* (*bottom grey diamonds*): amino acids 394–411 in the β subunit are determinants of gephyrin binding

MALOSIO et al. 1991a). Since the α4 subunit mRNA is not abundant in brain (MATZENBACH et al. 1994) these findings have led to speculation (MALOSIO et al. 1991a) that either additional GlyR α subunit genes remain to be identified, or that the β subunit forms part of another receptor complex. Additional diversity arises from alternative splicing, which generates variants of the α1 (MALOSIO et al. 1991b), α2 (KUHSE et al. 1991), and β subunits (HECK et al. 1997). A rat α2 subunit variant (α2*) has also been described (KUHSE et al. 1990a) which, in contrast to the human or rat α1, α2 and α3 subunits, displays only low affinity for strychnine and may represent the neonatal GlyR isoform described above.

III. The GlyR Ligand-Binding Domain

The first evidence that the GlyR ligand-binding site resides on α subunits came from photoaffinity labeling experiments using the GlyR antagonist strychnine. Peptide mapping of [^{3}H]strychnine-labeled GlyR preparations revealed covalent incorporation of this antagonist between amino acids 170 and 220 of the N-terminal domain of the rat GlyR α1 subunit (GRAHAM et al. 1983; RUIZ-GOMEZ et al. 1990). Further information about the location of ligand-binding site determinants came from functional expression studies using cloned GlyR cDNAs and *Xenopus laevis* oocytes or mammalian cells. In these systems, GlyR α subunits are capable of forming robust homo-oligomeric chloride channels which can be gated by micromolar concentrations of glycine, taurine, and β-alanine, and antagonized by nanomolar amounts of strychnine (SCHMIEDEN et al. 1989; SONTHEIMER et al. 1989). The GlyR β subunit does not form functional homomeric GlyRs (PRIBILLA et al. 1992; BORMANN et al. 1993), but on incorporation into heteromeric receptors alters several functional aspects of the ion channel (see below). By comparison of the functional receptors produced by different GlyR α subunit variants (KUHSE et al. 1990a,b, SCHMIEDEN et al. 1989, 1992, 1993) in combination with site-directed mutagenesis, it has become clear that several discontinuous domains of the α subunit extracellular domain are responsible for forming the ligand-binding pocket (Fig. 1). By comparing the pharmacology of the GlyR α2 and α2* variants, G167 (equivalent to G160 in the α1 subunit) was shown to be a crucial determinant of glycine and strychnine binding (KUHSE et al. 1990a). The two neighboring residues (F159 and Y161 in the α1 subunit) have also been found be involved in agonist selectivity and antagonist efficacy (SCHMIEDEN et al. 1993). Another domain in the GlyR α1 subunit, encompassing K200 and Y202, has been shown to be a determinant of the strychnine binding site (VANDENBERG et al. 1992). Substitution of residues I111 and A212 strongly affects the potency of the glycinergic agonists β-alanine and taurine (SCHMIEDEN et al. 1992).

Studies of naturally occurring mutations of the GlyR α1 subunit have revealed additional determinants of agonist binding (see Fig. 1 and below). In the mouse mutant *spasmodic*, a missense mutation (A52 to serine) results in a modest reduction of glycine affinity, but does not affect strychnine binding

(RYAN et al. 1994; SAUL et al. 1994). Point mutations in the human GlyR $\alpha 1$ subunit gene that underlie hereditary hyperekplexia have uncovered domains that are likely to link agonist binding and channel gating. Heterologous expression of mutants R271L, R271Q, K276E, or Y279C (residues found in the M2-M3 loop) results in GlyRs that exhibit a decreased sensitivity to glycine and a loss of β-alanine and taurine responses (LANGOSCH et al. 1994; RAJENDRA et al. 1994; LAUBE et al. 1995b; LYNCH et al. 1997). However, none of these mutations appears to affect receptor expression, as assessed by strychnine binding. There is evidence that some of these mutations (R271L/Q and K276E) reduce the single-channel conductance and/or the open channel probability of the expressed GlyRs (LANGOSCH et al. 1994; RAJENDRA et al. 1994; LEWIS et al. 1998), implying that the M2-M3 loop is vital for coupling signal transduction and ligand binding. Mutation of I244N within segment M1 also reduces channel gating (LYNCH et al. 1997), but additionally impairs the efficiency of GlyR expression. Taken together, these data point to a multi-site ligand-binding/signal transduction mechanism that involves distant segments of the large extracellular domain and residues between M2 and M3.

IV. Determinants of Ion Channel Function

Single-channel analysis has allowed a precise characterization of the anion selectivity of the GlyR chloride channel. In addition to Cl^-, the latter is also permeable to other halides as well as nitrate, bicarbonate, and small organic ions. Ion substitution studies have established a permeability sequence of $SCN^- > I^- > NO_3^- > Br^- > Cl^- > HCO_3^- >$ acetate $> F^- >$ propionate (BORMANN et al. 1987). The predicted membrane spanning segments M1 to M3 are highly conserved between GlyR and $GABA_A$ receptor subunits (BETZ 1990), suggesting their importance in chloride channel function. Segment M2 has a high content of uncharged polar amino-acid residues, and is generally thought to constitute the hydrophilic inner lining of the chloride channel. Indeed, a synthetic peptide corresponding to the M2 segment of the GlyR $\alpha 1$ subunit is capable of producing channel activity in liposomes and planar lipid bilayers (LANGOSCH et al. 1991; REDDY et al. 1993). Further evidence that the M2 segment of GlyR polypeptides contributes to ion channel formation came from a study that assigned determinants of resistance to channel blockade by the plant alkaloid picrotoxinin to residues E290 within the M2 segment, and E297 within the M2-M3 loop of the β subunit (PRIBILLA et al. 1992) (Fig. 1). Subsequently, residues within the carboxy–terminal half of the M2 segment in GlyR α and β subunits were shown (BORMANN et al. 1993) to regulate the main-state conductances of homo- and hetero-oligomeric GlyRs. GlyR α subunit homomeric receptors show distinct main-state conductances of 86 ($\alpha 1$), 111 ($\alpha 2$), and 105 ($\alpha 3$) pS, which are dependent on a single residue located within the M2 segment. Mutation of G254 in the rat GlyR $\alpha 1$ subunit (Fig. 1) to alanine (which is found in the equivalent position in $\alpha 2$ and $\alpha 3$ subunits) gave rise to a main-state conductance of 107 pS. Interestingly, the main-state conductances of heteromeric

$\alpha1\beta$, $\alpha2\beta$, and $\alpha3\beta$ GlyRs were significantly lower (44, 54, and 48 pS) than those of homomeric α subunit receptors and closely correspond to values recorded from spinal neurons (Takahashi et al. 1992; Bormann et al. 1993). This has been attributed to bulky side-chains within the M2 segment of the β subunit. Taken together, these studies indicate that native GlyRs are heteromeric, show the importance of the M2 segment for chloride conductance and underscore the role of the GlyR β subunit in determining the functional properties of the ion channel.

V. Clustering of GlyRs by the Anchoring Protein Gephyrin

GlyRs are densely clustered within postsynaptic specializations in spinal cord neurons. This ordered arrangement is thought to be mediated by gephyrin, a peripheral membrane protein of 93 kDa that co-purifies with GlyRs (Schmitt et al. 1987). Gephyrin is located at the cytoplasmic face of postsynaptic specializations containing GlyRs (Triller et al. 1985) and binds polymerized tubulin with nanomolar affinity (Kirsch et al. 1991). Gephyrin is also known to interact with GlyRs via an 18 amino acid motif (Fig. 1) that lies within the large intracellular loop of the GlyR β subunit (Meyer et al. 1995). Molecular cloning has elucidated several different isoforms of gephyrin that result from alternative splicing of four distinct exons (Prior et al. 1992). In situ hybridization (Kirsch et al. 1993a) and immunocytochemical (Araki et al. 1988; Kirsch and Betz 1993) studies have revealed a widespread expression of gephyrin in embryonic and adult rat brain and spinal cord. Antisense oligonucleotide treatment of cultured embryonic spinal cord neurons (Kirsch et al. 1993b; Kirsch and Betz 1995, 1998) indicates that gephyrin is required for the correct targeting of GlyRs to postsynaptic specializations (Kirsch et al. 1993b). Similarly, the addition of strychnine or L-type Ca^{2+} channel blockers has shown (Kirsch and Betz 1998) that the activation of embryonic GlyRs, resulting in Ca^{2+} influx, is crucial for the formation of gephyrin and GlyR clusters at the developing postsynaptic site. In addition, compounds that disrupt the integrity of microtubules (e.g., demecolcine) and microfilaments (e.g., cytochalsin D), affect the packing density of gephyrin and GlyR specializations in these cultures (Kirsch and Betz 1995). These cytoskeletal structures appear to operate antagonistically: microtubules condense GlyR clusters, while microfilaments disperse them. In conclusion, complex signaling mechanisms and membrane-cytoskeleton interactions recruit both gephyrin and GlyRs to postsynaptic sites.

C. Pharmacology of Glycine Receptors

I. Strychnine is a Selective GlyR Antagonist

The plant alkaloid strychnine (Fig. 2), derived from the Indian tree *Strychnos nux vomica,* is a potent convulsant that acts by antagonizing glycinergic

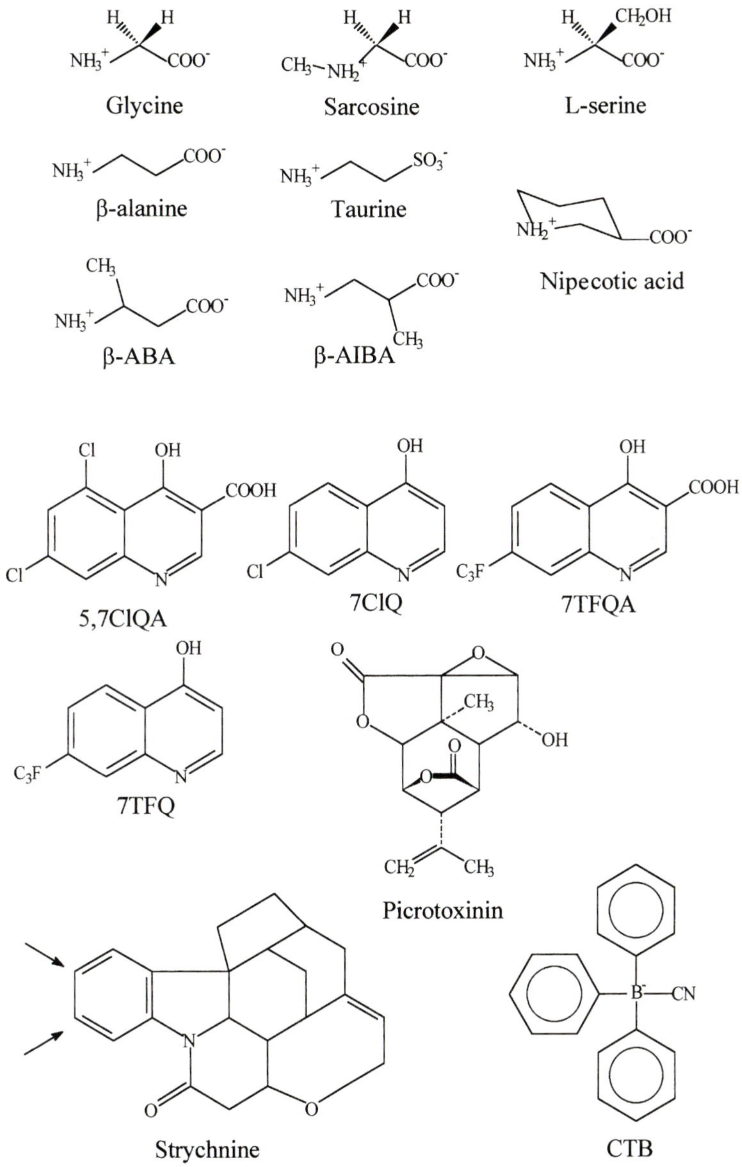

Fig. 2. Chemical structures of selected compounds which are active at GlyRs. *Top*: α-amino acids glycine, sarcosine, and serine, and β-amino acids β-alanine, taurine, β-aminobutyric acid (β-ABA), and β-aminoisobutyric acid (β-AIBA). *Middle*: the piperidine derivative nipecotic acid, and the quinolinic acid-based substances 5,7-dichloro-4-hydroxyquinoline-3-carboxylic acid (5,7ClQA), 7-chloro-4-hydroxyquinoline (7ClQ), 7-trifluoromethyl-4-hydroxyquinoline-3-carboxylic acid (7TFQA) and 7-trifluoromethyl-4-hydroxyquinoline (7TFQ). *Bottom*: the GlyR antagonists strychnine, picrotoxinin, and cyanotriphenylborate (CTB). Note that the aromatic ring positions indicated on the strychnine molecule (arrows) can be substituted without affecting toxicity; this has been exploited for both affinity purification and the synthesis of fluorescent derivatives

inhibition. Due to its high toxicity, strychnine has traditionally been used as a rat poison. Related alkaloids, such as brucine, act similarly as competitive glycine antagonists at the inhibitory GlyR, and extensive chemical modification studies have established detailed structure-function relationships (Becker 1992).

Strychnine constitutes a unique tool in the investigation of postsynaptic GlyRs. In electrophysiological studies, it is the most reliable antagonist to distinguish glycinergic from GABAergic inhibition. Glycine-displaceable [^{3}H]strychnine binding (Young and Snyder 1973) also constitutes the most reliable binding assay for this receptor system. Further, strychnine provides a natural photoaffinity label for the GlyR; upon UV illumination, [^{3}H]strychnine is incorporated into the ligand-binding α subunit (Graham et al. 1981, 1983). Lastly, substitutions at the aromatic ring have little effect on the toxicity of strychnine and have been exploited to generate affinity columns for GlyR purification (Pfeiffer et al. 1992) and to synthesize fluorescent derivatives for visualizing GlyR distribution on living neurons (St John and Stevens 1993).

The physiological symptoms of strychnine poisoning emphasize the importance of glycinergic inhibition in the control of both motor behavior and sensory processing. Consistent with the physiology of glycinergic synapses, sublethal strychnine poisoning leads to motor disturbances, increased muscle tonus, hyperacuity of sensory, visual and acoustic perception, and higher doses cause convulsions and death (Becker 1992). Surprisingly, the 'stimulation' of both motor and sensory pathways by strychnine and its derivatives has also been used therapeutically; for decades, strychnine was used as a tonic ingredient in several medical prescriptions.

II. Amino Acids and Piperidine Carboxylic Acid Compounds

In addition to glycine, the endogenous amino acids β-alanine and taurine (Fig. 2) display inhibitory activity when applied to neurons (e.g., Boehm et al. 1997). Nevertheless, until recently, neither of these amino acids had been firmly established as a neurotransmitter. Flint et al. (1998) demonstrated that GlyRs in the developing rodent neocortex are not only excitatory, but also activated by non-synaptically released taurine, which is stored in immature cortical neurons. Since foetal taurine deprivation has been linked with cortical dysgenesis, Flint and co-workers suggested that taurine may influence cortical development by activating GlyRs.

The agonist and antagonist actions of several α- and β-amino acids has been studied using recombinant GlyR α1 subunit homo-oligomers expressed in *Xenopus* oocytes (Schmieden and Betz 1995). This revealed that the *agonistic* activity of α-amino acids (e.g., glycine, sarcosine, alanine, and serine) (Fig. 2) exhibits marked stereoselectivity and is susceptible to substitutions at the C_α-atom. However, α-amino acid *antagonism* is neither influenced by C_α-atom substitutions nor enantiomer-dependent. β-Amino acids such as taurine,

β-aminobutyric acid (β-ABA), and β-aminoisobutyric acid (β-AIBA), which are partial agonists at GlyRs (Fig. 2), show competitive inhibition at low concentrations whereas high concentrations elicit a significant membrane current. This suggests that the partial agonist activity of a given β-amino acid at GlyRs may be determined by the relative amounts of the respective *cis/trans* isomers. In contrast, nipecotic acid (Fig. 2), and related compounds which contains a *trans*-β-amino acid configuration, behave as competitive GlyR antagonists.

III. Antagonism by Picrotoxinin, Cyanotriphenylborate, and Quinolinic Acid Compounds

Classically, the plant alkaloid picrotoxinin (Fig. 2) is considered as a use-dependent open-channel blocker of both $GABA_A$ receptor and GlyRs, and has been used as a pharmacological tool to discriminate homo-oligomeric from heteromeric GlyRs (PRIBILLA et al. 1992). Both native GlyRs and recombinantly expressed α,β hetero-oligomers are largely resistant to block by picrotoxinin, whereas α subunit homo-oligomers are sensitive to micromolar concentrations. Indeed, as stated above, studies using chimeric receptors have assigned determinants of resistance to open-channel blockade by picrotoxinin to residues G290 (within the M2 segment) and G297 (within the M2-M3 loop) of the β subunit (PRIBILLA et al. 1992). More recently, LYNCH et al. (1995) reported that, in contrast to its action at $GABA_A$ receptors, picrotoxinin antagonism of the GlyRs is competitive and not use-dependent, which is consistent with binding to extracellular domains of GlyR α subunits. One possible explanation for the apparent discrepancy between these two studies has been suggested from analyzing GlyRs composed of α1 subunits carrying the hyperekplexia mutations R271L or R271Q (LYNCH et al. 1995). These mutations apparently transform picrotoxin from an allosterically-acting competitive antagonist to an allosteric potentiator at low (0.01–3 μmol/l) concentrations and to a non-competitive antagonist at higher (~3 μmol/l) concentrations, suggesting that residues close to the membrane-spanning segments may be involved in coupling an extracellular picrotoxin binding site to channel gating. Alternatively, binding of this bulky antagonist to determinants located both at the channel mouth and on the extracellular domain may shift the equilibrium between active and desensitized receptor conformations, as suggested for the $GABA_A$ receptor (NEWLAND and CULL-CANDY 1992).

In contrast to picrotoxinin, antagonism by the organic anion cyanotriphenylborate (CTB) (Fig. 2) has been shown to be clearly non-competitive, use-dependent, and more pronounced at positive membrane potentials (RUNDSTRÖM et al. 1994) suggesting that it is an open-channel blocker. This compound can also be used to discriminate different GlyR subtypes: in contrast to GlyR α1 subunit receptors, homo-oligomeric α2 subunit receptors are resistant to CTB. Using site-directed mutagenesis this difference was traced to G254 in the M2 segment (RUNDSTRÖM et al. 1994). Novel derivatives of quinolinic acid compounds have also been developed as selective GlyR antagonists (SCHMIEDEN et al. 1996). These were based on 2-carboxy-4-hydroquinolines, which antagonize binding of the co-agonist glycine to the *N*-methyl-D-aspartate (NMDA) receptor. Closely related derivatives, 4-hydroxy-quinolines and 4-hydroxy-quinoline-3-carboxylic acids (Fig. 2) antagonize agonist responses of recombinantly expressed GlyRs (SCHMIEDEN et al. 1996). In *Xenopus* oocytes expressing GlyR $\alpha 1$ subunit homo-oligomers, the chloride-substituted derivatives 5,7-dichloro-4-hydroxyquinoline-3-carboxylic acid (5,7ClQA) and 7-chloro-4-hydroxyquinoline (7ClQ) inhibit glycine currents in a mixed high-affinity competitive and low-affinity non-competitive fashion. In contrast, the related compounds 7-trifluoromethyl-4-hydroxyquinoline-3-carboxylic acid (7TFQA) and 7-trifluoromethyl-4-hydroxyquinoline (7TFQ) show purely competitive antagonism. As well as providing new tools to study native and recombinant GlyRs, the latter results suggest that the GlyR agonist/antagonist binding pocket may show similarity to that proposed for the glycine-binding site of the NMDA receptor.

IV. Potentiation of GlyR Function by Anesthetics, Alcohol and Zn^{2+}

Electrophysiological studies have shown that GlyR function is enhanced by a number of volatile anesthetics and ethanol in both native systems (CELENTANO et al. 1988; AGUAYO and PANCETTI 1994) and *Xenopus* oocytes expressing homo-oligomeric $\alpha 1$ or $\alpha 2$ subunit GlyRs (MASCIA et al. 1996a,b). However, such compounds reduce the activity of receptors formed by the sequence-related $GABA_C$ receptor $\rho 1$ subunit (MIHIC and HARRIS 1996). By creating chimeric receptor constructs, MIHIC et al. (1997) identified a 45 amino-acid long region encompassing the M2 and M3 domains that is crucial for enhancement of GlyR function by such compounds. Mutation of a single amino acid (S267) (Fig. 1) within this region was shown to be sufficient to abolish enhancement of GlyR function by ethanol and the volatile anesthetic enflurane (MIHIC et al. 1997). Extending this study, YE et al. (1998) have shown that ethanol enhancement is inversely correlated with the molecular volume of the amino acid present at position 267.

The divalent cation Zn^{2+} exhibits biphasic effects on both native GlyRs on rat spinal cord neurons and on recombinantly expressed homo-oligomeric and heteromeric GlyRs (BLOOMENTHAL et al. 1994; LAUBE et al. 1995a). At low concentrations (nanomolar and low micromolar) Zn^{2+} potentiates glycine-induced currents, whereas at high micromolar concentrations Zn^{2+} decreases the glycine response (BLOOMENTHAL et al. 1994; LAUBE et al. 1995a). Dose-response analysis suggests that both the potentiating and inhibitory effects of Zn^{2+} result from changes in apparent agonist affinity. Using chimeric GlyR subunit cDNA constructs, LAUBE et al. (1995a) revealed that the positive and negative modulatory effects of Zn^{2+} are mediated by different regions of α subunits, and that determinants of the potentiating Zn^{2+} binding site are localized between amino acids 74–86 of the rat GlyR $\alpha 1$ subunit. This Zn^{2+} modulation of GlyRs is of potential physiological importance, since Zn^{2+} is stored in the synaptic vesicles of different neuronal populations and co-released with the transmitter upon stimulation.

D. Pathology of Glycine Receptors

In several mammalian species including mice, cattle, horses, and humans, defects in glycinergic neurotransmission have been implicated in complex motor disorders characterized by hypertonia and an exaggerated startle reflex (reviewed in BECKER 1995). Recently, mutations in GlyR subunit genes have been identified in the mouse mutants *spastic*, *spasmodic*, and *oscillator*, as well as in the human startle disease known as hereditary hyperekplexia.

I. Mouse Glycine Receptor Mutants: *Spastic*, *Spasmodic*, and *Oscillator*

The gene responsible for the recessive mouse mutant *spasmodic* ($Glra1^{spd}$) is located on mouse chromosome 11 (LANE et al. 1987) at 29.0cM, a region that exhibits synteny homology to human chromosome 5q31.3, where the human GlyR α1 subunit gene (GLRA1) has been mapped (SHIANG et al. 1993; BAKER et al. 1994). Homozygous *spasmodic* mice appear normal at rest, but around postnatal day 14 acquire an exaggerated acoustic startle reflex: when subjected to loud noises or handling, animals show rigidity, tremor, and an impaired righting reflex. This phenotype has been shown (RYAN et al. 1994) to be caused by a missense mutation in the mouse GlyR α1 subunit gene (*Glra1*) which results in an alanine to serine conversion at position 52 within the large N-terminal extracellular domain of the α1 subunit (Fig. 1). As discussed above, the A52S mutation lowers the agonist affinity of GlyRs containing the mutant subunit (RYAN et al. 1994; SAUL et al. 1994), but does not appear to affect receptor expression as monitored by strychnine binding.

The mouse mutant *oscillator* has been shown (BUCKWALTER et al. 1994) to be allelic to *spasmodic*, and hence a mutation in *Glra1*, by a direct breeding test. At two weeks of age *oscillator* ($Glra1^{ot}$) homozygotes begin to exhibit rapid, violent trembling, which increases in severity daily. At three weeks of age these mice show prolonged periods of rapid tremor, producing extreme rigor and stiffness, and normally die around this time. Western blot analysis employing GlyR subunit-specific antibodies, and [^{3}H]strychnine binding experiments have revealed that the spinal cord of *oscillator* homozygotes is totally devoid of the adult GlyR isoform (BUCKWALTER et al. 1994; KLING et al. 1997). This drastic loss of functional GlyRs, and the *oscillator* phenotype, result from a microdeletion of seven nucleotides within exon 8 of *Glra1* (BUCKWALTER et al. 1994). Depending on the use of an alternate splice acceptor site for exon 9, *oscillator* mice produce two mutant transcripts, neither of which encodes the large intracellular loop and M4 of the GlyR α1 subunit.

The *spastic* mutation ($Glrb^{spa}$) maps to mouse chromosome 3 at 38.5 cM (LANE and EICHER 1979; KINGSMORE et al. 1994b; MÜLHARDT et al. 1994). Homozygous *spastic* mice have a phenotype similar to that of *spasmodic* animals; at 14 days of age they suffer from muscle spasms, rapid tremor, stiffness of posture and difficulty in righting (WHITE and HELLER 1982). However, unlike *spasmodic* mice, GlyR levels in *spastic* homozygotes are drastically reduced (~20% of control levels) as assessed by strychnine binding (WHITE and HELLER 1982; BECKER et al. 1986). The *spastic* phenotype has been shown (KINGSMORE et al. 1994b; MÜLHARDT et al. 1994) to be due to the insertion of a LINE-1 transposable element into intron 5 of the mouse GlyR β subunit gene (*Glrb*). This LINE-1 element interferes with the correct splicing of the β subunit pre-mRNA, inducing 'skipping' of exons 4 and/or 5 and drastically reducing the level of full-length β subunit transcripts. However, low levels (20–30%) of correctly-spliced β subunit mRNAs are produced, and this appears to be sufficient to prevent the $Glrb^{spa/spa}$ genotype from being lethal. Introduction of a transgene encoding the rat GlyR β subunit into the $Glrb^{spa}$ genetic background has been found (HARTENSTEIN et al. 1996) to rescue the *spastic* phenotype, confirming the causal link between the LINE-1 element insertion in *Glrb* and the *spastic* phenotype. To date, no mutations have been identified in the mouse α2, α3, or α4 subunit genes, which map to chromosome X at 71.5 cM (DERRY et al. 1991), chromosome 8 at 25.0 cM (KINGSMORE et al. 1994a) and chromosome X at 56.0 cM (MATZENBACH et al. 1994), respectively.

II. Mutations in GLRA1 Underlie the Human Hereditary Disorder Hyperekplexia

Hereditary hyperekplexia (also called startle disease; symbol STHE) is a human autosomal neurological disorder that has symptoms which closely resemble sublethal strychnine poisoning (reviewed in BECKER 1995; RAJENDRA and SCHOFIELD 1995). Affected individuals commonly exhibit pronounced muscle rigidity in response to sudden stimuli, such as noise, light, or touch. In normal subjects, startle reactions include exaggerated jerks of the limbs, facial grimaces, and fist clenching. However, in affected patients much stronger responses are seen, which can result in general rigidity triggering loss of posture and unprotected falling. In some affected new-born babies and young infants, sudden noise or light stimuli may cause stiff baby syndrome, an excessive startle reaction involving strong muscle spasms. In severe cases, prolonged apnea can occur that may even be fatal. These symptoms normally ameliorate with age, and in most cases adults experience only a comparatively mild acoustic startle reaction.

Detailed genetic linkage analyses of two large families provided evidence that hyperekplexia maps to human chromosome 5q32 (RYAN et al. 1992), a region that is rich in neurotransmitter receptor genes, including that for the GlyR α1 subunit (GLRA1). Subsequently, point mutations were identified in exon 6 of GLRA1 which co-segregate with the disorder in affected families (SHIANG et al. 1993). This initial study identified two separate point mutations at arginine 271 (R271Q or R271L), which lies in the short extracellular loop linking M2 and M3. Since the pioneering work of SHIANG et al. (1993), further mutations of GLRA1 have been discovered that cause both dominant and recessive forms of startle disease (Fig. 1). These include other substitutions in the M2-M3 loop (K276E and Y279C) (SHIANG et al. 1995; ELMSLIE et al. 1996), within M1 (I244N) (REES et al. 1994) or M2 (Q266H) (MILANI et al. 1996). As discussed above, most of these mutations are thought to disrupt the coupling of ligand binding to signal transduction. A sporadic case of a recessive form of hyperekplexia has been found (BRUNE et al. 1996) that results from a genomic deletion which encompasses the first six exons of GLRA1. Since the symptoms of the affected child are ameliorated with age, mechanisms may exist that compensate the loss of GLRA1 gene function in man. In this regard, evidence has been obtained for a compensatory role of GABA, since drugs which potentiate $GABA_A$ receptor function, such as the benzodiazepine clonazepam, have proven efficient in the treatment of hyperekplexia patients (RYAN et al. 1992). Interestingly, more recent reports (e.g., VERGOUWE et al. 1997) have also uncovered families with hyperekplexia-like syndromes that do not have mutations in GLRA1, suggesting that mutations in other genes involved in glycinergic neurotransmission might also cause hyperekplexia. Candidates include the human GlyR β subunit gene (GLRB) which has been mapped to human chromosome 4q32 (HANDFORD et al. 1996) and the α2, α3 and α4 subunit genes, which have been localized to human chromosomes Xp21.2–p22.1 (GLRA2) (GRENNINGLOH et al. 1990), 4q33–q34 (GLRA3) (Nikolic and Becker, unpublished data) and Xq21–q22 (GLRA4) (Harvey and Betz, unpublished data).

E. Conclusions

The biochemical and molecular biology approaches outlined above have shown that GlyRs are heterogenous and widespread in the developing and adult central nervous system. The analysis of GlyR mutations in the mouse mutants *spastic*, *spasmodic*, and *oscillator* and in the human hereditary disorder hyperekplexia corroborate the pivotal role of GlyRs in the control of both motor and sensory functions. Although the analysis of these naturally-occurring and additional laboratory-designed GlyR mutants has allowed the correlation of structural features of GlyR subunits with receptor function, our picture of both the ligand-binding sites and the channel domain of these receptors is still incomplete. Similarly, although the interaction of the GlyR with gephyrin constitutes one of the most thoroughly studied model systems for postsynaptic membrane formation in the mammalian central nervous system, the mechanisms controlling the subcellular distribution of these proteins remains enigmatic. Furthermore, the pharmacology of GlyRs constitutes a largely uncharted terrain. Although recent years have seen the advent of new pharmacological compounds that selectively affect both native and recombinant GlyRs, potent agonists and/or positive modulators of GlyR channel function are not presently available. Such compounds have great promise as novel therapeutic agents for the treatment of spastic and convulsive motor disorders and peripheral pain syndromes, and may also serve as a new class of muscle relaxants. It is hoped that the coming years will see new attempts to develop such pharmaceuticals.

References

Aguayo LG, Pancetti FC (1994) Ethanol modulation of the γ-aminobutyric acid$_A$- and glycine-activated Cl^- current in cultured mouse neurons. J Pharmacol Exp Ther 270:61–69

Akagi H, Hirai K, Hishinuma F (1991) Cloning of a glycine receptor subtype expressed in rat brain and spinal cord during a specific period of neuronal development. FEBS Lett 281:160–166

Aprison MH (1990) The discovery of the neurotransmitter role of glycine. In: Ottersen OP, Storm-Mathiesen J (eds) Glycine neurotransmission. John Wiley and Sons Ltd, New York, pp. 1–23

Araki T, Yamano M, Murakami T, Wanaka A, Betz H, Tohyama M (1988) Localization of glycine receptors in the rat central nervous system: an immunocytochemical analysis using monoclonal antibody. Neuroscience 25:613–624

Baker E, Sutherland GR, Schofield PR (1994) Localization of the glycine receptor $\alpha 1$ subunit gene (GLRA1) to chromosome 5q32 by FISH. Genomics 40:396–400

Becker C-M (1992) Convulsants acting at the inhibitory glycine receptor. In: Herken H, Hucho F (eds) Handbook of experimental pharmacology, vol. 102. Springer Verlag, Berlin–Heidelberg, pp. 539–575

Becker C-M (1995) Glycine receptors: molecular heterogeneity and implications for disease. The Neuroscientist 1:130–141

Becker C-M, Hermans-Borgmeyer I, Schmitt B, Betz H (1986) The glycine receptor deficiency of the mutant mouse *spastic*: evidence for normal glycine receptor structure and localization. J Neurosci 6:1358–1364

Becker CM, Hoch W, Betz H (1988) Glycine receptor heterogeneity in rat spinal cord during postnatal development. EMBO J 7:3717–3726

Betz H (1990) Ligand-gated ion channels in the brain: the amino acid receptor superfamily. Neuron 5:383–392

Bloomenthal AB, Goldwater E, Pritchett DB, Harrison NL (1994) Biphasic modulation of the strychnine-sensitive glycine receptor by Zn^{2+}. Mol Pharmacol 46:1156–1159

Boehm S, Harvey RJ, von Holst A, Rohrer H, Betz H (1997) Glycine receptors in cultured chick sympathetic neurons are excitatory and trigger neurotransmitter release. J Physiol 504:683–694

Bormann J, Hamill OP, Sakmann B (1987) Mechanism of anion permeation through channels gated by glycine and γ-aminobutyric acid in mouse cultured spinal neurones. J Physiol 385:243–286

Bormann J, Rundström N, Betz H, Langosch D (1993) Residues within transmembrane segment M2 determine chloride conductance of glycine receptor homo- and hetero-oligomers. EMBO J 12:3729–3737

Brune W, Weber RG, Saul B, von Knebel Doeberitz M, Grond-Ginsbach C, Kellermann K, Meinck H-M, Becker C-M (1996) A *GLRA1* null mutation in recessive hyperekplexia challenges the functional role of glycine receptors. Am J Hum Genet 58:989–997

Buckwalter MS, Cook SA, Davisson MT, White WF, Camper S (1994) A frameshift mutation in the mouse $\alpha 1$ glycine receptor gene (*Glra1*) results in progressive neurological symptoms and juvenile death. Hum Mol Genet 3:2025–2030

Celentano JJ, Gibbs TT, Farb DH (1988) Ethanol potentiates GABA- and glycine-induced chloride currents in chick spinal cord neurons. Brain Res 455:377–380

Derry JM, Barnard PJ (1991) Mapping of the glycine receptor $\alpha 2$-subunit gene and the $GABA_A$ receptor $\alpha 3$-subunit gene on the mouse X chromosome. Genomics 10:593–597

Elmslie FV, Hutchings SM, Spencer V, Curtis A, Covanis T, Gardiner RM, Rees M (1996) Analysis of GLRA1 in hereditary and sporadic hyperekplexia: a novel mutation in a family co-segregating for hyperekplexia and spastic paraparesis. J Med Genet 33:435–436

Flint AC, Liu X, Kriegstein AR (1998) Nonsynaptic glycine receptor activation during early neocortical development. Neuron 20:43–53

Fujita M, Sato K, Sato M, Inoue T, Kozuka T, Tohyama M (1991) Regional distribution of the cells expressing glycine receptor β subunit mRNA in the rat brain. Brain Res 560:23–37

Graham D, Pfeiffer F, Betz H (1981) UV light–induced cross–linking of strychnine to the glycine receptor of rat spinal cord membranes. Biochem Biophys Res Commun 102:1330–1335

Graham D, Pfeiffer F, Betz H (1983) Photoaffinity-labelling of the glycine receptor of rat spinal cord. Eur J Biochem 131:519–525

Grenningloh G, Rienitz A, Schmitt B, Methfessel C, Zensen M, Beyreuther K, Gundelfinger ED, Betz H (1987) The strychnine–binding subunit of the glycine receptor shows homology with nicotinic acetylcholine receptors. Nature 328:215–220

Grenningloh G, Pribilla I, Prior P, Multhaup G, Beyreuther K, Taleb O, Betz H (1990a) Cloning and expression of the 58 kd β subunit of the inhibitory glycine receptor. Neuron 4:963–970

Grenningloh G, Schmieden V, Schofield PR, Seeburg PH, Siddique T, Mohandas TK, Becker C-M, Betz H (1990b) Alpha subunit variants of the human glycine receptor: primary structures, functional expression and chromosomal localisation of the corresponding genes. EMBO J 9:771–776

Handford CA, Lynch JW, Baker E, Webb GC, Ford JH, Sutherland GR, Schofield PR (1996) The human glycine receptor β subunit: primary structure, functional characterisation and chromosomal localisation of the human and murine genes. Mol Brain Res 25:211–219

Hartenstein B, Schenkel J, Kuhse J, Besenbeck B, Kling C, Becker C-M, Betz H, Weiher H (1996) Low level expression of glycine receptor β subunit transgene is sufficient for phenotype correction in *spastic* mice. EMBO J 15:1275–1282

Heck S, Enz R, Richter-Landsberg C, Blohm DH (1997) Expression and mRNA splicing of glycine receptor subunits and gephyrin during neuronal differentiation of P19 cells *in vitro*, studied by RT–PCR and immunocytochemistry. Dev Brain Res 98:211–220

Kingsmore SF, Suh D, Seldin MF (1994a) Genetic mapping of the glycine receptor $\alpha 3$ subunit on mouse chromosome 8. Mamm Genome 5:831–832

Kingsmore SF, Giros B, Suh D, Bieniarz M, Caron MG, Seldin MF (1994b) Glycine receptor β-subunit gene mutation in *spastic* mice associated with LINE-1 element insertion. Nature Genet 7:136–142

Kirsch J, Betz H (1993) Widespread expression of gephyrin, a putative glycine receptor-tubulin linker protein, in rat brain. Brain Res 621:301–310

Kirsch J, Betz H (1995) The postsynaptic localization of the glycine receptor-associated gephyrin is regulated by the cytoskeleton. J Neurosci 15:4148–4156

Kirsch J, Betz H (1998) Glycine-receptor activation is required for receptor clustering in spinal neurons. Nature 392:717–720

Kirsch J, Langosch D, Prior P, Littauer UZ, Schmitt B, Betz H (1991) The 93 kDa glycine receptor–associated protein binds to tubulin. J Biol Chem 266:22242–22245

Kirsch J, Malosio ML, Wolters I, Betz H (1993a) Distribution of gephyrin transcripts in the adult and developing rat brain. Eur J Neurosci 5:1109–1117

Kirsch J, Wolters I, Triller A, Betz H (1993b) Gephyrin antisense oligonucleotides prevent glycine receptor clustering in spinal neurons. Nature 366:745–748

Kling C, Koch M, Saul B, Becker C-M (1997) The frameshift mutation *oscillator* ($Glra1^{spd\text{-}ot}$) produces a complete loss of glycine receptor α1-polypeptide in mouse central nervous system. Neuroscience 78:411–417

Kuhse J, Schmieden V, Betz H (1990a) A single amino acid exchange alters the pharmacology of neonatal rat glycine receptor subunit. Neuron 5:867–873

Kuhse J, Schmieden V, Betz H (1990b) Identification and functional expression of a novel ligand binding subunit of the inhibitory glycine receptor. J Biol Chem 265:22317–22320

Kuhse J, Kuryatov A, Maulet Y, Malosio ML, Schmieden V, Betz H (1991) Alternative splicing generates two isoforms of the α2 subunit of the inhibitory glycine receptor. FEBS Lett 283:73–77

Lane PW, Ganser AL, Kerner AL, White WF (1987) *Spasmodic*, a mutation on chromosome 11 in the mouse. J Hered 78:353–356

Lane PW, Eicher EM (1979) Gene order in linkage group XVI of the house mouse. J Hered 70:239–244

Langosch D, Thomas L, Betz H (1988) Conserved quaternary structure of ligand gated ion channels: the postsynaptic glycine receptor is a pentamer. Proc Natl Acad Sci USA 85:7394–7398

Langosch D, Hartung K, Grell E, Bamberg E, Betz H (1991) Ion channel formation by synthetic transmembrane segments of the inhibitory glycine receptor – a model study. Biochim Biophys Acta 1063:36–44

Langosch D, Laube B, Rundström N, Schmieden V, Bormann J, Betz H (1994) Decreased agonist affinity and chloride conductance of mutant glycine receptors associated with human hereditary hyperekplexia. EMBO J 13:4223–4228

Laube B, Kuhse J, Rundström N, Kirsch J, Schmieden V, Betz H (1995a) Modulation by zinc ions of native rat and recombinant human inhibitory glycine receptors. J Physiol 483:613–619

Laube B, Langosch D, Betz H, Schmieden V (1995b) Hyperekplexia mutations of the glycine receptor unmask the inhibitory subsite for β–amino–acids. Neuroreport 6:897–900

Lewis TM, Sivilotti LG, Colquhoun D, Gardiner RM, Schoepfer R, Rees M (1998) Properties of human glycine receptors containing the hyperekplexia mutation α1(K276E), expressed in *Xenopus* oocytes. J Physiol 507:25–40

Lynch JW, Rajendra S, Barry PH, Schofield PR (1995) Mutations affecting the glycine receptor agonist transduction mechanism convert the competitive antagonist, picrotoxin, into an allosteric potentiator. J Biol Chem 270:13799–13806

Lynch JW, Rajendra S, Pierce KD, Handford CA, Barry PH, Schofield PR (1997) Identification of intracellular and extracellular domains mediating signal transduction in the inhibitory glycine receptor. EMBO J 16:110–120

Malosio ML, Marquèze-Pouey B, Kuhse J, Betz H (1991a) Widespread expression of glycine receptor subunit mRNAs in the adult and developing rat brain. EMBO J 10:2401–2409

Malosio ML, Grenningloh G, Kuhse J, Schmieden V, Schmitt B, Prior P, Betz H (1991b) Alternative splicing generates two variants of the α1 subunit of the inhibitory glycine receptor. J Biol Chem 266:2048–2053

Mascia MP, Machu TK, Harris RA (1996a) Enhancement of homomeric glycine receptor function by long–chain alcohols and anaesthetics. Br J Pharmacol 119:1331–1336

Mascia MP, Mihic SJ, Valenzuela CF, Schofield PR, Harris RA (1996b) A single amino acid determines differences in ethanol actions on strychnine-sensitive glycine receptors. Mol Pharmacol 50:402–406

Matzenbach B, Maulet Y, Sefton L, Courtier B, Avner P, Guénet J-L, Betz H (1994) Structural analysis of mouse glycine receptor α subunit genes: identification and chromosomal localization of a novel variant, α4. J Biol Chem 269:2607–2612

Meyer G, Kirsch J, Betz H, Langosch D (1995) Identification of a gephyrin binding motif on the glycine receptor β subunit. Neuron 15:563–572

Mihic SJ, Harris RA (1996) Inhibition of rho1 receptor GABAergic currents by alcohols and volatile anaesthetics. J Pharmacol Exp Ther 277:411–416

Mihic SJ, Ye Q, Wick MJ, Koltchine VV, Krasowski MD, Finn SE, Mascia MP, Valenzuela CF, Hanson KK, Greenblatt EP, Harris RA, Harrison NL (1997) Sites of alcohol and volatile anaesthetic action on $GABA_A$ and glycine receptors. Nature 389:385–389

Milani N, Dalprá L, Prete A, del Zanini R, Larizza L (1996) A novel mutation (Gln266→His) in the α1 subunit of the inhibitory glycine receptor gene (GLRA1) in hereditary hyperekplexia. Am J Hum Genet 58:420–422

Mülhardt C, Fischer M, Gass P, Simon Chazottes D, Guénet J-L, Kuhse J, Betz H, Becker C–M (1994) The *spastic* mouse: aberrant splicing of glycine receptor β subunit mRNA caused by intronic insertion of L1 element. Neuron 13:1003–1015

Nayeem N, Green TP, Martin IL, Barnard EA (1994) Quaternary structure of the native $GABA_A$ receptor determined by electron microscopic image analysis. J Neurochem 62:815–818

Newland CF, Cull–Candy SG (1992) On the mechanism of action of picrotoxin on GABA receptor channels in dissociated sympathetic neurones of the rat. J Physiol 447:191–213

Pfeiffer F, Graham D, Betz H (1982) Purification by affinity chromatography of the glycine receptor of rat spinal cord. J Biol Chem 257:9389–9393

Pribilla I, Takagi T, Langosch D, Bormann J, Betz H (1992) The atypical M2 segment of the β subunit confers picrotoxinin resistance to inhibitory glycine receptor channels. EMBO J 11:4305–4311

Prior P, Schmitt B, Grenningloh G, Pribilla I, Multhaup G, Beyreuther K, Maulet Y, Werner P, Langosch D, Kirsch J, Betz H (1992) Primary structure and alternative splice variants of gephyrin, a putative glycine receptor-tubulin linker protein. Neuron 8:1161–1170

Rajendra S, Schofield PR (1995) Molecular mechanisms of inherited startle syndromes. Trends Neurosci 18:80–82

Rajendra S, Lynch JW, Pierce KD, French CR, Barry PH, Schofield PR (1994) Startle disease mutations reduce the agonist sensitivity of the human inhibitory glycine receptor. J Biol Chem 269:18739–18742

Rajendra S, Lynch JW, Pierce KD, French CR, Barry PH, Schofield PR (1995) Mutation of an arginine residue in the human glycine receptor transforms β-alanine and taurine from agonists into competitive antagonists. Neuron 14:169–175

Reddy GL, Iwamoto T, Tomich JM, Montal M (1993) Synthetic peptides and four–helix bundle proteins as model systems for the pore–forming structure of channel proteins. II. Transmembrane segment M2 of the brain glycine receptor is a plausible candidate for the pore–lining structure. J Biol Chem 268:14608–14615

Rees MI, Andrew M, Jawad S, Owen MJ (1994) Evidence for recessive as well as dominant forms of startle disease (hyperekplexia) caused by mutations in the α1 subunit of the inhibitory glycine receptor. Hum Mol Genet 3:2175–2179

Reichling DB, Kyrozis A, Wang J, MacDermott AB (1994) Mechanisms of GABA and glycine depolarization-induced calcium transients in rat dorsal horn neurons. J Physiol 476:411–421

Ruiz-Gomez A, Morato E, Garcia-Calvo M, Valdivieso F, Mayor F Jr (1990) Localization of the strychnine binding site on the 48–kilodalton subunit of the glycine receptor. Biochemistry 29:7033–7040

Rundström N, Schmieden V, Betz H, Bormann J, Langosch D (1994) Cyanotriphenylborate: subtype-specific blocker of glycine receptor chloride channels. Proc Natl Acad Sci USA 91:8950–8954
Ryan SG, Sherman SL, Terry JC, Sparkes RS, Torres M, Mackey RW (1992) Startle disease, or hyperekplexia: response to clonazepam and assignment of the gene (STHE) to chromosome 5q by linkage analysis. Ann Neurol 31:663–668
Ryan SG, Buckwalter MS, Lynch JW, Handford CA, Segura L, Shiang R, Wasmuth JJ, Camper SA, Schofield P, O'Connell P (1994) A missense mutation in the gene encoding the α1 subunit of the inhibitory glycine receptor in the *spasmodic* mouse. Nature Genet 7:131–135
Sato K, Zhang JH, Saika T, Sato M, Tada K, Tohyama M (1991) Localization of glycine receptor α1 subunit mRNA-containing neurons in the rat brain: an analysis using *in situ* hybridization histochemistry. Neuroscience 43:381–395
Sato K, Kiyama H, Tohyama M (1992) Regional distribution of cells expressing glycine receptor α2 subunit mRNA in the rat brain. Brain Res 590:95–108
Saul B, Schmieden V, Kling C, Mülhardt C, Gass P, Kuhse J, Becker C-M (1994) Point mutation of glycine receptor α1 subunit in the *spasmodic* mouse affects agonist responses. FEBS Lett 350:71–76
Schmieden V, Betz H (1995) Pharmacology of the inhibitory glycine receptor: agonist and antagonist actions of amino acids and piperidine carboxylic acid compounds. Mol Pharmacol 48:919–927
Schmieden V, Grenningloh G, Schofield PR, Betz H (1989) Functional expression in *Xenopus* oocytes of the strychnine binding 48 kd subunit of the glycine receptor. EMBO J 8:695–700
Schmieden V, Kuhse J, Betz H (1992) Agonist pharmacology of neonatal and adult glycine receptor α-subunits: identification of amino acid residues involved in taurine activation. EMBO J 11:2025–2032
Schmieden V, Kuhse J, Betz H (1993) Mutation of glycine receptor subunit creates β-alanine receptor responsive to GABA. Science 262:256–258
Schmieden V, Jezequel S, Betz H (1996) Novel antagonists of the inhibitory glycine receptor derived from quinolinic acid compounds. Mol Pharmacol 50:1200–1206
Schmitt B, Knaus P, Becker C-M, Betz H (1987) The M_r 93,000 polypeptide of the postsynaptic glycine receptor complex is a peripheral membrane protein. Biochemistry 26:805–811
Shiang R, Ryan SG, Zhu Y-Z, Hahn AF, O'Connell P, Wasmuth JJ (1993) Mutations in the α1 subunit of the inhibitory glycine receptor cause the dominant neurologic disorder hyperekplexia. Nature Genet 5:351–358
Shiang R, Ryan SG, Zhu Y-Z, Fielder TJ, Allen RJ, Fryer A, Yamashita S, O'Connell P, Wasmuth JJ (1995) Mutational analysis of familial and sporadic hyperekplexia. Ann Neurol 38:85–91
Sontheimer H, Becker CM, Pritchett DB, Schofield PR, Grenningloh G, Kettenmann H, Betz H, Seeburg PH (1989) Functional chloride channels by mammalian cell expression of rat glycine receptor subunit. Neuron 2:1491–1497
St John PA, Stephens SL (1993) Adult-type glycine receptors form clusters on embryonic rat spinal cord neurons developing in vitro. J Neurosci 13:2749–2757
Takahashi T, Momiyama A, Hirai K, Hishinuma F, Akagi H (1992) Functional correlation of fetal and adult forms of glycine receptors with developmental changes in inhibitory synaptic receptor channels. Neuron 9:1155–1161
Triller A, Cluzeaud F, Pfeiffer F, Betz H, Korn H (1985) Distribution of glycine receptors at central synapses of the rat spinal cord. J Cell Biol 101:683–688
Vandenberg RJ, French CR, Barry PH, Shine J, Schofield PR (1992) Antagonism of ligand-gated ion channel receptors: two domains of the glycine receptor α subunit form the strychnine-binding site. Proc Natl Acad Sci USA 89:1765–1769
Vergouwe MN, Tijssen MAJ, Shiang R, van Dijk JG, Shahwan SA, Ophoff RA, Frants RR (1997) Hyperekplexia-like syndromes without mutations in the GLRA1 gene. Clin Neurol Neurosurg 99:172–178

Wang J, Reichling DB, Kyrozis A, MacDermott AB (1994) Developmental loss of GABA– and glycine–induced depolarization and Ca^{2+} transients in embryonic rat dorsal horn neurons in culture. Eur J Neurosci 6:1275–1280

Watanabe E, Akagi H (1995) Distribution patterns of mRNAs encoding glycine receptor channels in the developing rat spinal cord. Neurosci Res 23:377–382

White WF, Heller AH (1982) Glycine receptor alteration in the mutant mouse *spastic*. Nature 298:655–657

Ye Q, Koltchine VV, Mihic SJ, Mascia MP, Wick MJ, Finn SE, Harrison NL, Harris RA (1998) Enhancement of glycine receptor function by ethanol is inversely correlated with molecular volume at position α267. J Biol Chem 273:3314–3319

Young AB, Snyder SH (1973) Strychnine binding associated with glycine receptors of the central nervous system. Proc Natl Acad Sci USA 70:2832–2836

Zarbin MA, Wamsley JK, Kuhar MJ (1981) Glycine receptor: light microscopic autoradiographic localization with [^{3}H] strychnine. J Neurosci 1:532–547

Submission date May 7, 1998.

CHAPTER 19

$GABA_A$ Receptor Chloride Ion Channels

R.W. OLSEN and M. GORDEY

A. $GABA_A$ Receptors: Physiological Function, Molecular Structure, Pharmacological Subtypes

GABA is the major inhibitory neurotransmitter in the central nervous system (CNS). Virtually all neurons respond to GABA, while about 30% of them make and utilize it as a neurotransmitter. The $GABA_B$ receptors are defined as bicuculline-insensitive and baclofen-sensitive. They are members of the G protein-coupled, 7 membrane-spanning receptor family, and are associated with activation of K^+ channels or inhibition of Ca^{2+} channels; many times they are located on nerve endings and mediate presynaptic inhibition, including GABA autoreceptors (BOWERY 1993). The recent cloning of the $GABA_B$ receptor (KAUPMANN et al. 1997) raises hopes for the development of new drugs based on agonists or antagonists specific for $GABA_B$ receptor subtypes. $GABA_A$ receptors (GABAR) mediate rapid inhibitory synaptic transmission via chloride channel activation. They are defined by the antagonist bicuculline and the agonist muscimol. The GABA synapse is known to be site of action for many important clinical agents. GABAR appear to be modulated by several classes of CNS depressants, including benzodiazepines (BZ), barbiturates, neuroactive steroids, other general anesthetics including intravenous and volatile agents, and possibly ethanol. These drugs are used for anxiolytic, antiepileptic, sedative/hypnotic, and anesthetic applications. Antagonists of GABAR are generally convulsant, such as the competitive bicuculline and the noncompetitive picrotoxin, pentylenetetrazole, cage convulsants like TBPS, and benzyl penicillin. In addition, some agents acting at the BZ site on GABAR are antagonists of GABA function, and termed 'inverse agonists' (MACDONALD and OLSEN 1994; LUDDENS et al. 1995).

GABAR are the actual molecular targets in brain for many of these drugs that modulate GABA-mediated inhibition. Thus, binding of radiolabeled picrotoxin and benzodiazepines occurs on the same protein as the GABA binding site, i.e., the GABA receptor is the receptor for these other drugs. Further, allosteric modulation of GABA, picrotoxin, and benzodiazepine binding by nonradioactive barbiturates and related drugs and neuroactive steroids shows that binding sites for those drugs are also on the GABA receptor protein (OLSEN 1981). Indeed, all these binding sites co-purify on a single GABAR protein (SIGEL and BARNARD 1984; STAUBER et al. 1987; KING et al.

1987). The purified receptor protein produced partial sequence, leading to the cloning of cDNAs that code for a GABA-regulated chloride channel that is modulated by all these drugs, as shown by expression in heterologous cells (Schofield et al. 1987; Luddens et al. 1995).

GABAR are members of the ligand-gated ion channel superfamily of receptors, which includes nicotinic acetylcholine receptors, inhibitory glycine receptors, and 5HT3 receptors. These receptors share a homologous structure at the subunit level, with a common size of about 50 kDa, a long extracellular N-terminus, including the neurotransmitter binding site, four membrane-spanning domains (M1–M4), and a large intracellular domain between membrane-spanning regions M3 and M4. Five subunits form a pseudo-symmetric heterologous pentamer, with the five M2 domains forming the wall of the ion channel (Delorey and Olsen 1992; see Chap. 15).

GABAR are actually a family of pharmacological subtypes, resulting from differential region- and age-dependent expression of approximately 18 subunit genes, combined as different heteropentamers (Olsen and Tobin 1990; Burt and Kamatchi 1991; Vicini 1991; Macdonald and Olsen 1994; Luddens et al. 1995; Sieghart 1995; McKernan and Whiting 1996). The subunits identified to date by molecular cloning (α1–6, β1–4, γ1–4, δ, ε, ρ1–3) are grouped according to their degree of homology. Different Greek letters indicate subunits of about 30% identity, and then numbered subtypes within each subunit share about 70% identity. Most evidence is consistent with two copies of α, two copies of β, and one of γ, δ, or ε per pentamer. The exact stoichiometry and nearest-neighbor wheel arrangement of the pentamers is not specified. Most evidence also suggests that one major sort of pentamer is expressed for any given combination of subunits. On the other hand, most cells have more than one isoform, and the rules governing assembly of specific subtypes are only beginning to be unearthed. The γ subunit is required for BZ sensitivity, and the details of BZ selectivity vary with the nature of both γ and α subunit. Sensitivity to other drugs varies modestly with subunit composition (Vicini 1991; Macdonald and Olsen 1994; Luddens et al. 1995; Sieghart 1995; McKernan and Whiting 1996).

There are potentially thousands of subunit combinations, but about 20 isoforms appear to be reasonably abundant in nature. Figure 1 shows a cartoon of the basic GABAR structure and a summary of the drug binding sites present. Receptors of varying subunit composition show differential biological regulatory mechanisms, some variation in GABA affinity, and, concomitantly, variable sensitivity to modulatory drugs. This is reflected in a striking regional variation in modulatory drug effects on GABAR binding (Olsen et al. 1990; Olsen and Sapp 1995), as well as regional variation in GABAergic drug (e.g., anesthetic) actions on the central nervous system (Carlson et al. 1997). The regional variation in allosteric modulation of GABAR binding and function can be reconstituted in certain recombinant receptor subunit combinations expressed in heterologous cells. Differential sensitivity to GABAergic drugs for various GABAR subunits also allows the use of the chimeric and

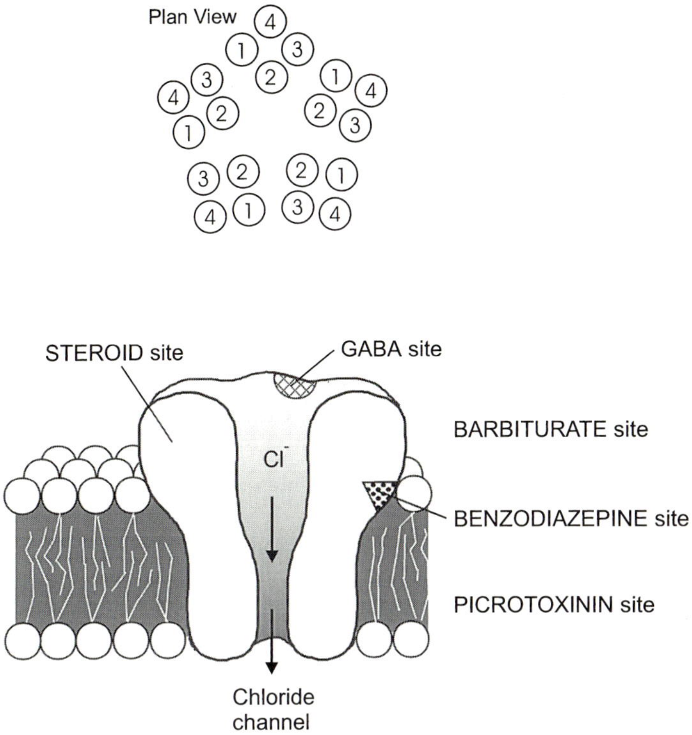

Fig. 1. Structure of the heteropentameric GABA$_A$ receptor indicating the membrane topological arrangment of each subunit, and a list of drugs binding to this protein (modified from OLSEN and DELOREY 1999)

site-directed mutagenesis approach in attempting to define domains of the protein which participate in the binding and actions of the modulatory drugs (DUNN et al. 1994; SMITH and OLSEN 1995; SIGEL and BUHR 1997).

Table 1 lists the major isoforms of GABAR that have been identified in nature and their approximate abundance (OLSEN 1998, modified from MCKERNAN and WHITING 1996). The general approach in obtaining this information is the localization of the subunit mRNAs and polypeptides and identification of co-localized subunits, plus co-immunoprecipitation or co-immunopurification of the various subunits. Also listed are the pharmacological properties of some of the more abundant subunit combinations studied in heterologous expression systems. It should also be admitted that the pharmacological specificity of GABAR in native cells is only approximately explained by the properties of recombinant subunit combinations studied to date. Furthermore, it has not been possible so far to relate the exact subunit expression pattern with the heterogeneity of allosteric drug modulation of GABAR binding reported above, although it appears that some such correspondence must exist.

Table 1. Naturally-occurring isoforms of GABAR, including pharmacological heterogeneity (OLSEN 1998, modified from MCKERNAN and WHITING 1996)

Isoform	Relative abundance	Location	Pharmacology/property
$\alpha1\beta2\gamma2$	40%	Most brain areas; hippocampal, cortical interneurons, cerebellar Purkinje cells	Common co-assembly, BZ-type I, Zn-insensitive
$\alpha2\beta3\gamma2$	15	Spinal cord motoneurons, hippocampal pyramidal cells;	BZ-type II, moderate Zn-sensitive
$\alpha3\beta\gamma2/3$	10	Cholinergic, monaminergic neurons	BZ-type II; Abecarnil-sensitive
$\alpha2\beta\gamma1$	10	Bergmann glia; thalamus; hypothalamus	BZ inverse agonist-enhanced
$\alpha5\beta3\gamma2/3$	3	Hippocampal pyramidal cells	BZ-type II; Zolpidem-insensitive, moderate Zn-sensitive
$\alpha6\beta\gamma2$	2	Cerebellar granule cells	BZ agonist-insensitive; moderate Zn-sensitive
$\alpha6\beta\delta$	3	Cerebellar granule cells	Insensitive to all BZ; GABA high affinity; high Zn-sensitivity; steroid-insensitive
$\alpha4\beta\gamma$	2	Cortical, hippocampal pyramidal cells, striatum	BZ agonist-insensitive; low steroid sensitivity
$\alpha4\beta2\delta$	4	Thalamus; dentate granule cells	Insensitive to all BZ; GABA high affinity; high Zn-sensitivity; steroid-insensitive
All other	11	Throughout CNS	

B. Activators and Inhibitors of $GABA_A$ Receptors

(see Table 2)

I. GABA Site

1. Agonists

Early studies on GABAR emphasized the sensitivity to the antagonist bicuculline. Besides GABA, analogues of equal or lower potency as agonists included 3-aminopropane sulfonate, β-hydroxy GABA, β-guanidino propionate, imidazole-acetic acid, *trans*-aminocrotonic acid, and some cyclic amino acids. β-Alanine was weak, taurine weaker, and glycine inactive. β-Chlorophenyl GABA (baclofen) was very weak. Then several more potent and specific naturally-occurring analogues were identified, primarily by GAR Johnston and P Krogsgaard-Larsen, such as isoguvacine from hypnotic plant

Table 2. List of compounds mentioned in this review

I. GABA Site
1. Agonists
GABA, 3-aminopropane sulfonate (3-APS), β-hydroxy GABA, β-guanidino propionate, imidazole-acetic acid, *trans*-aminocrotonic acid, β-alanine, taurine, glycine, β-chlorophenyl GABA (baclofen), isoguvacine, muscimol, 4,5,6,7-tetrahydroisoxazolo [4,5-*c*]pyridin-3-ol (THIP), piperidine-4-sulfonate (P4 S), ZAPA
2. Antagonists
Bicuculline, SR-95531, RU5135, δ-aminolevulinic acid, penicillin
II. The Picrotoxinin Site
Picrotoxinin, dieldrin, lindane, pentylenetetrazol, benzyl penicillin, *t*-butyl bicyclophosphorothionate (TBPS), pitrazepine, alklated butyrolactones, quinoxaline compounds (U-93631)
III. Benzodiazepine Site Ligands
Diazepam, quazepam, triazolopyridazines (Cl 218,872), β-carbolines (abecarnil), imidazopyridines (zolpidem), imidazolobenzodiazepines (Ro15–1788 = flumazenil, Ro15–4513, midazolam), triazolobenzodiazepines (triazolam), cyclopyrrolones (suriclone, zopiclone), pyrazoloquinolines (CSG8216), imidazoquinoxalines (U97775), imidazoquinolines (U101017), triazolo-benzoxazin-ones, indolyl-glyoxylyl-benzylamines, flavonoids (6,3'-dinitroflavone, 6-methyl flavone), furanocoumarins (phellopterin), yohimbines, Xenovulene A
IV. Barbiturates and Related Drugs
Barbiturates (phenobarbital, pentobarbital, thiopental, CHEB, DMBB, MPPB), pyrazolopyridines (etazolate), chlormethiazole, etomidate / lorecelezole, pyrazinones (U-92813)
V. Neuroactive Steroids
Alphaxalone, 3-α-hydroxy-5-α-pregnane-20-one, tetrahydrocorticosterone (THDOC), ganaxalone
VI. General Anesthetics: Propofol, Volatile Agents, and Alcohols
Propofol, halothane, ethanol, *n*-propanol through *n*-octanol
VII. Miscellaneous agents
Avermectin B_{1a}

extracts and muscimol from hallucinogenic mushrooms (Fig. 2) (Johnston 1996), and the synthetic analogues 4,5,6,7-tetrahydroisoxazolo[4,5-*c*]pyridin-3-ol (THIP) and piperidine-4-sulfonate (Falch et al. 1990). Recent additions to the list include the isothiouronium analogues like ZAPA (Allan et al. 1997). Analogues that do not pass the mammalian blood–brain barrier may have efficacy as veterinary anthelminthics, while GABA-active compounds that do cross the blood–brain barrier are potentially useful in the clinic. Most assays of rapid inhibitory synaptic transmission are more or less equivalently sensitive to all these GABA analogues, although some hints of heterogeneity with tissue have been observed. The relative potencies of most GABA analogues on subtypes of recombinant GABAR have been described (Ebert et al. 1994, 1997). Inhibitory actions of GABA that did not match this "GABA$_A$" pharmacology were ascribed to other receptor classes, GABA$_B$ (baclofen), or "non-A, non-B", or GABA$_C$ (Johnston 1996).

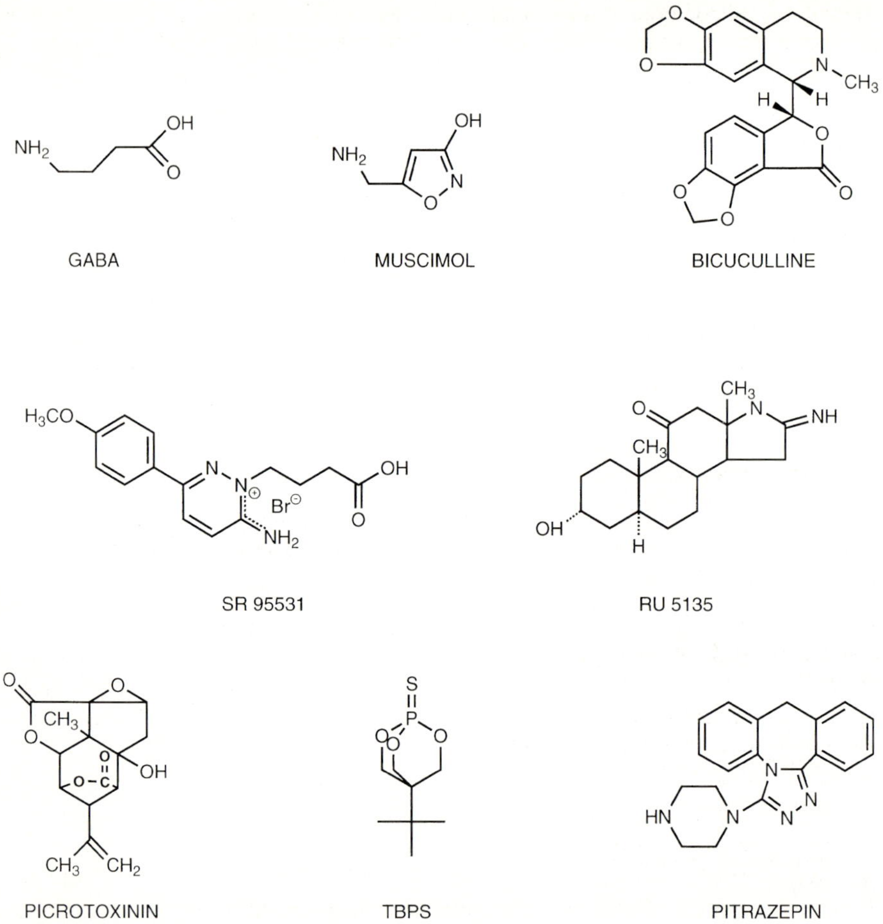

Fig. 2. Chemical structures of selected drugs that act on $GABA_A$ receptors

2. Antagonists

Competitive antagonists of the GABA site include the prototype bicuculline, a convulsant of plant origin (Johnston 1996). This molecule has some chemical similarity to GABA and blocks function and binding competitively. Site-directed mutagenesis of several residues in GABAR leads to simultaneous changes in affinity of GABA and bicuculline (Smith and Olsen 1995). Bicuculline crosses the blood–brain barrier, but its lactone moiety is sensitive to breakdown at pH7 (Olsen 1981); quaternary ammonium salts (bicuculline-methiodide, -methochloride) are chemically stable for in vitro work but do not cross the blood–brain barrier.

Very few chemical analogues of GABA with antagonist efficacy have been described, with the notable exception of certain aryl aminopyridazines typified by SR-95531 (CHAMBON et al. 1985). Another major antagonist of high potency is the amidine steroid RU5135 (OLSEN 1984), (Fig. 2). This compound contains a GABA-like moiety and appears to act at the GABA site, rather than binding to the neuroactive steroid site (see below). It is equipotent on GABAR and glycine receptors (HUNT and CLEMENTS-JEWERY 1981). The endogenous substance δ-aminolevulinic acid can inhibit GABAR, which can be dangerous if it enters the nervous system. Some compounds inhibit GABAR function with unknown mechanism, possibly at the GABA site or one of the many other allosteric and/or chloride channel sites. Certain penicillin antibiotics can be convulsant; this is generally regarded as channel block (OLSEN 1981; OLSEN and DELOREY 1999). Some quinolone antibacterial agents that show convulsant side effects are also GABAR antagonists (KAWAKAMI et al. 1997).

II. The Picrotoxin Site

Picrotoxin, a convulsant of plant origin, is a universal blocker of GABAR, and has been used for many years in many studies. Picrotoxin is a molecular 1:1 mixture of the less active picrotin and highly active picrotoxinin (Fig. 2). Its chemical structure bears no similarity to GABA, it contains no N atom and no charges, and has a lactone and some epoxides. Picrotoxinin's action to block GABAR chloride channels thus involves a site distinct from that for GABA and its functional antagonism is noncompetitive (OLSEN 1981). Its relatively low but not zero potency on ρ receptors parallels the insensitivity of those receptors to bicuculline and almost all of the modulatory drugs described here (JOHNSTON 1996).

Synthesis of a radiolabeled picrotoxinin analogue, [^{3}H]dihydropicrotoxinin, led to the ability to assay GABAR in vitro, and the demonstration of direct biochemical interaction with GABAR of several classes of drugs. The insecticides dieldrin and lindane appear to act like picrotoxinin at the same site, as do the experimental convulsants pentylenetetrazol and benzyl penicillin (OLSEN 1981; MAKSAY and TICKU 1985). Another class of compounds found to inhibit picrotoxinin binding were the 'cage convulsants (TICKU and OLSEN 1979), originally synthesized as potential insecticides but much more potent on mammals and noted as potential chemical warfare agents (CASIDA 1993). These trioxa-bicylo-octanes include bicyclo-phosphates, bicyclo-orthocarboxylates, and bicyclo-phosphorothionates, with three fused six-membered rings. One of these agents [^{35}S]*t*-butyl bicyclophosphorothionate (TBPS) (Fig. 2) was developed as a preferable radioligand for this site (SQUIRES et al. 1983). Another convulsant thought to act at this site is the triazolo-dibenzazepine, pitrazepine (Fig. 2) (GAHWILER et al. 1984). The barbiturates and related depressant agents were also found to inhibit picrotoxinin binding (TICKU and OLSEN 1978) and this was shown to be allosteric inhibition because, unlike competitive inhibitors, the barbiturates enhanced the dissociation rate of TBPS binding (MAKSAY and TICKU 1995). This allosteric mode of inhibition was also observed for GABA site ligands, steroids, and other anesthetics, and benzodiazepines, as well as chloride ions (Squires et al. 1983). Actually, benzodiazepine site ligands allosterically modulate TBPS binding with the same pharmacological specificity as they show at the BZ site, as expected, but also show a competitive inhibition with much lower potency and no correlation with BZ site interaction. Two compounds believed to inhibit GABAR function via this picrotoxin site interaction include the benzodiazepines Ro5–3663 (Leeb-Lundberg et al. 1981) and Ro5–4864 (Maksay and Ticku 1985; Gee et al. 1988a).

Picrotoxinin contains an essential conjugated butyrolactone moiety; the butyrolactone element was utilized in several synthetic alklated butyrolactones, some of which enhance, while most of them inhibit GABAR at the picrotoxin site or nearby (Holland et al. 1993). In the course of studying quinoxaline compounds for potential action at the BZ site (below), compounds with picrotoxin-like activity were also found (U-93631) (Dillon et al. 1995).

III. Benzodiazepine Site Ligands

The clinically important benzodiazepines (BZ), like diazepam (Fig. 3), act by enhancing GABAR function in the CNS (Haefely 1994). The receptor sites for BZ in the CNS (Squires and Braestrup 1977; Möhler and Okada 1977) are on GABAR proteins. The BZ are used as anxiolytics, sedative/hypnotics, and anticonvulsants (not anesthetics, although they may be useful adjuncts in induction of anesthesia). Side effects include sedation, intoxication, interaction with ethanol, paradoxical CNS stimulation, and addiction potential; they also exhibit tolerance, especially to the anticonvulsant efficacy.

BZs enhance GABAR function in many neurons. They do not directly activate the GABAR chloride channel in the absence of GABA. There are some subtypes of GABAR on which classical BZ agonists are inactive. Early receptor binding studies identified at least two sorts of BZ receptors based on differential affinity for some ligands. Although diazepam and classical BZ showed a single affinity, some binding sites had a higher (Type I) or lower (Type II) affinity for certain drugs (Fig. 3), like the triazolopyridazines (Cl 218,872) (Squires et al. 1979) and β-carbolines (Braestrup et al. 1984); other Type I selective drugs include imidazopyridines (zolpidem) (Arbilla et al. 1986) (Fig. 3) and certain atypical benzodiazepines (quazepam) (Billard et al. 1988).

The heterogeneity of BZ sites on GABAR results from the different receptor subtypes described above that are protein isoforms of differing subunit composition (Barnard et al. 1998). Type I was due to the isoforms containing the α1 subunit, while Type II is actually a mixture of isoforms containing the α2, α3, or α5 subunits. This group can be further differentiated by the drug zolpidem that binds α2 and α3 with moderate affinity, but has almost no affinity for α5-containing isoforms (Luddens et al. 1995). GABAR con-

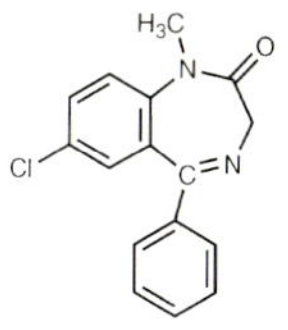

DIAZEPAM

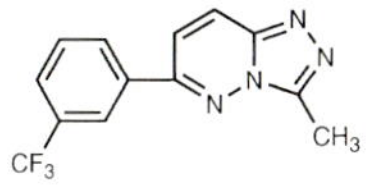

CI 218,872
(TRIAZOLOPYRIDAZINE)

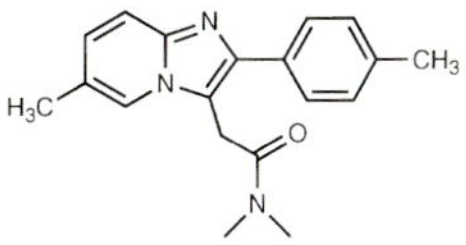

ZOLPIDEM
(IMIDAZOPYRIDINE)

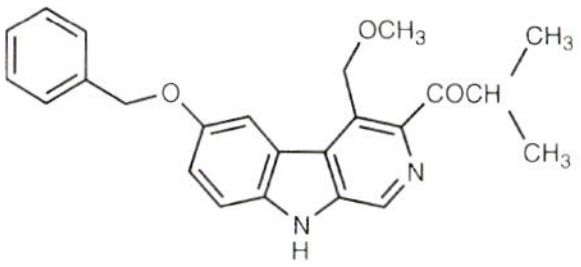

ABECARNIL
(β-CARBOLINE)

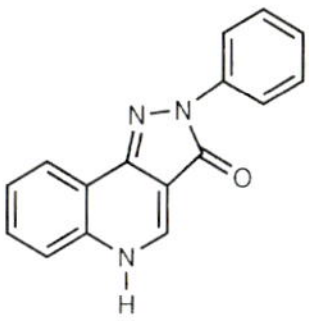

CGS 8216
(PYRAZOLOQUINOLINONE)

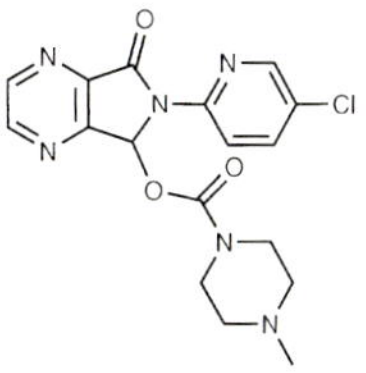

ZOPICLONE

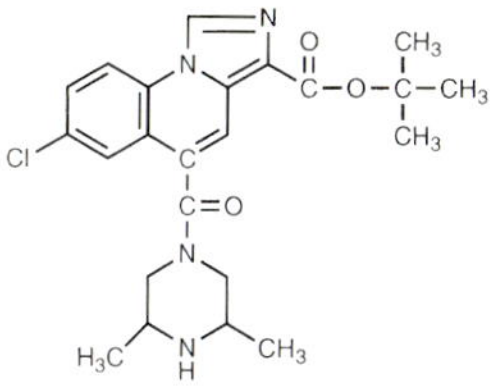

U-101017
(IMIDAZOQUINOLINE)

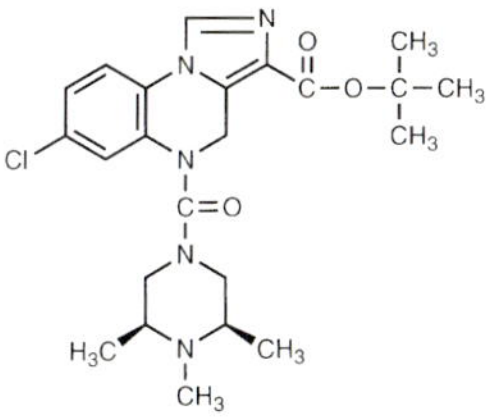

U-97775
(IMIDAZOQUINOXALINE)

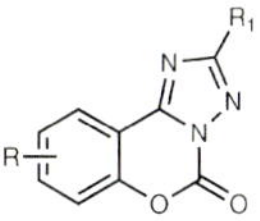

TRIAZOLOBENZOXAZINE

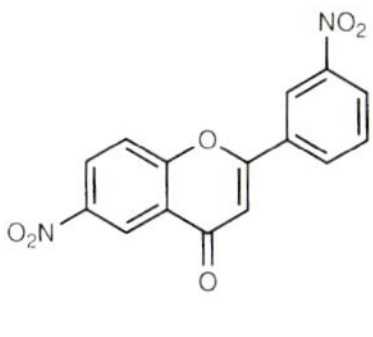

6, 3'-DINITROFLAVONE
(FLAVONOID)

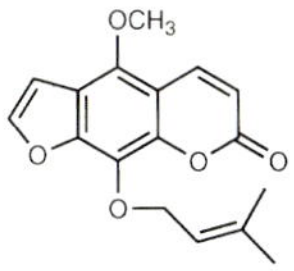

FURANOCUMARIN

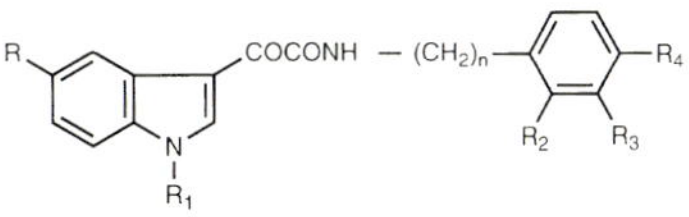

N-[(INDOL-3-YL)GLYOXYLYL]AMINE

Fig. 3. Chemical structures of selected drugs that act on the BZ site of GABA$_A$ receptors

taining $\alpha 4$ or $\alpha 6$ do not bind BZ agonists, but still bind antagonists and inverse agonists, with some differences in efficacy for these agents as well. There is hope that drugs with improved clinical profile may result from targeting receptor subtypes for specific efficacies, or compounds with partial agonist efficacy, as well as reduced tolerance potential (COSTA et al. 1995). Tolerance may result from down-regulation of GABAR by removal of the GABAR protein, post-translational modification, or a switch in expression of subunits, i.e., replacement of GABAR isoforms with new ones that do not respond to the drug given chronically. Understanding these mechanisms may also help to design more suitable BZ drugs.

Imidazolobenzodiazepines (e.g., Ro15–1788, below, and midazolam) (HESTER et al. 1980) and triazolobenzodiazepines (triazolam) (HESTER et al. 1980) of varying potency and efficacy were described early. Non-benzodiazepines active at the BZ site included cyclopyrrolone derivatives: suriclone and zopiclone (Fig. 3) have been described as hypnotics with clinical features superior to most BZ (JULOU et al. 1985). Analogs with anxiolytic efficacy and low sedative potential have also been developed (DOBLE et al. 1993). Several derivatives of pyrazoloquinolines (Fig. 3: see CGS8216 below) have antagonist or partial agonist activity at the BZ site (WANG et al. 1995).

Several other agents active at the BZ sites may show improved clinical profiles due to one of the factors mentioned: subtype selectivity, partial agonist efficacy, or reduced tolerance development. These include β-carbolines (abecarnil, see below). Several active compounds have been developed in the chemical category of imidazoquinoxalines (U97775) (IM et al. 1995) and the related imidazoquinolines (U101017: Fig. 3) (IM et al. 1996). Some of these analogues have nanomolar affinity and agonist or partial agonist efficacy at the BZ site. In addition, some of them have dual functionality, in that at high concentrations they also inhibit GABAR function at a second site which is not blocked by the BZ site antagonist flumazenil. The 'second site' may correspond to that for picrotoxin, barbiturates, or non-barbiturates of related pharmacology (below).

Thus several types of chemical structure help to define the pharmacophore at the BZ site(s). Two other types of chemicals that show activity at the BZ site include the triazolo-benzoxazin-ones (CATARZI et al. 1995) and indolylglyoxylyl-benzylamine derivatives (SETTIMO et al. 1996) (Fig. 3).

Antagonists for the BZ site have been described, such as the prototype imidazobenzodiazepine, flumazenil (Ro15–1788) (HUNKELER et al. 1981) and the pyrazoloquinoline CGS8216 (CZERNIK et al. 1982) . The BZ site is one of a few such drug receptors that has ligands of variable efficacy, not just partial agonists, but agents with the opposite efficacy as the classic agents. These are termed inverse agonists. Inverse agonists inhibit GABAR function at the cellular level, and are anxiogenic and proconvulsant in vivo. The first inverse agonists described were β-carbolines (BRAESTRUP et al. 1984), but other structures including BZ (Ro15–4513) can have this efficacy; conversely, β-carbolines can also be antagonists or agonists (e.g., abecarnil) (STEPHENS et al. 1990). The $\alpha 4$ and $\alpha 6$ subunit-containing receptors are insensitive to agonists like diazepam and still retain high affinity for inverse agonists, but many drugs that are inverse agonists on $\alpha 1$, $\alpha 2 \alpha 3$, and $\alpha 5$-containing receptors show agonist efficacy on these unusual $\alpha 4/\alpha 6$ receptors (LUDDENS et al. 1995). Pure antagonists have been tested clinically for treatment of BZ overdose, but also for alleviation of withdrawal symptoms in patients dependent on not only BZ but also ethanol and other drugs of abuse. Partial inverse agonists like the imidazobenzodiazepine Ro15–4513 have been proposed for therapy in alcohol withdrawal or even as nootropic agents, but they appear to be too excitatory for clinical use, due to proconvulsant activities and tremor induction (HINDMARCH and OTT 1988; HAEFELY 1994).

A number of naturally-occurring compounds have been identified as inhibitors of radioactive BZ binding to brain membranes. To date, no substance from brain has been shown to have any physiological activity at GABAR. Compounds in plants, animals, and microorganisms are used in folk medicines in several cultures for tranquilizers or analgesics. The major ingredients appear to be several flavonoids (Fig. 3). Most of these are agonists, in some cases anxioselective (6,3'-dinitroflavone) (MEDINA et al. 1997), while some antagonists also exist (6-methyl flavone) (AI et al. 1997a). Several potent partial agonists from plants used in Chinese and Japanese herbal medicine concoctions were identified as furanocumarins (phellopterin: Fig. 3) (DEKERMENDJIAN et al. 1996). Other active plant substances include certain yohimbines (AI et al. 1997b). A tetracylic microbial compound Xenovulene A has nanomolar affinity for the BZ site and antagonist/partial agonist efficacy (SUNDARAM et al. 1997).

IV. Barbiturates and Related Drugs

Barbiturates (short, medium, and long-lasting) that are active as CNS depressants all enhance GABAR-mediated inhibition. Long-acting barbiturates such as phenobarbital are used clinically for epilepsy; intermediate-acting barbiturates such as pentobarbital (Fig. 4) were previously used as sedative/hypnotics; and short-acting barbiturates such as thiopental are used as intravenous anesthetics. The structure-activity relationships for a series of barbiturates including stereoisomers agrees perfectly for allosteric modulation of GABAR binding and enhancement of GABA currents in neurons (LEEB-LUNDBERG et al. 1980; OLSEN 1981; SKOLNICK et al. 1981; WILLOW and JOHNSTON 1983; OLSEN et al. 1986, 1991). This is the major candidate mechanism for the pharmacological actions of barbiturates at the animal/patient level. The barbiturates, as well as other anesthetics described below, also directly activate the GABAR chloride channel in the absence of GABA. Some nonbarbiturates that appear to act exactly as barbiturates/neurosteroids and possibly at the same site(s) on GABAR include pyrazolopyridines (etazolate, an anxiolytic), clormethiazole (an anticonvulsant), etomidate (anesthetic)/lorecelezole (anxiolytic), and pyrazinones (U-92813) (IM et al. 1993). These drugs (Fig. 4) have varying degrees of antiepileptic, anxiolytic, and sedative/hypnotic/anesthetic efficacy (MAKSAY and TICKU 1985). These sites represent relatively under-utilized targets for clinical agents producing positive modulation of GABAR.

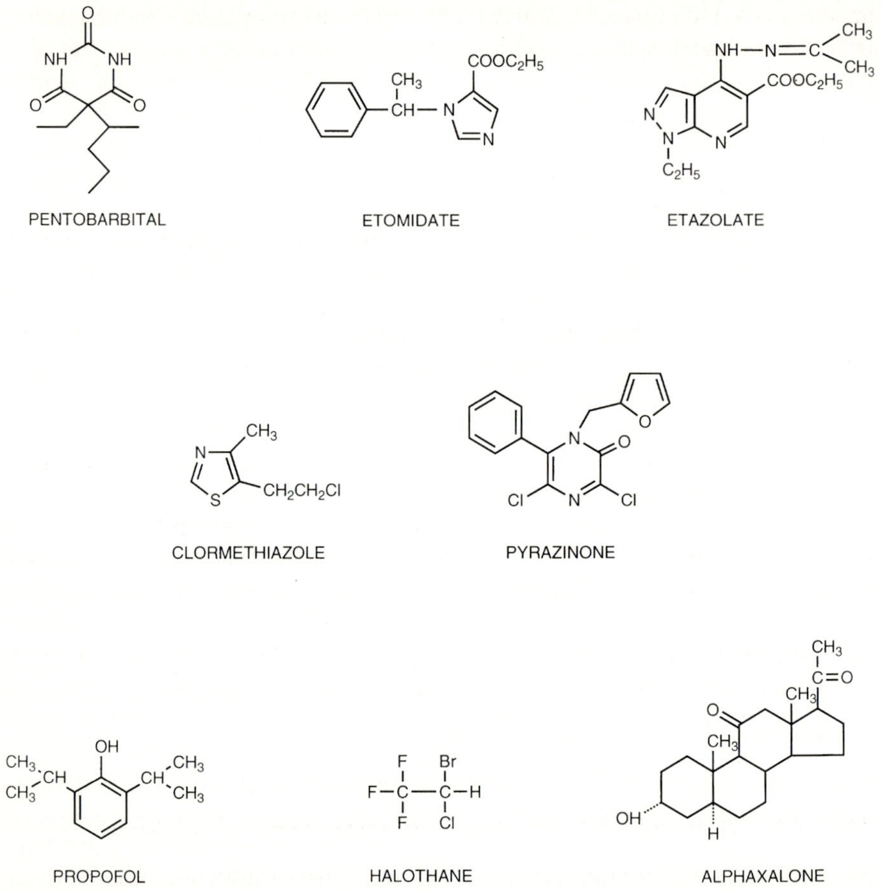

Fig. 4. Chemical structures of selected drugs that act on the barbiturate or anesthetics sites on $GABA_A$ receptors

Some barbiturates, e.g., cyclohexylidene-ethyl barbiturate (CHEB) and dimethyl dibutyl barbiturate, DMBB, are relatively potent convulsants. The (+) isomer of DMBB is more potent as a convulsant than (–). The excitatory activity is not due to inhibition of GABAR but rather action at some other target. Thus, both isomers of DMBB are enhancers of GABAR function and modulate binding accordingly. In this activity, the (–) isomer of DMBB is more potent than (+), as seen also with pentobarbital isomers. In the case of pentobarbital, the (–) isomer depressant effect to enhance GABA predominates over its (+) isomer convulsant activity at a non-GABA site; for DMBB, the (+) isomer excitatory action at a non-GABA site is predominant over its (–) isomer GABA enhancement (OLSEN 1981; OLSEN et al. 1986, 1991). Nevertheless, some isomers of barbiturates, e.g., (+) *N*-methyl, phenyl, propyl barbiturate (MPPB) may be antagonists of those barbiturates like (–)MPPB that enhance GABA. Thus, variable efficacy may be possible at this site (Maksay and Ticku 1995).

V. Neuroactive Steroids

The anesthetic steroid alphaxalone enhances GABAR function and modulates binding at the pharmacologically relevant concentration (Harrison et al. 1987; Gee et al. 1988b; Peters et al. 1988). This synthetic steroid is an analogue of a progesterone metabolite 3-α-hydroxy-5-α-pregnane-20-one that was shown to have sedative activity. Chemically related metabolites of corticosterone (tetrahydrocorticosterone, THDOC) and metabolites of testosterone are active as enhancers of GABAR. These agents have rapid and direct effects on the nervous system that are not related to steroid hormone action on cytoplasmic receptors to regulate gene expression. The endogenous neuroactive steroids are called neurosteroids; their actions may be physiological. The neurosteroids are inactive at the hormone receptors and the hormones are inactive at the neuroreceptors (Olsen and Sapp 1995; Lambert et al. 1995). New synthetic steroids and nonsteroid analogues have improved bioavailability and potential as anxiolytics and antiepileptics (Wittmer et al. 1996; Hogenkamp et al. 1997). Anticonvulsant efficacy was reported for these GABA enhancing steroids (Kokate et al. 1994), and at least one product, ganaxalone, protects against metrazole seizures and may be a candidate for epilepsy therapeutics (Beekman et al. 1998). GABA-active steroids may mediate, at least partly, the well-known antiepileptic action of glucocorticoids (Olsen and Sapp 1995).

VI. General Anesthetics: Propofol, Volatile Agents, and Alcohols

Exactly like barbiturates, etomidate, and steroid anesthetics, propofol (Fig. 4) also enhances GABAR function and modulates binding. Volatile anesthetics, such as halothane, and long chain alkyl alcohols, like *n*-propanol through *n*-octanol, enhance GABAR function and modulate binding in a similar manner, again implicating direct biochemical interaction with the GABAR protein (Moody et al. 1988; Narahashi et al. 1991; Longoni et al. 1993; Harris et al. 1995). The important alcohol ethanol (10–90 mmol/l) has been reported to enhance GABAR function in some but not all cases tested. This potent effect may be relevant to the mechanism of acute ethanol action. The tissue- (and investigator-) dependent nature of ethanol-GABAR interactions suggests that the interaction may be receptor subunit-dependent. However, because there is no interaction with GABAR binding at intoxicating concentrations, a tissue-dependent indirect action is most likely. Ethanol may affect GABAR binding at higher concentrations (150–500 mmol/l) but these are probably fatal levels in vivo (Narahashi et al. 1991; Mihic and Harris 1996).

These actions of anesthetics are consistent with the hypothesis that general anesthetics have a common mechanism of action which is to enhance GABAR-mediated inhibition. All of the general anesthetics of diverse chemical structure, except maybe ketamine which produces a unique sort of anesthesia, enhance GABA at the anesthetic concentration (Olsen 1998), while they do not affect any other neuronal targets (although at just slightly higher concentrations they interact with other ligand-gated ion channels, receptors for glycine, 5HT3, nicotinic acetylcholine, and glutamate). This specific interaction with GABAR includes stereoisomeric cyclobutane compounds that do not obey the Meyer-Overton correlation (Harris et al. 1995). The interaction with GABAR, like that of barbiturates and steroids, is directly on the receptor protein. Exactly how physical binding of the anesthetics to the GABAR increases receptor function by potentiating GABA responses and/or direct channel activation is under heavy study currently (Franks and Lieb 1994; Mihic et al. 1997; Olsen 1998).

VII. Miscellaneous Agents

Avermectin B_{1a}, a natural product from bacteria, is a cyclic polyene antihelminthic agent. It is active at a variety of ligand-gated ion channels operated by GABA, glutamate, and acetylcholine, especially in invertebrates. It has potent actions on mammalian GABAR (Olsen 1981; Sieghart 1995). Since it probably does not pass the blood–brain barrier, this is only of research interest. Novel azole derivatives inhibit GABAR function at an unknown site in nematodes (Bascal et al. 1996). Antispastic triazole drugs enhance GABAR function at an unknown site in rat brain (Miller et al. 1995).

C. Discussion

GABAR represent one of the richest targets for CNS-active drugs yet uncovered. With the benzodiazepines, their use for anxiolytic effects, as well as other stress syndromes and panic, reached legendary proportions. BZ as well as drugs of other chemical structures that act at this site have other applications in the clinic, including antiepileptic and sedative/hypnotic efficacy. The potential for GABAR subtype-selective agents, partial agonists, and agents with less tolerance development suggests that this BZ site can be tapped further for new therapeutic agents. In addition, at least four other allosteric sites on the GABAR/chloride ionophore complex, namely, the sites for barbiturates, neurosteroids, picrotoxin, and general anesthetics, are additional targets for which compounds of many classes of chemical structure may be designed, optimizing pharmacological profiles.

References

Ai J, Dekermendjian K, Wang X, Nielsen M, Witt MR (1997a) 6-Methylflavone, a benzodiazepine receptor ligand with antagonistic properties on rat brain and human recombinant $GABA_A$ receptors in vitro. Drug Devel Res 41:99–106

Ai J, Dekermendjian K, Nielsen M, Witt MR (1997b) The heteroyohimbine mayumbine binds with high affinity to rat brain benzodiazepine receptors in vitro. Natur Prod Lett 11:73–76
Arbilla S, Allen J, Wick A, Langer SZ (1986) High affinity [^{3}H]zolpidem binding in the rat brain: an imidazopyridine with agonist properties at central benzodiazepine receptors. Eur J Pharmacol 130:257–263
Barnard EA, Skolnick P, Olsen RW, Möhler H, Sieghart W, Biggio G, Braestrup C, Bateson AN, Langer SZ (1998) Sub-types of γ-aminobutyric acid$_A$ receptors: classification on the basis of subunit structure and receptor function. Int Union Pharmacol XV. Pharmacol Rev 50:291–313
Bascal Z, Holden-Dye L, Willis RJ, Smith SWG, Walker RJ (1996) Novel azole derivatives are antagonists at the inhibitory GABA receptor on the somatic muscle cells of the parasitic nematode Ascaris suum. Parasitology 112:253–259
Beekman M, Ungard JT, Gasior M, Carter RB, Kijksta D, Goldberg SR, Witkin JM (1998) Reversal of behavioral effects of pentylenetetrazol by the neuroactive steroid ganaxalone. J Pharmacol Exp Ther 284:868–877
Billard W, Crosby G, Iorio L, Chipkin R, Barnett A (1988) Selective affinity of the benzodiazepines quazepam and 2-oxoquazepam for BZ$_1$ binding sites and demonstration of [^{3}H]2-oxoquazepam as a BZ$_1$ selective radioligand. Life Sci 42:179–187
Bowery NG (1993) GABA$_B$ receptor pharmacology. Ann Rev Pharmacol Toxicol 33:109–147
Braestrup C, Honore T, Nielsen M, Petersen EN, Jensen LH (1984) Ligands for benzodiazepine receptors with positive and negative efficacy. Biochem Pharmacol 33:859–862
Burt DR, Kamatchi GL (1991) GABA$_A$ receptor subtypes: from pharmacology to molecular biology. FASEB J 5:2916–2923
Carlson BX, Hales TG, Olsen RW (1997) GABA$_A$ receptors and anesthesia, In Anesthesia, Biologic Foundations, Yaksh TL, Lynch C, Zapol WM, Maze M, Biebuyck JF, Saidman LJ (eds), Lippincott-Raven Publishers, New York, Ch 16, pp 259–275
Casida JE (1993) Insecticide action at the GABA-gated chloride channel: recognition, progress, and prospects. Arch Insect Biochem Physiol 22:13–23
Catarzi D, Cecchi L, Colotta V, Filacchioni G, Varano F (1995) Synthesis of some 2-aryl-1,2,4-triazolo[1,5-c][1,3]benzoxazin-5-ones as tools to define the essential pharmacophoric descriptors of a benzodiazepine receptor ligand. J Med Chem 38:2196–2201
Chambon JP, Feltz P, Heaulme M, Restle S, Schlichter R, Biziere K, Wermuth CG (1985) An aryl aminopyridazine derivative of GABA is a selective and competitive antagonist at the GABA$_A$ receptor site. Proc Natl Acad Sci USA 82:1832–1836
Costa E, Auta J, Caruncho H, Guidotti A, Impagnatiello F, Pesold C, Thompson DM (1995) A search for a new anticonvulsant and anxiolytic benzodiazepine devoid of side effects and tolerance liability. In GABA$_A$ Receptors in Anxiety, Biggio G, Sanna E, Serra M, Costa E (eds) Raven Press, NY, Adv Biochem Psychopharm 48:75–92
Czernik AJ, Petrack B, Kalinsky HJ, Psychoyos S, Cash WD, Tsai C, Rinehart RK, Granat FR, Lovell RA, Brundish DE, Wade R (1982) CGS-8216: receptor binding characteristics of a potent benzodiazepine antagonist. Life Sci 30:363–372
Dekermendjian K, Ai J, Nielsen M, Sterner O, Shan R, Witt MR (1996) Characterization of the furanocoumarin phellopterin as a rat brain benzodiazepine receptor partial agonist in vitro. Neurosci Lett 219:151–154
DeLorey TM, Olsen RW (1992) γ-Aminobutyric acid$_A$ receptor structure and function. J Biol Chem 267:16747–16750
Dillon GH, Im WB, Pregenzer JF, Carter DB, Hamilton BJ (1995) 4-Dimethyl-3-t-butylcarboxyl-4,5-dihydro(1,5-a) quinoxaline is a novel ligand to the picrotoxin site on GABA$_A$ receptors, and decreases single-channel open probability. J Pharmacol Exp Ther 272:597–603

Doble A, Canton T, Dreisler S, Piot O, Boireau A, Stutzmann JM, Bardone MC, Rataud J, Roux M, Roussel G, Bourzat JD, Cotrel C, Pauchet C, Zundel JL, Blanchard JC (1993) RP59037 and RP60503: Anxiolytic cyclopyrrolone derivatives with low sedative potential. Interaction with the $GABA_A$/benzodiazepine receptor complex and behavioral effects in the rodent. J Pharmacol Exp Ther 266:1213–1226

Dunn SMJ, Bateson AN, Martin IL (1994) $GABA_A$ receptors. Int Rev Neurobiol 36: 51–96

Ebert B, Wafford KA, Whiting PJ, Krogsgaard-Larsen P, Kemp JA (1994) Molecular pharmacology of $GABA_A$ receptor agonists and partial agonists in oocytes injected with different α, β, and γ receptor subunit combinations. Mol Pharmacol 46:957–963

Ebert B, Thompson SA, Saounatsou K, McKernan R, Krogsgaard-Larsen P, Wafford KA (1997) Differences in agonist/antagonist binding affinity and receptor transduction using recombinant human $GABA_A$ receptors. Mol Pharmacol 52:1150–1156

Falch E, Larsson OM, Schousboe A, Krogsgaard-Larsen P (1990) $GABA_A$ agonists and GABA uptake inhibitors. Drug Dev Res 21:169–188

Franks NP, Lieb WR (1994) Molecular and cellular mechanisms of general anaesthesia. Nature 367:607–614

Fritschy JM, Möhler H (1995) $GABA_A$ receptor heterogeneity in the adult rat brain: differential regional and cellular distribution of seven major subunits. J Comp Neurol 359:159–194

Gähwiler BH, Maurer R, Wüthrich HJ (1984) Pitrazepine, a novel GABA antgonist. Neurosci Lett 45:311–316

Gee KW, Brinton RE, McEwen BS (1988a) Regional distribution of a Ro5–4864 binding site that is functionally coupled to the GABA/benzodiazepine receptor complex in rat brain. J Pharmacol Exp Ther 244:379–383

Gee KW, Bolger MB, Brinton RE, Coirini H, McEwen BS (1988b) Steroid modulation of the chloride ionophore in rat brain: structure-activity requirements, regional dependence, and mechanism of action. J Pharmacol Exp Ther 246:803–812

Haefely WE (1994) Allosteric modulation of the $GABA_A$ receptor channel: a mechanism for interaction with a multitude of central nervous system functions. In The Challenge of Neuropharmacology (Möhler H, DaPrada M, eds) Editiones Roche, Basel, Switzerland, pp. 15–40

Harris RA, Mihic SJ, Dildy-Mayfield JE, Machu TK (1995) Actions of anesthetics on ligand-gated ion channels: role of receptor subunit composition. FASEB J 9:1454–1462

Harrison NL, Majewska MD, Harrington JW, Barker JL (1987) Structure activity relationships for steroid interaction with the $GABA_A$ receptor complex. J Pharmacol Expt Ther 241:346–353

Hester JB, Rudzik AD, von Voigtlander PF (1980) 1-(Aminoalkyl-6-aryl-4H-s-triazolo[4,3-a][1,4]benzodiazepines with antianxiety and antidepressant activity. J Med Chem 23:392–402

Hindmarch I, Ott H (eds) (1988) Benzodiazepine Receptor Ligands: Memory and Information Processing, Springer-Verlag, Berlin

Hogenkamp DJ, Tahir SH, Hawkinson JE, Upasani RB, Alauddin M, Kimbrough Cl, Acosta-Burruel M, Whittemore ER, Woodward RM, Lan NC, Gee KW, Bolger MB (1997) Synthesis and in vitro activity of 3β-substituted-3α-hydroxypregnan-20-ones: allosteric modulators of the $GABA_A$ receptors. J Med Chem 40, 61–72

Holland KD, Bouley MG, Covey DF, Ferrendelli JA (1993) Alkyl-substituted γ-butyrolactones act at a distinct site allosterically linked to the TBPS/picrotoxinin site on the $GABA_A$ receptor complex. Brain Res. 615:170–174

Hunkeler W, Möhler H, Pieri L, Polc P, Bonetti EP, Cumin R, Schaffner R, Haefely W (1981) Selective antagonists of benzodiazepines. Nature 290:514–516

Hunt P, Clements-Jewery S (1981) A steroid derivative R5135 antagonizes the GABA/benzodiazepine receptor interaction. Neuropharmacology 20:357–361

Im HK, Im WB, Judge TM, Gammill RB, Hamilton BJ, Carter DB, Pregenzer JF (1993) Substituted pyrazinones, a new class of allosteric modulators for $GABA_A$ receptors. Mol Pharmacol 44:468–472
Im HK, Im WB, Pregenzer JF, Carter DB, Jacobsen EJ, Hamilton BJ (1995) Characterization of U-97775 as a $GABA_A$ receptor ligand of dual functionality in cloned rat $GABA_A$ receptor subtypes. Br J Pharmacol 115:19–24
Im HK, Im WB, Von Voightlander PF, Carter DB, Murray BH, Jacobsen EJ (1996) Characterization of U-101017 as a $GABA_A$ receptor ligand of dual functionality. Brain Res 714:165–168
Johnston GAR (1996) $GABA_C$ receptors: relatively simple transmitter-gated ion channels? Trends Pharmacol Sci 17:319–323
Julou L, Blanchard JC, Dreyfus JF (1985) Pharmacological and clinical studies of cyclopyrrolones: zopiclone and suriclone. Pharmacol Biochem Behav 23:653–659
Kaupmann K, Huggel K, Heid J, Flor PJ, Bischoff S, Mickel SJ, McMaster G, Angst C, Bittiger H, Frostl W, Bettler B (1997) Expression cloning of $GABA_B$ receptors uncovers similarity to metabotropic glutamate receptors. Nature 386:239–246
Kawakami J, Yamamoto K, Asanuma A, Yanagisawa K, Sawada Y, Iga T (1997) Inhibitory effect of new quinolones on $GABA_A$ receptor-mediated responses and its potentiation with felbinac in Xenopus oocytes injected with mouse brain mRNA; correlation with convulsive potency in vivo. Tox App Pharmacol 145:246–254
King RG, Nielsen M, Stauber GB, Olsen RW (1987) Convulsant/ barbiturate activities on the soluble GABA/benzodiazepine receptor complex. Eur J Biochem 169:555–562
Kokate TG, Svensson BE, Rogawski MA (1994) Anticonvulsant activity of neurosteroids: correlation with GABA-evoked chloride current potentation. J Pharmacol Exp Ther 270:1223–1229
Lambert JJ, Belelli D, Hill-Venning C, Peters JA (1995) Neurosteroids and $GABA_A$ receptor function. Trends Pharmacol Sci 16:295–303
Leeb-Lundberg F, Snowman A, Olsen RW (1980) Barbiturate receptors are coupled to benzodiazepine receptors. Proc Natl Acad Sci USA 77:7468–7472
Leeb-Lundberg F, Napias C, Olsen RW (1981) Dihydropicrotoxinin binding sites in mammalian brain: interaction with convulsant and depressant benzodiazepines. Brain Res 216:399–408
Longoni B, Demontis G, Olsen RW (1993) Enhancement of $GABA_A$ receptor function and binding by the volatile anesthetic halothane. J Pharmacol Exp Ther 266:153–160
Lüddens H, Korpi ER, Seeburg PH (1995) $GABA_A$/benzodiazepine receptor heterogeneity: neurophysiological implications. Neuropharmacology 34:245–254
Macdonald RL, Olsen RW (1994) $GABA_A$ receptor channels. Ann Rev Neurosci 17:569–602
Maksay G, Ticku MK (1985) Dissociation of $[^{35}S]$t-butyl bicyclophosphorothionate binding differentiates convulsant and depressant drugs that modulate GABAergic transmission. J Neurochem 44:480–486
McKernan RM, Whiting PJ (1996) Which $GABA_A$-receptor subtypes really occur in the brain? Trends Neurosci 19:139–143
Medina JH, Viola H, Wolfman C, Marder M, Wasowski C, Calvo D, Paladini AC (1997) Overview– Flavonoids: a new family of benzodiazepine receptor ligands. Neurochem Res 22:419–425
Mihic SJ, Harris RA (1996) Alcohol actions at the $GABA_A$ receptor/chloride channel complex. In Deitrich RA, Erwin VG, Eds, Pharmacological effects of ethanol on the nervous system, CRC Press, Boca Raton, pp. 51–72
Mihic SJ, Ye Q, Wick MJ, Koltchine VV, Krasowski MD, Finn SE, Mascia MP, Valenzuela CF, Hanson KK, Greenblatt EP, Harris RA, Harrison NL (1997) Sites of alcohol and volatile anesthetic action on $GABA_A$ and glycine receptors. Nature 389:385–389

Miller JA, Braun D, Chmielewski PA, Kane JM (1995) The enhancement of muscimol-stimulated ^{36}Cl influx by the antispastic 5-aryl-3-(alkylsulfonyl)-4H-1,2,4-triazole (MDL 27 531) in rat brain membrane vesicles. Neurosci Lett 201:183–187

Möhler H, Okada T (1977) Benzodiazepine receptors– demonstration in the central nervous system. Science 198:849–851

Moody EJ. Suzdak PD, Paul SM, Skolnick P (1988) Modulation of the benzodiazepine/GABA receptor chloride channel complex by inhalational anesthetics. J Neurochem 51:1386–1393

Narahashi T, Arakawa O, Nakahiro M, Twombly DA (1991) Effects of alcohols on ion channels of cultured neurons, in Molecular and Cellular Mechanisms of Alcohol and Anesthetics, E Rubin, KW Miller, SH Roth, eds, Ann NY Acad Sci 625:26–36

Olsen RW (1981) GABA-benzodiazepine-barbiturate receptor interactions. J Neurochem 37:1–13

Olsen RW (1984) GABA receptor binding antagonism by the amidine steroid RU5135. Eur J Pharmacol 103:333–337

Olsen RW (1998) The molecular mechanism of action of general anesthetics: structural aspects of interactions with $GABA_A$ receptors. Toxicol Lett 101:193–201

Olsen RW, DeLorey TM (1999) GABA and Glycine, in Basic Neurochemistry, 6th edn, Siegel G, et al. eds. Lippincott-Raven, New York, pp 335–346

Olsen RW, Sapp DW (1995) Neuroactive steroid modulation of $GABA_A$ receptors, in $GABA_A$ Receptors and Anxiety: From Neurobiology to Treatment, G Biggio, E Sanna, M Serra, E Costa, eds, Raven Press, New York. Adv Biochem Psychopharm 48:57–74

Olsen RW, Tobin AJ (1990) Molecular biology of $GABA_A$ receptors. FASEB J 4:1469–1480

Olsen RW, Fischer JB, Dunwiddie TV (1986) Barbiturate enhancement of GABA receptor binding and function as a mechanism of anesthesia, in Molecular and Cellular Mechanisms of Anaesthetics, S Roth, K Miller, eds. Plenum Publishing Corporation, New York, pp 165–177

Olsen RW, McCabe RT, Wamsley JK (1990) $GABA_A$ receptor subtypes: autoradiographic comparison of GABA, benzodiazepine, and convulsant binding sites in the rat central nervous system. J Chem Neuroanat 3:59–76

Olsen RW, Sapp DW, Bureau MH, Turner DM, Kokka N (1991) Allosteric actions of CNS depressants including anesthetics on subtypes of the inhibitory $GABA_A$ receptor-chloride channel complex, in Molecular and Cellular Mechanisms of Alcohol and Anesthetics, E Rubin, KW Miller, SH Roth, eds, Ann NY Acad Sci 625:145–154

Peters JA, Kirkness EF, Callachan H, Lambert JL, Turner AJ (1988) Modulation of the $GABA_A$ receptor by depressant barbiturates and pregnane steroids. Br J Pharmacol 94: 1257–1269

Schofield PR, Darlison MG, Fujita N, Burt DR, Stephenson FA, Rodriguez H, Rhee LM, Ramachandran J, Reale V, Glencorse TA, Seeburg PH, Barnard EA (1987) Sequence and functional expression of the GABA-A receptor shows a ligand-gated receptor superfamily. Nature 328:221–227

Settimo ADa, Primofiore G, DaSettimo F, Marini AM, Novellino E, Greco G, Martini C, Giannaccini G, Lucacchini A (1996) Synthesis, structure-activity relationships, and molecular modeling studies of N-(indol-3-ylglyoxylyl)benzylamine derivatives acting at the benzodiazepine receptor. J Med Chem 39:5083–5091

Sieghart W (1995) Structure and pharmacology of $GABA_A$ receptor subtypes. Pharmacol Rev 47:181–234

Sigel E, Barnard EA (1984) A γ-aminobutyric acid/benzodiazepine receptor complex from bovine cerebral cortex. Improved purification with preservation of regulatory sites and their interactions. J Biol Chem 259:7219–7223

Sigel E, Buhr A (1997) The benzodiazepine binding site of $GABA_A$ receptors. Trends Pharmacol Sci 18:425–429

Skolnick P, Moncada V, Barker JL, Paul SM (1981) Pentobarbital: dual actions to increase brain benzodiazepine receptor affinity. Science 211:1448–1450

Squires RF, Braestrup C (1977) Benzodiazepine receptors in rat brain. Nature 266:732–734

Squires RF, Benson DI, Braestrup C, Coupet J, Klepner CA, Myers V, Beer B (1979) Some properties of brain specific benzodiazepine receptors: New evidence for multiple receptors. Pharmacol Biochem Behav 10:825–830

Squires RF, Casida JE, Richardson M, Saederup E (1983) [^{35}S]t–Butyl bicyclophosphorothionate binds with high affinity to brain-specific sites coupled to γ-aminobutyric acid-A and ion recognition sites. Mol Pharmacol 23:326–336

Stauber GB, Ransom RW, Dilber AI, Olsen RW (1987) The γ-aminobutyric acid–benzodiazepine receptor protein from rat brain: large-scale purification and preparation of antibodies. Eur J Biochem 167:125–133

Stephens DN, Schneider HH, Kehr W, Andrews JS, Rettig KJ, Turski L, Schmiechen R, Turner JD, Jensen LH, Petersen EN, Honore T, Hansen JB (1990) Abecarnil: a metabolically stable, anxioselective β-carboline acting at benzodiazepine receptors. J Pharmacol Exp Ther 253:334–343

Thomas P, Sundaram H, Krishek BJ, Chazot P, Xie X, Bevan P, Brocchini SJ, Latham CJ, Charlton P, Moore M, Lewis SJ, Thornton DM, Stephenson FA, Smart TG (1997) Regulation of neuronal and recombinant GABA$_A$ receptor ion channels by Xenovulene A, a natural product isolated from *Acremonium strictum*. J Pharmacol Expt Ther 282:513–520

Ticku MK, Olsen RW (1978) Interaction of barbiturates with dihydropicrotoxinin binding sites in mammalian brain. Life Sci 22:1643-1652

Ticku MK, Olsen RW (1979) Cage convulsants inhibit picrotoxinin binding. Neuropharmacology 18:315–318

Vicini, S (1991) Pharmacological significance of the structural heterogeneity of the GABA$_A$ receptor-chloride ion channel complex. Neuropsychopharmacology 4:9–15

Wang CG, Langer T, Kamath PG, Gu ZQ, Skolnick P, Fryer RI (1995) Computer-aided molecular modeling, synthesis, and biological evaluation of 8-(Benzyloxy)-2-phenyl-pyrazolo[4,3-c]quinoline as a novel benzodiazepine receptor agonist ligand. J Med Chem 38:950–957

Willow M, Johnston GAR (1983) Pharmacology of barbiturates: electrophysiological and neurochemical studies. Int Rev Neurobiol 24:15–36

Wittmer LL, Hu Y, Kalkbrenner M, Evers AS, Zorumski CF, Covey DF (1996) Enantioselectivity of steroid-induced GABA$_A$ receptor modulation and anesthesia. Mol Pharmacol 50:1581–1586

CHAPTER 20
P2X Receptors for ATP: Classification, Distribution, and Function

R.J. EVANS

A. Introduction

P2 receptors for ATP may be subclassified into ligand gated P2X receptor cation channels and metabotropic G-protein coupled P2Y receptors. ATP can be released from a variety of cell types including neurons (VON KUGELGEN and STARKE 1991; SILINSKY and REDMAN 1996) and endothelial cells (YANG et al. 1994; BODIN and BURNSTOCK 1996) or as a result of local tissue damage and cell lysis (BURNSTOCK 1996). P2X receptors mediate fast synaptic transmission between neurons in the periphery (EVANS et al. 1992; GALLIGAN and BERTRAND 1995), the central nervous system (EDWARDS et al. 1992), and between sympathetic nerves and smooth muscle (EVANS and SURPRENANT 1992). In addition P2X receptors are expressed in a variety of cell types ranging from immune cells to cochlear hair cells (BUELL et al. 1996a; RAYBOULD and HOUSLEY 1997). The aim of this chapter is to give an overview of the properties and distribution of cloned P2X receptors and how this corresponds to native P2X receptor phenotypes.

B. Molecular Biology of P2X Receptors

A major leap forward in the study of P2X receptors has come with the cloning of several P2X receptor genes. The first two P2X receptors were isolated by expression cloning in *Xenopus* oocytes using mRNA libraries generated from rat vas deferens ($P2X_1$) (VALERA et al. 1994) and rat phaeochromocytoma (PC12) cells ($P2X_2$) (BRAKE et al. 1994). Publication of the $P2X_1$ and $P2X_2$ amino acid sequences showed that P2X receptors define a new family of ligand gated cation channels with a novel molecular architecture (SURPRENANT et al. 1995). Additional members of the P2X receptor family have been isolated by screening of tissue libraries and/or using PCR. To date seven P2X receptors have been isolated; $P2X_1$ (VALERA et al. 1994, 1995; LONGHURST et al. 1996), $P2X_2$ (BRAKE et al. 1994; HOUSLEY et al. 1995; BRANDLE et al. 1997; SIMON et al. 1997), $P2X_3$ (CHEN et al. 1995; LEWIS et al. 1995; GARCIA-GUZMAN et al. 1997b), $P2X_4$ (BO et al. 1995; BUELL et al. 1996; SEGUELA et al. 1996; SOTO et al. 1996a; WANG et al. 1996; GARCIA-GUZMAN et al. 1997a), $P2X_5$ (COLLO et al. 1996; GARCIA-GUZMAN et al. 1996; LE et al. 1997), $P2X_6$ (COLLO et al. 1996;

Soto et al. 1996b), and P2X$_7$ (Surprenant et al. 1996; Rassendren et al. 1997b).

I. A New Structural Family of Ligand Gated Ion Channels

Analysis of the predicted amino acid sequence of P2X$_{1-7}$ receptors show that they are a homologous group of proteins (378 to 595 amino acids long) with an overall relatedness of 36–48% identity at the amino acid level between any pair of subunits (Collo et al. 1996; Surprenant et al. 1996). A similar degree of conservation is found in other families of ligand gated ion channels. Hydrophobicity analysis of the deduced amino acid sequence yields similar profiles for all seven cloned P2X receptors with only two hydrophobic domains of sufficient length to span the membrane (the first beginning at about 30 amino acids from the N terminus and the second beginning near residue 320). The lack of an N terminal signal peptide/secretion leader sequence suggests that the N terminal is intracellular. Therefore the predicted membrane topology of the channel is one with intracellular N and C termini, two transmembrane domains and a large extracellular loop (Fig. 1).

The predicted structure of P2X receptors is very different from other ligand gated ion channels which include two major families; nicotinic acetylcholine, 5-HT$_3$, glycine, and GABA channels which have extracellular N and C-termini and four transmembrane domains, and the glutamate receptor family which has an extracellular N terminus and three transmembrane domains. Ion channels with only two transmembrane domains include the mechanosensitive channels of *E. coli*, inwardly rectifying potassium channels, amiloride sensitive epithelial sodium channels (ENaC), and FMRFamide gated channels (FaNaC) from *Helix aspersa* (North 1996). However although P2X receptors are structurally similar to these channels they share no primary sequence homology.

II. The Extracellular Loop/Ligand Binding Site

The putative extracellular loop (about 270 amino acids) contains 67 amino acids that are completely conserved in all seven P2X receptors. The most abundant are glycine [14], cysteine [10] and lysine [6] residues. The large number of cysteines and their complete conservation between the P2X receptors may suggest the possibility of disulfide bridges forming to provide tertiary structure to the protein. As yet there is no information available regarding the folding/structural organization of the extracellular loop. Experimental evidence for this region forming the extracellular portion of the receptor comes from glycosylation studies. For the P2X$_1$ receptor consensus sequences for N-linked glycosylation NXT (where X is any amino acid) are only found in the proposed extracellular loop. In vitro translation of the human P2X$_1$ receptor shows that it can be glycosylated to yield a 60-kD product (un-glycosylated form 45 kD) (Valera et al. 1995). This correlates well with the molecular

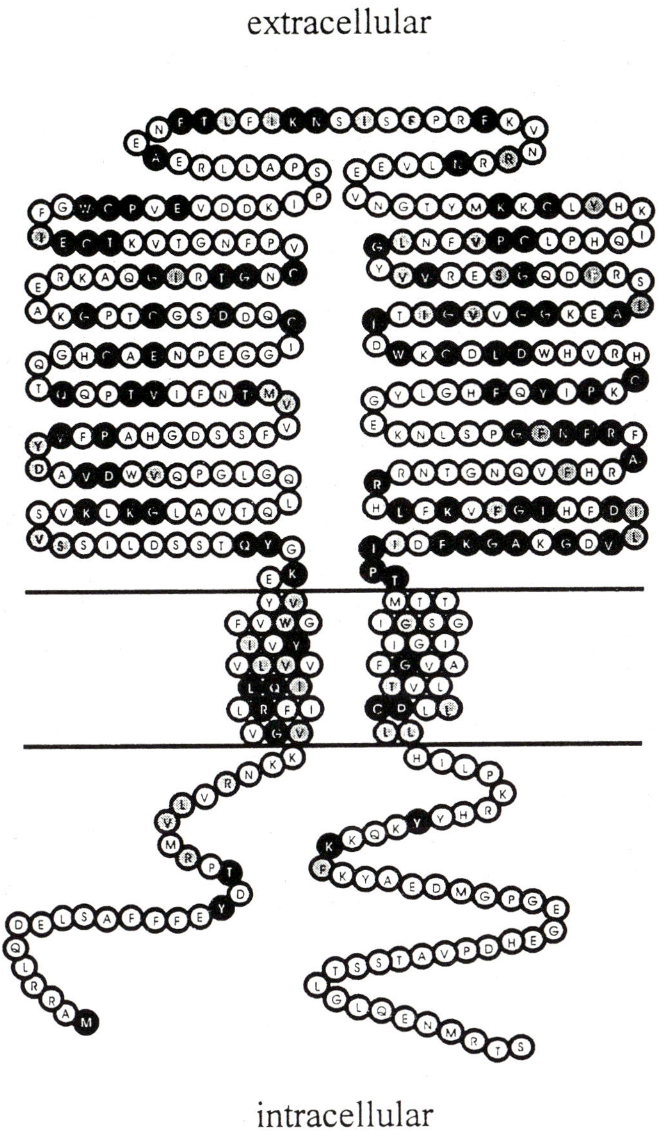

Fig. 1. The predicted membrane topology of the channel is one with intracellular N and C termini, two transmembrane domains and a large extracellular loop. Diagram show the amino acid sequence for the $P2X_1$ receptor residues in black are consemed throughout the P2X receptor family, shaded residues are similar

weight of smooth muscle P2X receptors extracted from the vas deferens by protein purification (Bo et al. 1992). The mutated receptor $P2X_1$ (N184S) produces only a 45-kD product following in vitro translation and suggests that asparagine at position 184 is glycosylated and therefore extracellular (G.

Buell, unpublished observations). Further support for the proposed topology comes from studies documenting the importance of the extracellular loop in determining the pharmacological properties of the receptor (Buell et al. 1996a; Garcia-Guzman et al. 1997a).

The nature of the nucleotide binding site of P2X receptors is unclear. P2X receptors appear to have no sequence homology with extracellular ATP binding proteins. The nucleotide binding sites of several proteins have been determined and it appears that lysine residues are important in ATP recognition, often co-ordinating the binding of the terminal phosphate (Traut 1994). For the metabotropic $P2Y_4$ (P2U) receptor a lysine residue has been shown to regulate the binding of ATP and the agonist selectivity (Erb et al. 1995) (although this residue is not conserved throughout all members of the P2Y receptor family). Given the large number of lysine residues conserved in the extracellular loop of P2X receptors a role for some of them in nucleotide binding seems likely.

III. Transmembrane Domains; Location of the Ionic Pore

The putative transmembrane domains are not considerably well conserved among the members of the P2X receptor family (although there is stronger conservation between $P2X_{1-6}$ with $P2X_7$ showing the greatest divergence). There are no obvious homologies between pore-lining sequences of known ion channels and regions of P2X receptors. Residues associated with the pore region of the $P2X_2$ receptor were investigated in work combining cysteine scanning mutagenesis and the channel blocking effects of sulfhydryl reagents (Rassendren et al. 1997a). These studies showed that the second transmembrane (TM2) domain contributes to the pore of the channel. These results in combination with structural predictions suggest that the TM2 region of P2X receptors has a β-sheeted secondary structure and is not formed as the polar face of an amphipathic helix as has been suggested for nicotinic receptors (Rassendren et al. 1997a). Further studies will be necessary to elucidate the fine structure of the pore region and to determine whether TM1 contributes to the pore of the channel.

The transmembrane domains also contribute to the functional properties of the channel and have been shown to be important in determining the time course of the response of P2X receptors. Replacement of either of the transmembrane regions of the $P2X_1$ (or $P2X_3$) receptor (desensitizing phenotypes) with the corresponding transmembrane region from the $P2X_2$ receptor (responses show a non-desensitizing phenotype) changes the kinetics of the response to ATP from a desensitizing phenotype to a non-desensitizing phenotype. However both transmembrane domains from $P2X_1$ (or $P2X_3$) receptors are necessary to change the $P2X_2$ receptor to one with a desensitizing phenotype. This suggests that the two transmembrane domains interact (Werner et al. 1996).

1. Intracellular N and C Termini

The P2X receptor family have a relatively short N terminal domain of 22–29 amino acids. The considerable variation in the length of P2X receptors is predominantly accounted for by variations in the C terminal tail (from 25–240 amino acids, $P2X_7$ is longest). Residues in the C-terminus of P2X receptors have been shown to contribute to their functional properties. For $P2X_2$ receptors splice variants of the C-terminus modify the time course of P2X responses (BRANDLE et al. 1997; SIMON et al. 1997) and the C terminus of the rat $P2X_7$ receptor have been shown to be important for the pore forming nature of this channel (SURPRENANT et al. 1996).

Analysis of the amino acid sequences of the P2X receptors reveals the presence of consensus motifs for kinase action on the intracellular domains of the P2X receptors. There is a conserved consensus sequence for protein kinase C (PKC) in the N terminal region (tyrosine 18 in $P2X_1$ receptors). For $P2X_1$, $P2X_4$, and $P2X_6$ receptors this is the sole PKC consensus sequence, there are additional sites in the C terminus for the other P2X receptors (1 for $P2X_3$ and $P2X_5$, 3 for $P2X_7$, and 4 for $P2X_2$ receptors). A consensus sequence for tyrosine kinase in $P2X_3$ receptors is found at position 398. However as yet there is no experimental evidence to support the modulation of P2X receptors by protein phosphorylation.

2. Genomic Organisation, Human P2X Receptors and Chromosomal Location

Studies on the genomic organisation of P2X receptors suggest that P2X receptor genes contain a number of introns (VALERA et al. 1994). The genomic sequence of the rat $P2X_2$ receptor gene has been determined (BRANDLE et al. 1997) and shown to contain ten introns (size range 78–320 b.p.) and a number of splice sites (BRANDLE et al. 1997; SIMON et al. 1997). Alternative splicing results in the expression of at least four alternative $P2X_2$ variants (HOUSLEY et al. 1995; BRANDLE et al. 1997; SIMON et al. 1997). A similar gene structure has been reported for the human $P2X_7$ receptor (BUELL et al. 1998).

The majority of human homologues of P2X receptors have been cloned and their chromosomal locations mapped. Certain P2X receptor iso-forms are co-localized/clustered, i.e., $hP2X_1$ and $hP2X_5$ at chromosome 17p13.3 (VALERA et al. 1995; LONGHURST et al. 1996, BUELL et al. 1998) and $hP2X_4$ and $hP2X_7$ at chromosome 12q24 (GARCIA-GUZMAN et al. 1997a, BUELL et al. 1998). The $hP2X_3$ receptor gene is found on the long arm of chromosome 11 locus q12 (GARCIA-GUZMAN et al. 1997b).

C. Distribution of P2X Receptors

Historically information on the localization of P2X receptors came from functional studies and radioligand experiments (MICHEL and HUMPHREY 1993; Bo and BURNSTOCK 1994; BO et al. 1994; BALCAR et al. 1995). Following the cloning of P2X receptors it is now possible to determine at the molecular level which P2X receptors are expressed in a given cell type. The distributions of $P2X_{1-7}$ receptor RNA transcripts have been characterized using northern analysis and PCR. More specific localization of P2X receptor transcripts has come from *in situ* hybridization and immunohistochemical studies with antibodies raised against specific P2X receptor subunits.

I. $P2X_1$ Receptors

$P2X_1$ receptors are expressed predominantly in smooth muscle including the vas deferens, bladder, and arteries (VALERA et al. 1994, 1995; COLLO et al. 1996; LONGHURST et al. 1996; VULCHANOVA et al. 1996). In addition, $P2X_1$ receptor transcripts have been localized in adult peripheral leukocytes, pancreas, spleen, placenta, prostate, testis, ovary, small intestine, colon, and liver (VALERA et al. 1994, 1995; LONGHURST et al. 1996).

$P2X_1$ receptors in neurons appear to be restricted to a discrete plexus of nerve fibers and terminals in the superficial horn of the spinal cord and in neurons of peripheral ganglia (COLLO et al. 1996; VULCHANOVA et al. 1996). In adult brain, $P2X_1$ receptors were below the level of detection by *in situ* hybridization, northern analysis, or immunohistochemical studies (VALERA et al. 1994; COLLO et al. 1996; VULCHANOVA et al. 1996). Using radiolabeled oligonucleotide probes $P2X_1$ receptor transcripts have been found in the cerebellum of neonatal brain (KIDD et al. 1995). In adult brain using this method $P2X_1$ receptor labeling was poorly detectable; however $P2X_1$ cDNAs were amplified by PCR from an adult rat brain (KIDD et al. 1995). Thus $P2X_1$ receptors may have a role in the development of the brain but are expressed at very low levels in the adult brain.

II. $P2X_2$ Receptors

$P2X_2$ receptor transcripts are expressed by a variety of neurons. The distribution of $P2X_2$ receptors in the brain is somewhat restricted when compared to the other P2X receptor subtypes that are expressed there (KIDD et al. 1995; COLLO et al. 1996). Strong staining for $P2X_2$ receptor transcripts is restricted to the medial septal nucleus of the subcortical telencephalon, the anterior nuclei and paraventricular nuclei of the thalamus, the hypothalamus, and in the hindbrain to the locus coeruleus, the dorsal motor vagal nucleus, and the area postrema (COLLO et al. 1996). These results in adult rat brain have been essentially confirmed using antibodies raised against the $P2X_2$ receptor (VULCHANOVA et al. 1996, 1997; FUNK et al. 1997). In the peripheral nervous system $P2X_2$ receptors have been detected in sensory ganglia (nodose and dorsal root ganglia (DRG) neurons; in the DRG some of the staining may be associated with supporting glial cells), superior cervical ganglia and in presynaptic nerve terminals in sub mucous and myenteric plexi (COLLO et al. 1996; VULCHANOVA et al. 1996, 1997).

$P2X_2$ receptor transcripts are expressed in vas deferens extracts (BRAKE et al. 1994; HOUSLEY et al. 1995). Subsequent immunohistochemical studies have shown that $P2X_2$ receptor immunoreactivity is confined to the nerve fibers that innervate the vas deferens and not the vas deferens smooth muscle cells (VULCHANOVA et al. 1996). This work illustrates an important point that analysis of tissue extracts does not give a direct indication of which particular cell types within a tissue express the receptor of interest, e.g., neuronal or smooth muscle, and care should be taken when drawing conclusions based on such data.

A number of splice variants of the $P2X_2$ receptors have been described (HOUSLEY et al. 1995; BRANDLE et al. 1997; SIMON et al. 1997). $P2X_{2-2}$ receptor transcripts are expressed in brain, spleen, kidney, intestine and organ of corti; indeed the $P2X_{2-2}$ splice variant is as highly expressed as the original $P2X_2$ sequence in various tissues. Another truncated form of $P2X_2$ receptor has also been reported (HOUSLEY et al. 1995) (confusingly named as $P2_XR1$–2, the $P2X_2$ receptor was initially named $P_{2X}R1$ by BRAKE et al. (1994) in their original Nature paper) which carries an 85 base pair insertion containing a stop codon towards the end of TM2 (this insertion is identical to intron x present in genomic structure of $P2X_2$). This truncated form was expressed in pituitary and cochlea cells but was not detected in brain.

III. $P2X_3$ Receptors

$P2X_3$ receptors are localized exclusively in sensory neurons including trigeminal, dorsal root, and nodose ganglia (CHEN et al. 1995; LEWIS et al. 1995; COLLO et al. 1996; COOK et al. 1997; VULCHANOVA et al. 1997). In the spinal cord, $P2X_3$ receptor immunoreactivity was restricted to the inner portion of lamina II and appeared to be of primary afferent origin (VULCHANOVA et al. 1997). Further studies have shown $P2X_3$ receptor immunoreactivity localized to nociceptive nerve fibers and endings in the tooth pulp (COOK et al. 1997) indicating they may play a role in pain transduction.

IV. $P2X_4$ Receptors

$P2X_4$ receptors are widely expressed in adult brain, spinal cord, and sympathetic and sensory ganglia (BO et al. 1995; BUELL et al. 1996; COLLO et al. 1996; SOTO et al. 1996a; TANAKA et al. 1996; WANG et al. 1996). This pattern is closely mirrored by that of $P2X_6$ receptors. The highest level of RNA expression is in Purkinge cells. In addition, $P2X_4$ receptor transcripts are also expressed in liver, kidney, spleen, lung, intestine, testis, adrenal, salivary gland, and many endocrine tissues and hormone secreting cell lines (SOTO et al. 1996a; WANG et al. 1996; BUELL et al. 1996b; COLLO et al. 1996a).

$P2X_4$ receptor transcripts are also expressed in the bladder, vas deferens, and aorta (BO et al. 1995; SOTO et al. 1996). The presence of $P2X_4$ receptor immunoreactivity in vascular smooth muscle has also been reported (BURNSTOCK 1997).

V. $P2X_5$ Receptors

In adult rat $P2X_5$ receptor transcripts are expressed at high levels in the heart (but not with associated blood vessels) (GARCIA-GUZMAN et al. 1996). For neurons the $P2X_5$ receptor is restricted to the mesencephalic nucleus of the trigeminal nerve and spinal cord and in sensory ganglia (COLLO et al. 1996). A truncated form of the human $P2X_5$ receptor ($hP2X_5R$) is expressed at high levels in foetal thymocytes and then down-regulated in adult and it has been suggested that the receptor could be involved in apoptosis of thymocytes during negative selection (LE et al. 1997).

VI. $P2X_6$ Receptors

$P2X_6$ receptor transcripts are expressed throughout the brain with the highest density in cerebellar Purkinge cells and strong signals in the olfactory bulb, cortex, hippocampus, some hypothlamic and thalamic nuclei, the mesencephalic nucleus of the trigeminal nerve, and the cranial nerve motor nuclei; no $P2X_6$ receptor RNA was detected in the cerebral white matter (COLLO et al. 1996b; SOTO et al. 1996b). This pattern of expression closely paralleled that of $P2X_4$ receptors with the exception of ependymal cells. In these cells, which constitute the epithelial layer surrounding the ventricles, $P2X_6$ receptors were the sole P2X receptor of those currently identified to be expressed. $P2X_6$ receptor transcripts are also expressed in the cervical spinal cord and in trigeminal, dorsal root, and coeliac ganglia. In peripheral tissues transcripts for $P2X_6$ receptors are found in the gland cells of the uterus, the granulosa cells of the ovary, and bronchial epithelia (COLLO et al. 1996). PCR analysis has also identified transcripts for $P2X_6$ receptors in brain, heart, spinal cord, adrenal, trachea, uterus, lung, testis, pituitary, and astrocytes (SOTO et al. 1996b).

VII. $P2X_7$ Receptors

$P2X_7$ receptors are expressed by microglia (FERRARI et al. 1997), the majority of bone marrow cells, including granulocytes, monocytes/macrophages, and B lymphocytes (COLLO et al. 1997). Adult tissues that express $P2X_7$ receptor transcripts include lung, spinal cord, spleen, salivary gland, and testis (SURPRENANT et al. 1996). The distribution of $P2X_7$ receptors in the brain appears restricted to the microglia (brain macrophages) and ependymal cells. $P2X_7$ receptors do not appear to be expressed in neurons in contrast to $P2X_{1-6}$ receptors (COLLO et al. 1996). Human $P2X_7$ receptor transcripts are expressed highly in the pancreas, liver, heart, and thymus with moderate to low levels in brain, skeletal muscle, lung, placenta, leukocytes, testis, prostate, and spleen (and shows a more widespread distribution than had previously been characterized for P2Z-like responses) (RASSENDREN et al. 1997b).

Receptor localization studies have contributed considerably to our understanding of the distribution of P2X receptors and their possible functional roles. In addition, they have indicated the presence of P2X receptors in cell types not previously known to express them. In some cell types a single P2X receptor subtype is expressed, for example $P2X_4$ receptors in acinar cells of the salivary gland (COLLO et al. 1996). However in the majority of cell types multiple P2X receptor subtypes are present. This raises the question of what is the subunit composition of the native P2X receptor(s) in tissues expressing multiple P2X receptor subunits? P2X receptor subunits can heteropolymerize to form functional channels (LEWIS et al. 1995). The evidence for this has come from the production of a composite phenotype following co-expression of $P2X_2$ and $P2X_3$ receptor subunits. To determine the native subunit composition of P2X receptors it will be necessary to use antibodies raised against different P2X receptor isoforms in co-immunoprecipitation studies (RADFORD et al. 1997). In addition this work may reveal the subunit stochiometry of these channels.

D. Functional Properties of P2X Receptors

I. General Features of P2X Receptors

Originally P2X receptors were characterized based on a rank order of potency of α,β-methylene ATP>>ATP based on contraction studies on smooth muscle preparations. There are two problems associated with using this classification system: (1) this apparent order of potency results from the differential breakdown of ATP by ectonucleotidases in whole tissue preparations – when breakdown is blocked then α,β-methylene ATP and ATP are roughly equipotent (EVANS and KENNEDY 1994; TREZISE et al. 1994); and (2) at a number of ligand gated P2X receptors, particularly neuronal ones, α,β-meATP is not an agonist. Previously some confusion has arisen on the classification of native P2 receptors using a system based on agonist potency. Now P2X receptors are classified based on their properties as ATP-gated cation channels and not based on their pharmacological profile.

Five P2X receptor phenotypes can be discriminated based on kinetic and pharmacological parameters (EVANS and SURPRENANT 1996). The key discriminating features are: (1) whether the response inactivates during short agonist applications; (2) sensitivity to methylated analogues of ATP; and (3) sensitivity to the antagonists suramin and PPADS.

$P2X_{1-7}$ receptors form homomeric channels when expressed in *Xenopus* oocytes or in mammalian cells (HEK 293 or CHO) (BRAKE et al. 1994; VALERA et al. 1994; COLLO et al. 1996; EVANS 1996); in addition $P2X_2$ and $P2X_3$ receptors may heteropolymerize to form a distinct channel phenotype. Under normal conditions (extracellular sodium (145 mmol/l) and intracellular potassium (145 mmol/l)) the reversal potential of P2X evoked currents is ~0 mV indicating a relatively non-selective cation current. Thus for cells at resting potential P2X receptor activation results in membrane depolarization. In neurons and smooth muscle this depolarization is sufficient to activate voltage dependent calcium channels and leads to calcium influx. In addition P2X receptor channels are permeable to calcium, and in normal calcium containing solutions it has been estimated that 5–10% of the inward current evoked by ATP is carried by calcium (Benham 1989; Schneider et al. 1991; Rogers and Dani 1995; Garcia-Guzman et al. 1997a).

The permeability of a number of monovalent organic cations through P2X receptors decreases as a function of the geometric mean diameter (Evans et al. 1996). The linear relationship between permeability of these organic cations and diameter indicates that the channel functions as a simple fluid filled pore with a minimum pore diameter of 9 Å (slightly larger than those for $5HT_3$ and nicotinic channels) (Evans et al. 1996).

1. $P2X_1$ Receptors

$P2X_1$ receptor mediated responses rapidly desensitize (EC_{50} for ATP ~1 μmol/l). Responses evoked during the continued application of ATP (or other purinergic agonists) decay in a relatively concentration independent manner which can be fitted by a single exponential function with a time constant of between 100 and 300 ms. Recovery from this desensitized state takes 3–5 min. Unitary currents flowing through homomeric $P2X_1$ receptors show the same rapid desensitization recorded at the whole cell level. Single channel studies have shown that $P2X_1$ receptors open in brief flickery bursts with a conductance of ~18 pS (Evans 1996). α,β-Methylene ATP is a full agonist at the receptor and has an EC_{50} value of ~3 μmol/l. The P2 purinoceptor antagonists suramin and PPADS are non-competitive inhibitors of $P2X_1$ mediated responses with IC_{50} values of 1–5 μmol/l (Evans et al. 1995).

2. $P2X_2$ Receptors

ATP evokes relatively sustained currents when applied to $P2X_2$ receptors (Brake et al. 1994). The EC_{50} value for ATP is ~10 μmol/l and α,β-methylene ATP is ineffective as an agonist or an antagonist at this receptor (Brake et al. 1994; Evans et al. 1995). Suramin and PPADS are non-competitive antagonists with IC_{50} values of 1–5 μmol/l (Evans et al. 1995). $P2X_2$ receptors have a single channel conductance of ~21 pS in low divalent cation containing solutions (in normal divalent cation solution 14 pS) (Evans 1996). These properties are very similar to those of native P2X receptors on PC12 cell (Neuhaus et al. 1991) from which the $P2X_2$ receptor was originally isolated (Brake et al. 1994).

Splice variants of the $P2X_2$ receptor have been reported (Housley et al. 1995; Liu and Sharom 1997; Simon et al. 1997). The $P2X_{2(b)}$ (Simon et al. 1997) or referred to as $P2X_{2\text{-}2}$ receptor (Brandle et al. 1997) forms functional channels that show a faster rate of inactivation (SIMON et al. 1997). There appears to be a similar difference in the time course of inactivation of native PC12 cell P2X receptors depending on whether the cells are cultured with nerve growth factor (NGF) (relatively slow inactivation (NAKAZAWA et al. 1991)) or without (faster inactivation (NAKAZAWA et al. 1990)). This raises the possibility that NGF treatment can modulate the splicing of $P2X_2$ receptor gene. Two other splice variants; one with a deletion in the first transmembrane domain and the second with an early truncation of sequence associated with intron X, fail to form functional channels (HOUSLEY et al. 1995; SIMON et al. 1997).

3. $P2X_3$ Receptors

The properties of recombinant $P2X_3$ receptors are essentially the same as those of $P2X_1$ receptors except for the inactivation of the currents and sensitivity to l-β,γ-methylene ATP. For $P2X_3$ receptors desensitization shows strong concentration dependence which is bi-exponential with decay time constants of <50 ms and about 1 s at maximum agonist concentrations (CHEN et al. 1995; LEWIS et al. 1995). In addition l-β,γ-methylene ATP is ineffective as an agonist at $P2X_3$ receptors (LEWIS 1998) or neuronal α,β-meATP sensitive P2X receptors (TREZISE et al. 1995). In contrast l-β,γ-methylene ATP is an effective agonist at $P2X_1$ receptors (EC_{50} ~2 μmol/l) (EVANS et al. 1995) and smooth muscle P2X receptors (TREZISE et al. 1995).

4. $P2X_2$/$P2X_3$ Heteromeric Receptors

Co-expression of $P2X_2$ and $P2X_3$ receptors results in the production of a novel heteromeric phenotype combining the properties of the constituent subunits; i.e., a non-desensitizing ($P2X_2$) α,β-methylene ATP sensitive ($P2X_3$) phenotype. The agonist sensitivity of the receptor seem to be determined by the $P2X_3$ receptor subunit (LEWIS et al. 1995). The $P2X_2$ receptor subunit does however contribute to ligand binding in terms of determining the pH sensitivity of the receptor (STOOP et al. 1997).

5. $P2X_4$ Receptors

Rat $P2X_4$ receptor ($rP2X_4$) mediated currents show a non-desensitizing phenotype (EC_{50} for ATP ~10 μmol/l) and are insensitive to α,β-methylene ATP and the P2 receptor antagonists suramin, PPADS, and pyridoxal 5-phosphate (BO et al. 1995; BUELL et al. 1996; SEGUELA et al. 1996; SOTO et al. 1996a; WANG et al. 1996). The sensitivity of the $rP2X_4$ receptor to ATP was increased by zinc (10 μmol/l) with no effect on the amplitude of the maximal response (SOTO et al. 1996a). $rP2X_4$ receptors have a unitary conductance of ~9 pS and channel openings occurred in flickery bursts (EVANS 1996).

Site directed mutagenesis studies on the $rP2X_4$ receptors showed that sensitivity to the antagonists PPADS and pyridoxal 5-phosphate can be restored by the single amino acid mutation E249 K. This suggests that the formation of a Schiff base between PPADS and the receptor is necessary for the slowly reversible antagonist actions of PPADS (BUELL et al. 1996b). In contrast the human homologue of the receptor $hP2X_4$ has a higher sensitivity to the P2 receptor antagonists suramin and PPADS than the $rP2X_4$ receptor (human and rat isoforms have 87% sequence identity). A combination of studies using chimeric receptors and mutagenesis showed a single point mutant of the $rP2X_4$ receptor Q78K was sufficient to increase suramin affinity (GARCIA-GUZMAN et al. 1997a). However interpretation of where the binding sites for suramin are located remains complicated as this region associated with suramin binding for $P2X_4$ receptors is deleted from suramin sensitive $P2X_2$ and $P2X_3$ receptors. Thus no simple, unifying model to explain antagonists sensitivity can be made which indicates that antagonist binding may be co-ordinated by a number of structural features.

6. $P2X_5$ Receptors

The functional responses of $P2X_5$ receptors (EC_{50} for ATP ~15 μmol/l) are essentially the same as those of the $P2X_2$ receptor. However, the level of expression of currents were ~5–10% of those observed with any of the other recombinant P2X receptors (COLLO et al. 1996; GARCIA-GUZMAN et al. 1996). This suggests that either the receptor is expressed at lower levels and/or has a lower conductance. A putative human homologue of the $P2X_5$ receptor has been cloned from fetal brain (LE et al. 1997) but does not form functional channels following heterologous expression. Analysis of the $hP2X_5$ receptor amino acid sequence shows that there is a deletion of 22 amino acids (region 328–349 of $rP2X_5$) that are thought to form part of the external vestibule of the channel and a portion of the second transmembrane domain associated with the pore region (RASSENDREN et al. 1997a).

7. $P2X_6$ Receptors

The pharmacological properties of $P2X_6$ receptors (EC_{50} for ATP ~12 μmol/l) are essentially the same as those of $P2X_4$ receptors and, like the $P2X_4$ receptor, PPADS sensitivity could be induced by the introduction of a lysine residue in the extracellular loop (position 246 for $P2X_4$ and the equivalent position 251 for the $P2X_6$ receptor) (COLLO et al. 1996). The amplitude of recombinant $P2X_6$ receptor whole cell currents expressed in HEK 293 cells was in the same order of magnitude as those for other P2X receptors; however they were recorded in only a low percentage ~5% of cells examined (COLLO et al. 1996). When expressed in *Xenopus* oocytes $P2X_6$ receptors failed to make functional channels (SOTO et al. 1996b).

8. $P2X_7$ Receptors

$P2X_7$ receptors are distinct from other recombinant P2X receptors in terms of their pharmacology and ability to form cytolytic pores. BzATP is the most potent agonist (EC_{50} ~7 μmol/l) and the potency of ATP is relatively low (EC_{50}

~100 μmol/l) (SURPRENANT et al. 1996). The amplitude of $P2X_7$ receptor mediated responses are sensitive to divalent cations. Reductions in magnesium and/or calcium from the external medium markedly potentiated the response to agonists (essentially no change in EC_{50} value) and prolonged the timecourse of response. Repeated or prolonged applications of agonist lead to the opening of a large "permeabilizing" pore in the membrane through which the propidium dye YO-PRO (629 daltons) could pass (SURPRENANT et al. 1996). In contrast to the other P2X receptor isoforms the $P2X_7$ receptor has a very long intracellular C terminal domain which is associated with the pore forming ability of the $P2X_7$ receptor (SURPRENANT et al. 1996). Subsequent studies have shown that the ion channel and pore forming properties of the channel may be separate because inward currents, but not dye uptake into cells expressing $rP2X_7$ receptors is blocked by calmidazolium (VIRGINIO et al. 1997).

The human $hP2X_7$ receptor has lower apparent potencies to ATP and BzATP (10- and 25-fold greater EC_{50} values respectively) than the $rP2X_7$ receptor (80% identity) and responses were more sensitive to changes in extracellular divalent cations. In addition the pore forming ability was considerably reduced.

II. Modulation of P2X Receptors

The binding of ATP to P2X receptors can be modulated by changes in extracellular pH (KING et al. 1996; STOOP et al. 1997). Changes in pH result in essentially parallel shifts in the concentration response curves for ATP with no effect on the maximal response; thus they change the apparent affinity of ATP for the receptor. For $P2X_1$, $P2X_3$, $P2X_4$, and $P2X_7$, receptors acidification reduces the potency of ATP (agonist was BzATP for $P2X_7$) (VIRGINIO et al. 1997), and for $P2X_2$ and $P2X_{2/3}$ heteromeric receptors acidification increases the apparent potency of ATP (KING et al. 1996; STOOP et al. 1997). Native P2X receptor mediated responses in rat nodose ganglion neurons are also potentiated by acidification (LI et al. 1996) providing further corroborative evidence that this native phenotype can be accounted for by the heteromeric expression of $P2X_{2/3}$ receptors (LEWIS et al. 1995). Marked changes in extracellular pH have been reported (both acidification and alkalinization) following nerve stimulation in the brain (CHESLER and KAILA 1992). This suggests that the level of neuronal activity may act to modulate the purinergic transmission.

Zinc can also modulate the potency of ATP at both native and recombinant P2X receptors. The concentration of zinc in the extracellular space can vary depending on the level of neuronal activity (ASSAF and CHUNG 1984). The modulation of P2X responses by zinc is concentration dependent, low concentrations (<10 μmol/l) potentiate the actions of ATP while higher concentrations (>30 μmol/l) inhibit the response (CLOUES et al. 1993a; GARCIA-GUZMAN et al. 1997a) (for $P2X_7$ receptors no potentiation is seen, only inhibition, threshold 1 μmol/l zinc) (VIRGINIO et al. 1997). Zinc potentiates $P2X_2$ (SCHACHTER and HARDEN 1997) and $P2X_4$ receptor mediated currents (SEGUELA et al. 1996; SOTO et al. 1996a; REN and BURNSTOCK 1997), and native

P2X receptor mediated responses in rat nodose (Li et al. 1993, 1996; Wright and Li 1995) and superior cervical ganglion neurons (Cloues et al. 1993; Cloues 1995). The mechanism associated with potentiation appears to be an increase in the opening frequency and burst duration of P2X channels in neurons (Cloues 1995; Wright and Li 1995). Other divalent cations affect ATP binding but they are required at much higher concentrations; calcium for example can decrease the potency of ATP at $P2X_2$ receptors (Evans et al. 1996) and PC12 cell P2X receptors (Nakazawa and Hess 1993). In addition divalent cations can reduce the single channel conductance of P2X receptor channels (Benham and Tsien 1987; Neuhaus et al. 1991; Nakazawa and Hess 1993; Evans 1996). Lanthanum and other trivalent cations (~100–300 μmol/l) have an inhibitory effect on P2X meditated currents through native PC12 cell P2X receptors and recombinant $P2X_1$ and $P2X_2$ receptors (Nakazawa et al. 1997). Similar effects of cations have been shown in binding studies at P2X receptors (Michel and Humphrey 1994; Michel et al. 1997).

Neuronal P2X receptor mediated responses can also be inhibited by ethanol (Li et al. 1998). These inhibitory effects of ethanol are thought to result from an allosteric action of ethanol to decrease the apparent agonist affinity of ATP at the receptor (Li et al. 1998).

Thus agonist binding at P2X receptors can be modulated by a variety of factors including changes in extracellular pH, cations and ethanol. The presence of a number of consensus sites for various protein kinases in the intracellular portions of the P2X receptors raises the possibility that these channels may be modulated following the activation of G-protein coupled receptors. Experiments on recombinant $P2X_2$ receptors in oocytes have shown that P2X receptor mediated responses can be potentiated by the application of enkephalin, substance P, calcitonin gene related peptide, and nerve growth factor (Wildman et al. 1997). Whether these effects are mediated through allosteric effects or the activation of G-protein coupled receptors and subsequent second messenger systems remains to be determined. However these results raise the interesting possibility that native P2X receptor function may be modulated by G-protein coupled receptors.

III. Native P2X Receptor Phenotypes; Molecular Correlates

The lack of potent and selective P2X receptor antagonists has frustrated the direct characterization of native P2X receptor phenotypes. Comparison of the native phenotype with the properties and distribution of cloned P2X receptors means that it is now possible to account for native P2X receptor phenotypes at the molecular level (see Table 1).

1. Smooth Muscle

P2X receptors play an important role in the response of smooth muscle preparations to sympathetic nerve stimulation. ATP is co-released with noradrena-

Table 1. Properties of cloned P2X receptors and their native correlates

Recept or	α,β-meATP-sensitive	Desensitisation	Antagonist-sensitive	Native receptor
$P2X_1$	Yes	Yes	Yes	Smooth muscle, platelets, HL60 cells, RBL cells
$P2X_2$	No	No	Yes	SCG neurons, PC12 cells, submucosal ganglia
$P2X_3$	Yes	Yes	Yes	Dorsal root ganglion neurons, a subset of trigeminal neurons
$P2X_{2/3}$	Yes	No	Yes	Nodose ganglia, subset of trigeminal neurons
$P2X_4$	No	No	No	Salivary gland
$P2X_5$	No	No	Yes	–
$P2X_6$	No	Yes	No	–
$P2X_7$	No	Yes	Partial	Microglia, macrophages

line from sympathetic nerves and acts through P2X receptors to mediate contraction (von Kugelgen et al. 1989). For arteries the P2X receptor mediated component of neurogenic contraction becomes progressively more important with the size of the artery. For example in submucosal arterioles, which contribute ~40% of the mesenteric-splanchnic resistance, sympathetic constriction is mediated solely by ATP acting at P2X receptors (Evans and Surprenant 1992).

The properties of smooth muscle P2X receptors (on arteries, bladder, and vas deferens) can be accounted for by the expression of $P2X_1$ receptors. In situ hybridization studies have suggested that $P2X_1$ receptors were the only currently identified P2X receptors expressed in smooth muscle preparations. Recent immunohistochemical studies have suggested that $P2X_4$ receptors may also be present in some arteries (Burnstock 1997). However the properties of native P2X receptors, in particular their sensitivity to α,βmeATP and the P2 receptor antagonists coupled with their desensitizing phenotype show that responses are dominated by the expression of the $P2X_1$ receptor subunits.

2. Sensory Neurons

The initial suggestion that P2X receptors could be involved in sensory processing came from Jahr and Jessel (1983). Subsequent studies have demonstrated the presence of P2X receptors on a variety of sensory neurons (Krishtal et al. 1988a,b; Khakh et al. 1995; Robertson et al. 1996; Gu and MacDermott 1997) and suggested that they may be involved in the sensation of pain (Driessen et al. 1994; Burnstock 1996; Bland-Ward and Humphrey 1997). Patch clamp studies on dissociated sensory neurons have revealed the presence of three sensory neuron phenotypes: (1) a rapidly inactivating α,β, meATP sensitive P2X receptor mediated response which can be accounted for by $P2X_3$ receptor expression – this phenotype is found in the majority of cultured dorsal root ganglion (DRG) neurons, both neonatal (ROBERTSON et al. 1996) and adult (Grubb and Evans, unpublished observations) and a subset of trigeminal ganglion neuron nociceptors (COOK et al. 1997); (2) a sustained α,β-meATP sensitive $P2X_{2/3}$ heteromeric receptor phenotype is found in nodose ganglia, a subset of trigeminal ganglion nociceptive neurons and a small percentage of DRG neurons (and amphibian DRG neurons) (LI et al. 1998); and (3) α,βmeATP insensitive, sustained $P2X_2$ receptor like responses in stretch receptors (COOK et al. 1997). These results in combination with distribution studies (CHEN et al. 1995; COLLO et al. 1996) have lead to the suggestion that $P2X_3$ receptors are expressed exclusively on nociceptive neurons and may provide novel targets for the development of analgesic drugs.

3. Peripheral Neurons

The properties of rat superior cervical ganglion neurons (KHAKH et al. 1995), rat phaeochromocytoma PC12 cells, and neurons from guinea-pig submucosal ganglia (BARAJAS-LOPEZ et al. 1994) show an α,β-meATP insensitive sustained P2X receptor phenotype that can be accounted for by the expression of $P2X_2$ receptors. However the properties of a number of peripheral neurons do not correspond to those of the cloned P2X receptors, for example guinea pig myenteric neurons (BARAJAS-LOPEZ et al. 1996; ZHOU and GALLIGAN 1996) and rat parasympathetic cardiac ganglia (FIEBER and ADAMS 1991), and suggest the existence of additional P2X receptor subunits.

4. Brain

P2X receptor mediated responses have been reported in several brain regions including the locus coerulus (SHEN and NORTH 1993), medial habenula (EDWARDS et al. 1992), mesencephalic nucleus (KHAKH et al. 1997), and the medial vestibular nucleus (CHESSEL et al. 1997). In all these preparations α,β-meATP was an agonist at the native P2X receptor. This suggests that there is an additional P2X receptor subunit yet to be identified in the brain that contributes to these properties as the currently identified α,β-meATP sensitive subunits $P2X_1$ and $P2X_3$ are not expressed in the brain (COLLO et al. 1996). The antagonist insensitive $P2X_4$ and $P2X_6$ receptors have been shown to have a widespread distribution in the brain; however, the fact that the majority of native brain P2X responses have been shown to be sensitive to suramin, and PPADS would suggest that the $P2X_4$ or $P2X_6$ receptors do not dominate these native brain P2X receptor phenotypes.

5. Immune/Blood Cells

$P2X_1$ receptors are expressed by and can account for the native P2X receptor phenotype of a variety of blood cells including HL60 cells (macrophage type lineage) (VALERA et al. 1995), rat basophilic leukaemia cells (RBL, granulocytic characteristics) human platelets (MacKenzie et al. 1996) and megakaryocytes (platelet progenitors) (Somasundaram and Mahaut-Smith 1994).

The properties of $P2X_7$ receptors are essentially the same as those for the pore forming P2Z receptor which has been described in a variety of immune cells, e.g., macrophages (Naumov et al. 1995), lymphocytes (Wiley et al. 1992). $P2X_7$ receptor expression in macrophages (Surprenant et al. 1996) formation of macrophage polykarions (Falzoni et al., 1995) and mitogenic stimulation of T-cells (Baricordi et al., 1996).

6. Salivary Gland

The phenotype of native salivary gland acinar cells corresponds to that of $P2X_4$ receptors, which are the only subtype currently identified at the molecular level to be expressed in these cells (Buell et al. 1996; Collo et al. 1996).

E. Future Directions

The development of specific subytpe selective P2X receptors agonists and antagonists and the production of transgenic mice deficient in P2X receptor subtype(s) should clarify the role of these receptors in physiological processes. In addition it is likely that there are still a number of P2X receptor subtypes, particularly in the brain, which have yet to be identified at the molecular level.

Acknowledgements. I would like to thank Carolyn Lewis, Daniel Gitterman, and Blair Grubb for their helpful comments on the manuscript.

References

Assaf SY, Chung S (1984) Release of endogenous Zn^{2+} from brain tissue during activity. Nature 308:734–736

Balcar VJ, Li Y, Killinger S, Bennett MR (1995) Autoradiography of P2X ATP receptors in rat brain. Br J Pharmacol 115:302–306

Barajas-Lopez C, Espinosa-Luna R, Gerzanich V (1994) ATP closes a potassium and opens a cationic conductance through different receptors in neurons of guinea pig submucous plexus. J Pharm Exp Ther 268:1396–1402

Barajas-Lopez C, Huizinga JD, Collins SM, Gerzanich V, Espinosa-Luna R, Peres AL (1996) P2X-purinoceptors of myenteric neurones from the guinea-pig ileum and their unusual pharmacological properties. Br J Pharmacol 119:1541–1548

Benham CD (1989) ATP-activated channels gate calcium entry in single smooth muscle cells from rabbit ear artery. J Physiol 419:689–701

Benham CD, Tsien RW (1987) A novel receptor-operated Ca^{2+}-permeable channel activated by ATP in smooth muscle. Nature 328:275–278

Bland-Ward PA, Humphrey PPA (1997) Acute nociception mediated by hindpaw P2X receptor activation in the rat. Br J Pharmacol 122:365–371

Bo X, Simon J, Burnstock G, Barnard EA (1992) Solubilization and molecular size determination of the P2X purinoceptor from rat vas deferens. J Biol Chem 25:17581–17587

Bo X, Fischer B, Maillard M, Jacobson KA, Burnstock G (1994) Comparative studies on the affinities of ATP derivatives for P2X-purinoceptors in rat urinary bladder. Br J Pharmacol 112:1151–1159

Bo X, Burnstock G (1994) Distribution of [3H] α,β-methyleneATP binding sites in rat brain and spinal cord. Neuroreport 5:1601–1604
Bo X, Zhang Y, Nassar M, Burnstock G, Schoepfer R (1995) A P2X purinoceptor cDNA conferring a novel pharmacological profile. FEBS Letts 375:129–133
Bodin P, Burnstock G (1996) ATP-stimulated release of ATP by human endothelial cells. J. Card Pharmacol 27:872–875
Brake AJ, Wagenbach MJ, Julius D (1994) New structural motif for ligand-gated ion channels defined by an ionotropic ATP receptor. Nature 371:519–523
Brandle U, Spielmanns P, Osteroth R, Sim J, Surprenant A, Buell G, Ruppersberg JP, Plinkert PK, Zenner H-P, Glowatzki E (1997) Desensitisation of the $P2X_2$ receptor controlled by alternative splicing. FEBS Letts 404:294–298
Buell G, Michel A, Lewis C, Collo G, Humphrey PPA, Surprenant A (1996a) $P2X_1$ receptor activation in HL60 cells. Blood 7:2659–2664
Buell G, Lewis C, Collo G, North RA, Surprenant A (1996b) An antagonist-insensitive P2X receptor expressed in epithelia and brain. EMBO J 15:55–62
Buell G, Talabot F, Gos A, Lorenz J, Lai E, Morris MA, Antonarakis SE (1998) Gene strucutre and chromosomal location of the human $P2X_7$ receptor. Receptors and Channels (in press)
Burnstock G (1996) A unifying purinergic hypothesis for the initiation of pain. Lancet 347:1604–1605
Burnstock G (1997) The past, present and future of purine nucleotides as signalling molecules. Neuropharmacol 36:1127–1139
Chen C, Akopian AN, Sivilotti L, Colquhoun D, Burnstock G, Wood JN (1995) A P2X purinoceptor expressed by a subset of sensory neurons. Nature 377:428–431
Chesler M, Kaila K (1992) Modulation of pH by neuronal activity. Tr Neurosci 15:396–402
Chessel IP, Michel AD, Humphrey PPA (1997) Functional evidence for multiple purinoceptor subtypes in the rat medial vestibular nucleus. Neurosci 77:783–791
Cloues R, Jones S, Brown DA (1993) Zn^{2+} potentiates ATP-activated currents in rat sympathetic neurons. Pflugers Arch – Eur J Physiol 424:152–158
Cloues R (1995) Properties of ATP-gated channels recorded from rat sympathetic neurons: voltage dependence and regulation by Zn^{2+} ions. J Neurophysiol 73:312–319
Collo G, North RA, Kawashima E, Merlo-Pich E, Neidhart S, Surprenant A, Buell G (1996) Cloning of $P2X_5$ and $P2X_6$ receptors and the distribution and properties of an extended family of ATP-gated ion channels. J Neurosci 16:2495–2507
Collo G, Neidhart S, Kawashima E, Kosco-Vilbois M, North RA, Buell G (1997) Tissue distribution of the $P2X_7$ receptor. Neuropharmacol 36:1277–1283
Cook S.P, Vulchanova L, Hargreaves KM, Elde RP, McCleskey EW (1997) Distinct ATP receptors on pain-sensing and stretch-sensing neurons. Nature 387:505–508
Driessen B, Reimann W, Selve N, Friderichs E, Bultmann R (1994) Antinociceptive effect of intrathecally administered P2-purinoceptor antagonists in rats. Brain Res 666:182–188
Edwards FA, Gibb AJ, Colquhoun D (1992) ATP receptor-mediated synaptic currents in the central nervous system. Nature 359:144–147
Erb L, Garrad R, Wang Y, Quinn T, Turner JT, Weisman G (1995) Site-directed mutagenesis of P2U purinoceptors. J Biol Chem 270:4185–4188
Evans RJ (1996) Single channel properties of ATP-gated cation channels (P2X receptors) heterologously expressed in Chinese hamster ovary cells. *Neurosci Letts* 212:212–214
Evans RJ, Surprenant A (1992) Vasoconstriction of guinea-pig submucosal arterioles following sympathetic nerve stimulation is mediated by the release of ATP. Br J Pharmacol 106:242–249
Evans RJ, Derkach V, Surprenant A (1992) ATP mediates fast synaptic transmission in mammalian neurons. Nature 357:503–505

Evans RJ, Kennedy C (1994) Characterization of P2-purinoceptors in the smooth muscle of the rat tail artery: a comparison between contractile and electrophysiological responses. Br J Pharmacol 113:853–860
Evans RJ, Lewis C, Buell G, Valera S, North RA, Surprenant A (1995) Pharmacological characterization of heterologously expressed ATP-gated cation channels (P2X purinoceptors). Mol Pharmacol 48:178–183
Evans RJ, Surprenant A (1996) P2X receptors in autonomic and sensory neurons. Sem Neurosi 8:217–223
Evans RJ, Lewis C, Virginio C, Lundstrom K, Buell G, Surprenant A, North RA (1996) Ionic permeability of, and divalent cation effects on, two ATP-gated cation channels (P2X receptors) expressed in mammalian cells. J Physiol 497:413–422
Ferrari D, Chiozzi P, Falzoni S, Dal Susio M, Collo G, Buell G, di Virgilio F (1997) ATP-mediated cytotoxicity in microglial cells. Neuropharmacol 36:1295–1301
Fieber LA, Adams DJ (1991) Adenosine triphosphate-evoked currents in cultured neurones dissociated from rat parasympathetic cardiac ganglia. J Physiol 434:239–256
Funk GD, Kanjhan R, Walsh C, Lipski J, Comer AM, Parkis MA, Housley GD (1997) P2 receptor excitation of rodent hypoglossal motoneuron activity in vitro and in vivo: a molecular physiological analysis. J Neurosci 17:6325–6337
Galligan JJ, Bertrand PP (1995) ATP mediates fast synaptic potentials in enteric neurons. J Neurosci 14:7563–7571
Garcia-Guzman M, Soto F, Laube B, Stuhmer W (1996) Molecular cloning and functional expression of a novel rat heart P2X purinoceptor. FEBS Letts 388:123–127
Garcia-Guzman M, Soto F, Gomez-Hernandez JM, Lund P-E, Stuhmer W (1997a) Characterization of recombinant human $P2X_4$ receptor reveals pharmacological differences to the rat homologue. Mol Pharmacol 51:109–118
Garcia-Guzman M, Stuhmer W, Soto F (1997b) Molecular characterization and pharmacological properties of the human $P2X_3$ purinoceptor. Mol Br Res 47:59–66
Gu JG, MacDermott AB (1997) Activation of ATP P2X receptors elicits glutamate release from sensory neuron synapses. Nature 389:749–753
Housley GD, Greenwood D, Bennett T, Ryan AF (1995) Identification of a short form of the P2XR1-purinoceptor subunit produced by alternative splicing in the pituitary and cochlea. BBRC 212:501–508
Jahr CE, Jessell TM (1983) ATP excites a subpopulation of rat dorsal horn neurones. Nature 304:730–733
Kanjhan R, Housley GD, Thorne PR, Christie DL, Palmer DJ, Luo L, Ryan AF (1996) Localization of ATP-gated ion channels in cerebellum using P2X2R subunit-specific antisera. Neuroreport 7:2665–2669
Khakh BS, Humphrey PPA, Surprenant A (1995) Electrophysiological properties of P2X-purinoceptors in rat superior cervical, nodose and guinea-pig coeliac neurones. J Physiol 484:385–396
Khakh BS, Humphrey PPA, Henderson G (1997) ATP-gated cation channels (P2X purinoceptors) in trigeminal mesencephalic nucleus neurones of the rat. J Physiol 498:709–715
Kidd EJ, Grahames CBA, Simon J, Michel AD, Barnard EA, Humphrey PPA (1995) Localization of P2X purinoceptor transcripts in the rat nervous system. Mol Pharmacol 48:569–573
King BF, Ziganshin LE, Pintor J, Burnstock G (1996) Full sensitivity of $P2X_2$ purinoceptor revealed by changing extracellular pH. Br J Pharmacol 117:1371–1373
Krishtal OA, Marchenko SM, Obukhov AG (1988a) Cationic channels activated by extracellular ATP in rat sensory neurons. Neurosci 3:995–1000
Krishtal OA, Marchenko SM, Obukhov AG, Volkova TM (1988b) Receptors for ATP in rat sensory neurones: the structure-function relationship for ligands. Br J Pharmacol 95:1057–1062

Le K, Paquet M, Nouel D, Babinski K, Seguela P (1997) Primary structure and expression of a naturally truncated human P2X ATP receptor subunit from brain and immune system. FEBS Letts 418:195–199

Lewis, C.J. (1998) Actions of ATP receptors and NPY receptors in the peripheral nervous system. Ph.D. thesis Flinders University, Australia.

Lewis C, Neidhart S, Holy C, North RA, Buell G, Surprenant A (1995) Coexpression of $P2X_2$ and $P2X_3$ receptor subunits can account for ATP-gated currents in sensory neurons. Nature 377:432–435

Li C, Peoples RW, Li Z, Weight FF (1993) Zn^{2+} potentiates excitatory action of ATP on mammalian neurons. Proc Natl Acad Sci USA 90:8264–8267

Li C, Peoples RW, Weight FF (1996) Proton potentiation of ATP-gated ion channel responses to ATP and Zn^{2+} in rat nodose ganglion neurons. J Neurophysiol 76:3048–3058

Li C, Peoples RW, Weight FF (1998) Ethanol-induced inhibition of a neuronal P2X purinoceptor by an allosteric mechanism. Br J Pharmacol 123:1–3

Liu R, Sharom FJ (1997) Fluorescence studies on the nucleotide binding domains of the P-glycoprotein multidrug transporter. Biochem 36:2836–2843

Longhurst PA, Schwegel T, Folander K, Swanson R (1996) The human $P2X_1$ receptor: molecular cloning, tissue distribution, and localization to chromosome 17. Biochem Biophys Acta 1308:185–188

MacKenzie AB, Mahaut-Smith MP, Sage SO (1996) Activation of receptor-operated cation channels via $P2X_1$ not P2T purinoceptors in human platelets. J Biol Chem 271:2879–2881

Michel A, Humphrey PPA (1993) Distribution and characterisation of $[^3H]\alpha,\beta$-methylene ATP binding sites in the rat. Naunyn Schmiedeberg's Arch Pharmacol 348:608–617

Michel AD, Humphrey PPA (1994) Effects of metal cations on $[^3H]\alpha,\beta$-methylene ATP binding in rat vas deferens. Naunyn Schmiedeberg's Arch Pharmacol 350:113–122

Michel AD, Miller KJ, Lundstrom K, Buell G, Humphrey PPA (1997) Radiolabeling of the rat $P2X_4$ purinoceptor: evidence for allosteric interactions of purinoceptor antagonists and monovalent cations with P2X purinoceptors. Mol Pharmacol 51:524–532

Nakazawa K, Fujimori K, Takanaka A, Inoue K (1990) An ATP-activated conductance in phaeochromocytoma cells and its suppression by extracellular calcium. J Physiol 428:257–272

Nakazawa K, Fujimori K, Takanaka A, Inoue K (1991) Comparison of adenosine triphosphate- and nicotine-activated inward currents in rat phaeochromocytoma cells. J Physiol 434:647–660

Nakazawa K, Hess P (1993) Block by calcium of ATP-activated channels in pheochromocytoma cells. J Gen Physiol 101:377–392

Nakazawa K, Liu M, Inoue K, Ohno Y (1997) Potent inhibition by trivalent cations of ATP-gated channels. Eur J Pharmacol 325:237–243

Naumov AP, Kaznacheyeva EV, Kiselyov KI, Kuryshev YA, Mamin AG, Mozhayeva GN (1995) ATP-activated inward current and calcium permeable channels in rat macrophage plasma membranes. J Physiol 486:323–337

Neuhaus R, Reber BFX, Reuter H (1991) Regulation of bradykinin- and ATP-activated Ca^{2+} permeable channels in rat pheochromocytoma (PC12) cells. J Neurosci 11:3984–3990

North RA (1996) Families of ion channels with two hydrophobic segments. Curr Opinions Cell Biol 8:474–483.

Radford KM, Virginio C, Surprenant A, North RA, Kawashima E (1997) Baculovirus expression provides direct evidence for heteromeric assembly of $P2X_2$ and $P2X_3$ receptors. J Neurosci 17:6529–6533

Rassendren F, Buell G, Newbolt A, North RA, Surprenant A (1997a) Identification of amino acid residues contributing to the pore of a P2X receptor. EMBO J 18:3446–3454

Rassendren F, Buell G, Virginio C, Collo G, North RA, Surprenant A (1997b) The permeabilizing ATP receptor, $P2X_7$. J Biol Chem 272:5482–5486
Raybould NP, Housley GD (1997) Variation in expression of the outer hair cell P2X receptor conductance along the guinea-pig cochlea. J Physiol 498:717–727
Ren L, Burnstock G (1997) Prominent sympathetic purinergic vasoconstriction in the rabbit splenic artery: potentiation by 2,2'-pyridylisatogen tosylate. Br J Pharmacol 120:530–536
Robertson SJ, Rae MG, Rowan EG, Kennedy C (1996) Characterization of a P2X-purinoceptor in cultured neurones of the rat dorsal root ganglia. Br J Pharmacol 118:951–956
Rogers M, Dani JA (1995) Comparison of quantitative calcium flux through NMDA, ATP, and ACh receptor channels. Biophys J 68:501–506
Schachter JB, Harden TK (1997) An examination of deoxyadenosine 5'(α-thio)triphosphate as a ligand to define P2Y receptors and its selectivity as a low potency partial agonist of the $P2Y_1$ receptor. Br J Pharmacol 121:338–344
Schneider P, Hopp HH, Isenberg G (1991) Ca^{2+} influx through ATP-gated channels increments $[Ca^{2+}]_i$ and activated I_{Ca} in myocytes from guinea-pig urinary bladder. J Physiol 440:479–496
Seguela P, Haghighi A, Soghomonian J, Cooper E (1996) A novel neuronal P2X ATP receptor ion channel with widespread distribution in the brain. J Neurosci 16:448–455
Shen K, North RA (1993) Excitation of rat locus coeruleus neurons by adenosine 5'-triphosphate: ionic mechanis and receptor characterization. J Neurosci 13:894–899
Silinsky E.M, Redman RS (1996) Synchronous release of ATP and neurotransmitter within milliseconds of a motor nerve impulse in the frog. J Physiol 492:815–822
Simon J, Kidd EJ, Smith FM, Chessel IP, Murrell-Lagnado R, Humphrey PPA, Barnard EA (1997) Localization and functional expression of splice variants of the $P2X_2$ receptor. Mol Pharmacol 52:237–248
Somasundaram B, Mahaut-Smith MP (1994) Three cation influx currents activated by purinergic receptor stimulation in rat megakaryocytes. J Physiol 480:225–231
Soto F, Garcia-Guzman M, Gomez-Hernandez JM, Hollmann M, Karschin C, Stuhmer W (1996a) $P2X_4$: an ATP-activated ionotropic receptor cloned from rat brain. Proc Natl Acad Sci USA 93:3684–3688
Soto F, Garcia-Guzman M, Karschin C, Stuhmer W (1996b) Cloning and tissue distribution of a novel P2X receptor from rat brain. BBRC 223:456–460
Stoop R, Surprenant A, North RA (1997) Different sensitivities to pH of ATP-induced currents at four cloned P2X receptors. *J Neurophysiol* 78:1837–1840
Surprenant A, Buell G, North RA (1995) P2X receptors bring new structure to ligand-gated ion channels. TINS 18:224–229
Surprenant A, Rassendren F, Kawashima E, North RA, Buell G (1996) The cytolytic P2Z receptor for extracellular ATP identified as a P2X receptor $P2X_7$. Science 272:735–738
Tanaka J, Murate M, Wang C, Seino S, Iwanaga T (1996) Cellular distribution of the $P2X_4$ ATP receptor mRNA in the brain and non-neuronal organs of rats. Arch Histol Cytol 59:485–490
Traut TW (1994) The functions and consensus motifs of nine types of peptide segments that form different types of nucleotide-binding sites. Eur J Biochem 222:9–19
Trezise DJ, Bell NJ, Kennedy I, Humphrey PPA (1994) Effects of divalent cations on the potency of ATP and related agonists in rat isolated vagus nerve: implications for P2 purinoceptor classification. Br J Pharmacol 113:463–470
Trezise DJ, Michel AD, Grahames CBA, Khakh BS, Surprenant A, Humphrey PPA (1995) The selective P2X purinoceptor agonist, beta,gamma-methylene-L-adenosine 5'-triphosphate, discriminates between smooth muscle and neuronal P2X purinoceptors. Naunyn Schmiedeberg's Arch Pharmacol 351:603–609

Valera S, Hussy N, Evans RJ, Adami N, North RA, Surprenant A, Buell G (1994) A new class of ligand-gated ion channel defined by P2X receptor for extraxcellular ATP. Nature 371:516–519

Valera S, Talabot F, Evans RJ, Gos A, Antonarakis SE, Morris MA, Buell G (1995) Characterization and chromosomal localization of a human P(2X) receptor from the urinary bladder. Receptors and Channels 3:283–289

Virginio C, Church D, North RA, Surprenant A (1997) Effects of divalent cations, protons and calmidazolium at the rat $P2X_7$ receptor. Neuropharmacol 36:1285–1294

von Kugelgen I, Bultmann R, Starke K (1989) Effects of suramin and α,β-methylene ATP indicate noradrenaline-ATP co-transmission in the response of the mouse vas deferens to single low frequency pulses. Naunyn Schmiedeberg's Arch Pharmacol 340:760–763

von Kugelgen I, Starke K (1991) Release of noradrenaline and ATP by electrical stimulation and nicotine in guinea-pig vas deferens. Naunyn Schmiedeberg's Arch Pharmacol 344:419–429

Vulchanova L, Arvidsson U, Riedl M, Buell G, Surprenant A, North RA, Elde RP (1996) Differential distribution of two ATP-gated ion channels (P2X receptors) determined by immunohistochemistry. Proc Natl Acad Sci USA 93:8063–8067

Vulchanova L, Riedl M, Shuster SJ, Buell G, Surprenant A, North RA, Elde RP (1997) Immunohistochemical study of the $P2X_2$ and $P2X_3$ receptor subunits in rat and monkey sensory neurons and their central terminals. Neuropharmaco*l* 36:1229–1242

Wang CW, Namba N, Gonoi T, Inagaki N, Seino S (1996) Cloning and pharmacological characterization of a fourth P2X receptor widely expressed in brain and peripheral tissues including various endocrine tissues. BBRC 220:196–202

Werner P, Seward EP, Buell G, North RA (1996) Domains of P2X receptors involved in desensitization. Proc Natl Acad Sci USA 93:15485–15490

Wildman SS, King BF, Burnstock G (1997) Potentiation of ATP-responses at recombinant $P2X_2$ receptor by neurotransmitters and related substances. Br J Pharmacol 120:221–224

Wiley JS, Chen R, Wiley MJ, Jamieson GP (1992) The ATP^{4-} receptor-operated ion channel of human lymphocytes: inhibition of ion fluxes by amiloride analogs and by extracellular sodium ions. Arch Biochem Biophys 292:411–418

Wright JM, Li C (1995) Zn^{2+} potentiated steady state ATP activated currents in rat nodose ganglion neurons by increasing the burst duration of a 35 pS channel. Neurosci Letts 193:177–180

Yang SY, Cheek DJ, Westfall DP, Buxton ILO (1994) Purinergic axis in cardiac blood vessels. Circ Res, 74:401–407

Zhou X, Galligan JJ (1996) P2X purinoceptors in cultured myenteric neurons of guinea-pig small intestine. J Physiol 496:719–729

CHAPTER 21

The 5-HT_3 Receptor Channel: Function, Activation and Regulation

J.L. Yakel

A. Introduction

The 5-HT_3 receptor (5-HT_3R) is a ligand-gated ion channel gated by the neurotransmitter serotonin (5-HT) and belonging to the superfamily of ligand-gated ion channels, a group that includes nicotinic acetylcholine (ACh), GABA, and glycine receptor channels (Maricq et al. 1991). The activation of the 5-HT_3R opens a cationic ion channel that depolarizes the membrane, thereby activating a rapid excitatory response in a variety of central and peripheral nervous system (CNS and PNS) preparations. Thus the 5-HT_3R is unique from the other classes of 5-HT receptors which all couple to GTP binding proteins.

The first clear evidence for distinct 5-HT receptor subtypes appeared more than 40 years ago (Gaddum and Picarelli 1957), which eventually led to the classification that first defined the 5-HT_3R (Bradley et al. 1986; Richardson and Engel 1986). The development of very potent and highly selective 5-HT_3R ligands in the mid-1980s (Richardson and Engel 1986) revolutionized the 5-HT_3R receptor field and led to a series of very important observations shortly thereafter. Kilpatrick et al. (1987) published the first radioligand binding data that showed the distribution of the 5-HT_3R in the rat brain. The distribution of the 5-HT_3R is widespread throughout the CNS and PNS where it is known to participate in a variety of physiological responses. Due to the rapid activation of 5-HT_3R-mediated responses and its lack of 'washout' in whole-cell patch-clamp electrophysiological recordings, it was proposed that the 5-HT_3R incorporated a ligand-gated ion channel (Yakel and Jackson 1988). Derkach et al. (1989) confirmed this by showing that 5-HT_3R-mediated single-channel currents could be elicited in outside-out membrane patches from guinea pig submucous plexus neurons. The 5-HT_3R was cloned by Maricq et al. (1991).

This review will cover the basic functional and molecular aspects of the 5-HT_3R. Although much has been learned about the functional role the 5-HT_3R may be playing in the nervous system, much has yet to be learned. As a variety of neuronal cell lines have functional 5-HT_3Rs, these cells often serve as a valuable model to study the molecular properties of this receptor (Jackson and Yakel 1995).

B. Receptor Distribution

The 5-HT_3R is widely distributed within the CNS and PNS. In the periphery, a variety of tissues have functional 5-HT_3Rs, including sympathetic, parasympathetic, and afferent nerves, the heart, and gastrointestinal tract (COHEN 1992; JACKSON and YAKEL 1995). Due to this widespread distribution, 5-HT_3R activation controls a diffuse array of diverse physiological responses in the periphery.

In the CNS, widespread distribution of 5-HT_3R binding sites have been reported by many groups (LAPORTE et al. 1992; JACKSON and YAKEL 1995). High or moderate density 5-HT_3R binding sites have been observed in the forebrain (e.g., cerebral cortex, hippocampus, and amygdala), hindbrain (e.g., entorhinal cortex), medulla oblongata (e.g., nucleus tractus solitarius, area postrema, dorsal motor nucleus of the vagus nerve, and the nucleus of the spinal tract of the trigeminal nerve), and spinal cord. Other regions of the brain, such as the nucleus accumbens, striatum, and substantia nigra, are sometimes reported to possess 5-HT_3R binding sites, although generally at a much lower density.

C. Molecular Structure

I. Sequence, Assembly, and Splice Variants

The 5-HT_3R was initially cloned from the NCB-20 neuroblastoma cell line by MARICQ et al. (1991) (this will be referred to as the 5-HT_3R-A subunit), and has a predicted amino acid length of 487 amino acids with a molecular weight of 56 kDa. Similar to the nicotinic, GABA, and glycine receptor channels, hydrophobicity analysis of the 5-HT_3R predicts that it contains four hydrophobic putative transmembrane domains (i.e., M1–M4), a large N-terminal extracellular domain, and a long cytoplasmic loop connecting M3 and M4 (Fig. 1). By analogy with the nAChR, the second transmembrane domain (M2) from each subunit is thought to line the pore of the channel, and the long linker region between M3 and M4 contains several putative phosphorylation sites (MARICQ et al. 1991; YAKEL et al. 1993). A variety of data (see Sect. E below) indicates that the 5-HT_3R ligand binding site is located on the N-terminal extracellular domain. The C-terminal domain has also been shown to be extracellular (MUKERJI et al. 1996). Also, like other members in this superfamily, the 5-HT_3R channel is likely to be a pentamer. BOESS et al. (1995) used electron microscopic techniques to show that the 5-HT_3R purified from NG108–15 cells was composed of five subunits arranged symmetrically around a central cavity (Fig. 1), with a length of approximately 11 nm, a diameter of approximately 8 nm, a closed end, and central cavity with a diameter of approximately 3 nm.

Apparent splice variants of the mouse and guinea pig 5-HT_3Rs have been cloned (HOPE et al. 1993; LANKIEWICZ et al. 1998) in which six amino acid residues located within the putative large intracellular loop between M3 and M4 were deleted; these variant subunits will be referred to as the long (i.e.,

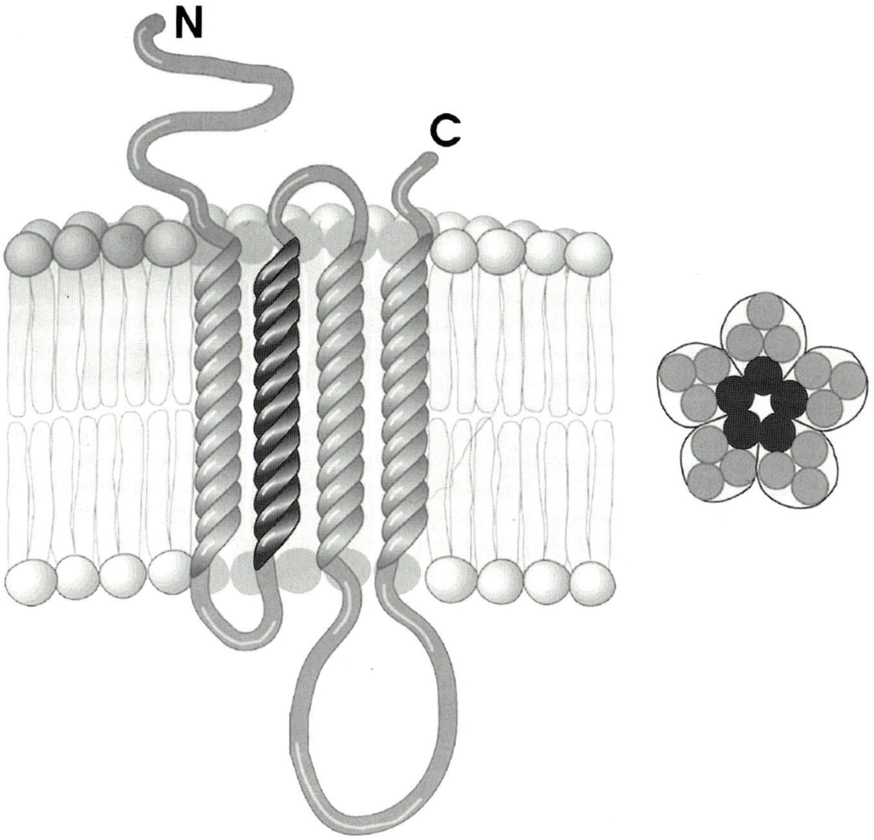

Fig. 1. Suspected topology of 5-HT$_3$R in the membrane. *Left*, the amino (*N*) and carboxy (*C*) termini are known to be extracellular, there are four predicted transmembrane domains, and the second (*darker gray*) is thought to line the pore of the channel. *Right*, birds' eye view of the 5-HT$_3$R. Functional 5-HT$_3$R channels are thought to be composed of five subunits, each surrounding a central pore

5-HT$_3$R-A$_L$ or 5-HT$_3$R-A) and short (5-HT$_3$R-A$_S$) subunits. In the rat, these two splice variants were also found; however the deleted region is composed of five rather than six amino acid residues (Miquel et al. 1995). Interestingly in the human, only the short form of the 5-HT$_3$R appears to exist (Werner et al. 1994; Belelli et al. 1995; Miyake et al. 1995).

Recently, a new 5-HT$_3$R subunit has been cloned (Davies et al. 1999); this will be referred to as the 5-HT$_3$R-B subunit. The 5-HT$_3$R-B subunit is 441 amino acid residues in length, and has 41% amino acid identity with the 5-HT$_3$R-A subunit. Interestingly, the 5-HT$_3$R-B subunit cannot form homo-oligomeric receptors on its own, but can only form hetero-oligomeric channels by co-assembling with the 5-HT$_3$R-A subunit.

II. Gene Structure

Analysis of the mouse 5-HT_3R gene (UETZ et al. 1994; WERNER et al. 1994) showed that the 5-HT_3R is most similar to the nicotinic ACh receptor (nAChR), in particular with the α7 nAChR subunit. Further evidence of the close structural similarity between the 5-HT_3R and α7 nAChR was demonstrated by formation of a functional recombinant chimeric α7–5-HT_3 receptor consisting of the N-terminal domain of the α7 nAChR and the remainder of the 5-HT_3R (EISELÉ et al. 1993). The coding region of the 5-HT_3R gene is interrupted by eight introns, three of which are conserved between the 5-HT_3R and vertebrate nAChRs. In addition the use of two alternative splice acceptor sites in intron 8 results in either the long or short form of the 5-HT_3R-A subunit (WERNER et al. 1994; UETZ et al. 1994). As the human gene does not contain the splice acceptor site that creates the long form of the 5-HT_3R-A subunit (WERNER et al. 1994), the fact that only the short form of the 5-HT_3R has been found in the human is consistent with this finding (BELELLI et al. 1995; MIYAKE et al. 1995). The 5-HT_3R gene was localized to human chromosome 11 (UETZ et al. 1994).

III. Developmental Regulation

Both the short and long forms of the 5-HT_3R-A subunit were found in a variety of neuronal tissues (e.g., SCG, hippocampus, and cortex) and mouse neuroblastoma cell lines (e.g., NCB-20, NG108–15), with the short form being approximately five times more abundant than the long form in each of these different preparations (WERNER et al. 1994). In the rat, both the short and long forms were also found in a variety of central and peripheral tissues, with the short form also much more abundant than the long; the relative amount of the long form in the adult rat (~10%) was consistent between these various tissues (MIQUEL et al. 1995). However the relative amount of the long form was developmentally regulated. At embryonic day 17, the relative percentage of the long form increased from ~10% to 30–35% in the hippocampus and cortex, and to 50–75% in the SCG and nodose ganglia (MIQUEL et al. 1995). In NG108–15 cells, the relative proportion of the two splice variants of the 5-HT_3R was also regulated in a similar fashion by differentiation (EMERIT et al. 1995).

IV. Homo-Oligomeric Vs Hetero-Oligomeric Assembly

Most members of this superfamily of ligand-gated ion channels are comprised of structurally different subunits and are therefore hetero-oligomeric receptor proteins. Functional 5-HT_3R-activated responses, with properties similar to those from natively expressed 5-HT_3Rs, can be obtained by expressing a single 5-HT_3R-A subunit in heterologous expression systems (e.g., HEK 293 cells or *Xenopus* oocytes); this suggests that functional native 5-HT_3Rs may be homo-oligomeric proteins (FLETCHER and BARNES 1998). However, diverse functional and pharmacological properties of 5-HT_3Rs have been reported that are not accounted for by the different 5-HT_3R-A splice variants (see below), which might suggest that other as yet unknown 5-HT_3R subunits exist. Therefore whether or not native 5-HT_3Rs are homo- or hetero-oligomeric proteins is currently unknown, and the search for other possible 5-HT_3R subunits (or subunits interacting with the 5-HT_3R) continues.

Data consistent with the idea that additional subunits may be interacting with the 5-HT_3R has appeared. 5-HT_3R purified from porcine brain contains both 5-HT_3R-A and non-5-HT_3R-A proteins (FLETCHER and BARNES 1997). HUSSY et al. (1994) suggested that nAChR subunits, or subunits from another ligand-gated ion channel, might associate with 5-HT_3R subunits to generate different 5-HT_3R subtypes. Interestingly, VAN HOOFT et al. (1998) and KRIEGLER et al. (1999) recently reported that the $\alpha 4$ nAChR subunit can co-assemble with the 5-HT_3R subunit in HEK 293 cells and *Xenopus* oocytes to form a functional channel with an enhanced permeability to calcium (Ca^{2+}); the pharmacological properties of this channel were similar to expressed homo-oligomeric 5-HT_3Rs. Whether or not co-assembly occurs *in vivo* has yet to be tested. Interestingly, NAYAK et al. (1998) have recently reported that the 5-HT_3R and $\alpha 4$ nAChR subunits co-localize on a subset of rat striatal and cerebellar synaptosomes. However, FLETCHER and BARNES (1998) recently reported that the non-5-HT_3R-A proteins purified from porcine brain along with the 5-HT_3R were not either the $\alpha 1$, $\alpha 3$, $\alpha 4$, $\alpha 5$, $\alpha 7$, or the $\beta 2$ nAChR subunits. The recent cloning of the 5-HT_3R-B subunit (DAVIES et al. 1999), whose co-expression with the 5-HT_3R-A subunit alters many of the functional properties of the 5-HT_3R channels, may help to explain in part some of the diverse functional and pharmacological properties of native 5-HT_3Rs. Clearly this issue awaits further study.

D. Function in the Nervous System

5-HT_3Rs are thought to be involved in a variety of physiological responses in the CNS and PNS, including cognition, pain reception, motor neuron activity, and sensory processing (JACKSON and YAKEL 1995). Clinically, 5-HT_3R ligands are powerful therapeutic agents in the control and treatment of emesis, drug and alcohol dependence, schizophrenia, anxiety, and cognitive dysfunction (GREENSHAW 1993; GRANT 1995). At the level of the synapse, the 5-HT_3R has been shown to function both at presynaptic sites to control the release of various neurotransmitters, and at postsynaptic sites where it participates in fast synaptic transmission.

I. Presynaptic Role and Neurotransmitter Release

5-HT_3Rs, probably located on presynaptic terminals, have been shown to regulate the release (from various brain regions) of dopamine, ACh, cholecystokinin (CCK), GABA, and glutamate (BARNES et al. 1989; BLANDINA et al. 1989; GLAUM et al. 1992; MAURA et al. 1992; PAUDICE and RAITERI 1991). Consistent with this, KIDD et al. (1993) previously reported that 5-HT_3Rs in the rat CNS appear mainly on presynaptic nerve terminals. The ability of the 5-HT_3R to regulate dopamine release in the mesolimbic pathway suggests a potentially important role in the reward pathway and drug abuse (GRANT 1995). Activation of 5-HT_3Rs in rat striatal slices increases dopamine release (BLANDINA et al. 1989), and direct evidence for the presence of functional presynaptic 5-HT_3Rs on striatal synaptosomes was recently reported (NICHOLS and MOLLARD 1996). Interestingly, over-expressing the 5-HT_3R in the mouse forebrain resulted in a decrease in ethanol consumption in these transgenic mice (ENGEL et al. 1998).

II. Postsynaptic Role

5-HT, via activation of the 5-HT_3R, can rapidly depolarize and excite neurons from many different neuronal tissues and cell lines, consistent with a postsynaptic role for 5-HT_3Rs in the nervous system (JACKSON and YAKEL 1995). Nevertheless, few examples of 5-HT_3R-mediated synaptic events have been directly observed; SUGITA et al. (1992) in the rat amygdala and ROERIG et al. (1997) in the ferret visual cortex have reported fast synaptic events mediated by the 5-HT_3R.

In the rat hippocampus, serotonergic projections from the raphe-hippocampal pathway selectively innervate and form multiple synaptic contacts with GABAergic interneurons, selectively onto the somata or dendritic trees of interneurons that contain calbindin (FREUND et al. 1990). Recently MORALES and BLOOM (1997) reported that 5-HT_3Rs were present in certain subpopulations of GABAergic neurons in the telencephalon (e.g., neocortex, olfactory cortex, hippocampus, and amygdala); these neurons were immunoreactive for cholecystokinin and the Ca^{2+}-binding proteins calbindin and calretinin, but not somatostatin and parvalbumin. ROPERT and GUY (1991) reported that the activation of 5-HT_3Rs in rat hippocampal slices increases the frequency of inhibitory synaptic events in CA1 pyramidal cells, and they suggested that this was due to the direct activation of GABAergic interneurons via the 5-HT_3R. To confirm this, direct electrical recordings were obtained from rat inhibitory hippocampal interneurons (mostly likely GABAergic), both in the stratum radiatum of the CA1 region (MCMAHON and KAUER 1997) and in the dentate gyrus (KAWA 1994), and the activation of functional 5-HT_3R channels was demonstrated. As described below (see section D/IV), a strong link between the function of the 5-HT_3R on these hippocampal GABAergic interneurons and learning and memory has been reported (STAUBLI and XU 1995; REZNIC and STAUBLI 1997). Nevertheless, the direct synaptic activation of these hippocampal 5-HT_3Rs has not yet been reported.

III. Physiological Properties

1. Receptor Activation

The activation of the 5-HT_3R by 5-HT and other ligands has been extensively studied (JACKSON and YAKEL 1995). Dose-response data have yielded estimates for a dissociation constant of ~1–5 μM, with a Hill coefficient significantly greater than 1. These data suggest that the 5-HT_3R has multiple agonist binding sites, and that the occupation of at least two of these sites is required for full activation of the receptor (i.e., co-operativity). Below (Sect. E) a more detailed description of the pharmacological properties of the 5-HT_3R ligand binding site will be discussed.

The rate of activation of 5-HT_3R-mediated responses is slower in comparison to other ligand-gated ion channels, where the rate of activation approaches the diffusion limit (JACKSON and YAKEL 1995). By rapidly (1 ms) applying a maximum dose (30–100 μM) of 5-HT to activate native 5-HT_3Rs in N1E-115 neuroblastoma cells, MIENVILLE (1991) reported that the activation was exponential with a time constant of ~24 ms. More recently in HEK 293 cells expressing 5-HT_3Rs, activation by 5-HT (100 μM applied in <0.5 ms) activated a response with a 10–90% risetime of 14 ms (TRAYNELIS and MOTT 1996). The reason and significance for the slow nature of 5-HT_3R activation have yet to be appreciated.

2. Single-Channel Properties

Some of the most intriguing data suggesting that additional 5-HT_3R subunits exist comes from studies investigating the diversity in single channel conductance levels; these levels have been estimated to range from 0.3 pS to 19 pS (JACKSON and YAKEL 1995; FLETCHER and BARNES 1998). In many neuronal preparations from guinea pig, rabbit, rat and mouse, observable single channel events up to 19 pS in conductance have been observed. However in various neuroblastoma cell lines or for homo-oligomeric 5-HT_3R channels expressed in mammalian cells, observable single channel events are not seen in most cases, suggesting that the single channel conductance is at the sub pS level (HUSSY et al. 1994; JACKSON and YAKEL 1995). In addition to the observable 5-HT_3R single channel conductances in neuronal preparations, the presence of a distinct and non-resolvable conductance level within the same cells was also seen, indicating the possibility of functional diversity (i.e., more than one type of functional 5-HT_3R) (DERKACH et al. 1989; YANG et al. 1992; HUSSY et al. 1994). This has led to the suggestions that functional homo-oligomeric 5-HT_3Rs can form and that the 5-HT_3Rs in these neuroblastoma cell lines are homo-oligomeric assemblies of 5-HT_3R-A subunits, and that the higher conductance level might be due to the participation of other subunits in the structure of native neuronal 5-HT_3Rs (HUSSY et al. 1994). Consistent with this idea is that the co-expression of the recently cloned 5-HT_3R-B subunit along with the 5-HT_3R-A subunit in HEK 293 cells resulted in relatively large single channel currents (16 pS) (DAVIES et al. 1999).

There have been two reports of discrete observable 5-HT_3R-gated single channel currents in neuroblastoma cell lines. In undifferentiated NG108–15 cells, observable single-channels of 9 pS and 13 pS were evident, whereas no observable single-channel currents were observed in differentiated NG108–15 cells (SHAO et al. 1991). In addition VAN HOOFT et al. (1994) observed single channel currents of ~6 pS in excised outside-out patches of differentiated N1E-115 cells when the driving force for Na^+ was enhanced. In cell-attached patches, single channels with a conductance of up to 27 pS were observed; this conductance level was thought to be dependent on the protein kinase C (PKC)-induced phosphorylation of the 5-HT_3R (VAN HOOFT and VIJVERBERG 1995).

3. Desensitization

Like other ligand-gated ion channels, the 5-HT_3R channel undergoes desensitization (i.e., a closed and/or inactivated state) in the continued presence of agonist. Even though the molecular mechanism of desensitization is currently unknown, several factors are known to regulate the kinetics of desensitization (JACKSON and YAKEL 1995). In NG108–15 cells, where the kinetics and regulation of desensitization have been extensively studied, the onset of desensitization is a biphasic process (with a fast time constant of decay of ~200–300 ms and a slow time constant of decay of ~2–5 s) with kinetics that are strongly dependent on voltage (JACKSON and YAKEL 1995). However, others have reported that the onset of desensitization is monophasic and voltage-independent (JACKSON and YAKEL 1995). Some of these differences might be explained by different laboratories using different methods of 5-HT application; however, the fact that differences exist under identical recording conditions, either between different cell types (YANG 1990; YANG et al. 1992) or identical cell types but under different developmental states (SHAO et al. 1991), may be another indication of molecular diversity in the 5-HT_3R.

The kinetics of desensitization were previously shown to be regulated by a phosphorylation process in NG108–15 cells (YAKEL et al. 1991; JACKSON and YAKEL 1995). More recently changes in the cytoplasmic Ca^{2+} concentration ($[Ca^{2+}]_i$) were also found to regulate the kinetics of desensitization in these cells (JONES and YAKEL 1998). The molecular mechanisms responsible for this regulation are currently unknown, but a Ca^{2+}-dependent signal transduction cascade is likely to be involved. Consistent with this, the function of the 5-HT_3R has been reported to be regulated by PKC (VAN HOOFT and VIJVERBERG 1995) and calcineurin (BODDEKE et al. 1996). However even though the 5-HT_3R contains several putative phosphorylation sites (MARICQ et al. 1991), no biochemical evidence has been published to date showing that the 5-HT_3R can be directly phosphorylated.

BARTRUP and NEWBERRY (1996) studied the kinetics of desensitization in NG108–15 cells to explore in more detail how the binding of ligand leads to receptor activation and desensitization. Their data is consistent with the cyclic model of receptor desensitization first proposed by KATZ and THESLEFF (1957), although a more complex co-operative model (NEIJT et al. 1989) cannot be ruled out. Even though relatively high concentrations of 5-HT (~1–5 μM) are required to activate the 5-HT_3R, low doses of 5-HT (~50–100 nM) or other 5-HT_3R agonists, which evoke little or no detectable current, can potently block the 5-HT_3R-activated responses (NEIJT et al. 1988; BARTRUP and NEWBERRY 1996). The concentration and use-dependence of this inhibitory effect suggests that it is the result of the high-affinity binding of 5-HT to the desensitized state of the 5-HT_3R, which prevents the subsequent recovery from receptor desensitization (BARTRUP and NEWBERRY 1996).

The molecular structure of the 5-HT_3R, like other members of this family, also contributes to the process of desensitization. YAKEL et al. (1993) reported that alterations in an amino acid residue thought to line the pore of the channel greatly altered the kinetics of desensitization, an effect similar to observations for nACh and GABA receptor channels (JACKSON and YAKEL 1995). The 5-HT_3R was recently cloned from the guinea pig and expressed in HEK 293 cells (LANKIEWICZ et al. 1998). Besides having markedly different pharmacological properties (see Sect. E), the kinetics of desensitization of the guinea pig 5-HT_3R were much slower than for either the human or mouse 5-HT_3R. These data also suggest that the molecular structure of the 5-HT_3R may be important in determining the properties of desensitization. There is also evidence for an allosteric regulatory site on the 5-HT_3R that controls desensitization; 5-hydroxyindole (and its analogs) can potentiate 5-HT_3R-mediated responses and decrease the rate of desensitization in N1E-115 neuroblastoma cells (VAN HOOFT et al. 1997a).

4. Ion Permeation and Pore Structure

The permeation of the 5-HT_3R has been extensively studied by many investigators in a variety of cell types (JACKSON and YAKEL 1995). Consistent with its molecular similarities with the nAChRs, the 5-HT_3R also appears to be a relatively non-selective cationic channel that discriminates poorly among the inorganic monovalent cations. The finite permeation by the large organic cation *N*-methyl-D-glucamine (NMDG) indicates that the 5-HT_3R has an effective minimum pore size estimated to be ~7.6–8.1 Å (YANG 1990; BROWN et al. 1998)

The main issue of contention is whether the 5-HT_3R channel is permeable to Ca^{2+} (HARGREAVES et al. 1994; JACKSON and YAKEL 1995; GILON and YAKEL 1995). This is important due to the role that $[Ca^{2+}]_i$ plays in various signal transduction cascades and synaptic plasticity. Initially based on electrophysiological reversal potential measurements, the Ca^{2+} permeability of native 5-HT_3Rs was thought to be very low, although data showing a relatively high Ca^{2+} permeability was reported (YANG 1990; YANG et al. 1992; HARGREAVES et al. 1994; BROWN et al. 1998). However in these later studies, the high Ca^{2+} permeability was associated with a very low external bath concentration of Na^+ and K^+;

Hargreaves et al. (1994) showed that the Ca^{2+} permeability of the 5-HT_3R was much lower in physiological ionic solutions (i.e., solutions high in Na^+ and/or K^+). Thus these data suggest that under physiological ionic conditions, the Ca^{2+} permeability of the native 5-HT_3R is very low or non-detectable. However it is possible that the Ca^{2+} permeability of the 5-HT_3R may be variable and/or regulated by factors other than the ionic composition of the solution. Using laser-scanning confocal microscopic techniques to measure $[Ca^{2+}]_i$ signals, the 5-HT_3R located on presynaptic rat striatal nerve terminals was found to be significantly Ca^{2+} permeant (Rondé and Nichols 1998), whereas the 5-HT_3R in NG108–15 cells did not seem to be (Rondé and Nichols 1997). In addition, as mentioned above, van Hooft et al. (1998) recently reported that co-assembly of the nicotinic $\alpha 4$ with the 5-HT_3R subunit in HEK 293 cells and *Xenopus* oocytes enhanced the permeability of the 5-HT_3R to Ca^{2+}, however whether such an interaction occurs *in vivo* has not yet been demonstrated. Furthermore co-expression of the 5-HT_3R-B subunit along with the 5-HT_3R-A subunit results in a channel with a lower permeability to Ca^{2+} as compared to the 5-HT_3R-A subunit alone (Davies et al. 1999).

Divalent cations (e.g., Ca^{2+} and Mg^{2+}) have been reported to block the function of the 5-HT_3R channel, both in a voltage-dependent and voltage-independent fashion (Jackson and Yakel 1995). The voltage-dependent block may be relevant to issues relating to synaptic plasticity. For example, McMahon and Kauer (1997) showed that Ca^{2+}, but not Mg^{2+}, resulted in a voltage-dependent block of the 5-HT_3R in rat inhibitory interneurons. The relative block was more significant at negative holding potentials, resulting in an I–V curve with a region of negative slope conductance. By analogy with the voltage-dependent Mg^{2+} block of the NMDA subtype of glutamate receptor channel, McMahon and Kauer (1997) suggested that the 5-HT_3R may also be serving a role as a coincident detector relating to synaptic plasticity and LTP. It should be noted however that such a voltage-dependent block is not always observed (Jackson and Yakel 1995). Since blockade of the 5-HT_3R by divalent ions appears to be different under different experimental conditions, it suggests the possibility that divalent ions may have multiple mechanisms for blocking the 5-HT_3Rs.

5. Rectification and Voltage-Dependence

The shape of the I–V curve of 5-HT_3R-mediated responses is generally non-linear (i.e., inwardly rectifies), such that the magnitude of the slope conductance at negative membrane potentials is generally several times greater than that at positive membrane potentials (Jackson and Yakel 1995). This non-linearity is not due to the voltage-dependent block by divalent cations, nevertheless the molecular basis relating to the non-linear I–V curve remains to be determined. This rectification seen in macroscopic current traces (i.e., whole-cell responses) is also observed at the microscopic single-channel current level. Brown et al. (1998) studied the functional properties of human 5-HT_3Rs expressed in HEK 293 cells and reported that the rectification of the whole-cell currents could be accounted for by the non-linearity in the single channel conductance estimated by noise analysis. A similar correlation was observed by others in different preparations (YANG et al. 1992; HUSSY et al. 1994). Furthermore BROWN et al. (1998) also showed that the voltage-*independent* block of the 5-HT_3R by divalent cations could mostly be accounted for by the ability of divalent ions to decrease the single channel conductance of the 5-HT_3R. In addition the co-expression of the 5-HT_3R-B subunit along with the 5-HT_3R-A subunit resulted in a more linear I-V curve as compared to expression of the 5-HT_3R-A subunit alone (DAVIES et al. 1999).

IV. Modulation, Synaptic Plasticity, and Learning and Memory

There is strong evidence linking the function of 5-HT_3Rs to long-term potentiation (LTP; a potential cellular model for learning and memory), and learning and memory. In the hippocampus, a region known to be important for certain forms of memory processing, patterns of neuronal activity that are known to be correlated with learning and LTP (e.g., theta rhythms) are controlled via GABAergic interneurons. As described above, GABAergic inhibitory interneurons in the rat hippocampus have functional 5-HT_3Rs (KAWA 1994; MCMAHON and KAUER 1997), and the activation of these receptors increased the frequency of GABAergic synaptic events in CA1 pyramidal cells (ROPERT and GUY 1991). In freely moving rats, the systemic injection of selective 5-HT_3R antagonists facilitates the induction of LTP in the CA1 subfield of the hippocampus, increased hippocampal theta rhythm, and enhanced the retention of memory in hippocampal-dependent tasks (STAUBLI and XU 1995). These effects of 5-HT_3R antagonists were shown to be due to a decrease in the firing activity of a subset of CA1 hippocampal interneurons, and a concomitant increase in the firing rate of the hippocampal pyramidal cells (REZNIC and STAUBLI 1997). 5-HT_3Rs have also been shown to play an important role in mediating the induction and maintenance of LTP in the rat superior cervical ganglion (SCG) (ALKADHI et al. 1996).

It is widely believed that the molecular/cellular mechanisms responsible for changes in neuronal activity and plasticity involve various intracellular signal transduction cascades. Like other ligand-gated ion channels, 5-HT_3Rs have been shown to be regulated by such processes. The activation of PKC enhanced the function of the 5-HT_3R channel in N1E-115 neuroblastoma cells (VAN HOOFT and VIJVERBERG 1995) and in *Xenopus* oocytes expressing the 5-HT_3R channel (ZHANG et al. 1995). In addition the Ca^{2+}-calmodulin regulated protein phosphatase, calcineurin, was also reported to regulate the function of the 5-HT_3R channel in NG108–15 cells (BODDEKE et al. 1996). Recently JONES and YAKEL (1998) have shown that $[Ca^{2+}]_i$ levels can have profound effects on the kinetics of desensitization of the 5-HT_3R. Although the molecular mechanism responsible to explain this action of Ca^{2+} is currently unknown, a Ca^{2+}-dependent enzymatic processes is suspected.

E. Pharmacological Properties

I. 5-HT$_3$R Ligands: Agonists and Antagonists

A variety of highly selective and potent 5-HT$_3$R ligands exist (KILPATRICK and TYERS 1992). One of the first and most often used 5-HT$_3$R-selective agonist is 2-methyl-5-HT. A much more potent and selective agonist, *m*-chlorophenylbiguanide (mCPBG), was identified by KILPATRICK et al. (1990). Even more potent biguanide derivatives exist (MORAIN et al. 1994). Both 2-methyl-5-HT and mCPBG can be either full or partial agonists under certain conditions (see below).

There are many selective and extremely potent 5-HT$_3$R antagonists (KILPATRICK and TYERS 1992). The most commonly used are tropisetron (ICS 205–930), MDL 72222, ondansetron (GR 38032), granisetron (BRL 43694), and zacopride. D-Tubocurarine (curare), although certainly not selective, is also a potent inhibitor of the 5-HT$_3$R (JACKSON and YAKEL 1995), and its blocking action has yielded important clues relating to the 5-HT$_3$R binding pocket (see below). In addition, tetraethylammonium (TEA), a classical blocker of voltage-gated K^+ channels, blocks the 5-HT$_3$R at an agonist recognition site and prevents desensitization (KOOYMAN et al. 1993b).

There is much pharmacological evidence indicating that there are species-specific differences (i.e., interspecies heterogeneity) in the properties of the 5-HT$_3$R (BUTLER et al. 1990; NEWBERRY et al. 1991; FLETCHER and BARNES 1998). For example, the human and guinea pig 5-HT$_3$Rs are distinct from the rat, mouse, and rabbit (see FLETCHER and BARNES 1998). The 5-HT$_3$R from the guinea pig was recently cloned (LANKIEWICZ et al. 1998) and expressed in HEK 293 cells, along with the human and mouse 5-HT$_3$Rs, to compare directly their pharmacological and physiological properties. Interestingly, the properties of the guinea pig receptor were markedly different from either the mouse or human. For example, mCPBG is a much less potent agonist for the guinea pig than for either the human or mouse 5-HT$_3$Rs, and tropisetron and metoclopramide are much less potent antagonists; all three ligands are more potent at mouse than at human 5-HT$_3$Rs. Furthermore the selective 5-HT$_3$R agonist 1-phenylbiguanide, which activates the human and mouse 5-HT$_3$Rs, neither binds to nor activates the guinea pig 5-HT$_3$R (LANKIEWICZ et al. 1998). These data for expressed guinea pig 5-HT$_3$Rs correspond to the data obtained in native neuronal guinea pig preparations (BUTLER et al. 1990). Therefore, as the overall sequence for the guinea pig receptor reveals >80% homology with the mouse and human 5-HT$_3$Rs, the different amino acid residues in the N-terminal region may help to determine the precise residues of interaction of these various ligands with the 5-HT$_3$R.

There is also somc pharmacological evidence supporting the case for intraspecies 5-HT$_3$R diversity (BONHAUS et al. 1993). In addition to these data, there is a variety of functional and molecular evidence suggesting that there is intraspecies and even intracellular 5-HT$_3$R diversity. Some differences in pharmacological properties can be explained by the properties of the different splice variants. For example, differences have been reported in 5-HT_3R-selective agonist properties between the splice variants expressed in various systems (Downie et al. 1994; Niemeyer and Lummis 1998); however such differences were not observed by others (Glitsch et al. 1994; Werner et al. 1994; van Hooft et al. 1997b). Even though similarities in the properties of native 5-HT_3Rs in neuroblastoma cell lines and expressed receptors led to the suggestion that functional homo-oligomeric 5-HT_3Rs can form, and that the 5-HT_3Rs in these neuroblastoma cell lines are homo-oligomeric 5-HT_3R-As (Hussy et al. 1994; see Sect. D.II.2 above), differences exist between these two preparations. This has led to the suggestion that the native 5-HT_3Rs in neuroblastoma cell lines are distinct from expressed receptors, and provides further evidence that other, as yet unknown, 5-HT_3R subunits exist (Gill et al. 1995; van Hooft et al. 1997b). Interestingly, co-expressing the recently cloned 5-HT_3R-B subunit along with the 5-HT_3R-A subunit resulted in alterations in pharmacological properties; for example the potency of both 5-HT to activate and curare to block co-assembled hetero-oligomeric 5-HT_3R channels was reduced (Davies et al. 1999).

II. 5-HT_3R Ligand Binding Site

Miquel et al. (1991) first reported that tryptophan residues were involved in the binding of ligands to the 5-HT_3R. Evidence for the involvement of a specific tryptophan residue, position 89 (W89), was reported by Schulte et al. (1995) (originally reported as W66). More recently, Yan et al. (1999) have extended these findings to show that W89 appears to be important for antagonist (i.e., curare and granisetron) but not agonist (i.e., 5-HT) binding, that position R91 affects 5-HT and granisetron binding but not curare, and that position Y93 affects granisetron but neither curare nor 5-HT binding. These data clearly show that different ligands have different points of interaction with the 5-HT_3R, and that the periodicity of the effect on granisetron binding (i.e., involvement of positions 89, 91, and 93) suggests that this region is in a β-strand configuration (Yan et al. 1999).

In a comprehensive study of the role played by all N-terminal domain tryptophan residues in the binding of ligands to the 5-HT_3R (Spier and Lummis 2000), the likely involvement of W90 (equivalent to W89 of Yan et al. 1999) was confirmed, the importance of W183, W195, and W214 was suggested, and the involvement of W60 was found to be unlikely. Mutating the tryptophan residues at positions 95, 102, and 121 to tyrosine or serine resulted in no ligand binding nor functional expression, suggesting that these residues may be important for ligand binding. However other explanations are possible. For example, perhaps the expression of these non-functional mutants was disrupted due to the lack of correct subunit assembly and/or proper insertion of functional channels into the membrane (Green and Millar 1995). Residues W90 and W183 were also found to be important for ligand binding for the nACh, GABA, and glycine receptors (Dennis et al. 1988, Amin and Weiss

1993; SCHMIEDEN et al. 1993; VANDENBERG et al. 1992), confirming the homology between structure and function among these different receptors.

Other residues appear to be involved in ligand binding to the 5-HT_3R. The glutamate at position 129 (E129; originally reported as E106) appears to contribute (BOESS et al. 1997), as well as the phenylalanine at position 130 (F130; originally reported as F107) (STEWARD et al. 1996).

F. Allosteric Regulation

Besides the recognition sites for agonists and antagonists, the 5-HT_3R is thought to have other sites that can be regulated by a variety of allosteric agents (PARKER et al. 1996). These agents include alcohols, anesthetic agents, and 5-hydroxyindole.

I. Alcohols

Alcohols can enhance the function of the 5-HT_3R and other members of the superfamily of ligand-gated ion channels at concentrations that are known to produce intoxication and/or anesthesia (GRANT 1995; PARKER et al. 1996). In general alcohols shift the dose-response curve for 5-HT_3R activation to lower 5-HT concentrations; thus the maximum alcohol-induced potentiation was observed at lower agonist concentrations. Alcohols have also been shown to alter the gating kinetics of the 5-HT_3R, suggesting that alcohols either increase the rate of channel activation and/or enhance the rate of desensitization (PARKER et al. 1996). Recently ZHOU et al. (1998) reported that alcohols potentiate the function of the 5-HT_3R by stabilizing the open channel state. There is strong evidence to suggest that there are multiple alcohol regulatory sites. For example, ZHOU and LOVINGER (1996) reported both positive and negative allosteric actions when different concentrations of different alcohols were co-applied, and they suggested that alcohols may interact with several hydrophobic sites associated with the 5-HT_3R.

In addition to enhancing the function of the 5-HT_3R, higher *n*-alcohols have been reported to inhibit the function of the 5-HT_3R (JENKINS et al. 1996). In addition, differences have been observed in different preparations, suggesting the possibility yet again for molecular diversity in the 5-HT_3R (PARKER et al. 1996).

II. Anesthetics

Similar to alcohols, general anesthetics are thought to enhance the function of the 5-HT_3R, although some inhibitory and/or differential actions have been observed (PARKER et al. 1996). PETERS et al. (1991) reported that ketamine, unlike the inhibitory effect reported for nACh and the NMDA-subtype of glutamate receptor channels, potentiated the function of the 5-HT_3R in rabbit nodose ganglion neurons. Halothane and isoflurane potentiated 5-

HT$_3$R channel function in N1E-115 cells (JENKINS et al. 1996) and in *Xenopus* oocytes expressing 5-HT$_3$Rs (MACHU and HARRIS 1994). Similar to the alcohols, these volatile anesthetics shifted the dose-response curve for the activation of the response to lower 5-HT concentrations (JENKINS et al. 1996; MACHU and HARRIS 1994). There is also evidence for the anesthetics (as with the alcohols) for differential effects on the function of the 5-HT$_3$R depending on species and/or preparation (PARKER et al. 1996). In addition evidence suggests that anesthetics interact with a site on the 5-HT$_3$R different than alcohols (MACHU and HARRIS 1994; ZHOU and LOVINGER 1996).

III. 5-Hydroxyindole

5-Hydroxyindole (5-OHi; 1 mM) was reported to potentiate the amplitude and slow the kinetics of desensitization of native 5-HT$_3$R-mediated responses in N1E-115 cells, without producing a response on its own (KOOYMAN et al. 1993a). At higher concentrations, 5-OHi reduced the amplitude of the response while still slowing the kinetics of desensitization. It was concluded that the blocking effect of 5-OHi was due to a competitive interaction at an antagonist recognition site on the 5-HT$_3$R, whereas the potentiating effect of low doses of 5-OHi was mediated by a non-competitive interaction (KOOYMAN et al. 1994). More recently VAN HOOFT et al. (1997a) studied the actions of 5-OHi and various analogs and determined that these compounds were acting as allosteric modulators of the 5-HT$_3$R.

G. Conclusion

The 5-HT$_3$R is now a firmly established member of the superfamily of ligand-gated ion channels that includes the nACh, GABA, and glycine receptor channels. Besides having multiple sites at which 5-HT$_3$R ligands (both agonists and antagonists) can bind, several possible allosteric modulatory sites also appear to exist. In addition, cytoplasmic signal transduction cascades, including those regulated by Ca^{2+} and possibly involving phosphorylation, regulate the function of the 5-HT$_3$R. Therefore there are a variety of ways in which the function of the 5-HT$_3$R can be modulated. Molecular diversity of the 5-HT$_3$R, both between and within species, strongly indicates that other 5-HT$_3$R subunits, or subunits that interact intimately with the 5-HT$_3$R, exist. Although we know that the 5-HT$_3$R is involved in a variety of physiological processes, we still have much to learn about the precise role that it plays, in particular in the CNS. In conclusion, it is clear that many pathways converge to regulate the function of the 5-HT$_3$R, both extra- and intracellularly. Therefore the overall regulation of neuronal excitability via the 5-HT$_3$R will be a complex and multifaceted process.

Acknowledgments. I would like to thank Sterling Sudweeks for critical reading of this manuscript, and Andrea Allan, Nicholas Barnes, Robert Nichols, Avi Spier, and Michael White for access to unpublished work.

References

Alkadhi KA, Salgado-Commissariat D, Hogan YH, Akpaudo SB (1996) Induction and maintenance of ganglionic long-term potentiation require activation of 5-hydroxytryptamine (5-HT_3) receptors. J Physiol 496:479–489

Amin J, Weiss DS (1993) GABAA receptor needs two homologous domains of the beta-subunit for activation by GABA but not by pentobarbital. Nature 366:565–569

Barnes JM, Barnes NM, Costall B, Naylor RJ, Tyers MB (1989) 5-HT_3 receptors mediate inhibition of acetylcholine release in cortical tissue. Nature 338:762–763

Bartrup JT, Newberry NR (1996) Electrophysiological consequences of ligand binding to the desensitized 5-HT_3 receptor in mammalian NG108–15 cells. J Physiol 490:679–690

Belelli D, Balcarek JM, Hope AG, Peters JA, Lambert JJ, Blackburn TP (1995) Cloning and functional expression of a human 5-hydroxytryptamine type $3A_S$ receptor subunit. Mol Pharmacol 48:1054–1062

Blandina P, Goldfarb J, Craddock-Royal B, Green JP (1989) Release of endogenous dopamine by stimulation of 5-$hydroxytryptamine_3$ receptors in rat striatum. J Pharmacol Exp Ther 251:803–809

Boddeke HWGM, Meigel I, Boeijinga P, Arbuckle J, Docherty RJ (1996) Modulation by calcineurin of 5-HT_3 receptor function in NG108–15 neuroblastoma X glioma cells. Brit J Pharmacol 118:1836–1840

Boess FG, Beroukhim R, Martin IL (1995) Ultrastructure of the 5-$hydroxytryptamine_3$ receptor. J Neurochem 64:1401–1405

Boess FG, Steward LJ, Steele JA, Liu D, Reid J, Glencorse TA, Martin IL (1997) Analysis of the ligand binding site of the 5-HT_3 receptor using site directed mutagenesis: Importance of glutamate 106. Neuropharmacol 36:637–647

Bonhaus DW, Wong EH, Stefanich E, Kunysz EA, Eglen RM (1993) Pharmacological characterization of 5-$hydroxytryptamine_3$ receptors in murine brain and ileum using the novel radioligand [3H]RS-42358–197: evidence for receptor heterogeneity. J Neurochem 61:1927–1932

Bradley PB, Engel G, Feniuk W, Fozard JR, Humphrey PP, Middlemiss DN, Mylecharane EJ, Richardson BP, Saxena PR (1986) Proposals for the classification and nomenclature of functional receptors for 5-hydroxytryptamine. Neuropharmacol 25:563–576

Brown AM, Hope AG, Lambert JJ, Peters JA (1998) Ion permeation and conduction in a human recombinant 5-HT_3 receptor subunit (h5-HT_{3A}). J Physiol 507:653–665

Butler A, Elswood CJ, Burridge J, Ireland SJ, Bunce KT, Kilpatrick GJ, Tyers MB (1990) The pharmacological characterization of 5-HT_3 receptors in three isolated preparations derived from guinea-pig tissues. Br J Pharmacol 101:591–598

Cohen ML (1992) 5-HT_3 receptors in the periphery. In: Hamon M (ed) Central and peripheral 5-HT_3 receptors, Academic Press, London, England, pp. 19–32

Davies PA, Pistis M, Hanna MC, Peters JA, Lambert JJ, Hales TG, Kirkness EF (1999) The 5-HT_{3B} subunit is a major determinant of serotonin-receptor function. Nature 397:359–363

Dennis M, Giraudat J, Kotzyba-Hibert F, Goeldner M, Hirth C, Chang JY, Lazure C, Chretien M, Changeux JP (1988) Amino acids of the Torpedo marmorata acetylcholine receptor alpha subunit labeled by a photoaffinity ligand for the acetylcholine binding site. Biochem 27:2346–2357

Derkach V, Surprenant A, North RA (1989) 5-HT_3 receptors are membrane ion channels. Nature 339:706–709

Downie DL, Hope AG, Lambert JJ, Peters JA, Blackburn TP, Jones BJ (1994) Pharmacological characterization of the apparent splice variants of the murine 5-HT_3 R-A subunit expressed in *Xenopus laevis* oocytes. Neuropharmacol 33:473–482

Eiselé JL, Bertrand S, Galzi JL, Devillers-Thiéry A, Changeux JP, Bertrand D (1993) Chimaeric nicotinic-serotonergic receptor combines distinct ligand binding and channel specificities. Nature 366:479–483
Emerit MB, Martres MP, Miquel MC, El Mestikawy S, Hamon M (1995) Differentiation alters expression of the two splice variants of the serotonin 5-HT_3 receptor-A mRNA in NG108–15 cells. J Neurochem 65:1917–1925
Engel SR, Lyons CR, Allan AM (1998) 5-HT_3 receptor over-expression decreases ethanol self administration in transgenic mice. Psychopharmacology (Berl) 140: 243–248
Fletcher S, Barnes NM (1997) Purification of 5-hydroxytryptamine$_3$ receptors from porcine brain. Brit J Pharmacol 122:655–662
Fletcher S, Barnes NM (1998) Desperately seeking subunits: Are native 5-HT_3 receptors really homomeric complexes? TIPS 19:212–215
Freund TF, Gulyás AI, Acsády L, Görcs T, Tóth K (1990) Serotonergic control of the hippocampus via local inhibitory interneurons. PNAS 87:8501–8505
Gaddum JHR, Picarelli ZP (1957) Two kinds of tryptamine receptor. Brit J Pharmacol 12:323–328
Gill CH, Peters JA, Lambert JJ (1995) An electrophysiological investigation of the properties of a murine recombinant 5-HT_3 receptor stably expressed in HEK 293 cells. Brit J Pharmacol 114:1211–1221
Gilon P, Yakel JL (1995) Activation of 5-HT_3 receptors expressed in *Xenopus* oocytes does not increase cytoplasmic Ca^{2+} levels. Receptors Channels 3:83–88
Glaum SR, Brooks PA, Spyer KM, Miller RJ (1992) 5-Hydroxytryptamine-3 receptors modulate synaptic activity in the rat nucleus tractus solitarius in vitro. Brain Res 589:62–68
Glitsch M, Wischmeyer E, Karschin A (1996) Functional characterization of two 5-HT_3 receptor splice variants isolated from a mouse hippocampal cell line. Pflügers Arch 432:134–143
Grant KA (1995) The role of 5-HT_3 receptors in drug dependence. Drug Alcohol Depen 38:155–171
Green WN, Millar NS (1995) Ion-channel assembly. TINS 18:280–287
Greenshaw AJ (1993) Behavioural pharmacology of 5-HT_3 receptor antagonists: a critical update on therapeutic potential. TIPS 14:265–270
Hargreaves AC, Lummis SCR, Taylor CW (1994) Ca^{2+} permeability of cloned and native 5-hydroxytryptamine type 3 receptors. Mol Pharmacol 46:1120–1128
Hope AG, Downie DL, Sutherland L, Lambert JJ, Peters JA, Burchell B (1993) Cloning and functional expression of an apparent splice variant of the murine 5-HT_3 receptor A subunit. Europ J Pharmacol 245:187–192
Hussy N, Lukas W, Jones KA (1994) Functional properties of a cloned 5-hydroxytryptamine ionotropic receptor subunit: comparison with native mouse receptors. J Physiol 481:311–323
Jackson MB, Yakel JL (1995) The 5-HT_3 receptor channel. Annu Rev Physiol 57:447–468
Jenkins A, Franks NP, Lieb WR (1996) Actions of general anaesthetics on 5-HT_3 receptors in N1E-115 neuroblastoma cells. Brit J Pharmacol 117:1507–1515
Jones S, Yakel JL (1998) Ca^{2+} influx through voltage-gated Ca^{2+} channels regulates 5-HT_3 receptor channel desensitization in rat glioma X mouse neuroblastoma hybrid NG108–15 cells. J Physiol 510:361–370
Katz B, Thesleff S (1957) A study of the 'desensitization' produced by acetylcholine at the motor end-plate. J Physiol 138:63–80
Kawa K (1994) Distribution and functional properties of 5-HT_3 receptors in the rat hippocampal dentate gyrus: A patch-clamp study. J Neurophysiol 71:1935–1947
Kidd EJ, Laporte AM, Langlois X, Fattaccini CM, Doyen C, Lombard MC, Gozlan H, Hamon M (1993) 5-HT_3 receptors in the rat central nervous system are mainly located on nerve fibres and terminals. Brain Res 612:289–298

Kilpatrick GJ, Jones BJ, Tyers MB (1987) Identification and distribution of 5-HT_3 receptors in rat brain using radioligand binding. Nature 330:746–748
Kilpatrick GJ, Butler A, Burridge J, Oxford AW (1990) 1-(*m*-Chlorophenyl)-biguanide, a potent high affinity 5-HT_3 receptor agonist. Eur J Pharmacol 182:193–197
Kilpatrick GJ, Tyers MB (1992) The pharmacological properties and functional roles of central 5-HT_3 receptors. In: Hamon M (ed) Central and peripheral 5-HT_3 receptors, Academic Press, London, England, pp. 33–57
Kooyman AR, van Hooft JA, Vijverberg HPM (1993a) 5-Hydroxyindole slows desensitization of the 5-HT_3 receptor-mediated ion current in N1E-115 neuroblastoma cells. Brit J Pharmacol 108:287–289
Kooyman AR, Zwart R, Vijverberg HPM (1993b) Tetraethylammonium ions block 5-HT_3 receptor-mediated ion current at the agonist recognition site and prevent desensitization in cultured mouse neuroblastoma cells. Europ J Pharmacol 246:247–254
Kooyman AR, van Hooft JA, Vanderheijden PML, Vijverberg HPM (1994) Competitive and non-competitive effects of 5-hydroxyindole on 5-HT_3 receptors in N1E-115 neuroblastoma cells. Brit J Pharmacol 112:541–546
Kriegler S, Sudweeks S, Yakel JL (1999) Communication: The nicotinic $\alpha 4$ receptor subunit contributes to the lining of the ion channel pore when expressed with the 5-HT_3 receptor subunit.: 3934–3936. J Biol Chem 274: 3934–3936
Lankiewicz S, Lobitz N, Wetzel CHR, Rupprecht R, Gisselmann G, Hatt H (1998) Molecular cloning, functional expression, and pharmacological characterization of 5-hydroxytryptamine$_3$ receptor cDNA and its splice variants from guinea pig. Mol Pharmacol 53:202–212
Laporte AM, Kidd EJ, Vergé D, Gozlan H, Hamon M (1992) Autoradiographic mapping of central 5-HT_3 receptors. In: Hamon M (ed) Central and peripheral 5-HT_3 receptors, Academic Press, London, England, pp. 157–187
Machu TK, Harris RA (1994) Alcohols and anesthetics enhance the function of 5-hydroxytryptamine$_3$ receptors expressed in *Xenopus laevis* oocytes. J Pharmacol Exp Ther 271:898–905
Maricq AV, Peterson AS, Brake AJ, Myers RM, Julius D (1991) Primary structure and functional expression of the 5-HT_3 receptor, a serotonin-gated ion channel. Science 254:432–437
Maura G, Andrioli GC, Cavazzani P, Raiteri M (1992) 5-Hydroxytryptamine$_3$ receptors sited on cholinergic axon terminals of human cerebral cortex mediate inhibition of acetylcholine release. J Neurochem 58:2334–2337
McMahon LL, Kauer JA (1997) Hippocampal interneurons are excited via serotonin-gated ion channels. J Neurophysiol 78:2493–2502
Mienville JM (1991) Comparison of fast responses to serotonin and 2-methyl-serotonin in voltage-clamped N1E-115 neuroblastoma cells. Neurosci Letts 133:41–44.
Miquel MC, Emerit MB, Gozlan H, Hamon M (1991) Involvement of tryptophan residue(s) in the specific binding of agonists/antagonists to 5-HT_3 receptors in NG108–15 clonal cells. Biochem Pharmacol 42:1453–1461
Miquel MC, Emerit MB, Gingrich JA, Nosjean A, Hamon M, El Mestikawy S (1995) Developmental changes in the differential expression of two serotonin 5-HT_3 receptor splice variants in the rat. J Neurochem 65:475–483
Miyake A, Mochizuki S, Takemoto Y, Akuzawa S (1995) Molecular cloning of human 5-hydroxytryptamine$_3$ receptor: heterogeneity in distribution and function among species. Mol Pharmacol 48:407–416
Morain P, Abraham C, Portevin B, de Nanteuil G (1994) Biguanide derivatives: Agonist pharmacology at 5-hydroxytryptamine type 3 receptors *in vitro*. Mol Pharmacol 46:732–742
Morales M, Bloom FE (1997) The 5-HT_3 receptor is present in different subpopulations of GABAergic neurons in the rat telencephalon. J Neurosci 17:3157–3167
Mukerji J, Haghighi A, Séguéla P (1996) Immunological characterization and transmembrane topology of 5-hydroxytryptamine$_3$ receptors by functional epitope tagging. J Neurochem 66:1027–1032

Nayak SV, Rondé P, Spier AD, Lummis SCR, Nichols RA (1998) Co-localization of 5-HT_3 serotonin and alpha4 nicotinic receptor subunits on mammalian brain nerve terminals. Soc Neuro Abstr 24:1819
Neijt HC, te Duits IJ, Vijverberg HPM (1988) Pharmacological characterization of serotonin 5-HT_3 receptor-mediated electrical response in cultured mouse neuroblastoma cells. Neuropharmacol 27:301–307
Neijt HC, Plomp JJ, Vijverberg HPM (1989) Kinetics of the membrane current mediated by serotonin 5-HT_3 receptors in cultured mouse neuroblastoma cells. J Physiol 411:257–269
Newberry NR, Cheshire SH, Gilbert MJ (1991) Evidence that the 5-HT_3 receptors of the rat, mouse and guinea-pig superior cervical ganglion may be different. Brit J Pharmacol 102:615–620
Nichols RA, Mollard P (1996) Direct observation of serotonin 5-HT_3 receptor-induced increases in calcium levels in individual brain nerve terminals. J Neurochem 67:581–592
Niemeyer MI, Lummis SCR (1998) Different efficacy of specific agonists at 5-HT_3 receptor splice variants: the role of the extra six amino acid segment. Brit J Pharmacol 123:661–666
Parker RMC, Bentley KR, Barnes NM (1996) Allosteric modulation of 5-HT_3 receptors: Focus on alcohols and anaesthetic agents. TIPS 17:95–99
Paudice P, Raiteri M (1991) Cholecystokinin release mediated by 5-HT_3 receptors in rat cerebral cortex and nucleus accumbens. Brit J Pharmacol 103:1790–1794
Peters JA, Malone HM, Lambert JJ (1991) Ketamine potentiates 5-HT_3 receptor-mediated currents in rabbit nodose ganglion neurones. Brit J Pharmacol 103: 1623–1625
Reznic J, Staubli U (1997) Effects of 5-HT_3 receptor antagonism on hippocampal cellular activity in the freely moving rat. J Neurophysiol 77:517–521
Richardson BP, Engel G (1986) The pharmacology and function of 5-HT_3 receptors. TINS 9:424–428
Roerig B, Nelson DA, Katz LC (1997) Fast synaptic signaling by nicotinic acetylcholine and serotonin 5-HT_3 receptors in developing visual cortex. J Neurosci 17:8353–8362
Rondé P, Nichols RA (1997) 5-HT_3 receptors induce rises in cytosolic and nuclear calcium in NG108–15 cells via calcium-induced calcium release. Cell Calcium 22:357–365
Rondé P, Nichols RA (1998) High calcium permeability of serotonin 5-HT_3 receptors on presynaptic nerve terminals from rat striatum. J Neurochem 70:1094–1103
Ropert N, Guy N (1991) Serotonin facilitates GABAergic transmission in the CA1 region of rat hippocampus *in vitro*. J Physiol 441:121–136
Schmieden V, Kuhse J, Betz H (1993) Mutation of glycine receptor subunit creates beta-alanine receptor responsive to GABA. Science 262:256–258
Schulte MK, Bloom KE, White MM (1995) Evidence for the involvement of tryptophan in the binding of curare to 5HT-3 receptors. Soc Neuro Abstr 21:55
Shao XM, Yakel JL, Jackson MB (1991) Differentiation of NG108–15 cells alters channel conductance and desensitization kinetics of the 5-HT_3 receptor. J Neurophysiol 65:630–638
Spier AD, Lummis SCR (2000) The role of tryptophan residues in the 5-Hydroxytryptamine$_3$ receptor ligand binding domain. J Bol Chem 275:5620–5625
Stäubli U, Xu FB (1995) Effects of 5-HT_3 receptor antagonism on hippocampal theta rhythm, memory, and LTP induction in the freely moving rat. J Neurosci 15:2445–2452
Steward LJ, Boess FG, Steele JA, Phipps BP, Liu D, Martin IL (1996) The importance of the amino acid phenylalanine 107 for function and ligand recognition at the 5-HT_3 receptor. Brit J Pharamacol 119:290P
Sugita S, Shen KZ, North RA (1992) 5-Hydroxytryptamine is a fast excitatory transmitter at 5-HT_3 receptors in rat amygdala. Neuron 8:199–203

Traynelis SF, Mott DD (1996) Recombinant 5-HT3 receptor activation and desensitization. Soc Neuro Abstr 22:1781
Uetz P, Abdelatty F, Villarroel A, Rappold G, Weiss B, Koenen M (1994) Organisation of the murine 5-HT_3 receptor gene and assignment to human chromosome 11. FEBS Letts 339:302–306
van Hooft JA, van der Haar E, Vijverberg HPM (1997a) Allosteric potentiation of the 5-HT_3 receptor-mediated ion current in N1E-115 neuroblastoma cells by 5-hydroxyindole and analogues. Neuropharmacol 36:649–653
van Hooft JA, Kooyman AR, Verkerk A, van Kleef RGDM, Vijverberg HPM (1994) Single 5-HT_3 receptor-gated ion channel events resolved in N1E-115 mouse neuroblastoma cells. Biochem Biophys Res Comm 199:227–233
van Hooft JA, Kreikamp AP, Vijverberg HPM (1997b) Native serotonin 5-HT_3 receptors expressed in *Xenopus* oocytes differ from homopentameric 5-HT_3 receptors. J Neurochem 69:1318–1321
van Hooft JA, Spier AD, Yakel JL, Lummis SCR, Vijverberg HPM (1998) Promiscuous coassembly of serotonin 5-HT_3 and nicotinic $\alpha4$ receptor subunits into Ca^{2+} permeable ion channels. PNAS 95:11456–11461
van Hooft JA, Vijverberg HPM (1995) Phosphorylation controls conductance of 5-HT_3 receptor ligand-gated ion channels. Receptors Channels 3:7–12
Vandenberg RJ, Handford CA, Schofield PR (1992) Distinct agonist- and antagonist-binding sites on the glycine receptor. Neuron 9:491–496
Werner P, Kawashima E, Reid J, Hussy N, Lundström K, Buell G, Humbert Y, Jones KA (1994) Organization of the mouse 5-HT_3 receptor gene and functional expression of two splice variants. Mol Brain Res 26:233–241
Yakel JL, Jackson MB (1988) 5-HT_3 receptors mediate rapid responses in cultured hippocampus and a clonal cell line. Neuron 1:615–621
Yakel JL, Shao XM, Jackson MB (1991) Activation and desensitization of the 5-HT_3 receptor in a rat glioma x mouse neuroblastoma hybrid cell. J Physiol 436:293–308
Yakel JL, Lagrutta A, Adelman JP, North RA (1993) Single amino acid substitution affects desensitization of the 5-HT_3 receptor expressed in *Xenopus* oocytes. PNAS 90:5030–5033
Yan D, Schulte MK, Bloom KE, White MM (1999) Structural features of the ligand-binding domain of the serotonin 5HT3 receptor. J Biol Chem 274:5537–5541
Yang J (1990) Ion permeation through 5-hydroxytryptamine-gated channels in neuroblastoma N18 cells. J Gen Physiol 96:1177–1198
Yang J, Mathie A, Hill B (1992) 5-HT_3 receptor channels in dissociated rat superior cervical ganglion neurons. J Physiol 448:237–256
Zhang L, Oz M, Weight FF (1995) Potentiation of 5-HT_3 receptor-mediated responses by protein kinase C activation. Neuroreport 6:1464–1468
Zhou Q, Lovinger DM (1996) Pharmacologic characteristics of potentiation of 5-HT_3 receptors by alcohols and diethyl ether in NCB-20 neuroblastoma cells. J Pharmacol Exp Ther 278:732–740
Zhou Q, Verdoorn TA, Lovinger DM (1998) Alcohols potentiate the function of 5-HT_3 receptor-channels on NCB-20 neuroblastoma cells by favouring and stabilizing the open channel state. J Physiol 507:335–352

CHAPTER 22

Cyclic Nucleotide-Gated Channels: Classification, Structure and Function, Activators and Inhibitors

M.E. GRUNWALD, H. ZHONG, and K.-W. YAU

A. Introduction

Cyclic nucleotide-gated channels are ion channels the gating of which is directly controlled by cyclic nucleotides. The most well-known of these are non-selective cation channels that require the binding of cyclic nucleotide in order to open (cyclic nucleotide-activated channels). These channels pass not only monovalent cations, but divalent cations even better. They are prominently present in retinal photoreceptors and olfactory receptor neurons, where they play crucial roles in visual and olfactory transductions (for reviews, see YAU and BAYLOR 1989; BIEL et al. 1995; KAUPP 1995; ZIMMERMAN 1995; FINN et al. 1996; ZAGOTTA and SIEGELBAUM 1996). These channels now appear to be widely present in both neural and non-neural tissues (FINN et al. 1996). Their exact functions in cells other than sensory receptors are still mostly unclear, though they probably serve as a pathway for Ca^{2+} influx that is controlled directly by an intracellular second messenger. For example, such a channel is present at the synaptic terminal, where it is involved in neurotransmitter release (RIEKE and SCHWARTZ 1994; SAVCHENKO et al. 1997).

There are also cyclic nucleotide-activated channels that are selective for K^+, such as that mediating the hyperpolarizing response in the scallop hyperpolarizing photoreceptor (GOMEZ and NASI 1995) and the molluscan extraocular photoreceptor (GOTOW et al. 1994), and that mediating the inhibitory response of primary olfactory receptor neurons in the lobster (HATT and ACHE 1994). It is also present on larval *Drosophila* skeletal muscle (DELGADO et al. 1991). The first of these is activated by cGMP and the other two by cAMP. There is also report of a cAMP-activated Na^+ channel (SUDLOW et al. 1993), and even a cAMP-activated Cl^- channel (DELAY et al. 1997).

Besides cyclic nucleotide-activated channels, there are channels that do not require the binding of cyclic nucleotide in order to open, but their open probability is modulated by the binding of cyclic nucleotide (cyclic nucleotide-modulated channels; see FINN et al. 1996). A well-known example is I_f (also called I_h), a channel activated by hyperpolarization which, upon binding cAMP, has its activation curve shifted to less negative voltages so that it is more readily activated by hyperpolarization (DIFRANCESCO and TORTORA 1991; BOIS et al. 1997); in other words, it is dually controlled by voltage and cyclic nucleotide. This channel is important for pacemaker activity in the heart, but is also present in the brain (PAPE 1996). Related members of this channel type have recently been cloned, with their amino-acid sequences showing a consensus cyclic nucleotide-binding site on the cytoplasmic C-terminus (GAUSS et al. 1998; LUDWIG et al. 1998; SANTORO et al. 1998). Another example is a cation channel in the kidney inner medullary collecting duct, which is constitutively open but its open probability is reduced when cGMP binds (LIGHT et al. 1989). A third example is a cation channel on vertebrate taste sensory cells that is inhibited by cyclic nucleotides (KOLESNIKOV and MARGOLSKEE 1995). None of these channels has been cloned. Finally, members of the Eag K^+ channel family (WARMKE et al. 1994), as well as the plant AKT1/KAT1 channels (ANDERSON et al. 1992; SENTENAC et al. 1992), also show a consensus cyclic nucleotide-binding site at the C-termini of their sequences, but the nature of their modulation by cyclic nucleotides is still unknown (see SATLER et al. 1996; FRINGS et al. 1998).

In this chapter, we shall focus exclusively on the cyclic-nucleotide-activated, non-selective cation channels first identified in primary sensory receptor neurons as mentioned above. Not only are the amino-acid sequences of these channels known, but there is a wealth of information about their structure-function relations. Details of their properties can be found in reviews (for example YAU and BAYLOR 1989; FINN et al. 1996; ZAGOTTA and SIEGELBAUM 1996).

B. Structure

These channels are composed of α- and β-subunits, apparently as tetrameric complexes (GORDON and ZAGOTTA 1995c; LIU et al. 1996; VARNUM et al. 1996). The α-subunit, but not the β-subunit, can form homomeric channels that are activated by cyclic nucleotide (CHEN et al. 1993; KÖRSCHEN et al. 1995; LIMAN and BUCK 1994; BRADLEY et al. 1994). When co-assembled, however, the β-subunit modifies the functional properties of the α-subunit (see below). The stoichiometries of α- and β-subunits in the tetrameric native rod, cone, and olfactory channels are not yet clear. So far, three distinct α-subunits have been identified in vertebrates, designated here as CNGα1, CNGα2, and CNGα3, with the number indicating the chronological order of cloning. They are the α-subunits present in the cyclic nucleotide-activated channels of retinal rods, the olfactory receptor cells, and retinal cones, respectively. Another designation for these α-subunits found in the literature is RCNC1 (**R**od **C**yclic-**N**ucleotide **C**hannel subunit **1**), OCNC1 (**O**lfactory **C**yclic-**N**ucleotide **C**hannel subunit **1**) and CCNC1 (**C**one **C**yclic-**N**ucleotide **C**hannel subunit **1**) respectively. Two distinct β-subunits have been identified in vertebrates, designated here as CNGβ1 and CNGβ2, with the number again indicating the chronological order of cloning (also referred to in some literature as RCNC2 and OCNC2, respectively). CNGβ2 actually shows stronger structural homology to the α-subunits than to CNGβ1. However, with the β-subunit defined as being unable to form homomeric channels that can be activated by cyclic nucleotide, the classification of CNGβ2 is still appropriate. It now appears that different gene-splice variants of CNGβ1 are present in the native rod and olfactory channels (SAUTTER et al. 1998; FRINGS et al. 1998), while CNGβ2 is present only in the native olfactory channels (LIMAN and BUCK 1994; BRADLEY et al. 1994). Thus, one α- and two β-subunits are apparently present in the native olfactory channel. The identity of the β-subunit present in the native cone channel is still unclear, though there is indication that it may be similar, or related, to CNGβ1 (YU et al. 1996). The α- and β-subunits from all vertebrate and invertebrate species studied so far can all freely cross-assemble, suggesting high evolutionary conservation (FINN et al. 1998).

The amino-acid sequences of both α- and β-subunits suggest that they have a similar topology in the membrane as the Shaker K^+ channel family, with cytoplasmic N- and C-terminal regions and six putative transmembrane domains, designated S1 through S6 respectively (Fig. 1a). The S4 domain resembles the corresponding domain in voltage-activated channels by having regularly spaced, positively charged residues (Fig. 1b) (see JAN and JAN 1990, 1992), the latter thought to be the voltage-sensor. While the cyclic nucleotide-activated channels cannot be activated by voltage alone, their S4 domain can functionally replace that of a voltage-activated channel (TANG and PAPAZIAN 1997). There is a pore-region between S5 and S6 that dips as a loop into the membrane, again similar to voltage-activated channels (Fig. 1c) (see GUY et al. 1991). These common features suggest that cyclic nucleotide-activated channels share an ancient ancestor with voltage-gated channels (see above references, and HEGINBOTHAM et al. 1992; GOULDING et al. 1993; KRAMER et al. 1994). The distinctive feature of the cyclic nucleotide-activated channel subunits, however, is the presence of a consensus cyclic nucleotide-binding site in the C-terminal region. This cyclic nucleotide-binding site is homologous to those present in cGMP- and cAMP-dependent protein kinases, and in the catabolite gene activating protein (CAP) in *Escherichia coli* (KAUPP et al. 1989).

C. Ion Permeation Properties

As pointed out above, these channels are non-selective among cations. They pass Na^+ and K^+ about equally well, and Ca^{2+} even better (HAYNES 1995; FRINGS et al. 1995; PICONES and KORENBROT 1995; FINN et al. 1997; for early literature, see YAU and BAYLOR 1989; FINN et al. 1996). In addition to being permeant, divalent cations also block the channels (see YAU and BAYLOR 1989; FINN et al. 1996 for early literature). The selectivity of the channels among extracellular divalent cations, as well as their blockage by these cations, are due to divalent cations binding to a glutamate residue in the pore region of the α-subunit (Glu^{363} for bovine CNGα1; see Fig. 1c), presumably located near the extracellular side of the pore region (ROOT and MACKINNON 1993; EISMANN et al. 1994;

a

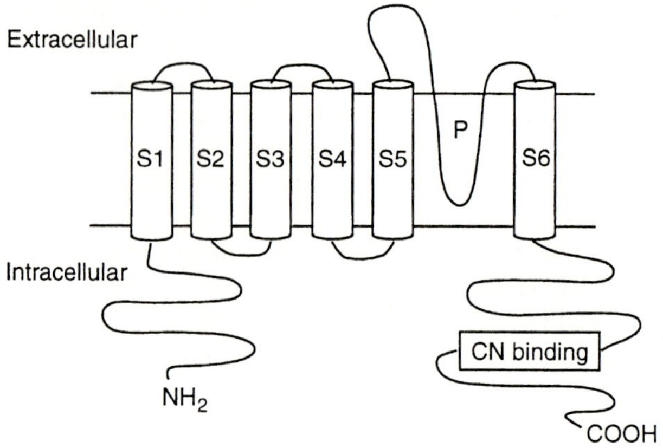

b

S4 region

Shaker	LAIL**R**VI**RL**V**R**VF**R**I**F**KLSRHSKG
CNGα1	YPEI**R**LN**RL**L**R**IS**R**M**F**EFFQRTET

c

Pore region

CNGα1	LY**W**S**TLT**L**TT**I**G**--**E**TP
Shaker	FW**W**AVV**T**M**TT**V**G**YGDMT
CaB1 I	FAVL**T**VFQCITM--**E**GW
II	AAIM**T**VFQILT**G**--**E**DW
III	WALL**TL**FTVST**G**--**E**GW
IV	QALML**L**FRSAT**G**--**E**AW

Fig. 1. **a** Putative folding pattern of a cyclic nucleotide-activated channel. S1–6 are the putative transmembrane domains. P indicates the pore region, and CN binding indicates the cyclic nucleotide-binding domain. Lengths of N- and C-termini are drawn roughly to scale. **b** Alignment of the amino-acid sequences of the S4 region in the Shaker K^+ channel (TEMPEL et al. 1987) and bovine CNGα1 (KAUPP et al. 1989). Identical amino acids are shown in boldface. *Dots* mark the repeated basic residues characteristic of the S4 region. **c** Alignment of the sequences of the putative pore region in bovine CNGα1, Shaker, and each of the four repeats (I–IV) in the calcium channel CaB1 (MORI et al. 1991). Amino acids identical to the CNGα1 sequence are indicated in BOLDFACE. Adapted from HEGINBOTHAM et al. (1992) and FINN et al. (1996)

PARK and MACKINNON 1995; see also SESTI et al. 1995). The same residue is present at the corresponding position in Ca^{2+} channels, with similar functional characteristics (MORI et al. 1991; YANG et al. 1993; see Fig. 1c). For a homotetrameric channel complex, there should be four such glutamate residues, but these appear to interact to produce effectively two identical and independent binding sites for metal cations or protons (ROOT and MACKINNON 1994). Because CNGβ1 does not have a negatively charged residue at the corresponding position in the pore region, a channel complex composed of both α- and β-subunits, such as the native rod or olfactory channel, is expected to show a weaker divalent cation permeability and blockage compared to homomeric channels formed by the α-subunit alone, which is indeed the case (CHEN et al. 1993). Finally, divalent cations also block the channels from the cytoplasmic side, but the binding site mediating this effect has not been identified (ROOT and MACKINNON 1993).

D. Cyclic-Nucleotide Binding and Channel Gating

It is difficult to separate the discussions of ligand binding and channel gating because the two steps are kinetically linked to each other. Thus, the relative effectiveness of a cyclic nucleotide or cyclic-nucleotide analog in activating a channel reflects both the absolute affinity of the ligand for the binding site and the ease of opening (open probability) of the liganded channel. Likewise, an observed change in a channel's apparent affinity for cyclic nucleotide often results from a change in the gating step instead of the binding step (see below).

The cyclic nucleotide-binding sites on both α- and β-subunits bear homology at the amino-acid level to those on cAMP- and cGMP-dependent protein kinases, and to those on the catabolite gene activator protein (CAP) (KAUPP et al. 1989; CHEN et al. 1993; KUMAR and WEBER 1992) (see Fig. 2b). The structure of the CAP protein with cAMP bound has been solved, which suggests that each cAMP-binding site is composed of an 8-stranded β-barrel and three α-helices (MCKAY and STEITZ 1981; MCKAY et al. 1982) (see Fig. 2a). The cAMP molecule is stabilized in the pocket by hydrogen bonds and non-polar interactions with the protein. Key arginine, glutamate, and glycine residues have been identified in CAP that interact with the cyclic nucleotide ribose phosphate moiety (MCKAY et al. 1982); the same residues are present in the cAMP- and cGMP-dependent protein kinases (WEBER et al. 1987, 1989), as well as the cyclic nucleotide-activated channels (marked by dots in Fig. 2b). The CNGα1 residue Thr^{560} (bovine) in the β7 strand (ALTENHOFEN et al. 1991) and Val^{523}, Val^{524} and Ala^{525} (bovine) in the β4 strand (BROWN et al. 1995) may form additional contact points with the ligand; in particular, Thr^{560} appears to help the channel prefer cGMP over cAMP by providing an additional hydrogen bond with cGMP but not cAMP. Asp^{604} in α-helix C also has a key role in a channel's preference for cGMP over cAMP, in this case because its presence generates

a

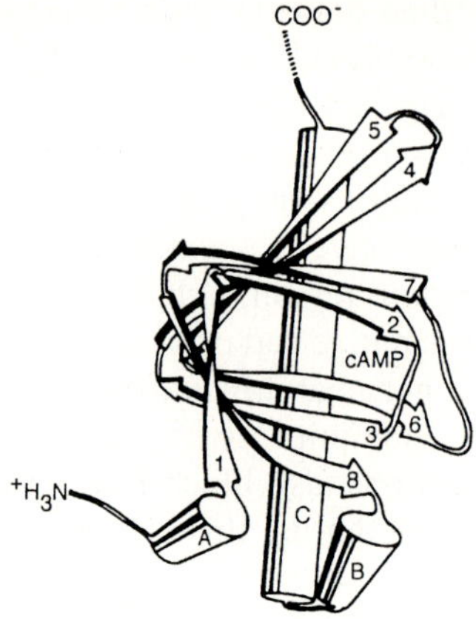

b

	αA β1 β2 β3
CAP	T**L**EWFLSHCHIHK**Y**PSKSTLIHQ**G**EKAETL**Y**Y**I**VK
CNGα1	L**L**VELVLKLQPQV**Y**SPGDYICKK**G**DIGREM**Y**I**I**KE
	β4 β5 β6
CAP	**G**SV**AV**LIK**D**EEGKEMILSY**L**NQ**G**DFI**GE**LGLFEE-
CNGα1	**G**KL**AV**VAD**D**---GITQFVV**L**SD**G**SYF**GE**ISILNIK
	β7 β8 αB
CAP	----**G**QE**R**S**A**WVRAKTACEVAEI**S**YKKFRQLIQVN
CNGα1	GSKA**G**NR**R**T**A**NIKSIGYSDLFCL**S**KDDLMEALTEY
	αC
CAP	**PD**ILMR**L**SAQMARR**L**QVTSEKVG**N**
CNGα1	**PD**AKGM**L**EEKGKQI**L**MEDGLLDI**N**

Fig. 2. **a** A three-dimensional, schematic drawing of the cyclic nucleotide-binding site of the CAP monomer in the closed form (i.e., with cAMP bound). The α-helices, shown as *tubes* are *lettered* A through C. The β-sheets, represented by *arrows*, are *numbered* 1 through 8. The approximate position of the cAMP in the β roll is shown. (McKAY et al. 1982). **b** Alignment of the amino-acid sequences of the cyclic nucleotide-binding domains of CAP and bovine CNGα1 (KUMAR and WEBER 1992). Identical residues are indicated in boldface. *Dots* mark some of the key residues for cyclic nucleotide-binding. The elements of secondary structure are *overlined* and *marked* αA through αC for α-helices and β1 through β8 for β-sheets (see panel a). Adapted from FINN et al. (1996)

an open probability that is higher when the channel is liganded with cGMP than with cAMP (VARNUM et al. 1995, 1996; see also GOULDING et al. 1994). It has been suggested that the β-roll of the binding site is more important for ligand stabilization in a state-independent manner, whereas α-helix C (the "C-helix") selectively stabilizes the ligand in the open state of the channel (TIBBS et al. 1998).

Despite the homology between the α- and β-subunits, their effective cyclic nucleotide-binding characteristics may not be identical (KARPEN and BROWN 1996). Indeed, the presence or absence of the β-subunit in a channel complex can affect both the channel's half-activation constant ($K_{1/2}$) for a particular ligand and whether the ligand is a partial or full agonist (see, for example, LIMAN and BUCK 1994; BRADLEY et al. 1994; FODOR and ZAGOTTA 1996).

A number of agonistic and antagonistic cyclic nucleotide analogs have been synthesized and/or studied (BROWN et al. 1993a,b; SCOTT and TANAKA 1995; KRAMER and TIBBS 1996; WEI et al. 1996). In particular, polymer-linked dimers of cGMP have recently been synthesized that are up to 1000-fold more potent than cGMP in activating cyclic nucleotide-activated channels (KRAMER and KARPEN 1998). The dependence of dimer potency on the polymer chain length allows one to estimate roughly the distance between two cyclic nucleotide-binding sites in an oligomeric complex.

Some regions of the channels involved in gating have been identified. One of these, not surprisingly, is the cytoplasmic C-terminal region containing the cyclic nucleotide-binding site: not just the binding site, however, but also the region between the binding site and the end of the last transmembrane domain, S6 (also called the linker region) (GOULDING et al. 1994; GORDON and ZAGOTTA 1995b; ZONG et al. 1998; see also following sections on modulations by transition metals, protons and sulfhydryl reagents). Another important region is the cytoplasmic N-terminus and the first two transmembrane domains, S1 and S2 (GOULDING et al. 1994; GORDON et al. 1997; see also following sections on modulations by calmodulin and phosphorylation). Finally, the interior of the pore region also appears to be coupled to the gating process (ROOT and MACKINNON 1993; BUCOSSI et al. 1996, 1997).

The cyclic nucleotide-activated channels can open spontaneously in the absence of cyclic nucleotide, though the open probability is very low under these circumstances (PICONES and KORENBROT 1995; TIBBS et al. 1997). The open probability increases sharply as the number of binding sites (altogether four in a tetrameric channel complex) occupied by cyclic nucleotide increases. The dose-response relation between the macroscopic activated current and cyclic nucleotide concentration has a Hill coefficient typically between 2 and 3 (see YAU and BAYLOR 1989; FINN et al. 1996; ZAGOTTA and SIEGELBAUM 1996). An empirical fit of the Hill equation to the dose-response relation belies the underlying channel kinetics, which now appear to be highly complex (RUIZ and KARPEN 1999). They involve multiple conductance levels, each of which has more than one kinetically distinguishable state, depending on the number of occupied binding sites (RUIZ and KARPEN 1997, 1999; see also ILDEFONSE and BENNETT 1991; ILDEFONSE et al. 1992; BUCOSSI et al. 1997; LI et al. 1997; LIU et al. 1998).

E. Modulations

I. $Ca^{2\pm}$-Calmodulin

One modulation of known physiological importance is by Ca^{2+}-calmodulin, which directly binds to CNGα2 (CHEN and YAU 1994) and CNGβ1 (CHEN et al. 1994; KÖRSCHEN et al. 1995; see also HSU and MOLDAY 1993, 1994). The modulation of CNGα2 by Ca^{2+}-calmodulin is strong, increasing the cGMP $K_{1/2}$ of the homomeric channel by ca. 10-fold, and the cAMP $K_{1/2}$ by about 15-fold (CHEN and YAU 1994). Ca^{2+}-calmodulin binds to a site on the cytoplasmic N-terminal region of this α-subunit (LIU et al. 1994). The site has a consensus 1–8-14 motif and an amphiphilic structure, and binds Ca^{2+}-calmodulin with a K_d of ca. 4 nmol/l based on binding assays with dansyl-calmodulin and a peptide corresponding to the binding-site (LIU et al. 1994). The increase in $K_{1/2}$ appears to result from the channel favoring the closed state when Ca^{2+}-calmodulin binds, suggested by the change in cAMP $K_{1/2}$ being accompanied by a decrease in the saturated cAMP-induced current but no change in the single-channel current (CHEN et al. 1994; LIU et al. 1994). As a mechanism for this modulation, the following has been suggested. An N-terminal domain that overlaps with the Ca^{2+}-calmodulin binding site normally promotes a high open probability of the liganded channel; upon binding to the N-terminus, however, Ca^{2+}-calmodulin disrupts the influence of this domain on channel gating (LIU et al. 1994). Consistent with this idea, the same increase in $K_{1/2}$ is observed in the absence of Ca^{2+}-calmodulin when the calmodulin-binding site is removed, presumably by abolishing the integrity of the domain influencing gating (LIU et al. 1994). Subsequent biochemical experiments with fusion proteins have indicated that the N- and C-termini of αCNG2 physically interact with each other, but this interaction is disrupted by the binding of Ca^{2+}-calmodulin to the N-terminus (VARNUM and ZAGOTTA 1997). It now appears that this N-terminal domain important for gating is quite diffuse, stretching from the calmodulin-binding site to the beginning of the first transmembrane domain, so that mutations throughout this region can disrupt this interaction (Grunwald and Yau, manuscript in preparation).

The relevant Ca^{2+}-calmodulin binding site on CNGβ1 corresponds roughly in position to that on CNGα2 (GRUNWALD et al. 1998b; WEITZ et al. 1998). This binding site is less readily recognizable than that on CNGα2, and, from peptide-binding experiments, it appears to be of low affinity, with a K_d of a few μmol/l Ca^{2+}-calmodulin. In the context of the entire β-subunit, however, the affinity increases to the nanomolar range (HSU and MOLDAY 1993, 1994; GRUNWALD et al. 1998b), possibly suggesting that other regions of the protein contribute to the high affinity (GRUNWALD et al. 1998b; WEITZ et al. 1998). The effect of Ca^{2+}-calmodulin imparted by CNGβ1 is quite weak. For the native rod channel (HSU and MOLDAY 1993, 1994), or when CNGβ1 is co-expressed with CNGα1 (CHEN et al. 1994), Ca^{2+}-calmodulin increases the cGMP $K_{1/2}$ by only twofold or less. Nonetheless, this small effect imparted by CNGβ1 may still add to the strong effect imparted by CNGα2 when both subunits are co-assembled in the native olfactory channel (see earlier), where a 20-fold increase in the cAMP $K_{1/2}$ due to Ca^{2+}-calmodulin has been observed (CHEN and YAU 1994).

CNGα3 also has a high-affinity binding site for calmodulin at a position corresponding to that on the N-terminus of CNGα2 (GRUNWALD et al. 1998a; Grunwald and Yau, manuscript in preparation). However, homomeric channels formed by this subunit do not show any sensitivity to Ca^{2+}-calmodulin (Grunwald and Yau, manuscript in preparation; see also YU et al. 1996; HAYNES and STOTZ 1997; HACKOS and KORENBROT 1997). It is possible that the N- and C-termini of CNGα3 do not interact with each other, so that no change in channel gating results even when Ca^{2+}-calmodulin binds. Alternatively, additional domains may be necessary for a modulation by calmodulin.

Neither CNGα1 nor CNGβ2 appear to impart any Ca^{2+}-calmodulin modulation, nor has any bona fide binding site been identified (CHEN et al. 1994; FINN et al. 1998).

Finally, there are experiments to indicate that one or more still-unknown Ca^{2+}-binding proteins may modulate the native rod, olfactory and cone channels in the same way as calmodulin, possibly by binding to the same binding site (GORDON et al. 1995b; BALASUBRAMANIAN et al. 1996; HACKOS and KORENBROT 1997)

II. $Ca^{2\pm}$

Ca^{2+} has been reported to modulate directly the open probability of the native olfactory channel (ZUFALL et al. 1991), but this finding has not been verified by others (BALASUBRAMANIAN et al. 1996).

III. Phosphorylation

There was an early report of a decrease in the cGMP $K_{1/2}$ of the native rod cGMP-activated channel due to serine/threonine phosphatases (GORDON et al. 1992), but the phosphorylation site involved has not been identified. However, tyrosine phosphorylation of CNGα1 has recently been found to increase the cGMP $K_{1/2}$ by about a factor of two, apparently through influencing channel gating, and the same was observed for the native rod channel (MOLOKANOVA et al. 1997). The phosphorylation site has been localized to Tyr^{498} for bovine CNGα1 (MOLOKANOVA et al. 1998), situated in the β1 strand of the cyclic nucleotide-binding site (see Fig. 2). Interestingly, this modulation is state-dependent, with phosphorylation requiring the closed state and dephosphorylation requiring the open state (MOLOKANOVA et al. 1998). The physiological significance of this phosphorylation, apparently activated by growth factors, is unclear. As for CNGα2, the rat protein does not have this tyrosine residue, but the catfish does, making the latter also a likely candidate for tyrosine phosphorylation (MOLOKANOVA et al. 1998).

Homomeric rat CNGα2 channels, on the other hand, are modulated by protein kinase C, producing a decrease in the cAMP $K_{1/2}$ by about fourfold (MÜLLER et al. 1998). The target is Ser^{93}, situated immediately downstream of the calmodulin-binding site on the N-terminal region described above. Phosphorylation of this serine does not affect the binding and action of Ca^{2+}-calmodulin (MÜLLER et al. 1998), with the result that Ca^{2+}-calmodulin produces a larger decrease in cAMP sensitivity when the channel is in the phosphorylated than in the unphosphorylated state. Thus, phosphorylation has the potential to enhance the channel's sensitivity to cyclic nucleotide, and make it more dramatically modulated by Ca^{2+}-calmodulin. Because an effect of protein kinase C on the native olfactory channel has not been detected (MÜLLER et al. 1998), however, the physiological interpretation of this modulation remains to be established.

IV. Transition Metals

The cGMP sensitivity of the native rod channel is increased in the presence of micromolar concentrations of transition metals, such as Ni^{2+} (ILDEFONSE and BENNETT 1991; ILDEFONSE et al. 1992; KARPEN et al. 1993). The target of Ni^{2+} has been localized to a histidine residue upstream of the cyclic nucleotide-binding site on the cytoplasmic C-terminus of CNGα1 (H^{418} for the human and H^{420} for the bovine proteins) (FINN et al. 1995; GORDON and ZAGOTTA 1995a). The underlying mechanism appears to involve a coordination of this residue in two adjacent subunits by Ni^{2+}, resulting in a stabilization of the open state (GORDON and ZAGOTTA 1995c). CNGα2 does not have the correspondent histidine, but has a histidine at a position four residues upstream (H^{396} for the rat protein), which upon binding Ni^{2+} leads to a small reduction in open probability (GORDON and ZAGOTTA 1995b). Thus, the region containing these histidine residues appears to be intimately involved in the gating of these channels.

V. Sulfhydryl Reagents

The cGMP sensitivity of the native rod channel is increased by sulfhydryl reagents, such as *N*-ethylmaleimide (NEM) (BALAKRISHNAN et al. 1990; DONNER et al. 1990; SUN et al. 1993; SERRE et al. 1995). This enhancement of sensitivity likewise involves an increase in open probability, and the target of NEM has been localized to a cysteine residue, C^{479} (human), also upstream of the cyclic nucleotide-binding site on the cytoplasmic C-terminus of CNGα1 (FINN et al. 1995; BROWN et al. 1998; see also GORDON et al. 1997). CNGα2 likewise has a histidine in the corresponding position (C^{460} for the rat protein), and is similarly modulated by NEM (FINN et al. 1995).

Nitric oxide-generating compounds such as *S*-nitroso-cysteine and 3-morpholino-sydnonomine have been reported to directly open the native olfactory channel in the absence of cyclic nucleotide (Broillet and Firestein 1996). This effect was proposed to involve the nitrosonium ion (a redox state of nitric oxide) transnitrosylating a sulfhydryl group on a cysteine residue, based on the observation that sulfhydryl reagents have the same action (Broillet and Firestein 1996). The same authors have reported that nitric oxide can activate not only homomeric CNGα2 channels, but also homomeric CNGβ2 channels (Broillet and Firestein 1997), even though the latter is known not to be activated by cyclic nucleotide as mentioned earlier. This unusual finding is exciting, but should also await verification by others.

VI. Protons

Apart from having a blocking effect on cyclic nucleotide-activated channels (see above), protons also modulate them. Specifically, protons on the cytoplasmic side enhance the sensitivity of the native rod channel, as well as homomeric CNGα1 and CNGα2 channels, to cyclic nucleotides (Picco et al. 1996; Gordon et al. 1996; Gavazzo et al. 1997). At least for CNGα1, it appears that protons act in two distinct ways (Gordon et al. 1996). First, protonation of residue Asp^{604} in the cyclic nucleotide-binding site alluded to earlier in connection with channel gating removes the unfavorable electrostatic interaction between the carboxylate of this residue and the purine ring of cAMP, allowing cAMP to act as a nearly full agonist. This effect is cAMP-specific. Second, protonation of His^{468} (bovine) on the cytoplasmic C-terminus upstream of the cyclic nucleotide-binding site favors the open state of the channel. This effect is not specific to a particular cyclic nucleotide. Because CNGβ1 does not have protonatable residues at positions corresponding to His^{468} and Asp^{604}, native rod channels (composed of both CNGα1 and CNGβ1) show a smaller pH-dependent increase in sensitivity to cyclic nucleotide compared to homomeric CNGα1 channels (Gordon et al. 1996).

VII. Other Modulators

The native rod channel has also been reported to be modulated negatively by diacylglycerol analogs (Gordon et al. 1995a) and nicotine (McGeoch et al. 1995).

F. Blockers

Apart from divalent cations and protons (see above), several chemicals have been found to inhibit these channels.

The most well known among these is L-*cis*-diltiazem, which at physiological pH blocks the native rod and cone channels from the cytoplasmic side at micromolar concentrations (KOCH and KAUPP 1985; STERN et al. 1986; QUANDT et al. 1991; HAYNES 1992; MCLATCHIE and MATTHEWS 1992, 1994). Its blocking effect on the native olfactory channel is weaker by perhaps an order of magnitude (FRINGS et al. 1992). The blockage increases with increasing depolarization, suggesting a binding site situated within the transmembrane electric field. The effectiveness of the drug requires the presence of CNGβ1, which when co-assembled with CNGα1 increases the latter's sensitivity to L-*cis*-diltiazem by almost 100-fold at +60 mV (CHEN et al. 1993), and when co-assembled with CNGα2 increases the latter's sensitivity by about 15-fold at the same voltage (FINN et al. 1998). The chemical can also block these channels from the extracellular side of the membrane, but requires concentrations perhaps 100-fold higher (GOMEZ and NASI 1997; XIONG et al. 1998). D-*cis*-Diltiazem, which blocks Ca^{2+} channels, is much less effective than the L-*cis*-isomer in blockage (KOCH and KAUPP 1985).

Several Ca^{2+}-channel blockers have an effect from the cytoplasmic side. Pimozide blocks the native rod channel with a potency similar to that of L-*cis*-diltiazem (NICOL 1993), while D-600 and nifedipine block the native olfactory channel in the 20–50 μmol/l range at positive voltages (FRINGS et al. 1992; ZUFALL and FIRESTEIN 1993). The Na^+-channel blocker amiloride also inhibits the olfactory channel (FRINGS et al. 1992), and its derivative 3′,4′-dichlorobenzamil inhibits the rod channel (NICOL et al. 1987), at micromolar concentrations and positive voltages. The rod channel, and also homomeric CNGα1 and CNGα2 channels, are blocked by the local anesthetic tetracaine at 100 μmol/l or lower from the cytoplasmic side (ILDEFONSE and BENNETT 1991; SCHNETKAMP 1987, 1990; QUANDT et al. 1991; FODOR et al. 1997a). Tetracaine binds more tightly to the closed state of the channel, apparently through electrostatic interaction with Glu^{363}, the same residue that binds divalent cations and protons as mentioned earlier. In the open state of the channel, this electrostatic interaction with Glu^{363} somehow disappears, thus weakening the tetracaine binding and effect (FODOR et al. 1997b).

Finally, certain calmodulin inhibitors such as W-7, calmidazolium, and trifluoperazine (KLEENE 1994), the guanylyl cyclase inhibitor LY83583 (LEINDERS-ZUFALL and ZUFALL 1995) and the non-specific inhibitor of cyclic nucleotide-dependent protein kinase, H-8 (WEI et al. 1997) have all been found to inhibit the native rod and olfactory channels.

G. Conclusions

Ever since the discovery of the first cyclic nucleotide-activated channel (FESENKO et al. 1985) and its subsequent molecular cloning (KAUPP et al. 1989), our knowledge about this family of ion channels has increased by leaps and bounds, and at the same time has expanded into related channels that are modulated by the binding of cyclic nucleotides (see above and FINN et al. 1996). The fascinating kinship of these ion channels to voltage-activated channels has also shed unexpected but important light on the voltage-gating and ion-permeation properties of ion channels in general.

References

Altenhofen W, Ludwig J, Eismann E, Kraus W, Bönigk W, Kaupp UB (1991) Control of ligand specificity in cyclic nucleotide-gated channels from rod photoreceptors and olfactory epithelium. Proc Natl Acad Sci USA 88:9868–9872

Anderson JA, Huprikar SS, Kochian LV, Lucas WJ, Gaber RF (1992) Functional expression of a probable *Arabidopsis thaliana* potassium channel in *Saccharomyces cerevisiae*. Proc Natl Acad Sci USA 89:3736–3740

Balakrishnan K, Padgett J, Cone RA (1990) Calcium flux in rod outer segment membranes: NEM potentiates the effects of cGMP. Biophy J 57:A371 (Abstr)

Balasubramanian S, Lynch JW, Barry PH (1996) Calcium-dependent modulation of the agonist affinity of the mammalian olfactory cyclic nucleotide-gated channel by calmodulin and a novel endogenous factor. J Membr Biol 152:13–23

Biel M, Zong X, Hofmann F (1995) Molecular diversity of cyclic nucleotide-gated cation channels, Naunyn Schmiedebergs Arch Pharmacol 353:1–10

Bois P, Renaudon B, Baruscotti M, Lenfant J, DiFrancesco D (1997) Activation of f-channels by cAMP analogues in macropatches from rabbit sino-atrial node myocytes. J Physiol (Lond) 501:565–571

Bradley J, Li J, Davidson N, Lester HA, Zinn K (1994) Heteromeric olfactory cyclic nucleotide-gated channels: A new subunit that confers increased sensitivity to cAMP. Proc Natl Acad Sci USA 91:8890–8894

Broillet MC, Firestein S (1996) Direct activation of the olfactory cyclic nucleotide-gated channel through modification of sulfhydryl groups by NO compounds. Neuron 16:377–385

Broillet M-C, Firestein S (1997) β subunits of the olfactory cyclic nucleotide-gated channel form a nitric oxide activated Ca^{2+} channel. Neuron 18:951–958

Brown RL, Bert RJ, Evans FE, Karpen JW (1993b) Activation of retinal rod cGMP-gated channels: What makes for an effective 8-substituted derivative of cGMP? Biochemistry 32:10089–10095

Brown RL, Gerber WV, Karpen JW (1993a) Specific labeling and permanent activation of the retinal rod cGMP-activated channel by the photoaffinity analog 8-p-azidophenacylthio-cGMP. Proc Natl Acad Sci USA 90:5369–5373

Brown RL, Gramling R, Bert RJ, Karpen JW (1995) Cyclic GMP contact points within the 63-kDa subunit and a 240-kDa associated protein of retinal rod cGMP-activated channels. Biochemistry 34:8365–8370

Brown RL, Snow SD, Haley TL (1998) Movement of gating machinery during the activation of rod cyclic nucleotide-gated channels. Biophys J 75:825–833

Bucossi G, Eismann E, Sesti F, Nizzari M, Seri M, Kaupp UB, Torre V (1996) Time-dependent current decline in cyclic GMP-gated bovine channels caused by point mutations in the pore region expressed in Xenopus oocytes. J Physiol (Lond) 493:409–418

Bucossi G, Nizzari M, Torre V (1997) Single-channel properties of ionic channels gated by cyclic nucleotides. Biophys J 72:1165–1181

Chen T-Y, Illing M, Molday LL, Hsu Y-T, Yau K-W, Molday RS (1994) Subunit 2 (or β) of retinal rod cGMP-gated cation channel is a component of the 240-kDa channel-associated protein and mediates Ca^{2+}-calmodulin modulation. Proc Natl Acad Sci USA 91:11757–11761

Chen T-Y, Peng Y-W, Dhallan RS, Ahamed B, Reed RR, Yau K-W (1993) A new subunit of the cyclic nucleotide-gated channel of rat olfactory receptor neurons. Nature 362:764–767

Chen T-Y, Yau K-W, (1994) Direct modulation by Ca^{2+}-calmodulin of cyclic nucleotide-activated channel of rat olfactory receptor neurons. Nature 368:545–548

Delay RJ, Dubin AE, Dionne VE (1997) A cyclic nucleotide-dependent chloride conductance in olfactory receptor neurons. J Membr Biol 159:53–60

Delgado R, Hidalgo P, Diaz F, Latorre R, Labarca P (1991) A cyclic AMP-activated K^+ channel in *Drosophila* larval muscle is persistently activated in dunce. Proc Natl Acad Sci USA 88:557–560

DiFrancesco D, Tortora P (1991) Direct activation of cardiac pacemaker channels by intracellular cyclic AMP. Nature 351:145–147

Donner K, Hemila S, Kalamkarov G, Koskelainen A, Pogozheva I, Rebrik T (1990) Sulfhydryl binding reagents increase the conductivity of the light-sensitive channel and inhibit phototransduction in retinal rods, Exp Eye Res 51:97–105

Eismann E, Muller F, Heinemann SH, Kaupp UB (1994) A single negative charge within the pore region of a cGMP-gated channel controls rectification, Ca^{2+} blockage, and ionic selectivity. Proc Natl Acad Sci USA 91:1109–1113

Fesenko EE, Kolesnikov SS, Lyubarsky AL (1985) Induction by cyclic GMP of cationic conductance in plasma membrane of retinal rod outer segment. Nature 313:310–313

Finn JT, Grunwald ME, Yau K-W (1996) Cyclic nucleotide-gated ion channels: an extended family with diverse functions, Annu Rev Physiol. 58:395–426

Finn JT, Li J, Yau K-W (1995) C-terminus involvement in the gating of cyclic nucleotide-activated channels as revealed by Ni^{2+} and NEM. Biophys J 68:A385 (Abstr)

Finn JT, Solessio EC, Yau KW (1997) A cGMP-gated cation channel in depolarizing photoreceptors of the lizard parietal eye. Nature 385:815–819

Finn JT, Krautwurst D, Schroeder JE, Chen T-Y, Reed RR, Yau K-W (1998) Functional co-assembly among subunits of cyclic-nucleotide-activated nonselective cation channels, and across species from nematode to human. Biophys J 74:1333–1345

Fodor AA, Black KD, Zagotta WN (1997b) Tetracaine reports a conformational change in the pore of cyclic nucleotide-gated channels. J Gen Physiol 110:591–600

Fodor AA, Gordon SE, Zagotta WN (1997a) Mechanism of tetracaine block of cyclic nucleotide-gated channels. J Gen Physiol 109:3–14

Fodor AA, Zagotta WN (1996) Subunit 2 alters ligand specificity of rod CNG channels. Biophys J 70:368 (Abstr)

Frings S, Brull N, Dzeja C, Angele A, Hagen V, Kaupp UB, Baumann A (1998) Characterization of ether-a-go-go channels present in photoreceptors reveals similarity to IKx, a K^+ current in rod inner segments. J Gen Physiol 111:583–599

Frings S, Dzeja C, Bönigk W, Müeller F, Bradley J, Sesti F, Kaupp UB (1998) cAMP-gated channels of olfactory sensory neurons: subunit composition; Ca^{2+} permeation, and modulation of ligand sensitivity. Eur J Neurosci 10:358 (Abstr)

Frings S, Lynch JW, Lindemann B (1992) Properties of cyclic nucleotide-gated channels mediating olfactory transduction. Activation, selectivity, and blockage. J Gen Physiol 100:45–67

Frings S, Seifert R, Godde M, Kaupp UB (1995) Profoundly different calcium permeation and blockage determine the specific function of distinct cyclic nucleotide-gated channels. Neuron 15:169–179

Gauss R, Siefert R, Kaupp UB (1998) Molecular identification of a hyperpolarization-activated channel in sea urchin sperm. Nature 393:583–587

Gavazzo P, Picco C, Menini A (1997) Mechanisms of modulation by internal protons of cyclic nucleotide-gated channels cloned from sensory receptor cells. Proc R Soc Lond B Biol Sci 264:1157–1165

Godde M, Molday L, Kaupp UB, Molday RS (1995) A 240 kDA protein represents the complete β-subunit of cyclic nucleotide-dated channel from rod photoreceptors. Neuron 15:627–636

Gomez MD, Nasi E (1995) Activation of light-dependent potassium channels in ciliary invertebrate photoreceptors involves cGMP but not the IP_3/Ca^{2+} cascade. Neuron 15:607–618

Gomez MP, Nasi E (1997) Antagonists of the cGMP-gated conductance of vertebrate rods block the photocurrent in scallop ciliary photoreceptors. J Physiol (Lond) 500:367–78
Gordon SE, Brautigan DL, Zimmerman AL (1992) Protein phosphatases modulate the apparent agonist affinity of the light-regulated ion channel in retinal rods. Neuron 9:739–748
Gordon SE, Downing-Park J, Tam B, Zimmerman AL (1995a) Diacylglycerol analogs inhibit the rod cGMP-gated channel by a phosphorylation-independent mechanism. Biophys J 69:409–417
Gordon SE, Downing-Park J, Zimmerman AL (1995b) Modulation of the cGMP-gated ion channel in frog rods by calmodulin and an endogenous inhibitory factor. J Physiol (Lond) 486:533–546
Gordon SE, Oakley JC, Varnum MD, Zagotta WN (1996) Altered ligand specificity by protonation in the ligand binding domain of cyclic nucleotide-gated channels. Biochem 35:3994–4001
Gordon SE, Varnum MD, Zagotta WN (1997) Direct interaction between amino- and carboxyl-terminal domains of cyclic nucleotide-gated channels. Neuron 19:431–441
Gordon SE, Zagotta WN (1995a) A histidine residue associated with the gate of the cyclic nucleotide-activated channels in rod photoreceptors. Neuron 14:177–183
Gordon SE, Zagotta WN (1995b) Localization of regions affecting an allosteric transition in cyclic nucleotide-activated channels. Neuron 14:857–864
Gordon SE, Zagotta WN (1995c) Subunit interactions in coordination of Ni^{2+} in cyclic nucleotide-gated channels. Proc Natl Acad Sci USA 92:10222–10226
Gotow T, Nishi T, Kijima H (1994) Single K^+ channels closed by light and opened by cyclic GMP in molluscan extraocular photoreceptor cells. Brain Res 662:268–272
Goulding EH, Tibbs GR, Liu D, Siegelbaum SA (1993) Role of H5 domain in determining pore diameter and ion permeation through cyclic nucleotide-gated channels. Nature 364:61–64
Grunwald ME, Lai J, Yau K-W (1998a) Molecular determinants of the calmodulin-mediated inhibition and the gating of cyclic nucleotide-gated ion channels. Biophys J 74:A125 (Abstr)
Grunwald ME, Yu W-P, Yu H-H, Yau K-W (1998b) Identification of a domain on the β-subunit of the rod cGMP-gated cation channel that mediates inhibition by calcium-calmodulin. J Biol Chem 273:9148–9157
Guy Hr, Durell SR, Warmke J, Drysdale R, Ganetzky B (1991) Similarities in amino acid sequences of *Drosophila* eag and cyclic nucleotide-gated channels. Science 254:730
Hackos DH, Korenbrot JI (1997) Calcium modulation of ligand affinity in the cyclic GMP-gated ion channels of cone photoreceptors. J Gen Physiol 110:515–528
Hatt H, Ache BW (1994) Cyclic nucleotide- and inositol phosphate-gated ion channels in lobster olfactory receptor neurons. Proc Natl Acad Sci USA 91:6264–6268
Haynes LW (1992) Block of the cyclic GMP-gated channel of vertebrate rod and cone photoreceptors by l-cis-diltiazem. J Gen Physiol 100:783–801
Haynes LW (1995) Permeation and block by internal and external divalent cations of the catfish cone photoreceptor cGMP-gated channels. J Gen Physiol 106:507–523
Haynes LW, Stotz SC (1997) Modulation of catfish rod, but not cone, cGMP-gated photoreceptor channels by calcium-calmodulin. Vis Neurosci 14:233–239
Heginbotham L, Abramson T, MacKinnon R (1992) A functional connection between the pores of distantly related ion channels as revealed by mutant K^+ channels. Science 258:1152–1155
Hsu YT, Molday RS (1993) Modulation of the cGMP-gated channel of rod photoreceptor cells by calmodulin. Nature 361:76–79
Hsu YT, Molday RS (1994) Interaction of calmodulin with the cyclic GMP-gated channel of rod photoreceptor cells. Modulation of activity, affinity purification, and localization. J Biol Chem 269:29765–29770

Ildefonse M, Bennett N (1991) Single-channel study of the cGMP-dependent conductance of retinal rods from incorporation of native vesicles into planar lipid bilayers. J Membr Biol 123:133–147
Ildefonse M, Crouzy S, Bennett N (1992) Gating of retinal rod cation channel by different nucleotides: comparative study of unitary currents. J Membr Biol 130:91–104
Jan LY, Jan YN (1990) A super family of ion channels. Nature 345:672
Jan LY, Jan YN (1992) Tracing the roots of ion channels. Cell 69:715–718
Karpen JW, Brown RL (1996) Covalent activation of retinal rod cGMP-gated channels reveals a functional heterogeneity in the ligand binding sites. J Gen Physiol 107:169–181
Karpen JW, Brown RL, Stryer L, Baylor DA (1993) Interactions between divalent cations and the gating machinery of cyclic GMP-activated channels in salamander retinal rods. J Gen Physiol 101:1–25
Kaupp UB (1995) Family of cyclic nucleotide-gated ion channels. Curr Opin Neurobiol 5:434–442
Kaupp UB, Niidome T, Tanabe T, Terada S, Bönigk W, Stuhmer W, Cook NJ, Kangawa K, Matsuo H, Hirose T, Miyata T, Numa S (1989) Primary structure and functional expression from complementary DNA of the rod photoreceptor cyclic GMP-gated channel. Nature 342:762–766
Kleene SJ (1994) Inhibition of olfactory cyclic nucleotide-activated current by calmodulin antagonists. Br J Pharmacol 111:469–472
Koch KW, Kaupp UB (1985) Cyclic GMP directly regulates a cation conductance in membranes of bovine rods by a cooperative mechanism. J Biol Chem 260:6788–6800
Kolesnikov SS, Margolskee RF (1995) A cyclic-nucleotide-suppressible conductance activated by transducin in taste cells, Nature 376:85–88
Körschen HG, Illing M, Seifert R, Sesti F, Williams A, Gotzes S, Colville C, Müller F, Dosé A, Godde M, Molday L, Kaupp UB, Molday RS (1995) A 240 kDA protein represents the complete β-subunit of cyclic nucleotide-dated channel from rod photoreceptors, Neuron 15:627–636
Kramer RH, Goulding E, Siegelbaum SA (1994) Potassium channel inactivation peptide blocks cyclic nucleotide-gated channels by binding to the conserved pore domain. Neuron 12:655–662
Kramer RH, Karpen JW (1998) Spanning binding sites on allosteric proteins with polymer-linked ligand dimers. Nature 395:710–713
Kramer RH, Tibbs GR (1996) Antagonists of cyclic nucleotide-gated channels and molecular mapping of their site of action. J Neurosci 16:1285–1293
Kumar VD, Weber IT (1992) Molecular model of the cyclic GMP-binding domain of the cyclic GMP-gated ion channel. Biochemistry 31:4643–4649
Leinders-Zufall T, Zufall F (1995) Block of cyclic nucleotide-gated channels in salamander olfactory receptor neurons by the guanylyl cyclase inhibitor LY83583. J Neurophysiol 74:2759–2762
Li P, Zagotta WN, Lester HA (1997) cyclic nucleotide-gated channels: structural basis of ligand efficacy and allosteric modulation. Q Rev Biophys 30:177–193
Light DB, Schwiebert EM, Karlson KH, Stanton BA (1989) Atrial natriuretic peptide inhibits a cation channel in renal inner medullary collecting duct cells. Science 243:383–385
Liman ER, Buck LB (1994) A second subunit of the olfactory cyclic nucleotide-gated channel confers high sensitivity to cAMP. Neuron 13:611–621
Liu DT, Tibbs GR, Siegelbaum SA (1996) Subunit stoichiometry of cyclic nucleotide-gated channels and effects of subunit order on channel function. Neuron 16:983–990
Liu DT, Tibbs GR, Paoletti P, Siegelbaum SA (1998) Constraining ligand-binding site stoichiometry suggests that a cyclic nucleotide-gated channel is composed of two functional dimers. Neuron 21:235–248

Liu M, Chen T-Y, Ahamed B, Li J, Yau K-W (1994) Calcium-calmodulin modulation of the olfactory cyclic nucleotide-gated cation channel. Science 266:1348–1354
Ludwig A, Zong X, Jeglitsch M, Hofmann F, Biel M (1998) A family of hyperpolarization-activated mammalian cation channels. Nature 393:587–591
McGeoch JE, McGeoch MW, Guidotti G (1995) Eye CNG channel is modulated by nicotine. Biochem Biophys Res Commun 214:879–887
McKay DB, Steitz TA (1981) Structure of catabolite gene activator protein at 2.9-Å resolution suggests binding to left-handed B-DNA. Nature 290:744–749
McKay DB, Weber IT, Steitz TA (1982) Structure of catabolite gene activator protein at 2.9-Å resolution. J Biol Chem 257:9518–9524
McLatchie LM, Matthews HR (1992) Voltage-dependent block by L-*cis*-diltiazem of the cyclic GMP-activated conductance of salamander rods. Proc R Soc Lond B biol Sci 247:113–119
McLatchie LM, Matthews HR (1994) The effect of pH on the block by L-cis-diltiazem and amiloride of the cyclic GMP-activated conductance of salamander rods. Proc R Soc Lond B Biol Sci 255:231–236
Molokanova E, Maddox F, Luetje CW, Kramer RH (1999) Activity-dependent modulation of rod photoreceptor cyclic nucleotide-gated channels mediated by phosphorylation of a specific residue. J Neurosci 19:4786–4795
Molokanova E, Trivedi B, Savchenko A, Kramer RH (1997) Modulation of rod photoreceptor cyclic nucleotide-gated channels by tyrosine phosphorylation. J Neurosci 17:9068–9076
Mori Y, Friedrich T, Kim MS, Mikami A, Nakai J, Ruth P, Bosse E, Hofmann F, Flockerzi V, Foruichi T, Mikoshiba K, Imoto K, Tanabe T, Numa S (1991) Primary structure and functional expression from complementary DNA of a brain calcium channel. Nature 350:398–402
Müller F, Bönigk W, Sesti F, Frings S (1998) Phosphorylation of mammalian olfactory cyclic nucleotide-gated channels increases ligand sensitivity, J Neurosci 18:164–173
Nicol GD (1993) The calcium channel antagonist, pimozide, blocks the cyclic GMP-activated current in rod photoreceptors. J Pharmacol Exp Ther 265:626–32
Nicol GD, Schnetkamp PP, Saimi Y, Gragoe EJ Jr, Bownds MD (1987) A derivative of amiloride blocks both the light-regulated and cyclic GMP-regulated conductances in rod photoreceptors. J Gen Physiol 90:651–669
Pape HC (1996) Queer current and pacemaker: the hyperpolarization-activated cation current in neurons. Annu Rev Physiol 58:299–327
Park CS, MacKinnon R (1995) Divalent cation selectivity in a cyclic nucleotide-gated ion channel. Biochemistry 34:13328–13333
Picco C, Sanfilippo C, Gavazzo P, Menini A (1996) Modulation by internal protons of native cyclic nucleotide-gated channels from retinal rods. J Gen Physiol 108:265–276
Picones A, Korenbrot JI (1995) Permeability and interaction of Ca^{2+} with cGMP-gated ion channels differ in retinal rod and cone photoreceptors. Biophys J 69:120–127
Quandt FN, Nicol GD, Schnetkamp PP (1991) Voltage-dependent gating and block of the cyclic-GMP-dependent current in bovine rod outer segments. Neurosci 42:629–638
Rieke F, Schwartz EA (1994) A cGMP-gated current can control exocytosis at cone synapses. Neuron 13:863–873
Root MJ, MacKinnon R (1993) Identification of an external divalent cation-binding site in the pore of a cGMP-activated channel. Neuron 11:459–466
Root MJ, MacKinnon R (1994) Two identical noninteracting sites in an ion channel revealed by proton transfer. Science 265:1852–1856
Ruiz ML, Karpen JW (1997) Single cyclic nucleotide-gated channels locked in different ligand-bound states. Nature 389:389–392
Ruiz ML and Karpen, JW (1999) Opening mechanism of a cyclic nucleotide-gated channel based on analysis of single channels locked in each liganded state. J Gen Physiol 113:873–895

Santoro B, Liu DT, Yao H, Bartsch D, Kandel ER, Siegelbaum SA, Tibbs GR (1998) Identification of a gene encoding a hyperpolarization-activated pacemaker channel of brain. Cell 93:717–729
Satler CA, Walsh EP, Vesely MR, Plumer MH, Ginsburg GS, Jacob HJ (1996) Novel missense mutation in the cyclic nucleotide-binding domain of HERG causes long QT syndrome. Am J Med Genet 65:27–35
Sautter A, Zong X, Hofmann F, Biel M (1998) an isoform of the rod photoreceptor cyclic nucleotide-gated channel β subunit expressed in olfactory neurons. Proc Natl Acad Sci USA 95:4696–4701
Savchenko A, Barnes S, Kramer RH (1997) Cyclic-nucleotide-gated channels mediated synaptic feedback by nitric oxide. Nature 390:694–698
Schnetkamp PPM (1987) Sodium ions selectively eliminate the fast component of guanosine cyclic 3′,5′-phosphate induced Ca^{2+} release from bovine rod outer segment disks. Biochemistry 26:3249–3253
Schnetkamp PPM (1990) Cation selectivity of and cation binding to the cGMP-dependent channel in bovine rod outer segment membranes. J Gen Physiol 96:517–34
Scott S-P, Tanaka JC (1995) Molecular interactions of 3’5’-cyclic purine analogues with the binding site of retinal rod ion channels. Biochemistry 34:2338–2347
Sentenac H, Bonneaud N, Minet M, Lacroute F, Salmon J-M, Gaymard F, Grignon C (1992) Cloning and expression in yeast of a plant potassium ion transport system. Science 256:663–65
Serre V, Ildefonse M, Bennett N (1995) Effects of cysteine modification on the activity of the cGMP-gated channel from retinal rods. J Membr Biol 146:145–162
Sesti F, Eismann E, Kaupp UB, Nizarri M, Torre V (1995) The multi-ion nature of the cGMP-gated channel from vertebrate rods. J Physiol 487:17–36
Stern JH, Kaupp UB, MacLeish PR (1986) Control of the light-regulated current in rod photoreceptors by cyclic GMP, calcium and L-*cis*-diltiazem. Proc Natl Acad Sci USA 83:1163–1167
Sudlow LC, Huang R-C, Green DJ, Gillette R (1993) cAMP-activated Na^+ current of molluscan neurons is resistant to kinase inhibitors and is gated by cAMP in the isolated patch. J Neurosci 13:5188–5193
Sun Z-P, Akabas MH, Karlin A, Siegelbaum SA (1993) Covalent modification of cysteines by MTSEA changes cGMP responses of a cyclic nucleotide-gated channel. Soc Neurosci Abstr 19:713
Tang CY, Papazian DM (1997) Transfer of voltage independence from a rat olfactory channel to the Drosophila ether-a-go-go K^+ channel. J Gen Physiol 109:301–311
Tempel BL, Papazian DM, Schwarz TL, Jan YN, Jan LY (1987) Sequence of a probable potassium channel component encoded at Shaker locus of Drosophila. Science 237:770–775
Tibbs GR, Goulding EH, Siegelbaum SA (1997) Allosteric activation and tuning of ligand efficacy in cyclic-nucleotide-gated channels. Nature 368:612–615
Tibbs GR, Liu DT, Leypold BG, Siegelbaum SA (1998) A state-independent interaction between ligand and a conserved arginine residue in cyclic nucleotide-gated channels reveals a functional polarity of the cyclic nucleotide binding site. J Biol Chem 273:4497–4505
Varnum MD, Black KD, Zagotta WN (1995) Molecular mechanism for ligand discrimination of cyclic nucleotide-gated channels. Neuron 15:619–625
Varnum MD, Zagotta WN (1996) Subunit interactions in the activation of cyclic nucleotide-gated ion channels. Biophys J 70:2667–2679
Varnum MD, Zagotta WN (1996) Activation of cyclic nucleotide-gated channels examined by tandem linkage of subunits. Biophys J 70:368 (Abstr)
Varnum MD, Zagotta WN (1997) Interdomain interactions underlying activation of cyclic nucleotide-gated channels. Science 278:110–113
Warmke JW, Ganetzky B (1994) A family of potassium channel genes related to *eag* in *Drosophila* and mammals. Proc Natl Acad Sci USA 91:3438–3442

Weber IT, Shabb JB, Corbin JD (1989) Predicted structures of the cGMP binding domains of the cGMP-dependent protein kinase: a key alanine/threonine difference in evolutionary divergence of cAMP and cGMP binding sites. Biochemistry 28:6122–6127

Weber IT, Steitz TA, Bubis J, Taylor SS (1987) Predicted structures of cAMP binding domains of type I and II regulatory subunits of cAMP-dependent protein kinase. Biochemistry 26:343–351

Wei JY, Cohen ED, Barnstable CJ (1997) Direct blockage of both cloned rat rod photoreceptor cyclic nucleotide-gated non-selective cation (CNG) channel α-subunit and native CNG channels from Xenopus rod outer segments by H-8, a non-specific nucleotide-dependent protein kinase inhibitor. Neurosci Lett 233:37–40

Wei J-Y, Cohen ED, Yan Y-Y, Genieser H-G, Barnstable CJ (1996) Identification of competitive antagonists of the rod photoreceptor cGMP-gated cation channel: β-phenyl-1,N^2-etheno-substituted cGMP analogues as probes of the cGMP-binding site. Biochemistry 35:16815–16823

Weitz D, Zoche M, Müller F, Beyermann M, Körschen HG, Kaupp UB, Koch K-W (1998) Calmodulin controls the rod photoreceptor CNG channel through an unconventional binding site in the N-terminus of the β-subunit. EMBO J 17:2273–2284

Xiong W-H, Solessio EC, Yau K-W (1998) An unusual cGMP pathway underlying depolarizing light response of the vertebrate parietal-eye photoreceptor. Nat Neurosci 1:359–365

Yang J, Ellinor PT, Sather WA, Zhang JF, Tsien RW (1993) Molecular determinants of Ca^{2+} selectivity and ion permeation in L-type Ca^{2+} channels. Nature 366:158–161

Yau KW, Baylor DA (1989) Cyclic GMP-activated conductance of retinal photoreceptor cells. Annu Rev Neurosci 12:289–327

Yu W-P, Grunwald ME, Yau K-W (1996) Molecular cloning, functional expression and chromosomal localization of a human homolog of the cyclic nucleotide-gated ion channel of retinal cone photoreceptors. FEBS Lett 393:211–215

Zagotta WN, Siegelbaum SA (1996) Structure and function of cyclic nucleotide-gated channels. Annu Rev Neurosci 19:235–263

Zimmerman AL (1995) Cyclic nucleotide gated channels. Curr Opin Neurobiol 5:296–303

Zong, X, Zucker H, Hofmann F, Biel M (1998) Three amino acids in the C-linker are major determinants of gating in cyclic nucleotide-gated channels. EMBO J 17:353–362

Zufall F, Firestein S (1993) Divalent cations block the cyclic nucleotide-gated channel of olfactory receptor neurons. J Neurophysiol 69:1758–1768

Zufall F, Shepherd GM, Firestein S (1991) Inhibition of the olfactory cyclic nucleotide gated ion channel by intracellular calcium. Proc R Soc Lond B Biol Sci 246:225–230

Section III
Miscellaneous Ion Channels – Intracellular Ca Release Channels

CHAPTER 23

Regulation of Ryanodine Receptor Calcium Release Channels

M. ENDO and T. IKEMOTO

A. Introduction

Ryanodine receptor (RyR)/calcium release channel is one of the two major classes of intracellular calcium release channels, the other being IP_3 receptor described in the Chap. 24. RyRs are present in the membrane of the intracellular calcium store, usually endoplasmic reticulum, in almost every kind of cell and, by its opening, it provides a path through which calcium ion is supplied from the store to the cytoplasm. The calcium ion in turn triggers a series of events that lead to various cellular responses. While the presence of RyRs is thus ubiquitous, the study of this type of calcium release channel started with the skeletal muscle, and much of the important information about RyRs has been obtained with this tissue.

The problem as to how an action potential of skeletal muscle cells can lead to the cellular contractile response was one of the main themes of life science, and the framework of our present understanding about this problem was established in the early 1960s by discoveries of the following important facts (cf. EBASHI and ENDO 1968):

1. Contractile response of the native contractile protein system (myosin-actin-tropomyosin-troponin system) by MgATP requires a minute amount of calcium ion.
2. The active principle of "relaxing factor," which was obtained from muscle homogenate and had the ability to bring about relaxation of glycerinated fiber in the presence of MgATP, is fragments of the sarcoplasmic reticulum (SR).
3. The relaxing factor can strongly accumulate calcium ion from the medium in the presence of MgATP and resulting removal of calcium ion from the medium surrounding glycerinated fibers is the mechanism of relaxation.

These discoveries led to the present idea about excitation–contraction coupling (E–C coupling), the reverse of relaxation, whereby the SR accumulates and holds calcium ion during the resting state and action potentials can somehow cause a release of calcium ion from the SR. Since then, the mechanism of calcium release from the SR was one of the main targets in the studies of E–C coupling.

FORD and PODOLSKY (1970) and ENDO et al. (1970) reported independently on somewhat different grounds that calcium ion itself can cause calcium release from the SR. This release process was named "calcium-induced calcium release (CICR)" and its properties have been extensively studied (cf. ENDO 1985). However, CICR was then shown not to be the physiological mechanism of calcium release during E–C coupling of skeletal muscle, as described later in detail. (Although this problem is still somewhat controversial, the point of dispute is not on the essential release mechanism but as to whether or not CICR plays a secondary role as an amplification mechanism.) The question arises then as to whether the physiological calcium release (PCR) during E–C coupling of skeletal muscle is through the same channel as CICR or a different channel? The former is true as described later. The answer was obtained only after the calcium release channel of the SR was cloned and its knockout mice were made.

In the late 1980s, the calcium release channel protein of SR of skeletal muscle was isolated and purified utilizing ryanodine, a plant alkaloid, that specifically binds to the channel protein (INUI et al. 1987; LAI et al. 1988). This is the origin of the name of this type of calcium release channel, ryanodine receptor. The primary structure of the RyR was then determined (TAKESHIMA et al. 1989) and successive studies from many laboratories, especially with molecular biological techniques, revealed the presence of similar calcium release channels in many cells, their distribution in the body, their relation in evolution to another kind of intracellular calcium release channel, the IP_3 receptor, and more about the properties of RyR channels.

B. Molecular Structure and Function of RyR

Ryanodine receptor is a large protein consisting of about 5000 amino acid residues, with a molecular weight of about 560kDa (TAKESHIMA et al. 1989). Towards the C-terminal end, it has the transmembrane segments that are thought to be embedded in the SR membrane. The big N-terminal region is exposed to the cytoplasm. Four of the 560kDa proteins assemble to form a tetramer with a characteristic quatrefoil appearance under the electron microscope (INUI et al. 1987; LAI et al. 1988; SAITO et al. 1988). The cytoplasmic part of the tetramer is identified as the foot structure that spans the gap between the SR and the T-tubule in the skeletal muscle (INUI et al. 1987; BLOCK et al. 1988; LAI et al. 1988; TAKEKURA et al. 1995). The tetramer forms an ion channel which, when it is open, allows calcium and other ions to pass with a conductance of about 100–150pS for calcium ion or 550–750pS for monovalent ion (SMITH et al. 1986; ANDERSON et al. 1989). As already mentioned, it specifically binds ryanodine, 1mol of ryanodine being bound to 1mol of tetramer with a KD of 2–7nmol/l (MCGREW et al. 1989).

In mammals, three types of RyR – RyR1, RyR2, and RyR3 – are known, homologies between any two types being about 70% (TAKESHIMA 1993). RyR1

Table 1. Comparison among mammalian RyR

	RyR1	RyR2	RyR3
Amino acid residues	5037	4976	4872
Calculated M_w (kDa)	565	565	552
Mobility on SDS-PAGE	Smallest	Intermediate	Largest
Effective stimulus for opening			
Calcium ion (sensitivity)	High	High	Low
Mechanical (protein-protein interaction)	+	–?	–?
SR "depolarization"	+	–?	–
Inhibition by calcium ion	++	+	+

++ and +: positive effect depending on magnitudes of the action.
–?: probably negative but not certain.
– : negative.

and RyR2 exist mainly in skeletal and cardiac muscle, respectively, while RyR3 is present in many tissues more or less ubiquitously (SUTKO and AIREY 1996). A small amount of RyR3 is present in skeletal muscle as well (usually less than 1% of RyR1 (MURAYAMA and OGAWA 1997) but 5% or more in a certain muscle or at different developmental stages (JEYAKNMAR et al. 1997; FLUCHER et al. 1999)), but not in cardiac muscle. In brain, all three kinds of RyRs are expressed. In Table 1, some properties of the three types of mammalian RyRs are summarized.

In non-mammalian vertebrates, such as chicken and bullfrog, skeletal muscle contains two types of RyRs – α-RyR and β-RyR – in about equal amounts (SUTKO and AIREY 1996). α-RyR is homologous with RyR1 in mammals, and β-RyR with RyR3 (OYAMADA et al. 1994; OTTINI et al. 1996). These two types of RyRs form only *homo*tetramers (AIREY et al. 1990; MURAYAMA and OGAWA 1992). The same is true in the case of mammalian cells, in which different types of RyRs are coexistent (MURAYAMA and OGAWA 1997). This property of forming only homotetramer is in sharp contrast to the case of IP_3 receptors, which also have three genetically different types and form tetramers to make ion channels. In the case of IP_3 receptor, *hetero*tetramers can be formed, which probably provides the basis of functional diversities of this kind of calcium release channels in a variety of cells (for references concerning IP_3 receptor, see Chap. 24).

C. Different Modes of Opening of RyR1 Calcium Release Channel

The difference between PCR and CICR in skeletal muscle should be fully described first. If PCR is mediated by calcium ion that in turn opens the calcium release channel of the SR, i.e., CICR channel, an inhibitor of CICR should also inhibit PCR. However, two different kinds of inhibitors of CICR, procaine and adenine, were shown not to inhibit PCR at all. *Procaine*, a well-known inhibitor of CICR, in fact strongly inhibits contracture of a single skeletal muscle fiber induced by activation of CICR through caffeine application, but does not at all inhibit contracture of the same fiber induced by potassium depolarization of the surface membrane (Thorens and Endo 1975). *Adenine* is another inhibitor of CICR in the presence of ATP, although it is an activator by itself. Since ATP is present in millimolar concentration in living cells, adenine acts as an inhibitor of CICR there (Ishizuka and Endo 1983). Indeed, adenine was shown to inhibit caffeine contracture of living skeletal muscle fiber. However, adenine does not inhibit twitch of the same fiber at all (Ishizuka et al. 1983).

An example of the opposite is dantrolene. This drug was well known to inhibit twitch contractions of skeletal muscle by suppressing PCR at any temperature (Ellis and Bryant 1972; Ellis and Carpenter 1972), but its inhibition on CICR and hence on caffeine contracture was shown to be negligible at room temperature, although at 37 °C the inhibition is rather strong with the same potency as its inhibition on PCR (Ohta and Endo 1986; Kobayashi and Endo 1986). A screening study of derivatives of dantrolene revealed several more specific inhibitors of PCR with no effect on CICR (Ikemoto T. et al., in preparation).

Thus, PCR in skeletal muscle is not activated by calcium ion. It is generally considered to be activated by direct protein-protein interaction with a voltage-sensing molecule, the dihydropyridine receptor (DHPR), in the membrane of the transverse tubule (T-tubule).

These results at first appeared to suggest that two different calcium release channels are responsible for PCR and CICR, respectively. However, it turned out that PCR and CICR are the function of one and the same kind of channel with different opening modes. When purified RyR proteins of any kinds were incorporated into lipid bilayer, they behaved as CICR channels, the open probability of which increased with an increase in free calcium ion concentration on the *cis* side (Smith et al. 1988). All the other properties of CICR, inhibitory effect of calcium ion at high concentrations, inhibition by magnesium ion, stimulation by adenine compounds, and so on, were also reproduced in bilayer experiments of RyRs. These results clearly indicate that RyRs are proteins that form CICR channel. This conclusion is also supported by the result of lack of CICR in RyR1-RyR3-double knockout mice described below.

Evidence that RyR1 is also responsible for PCR of skeletal muscle came from experiments on knockout mice. So far, knockout mice have been made for each of all three kinds of RyRs, RyR1 (Takeshima et al. 1994), RyR2 (Takeshima et al. 1998), and RyR3 (Takeshima et al. 1996), as well as both RyR1 and RyR3 double knockout mice (Ikemoto et al. 1997). Skeletal muscle cells of RyR1-knockout mice cannot contract in response to electrical stimulation whereas caffeine could still evoke contracture of these cells (Takeshima et al. 1994). The animals, therefore, cannot survive after birth due to respiratory failure. The preservation of caffeine response was found to be because of the presence of RyR3, which is also expressed normally in skeletal muscle cells although in a small amount. In fact, skinned fibers from skeletal muscle of RyR1-RyR3-double knockout mice showed neither calcium nor caffeine responses while the SR retained an ability to accumulate calcium ion (IKEMOTO et al. 1997). The failure of E–C coupling of skeletal muscle of RyR1 knockout mice, which still have RyR3 calcium release channels, clearly indicates that RyR1 that can form CICR channel as described in the previous paragraph also forms PCR channels, and the latter function cannot be substituted by RyR3. RyR2 can neither replace RyR1 in this sense, which was demonstrated by reconstitution experiments: E–C coupling of defective myotubes from RyR1 knockout mice was recovered with the incorporation of RyR1 cDNA but not that of RyR2 cDNA (YAMAZAWA et al. 1996).

RyR1 appears to be able operate in still other mode of opening which is pharmacologically different from both CICR and PCR. "Depolarization" of the SR membrane, which is defined as a potential change in such a direction that the luminal side becomes more negative, was shown to cause calcium release (ENDO and NAKAJIMA 1973). Since these authors induced "depolarization" by changing ionic composition of the medium surrounding SR, it was argued that some effects of the ionic replacement other than "depolarization," for example, swelling of the SR, specific effects of the ion introduced, and so on, might be responsible for calcium release. However, ENDO (1985) put forward the reasons to believe "depolarization" is the cause of calcium release:

1. "Depolarization" induced by cationic replacement or ionic replacement by keeping [K]·[Cl] product constant should not be accompanied by swelling of the SR but still causes calcium release.
2. In this type of calcium release "inactivation" is demonstrated to occur, which is most conveniently interpreted by "depolarization."

Therefore, in the present article this type of calcium release is called "depolarization"-induced calcium release (DICR).[1] The pharmacological profile of DICR is different from both that of CICR and of PCR (ENDO 1985). DICR was absent in RyR1 knockout and RyR1-RyR3-double knockout mice, but it is present in RyR3 knockout mice (IKEMOTO et al. 1997). These results indicate that DICR is carried by RyR1, as its third mode of opening. Unfortunately there have been no proper studies of DICR in RyR1 channels incorporated into lipid bilayer (CICR should be completely inhibited), although "depolarization"-induced inactivation of CICR was reported (PERCIVAL et al. 1994; ZAHRADNIKOVA and MESZAROS 1998). Since proper bilayer studies with applied voltages are free from the objections to ionic replacement studies mentioned

---

[1] DICR is sometimes used to stand for the SR calcium release induced by depolarization of the T-tubule membrane, which, therefore, corresponds to PCR in this article. This terminology should clearly be distinguished from DICR as used in this article

above, we hope these studies will soon be conducted to make the pure effect of potential across the SR membrane clearer.

Thus, RyR1 operates in three different modes,[2] PCR, CICR, and DICR, depending on stimulus given. Therefore, activators and inhibitors of RyR1 should be considered for each mode separately. It is not known whether or not RyR2 and RyR3 can also operate in a mode different from CICR. Unlike skeletal muscle, CICR is considered to be the physiological calcium release mechanism in cardiac muscle where RyR2 is the responsible channel. The presence of DICR in RyR2 is questionable because FABIATO and FABIATO (1977) reported that the ionic replacement which caused calcium release in skinned skeletal muscle fibers was ineffective in cardiac preparation. In the case of RyR3 what is the physiological stimulus is still unknown. The fact that skeletal muscles from RyR1 knockout mice in which RyR3 was still expressed were deprived of DICR activity (IKEMOTO et al. 1997) suggests that RyR3 does not operate in DICR mode. Therefore, for RyR2 and RyR3 only CICR mode will be discussed.

D. Activators of RyRs

As discussed in the previous section, activators and inhibitors of RyRs should be discussed for three different modes of opening separately, at least as far as RyR1 is concerned. The analysis of effects of activators on PCR and DICR should be made under the condition that CICR is completely inhibited, otherwise calcium released by PCR or DICR might activate secondarily RyR channel further to open in CICR mode. In this situation the secondary CICR must be modulated by such a modulator of CICR that has no action on PCR and DICR, which might mislead the experimenter into a wrong conclusion that the modulator examined directly affects PCR or DICR. However, this kind of careful study has not been done for most of modulators and, therefore, reliable information about PCR and DICR is rather scanty. In the sections below, discussion of modulators are mainly confined to those of CICR and effects on PCR and DICR are mentioned only when information is available.

In addition, there is no guarantee that effects of modulators of CICR are common among RyR1, 2, and 3, and/or α- and β-RyR, even qualitatively. For example, dantrolene inhibits CICR of skeletal muscle under appropriate conditions but not that of cardiac muscle. However, since most CICR modulators appear to show similar actions on every kind of RyRs, mention will be made only when the differences among different kinds of RyRs are clearly known.

[2] Single channel studies on RyRs incorporated into lipid bilayer demonstrated the presence of different level of activities of CICR, and they are called L-(low activity), H-(high activity) and I-(inactivation) mode. These modes are of course different from those used in the present article and these two terminology should clearly be distinguished (Percival et al. 1994; Zahradnikova and Meszaros 1998)

The action of some activators of CICR are complex and in a different condition, e.g., at higher concentrations, their action on CICR turns into inhibition. For the sake of convenience, inhibitory actions of such activators will be included under the section on activators.

I. Calcium, Strontium, and Barium Ions

Calcium ion activates CICR at lower concentrations but inhibits at higher concentrations, thus showing a bell-shaped concentration dependence (ENDO 1981, 1985; MEISSNER 1994). This is true for all types of RyRs. Therefore, RyRs are thought to have two calcium-binding sites, a high affinity activating site and a low affinity inhibitory site. Calcium sensitivity of RyR1 for activation is higher than that of RyR3 (TAKESHIMA et al. 1995; MURAYAMA and OGAWA 1997). However, α-RyR and β-RyR of the frog were reported to have similar calcium sensitivities for CICR activation (MURAYAMA and OGAWA 1996). Mammalian skeletal and cardiac skinned fibers showed a similar calcium sensitivity for CICR activation (ENDO 1985), which is consistent with a recent determination of that of RyR2 compared with that of RYR1 (LIU et al. 1998), although it was once believed that RyR2 has a higher calcium sensitivity for CICR activation than RyR1. The affinity of calcium ion for inhibitory site is similar in RyR1 and RyR3 but lower in RyR2 (LIU et al. 1998).

Early skinned fiber experiments showed that at the lowest activating concentrations CICR activity was proportional to the square of calcium ion concentration, suggesting two calcium ions are necessary to activate the channel (ENDO 1985). This should be examined in the light of molecular structure.

Calcium ion cannot fully open RyR channels by itself. Even at an optimal concentration of calcium ion, calcium release rate of the SR or open probability of RyR channels remains at a low level and maximal response requires the presence of physiological and/or pharmacological potentiators such as ATP and/or caffeine.

Strontium ion is similar to calcium ion in that it can activate CICR at lower but inhibit it at higher concentrations. The activating concentrations of strontium ion is about one order of magnitude higher, and maximal level of activation achieved is much lower than calcium, while the inhibitory effect is exerted at a concentration range similar to calcium (ENDO 1981; HORIUTI 1986; MEISSNER 1994). Barium ion can stimulate CICR in the presence of other potentiators such as caffeine with a potency similar to that of strontium (ENDO 1981), but it was reported that barium ion by itself cannot activate CICR in the absence of other potentiators (NAGASAKI and KASAI 1984; ROUSSEAU et al. 1992).

In the presence of a high concentration of calcium chelators such as fura-2 (2–3 mmol/l), PCR is not inhibited but enhanced (BAYLOR and HOLLINGWORTH 1988; HOLLINGWORTH et al. 1992; PAPE et al. 1993). Thus, calcium ion is not required for PCR activation, but it only inhibits PCR. At a very high concentration of fura-2 (>6 mmol/l), apparent PCR is inhibited, which is interpreted either as a result of inhibition of secondary CICR or due to pharmacological effects of fura-2.

DICR can be evoked in the practical absence of calcium ion with a high concentration of a calcium chelator, EGTA (THORENS and ENDO 1975).

II. Adenine Compounds

The fact that ATP potentiates caffeine-induced calcium release was first shown by ENDO and KITAZAWA (1976) and it was then demonstrated that non- or much less hydrolyzable analogues of ATP, AMPPCP, or AMPCPP, and other adenine compounds, ADP, AMP, cAMP, adenosine, and adenine, also potentiate CICR (ENDO et al. 1981; KAKUTA 1984; ISHIZUKA and ENDO 1983; MEISSNER 1984). The potency of this action is in the sequence of ATP, AMPPCP>ADP>AMP>adenosine, adenine. These agents potentiate CICR without altering the calcium-dependence of CICR, which is in marked contrast with that of caffeine that strongly increased the calcium sensitivity of CICR. The extent of potentiation by ATP, i.e., increase in the maximum calcium release rate or maximum open probability at the optimal calcium concentration, is much greater than that of caffeine.

In the absence of magnesium ion, ATP strongly potentiates calcium release even in the practical absence of calcium ion (ISHIZUKA and ENDO 1983). This opening of RyR channels by ATP is still considered as an operation in the CICR mode, because inhibitors of CICR such as procaine inhibit this channel opening.

Adenine moiety is important for this potentiating action because all the adenine compounds have this action and because nucleotides other than adenine nucleotides such as ITP, UTP, CTP, and GTP have only a very weak action, if any (ENDO et al. 1981; KAKUTA 1984; MEISSNER 1984). A weak agonist such as adenine acts as an inhibitor of CICR in the presence of a strong agonist, ATP.

Quite different from CICR, ATP does not potentiate PCR. This was shown in cut fiber experiments. When ATP was removed so that CICR activity was strongly attenuated, PCR was practically unchanged (M. IINO, personal communication). This is consistent with the fact that adenine, a competitive inhibitor against ATP as a potentiator of CICR, did not inhibit PCR. Alternative interpretation of the latter result is that ATP and adenine have a potentiating action on PCR to about the same extent, but this seems very unlikely.

DICR is potentiated by ATP and other adenine compounds like CICR, but the magnitude of potentiation appears smaller than that of CICR (ENDO and KITAZAWA 1976; T. IKEMOTO, unpublished observation).

III. Caffeine and Related Compounds

Caffeine strongly potentiates CICR by increasing its calcium sensitivity and by increasing its maximal response at the optimum calcium ion concentration

(ENDO 1975, 1985). In this sense the mode of action of caffeine on CICR is different from that of ATP and other adenine compounds as described in the previous section. Potentiation by caffeine is exerted similarly in the absence and in the presence of adenine compounds, and vice versa, indicating that these two kinds of potentiators act on different sites (ENDO 1985; MEISSNER 1994). This action of caffeine is the mechanism of its well-known contracture-inducing action on skeletal muscle: caffeine enhances calcium sensitivity of CICR so much that the resting calcium ion concentration can induce calcium release from the SR.

Caffeine has been utilized as a tool in such a way that caffeine-induced calcium release is almost equivalent to CICR itself. This is reliable if one compares RyRs with IP_3-receptors, because IP_3-induced calcium release is *inhibited* by caffeine (HIROSE et al. 1993). However, if PCR or DICR is to be distinguished from CICR, caffeine may not be a good tool, since direct effects of caffeine on PCR or DICR is quite possible but has not so far been properly examined because it is very difficult to establish such direct effects due to the inevitable presence of secondary CICR.

Other xanthine derivatives, including theophylline and theobromine as well as xanthine itself, exert similar action on CICR with more or less similar potencies (ROUSSEAU et al. 1988; LIU and MEISNER 1997). The potency of xanthine derivatives on intact muscle, however, is quite different (caffeine> theophylline>theobromine>>xanthine) because of the different permeability of the muscle cell membrane to these compounds.

9-Methyl-7-bromoeudistomin D (MBED), a derivative of eudistomin D isolated from a marine tunicate, was shown to have a strong caffeine-like calcium releasing action with a potency 1000 times of that of caffeine (SEINO et al. 1991). [^{3}H]MBED was found to bind to terminal cisternae of skeletal muscle SR with a K_D of 40 nmol/l. The binding was competitively inhibited by caffeine with an IC_{50} value of 0.8 mmol/l, in accordance with its potency of CICR activation, indicating that MBED shares the same binding site with caffeine. It is interesting that [^{3}H]MBED binding was not affected by calcium and magnesium ions and ryanodine, although it was enhanced by AMPPCP and inhibited by procaine (FANG et al. 1993). Further studies along this line should be useful to elucidate the nature of the caffeine and MBED binding site.

IV. Ryanodine and Ryanoid

Ryanodine, a neutral plant alkaloid from *Ryana speciosa*, specifically binds to all kinds of RyRs with a dissociation constant of 2–7 nmol/l (FLEISCHER et al. 1985; PESSAH et al. 1985; LAI et al. 1989). The binding occurs only when the channel is open, and the alkaloid-bound channels are fixed in an open but low conducting state (ROUSSEAU et al. 1987; SMITH et al. 1988). By this effect ryanodine abolishes the ability of SR to store calcium ion in the lumen. The activators or inhibitors of RyRs potentiate or inhibit, respectively, the binding of ryanodine. Conversely, modulation of ryanodine binding by some agent indicates the modulation of RyR channel opening by the agent. Thus, calcium ion enhances the binding at micromolar range but inhibits at millimolar range. The effect of all the modulators of CICR is also in parallel with that on ryanodine binding (CHU et al. 1990; OYAMADA et al. 1993), indicating clearly that ryanodine binds RyR channels that are open in CICR mode. Ryanodine also appears to bind RyR open in PCR mode, because time required for ryanodine to induce contracture of frog skeletal muscle is much reduced if electrical stimulation to cause twitch is given in the condition that the secondary CICR must have been appreciably suppressed by adenine (M. ENDO et al., unpublished results).

As already mentioned, the stoichiometry of the specific binding of ryanodine is one ryanodine molecule per one channel, i.e., four RyR molecules. With the very high affinity of binding, ryanodine bound to the channel does not dissociate from the site within the time frame of usual experiments. TANNA et al. (1998) found that a synthetic ryanoid, 21-amino-9α hydroxy-ryanodine interacted with the RyR channel in a similar way to ryanodine except that the interaction was reversible. They also found that the ryanoid binding was strongly influenced by transmembrane voltage, probably because of voltage-driven conformational alteration.

At high concentrations, ryanodine occludes the RyR channel in contrast to its activating effect at a low concentration. RyR channel consisting of four monomers has four binding sites for ryanodine (one on each monomer) and when ryanodine molecules occupy an increasing number of sites, affinity of the remaining sites becomes increasingly lower (PESSAH et al. 1985; MCGREW et al. 1989). When all the four sites are occupied at a high concentration of the alkaloid, the channel is totally blocked (PESSAH and ZIMANYI 1991; CARROLL et al. 1991).

V. Halothane and Other Inhalation Anesthetics

Malignant hyperthermia (MH) is a human hereditary disorder in which susceptible individuals respond to halothane or other inhalation anesthetics with high fever and skeletal muscle contracture (DENBOROUGH and LOVELL 1960; LOKE and MACLENNAN 1998). Halothane shows a caffeine-like calcium releasing action due to potentiation of CICR (TAKAGI et al. 1976; ENDO et al. 1983). Other inhalation anesthetics also potentiate CICR (MATSUI and ENDO 1986). The calcium sensitivity of CICR in skeletal muscle of MH patients was shown to be higher than that of normal individuals (ENDO et al. 1983; KAWANA et al. 1992) so that potentiators of CICR can more easily induce contracture of skeletal muscle in MH patients, which essentially explains the disease. A similar disorder exists in swine with a similarly greater-than-normal calcium sensitivity of CICR (OHTA et al. 1989), and a point mutation on RyR1 of animals in disorder was found and proven to cause the greater calcium sensitivity of CICR (MACLENNAN and PHILLIPS 1992). Similar molecular biological studies in human disorder have been made and are still being continued. Many different and quite heterogeneous mutations appear to be involved in the human disorder (LOKE and MACLENNAN 1998).

VI. Oxidizing Agents and Doxorubicin

Several agents that oxidize reactive SH groups on RyRs, such as heavy metals like Ag^{+} (SALAMA and ABRAMSON 1984), thimerosal (ABRAMSON et al. 1995), 4,4'-dithiodipyridine (EAGER et al. 1997), *N*-ethylmaleimide, diamide (AGHDASI et al. 1997), H_2O_2 (OBA et al. 1998), and so on, were reported to activate or inhibit calcium release through RyR1 and RyR2 channels depending on which SH group is affected. The effects of these oxidizing agents are prevented or reversed by reducing agents, dithiothreitol, β-mercaptoethanol, or gluthathione. The oxidizing agent-activated channel is inhibited by magnesium ion and procaine and potentiated by ATP (SALAMA and ABRAMSON 1984; EAGER and DULHUNTY 1998), showing the properties of CICR mode of opening.

Doxorubicin, a widely used chemotherapeutic agent for malignant tumors, first activates and then inhibits CICR channel of skeletal and cardiac muscle (ABRAMSON et al. 1988; ONDRIAS et al. 1990). The inhibition can be prevented by dithiothreitol, indicating the involvement of SH groups. However, the activation, which is modulated by ATP, procaine, and other modulators of CICR as in the case of caffeine, is not prevented by the reducing agent, which suggests a molecular mechanism of action for activation different from that with involvement of SH group. These effects of doxorubicin on RyR2 probably contribute to the cardiotoxicity, a well-known clinical side-effect of this agent.

VII. Cyclic ADP-Ribose

A cyclic nucleotide derived from NAD, cyclic ADP-ribose, was claimed to enhance CICR in sea urchin eggs (LEE 1991; GALIONE et al. 1991). In higher animals the activating effect of cyclic ADP-ribose on CICR has been reported in nonmuscle cells such as pancreatic β cell (TAKASAWA et al. 1993) and sympathetic neurons (HUA et al. 1994). While the effect of cyclic ADP-ribose on RyR2 is controversial [activation (MESZAROS et al. 1993), modulation (CUI et al. 1999), or no effect (OGAWA 1999)], the effect on RyR1 appears null (MESZAROS et al. 1993). Although there is a report of an activation of skeletal muscle SR calcium release channel by cyclic ADP-ribose (SITSAPESAN and WILLIAMS 1995), it could be interpreted as the effect on coexisting RyR3 (SONNLEITNER et al. 1998).

VIII. Calmodulin and Other Endogenous Modulatory Proteins

Calmodulin (CaM) potentiates or inhibits CICR depending on experimental conditions (MEISSNER 1986; SMITH et al. 1989; IKEMOTO et al. 1995; TRIPATHY et al. 1995). Potentiation and inhibition tend to dominate at lower and higher calcium ion concentrations respectively. While potentiation is the main effect of CaM in RyR1, potentiation in RyR3 at low calcium is very weak and inhibition at high calcium strong (IKEMOTO et al. 1998). Inhibition by CaM on RyR2 was also reported (MEISSNER and HENDERSON 1987; SMITH et al. 1989). Greater concentrations of CaM are required for potentiation than for inhibition, suggesting multiple CaM-binding sites on RyRs. These effects of CaM are not mediated by phosphorylation of RyRs because the effects are obtained in the absence of ATP and also because they are not affected by CaM antagonists (IKEMOTO et al. 1996), although this is not to deny the possible presence of modulatory influence of phosphorylation.

Several other proteins, such as FK506-binding protein (cf. MARKS 1996), triadin (FAN et al. 1995), calsequestrin (IKEMOTO et al. 1989) and so on, were reported to have implications in the regulation of RyR1 channels in skeletal muscle. Therefore, in considering mechanisms of actions of activators and inhibitors of physiological RyR channels, their effects on these proteins should be kept in mind as a possibility.

IX. Imperatoxin Activator

Imperatoxin activator ($IpTx_a$), a 33 amino acid peptide isolated from the venom of the African scorpion *Pandinus imperator*, activated RyR1 channels incorporated into lipid bilayers, and enhanced calcium-dependent [^{3}H]ryanodine binding of RyR1, but exerted little or no effect on RyR2 (VALDIVIA et al. 1991; EL-HAYEK et al. 1995). GURROLA et al. (1999) noticed that $IpTx_a$ has a sequence similar to a segment of II–III loop of α subunit of DHPR, a putative site for interaction with RyR1 in E–C coupling of skeletal muscle. Although modulators of CICR similarly modulate $IpTx_a$-stimulated ryanodine binding, at present it is not very certain that activation of RyR1 by $IpTx_a$ is in CICR mode because the modulators used are mostly not very specific to CICR. Especially in the case of AMPPCP, a strong potentiator of CICR, the effect on $IpTx_a$-induced ryanodine binding is very weak (EL-HAYEK et al. 1995). Further studies are awaited on the problem as to whether $IpTx_a$-activated channel is in PCR mode, as it should be if the toxin indeed closely mimics the physiological activation.

X. Clofibric Acid

Clofibric acid, a hypolipidemic-related agent, is reported to have an ability to increase open probability of SR calcium release channel incorporated into lipid bilayers (SUKHAREVA et al. 1994). When applied to the SR of skinned skeletal muscle fibers, it causes strong calcium release in the complete absence of calcium ion, although a rather high concentration such as 10 mmol/l is required. This calcium release is considered to be due to an opening of RyR, because the effect of ryanodine to fix RyR channel in an open state is very much enhanced under the condition. Unlike CICR activation, the clofibric acid-induced calcium release is not inhibited by procaine and inhibited rather than potentiated by AMP. In addition, it is inhibited by twitch inhibitors of dantrolene derivatives with no inhibitory action on CICR, suggesting that clofibric acid might be the first agent to activate RyR channel in the PCR mode (T. Ikemoto et al., in preparation). Further studies are awaited.

XI. Miscellaneous Activators

4-Chloro-*m*-cresol is a caffeine-like, more potent potentiator of CICR, effective in micromolar concentrations (Zorzato et al. 1993; Herrmann-Frank et al. 1996). Amentoflavone was recently reported to be a similar CICR potentiator (Suzuki et al. 1999).

Some anion transport inhibitors, 4,4′-diisothiocyanostilbene-2,2′-disulfonic acid (DIDS) and 4-acetoamido-4′-isothiocyanostilbene-2,2′-disulfonic acid (SITS) were reported to open RyR channel operating in CICR mode and lock it almost irreversibly at open state (Kawasaki and Kasai 1989; Zahradnikova and Zahradnik 1993). Another anion transport inhibitor, niflumic acid, was also reported to increase open probability of RyR channels (Oba et al. 1996). DIDS binds not the calcium release channel but the 30-kDa protein (Yamaguchi et al. 1995) to which another activator, myotoxin, a polypeptide toxin isolated from the venom of prairie rattlesnake *Crotalus viridis viridis*, also binds (Hirata et al. 1999).

E. Inhibitors of RyRs

I. Magnesium Ion

Magnesium ion exerts an inhibitory effect on CICR. In the presence of magnesium ion, calcium-dependence of CICR is shifted to a higher concentration range, indicating that magnesium competes with calcium for the high affinity activating calcium site (Endo 1985; Meissner 1994). However, the inhibitory effect of magnesium does not diminish even at a very high calcium concentration as a simple competitive antagonism predicts. This is because magnesium ion also acts on the low affinity inhibitory calcium site as an agonist (Meissner 1994).

Magnesium ion appears to inhibit PCR as well: In skinned fibers of the toad and the rat, T-tubule depolarization is reported to cause calcium release from the SR in the low but not high concentrations of magnesium ion (Lamb and Stephenson 1991, 1994). DICR is reported to be not inhibited in an amphibian skeletal muscle (Thorens and Endo 1975), but inhibited in mammalian skeletal muscle at least at a high concentration (T. Ikemoto, unpublished results).

II. Procaine and Other Local Anesthetics

Procaine inhibits CICR, but unlike magnesium ion it does not change the dependence of CICR on calcium ion concentration appreciably (Ford and Podolsky 1972; Thorens and Endo 1975; Endo 1985). This must be the mechanism of long known procaine effect to antagonize the contracture-inducing action of caffeine. Tetracaine also inhibits CICR with a higher potency than procaine (Ohnishi 1979). While procaine does not inhibit PCR at all, tetracaine inhibits it (Almers and Best 1976).

Some other local anesthetics such as cocaine and lidocaine inhibit neither caffeine contracture nor CICR (Sakai 1965; Bianchi and Bolton 1967).

III. Ruthenium Red

Ruthenium red completely blocks CICR (Ohnishi 1979; Miyamoto and Racker 1981; Smith et al. 1986) and is often used to check whether the calcium transport is through CICR. This certainly distinguishes CICR from IP_3-induced calcium release. However, other modes of opening of RyR1, PCR, and DICR, may also be inhibited.

IV. Dantrolene

Dantrolene inhibits CICR of RyR1 at 37 °C, but the inhibition is very weak or negligible at 20 °C. In the case of frog muscle, a similar tendency exists: dantrolene inhibits caffeine-induced calcium release to a small but definite extent at 15 °C, but not at all at 0 °C (Yagi and Endo 1976). Since dantrolene inhibits twitch contraction to the same extent at any temperature, the drug can somehow distinguish CICR and PCR at low temperature. Further study on the mechanism of this difference in inhibitory action might give some clue for the understanding of the nature of different modes of opening of RyR1. A search for more specific inhibitors that distinguish between CICR and PCR utilizing derivatives of dantrolene is now being conducted in our laboratory.

Dantrolene appears to distinguish between RyR1 and RyR2, because cardiac contraction that is thought to be mediated by CICR of RyR2 is not at all inhibited even at 37 °C (Ellis et al. 1976).

DICR of amphibian skeletal muscle was not inhibited by dantrolene (Yagi and Endo 1976), but that of mammalian muscle was inhibited (T. Ikemoto, unpublished results).

F. Closing Remarks

A comparison between activators and inhibitors of RyRs and those of IP_3 receptors is tabulated in Chap. 24.

Important similarities and differences among activators and inhibitors of CICR and those of PCR or of DICR so far known are summarized in

Table 2. Comparison of activators and inhibitors of RyRl among three different modes of opening

Mode of opening	Stimulus to cause Ca^{2+} release	Ca^{2+} is required	Activators			Inhibitors		
			Caffeine	Adenine compounds	Ryanodine	Mg^{2+}	Procaine	Dantrolene
PCR	Depolarization of T-tubule	–	?	–	+?	+	–	+
CICR	Increase in cytoplasmic Ca^{2+} concentration	+	++	++	++	+	+	37 °C+ 20 °C–
DICR	"Depolarization" of SR membrane	–	?	+	?	±	–	+

++ and +: positive effect depending on magnitudes of the action.
+?: probably positive but not certain.
–: no effect.
?: unknown.
For the details, see text.

Table 2. At present, the fact that RyR1 operates in multiple (at least more than one) modes of opening is not well recognized. Activators and inhibitors more specific to any one mode will be very useful for the study of physiological functions, and development of such specific agents is awaited.

Also, activators and inhibitors specific to any one type of RyRs are undoubtedly useful, and in this case information from molecular biological studies must fully be utilized. When one thinks about physiological functions of RyRs, associated proteins briefly described in Sect. D.VIII must also be taken into consideration.

Although much information has been accumulated recently, we still know only a very little at present. Further development is anticipated.

References

Abramson JJ, Buck E, Salama G, Casida JE, Pessah IN (1988) Mechanism of anthraquinone-induced calcium release from skeletal muscle sarcoplasmic reticulum. J Biol Chem 263:18750–18758

Abramson JJ, Zable AC, Favero TG, Salama G (1995) Thimerosal interacts with the Ca^{2+} release channel ryanodine receptor from skeletal muscle sarcoplasmic reticulum. J Biol Chem 270:29644–29647

Aghdasi B, Zhang JZ, Wu Y, Reid MB, Hamilton SL (1997) Multiple classes of sulfhydryls modulate the skeletal muscle Ca^{2+} release channel. J Biol Chem 272:3739–3748

Airey JA, Beck CF, Murakami K, Tanksley SJ, Deerinck TJ, Ellisman MH, Sutko JL (1990) Identification and localization of two triad junctional foot protein isoforms in mature avian fast twitch skeletal muscle. J Biol Chem 265:14187–14194

Almers W, Best PM (1976) Effects of tetracaine on displacement currents and contraction of frog skeletal muscle. J Physiol 262:583–611

Anderson K, Lai FA, Liu Q-Y, Rousseau E, Erickson HP, Meissner G (1989) Structural and functional characterization of purified cardiac ryanodine receptor-Ca^{2+} release channel complex. J Biol Chem 264:1329–1335

Baylor SM, Hollingworth S (1988) Fura-2 calcium transients in frog skeletal muscle fibres. J Physiol 403:151–192

Bianch CP, Bolton TC (1967) Action of local anesthetics on coupling system in muscle. J Pharmacol Exp Ther 157:388–405

Block BA, Imagawa T, Campbell KP, Franzini-Armstrong C (1988) Structural evidence for direct interaction between the molecular components of the transverse tubule/sarcoplasmic reticulum junction in skeletal muscle. J Cell Biol 107:2587–2600

Carroll S, Skarmeta JG, Yu X, Collins KD, Inesi G (1991) Interdependence of ryanodine binding, oligomeric receptor interactions, and Ca^{2+} release regulation in junctional sarcoplasmic reticulum. Arch Biochem Biophys 290:239–247

Chu A, Diaz-Munoz M, Hawkes MJ, Brush K, Hamilton SL (1990) Ryanodine as a probe for the functional state of the skeletal muscle sarcoplasmic reticulum calcium release channel. Mol Pharmacol 37:735–741

CuiY, Galione A, Terrar DA (1999) Effects of photoreleased cADP-ribose on calcium transients and calcium sparks in myocytes isolated from guinea-pig and rat ventricle. Biochem J 342:269–73

Denborough MA, Lovell RRH (1960) Anaesthetic death in a family. Lancet II, 45

Eager KR, Dulhunty AF (1998) Activation of the cardiac ryanodine receptor by sulfhydryl oxidation is modified by Mg^{2+} and ATP. J Membr Biol 163:9–18

Eager KR, Roden LD, Dulhunty AF (1997) Actions of sulfhydryl reagents on single ryanodine receptor Ca^{2+}-release channels from sheep myocardium. Am J Physiol 272:C1908–C1918

Ebashi S, Endo M (1968) Ca ion and muscle contraction. Prog Biophys Mol Biol 18: 123–183

El-Hayek R, Lokuta AJ, Arevalo C, Valdivia HH (1995) Peptide probe of ryanodine receptor function. Imperatoxin A, a peptide from the venom of the scorpion *Pandinus imperator*, selectively activates skeletal-type ryanodine receptor isoforms. J Biol Chem 270:28696–28704

Ellis KO, Bryant SH (1972) Excitation-contraction uncoupling in skeletal muscle by dantrolene sodium. Naunyn-Schmiedeberg's Arch Pharmacol 274:107–109

Ellis KO, Carpenter JF (1972) Studies on the mechanism of action of dantrolene sodium. A skeletal muscle relaxant. Naunyn-Schmiedeberg's Arch Pharmacol 275:83–94

EllisK O, Butterfield JL, Wessels FL, Carpenter JF (1976). A comparison of skeletal, cardiac, and smooth muscle actions of dantrolene sodium -a skeletal muscle relaxant. Arch Int Pharmacodyn Ther 224:118–132

Endo M (1975) Mechanism of action of caffeine on the sarcoplasmic reticulum of skeletal muscle. Proc Jpn Acad 51:479–484

Endo M (1981) Mechanism of calcium-induced calcium release in the SR membrane. In The mechanism of Gated Calcium Transport across Biological Membranes, ed. Ohnishi ST, Endo M, pp 257–264. Academic Press, New York

Endo M (1985) Ca^{2+} release from sarcoplasmic reticulum. Curr Top Membr Transp 25:181–230

Endo M, Kitazawa T (1976) The effect of ATP on calcium release mechanisms in the sarcoplasmic reticulum of skinned muscle fibers. Proc Jpn Acad, 52:599–602

Endo M, Nakajima Y (1973) Release of Calcium induced by "depolarisation" of the sarcoplasmic reticulum membrane. Nature New Biol, 246:216–218

Endo M, Kakuta Y, Kitazawa T (1981) A further study of the Ca-induced Ca release mechanism. In The Regulation of Muscle Contraction: Excitation-Contraction Coupling, ed. Grinnell A D, Brazier MA, pp 181–193. Academic Press, New York

Endo M, Tanaka M, Ogawa Y (1970) Calcium-induced release of calcium from the sarcoplasmic reticulum of skinned skeletal muscle fibres. Nature 228:34–36

Endo M, Yagi S, Ishizuka T, Horiuti K, KogaY, Amaha K (1983) Changes in the Ca-induced Ca release mechanism in the sarcoplasmic reticulum of muscle from patient with malignant hyperthermia. Biomed Res 4:83–92

Fabiato A, Fabiato F (1977) Calcium release from the sarcoplasmic reticulum. Circ Res 40:19–29

Fan H, Brandt NR, Caswell AH (1995) Disulfide bonds, N-glycosylation and transmembrane topology of skeletal muscle triadin. Biochemistry 34:14902–14908

Fang YI, Adachi M, Kobayashi J, Ohizumi Y (1993) High affinity binding of 9-[3H]methyl-7-bromoeudistomin D to the caffeine-binding site of skeletal muscle sarcoplasmic reticulum. J Biol Chem 268:18622–18625

Fleischer S, Ogunbunmi EA, Dixon MC, Fleer EAM (1985) Localization of Ca^{2+} release channels with ryanodine in junctional terminal cisternae of sarcoplasmic reticulum of fast skeletal muscle. Proc Natl Acad Sci USA 82:7256–7259

Flucher BE, Conti A, Takeshima H, Sorrentino V (1999) Type 3 and type 1 ryanodine receptor are localized in triads of the same mammalian skeletal muscles. J Cell Biol 146:621–629

Ford LE, Podolsky RJ (1970) Regenerative Ca release within muscle cells. Science 167:58–59

Ford LE, Podolsky RJ (1972) Calcium uptake and force development by skinned muscle fibres in EGTA buffered solutions. J Physiol 223:1–19

Galione A, Lee HC, Busa WB (1991) Ca^{2+}-induced Ca^{2+} release in sea urchin egg homogenates: modulation by cyclic ADP-ribose. Science 253:1143–1146

Gurrola GB, Arevalo C, Sreekumar R, Lokuta AJ, Walker JW, Valdivia HH (1999). Activation of ryanodine receptors by imperatoxin A and a peptide segment of the II-III loop of the dihydropyridine receptor. J Biol Chem 274:7879–7886

Herrmann-Frank A, Richter M, Sarkozi S, Mohr U, Lehmann-Horn F (1996) 4-Chloro-*m*-cresol, a potent and specific activator of the skeletal muscle ryanodine receptor. Biochim Biophys Acta 1289:31–40

HirataY, Nakahata N, Ohkura M, Ohizumi Y (1999) Identification of 30kDa protein for Ca^{2+} releasing action of myotoxin a with a mechanism common to DIDS in skeletal muscle sarcoplasmic reticulum. Biochim Biophys Acta 1451: 132–140

Hirose K, Iino M, Endo M (1993) Caffeine inhibits Ca^{2+}-mediated potentiation of inositol 1,4,5-trisphosphate-induced Ca^{2+} release in permeabilized vascular smooth muscle cells. Biochem Biophys Res Commun 94:726–732,

Hollingworth S, Harkins AB, Kurebayashi N, Konishi M, Baylor SM (1992) Excitation-contraction coupling in intact frog skeletal muscle fibers injected with mmolar concentrations of fura-2. Biophys J 63:224–234

Horiuti K (1986) Some properties of the contractile system and sarcoplasmic reticulum of skinned slow fibres from *Xenopus* muscle. J Physiol 373:1–23

Hua SY, Tokimasa T, Takasawa S, Furuya Y, Nohmi M, Okamoto H, Kuba K (1994) Cyclic ADP-ribose modulates Ca^{2+} release channels for activation by physiological Ca^{2+} entry in bullfrog sympathetic neurons. Neuron 12:1073–1079

Ikemoto N, Ronjat M, Meszaros LG, Koshita M (1989) Postulated role of calsequestrin in the regulation of calcium release from sarcoplasmic reticulum. Biochemistry 28:6764–6771

Ikemoto T, Iino M, Endo M (1995) Enhancing effect of calmodulin on Ca^{2+}-induced Ca^{2+} release in the sarcoplasmic reticulum of skeletal muscle fibres. J Physiol 487: 573–582

Ikemoto T, Iino M, Endo M (1996) Effect of calmodulin antagonists on calmodulin-induced biphasic modulation of Ca^{2+}-induced Ca^{2+} release. Br J Pharmacol 118: 690–694

Ikemoto T, Komazaki S, Takeshima H, Nishi M, Noda T, Iino M, Endo M (1997) Functional and morphological features of skeletal muscle from mutant mice lacking both ryanodine receptors type 1 and Type 3. J Physiol 501:305–312

Ikemoto T, Takeshima H, Iino M, Endo M (1998) Effect of calmodulin on Ca^{2+}-induced Ca^{2+} release of skeletal muscle from mutant mice expressing either ryanodine receptor type 1 or type 3. Pflügers Archiv 437:43–48

Inui M, Saito A, Fleischer S (1987) Purification of ryanodine receptor and identity with feet structures of junctional terminal cisternae of sarcoplasmic reticulum from fast skeletal muscle. J Biol Chem 262:1740–1747

Ishizuka T, Endo M (1983) Effects of adenine on skinned fibers of amphibian fast skeletal muscle. Proc Jpn Acad 59:93–96

Ishizuka T, Iijima T, Endo M (1983) Effects of adenine on twitch and other contractile response of single fibers of amphibian fast skeletal muscle. Proc Jpn Acad 59:97–100

Jeyaknmar LH, Copello JA, O'Malley AM, Grassucci R, Wagenknecht T, Fleischer S (1997) Purification and characterization of ryanodine receptor 3 from mammalian tissue. J Biol Chem 273:16011–16020

Kakuta Y (1984) Effects of ATP and related compounds on the Ca-induced Ca release mechanism of the *Xenopus* SR. Pflügers Archiv 400:72–79

Kawasaki T, Kasai M (1989) Disulfonic stilbene derivatives open the Ca^{2+} release channel of sarcoplasmic reticulum. J Biochem 106:401–405

Kawana Y, Iino M, Horiuti K, Matsumura N, Ohta T, Matsui K, Endo M (1992) Acceleration in calcium-induced calcium release in the biopsied muscle fibers from patients with malignant hyperthermia. Biomed Res 13:287–297

Kobayashi T, Endo M (1986) Temperature-dependent inhibition of caffeine contracture of mammalian skeletal muscle by dantrolene. Proc Jpn Acad 62:329–332

Lai FA, Erickson HP, Rousseau E, Liu Q-Y, Meissner G (1988) Purification and reconstitution of calcium release channel from skeletal muscle. Nature 331:315–319

Lai FA, Misra M, Xu L, Smith HA, Meissner G (1989) The ryanodine receptor-Ca^{2+} release channel complex of skeletal muscle sarcoplasmic reticulum. Evidence for a cooperatively coupled, negatively charged homotetramer. J Biol Chem 264: 16776–16785

Lamb GD, Stephenson DG (1991) Effect of Mg^{2+} on the control of Ca^{2+} release in skeletal muscle fibres of the toad. J Physiol 434:507–528

Lamb GD, Stephenson DG (1994) Effects of intracellular pH and [Mg^{2+}] on excitation-contraction coupling in skeletal muscle fibres of the rat. J Physiol 478:331–339

Lee HC (1991) Specific binding of cyclic ADP-ribose to calcium-storing microsomes from sea urchin eggs. J Biol Chem 266:2276–2281

Liu W, Meissner G (1997) Structure-activity relationship of xanthines and skeletal muscle ryanodine receptor/Ca^{2+} release channel. Pharmacology 54:135–143

Liu W, Pasek DA, Meissner G (1998) Modulation of Ca^{2+}-gated cardiac muscle Ca^{2+}-release channel (ryanodine receptor) by mono- and divalent ions. Am J Physiol 274:C120–C128

Loke J, MacLennan DH (1998) Malignant hyperthermia and central core disease: disorders of Ca^{2+} release channels. Am J Med 104:470–486

MacLennan DH, Phillips MS (1992) Malignant hyperthermia. Science 256:789–794

Marks AR (1996) Cellular functions of immunophilins. Physiol Rev 76:631–649

Matsui K, Endo M (1986) Effect of inhalation anesthetics on the rate of Ca release from the sarcoplasmic reticulum of skeletal muscle in the guinea pig. Jpn J Pharmacol 40:245P

McGrew SG, Wolleben C, Siegl P, Inui M, Fleischer S (1989) Positive cooperativity of ryanodine binding to the calcium release channel of sarcoplasmic reticulum from heart and skeletal muscle. Biochemistry 28:1686–1691

Meissner G (1984) Adenine nucleotide stimulation of Ca^{2+}-induced Ca^{2+} release in sarcoplasmic reticulum. J Biol Chem 259:2365–2374

Meissner G (1986) Evidence of a role for calmodulin in the regulation of calcium release from skeletal muscle sarcoplasmic reticulum. Biochemistry 25:244–251

Meissner G (1994) Ryanodine receptor/Ca^{2+} release channels and their regulation by endogenous effector. Annu Rev Physiol 56:485–508

Meissner G, Henderson JS (1987) Rapid calcium release from cardiac sarcoplasmic reticulum vesicles is dependent on Ca^{2+} and is modulated by Mg^{2+}, adenine nucleotide, and calmodulin. J Biol Chem 262:3065–3073

Meszaros LG, Bak J, Chu A (1993) Cyclic ADP-ribose as an endogenous regulator of the non-skeletal type ryanodine receptor Ca^{2+} channel. Nature 364:76–79
MiyamotoH, Racker E (1981) Calcium-induced calcium release at terminal cisternae of skeletal sarcoplasmic reticulum. FEBS Lett 133:235–238
Murayama T, Ogawa Y (1992) Purification and characterization of two ryanodine-binding protein isoforms from sarcoplasmic reticulum of bullfrog skeletal muscle. J Biochem 112:514–522
Murayama T, Ogawa Y (1996) Similar Ca^{2+} dependences of [^{3}H]ryanodine binding to alpha- and beta-ryanodine receptors purified from bullfrog skeletal muscle in an isotonic medium. FEBS Lett 380:267–271
Murayama T, Ogawa Y (1997) Characterization of type 3 ryanodine receptor (RyR3) of sarcoplasmic reticulum from rabbit skeletal muscles. J Biol Chem 272: 24030–24037
Nagasaki K, Kasai M (1984) Channel selectivity and gating specificity of calcium-induced calcium release channel in isolated sarcoplasmic reticulum. J Biochem 96: 1769–1775
Oba T, Ishikawa T, Yamaguchi M (1998) Sulfhydryls associated with H_2O_2-induced channel activation are on luminal side of ryanodine receptors. Am J Physiol 274: C914–C921
Oba T, Koshita M, Van Helden DF (1996) Modulation of frog skeletal muscle Ca^{2+} release channel gating by anion channel blockers. Am J Physiol 271: C819–C824
Ogawa Y (1999) Ryanodine receptor isoforms as putative sites for cyclic ADP-ribose. Jpn J Pharmacol 79:10P
Ohnishi ST (1979) Interaction of metallochromic indicators with calcium sequestering organelles. Biochim Biophys Acta 585:15–19
Ohta T, Endo M (1986) Inhibition of calcium-induced calcium release by dantrolene at mammalian body temperature. Proc Jpn Acad 64:76–79
OhtaT, Endo M, Nakano T, Morohoshi Y, Wanikawa K, Ohga A (1989) Ca-induced Ca release in malignant hyperthermia-susceptible pig skeletal muscle. Am J Physiol 256:C358–C367
Ondrias K, Borgatta L, Kim DH, Ehrlich BE (1990) Biphasic effects of doxorubicin on the calcium release channel from sarcoplasmic reticulum of cardiac muscle Circ Res 67:1167–1174
Ottini L, Marziali G, Conti A, Charlesworth A, Sorrentino V (1996) α and β isoforms of ryanodine receptor from chicken skeletal muscle are the homologues of mammalian RyR1 and RyR3. Biochem J 315:207–216
Oyamada H, Iino M, Endo M (1993) Effects of ryanodine on the properties of Ca^{2+} release from sarcoplasmic reticulum in skinned skeletal muscle fibres of the frog. J Physiology 470:335–348
Oyamada H, Murayama T, Takagi T, Iino M, Iwabe N, Miyata T, Ogawa Y, Endo M (1994) Primary structure and distribution of ryanodine-binding protein isoforms of bullfrog skeletal muscle. J Biol Chem 269:17206–17214
Pape PC, Jong DS, Chandler WK, Baylor SM (1993) Effect of fura-2 on action potential-stimulated calcium release in cut twitch fibers from frog muscle. J Gen Physiol 102:295–332
Percival AL, Williams AJ, Kenyon JL, Grinsell MM, Airey JA, Sutko JL (1994) Chicken skeletal muscle ryanodine receptor isoforms: ion channel properties. Biophys J 67:1834–1850
Pessah IN, Zimanyi I (1991) Characterization of multiple [^{3}H]ryanodine binding sites on the Ca^{2+} release channel of sarcoplasmic reticulum from skeletal and cardiac muscle: evidence for a sequential mechanism in ryanodine action. Mol Pharmacol 39:679–689
Pessah IN, Stambuk RA, Casida JE (1987) Ca^{2+}-activated ryanodine binding: mechanisms of sensitivity and intensity modulation by Mg^{2+}, caffeine, and adenine nucleotides. Mol Pharmacol 31:232–238

Pessah IN, Waterhouse AL, Casida JE (1985) The calcium-ryanodine receptor complex of skeletal and cardiac muscle. Biochem Biophys Res Commun 128:449–456

Rousseau E, Ladine J, Liu QY, Meissner G (1988) Activation of the Ca^{2+} release channel of skeletal muscle sarcoplasmic reticulum by caffeine and related compounds. Arch Biochem Biophys 267:75–86

Rousseau E, Pinkos J, Savaria D (1992) Functional sensitivity of the native skeletal muscle Ca^{2+} release channel to divalent cations and Mg-ATP complex. Can J Physiol Pharmacol 70:394–402

Rousseau E, Smith JS, Meissner G (1987) Ryanodine modifies conductance and gating behavior of single Ca^{2+} release channel. Am J Physiol 53:C364–C368

Saito A, Inui M, Radermacher M, Frank J, Fleischer S (1988) Ultrastructure of calcium release channel of sarcoplasmic reticulum. J Cell Biol 107:211–219

Sakai T (1965) The effect of temperature and caffeine on activation of the contractile mechanism in the striated muscle fibres. Jikeikai Med J 12:88–102

Salama G, Abramson J (1984) Silver ions trigger Ca^{2+} release by acting at the apparent physiological release site in sarcoplasmic reticulum. J Biol Chem 259: 13363–13369

Seino A, Kobayashi M, Kobayashi J, Fang YI, Ishibashi M, Nakamura H, Momose K, Ohizumi Y (1991) 9-Methyl-7-bromoeudistomin D, a powerful radio-labelable Ca^{++} releaser having caffeine-like properties, acts on Ca^{++}-induced Ca^{++} release channels of sarcoplasmic reticulum. J Pharmacol Exp Ther 256:861–867

Sitsapesan R, Williams AJ (1995) Cyclic ADP-ribose and related compounds activate sheep skeletal sarcoplasmic reticulum Ca^{2+} release channel. Am J Physiol 268: C1235–C1240

Smith JS, Coronado R, Meissner G (1986) Single channel measurements of the calcium release channel from skeletal muscle sarcoplasmic reticulum: activation by Ca^{2+} and ATP and modulation by Mg^{2+}. J Gen Physiol 88:573–588

Smith JS, Imagawa T, Ma J, Fill M, Campbell K P, Coronado R (1988) Purified ryanodine receptor from rabbit skeletal muscle is the calcium-release channel of sarcoplasmic reticulum. J Gen Physiol 92:1–26

Smith JS, Rousseau E, Meissner G (1989) Calmodulin modulation of single sarcoplasmic reticulum Ca^{2+}-release channels from cardiac and skeletal muscle. Circ Res 64:352–359

Sonnleitner A, Conti A, Bertocchini F, Schindler H, Sorrentino V (1998) Functional properties of the ryanodine receptor type 3 (RyR3) Ca^{2+} release channel. EMBO J 17:2790–2798

Sukhareva M, Morrissette J, Coronado R (1994) Mechanism of chloride-dependent release of Ca^{2+} in the sarcoplasmic reticulum of rabbit skeletal muscle. Biophys J 67:751–765

Sutko JL, Airey JA (1996) Ryanodine receptor Ca^{2+} release channels; Does diversity in form equal diversity in function? Physiol. Rev 76:1027–1071

Suzuki A, Matsunaga K, Mimaki Y, Sashida Y, Ohizumi Y (1999) Properties of amentoflavone, a potent caffeine-like Ca^{2+} releaser in skeletal muscle sarcoplasmic reticulum. Eur J Pharmacol 372:97–102

Takagi A, Sugita H, Toyokura Y, Endo M (1976) Malignant hyperpyrexia; effect of halothane on single skinned muscle fibers. Proc Jpn Acad 52:603–606

Takasawa S, Nata K, Yonekura H, Okamoto H (1993) Cyclic ADP-ribose in insulin secretion from pancreatic β cells. Science 259:370–373

Takekura H, Nishi M, Noda T, Takeshima H, Franzini-Armstrong C (1995) Abnormal junctions between surface membrane and sarcoplasmic reticulum in skeletal muscle with a mutation targeted to the ryanodine receptor. Proc Natl Acad Sci USA 92:3381–3385

Takeshima H (1993) Primary structure and expression from cDNAs of the ryanodine receptor. Ann N Y Acad Sci 707:165–177

Takeshima H, Iino M, Takekura H, Nishi M, Kuno J, Minowa O, Takano H, Noda T (1994) Excitation-contraction uncoupling and muscular degeneration in mice lacking functional skeletal muscle ryanodine receptor gene. Nature 369:556–559

Takeshima H, Ikemoto T, Nishi M, Nishiyama N, Shimuta M, Kuno J, Saito I, Saito H, Endo M, Iino M, Noda T (1996) Generation and characterization of mutant mice lacking ryanodine receptor type 3. J Biol Chem 271:19649–19652

Takeshima H, Komazaki S, Hirose K, Nishi M, NodaT, Iino M (1998) Embryonic lethality and abnormal cardiac myocytes in mice lacking ryanodine receptor type 2. EMBO J 17:3309–3316

Takeshima H, Nishimura S, Matsumoto T, Ishida H, Kangawa K, Minamino N, Matsuo H, Ueda M, Hanaoka M, HiroseI T, Numa S (1989) Primary structure and expression from complementary DNA of skeletal muscle ryanodine receptor. Nature 339:439–445

Takeshima H, Yamazawa T, Ikemoto T, Takekura H, Nishi M, Noda T, Iino, M (1995) Ca^{2+}-induced Ca^{2+} release in myocytes from dyspedic mice lacking type-1 ryanodine receptor. EMBO J 14:2999–3006

Tanna B, Welch W, Ruest L, Sutko JL, Williams AJ (1998) Interactions of a reversible ryanoid (21-amino-9α-hydroxy-ryanodine) with single sheep cardiac ryanodine receptor channels. J Gen Physiol 112:55–69

Thorens S, Endo M (1975) Calcium-induced calcium release and "depolarization"-induced calcium release: Their physiological significance. Proc Jpn Acad 51:473–478

Tripathy A, Xu L, Mann G, Meissner G (1995) Calmodulin activation and inhibition of skeletal muscle Ca^{2+} release channel (ryanodine receptor). Biophys J 69:106–119

Valdivia HH, Fuentes O, El-Hayek R, Morrissette J, Coronado R (1991) Activation of the ryanodine receptor Ca^{2+} release channel of sarcoplasmic reticulum by a novel scorpion venom. J Biol Chem 266:19135–19138

Yagi S, Endo M (1976) Effect of dantrolene on excitation-contraction coupling of skeletal muscle. Jpn J Pharmacol 26:164P

Yamaguchi N, Kawasaki T, Kasai M (1995) DIDS binding 30-kDa protein regulates the calcium release channel in the sarcoplasmic reticulum. Biochem Biophys Res Commun 210:648–653

Yamazawa T, Takeshima H, Sakurai T, Endo M, Iino M (1996) Subtype specificity of the ryanodine receptor for Ca^{2+} signal amplification in excitation-contraction coupling. EMBO J 15:6172–6177

Zahradnikova A, Meszaros LG (1998) Voltage change-induced gating transitions of the rabbit skeletal muscle Ca^{2+} release channel. J Physiol 509:29–38

Zahradnikova A, Zahradnik I (1993) Modification of cardiac Ca^{2+} release channel gating by DIDS. Pflügers Archiv 425:555–557

Zorzato F, Scutari E, Tegazzin V, Clementi E, Treves S (1993) Chlorocresol: an activator of ryanodine receptor-mediated Ca^{2+} release. Mol Pharmacol 44:1192–1201

CHAPTER 24

Regulation of IP_3 Receptor Ca^{2+} Release Channels

M. IINO

A. Introduction

Many types of cell surface receptors transmit signals into the cell using inositol 1,4,5-trisphosphate (IP_3) as the intracellular messenger (BERRIDGE 1993). Upon binding of agonists to G-protein coupled receptors or to tyrosine kinase receptors, phospholipase C-β (PLC-β) or PLC-γ, respectively, is activated to induce production of IP_3 via hydrolysis of phosphatidylinositol 4,5-bisphosphate. IP_3 then diffuses freely within the cell and binds to the IP_3 receptor (IP_3R) that functions as a Ca^{2+} release channel on the endoplasmic reticulum (ER), the major intracellular Ca^{2+} storage site. Binding of IP_3 to the IP_3R induces opening of the channel and release of Ca^{2+} from the stores. The resulting Ca^{2+} signal controls a variety of important cell functions, such as contraction, secretion, fertilization, synaptic plasticity, and gene expression. In this chapter I will review recent research on the structure and function of IP_3R with reference to compounds that interact with the IP_3R-mediated signalling pathway.

B. Molecular Structure and Function of IP_3R

The IP_3R is a family of proteins with some 2700 amino acid residues and a molecular mass of about 304–313 kDa. The membrane spanning region that forms the Ca^{2+} channel is assumed to be located near the C-terminus, while the large N-terminal segment forms the cytoplasmic structure that contains an IP_3 binding domain. The number of membrane spanning segments in the channel domain was proposed to be six with similarity to voltage-sensitive cation channels on the plasma membrane (MICHIKAWA et al. 1994). Up to about 580 amino acid residues from the N-terminus form the IP_3 binding domain (MIGNERY and SÜDHOF 1990; MIYAWAKI et al. 1991; NEWTON et al. 1994; YOSHIKAWA et al. 1996). The functional IP_3R consists of a tetramer of the proteins with a total molecular mass exceeding 1.2 MDa.

There are at least three subtypes of IP_3Rs that are derived from separate genes. The subtypes show 60–70% amino acid sequence identity (FURUICHI et al. 1989; MIGNERY et al. 1990; SÜDHOF et al. 1991; BLONDEL et al. 1993). Although the presence of other subtypes was suggested by partial cDNA cloning (ROSS et al. 1992; DE SMEDT et al. 1994), whether or not they are derived from independent genes remains to be confirmed. Indeed, only three subtypes have been observed in several cell types (De Smedt et al. 1997).

IP_3Rs are expressed throughout the body, both in excitable and non-excitable cells. Multiple subtypes are coexpressed in various tissues (Newton et al. 1994; Wojcikiewicz 1995; Joseph et al. 1995; De Smedt et al. 1997). Furthermore, different IP_3R subtypes form heterotetramers to make up a Ca^{2+} release channel (Monkawa et al. 1995; Joseph et al. 1995). The heterogeneity of the subunit structure of the IP_3R has been assumed to generate diversity in the patterns of IP_3-mediated Ca^{2+} signaling. However, the functional characterization of each subtype is yet to be carried out.

Type 1 IP_3R (IP_3R-1)-deficient mice have been generated by gene targeting (Matsumoto et al. 1996). Most of the knockout mice died in utero, and those that were born had severe ataxia and tonic or tonic-clonic seizures and died by the weaning period. Therefore, IP_3R-1 is essential at least for development and proper brain function, albeit specific cellular functions compromised by IP_3R-1-knockout are not yet known. Gene targeting of IP_3R-2 and IP_3R-3 may also provide insight into the functional roles of these IP_3R subtypes.

The IP_3R is primarily localized on the ER membrane. The presence of the IP_3R on the outer nuclear membrane, which is continuous with the ER membrane, was suggested based on the detection of IP_3-activated single channel currents on the surface of isolated nucleus using the patch-clamp method (Mak and Foskett 1994; Stehno-Bittel et al. 1995). It was proposed that the IP_3R is also located on the plasma membrane (DeLisle et al. 1996), inner nuclear membrane (Gerasimenko et al. 1995), and secretory vesicles of both endocrine and exocrine pancreas (Blondel et al. 1994; Gerasimenko et al. 1996). The extra-ER localization of the IP_3R and its physiological roles require further study.

The negatively stained image of purified IP_3R shows a pinwheel-like structure having surface dimensions of approximately $25 \times 25\,nm$ (as viewed perpendicular to the sample grid or ER membrane) in accordance with the tetrameric structure (Chadwick et al. 1990; Maeda et al. 1990). The size of the IP_3R thus observed is only slightly smaller than that of the ryanodine receptor (RyR), which is another type of intracellular Ca^{2+} release channel (see below and Chap. 25) that is square-shaped in appearance with each side ~29 nm in length (Radermacher et al. 1994; Orlova et al. 1996). However, a much more compact IP_3R structure was observed by quick-freeze deep-etch replica electron microscopy using bovine cerebellar microsomes, on which IP_3Rs are arranged in a nearly two-dimensional crystalline manner (Katayama et al. 1996). Each IP_3R was square-shaped in appearance, each side being ~12 nm in length. The difference in the size of the IP_3R in different studies was ascribed by Katayama et al. (1996) to the possible flattening in the negatively stained images and binding of detergent molecules to the purified IP_3R.

Single channel recording in a lipid planar bilayer in which cerebellar microsomes that contain predominantly IP_3R-1 were incorporated showed that the single channel conductance of the IP_3R was 53 pS (55 mmol/l Ca^{2+} on the luminal side) and the maximum open channel probability was less than 0.1 in the presence of 2–40 μmol/l IP_3, 200 nmol/l Ca^{2+} and 0.5–1.0 mol/l ATP at pH 7.35 on the cytoplasmic side (Bezprozvanny and Ehrlich 1994; Lupu et al. 1998). The corresponding values obtained from RyR-1 (rabbit skeletal muscle) were greater than those from IP_3R. The single channel conductance was 91 or 110 pS (50 mol/l or 54 mmol/l Ca^{2+} on the luminal side) (Lai et al. 1988; Smith et al. 1988). The open probability, which was dependent on Ca^{2+} concentration, approached 1.0 depending on the activating conditions and was 0.93 in the presence of 10 μmol/l Ca^{2+}, 10 mmol/l ATP at pH 7.4 on the cytoplasmic side (Smith et al. 1988).

C. Physiological Agonists and Modulators of IP_3R

I. IP_3

The binding of IP_3 (Fig. 1) is prerequisite, but not sufficient, for IP_3R channel opening. Isotope labeled IP_3 was used to determine the relationship between IP_3 concentration and IP_3 binding to the IP_3R. The dissociation constant (K_d) of IP_3 thus obtained was of the order of 1–100 nmol/l. The affinity of IP_3R subtypes to IP_3 followed the order type 2 > type 1 > type 3 which was determined by an expression experiment in COS cells and fusion peptides produced in *E. coli* (Newton et al. 1994).

The relationship between IP_3 binding and IP_3 dependence of Ca^{2+} release is not straightforward, and the K_d values of binding are often much lower than EC_{50} estimated in many functional assays of Ca^{2+} release. IP_3 binding has strong pH dependence and the affinity increases with increasing pH (Worley et al. 1987). IP_3 binding also depends on Ca^{2+} concentration and temperature (Kaftan et al. 1997). Arginine 265, lysine 508, and arginine 511 within the IP_3 binding domain of the IP_3R-1 were shown to be critical for the binding of IP_3 (Yoshikawa et al. 1996), indicating the importance of the ionic interaction between the basic amino acid residues of the IP_3R and the phosphate groups of IP_3. It is, therefore, expected that ionic strength is also an important factor for the determination of ligand binding affinity. In fact, many binding studies were carried out at high pH (pH ~8) and low ionic strength at 4 °C in the presence of a divalent cation chelator, EDTA, to maximize IP_3 binding. On the other hand, functional assays of Ca^{2+} release were usually carried out at neutral pH, at a physiological ionic strength, at room temperature and at Ca^{2+} concentrations of 0.1–1 μmol/l. Indeed, an increase in pH shifted the dose response relationship and decreased EC_{50} (Tsukioka et al. 1994), and an increase in the Ca^{2+} concentration increased EC_{50} (Kaftan et al. 1997; Hirose et al. 1998). Therefore, a comparison of IP_3 binding with Ca^{2+} release under comparable conditions would give better matching between K_d and EC_{50}.

In most binding studies, the Hill coefficient of the IP_3 concentration-binding relationship is close to unity, which suggests that there is no obvious

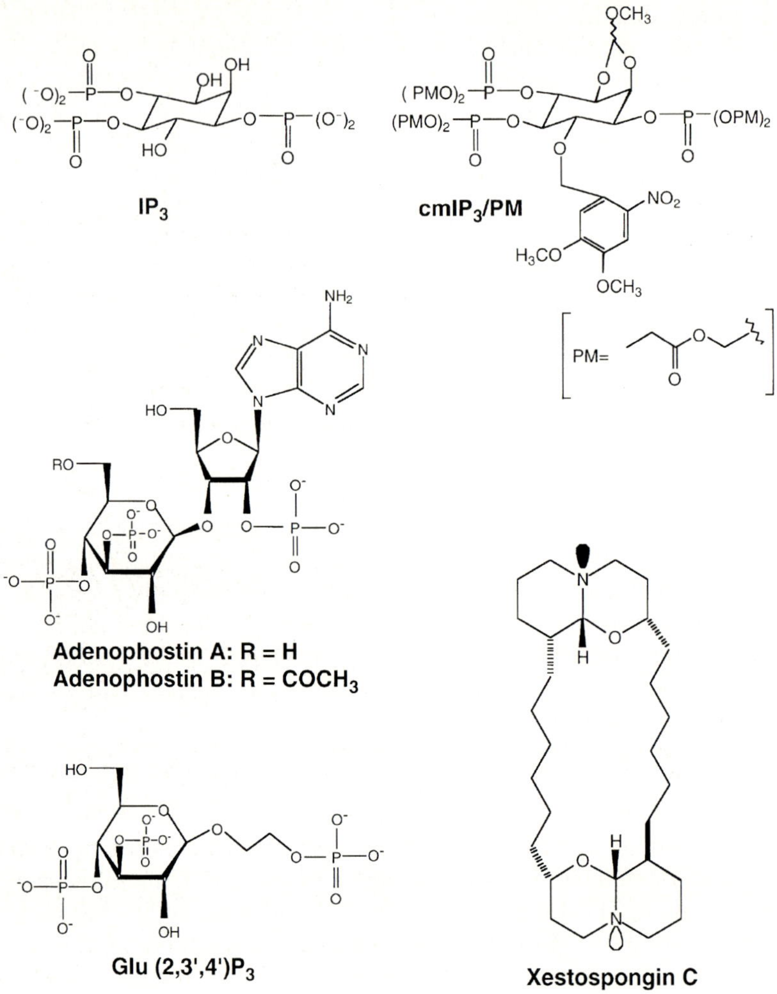

Fig. 1. Structures of IP_3, its analogues and an IP_3R inhibitor. Adenophostin (TAKAHASHI et al. 1994), Glu(2,3′,4′)P_3 (WILCOX et al. 1995), membrane-permeable caged IP_3 (LI et al. 1998), and xestospongin C (NAKAGAWA and ENDO 1984)

cooperativity in IP_3 binding among the IP_3R subunits within the tetrameric structure. However, in functional assays, Hill coefficients between 1 and 4 were reported for the IP_3 concentration-Ca^{2+} release channel activity relationship (MEYER et al. 1988; WATRAS et al. 1991; IINO and ENDO 1992; KAFTAN et al. 1997; HIROSE et al. 1998). Since IP_3-mediated Ca^{2+} release depends on cytosolic Ca^{2+} concentration and makes the Ca^{2+} release regenerative (see below), the apparent cooperativity in Ca^{2+} release may be spurious and be brought about by the regenerativity. Indeed, in experiments where cytosolic Ca^{2+} concentration was well buffered, the Hill coefficient for Ca^{2+} release approached unity (WATRAS et al. 1991; HIROSE et al. 1998; LUPU et al. 1998).

Single channel recordings of IP_3R Ca^{2+} release channels incorporated into a planar lipid bilayer showed four conductance levels (WATRAS et al. 1991). This may reflect the tetrameric subunit structure of the IP_3R Ca^{2+} release channel. Since each subunit has one IP_3 binding site, there are four IP_3 binding sites in a Ca^{2+} release channel molecule. The relationship between the number of IP_3 molecules bound to the tetramer and the probability of channel opening to the full and subconductance states is not yet known. The relationship between the number of bound agonist molecules and the open probability was shown to be extremely nonlinear in the cGMP-gated channel, which also comprises four subunits (RUIZ and KARPEN 1997). If multiple occupancy of the IP_3 binding sites per tetramer were required, dissociation between K_d of IP_3 binding and EC_{50} of Ca^{2+} release would be expected.

II. Ca^{2+}

Ca^{2+} is a coactivator of the IP_3R Ca^{2+} release channel. The potentiating effect of Ca^{2+} on IP_3-induced Ca^{2+} release was first observed in permeabilized smooth muscle cells (IINO 1987). It was subsequently shown that Ca^{2+} exerts a biphasic effect on the rate of IP_3-induced Ca^{2+} release with the peak effect obtained near 300 nmol/l (IINO 1990). The biphasic effect of Ca^{2+} was later found in other cell types including cerebellum (BEZPROZVANNY et al. 1991), rat brain (FINCH et al. 1991), *Xenopus* oocytes (PARYS et al. 1992; STEHNO-BITTEL et al. 1995), hepatocytes (MARSHALL and TAYLOR 1993), smooth muscle cell line (BOOTMAN et al. 1995), renal epithelial cell line (TSHIPAMBA et al. 1993), and bronchial mucosal cell line (MISSIAEN et al. 1998). The Ca^{2+} dependence is regarded to be important for both regenerative Ca^{2+} release and Ca^{2+} waves that are observed in many cell types (LECHLEITER and CLAPHAM 1992; IINO and ENDO 1992; IINO et al. 1993; HORNE and MEYER 1997; BOOTMAN et al. 1997).

Functional analyses of Ca^{2+} dependence of IP_3-induced Ca^{2+} release suggested that the binding of one to two Ca^{2+} ions on the IP_3R facilitates the opening of the IP_3R Ca^{2+} release channel (BEZPROZVANNY et al. 1991; KAFTAN et al. 1997; HIROSE et al. 1998). Ca^{2+} exhibits an immediate potentiating effect on the IP_3R Ca^{2+} channel (IINO and ENDO 1992). At higher Ca^{2+} concentrations, Ca^{2+} exhibits an inhibitory effect. Whether or not onset of and recovery from the high Ca^{2+} concentration-induced inhibition is instantaneous is a matter of controversy. In a rapid mixing experiment, the onset of $10\,\mu$mol/l Ca^{2+}-mediated inhibition took place with a time constant of ~0.58 s (FINCH et al. 1991), while in an experiment with a Ca^{2+} concentration jump using caged Ca^{2+}, an immediate inhibition was observed (IINO and ENDO 1992). It was shown that the peak size of Ca^{2+} release induced by photolysis of caged IP_3 (see below) depends on the interval between two successive photolyses (ILYIN and PARKER 1994; OANCEA and MEYER 1996). This was taken to suggest that the IP_3R assumes an inactivated state as cytosolic Ca^{2+} concentration increases, and the recovery from this state takes time (10–70 s). However, in another experimental paradigm, where permeabilized smooth muscle cells and a rapid solution exchange system were used, recovery of Ca^{2+} release following an increase in Ca^{2+} took place within 5 s (IINO and TSUKIOKA 1994). Further studies are required to clarify the kinetics of high Ca^{2+}-induced inhibition of the Ca^{2+} release channel.

The molecular mechanism of Ca^{2+} dependence of IP_3R activity is not yet known. It is assumed that the IP_3R molecule has Ca^{2+} binding site(s) that regulate the gating property of the Ca^{2+} release channel. Indeed, the presence of high-affinity Ca^{2+} binding site(s) on the IP_3R was suggested experimentally (MIGNERY et al. 1992; SIENAERT et al. 1996), although whether these Ca^{2+} binding sites actually regulate channel gating was not demonstrated. The presence of an accessory protein that inhibits $[^3H]IP_3$ binding in a Ca^{2+}-dependent manner was proposed (DANOFF et al. 1988); however, it was challenged by a subsequent study that showed that contaminant PLC activity in the sample may have generated "cold" IP_3 which competed with $[^3H]IP_3$ binding in a Ca^{2+}-dependent manner (MIGNERY et al. 1992).

III. ATP

ATP exhibits dual effects on the function of IP_3R. At submillimolar concentrations, it potentiates the opening of the IP_3R Ca^{2+} release channel (FERRIS et al. 1990; IINO 1991; BEZPROZVANNY and EHRLICH 1993). The effect of ATP is not mediated by protein kinase activity, because it is observed in the absence of Mg^{2+} and non-hydrolyzable ATP analogues also exert the same effect. Thus, it is likely that there is an ATP binding site on the IP_3R that allosterically regulates channel gating. At millimolar concentrations, however, ATP seems to inhibit IP_3 binding to the IP_3R as well as channel gating. These effects of ATP on IP_3R activity were reported in smooth muscle cells and cerebellar microsomes (IINO 1991; BEZPROZVANNY and EHRLICH 1993), where the dominantly expressed subtype is IP_3R-1. The potentiating effect of ATP was also observed in 16HBE14o-bronchial mucosal cells that dominantly express IP_3R-3 (MISSIAEN et al. 1998).

IV. Phosphorylation

The IP_3R is phosphorylated by various protein kinases. The functional effects of phosphorylation have not yet been firmly established, although possible modulator effects have been proposed. The purified IP_3 receptor (IP_3R-1) can be phosphorylated using either cyclic AMP- or cyclic GMP-dependent protein kinase (PKA or PKG) *in vitro*. Phosphorylation was time-dependent and stoichiometric using both kinases. Serines 1755 and 1589 were identified as the sites of phosphorylation by PKA (FERRIS et al. 1991) and in addition serine 1755 was phosphorylated by PKG (KOMALAVILAS and LINCOLN 1994). In an earlier report, IP_3-induced Ca^{2+} release from cerebellar microsomes was inhibited by phosphorylation by PKA (SUPATTAPONE et al. 1988). However, in a later study, phosphorylation of the IP_3R-1 by PKA resulted in a 20% increase in passive Ca^{2+} influx into proteoliposomes prepared from immunoaffinity-purified IP_3R (NAKADE et al. 1994). Membrane permeable cGMP analogues induced Ca^{2+} oscillations in hepatocytes, and it was suggested that phosphorylation of IP_3Rs by PKG may cause them to be more labile to release Ca^{2+} (ROONEY et al. 1996). The IP_3 receptor, reconstituted in liposomes, was stoichiometrically phosphorylated by protein kinase C (PKC) and Ca^{2+} calmodulin-dependent protein kinase II (CaM kinase II) as well as by PKA. The phosphorylation by the three enzymes was additive and involved different peptide sequences (FERRIS et al. 1991). Tyrosine phosphorylation of the IP_3R by Fyn was reported to activate IP_3-gated calcium channels in vitro (JAYARAMAN et al. 1996).

D. Activators of IP_3R

I. IP_3 Analogues

Inositol 2,4,5-trisphosphate also binds and activates the IP_3R but at ~10-fold lower affinity than IP_3 (IRVINE et al. 1984). Since it is a poor substrate of the IP_3 5-phosphatase that inactivates IP_3, it was used as a metabolically stable IP_3 analogue. Inositol 1,4,5-trisphosphorothioate (IPS_3) (STRUPISH et al. 1988; TAYLOR et al. 1989) and glycerophosphoryl-myo-inositol 4,5-bisphosphate ($GPIP_2$) (IRVINE et al. 1984) were also used as metabolically stable IP_3 analogues. Inositol 1,3,4,5-tetrakisphosphate (IP_4) is the product of IP_3 3-kinase that phosphorylates the 3-position of IP_3. IP_4 binds to the IP_3R with 250-fold lower affinity than IP_3 (WORLEY et al. 1987). IP_4 was reported to activate Ca^{2+} release via the IP_3R (WILCOX et al. 1993), but in other studies IP_4 was reported not to induce Ca^{2+} release (STEHNO-BITTEL et al. 1995; BIRD and PUTNEY 1996). This discrepancy might be due to the difference in subtypes of IP_3R expressed in the cells used in the different studies.

Although IP_3 is membrane-impermeable, membrane-permeable derivatives of IP_3 were synthesized by esterifying the phosphate groups with either acetoxymethyl, propionyloxymethyl, or butyryloxymethyl groups (LI et al. 1997). While IP_3 hexakis(acetoxymethyl) ester (IP_3/AM) was not sufficiently membrane-permeable, both IP_3 hexakis(propionyloxymethyl) ester (IP_3/PM) and IP_3 hexakis(butyryloxymethyl) ester (IP_3/BM) showed sufficient lipophilicity for entry into the cells. Extracellular application of IP_3/PM or IP_3/BM at ≥ 20 or ≥ 2 μmol/l concentrations, respectively, to 1321N1 astrocytoma cells induced the mobilization of the intracellular Ca^{2+} stores with 1–5 min delay, which was required for permeation through the cell membrane and subsequent de-esterification by intrinsic esterase activity within the cells (LI et al. 1997).

Adenophostins A and B (Fig. 1), which were isolated from fungal products, activate the IP_3R with about 100 times higher potency than that of IP_3, and are metabolically stable (TAKAHASHI et al. 1994; MARCHANT et al. 1997). Adenophostins are highly charged and do not permeate cell membranes. Therefore, microinjection of adenophostin into the cells or permeabilization of the cell membrane is required to allow the compound to reach the IP_3R within the cells. A synthetic analogue of adenophostin A, 2-hydroxyethyl-α-D-glucopyranoside-2,3',4'-trisphosphate (Glu (2,3′,4′)P_3) (Fig. 1), in which most of the adenosine moiety of adenophostin A is excised, has about 500-fold lower affinity than adenophostin A, although it is a full agonist of the IP_3R (WILCOX et al. 1995). The adenosine component of adenophostin A was argued by these authors to be important for increasing the affinity to the IP_3 binding site by keeping a distance between the two vicinal ring phosphates from the remaining phosphate. It was reported that adenophostin may directly activate Ca^{2+} influx (without depleting the Ca^{2+} stores) in *Xenopus* oocytes (DELISLE et al. 1997).

II. Caged IP_3

Caged IP_3 is an IP_3 molecule with its phosphate at either 4- or 5-position being esterified with a nitrobenzyl, or a caging, group. Such a compound does not bind to the IP_3R and hence has no Ca^{2+} releasing activity. Upon illumination with UV light near 360 nm, the caging group is released, resulting in the rapid formation of an active form of IP_3. Caged IP_3 was used in the studies of the kinetics of IP_3-induced Ca^{2+} release (e.g., SOMLYO and SOMLYO 1990; PARKER and IVORRA 1990; LECHLEITER and CLAPHAM 1992; IINO and ENDO 1992; KHODAKHAH and OGDEN 1993). The caged compound of a metabolically stable IP_3 analogue, caged $GPIP_2$ (1-(α-glycerophosphoryl)-myo-inositol 4,5-bisphosphate P4(5)-1-(2-nitrophenyl)ethyl ester), was also used to study the effect of prolonged activation of the IP_3R (BIRD et al. 1992).

Caged IP_3 is highly charged, hence, it is membrane-impermeable. Therefore, the introduction of caged IP_3 into the cytoplasm requires microinjection of the compound or the permeabilization of the surface membrane. Recently, a membrane-permeable caged IP_3 analogue (cmIP_3/PM) (Fig. 1) was synthesized (LI et al. 1998). In this molecule, the three charged phosphate groups are masked by esterification with propionyloxymethyl groups and 6-hydroxyl is protected by a photolabile caging group, 4,5-dimethoxy-2-nitrobenzyl ester. It is membrane-permeable and once inside the cell, the ester groups on the phosphates are hydrolyzed by the intrinsic esterase activity. Photolysis of the hydrolyzed compound yields an active IP_3 analogue and mobilization of the intracellular Ca^{2+} stores. This compound will be an invaluable tool to study the role of IP_3 not only in isolated or cultured cells but also in tissue preparations.

III. Thimerosal

Thimerosal, a thiol reagent, increases the affinity of the IP_3R to IP_3 (BOOTMAN et al. 1992). Thimerosal induces the spontaneous release of Ca^{2+} and often Ca^{2+} oscillations. The effect of thimerosal can be reversed by a sulfhydryl-reducing agent such as dithiothreitol. At high concentrations (>10 μmol/l), thimerosal exhibits an inhibitory effect on IP_3-induced Ca^{2+} release (PARYS et al. 1993).

IV. Immunophilin Ligands

The immunophilin FK506 binding protein (FKBP12) has been implicated in the regulation of IP_3R. It was reported that calcineurin is physiologically anchored to the IP_3R via FKBP12 and regulates the phosphorylation status of the receptor, resulting in a dynamic Ca^{2+}-sensitive regulation of IP_3-mediated Ca^{2+} flux (CAMERON et al. 1995). The interaction between IP_3R and FKBP12 can be disrupted by FK506 or rapamycin, and IP_3-induced Ca^{2+} release is enhanced (CAMERON et al. 1995).

FKBP12 also modulates channel gating of the RyR by increasing the number of channels with full conductance levels (by >400%), decreasing the open probability after caffeine activation, and increasing the mean open time. FK506 or rapamycin displaces FKBP12 from the RyR and reverses these stabilizing effects (BRILLANTES et al. 1994).

V. Mn^{2+}

Mn^{2+}, similarly to Ca^{2+}, exhibits a biphasic effect (with a peak near 1 μmol/l Mn^{2+}) on the IP_3R Ca^{2+} release channel activity (STRIGGOW and EHRLICH 1996). The effect of Mn^{2+} probably has no physiological significance, but may be important for the evaluation of the results in a certain type of experiment often used in the study of Ca^{2+} signaling. Since Mn^{2+} quenches the fluorescence of Fura-2, a fluorescent Ca^{2+} indicator, Mn^{2+} influx via either the cytoplasmic Ca^{2+} influx pathway or IP_3R itself can be estimated by the decrease in the fluorescence intensity of Fura-2 in the cytoplasm or in the Ca^{2+} stores at its isosbestic point (~365 nm excitation). However, the introduction of Mn^{2+} would alter IP_3R activity, which would potentially interfere with the experiments involved. Furthermore, since Mn^{2+} may also compete with Ca^{2+} for both intrinsic and extrinsic Ca^{2+} buffering systems, its introduction may result in changes in the free Ca^{2+} concentration, which would again influence the experimental system. Therefore, it is important to keep these points in mind during interpretation of results of experiments using Mn^{2+}.

E. Inhibitors of IP_3R

We are awaiting the introduction of subtype-specific inhibitors of IP_3Rs. Presented here are currently available inhibitors with some limitations for use.

I. Heparin

Heparin acts as a competitive inhibitor by blocking the binding of IP_3 to the IP_3R (WORLEY et al. 1987) and inhibiting IP_3-mediated Ca^{2+} release (HILL et al. 1987; KOBAYASHI et al. 1988). Since heparin is membrane-impermeable due to its polyanionic structure, its delivery to the vicinity of the IP_3R requires either microinjection or permeabilization of the cell membrane. Heparin also exhibits effects such as activation of the RyR (BEZPROZVANNY et al. 1993), and careful interpretation of the results will be required.

II. Xestospongin

Xestospongins, which are macrocyclic 1-oxaquinolizidines, purified from a marine sponge (NAKAGAWA and ENDO 1984), were shown to inhibit Ca^{2+} flux through the IP_3R (GAFNI et al. 1997). Since xestospongins exerts a minimal effect on the binding of IP_3 at concentrations where inhibition of channel activity is observed, they seem to inhibit either the Ca^{2+} channel itself or the transducer that transmits allosteric signals from the IP_3 binding site to the ion channel. The structure of xestospongins, where two charged nitrogen groups are connected by a lipophylic core (Fig. 1), suggests that they may function to plug the channel pore (GAFNI et al. 1997). Among the structurally related derivatives, such as xestospongin A, xestospongin C, xestospongin D, demethylxestopsongin B, and araguspongine B, xestospongin C, with an IC_{50} of 360 nmol/l, seems to be the most potent for blocking IP_3-induced Ca^{2+} release from cerebellar microcosms. Xestospongin C seems to only slightly interfere with the RyR, because it exhibited little effect on the ryanodine- and caffeine-induced Ca^{2+} release. Xestospongins are membrane-permeable and were shown to block agonist-induced Ca^{2+} transients in PC12 cells and primary cultured astrocytes (GAFNI et al. 1997).

III. Caffeine

Caffeine, a well-known activator of the RyR (ENDO 1977), exhibits a direct inhibitory effect on the IP_3-induced Ca^{2+} release at millimolar concentrations (PARKER and IVORRA 1991; HIROSE et al. 1993; BEZPROZVANNY et al. 1994). Although caffeine is membrane-permeable, multiple effects of methylxanthine, such as inhibition of adenosine receptors and phosphodiesterases as well as activation of RyRs, preclude its use as a specific inhibitor of IP_3R.

IV. Cyclic ADP-Ribose

Cyclic ADP-ribose (cADPR) was first shown to induce Ca^{2+} release via the RyR in sea urchin eggs (LEE et al. 1989; GALIONE et al. 1991). It was subsequently suggested to have the same effects on vertebrate RyRs (see Chap. 23). cADPR was shown to inhibit IP_3-induced Ca^{2+} release with an IC_{50} of 20–30 μmol/l in smooth muscle and bronchial mucosal cell lines (MISSIAEN et al. 1998). The effect of cADPR seems to be exerted by an allosteric site that differs from the IP_3 binding site. Since the inhibitory effect of 40 μmol/l cADPR is suppressed by the addition of an equimolar concentration of

Table 1. Pharmacology of IP_3R and RyR

	IP_3R	RyR
IP_3	⇑	⇔
Ca^{2+}	⇑ (<300 nmol/l)	⇑ (<100 μ mol/l[a])
	⇓ (>300 nmol/l)	⇓ (>300 μ mol/l[a])
ATP	⇑ (<5 mmol/l)	⇑
	⇓ (>5 mmol/l)	
Caffeine	⇓	⇑
Ryanodine	⇔	⇑ (open-lock[b])
Cyclic ADP-ribose	⇓	⇑
Heparin	⇓	⇑
FK506	⇑	⇑
Xestospongin	⇓	⇔
Thimerosal	⇑	⇑

⇑: potentiation; ⇓: inhibition; ⇔: no effect.
[a] Approximate values, and the transition concentrations depend on the subtype of RyR.
[b] Binding of ryanodine to the RyR locks the channel in an open state with about half of the maximum conductance.

nucleotide triphosphates, it probably plays no physiological role in intact cells that contain millimolar concentrations of ATP.

F. Comparison of Pharmacology Between IP_3R and RyR

Both IP_3R and RyR are intracellular Ca^{2+} release channels. Although both receptors have three subtypes, RyR is thought to form homotetramers whereas IP_3R can form heterotetramers (Monkawa et al. 1995; Joseph et al. 1995). Phylogenetic analysis indicated that they are members of a super gene family (Furuichi et al. 1989; Mignery et al. 1989; Takeshima et al. 1994). Furthermore, both Ca^{2+} release channels are activated by cytoplasmic Ca^{2+} concentration (see pages 605 and 607). Table 1 compares the pharmacological interventions that modulate IP_3R and RyR.

G. Spatio-Temporal Patterns of IP_3R-Mediated Ca^{2+} Signals

The advent of both fluorescent Ca^{2+} indicators and digital imaging techniques has enabled visualization of $[Ca^{2+}]_i$ change within a cell with sub-second time resolution. Recent studies demonstrated the complex spatio-temporal patterns of IP_3-induced Ca^{2+} release, such as Ca^{2+} oscillations, Ca^{2+} waves and Ca^{2+} puffs (Berridge and Irvine 1989; Lechleiter and Clapham 1992; Kasai and Petersen 1994; Yao et al. 1995). These complex patterns of Ca^{2+} signals may have strong impact on cell function. For example, Ca^{2+} waves may be important for generating a time delay in the rise of Ca^{2+} concentration at different locations within a cell to induce a polarized flux of ions (Kasai and Augustine 1990). The frequency of Ca^{2+} oscillations may be important for the specificity and efficiency of gene expression (Dolmetsch et al. 1998; Li et al. 1998) or protein kinase activity (De Koninck and Schulman 1998).

For the spatial patterning of the Ca^{2+} release, Ca^{2+}-mediated control of the IP_3R activity seems important. Ca^{2+} puffs are localized transient increases in Ca^{2+} concentration probably due to Ca^{2+} release through a few IP_3Rs (Yao et al. 1995). The synchronization of IP_3Rs within a Ca^{2+} puff may be generated by a Ca^{2+}-mediated activation mechanism, i.e., Ca^{2+} release from an IP_3R may activate the opening of adjacent IP_3Rs. In Ca^{2+} waves, the Ca^{2+}-mediated activation of the IP_3R takes place on a more generalized scale. Ca^{2+} release will successively activate the adjacent IP_3Rs and often propagate throughout the cell (Berridge 1993). The velocity of the waves is usually 20–100 μm/s (Jaffe 1991).

A generalized increase in intracellular Ca^{2+} concentration may take place repetitively to generate Ca^{2+} oscillations. It was postulated that Ca^{2+} oscillations can be evoked at a constant IP_3 concentration (Wakui et al. 1989) or that they require oscillatory changes in the IP_3 concentration (Harootunian et al. 1991). However, the molecular mechanism for repetitive increases in either Ca^{2+} or IP_3 concentration is not fully understood.

The time course of IP_3-induced Ca^{2+} release often shows a complex pattern. The rate of Ca^{2+} release is greater for a transient period immediately after an increase in IP_3 concentration than the steady-state rate of release. A further increase in IP_3 concentration induces another phasic increase in the rate of Ca^{2+} release, which again decays with time to the slow steady-state level. This aspect of IP_3-induced Ca^{2+} release is referred to as "quantal Ca^{2+} release" or "incremental detection" (Muallem et al. 1989; Meyer and Stryer 1990). It was proposed that Ca^{2+} concentration not only on the cytosolic side (see page 609) but also on the luminal side of the Ca^{2+} stores controls the activity of the IP_3R, and that the decrease in luminal Ca^{2+} concentration deactivates the IP_3R, resulting in a decreased Ca^{2+} release rate with the progress of Ca^{2+} release (Missiaen et al. 1992). However, such luminal Ca^{2+} dependence was not observed in the Ca^{2+} stores of smooth muscle cells, which also showed a biphasic Ca^{2+} release time course (Hirose and Iino 1994). Neither did single channel recording in bilayer experiments support the luminal Ca^{2+} dependence of the IP_3R activity (Bezprozvanny and Ehrlich 1994). Another possible mechanism for quantal Ca^{2+} release is that the IP_3-sensitive stores are separated into many compartments, each of which has different sensitivities to IP_3 (Muallem et al. 1989). Therefore, at a certain IP_3 concentration, only a fraction of the compartments with high IP_3 sensitivities would respond to the IP_3 concentration, thereby inducing a quantal Ca^{2+} release pattern. This hypothesis received some support from experiments that used artificial proteoliposomes (Ferris et al. 1992; Hirota et al. 1995). However, no such compartments with different IP_3 sensitivities were observed in smooth muscle cells (Hirose and Iino 1994).

Based on current knowledge on the functions of the IP_3R molecule, it seems difficult to ascribe the cause of quantal Ca^{2+} release to the IP_3R alone. Since IP_3R molecules interact closely due to Ca^{2+}-mediated activation and inhibition mechanisms, the arrangement of the IP_3Rs or the structure of the Ca^{2+} stores is important for the generation of spatio-temporal patterns of Ca^{2+} release (Bootman and Berridge 1995; Golovina and Blaustein 1997). Indeed, the biphasic time course of IP_3-induced Ca^{2+} release can be explained by the presence of compartments with different IP_3R densities (Hirose and Iino 1994).

H. Perspectives

IP_3-mediated Ca^{2+} signaling controls a multitude of important cell functions, generating complex spatio-temporal patterns. We have only a limited number of useful pharmacological tools to study these interesting processes, although a very promising caged IP_3 analogue has been introduced recently. Similarly, were it for a membrane-permeable caged IP_3R antagonist, such agent would be a powerful tool to study Ca^{2+} signaling mechanisms. Since there are subtypes of IP_3R which are expressed in a tissue-specific manner (Newton et al. 1994; Wojcikiewicz 1995; De Smedt et al. 1997), it will be of use to look for agents that discriminate the IP_3R subtypes. Functional diversity among IP_3R subtypes has been demonstrated in genetically engineered DT40 B cells (Miyakawa et al. EMBO J. 18:1303–1308, 1999).

Acknowledgements. The author wish to thank Drs. M. Endo, T. Yamazawa, and K. Hirose for comments on the manuscript.

References

Berridge MJ (1993) Inositol trisphosphate and calcium signalling. Nature 361:315–325

Berridge MJ, Irvine RF (1989) Inositol phosphates and cell signalling. Nature 341: 197–205

Bezprozvanny I, Bezprozvannaya S, Ehrlich BE (1994) Caffeine-induced inhibition of inositol(1,4,5)-trisphosphate-gated calcium channels from cerebellum. Mol Biol Cell 5:97–103

Bezprozvanny I, Ehrlich B (1993) ATP modulates the function of inositol 1,4,5-trisphosphate-gated channels at two sites. Neuron 10:1175–1184

Bezprozvanny I, Ehrlich BE (1994) Inositol (1,4,5)-trisphosphate ($InsP_3$)-gated Ca channels from cerebellum: conduction properties for divalent cations and regulation by intraluminal calcium. J Gen Physiol 104:821–856

Bezprozvanny I, Watras J, Ehrlich BE (1991) Bell-shaped calcium-response curve of Ins(1,4,5)P_3- and calcium-gated channels from endoplasmic reticulum of cerebellum. Nature 351:751–754

Bezprozvanny IB, Ondrias K, Kaftan E, Stoyanovsky DA, Ehrlich BE (1993) Activation of the calcium release channel (ryanodine receptor) by heparin and other polyanions is calcium dependent. Mol Biol Cell 4:347–352

Bird GS, Obie JF, Putney JW, Jr. (1992) Sustained Ca^{2+} signaling in mouse lacrimal acinar cells due to photolysis of "caged" glycerophosphoryl-myo-inositol 4,5-bisphosphate. J Biol Chem 267:17722–17725

Bird GS, Putney JW, Jr. (1996) Effect of inositol 1,3,4,5-tetrakisphosphate on inositol trisphosphate-activated Ca^{2+} signaling in mouse lacrimal acinar cells. J Biol Chem 271:6766–6770

Blondel O, Moody MM, Depaoli AM, Sharp AH, Ross CA, Swift H, Bell GI (1994) Localization of inositol trisphosphate receptor subtype 3 to insulin and somatostatin secretory granules and regulation of expression in islets and insulinoma cells. Proc Natl Acad Sci USA 91:7777–7781

Blondel O, Takeda J, Janssen H, Seino S, Bell GI (1993) Sequence and functional characterization of a third inositol trisphosphate receptor subtype, IP_3R-3, expressed in pancreatic islets, kidney, gastrointestinal tract, and other tissues. J Biol Chem 268:11356–11363

Bootman MD, Berridge MJ (1995) The elemental principles of calcium signaling. Cell 83:675–678

Bootman MD, Berridge MJ, Lipp P (1997) Cooking with calcium: the recipes for composing global signals from elementary events. Cell 91:367–373

Bootman MD, Missiaen L, Parys JB, De Smedt H, Casteels R (1995) Control of inositol 1,4,5-trisphosphate-induced Ca^{2+} release by cytosolic Ca^{2+}. Biochem J 306:445–451

Bootman MD, Taylor CW, Berridge MJ (1992) The thiol reagent, thimerosal, evokes Ca^{2+} spikes in HeLa cells by sensitizing the inositol 1,4,5-trisphosphate receptor. J Biol Chem 267:25113–25119

Brillantes AB, Ondrias K, Scott A, Kobrinsky E, Ondriasova E, Moschella MC, Jayaraman T, Landers M, Ehrlich BE, Marks AR (1994) Stabilization of calcium release channel (ryanodine receptor) function by FK506-binding protein. Cell 77:513–523

Cameron AM, Steiner JP, Roskams AJ, Ali SM, Ronnett GV, Snyder SH (1995) Calcineurin associated with the inositol 1,4,5-trisphosphate receptor-FKBP12 complex modulates Ca^{2+} flux. Cell 83:463–472

Cameron AM, Steiner JP, Sabatini DM, Kaplin AI, Walensky LD, Snyder SH (1995) Immunophilin FK506 binding protein associated with inositol 1,4,5-trisphosphate receptor modulates calcium flux. Proc Natl Acad Sci USA 92:1784–1788

Chadwick CC, Saito A, Fleischer S (1990) Isolation and characterization of the inositol trisphosphate receptor from smooth muscle. Proc Natl Acad Sci USA 87:2132–2136

Danoff SK, Supattapone S, Snyder SH (1988) Characterization of a membrane protein from brain mediating the inhibition of inositol 1,4,5-trisphosphate receptor binding by calcium. Biochem J 254:701–705

De Koninck P, Schulman H (1998) Sensitivity of CaM kinase II to the frequency of Ca^{2+} oscillations. Science 279:227–230

De Smedt H, Missiaen L, Parys JB, Bootman MD, Mertens L, Van Den Bosch L, Casteels R (1994) Determination of relative amounts of inositol trisphosphate receptor mRNA isoforms by ratio polymerase chain reaction. J Biol Chem 269: 21691–21698

De Smedt H, Missiaen L, Parys JB, Henning RH, Sienaert I, Vanlingen S, Gijsens A, Himpens B, Casteels R (1997) Isoform diversity of the inositol trisphosphate receptor in cell types of mouse origin. Biochem J 322:575–583

DeLisle S, Blondel O, Longo FJ, Schnabel WE, Bell GI, Welsh MJ (1996) Expression of inositol 1,4,5-trisphosphate receptors changes the Ca^{2+} signal of Xenopus oocytes. Am J Physiol 270:C1255–1261

DeLisle S, Marksberry EW, Bonnett C, Jenkins DJ, Potter BV, Takahashi M, Tanzawa K (1997) Adenophostin A can stimulate Ca^{2+} influx without depleting the inositol 1,4,5-trisphosphate-sensitive Ca^{2+} stores in the Xenopus oocyte. J Biol Chem 272:9956–9961

Dolmetsch RE, Xu K, Lewis RS (1998) Calcium oscillations increase the efficiency and specificity of gene expression. Nature 392:933–936
Endo M (1977) Calcium release from the sarcoplasmic reticulum. Physiol Rev 57:71–108
Ferris C, Huganir R, Snyder S (1990) Calcium flux mediated by purified inositol 1,4,5-trisphosphate receptor in reconstituted lipid vesicles is allosterically regulated by adenine nucleotides. Proceedings of the National Academy of Science U.S.A 87:2147–2151
Ferris CD, Cameron AM, Bredt DS, Huganir RL, Snyder SH (1991) Inositol 1,4,5-trisphosphate receptor is phosphorylated by cyclic AMP-dependent protein kinase at serines 1755 and 1589. Biochem Biophys Res Commun 175:192–198
Ferris CD, Cameron AM, Huganir RL, Snyder SH (1992) Quantal calcium release by purified reconstituted inositol 1,4,5-trisphosphate receptor. Nature 356:350–352
Ferris CD, Huganir RL, Bredt DS, Cameron AM, Snyder SH (1991) Inositol trisphosphate receptor: phosphorylation by protein kinase C and calcium calmodulin-dependent protein kinases in reconstituted lipid vesicles. Proc Natl Acad Sci USA 88:2232–2235
Finch EA, Turner TJ, Goldin SM (1991) Calcium as a coagonist of inositol 1,4,5-trisphosphate-induced calcium release. Science 252:443–252
Furuichi T, Yoshikawa S, Miyawaki A, Wada K, Maeda N, Mikoshiba K (1989) Primary structure and functional expression of the inositol 1,4,5-trisphosphate-binding protein P_{400}. Nature 342:32–38
Gafni J, Munsch JA, Lam TH, Catlin MC, Costa LG, Molinski TF, Pessah IN (1997) Xestospongins: potent membrane permeable blockers of the inositol 1,4,5-trisphosphate receptor. Neuron 19:723–733
Galione A, Lee HC, Busa WB (1991) Ca^{2+}-induced Ca^{2+} release in sea urchin egg homogenates: Modulation by cyclic ADP-ribose. Science 253:1143–1146
Gerasimenko OV, Gerasimenko JV, Belan PV, Petersen OH (1996) Inositol trisphosphate and cyclic ADP-ribose-mediated release of Ca^{2+} from single isolated pancreatic zymogen granules. Cell 84:473–480
Gerasimenko OV, Gerasimenko JV, Tepikin AV, Petersen OH (1995) ATP-dependent accumulation and inositol trisphosphate- or cyclic ADP-ribose-mediated release of Ca^{2+} from the nuclear envelope. Cell 80:439–444
Golovina VA, Blaustein MP (1997) Spatially and functionally distinct Ca^{2+} stores in sarcoplasmic and endoplasmic reticulum. Science 275:1643–1648
Harootunian AT, Kao JPY, Paranjape S, Tsien RY (1991) Generation of calcium oscillations in fibroblasts by positive feedback between calcium and IP_3. Science 251:75–78
Hill TD, Berggren PO, Boynton AL (1987) Heparin inhibits inositol trisphosphate-induced calcium release from permeabilized rat liver cells. Biochem Biophys Res Commun 149:897–901
Hirose K, Iino M (1994) Heterogeneity of channel density in inositol 1,4,5-trisphosphate-sensitive Ca^{2+} stores. Nature 372:791–794
Hirose K, Iino M, Endo M (1993) Caffeine inhibits Ca^{2+}-mediated potentiation of inositol 1,4,5-trisphosphate-induced Ca^{2+} release in permeabilized vascular smooth muscle cell. Biochem Biophys Res Commun 194:726–732
Hirose K, Kadowaki S, Iino M (1998) Allosteric regulation by cytoplasmic Ca^{2+} and IP_3 of the gating of IP_3 receptors in permeabilized guinea-pig vascular smooth muscle cells. J Physiol 506:407–414
Hirota J, Michikawa T, Miyawaki A, Furuichi T, Okura I, Mikoshiba K (1995) Kinetics of calcium release by immunoaffinity-purified inositol 1,4,5-trisphosphate receptor in reconstituted lipid vesicles. J Biol Chem 270:19046–19051
Horne JH, Meyer T (1997) Elementary calcium-release units induced by inositol trisphosphate. Science 276:1690–1693
Iino M (1987) Calcium dependent inositol trisphosphate-induced calcium release in the guinea-pig taenia caeci. Biochem Biophys Res Commun 142:47–52

Iino M (1990) Biphasic Ca^{2+} dependence of inositol 1,4,5-trisphosphate-induced Ca release in smooth muscle cells of the guinea pig taenia caeci. J Gen Physiol 95:1103–1122
Iino M (1991) Effects of adenine nucleotides on inositol 1,4,5-trisphosphate-induced calcium release in vascular smooth muscle cells. J Gen Physiol 98:681–698
Iino M, Endo M (1992) Calcium-dependent immediate feedback control of inositol 1,4,5-trisphosphate-induced Ca^{2+} release. Nature 360:76–78
Iino M, Tsukioka M (1994) Feedback control of inositol trisphosphate signalling by calcium. Mol Cell Endocrinol 98:141–146
Iino M, Yamazawa T, Miyashita Y, Endo M, Kasai H (1993) Critical intracellular Ca^{2+} concentration for all-or-none Ca^{2+} spiking in single smooth muscle cells. EMBO J 12:5287–5291
Ilyin V, Parker I (1994) Role of cytosolic Ca^{2+} in inhibition of $InsP_3$-evoked Ca^{2+} release in Xenopus oocytes. J Physiol 477:503–509
Irvine RF, Brown KD, Berridge MJ (1984) Specificity of inositol trisphosphate-induced calcium release from permeabilized Swiss-mouse 3T3 cells. Biochem J 222:269–272
Jaffe LF (1991) The path of calcium in cytosolic calcium oscillations: A unifying hypothesis. Proceedings of the National Academy of Sciences of U.S.A. 88:9883–9887
Jayaraman T, Ondrias K, Ondriasova E, Marks AR (1996) Regulation of the inositol 1,4,5-trisphosphate receptor by tyrosine phosphorylation. Science 272:1492–1494
Joseph SK, Lin C, Pierson S, Thomas AP, Maranto AR (1995) Heteroligomers of type-I and type-III inositol trisphosphate receptors in WB rat liver epithelial cells. J Biol Chem 270:23310–23316
Kaftan EJ, Ehrlich BE, Watras J (1997) Inositol 1,4,5-trisphosphate ($InsP_3$) and calcium interact to increase the dynamic range of $InsP_3$ receptor-dependent calcium signaling. J Gen Physiol 110:529–538
Kasai H, Augustine GJ (1990) Cytosolic Ca^{2+} gradients triggering unidirectional fluid secretion from exocrine pancreas. Nature 348:735–738
Kasai H, Petersen OH (1994) Spatial dynamics of second messengers: IP_3 and cAMP as long-range and associative messengers. TINS 17:95–101
Katayama E, Funahashi H, Michikawa T, Shiraishi T, Ikemoto T, Iino M, Mikoshiba K (1996) Native structure and arrangement of inositol-1,4,5-trisphosphate receptor molecules in bovine cerebellar Purkinje cells as studied by quick-freeze deep-etch electron microscopy. EMBO J 15:4844–4851
Khodakhah K, Ogden D (1993) Functional heterogeneity of calcium release by inositol trisphosphate in single Purkinje neurons, cultured cerebellar astrocytes, and peripheral tissues. Proc Natl Acad Sci USA 90:4976–4980
Kobayashi S, Somlyo AV, Somlyo AP (1988) Heparin inhibits the inositol 1,4,5-trisphosphate-dependent, but not the independent, calcium release induced by guanine nucleotide in vascular smooth muscle. Biochem Biophys Res Commun 153:625–631
Komalavilas P, Lincoln TM (1994) Phosphorylation of the inositol 1,4,5-trisphosphate receptor by cyclic GMP-dependent protein kinase. J Biol Chem 269:8701–8707
Lai FA, Erickson HP, Rousseau E, Liu QY, Meissner G (1988) Purification and reconstitution of the calcium release channel from skeletal muscle. Nature 331:315–319
Lechleiter JD, Clapham DE (1992) Molecular mechanisms of intracellular calcium excitability in X. laevis oocytes. Cell 69:283–294
Lee HC, Walseth TF, Bratt GT, Hayes RN, Clapper DL (1989) Structural determination of a cyclic metabolite of NAD^+ with intracellular Ca^{2+}-mobilizing activity. J Biol Chem 264:1608–1615
Li W, Llopis J, Whitney M, Zlokarnik G, Tsien RY (1998) Cell-permeant caged $InsP_3$ ester shows that Ca^{2+} spike frequency can optimize gene expression. Nature 392:936–941
Li W, Schultz C, Llopis J, Tsien RY (1997) Membrane-permeant esters of inositol polyphosphates, chemical synthesis and biological applications. Tetrahedron 53:12017–12040

Lupu VD, Kaznacheyeva E, Krishna UM, Falck JR, Bezprozvanny I (1998) Functional coupling of phosphatidylinositol 4,5-bisphosphate to inositol 1,4,5-trisphosphate receptor. J Biol Chem 273:14067–14070

Maeda N, Niinobe M, Mikoshiba K (1990) A cerebellar Purkinje cell marker P400 protein is an inositol 1,4,5-trisphosphate ($InsP_3$) receptor protein. Purification and characterization of $InsP_3$ receptor complex. EMBO J 9:61–67

Mak DO, Foskett JK (1994) Single-channel inositol 1,4,5-trisphosphate receptor currents revealed by patch clamp of isolated Xenopus oocyte nuclei. J Biol Chem 269:29375–29378

Marchant JS, Beecroft MD, Riley AM, Jenkins DJ, Marwood RD, Taylor CW, Potter BV (1997) Disaccharide polyphosphates based upon adenophostin A activate hepatic D-myo-inositol 1,4,5-trisphosphate receptors. Biochemistry 36:12780–12790

Marshall ICB, Taylor CW (1993) Biphasic effects of cytosolic Ca^{2+} on Ins(1,4,5)P_3-stimulated Ca^{2+} mobilization in hepatocytes. J Biol Chem 268:13214–13220

Matsumoto M, Nakagawa T, Inoue T, Nagata E, Tanaka K, Takano H, Minowa O, Kuno J, Sakakibara S, Yamada M, Yoneshima H, Miyawaki A, Fukuuchi Y, Furuichi T, Okano H, Mikoshiba K, Noda T (1996) Ataxia and epileptic seizures in mice lacking type 1 inositol 1,4,5-trisphosphate receptor. Nature 379:168–171

Meyer T, Holowka D, Stryer L (1988) Highly cooperative opening of calcium channels by inositol 1,4,5-trisphosphate. Science 240:653–656

Meyer T, Stryer L (1990) Transient calcium release induced by successive increments of inositol 1,4,5-trisphosphate. Proc Natl Acad Sci USA 87:3841–3845

Michikawa T, Hamanaka H, Otsu H, Yamamoto A, Miyawaki A, Furuichi T, Tashiro Y, Mikoshiba K (1994) Transmembrane topology and sites of N-glycosylation of inositol 1,4,5-trisphosphate receptor. J Biol Chem 269:9184–9189

Mignery GA, Johnston PA, Südhof TC (1992) Mechanism of Ca^{2+} inhibition of inositol 1,4,5-trisphosphate ($InsP_3$) binding to the cerebellar $InsP_3$ receptor. J Biol Chem 267:7450–7455

Mignery GA, Newton CL, Archer BT, Südhof TC (1990) Structure and expression of the rat inositol 1,4,5-trisphosphate receptor. J Biol Chem 265:12679–12685

Mignery GA, Südhof TC (1990) The ligand binding site and transduction mechanism in the inositol-1,4,5-triphosphate receptor. EMBO J 9:3893–3898

Mignery GA, Südhof TC, Takei K, De Camilli P (1989) Putative receptor for inositol 1,4,5-trisphosphate similar to ryanodine receptor. Nature 342:192–195

Missiaen L, De Smedt H, Droogmans G, Casteels R (1992) Ca^{2+} release induced by inositol 1,4,5-trisphosphate is a steady-state phenomenon controlled by luminal Ca^{2+} in permeabilized cells. Nature 357:599–602

Missiaen L, Parys JB, De Smedt H, Sienaert I, Sipma H, Vanlingen S, Maes K, Kunzelmann K, Casteels R (1998) Inhibition of inositol trisphosphate-induced calcium release by cyclic ADP-ribose in A7r5 smooth-muscle cells and in 16HBE14o- bronchial mucosal cells. Biochem J 329:489–495

Missiaen L, Parys JB, Sienaert I, Maes K, Kunzelmann K, Takahashi M, Tanzawa K, De Smedt H (1998) Functional properties of the type-3 $InsP_3$ receptor in 16HBE14o– bronchial mucosal cells. J Biol Chem 273:8983–8986

Miyawaki A, Furuichi T, Ryou Y, Yoshikawa S, Nakagawa T, Saitoh T, Mikoshiba K (1991) Structure-function relationships of the mouse inositol 1,4,5-trisphosphate receptor. Proc Natl Acad Sci USA 88:4911–4915

Monkawa T, Miyawaki A, Sugiyama T, Yoneshima H, Yamamoto-Hino M, Furuichi T, Saruta T, Hasegawa M, Mikoshiba K (1995) Heterotetrameric complex formation of inositol 1,4,5-trisphosphate receptor subunits. J Biol Chem 270:14700–14704

Muallem S, Pandol SJ, Beeker TG (1989) Hormone-evoked calcium release from intracellular stores is a quantal process. J Biol Chem 264:205–212

Nakade S, Rhee SK, Hamanaka H, Mikoshiba K (1994) Cyclic AMP-dependent phosphorylation of an immunoaffinity-purified homotetrameric inositol 1,4,5-trisphosphate receptor (type I) increases Ca^{2+} flux in reconstituted lipid vesicles. J Biol Chem 269:6735–6742

Nakagawa M, Endo M (1984) Structure of xestospongin A, B, C and D, novel vasodilative compounds from marine sponge, *Xestospongia exigua*. Tetrahedron Lett 25:3227–3230
Newton CL, Mignery GA, Südhof TC (1994) Co-expression in vertebrate tissues and cell lines of multiple inositol 1,4,5-trisphosphate ($InsP_3$) receptors with distinct affinities for $InsP_3$. J Biol Chem 269:28613–28619
Oancea E, Meyer T (1996) Reversible desensitization of inositol trisphosphate-induced calcium release provides a mechanism for repetitive calcium spikes. J Biol Chem 271:17253–17260
Orlova EV, Serysheva, II, van Heel M, Hamilton SL, Chiu W (1996) Two structural configurations of the skeletal muscle calcium release channel. Nature Str Biol 3:547–552
Parker I, Ivorra I (1990) Localized all-or-none calcium liberation by inositol trisphosphate. Science 250:977–979
Parker I, Ivorra I (1991) Caffeine inhibits inositol trisphosphate-mediated liberation of intracellular calcium in Xenopus oocytes. J Physiol 433:229–240
Parys JB, Missiaen L, De Smedt H, Droogmans G, Casteels R (1993) Bell-shaped activation of inositol-1,4,5-trisphosphate-induced Ca^{2+} release by thimerosal in permeabilized A7r5 smooth-muscle cells. Pflugers Archiv 424:516–522
Parys JB, Sernett SW, DeLisle S, Snyder PM, Welsh MJ, Campbell KP (1992) Isolation, characterization, and localization of the inositol 1,4,5-trisphosphate receptor protein in Xenopus laevis oocytes. J Biol Chem 267:18776–18782
Radermacher M, Rao V, Grassucci R, Frank J, Timerman AP, Fleischer S, Wagenknecht T (1994) Cryo-electron microscopy and three-dimensional reconstruction of the calcium release channel/ryanodine receptor from skeletal muscle. J Cell Biol 127:411–423
Rooney TA, Joseph SK, Queen C, Thomas AP (1996) Cyclic GMP induces oscillatory calcium signals in rat hepatocytes. J Biol Chem 271:19817–19825
Ross CA, Danoff SK, Schell MJ, Snyder SH, Ullrich A (1992) Three additional inositol 1,4,5-trisphosphate receptors: molecular cloning and differential localization in brain and peripheral tissues. Proc Natl Acad Sci USA 89:4265–4269
Ruiz ML, Karpen JW (1997) Single cyclic nucleotide-gated channels locked in different ligand-bound states. Nature 389:389–392
Sienaert I, De Smedt H, Parys JB, Missiaen L, Vanlingen S, Sipma H, Casteels R (1996) Characterization of a cytosolic and a luminal Ca^{2+} binding site in the type I inositol 1,4,5-trisphosphate receptor. J Biol Chem 271:27005–27012
Smith JS, Imagawa T, Ma J, Fill M, Campbell KP, Coronado R (1988) Purified ryanodine receptor from rabbit skeletal muscle is the calcium-release channel of sarcoplasmic reticulum. J Gen Physiol 92:1–26
Somlyo AP, Somlyo AV (1990) Flash photolysis studies of excitation-contraction coupling, regulation, and contraction in smooth muscle. Annu Rev Physiol 52:857–874
Stehno-Bittel L, Lückhoff A, Clapham D (1995) Calcium release from the nucleus by $InsP_3$ receptor channels. Neuron 14:163–167
Striggow F, Ehrlich BE (1996) The inositol 1,4,5-trisphosphate receptor of cerebellum. Mn^{2+} permeability and regulation by cytosolic Mn^{2+}. J Gen Physiol 108:115–124
Strupish J, Cooke AM, Potter BV, Gigg R, Nahorski SR (1988) Stereospecific mobilization of intracellular Ca^{2+} by inositol 1,4,5-triphosphate. Comparison with inositol 1,4,5-trisphosphorothioate and inositol 1,3,4-trisphosphate. Biochem J 253:901–905
Supattapone S, Danoff SK, Theibert A, Joseph SK, Steiner J, Snyder SH (1988) Cyclic AMP-dependent phosphorylation of a brain inositol trisphosphate receptor decreases its release of calcium. Proc Natl Acad Sci USA 85:8747–8750
Südhof TC, Newton CL, Archer BT, Ushkaryov YA, Mignery GA (1991) Structure of a novel $InsP_3$ receptor. EMBO J 10:3199–3206
Takahashi M, Tanzawa K, Takahashi S (1994) Adenophostins, newly discovered metabolites of Penicillium brevicompactum, act as potent agonists of the inositol 1,4,5-trisphosphate receptor. J Biol Chem 269:369–372

Takeshima H, Nishi M, Iwabe N, Miyata T, Hosoya T, Masai I, Hotta Y (1994) Isolation and characterization of a gene for a ryanodine receptor/calcium release channel in *Drosophila melanogaster*. FEBS Lett 337:81–87

Taylor CW, Berridge MJ, Cooke AM, Potter BV (1989) Inositol 1,4,5-trisphosphorothioate, a stable analogue of inositol trisphosphate which mobilizes intracellular calcium. Biochem J 259:645–650

Tshipamba M, De Smedt H, Missiaen L, Himpens B, Van Den Bosch L, Borghgraef R (1993) Ca^{2+} dependence of inositol 1,4,5-trisphosphate-induced Ca^{2+} release in renal epithelial LLC-PK1 cells. J Cell Physiol 155:96–103

Tsukioka M, Iino M, Endo M (1994) pH dependence of inositol 1,4,5-trisphosphate-induced Ca^{2+} release in permeabilized smooth muscle cells of the guinea-pig. J Physiol 485:369–375

Wakui M, Potter BVL, Petersen OH (1989) Pulsatile intracellular calcium release does not depend on fluctuations in inositol trisphosphate concentration. Nature 339:317–320

Watras J, Bezprozvanny I, Ehrlich BE (1991) Inositol 1,4,5-trisphosphate-gated channels in cerebellum: presence of multiple conductance states. J Neurosci 11:3239–3245

Wilcox RA, Challiss RA, Liu C, Potter BV, Nahorski SR (1993) Inositol-1,3,4,5-tetrakisphosphate induces calcium mobilization via the inositol-1,4,5-trisphosphate receptor in SH-SY5Y neuroblastoma cells. Mol Pharmacol 44: 810–817

Wilcox RA, Erneux C, Primrose WU, Gigg R, Nahorski SR (1995) 2-Hydroxyethyl-alpha-D-glucopyranoside-2,3',4'-trisphosphate, a novel, metabolically resistant, adenophostin A and myo-inositol-1,4,5-trisphosphate analogue, potently interacts with the myo-inositol-1,4,5-trisphosphate receptor. Mol Pharmacol 47:1204–1211

Wojcikiewicz RJ (1995) Type I, II, and III inositol 1,4,5-trisphosphate receptors are unequally susceptible to down-regulation and are expressed in markedly different proportions in different cell types. J Biol Chem 270:11678–11683

Worley PF, Baraban JM, Supattapone S, Wilson VS, Snyder SH (1987) Characterization of inositol trisphosphate receptor binding in brain. Regulation by pH and calcium. J Biol Chem 262:12132–12136

Yao Y, Choi J, Parker I (1995) Quantal puffs of intracellular Ca^{2+} evoked by inositol trisphosphate in Xenopus oocytes. J Physiol 482:533–553

Yoshikawa F, Morita M, Monkawa T, Michikawa T, Furuichi T, Mikoshiba K (1996) Mutational analysis of the ligand binding site of the inositol 1,4,5-trisphosphate receptor. J Biol Chem 271:18277–18284

CHAPTER 25

Ca^{2+}-Activated Non-Selective Cation Channels

J. TEULON

A. Introduction

Non-selective cation channels form a mixed group of ion channels including ligand-gated, hyperpolarization-activated, mechanosensitive channels, as well as channels activated by noxious stimuli or involved in capacitative Ca^{2+} entry (see SIEMEN and HESCHELER 1993; ZHU et al. 1996; CATERINA et al. 1997; GAUSS et al. 1998; LUDWIG et al. 1998). Non-selective cation channels of the Ca^{2+}-activated type (NSC_{Ca}) were described very shortly after the development of the patch-clamp technique (HAMILL et al. 1981), probably because these channels, usually closed in intact cells, activate spontaneously upon excision when the internal side of the channel comes into contact with millimolar concentrations of calcium. Many of their biophysical and regulatory properties have been described, but these channels remain something of an enigma because their molecular sequence has not been elucidated (see however SUZUKI et al. 1998), and channel function is unknown in many cases. Reviews on NSC_{Ca} channels appeared elsewhere (PARTRIDGE and SWANDULLA 1988; SIEMEN and HESCHELER 1993; CONLEY 1996). A different type of non-selective, Ca^{2+}-activated cation channel, which depends on nicotinamide-adenine-dinucleotide, has been reported in insulinoma cells (HERSON et al. 1997).

B. Tissue Distribution

With the discovery of NSC_{Ca} channels in cardiac muscle cells (COLQHOUN et al. 1981), neuroblastoma (YELLEN 1982) and pancreatic acinar cells (MARUYAMA and PETERSEN 1982), it rapidly became clear that this type of channel has a very wide distribution. Table 1 gives an overview of the various native tissues and cultured cells in which the channel is present[1]. The NSC_{Ca} channel has been detected in a variety of neuron types, in exocrine tissues

[1] Studies on channel location in cultured cells should be interpreted with caution because culture may affect the channels. A well known example is the appearance of maxi K^+ channels in almost all cultured renal cell whereas this channel is almost completely restricted to the intercalated cells of the collecting tubule in native tissue (see TEULON et al. 1992; GÖGELEIN 1997; WANG et al. 1997)

Table 1. Cell distribution of the Ca^{2}-activated non-selective cation channel

Tissue	g (pS)	Activ. by Ca^{2+} (μM)	Voltage	Inhibition by 1 mM ATP	References
Adipous and connective tissues					
Brown adipocytes (rat, prim. culture)	30	≥10	?	100%	Siemen et al. 1987; Koivisto et al. 1992,1993
Fibroblasts (human skin, prim. culture)	19	>200	↑ dep.	?	Galietta et al. 1989
Airways					
Adult alveolar cells (rat, prim. culture)	20	≥10	?	?	Feng et al. 1993
Fetal distal lung (rat, prim. culture)	25	>1	?	?	Tohda et al. 1994; Orser et al. 1991
Nasal polyps (human, prim. culture)	21	≥100	↑ dep.		Jorissen et al. 1990;
Blood cells					
Neutrophils (human)[a]	18–25	>0.1	↑ dep.	?	von Tscharner et al. 1986
Mast cells (rat)	18–25	>0.1	?	?	Lindau and Fernandez 1986
Cardiac muscle					
Ventricular myocyte (rat, prim. culture)	30	<1	No	?	Colqhoun et al. 1981
Ventricular cell (guinea pig)[b]	28	>100	↑ dep.	?	Ehara et al. 1988
Endocrine cells					
Insulinoma cells (rat, CRI-G1 cell line)	25	>100	↑ dep.	100%	Sturgess et al. 1987
Thyroid gland (rat)	35	≥1	No	?	Maruyama et al. 1985
Endothelia					
Coronary endothelium (pig, prim. culture)[a]	44	>0.1	No	?	Baron et al. 1996
Umbilical vein (human, prim. culture)[a]	20	>0.1	?	?	Bregestovski et al. 1988
Cerebral capillary (rat)	31	>1	No	100%	Popp and Gögelein 1992
Exocrine glands					
Pancreas acinus (mouse)	?	≥0.1	?	100% (2 mM)	Maruyama and Petersen 1984; Thorn and Petersen 1993
Pancreas acinus (rat)	27	>0.1	No	?	Gögelein et al. 1989 1990; Maruyama et al. 1982
Pancreas acinus (guinea pig)	26	>0.1	?	100%	Suzuki and Petersen 1988
Lacrimal gland (rat)	26	≥1	?	?	Marty et al. 1984
Salivary cells (mouse, ST 885 cell line)	25	>1	↑ dep.	100% (0.1 mM)	Cook et al. 1990
Pancreatic duct (rat, prim. culture)	25	≥1	↑ dep.	100%	Gray and Argent 1990
Gastro-intestinal tract and liver					

Jejunum (mouse)	20	≥10	↑ dep.	100%	Nonaka et al. 1995
Distal colon (rat)[c]	27	≥10	?	90% (0.7 mM)	Siemer et al. 1992; Bleich et al. 1996
HT29D4 and T84 cell lines (human)	20	>100	↑ dep.	100%	Champigny et al. 1991; Braun and Schulman 1995
Liver (rat, HTC cell line)[a]	28	>0.1	No	45% (0.2 mM)	Fitz et al. 1994; Lidofsky et al. 1997
Kidney					
Mesangial cells (rat, prim. culture)	25	>1	?	?	Matsunaga et al. 1991
Straight proximal tubule (rabbit)[d]	28	≥1	↑ dep.	?	Gögelein 1990; Gögelein and Greger 1986
Thick ascending limb (mouse)	27	≥1	↑ dep.	93%	Teulon et al. 1987; Paulais et al. 1989; Chraibi et al. 1994
Collecting tubule (mouse)	27	≥10	↑ dep.	82%	Chraibi et al. 1994
Collecting tubule (mouse, M-1 cell line)	23	>0.1	↑ dep.	69%	Korbmacher et al. 1995
Inner medulla (rat, mouse, cultures)	24–28	≥1	?	100%	Nonaka et al. 1995; Ono et al. 1994
Neurons and related tissues					
Neuroblastoma cells (N1E-115)	22	≥1	No		(Yellen 1982
Neurons (*Helix pomatia*)	30	>0.1	?	?	Partridge and Swandulla 1987
Dorsal root ganglion cells (mouse, culture)	20	>0.1	No	?	
Embryonic sensory neurons (chick)	34	>0.01	?		Razani-Boroujerdi and Partridge 1993
Schwann cells (rat sciatic nerve, culture)	32	>10	↑ dep.	?	Bevan et al. 1984
Sensory organs					
Corneal endothelium (rabbit)	22	≥100	↑ dep.	100%	Rae et al. 1990
Lens (human, prim. culture)	35	yes	?	?	Cooper et al. 1990
Vestibular dark cells (gerbil)	28	>0.1	?	?	Marcus et al. 1992
Stria vascularis (apical, guinea pig)	27	≥1	?	96%	Sunose et al. 1993; Takeuchi et al. 1992
Outer hair cells (cochlea, guinea pig)	26	≥1	No	58%	Van den Abbeele et al. 1994

↑ dep., *Po* increases towards $V > 0$; ?, not tested or contradictory results.

[a,b] This table only includes channels for which non-selectivity among cations, and Ca^{2+}-dependence have been demonstrated. These channels usually have low permeability for Ca^{2+}, the few exceptions are marked with "a" and "b" indicates when a second NSC_{Ca} channel with 15-pS conductance and higher Ca^{2+}-sensitivity was recorded for the same cell type. Unit conductances were measured at room temperature.

[c] Conductance at 37°C was 44 pS.

[d] A P_{Cl}/P_{Na} of 0.5 attributed to this cation channel was revised in a later study, which also reported Ca^{2+}-sensitivity.

(pancreas, salivary and lacrimal glands), renal tubules, intestine, and dissociated cells from sensory organs (cochlea, vestibule). It is also found in a variety of cultured cells including those from the respiratory system, liver, endocrine cells, and other cell types such as adipocytes and fibroblasts (see Table 1 for references). A classical NSC_{Ca} channel has been described in endothelial cells from rat cerebral capillaries (POPP and GÖGELEIN 1992). The channel was initially discovered in cultured ventricular myocytes (COLQHOUN et al. 1981), but it does not seem to be present in other muscle cells. I have found no report of NSC_{Ca} channels in skeletal muscle and, according to ISENBERG (1993), there is no NSC_{Ca} channel in smooth muscle cells. However, an unusual form of Ca^{2+}-dependent non-selective cation channel was found by LOIRAND et al. (1991) in the portal vein.

The high frequency of NSC_{Ca} channels in polarized epithelial cells raises the question as to whether this channel has a particular apical (mucous) or basolateral (serous) membrane distribution in native epithelial cells. We have shown that the NSC_{Ca} channel is present in the basolateral membranes of renal cells throughout the mouse renal tubule (CHRAIBI et al. 1994; TEULON et al. 1987). It is absent in the apical membranes, at least in the cortical thick ascending limb (Guinamard and Teulon, unpublished results) and in the cortical collecting tubule, in which Ca^{2+}-independent cation channels have been found (PALMER and FRINDT 1992; LING et al. 1991). The NSC_{Ca} channel has also been detected on the basolateral side of pancreatic acini (MARUYAMA and PETERSEN 1984; SUZUKI and PETERSEN 1988; GÖGELEIN and PFANNMULLER 1989), salivary ducts (DINUDOM et al. 1994), distal colon (SIEMER and GÖGELEIN 1992; BLEICH et al. 1996), jejunum (BUTT and HAMILTON 1998), and outer hair cells of the guinea pig cochlea (VAN DEN ABBEELE et al. 1994). It has been found more rarely in the apical membranes of native tissues: vestibular dark cells of the gerbil (MARCUS et al. 1992), Reissner membrane (YEH et al. 1998). There are also reports of NSC_{Ca} channels in both membranes in the vestibular dark cells of the gerbil stria vascularis (SUNOSE et al. 1993; TAKEUCHI et al. 1992, 1995) and rat pancreatic duct cells (GRAY and ARGENT 1990).

The specific subcellular location of NSC_{Ca} channels is not restricted to epithelia. A report by PARTRIDGE and SWANDULLA (1987) suggests that the NSC_{Ca} channel is present in the soma rather than in the axons of the bursting neurons of *Helix pomatia*.

C. Conductive Properties

I. Unit Conductance and Voltage Dependence

The NSC_{Ca} channel has a unit conductance of 18–34 pS at room temperature[2] but channels with similar properties have been reported that have lower

[2] Typically, the unit conductance at 37 °C is close to 40 pS (BLEICH et al. 1996; SIEMER and GÖGELEIN 1992)

Table 2. Cation selectivity of the Ca^{2}-activated non-selective cation channel

Cell model	NH_4^+	K^+	Na^+	Li^+	Rb^+	Cs^+	Ba^{2+}	Ca^{2+}
Adipocytes[a]	1.6	0,8	1.0	0.9	0.8	0.8	<0.02	<0.02
Salivary cells[b]	1.9	1.1	1.0	1.0	0.8	–	–	0.002
Distal colon[c]	1.6	1.0	1.0	0.9	0.9	–	–	0.14
Outer hair cells[d]	–	0.9	1.0	–	–	–	0.06	0.1
Renal tubule[e]	1.5	1.0	1.0	1.0	–	0.8	<0.01	0.09
Renal cells[f]	1.7	1.0	1.0	0.8	1.1	1.0	0	0
Renal cells[g]	–	1.0	1.0	1.0	1.0	1.0	0	0

[a] Weber and Siemen 1989.
[b] Cook et al. 1990.
[c] Siemer and Gögelein 1992.
[d] Van den Abbeele et al. 1994.
[e] Teulon et al. 1987; Chraibi et al. 1994.
[f] Nonaka et al. 1995.
[g] Korbmacher et al. 1995.

(Ehara et al. 1988) or higher (Loirand et al. 1991) unit conductances, or have conductance sub-states (Lipton 1986). In the absence of molecular biology data, it is unclear whether these channels are of the same type or whether they are different. This is the more true because our current knowledge of their properties does not allow a complete comparison to be made (in particular, the effects of ATP, pH, and blockers are unknown).

Voltage dependence is not a major property of this channel but it has generally been found that the probability of the channel being open increases with depolarization (Table 1). It is not always clear whether the voltage-independence encountered in some cases is due to experimental conditions, or whether it is an intrinsic property of some NSC_{Ca} channels.

II. Ion Selectivity

The selectivity pattern for monovalent cations was determined in detail by Cook et al. (1990) and Siemen et al. (1987), and has been confirmed by a number of studies (Table 2): (i) the channel is more permeable to NH_4^+ than to Na^+ (1.5–1.9)[3]; (ii) there is little if any difference in permeability to K^+, Na^+, Li^+, Cs^+, and Rb^+ (permeability range: 0.8–1.1); and (iii) large cations like piperazine (0.5), Tris (0.2) and *N*-methylglucamine (0.1) permeate the channel (Siemer and Gögelein 1992; Cook et al. 1990) to various extent. Several cation channels are similar to the NSC_{Ca} channel in being equally permeable (except for NH_4^+) to small monovalent cations but excluding anions. Such channels include the serotonin-gated cation channel [Cs^+ (1.2)>K^+ (1.1) > Rb^+ (1.0) = Na^+ (1.0) = Li^+ (1.0); Yang 1990], the endplate channel [NH_4^+ (1.8) > Cs^+ (1.4) > Rb^+ (1.3) > K^+ (1.1) > Na^+ (1.0) > Li^+ (0.9); Adams et al. 1980; Hille 1998],

[3] The numbers in brackets indicate the permeability of the named ion relative to Na^+

and the CGMP-dependent cation channel [Li^+ (1.1) > Na^+ (1.0)>K^+ (0.7)>Rb^+ (0.5)>Cs^+ (0.3); BARNSTABLE 1998].

Permeability to calcium is a key issue because it may indicate the function of the channel. In most cases, permeability to Ca^{2+} is low or undetectable (P_{Ca}/P_{Na} in the range 0–0.14, see Table 2). This is consistent with early results from YELLEN (1982). Similar values have been obtained for Ba^{2+} (Table 2) and Mn^{2+} (KORBMACHER et al. 1995). The low Ca^{2+} permeability of the NSC_{Ca} channel distinguishes this cation channel from most other types of cation channels such as endplate channels (P_{Ca}/P_{Na} about 0.2, ADAMS et al. 1980), CGMP-dependent cation channels (P_{Ca}/P_{Na} about 10 under physiological conditions, BARNSTABLE 1998) and serotonin-gated cation channels (P_{Ca}/P_{Na} about 1, YANG 1990). Mechanosensitive channels are also permeable to Ca^{2+}.

III. Ca-Permeable, Ca-Dependent Cation Channels: A Subtype of the NSC_{Ca} Channel?

Excluded from the preceding section are a few cases in which NSC_{Ca} channels are significantly permeable to Ca^{2+}: cation channels in endothelial cells (BREGESTOVSKI et al. 1988; NILIUS 1990; BARON et al. 1996), hepatocytes (FITZ et al. 1994; LIDOFSKY et al. 1997), and neutrophils (VON TSCHARNER et al. 1986). These channels otherwise have typical features of NSC_{Ca} channels, being sensitive to calcium and to ATP (hepatocytes only). Thus Ca^{2+}-permeable, Ca^{2+}-sensitive cation channels may constitute a subclass of NSC_{Ca} channels.

D. Blockers and Pharmacological Stimulators

I. Blockers

GÖGELEIN and PFANMÜLLER (1989) investigated the blocking effects of a variety of compounds in a series of experiments conducted with the well characterized NSC_{Ca} channel of the rat pancreas acinus. Following initial observation in rabbit proximal tubule (GÖGELEIN and GREGER 1986), diphenylamine-2-carboxylic acid (DPC) and related substances known to block Cl^- conductance in the thick ascending limb and various Cl^- channels (WANGEMANN et al. 1986) were also found to efficiently block the NSC_{Ca} channel (GÖGELEIN and PFANNMULLER 1989). These compounds acted by reducing the open state probability (*Po*) in a voltage-independent manner without changing unit current amplitude (slow block). All effects were fully reversible. 3′,5-Dichlorodiphenylamine-2-carboxylic acid (DCDPC) was the most potent, completely blocking channel activity at a concentration of 0.1 mmol/l (50% inhibitory concentration about 10^{-5} mol/l) whereas DPC and 5-nitro-2-(3-phenylpropyl-amino)-benzoic acid (NPPB) had no effect at a concentration of 10^{-5} mol/l and caused 50–90% inhibition at 10^{-4} mol/l. GÖGELEIN et al. (1990) then tested the effects of mefenamic acid, flufenamic acid and niflumic acid on the NSC_{Ca} channel. All these substances cause a reduction in *Po* which was reversible by a prolonged wash-out. The 50% inhibitory concentrations were not precisely determined but were estimated to be 10 μmol/l for mefenamic and flufenamic acids and about 50 μmol/l for niflumic acid. Anti-inflammatory drugs, with chemical structures different from those of these three acids, including indomethacin, aspirin, diltiazem, and ibuprofen, had no effect on the channel at a concentration of 0.1 mmol/l.

The effects of several blockers of K^+ and Na^+ channels have also been investigated in attempts to find specific blockers of the NSC_{Ca} channel: internal Ba^{2+} (70 mmol/l) (GÖGELEIN and PFANNMULLER 1989), internal and external TEA (up to 20 mmol/l) (GÖGELEIN and PFANNMULLER 1989; STURGESS et al. 1987), internal and external TTX (10–100 nmol/l) cause no change in channel activity or unit current amplitude. In contrast, the addition of quinine in the bath causes flickering indicative of an intermediate channel block. The effects of quinine (10 μmol/l–1 mmol/l) are observed in both the outside-out and inside-out configurations in insulinoma cells (STURGESS et al. 1987) and salivary cells (COOK et al. 1990). They are reversible and more intense at positive voltages. Similar results have been reported for distal colon cells (GÖGELEIN and CAPEK 1990). 4-Aminopyridine (2–10 mmol/l) added on the inner but not the outer side of the membrane patch decreases the mean duration of the openings and increases the mean duration of the closings (COOK et al. 1990; STURGESS et al. 1987).

Conflicting results have been obtained concerning the use of amiloride as a blocker of cation channels. Amiloride has often been tested for effects on the NSC_{Ca} channel because the epithelial amiloride-sensitive Na^+ channel may be present in a non-selective form in some tissues (GARTY and PALMER 1997). Amiloride is also a common blocker of stretch-activated cation channels (HAMILL and MCBRIDE Jr 1996) which have similar unit conductances. Several studies have tested this agent by including it in the pipette and comparing activity in separate patches in the presence and absence of amiloride. In such conditions, amiloride (0.5–25 μmol/l) was found to block the NSC_{Ca} channel in several respiratory cell models and in a renal cell line (DUSZIK et al. 1991; FENG et al. 1993; ONO et al. 1994; MARUNAKA 1996). However, amiloride (0.1–100 μmol/l) had no effect in several other renal cell lines and in human nasal cells in primary culture (HAMILTON and BENOS 1990; JORISSEN et al. 1990; KORBMACHER et al. 1995; NONAKA et al. 1995). Other studies have investigated the effects of external amiloride (10–1000 μmol/l) using the outside-out configuration of the patch-clamp method to compare channel activities in the presence and absence of amiloride in the same membrane patch (STURGESS et al. 1987; CHAMPIGNY et al. 1991; TAKEUCHI et al. 1992). No blocking effect of amiloride has been detected. The latter studies show that external amiloride is probably not an efficient blocker of NSC_{Ca} channels. Internal amiloride (10–100 μmol/l) is not a blocker of the NSC_{Ca} channel either (KORBMACHER et al. 1995; NONAKA et al. 1995).

The NSC_{Ca} channel is sensitive to internal gadolinium at relatively high concentrations (10^{-5} mmol/l to 10^{-3} mmol/l, POPP and GÖGELEIN 1992; NONAKA et al. 1995), but unfortunately the effects of gadolinium applied to the outer side of the NSC_{Ca} channel have not been studied. Thus, we do not know whether gadolinium could be used to distinguish the NSC_{Ca} channel from stretch-activated cation channels (HAMILL and MCBRIDE Jr 1996). Internal lanthanum (10 μmol/l) does not block NSC_{Ca} channels in rabbit distal cells in primary culture (which are blocked by 4'-methyl-2-diphenylamine carboxylic acid), while it does block a Ca^{2+}-specific channel at a concentration of 1 μmol/l (PONCET et al. 1992).

Thus, DCDPC, mefenamic acid, and related compounds are the best characterized blockers of the NSC_{Ca} channel. It is clear, given the concentrations required for a complete block, that these blockers are rather non-specific, because they also block chloride channels at similar concentrations. They may, however, be of some value for distinguishing between NSC_{Ca} channels and other cation channels. For example, in the Reissner membrane, there is a Ca^{2+}-independent, stretch-activated cation channel which is highly sensitive to external gadolinium (about 50% block with 1 μmol/l) but insensitive to 1 mmol/l internal flufenamic acid (YEH et al. 1998).

II. Pharmacological Stimulators

GÖGELEIN and PFANMÜLLER (1989) discovered that some stilbene sulfonates were able to increase NSC_{Ca} channel *Po* in rat pancreas. 4-Acetamido-4'-isothiocyanatostilbene-2,2'-disulfonic acid (SITS), added to the intracellular side of excised, inside-out patches at concentrations of 10–100 μmol/l, increases *Po* to values close to one, "locking" the NSC_{Ca} channel in the open state. This effect of SITS was hardly reversible since a long wash-out of 2 minutes was not enough to bring *Po* back to its control value. SITS does not induce additional current levels – instead it increases the channel mean open time and decreases the channel mean closed time. Studies in cultured salivary gland cells (COOK et al. 1990) and in cortical collecting duct (CHRAIBI et al. 1995) produced similar observations. A preliminary kinetics study (CHRAIBI et al. 1995) showed that SITS decreased the access to a long closed state. SITS has no effect when applied externally (GÖGELEIN and PFANNMULLER 1989). Similar effects were observed with 4,4'-diisothiocyanatostilbene-2,2'-disulfonic acid (DIDS) and 4,4'-dinitro-2,2'-stilbenedisulfonate (DNDS) added to the internal side of the patches at a concentration of 100 μmol/l (GÖGELEIN and PFANNMULLER 1989).

SITS does not interfere with the mechanisms responsible for Ca^{2+} sensitivity: channels "locked" open by SITS in the condition of a high Ca^{2+} concentration, can still close if the Ca^{2+} concentration is lowered to 10^{-7} mol/l, and SITS does not open NSC_{Ca} channels if Ca^{2+} concentration is below the threshold for activation of the channel (GÖGELEIN and PFANNMULLER 1989). CHRAÏBI et al. (1995) have shown that SITS largely reduces NSC_{Ca} channel inhibition caused by alkaline pH. This is probably because alkaline pH acts mainly by increasing the sojourn of the channel in a long closed state which no longer exists in the presence of SITS.

E. Intracellular Regulatory Elements

I. Calcium Sensitivity

All channels in this group are activated when the internal concentration of calcium increases, but sensitivity is very variable with an activation threshold ranging from 10^{-7} mol/l to 10^{-4} mol/l (Table 1). Sensitivity is usually high in excitable tissues (neurons and cardiac muscle), exocrine glands, and some epithelia. It is particularly low in many cultured cells.

There is a debate as to whether the Ca^{2+}-sensitivity determined in excised, isolated patches represents the true sensitivity of the channel in intact cells. Maruyama and Petersen (1984) showed that initially, just after the isolation of the membrane patch, sensitivity to calcium is high. The NSC_{Ca} channel is active at 10^{-7} mol/l in pancreatic acini but its activity declines with time. In other experiments Maruyama and Petersen (1984) added saponin to the bath to disrupt the cell membrane outside of the membrane patch and recorded NSC_{Ca} channel currents when the Ca^{2+} concentration in the bath was as low as 5×10^{-8} mol/l. It was concluded that a cytoplasmic factor was lost (or a channel state modified) by isolation of the membrane patch from the cell body (Maruyama and Petersen 1984).

However, the range of sensitivity observed is probably too large to be explained in these terms alone. For instance, the NSC_{Ca} channel has very different sensitivities to calcium in various parts of the renal tubule (from 10^{-7} mol/l to 10^{-4} mol/l), for similar experimental conditions (Chraibi et al. 1995; Teulon et al. 1987). This suggests that (i) sensitivity to calcium is regulated by other agents (phosphorylation?) or (ii) calcium per se is not the only activator of the channel in vivo (see later).

II. Inhibition by Intracellular Nucleotides

Sturgess et al. (1986) showed that the NSC_{Ca} channel is inhibited on the inside of the membrane by ATP and other adenine nucleotides which decrease *Po* without changing the unit conductance. Subsequent studies have demonstrated that inhibition by nucleotides is a general property of this channel (see Table 1), and that: (i) all inhibitory effects are reversible and largely voltage independent (Sturgess et al. 1986; Takeuchi et al. 1995; Paulais and Teulon 1989; but see Gray and Argent 1990); (ii) external ATP has no inhibitory effect (Sturgess et al. 1987; Takeuchi et al. 1995); and (iii) non-hydrolyzable ATP analogs also inhibit the channel, indicating that the effect of ATP is independent of phosphorylation (Paulais and Teulon 1989; Sturgess et al. 1987, 1997; Fitz et al. 1994; Rae et al. 1990). The quantitative characteristics of the inhibition have been determined in some studies (Table 3). In general, the 50% inhibitory concentration for ATP (8–400 μmol/l) is higher than that for ADP (3.5–21 μmol/l) or AMP (0.4–2.5 μmol/l) and Hill coefficients are around 1 (0.8–1.6). It is not known whether the channel is inhibited by the MgATP complex or by free ATP. The order of potency (AMP>ADP>ATP) contrasts

Table 3. Inhibition by adenosine-based nucleotides

Cell model	ATP IC_{50} (μmol/l)	ADP IC_{50} (μmol/l)	AMP IC_{50} (μmol/l)
Thick ascending limb[a]	20	21	2.5
Insulinoma cells[b]	8	3.5	0.4
Collecting tubule[c]	20	–	1.2
Stria vascularis[d]	400	15	–
Pancreatic duct[e]	200–400	–	–

IC_{50}: 50% inhibitory concentration.
[a] PAULAIS and TEULON 1989.
[b] STURGESS et al. 1987; REALE et al. 1994.
[c] CHRAIBI et al. 1994, 1995.
[d] TAKEUCHI et al. 1995.
[e] GRAY and ARGENT 1990.

with that observed for the ATP-dependent K^+ channel (ATP > ADP >> AMP; NOMA 1983).

The pharmacological profile of inhibition has been investigated in a number of studies. Nicotinamide-adenine nucleotides, β-NAD^+, β-NADH, β-$NADP^+$, and β-NADPH inhibit the NSC_{Ca} channel to similar extent (about 80% at a concentration of 0.1 mmol/l, REALE et al. 1994). In contrast, while 10–100 μmol/l adenine nucleotide completely inhibits the channel, 1 mmol/l adenosine is required to reduce *Po* by 90% in insulinoma cells (STURGESS et al. 1986) and by 17% in the thick ascending limb (PAULAIS and TEULON 1989). Nucleotides containing guanosine, inosine, or uridine are less efficient inhibitors than adenine nucleotides (STURGESS et al. 1987; PAULAIS and TEULON 1989; REALE et al. 1995), as are cyclic nucleotides, which have inhibitory effects at high concentrations only. Cyclic AMP only partially inhibits the NSC_{Ca} channel, reducing *Po* to 22–35% of control at a concentration of 1 mmol/l (PAULAIS and TEULON 1989; REALE et al. 1994). REALE et al. (1994a) demonstrated clear dose-dependent inhibition with a 50% inhibitory concentration of 12 μmol/l and a Hill coefficient of about 0.5. The dose-dependence of cyclic AMP effects was less obvious in the the study carried out by PAULAIS and TEULON (1989). REALE et al. (1994a) also investigated the inhibitory effects of other cyclic nucleotides and reported the order of potency: cyclic AMP > cyclic UMP > cyclic GMP > cyclic CMP > cyclic IMP. In particular, cyclic GMP (0.01–1 mmol/l) inhibited the NSC_{Ca} channel by only 0–27% (REALE et al. 1994a; PAULAIS and TEULON 1989; KORBMACHER et al. 1995; NONAKA et al. 1995), and 0–12% inhibition was obtained with 0.1 mmol/l 8-Br-cGMP (KORBMACHER et al. 1995; ONO et al. 1994). The NSC_{Ca} channel therefore differs from another cation channel found in cultured cells of the inner medulla collecting tubule, the other channel being inhibited by cyclic GMP and having no clear Ca^{2+} dependence (LIGHT et al. 1988). The inhibitory effects of cyclic AMP, and of other cyclic nucleotides, require such concentrations that they can have no physiological significance. However, there is evidence that cyclic nucleotides may positively modulate channel activity.

III. Tonic Influence of Intracellular ATP

The massive inhibition of the NSC_{Ca} channel activity caused by physiological concentrations of ATP (about 1 mmol/l) in the presence of high internal Ca^{2+} concentration (10^{-5} mol/l to 10^{-3} mol/l) raises the question of how the channel can open in situ. By analogy with ATP-dependent K^+ channels, in which channel activity depends on the ATP/ADP ratio rather than solely on ATP (see PETERSEN 1992), it was investigated whether ATP-evoked inhibition was affected by ADP. No such effect has been reported (THORN and PETERSEN 1992). However, as for ATP-dependent K^+ channels, ATP was found to "refresh" the activity of the NSC_{Ca} channel. The activity of NSC_{Ca} channel in the control solution is higher after exposure to ATP than it was before (THORN and PETERSEN 1992). This effect is not reproduced with ADP or AMP-PNP, a non-hydrolyzable ATP derivative. THORN and PETERSEN (1992) also noticed that ATP had a smaller inhibitory effect shortly after excision, when opening of the NSC_{Ca} channel was observed in the presence of 2 mmol/l ATP, even when the Ca^{2+} concentration was low (5 μmol/l). This quite interesting observation suggests that, as for Ca^{2+} dependence, patch isolation results in a different, higher, sensitivity to ATP, and that ATP may not prevent channel activation in situ.

IV. Stimulatory Effects of Intracellular Cyclic Nucleotides

REALE et al. (1994a) reported that ADP and AMP at low concentrations (0.1–5 μmol/l) activate the the NSC_{Ca} channel in 20/30% of patches. Cyclic nucleotides also have biphasic effects on channel activity, with low concentrations activating rather than inhibiting the channel. Cyclic AMP (0.1–1.0 μmol/l) increased *Po* by about 75% in 70% of patches. In contrast to the situation for inhibition by cyclic nucleotides, there is no base specificity for the activation because cyclic GMP, cyclic UMP, cyclic CMP, and cyclic IMP have similar effects to cyclic AMP. This may indicate that there is an alternative pathway to Ca^{2+} for activating the NSC_{Ca} channel, although it is unknown whether cyclic nucleotides can overcome nucleotide-induced inhibition. REALE et al. (1994b) have investigated in detail the structural requirements of cyclic nucleotides effects.

V. Other Regulators: Internal pH and Oxidation

CHRAÏBI et al. (1995) extended the original work of GRAY and ARGENT (1990) showing that the NSC_{Ca} channel is inhibited at low pH, by demonstrating that *Po* is a bell-shaped function of internal pH with maximum activity at pH 6.8–7.0. The two halves of the curve were not symmetrical with Hill coefficients of about 1 at high pH, and 3 at low pH. The inhibitory effects of low pH were more pronounced than those of high pH since *Po* at pH 6.0 was 11%, and *Po* at pH 8.0, 32% of Po at pH 7.2. Two independent observations strongly suggest that the effects of low and high pH are due to different mechanisms.

First, low pH decreases open times while high pH increases long closed times without affecting open times. Second, SITS, which locks the channel open by eliminating long closed times, prevents inhibition by high pH whereas it has no effect on the inhibition mediated by low pH. Another result from the same study (Chraibi et al. 1995) demonstrates that high pH changes the inhibition profile of AMP. The 50% inhibitory concentration is not affected by raising internal pH from 6.6 to 8.0, but the Hill coefficient decreases from 1 at pH 6.6 to 0.6 at pH 7.2 and 0.2 at pH 8.0. We found no evidence for similar regulation of the inhibition by ATP

Koivisto et al. (1993) used the inside-out configuration of the patch-clamp technique to test the effects of mercury and thimerosal on the NSC_{Ca} channel in rat brown fat cells. These two agents inhibited channel activity in a dose-dependent manner with 50% inhibitory concentrations of 0.2 μmol/l for mercury and 1.5 μmol/l for thimerosal. Inhibition was not reversible on wash-out, but was partially reversed by perfusing with the disulfide-reducing agent dithiothreitol (2 mmol/l). Inhibition was not dependent on voltage and there was no effect on the unit current amplitude. Koivisto and Nedergaard (1995) also reported that substances releasing nitric oxide (e.g., sodium nitroprusside, nitroglycerine) had a similar inhibitory effect, also reversible by dithiothreitol. These results were interpreted as resulting from the oxidation of sulfhydryl groups.

In contrast, a study using the whole-cell configuration provided evidence that extracellular oxygen-derived free radicals caused a non-selective cation current in guinea pig ventricular myocytes via a Ca^{2+}-independent pathway (Jabr and Cole 1995). Thimerosal had a similar activating effect while dithiothreitol prevented induction of the cation current. The authors concluded that although the activation was independent of Ca^{2+}, the current could be attributed to a Ca^{2+}-dependent non-selective cation channel because it was blocked by the same blockers as the cation current induced by an elevation in intracellular Ca^{2+} concentration. Oxygen-derived free radicals, when applied intracellularly, seem to cause a similar Ca^{2+}-dependent non-selective cation current via a different, indirect mechanism (Ca^{2+} increase; Jabr and Cole 1993). Oxidant stress caused by *tert*-butylhydroperoxide stimulated a Ca^{2+}-permeable non-selective cation channel in endothelial cells. In this latter case, the oxidant-induced cation channel was different from the Ca^{2+}-permeable NSC_{Ca} channel described in endothelial cells (Table 1) because it was Ca^{2+}-insensitive (Koliwad et al. 1996).

F. Phosphorylation-Dependent Regulation

I. Regulation via Protein Kinase A

The NSC_{Ca} channel is down-regulated via cAMP-dependent phosphorylation. A study of bursting neurons of *Helix pomatia* (Partridge and Swandulla 1990) showed that external application of the adenylylcyclase activator, forskolin, or of the phosphodiesterase inhibitor, 3-isobutyl-1-methylxanthine, or membrane-permeable analogs of cAMP, reversibly reduces NSC_{Ca} currents activated by injection of small quantities of Ca^{2+}. Injection of the catalytic subunit of the cAMP-dependent protein kinase A (PKA) had the same effect. Similar down-regulation has been demonstrated following application of serotonin, the action of which is mediated by cAMP. This effect was antagonized by an inhibitor of cyclic AMP-dependent protein kinase. Cyclic AMP-dependent regulation has been reported at the single-channel level in embryonic chick sensory neurons (Razani-Boroujerdi and Partridge 1993): PKA reduces the *Po* by decreasing open time durations. A similar down-regulation of channel activity by PKA has been reported for the outer hair cells of the guinea pig cochlea (Van den Abbeele et al. 1996): the application of the catalytic subunit of PKA reduced channel activity in maximally-activated inside-out patches in the presence of 1 mmol/l Ca^{2+}. In cell-attached patches, adenosine and forskolin had similar effects, reducing the channel activity elicited by the agonist external ATP.

In contrast, Sasaki and Gallacher (1992) observed that the cAMP-dependent transduction pathway potentiated an ATP-induced cation current in mouse lacrimal glands. However, although the channel involved has a unit conductance of 30 pS (Sasaki and Gallacher 1990), it is not clear whether the channel is depending on calcium and internal ATP. Another feature is not typical of classical NSC_{Ca} channels: permeability to Ca^{2+} was half that to sodium. Further studies are then necessary to ascertain whether this channel belongs to the same class of NSC_{Ca} channels or should be considered as a different type of channel.

II. Effects of Other Protein Kinases

Braun and Schulman (1995) reported that the calcium-mediated activation of NSC_{Ca} channels in the T84 cell line may involve a phosphorylation step mediated by a calmodulin-dependent kinase because intracellular dialysis with inhibitors of this kinase blocked the activation of cation currents recorded in the whole-cell configuration. However, the effect of the Ca^{2+}-calmodulin-dependent protein kinase is probably indirect. Local perfusion of purified caldmodulin-dependent protein kinase onto the internal face of inside-out patches does not activate NSC_{Ca} channels. Korbmacher et al. (1995) showed that the NSC_{Ca} channel in a mouse renal cell line is not modulated by cGMP-dependent kinase, and Fitz et al. (1994) demonstrated that NSC_{Ca} channels in liver cells are down-regulated by PKC in inside-out patches.

G. Dependence on Hypertonicity

Volk et al. (1995) investigated the effects of hypertonicity on cultured collecting duct cells in the whole-cell configuration and demonstrated that the inward currents elicited in hypertonic medium (140 mmol/l NaCl+100 mmol/l sucrose) were due the activation of the NSC_{Ca} channel. The currents induced in hypertonic conditions were not affected by replacing the Na^+ in the bath with K^+, Cs^+, Li^+, or Rb^+, but they are almost completely abolished in the presence of *N*-methyl-D-glucamine. The current is blocked by flufenamic acid (0.1 mmol/l) or gadolinium (0.1 mmol/l) but not by amiloride (0.1 mmol/l). ATP in the pipette (1 mmol/l or 10 mmol/l) decreases hypertonicity-induced currents in a dose-dependent manner. Surprisingly, these currents are independent of the concentration of internal Ca^{2+}. In subsequent experiments, the authors could record single-channel openings in the whole-cell configuration during the onset of the currents activated by hypertonicity, in conditions in which there were no K^+ currents. The channel activated by hypertonicity has a unit conductance of about 26 pS and is cation-selective. However, the current is not dependent on internal Ca^{2+}. Previous studies by the same authors on the same cell model showed the presence of a classical NSC_{Ca} channel (KORBMACHER et al. 1995), so the authors have suggested that the NSC_{Ca} channel is responsible for hypertonicity-induced current. This study is particularly important because it reported for the first time Ca^{2+}-independent activation of the NSC_{Ca} channel, which may be important when the Ca^{2+} sensitivity of the channel is too low to be physiologically significant. Indeed, this is the case in the mouse collecting duct (CHRAIBI et al. 1995). Activation of non-selective cation channels by hypertonicity occurs in various types of cultured cells (KOCH and KORBMACHER 1999) but not in rat colonic crypts (WEYAND et al. 1998).

While the NSC_{Ca} channel appears to be sensitive to hypertonicity, hypotonicity is apparently ineffective (but see ONO et al. 1994). Two studies on tissues containing both stretch-activated cation channels and NSC_{Ca} channels have shown that the NSC_{Ca} channel is not stimulated by application of a negative pressure onto the membrane patch via the electrode (POPP et al. 1993; YEH et al. 1998).

H. Agonist-Mediated Control of NSC_{Ca} Channels

NSC_{Ca} channels are usually, but not always, closed in basal conditions. Channel activity has been detected in a low percentage of cell-attached patches (4–25%) in kidney, nasal, and cochlea cells (TEULON et al. 1987; CHRAIBI et al. 1994; MATSUNAGA et al. 1991; JORISSEN et al. 1990; VAN DEN ABBEELE et al. 1994; MARUNAKA et al. 1992) whereas it has never been recorded in pancreas (GRAY and ARGENT 1990; PETERSEN 1992) or intestine cells (BUTT and HAMILTON 1998; SIEMER and GÖGELEIN 1992). As mentioned above, the sensitivities to calcium and ATP determined in inside-out patches seemed at first to preclude any Ca^{2+}-dependent activation of the channel in intact cells. However, it was later recognized that sensitivity to these agents were affected by excision. It has also been shown on several occasions that an increase in intracellular Ca^{2+} caused by an injection of Ca^{2+} or by superfusion of a Ca^{2+} ionophore is enough to cause the opening of the NSC_{Ca} channel. Therefore, intracellular calcium is currently considered to be the predominant, although perhaps not the only (cf. hypertonicity or oxidation), agent mediating activation of the NSC_{Ca} channel in situ.

The effects of agonists are most studied for the pancreatic acinus, in which cholecystokinin and acetylcholine stimulate the NSC_{Ca} channel at high (MARUYAMA and PETERSEN 1982; KASAI and AUGUSTINE 1990; RANDRIAMAMPITA et al. 1988) and physiological concentrations (THORN and PETERSEN 1993) via an increase in Ca^{2+} concentration. Carbamylcholine also causes channel activation in rat lacrimal glands (MARTY et al. 1984), a process which can be mimicked by a Ca^{2+} ionophore. Other Ca^{2+} mobilizing agonists such as ATP (VAN DEN ABBEELE et al. 1996; FITZ et al. 1994), histamine (BREGESTOVSKI et al. 1988; NILIUS et al. 1993), or bradykinin (BARON et al. 1996) activate the NSC_{Ca} channel in various cell types. TOHDA et al. (1994) reported that terbutaline, a β_2 agonist, increases intracellular Ca^{2+} up to 1.5 μmol/l in fetal distal lung epithelium and stimulates NSC_{Ca} channel activity. The posssible effects of cyclic AMP accumulation on channel activity were not discussed. In this case however, the channel activation is not due solely to an increase in Ca^{2+} but is also due to terbutaline apparently increasing channel sensitivity to Ca^{2+}. In addition, according to the authors, channel activation is facilitated by a decrease in intracellular Cl^- concentration following exposure to terbutaline, which seems per se to increase *Po*.

A different mechanism of activation has also been proposed. Using the perforated patch-clamp method on rat distal colon, SIEMER and GÖGELEIN (1992, 1993) established that forskolin and PGE_2 caused membrane depolarization of the crypt cells, which they attributed to stimulation of Cl^- conductance in the upper part of the crypts, and to the onset of a cation conductance in the base of the crypts. A single-channel current recording using the cell-attached and inside-out configurations of the patch-clamp technique provided evidence for the activation of a Ca^{2+}-impermeable, non-selective cation channel blocked by mefenamic acid and flufenamic acids (100 μmol/l). Channel sensitivity to Ca^{2+} was not investigated. These studies, which suggest that the NSC_{Ca} channel is stimulated via a cyclic AMP-dependent pathway, conflict with the work of ECKE et al. (1996) who found no evidence for forskolin activation of NSC_{Ca} channels. The most frequently reported influence of the cyclic AMP-dependent pathway on NSC_{Ca} channel activity is negative regulation, not activation. Inhibition has been demonstrated on inside-out patches using PKA and also on intact cells, using forskolin, 8-bromo-cyclic AMP, and agonists such as serotonin and adenosine (PARTRIDGE et al. 1990; VAN DEN ABBEELE et al. 1996).

I. Physiological Role

Paradoxically, given its broad tissue distribution, the NSC_{Ca} channel has rarely been attributed a clear physiological function. The search for physiological function is hampered by the absence of an identified agonist in many preparations[4], and by the lack of a specific blocker for physiological studies in organs. The effects of activation of this type of channel, membrane depolarization and Na^+ entry suggest possible functions. The channel is probably involved in Ca^{2+} signaling when permeability to Ca^{2+} is high, as in neutrophils or endothelial cells (Table 1), but the specific role NSC_{Ca} channels is difficult to distinguish from that of other Ca^{2+}-permeable channels in the same cells (NILIUS et al. 1997).

I. Excitable Cells: "Voltage Signal"

The possible function of NSC_{Ca} channels has been most thoroughly explored in neurons, in which NSC_{Ca} channels are thought to cause a sustained depolarization during bursts of action potential. Neurons of *Helix pomatia* isolated from the circumesophagal ganglion complex have a spontaneous bursting activity which is induced by inward currents carried by cations (HOFMEIER and LUX 1981; SWANDULLA and LUX 1985). Several studies have identified the NSC_{Ca} channel as being responsible for the long lasting depolarization that causes bursts after Ca^{2+} injection (PARTRIDGE and SWANDULLA 1987; PARTRIDGE et al. 1990). The NSC_{Ca} channel of *Helix pomatia* neurons has not been tested for Ca^{2+} permeability, but the non-selective cation current can be carried by Ca^{2+} (PARTRIDGE and SWANDULLA 1988). A similar Ca^{2+}-activated cation current has been described in bursting neurons of *Aplysia* (KRAMER and ZUCKER 1985; LEWIS 1984), which conducts both Ca^{2+} and Na^+ (KNOX et al. 1996; KRAMER and ZUCKER 1985). A recent study has also reported a Ca^{2+}-activated non-selective cation current in pyramidal cells of the rat cerebral cortex, that is thought to be involved in a particular pattern of spike firing (HAJ-DAHMANE and ANDRADE 1997). Given the diversity of cation channels, other, calcium-independent, cation currents may be responsible for long-lasting depolarization: this is the case, for instance, for the cation current induced by neurotensin in midbrain dopaminergic neurons (FARKAS et al. 1996; CHIEN et al. 1996).

The effect of NSC_{Ca} channel opening on cardiac cell function is unclear. Membrane depolarization associated with oscillations in the intracellular Ca^{2+} has been observed in cardiac cells under various pathological conditions (see KASS et al. 1978; SIPIDO et al. 1995; JABR and COLE 1993) and have been attributed to a transient inward current. The waves of membrane depolarization may propagate and induce arrhythmia. It is widely thought that the Na^+/Ca^{2+} exchanger is largely responsible for the inward current but there is much debate as to whether Ca^{2+}-activated non-selective cation currents are also involved. In ventricular cells of the guinea pig in which NSC_{Ca} channels have been described in detail (EHARA et al. 1988), neither NSC_{Ca} channels nor Ca^{2+}-activated Cl^- channels are involved in the membrane depolarization caused

[4] The case for the renal tubule, in which all attempts to activate the NSC_{Ca} channel via Ca^{2+}-mobilizing agonists have been unsuccessful (Teulon, unpublished observations)

by Ca^{2+} oscillations. This depolarization is therefore due only to the Na^+/Ca^{2+} exchanger (SIPIDO et al. 1995). In contrast, another study demonstrated that oxygen-derived free radicals induce a Ca^{2+}-activated non-selective cation current in ventricular cells of the guinea pig (JABR and COLE 1993, 1995). The authors suggested then that NSC_{Ca} channels may be implicated in disorders associated with myocardial injury during reperfusion after ischemia. NSC_{Ca} channels and Ca^{2+}-activated Cl^- channels may be responsible for inward currents in Purkinje fibers (KASS et al. 1978; CANNELL and LEDERER 1986; SIPIDO et al. 1993).

In the above examples, the effects of NSC_{Ca} channel activation derive from membrane depolarization which has a self-evident importance in excitable cells, but also in various non-excitable cell types, including the outer hair cells of the mammalian cochlea. These cells amplify the movements of the cochlear partition by generating motile responses and, through this process, are thought to be responsible for fine discrimination between sound frequencies. Membrane depolarization, like other agents, can cause slow motile response, suggesting the probable involvement of NSC_{Ca} channels (VAN DEN ABBEELE et al. 1996). It is also suggested that the channel modulates the process of insulin release in β cells (REALE et al. 1994). The glucose-induced closure of the ATP-dependent K^+ channel can depolarize the membrane only in the presence of an inward current, which may be partly due to NSC_{Ca} channel activity (ROE et al. 1998; LEECH and HABENER 1998).

II. Exocrine Glands: Participation in Cl^- Transport

Secretagogues like acetylcholine and cholecystokinin cause NSC_{Ca} channel opening in pancreatic and lacrimal acinar cells (see Sect. H), suggesting that this channel is involved in electrolyte secretion. Functions in NaCl secretion via modulation of the membrane potential, or in sustained NaCl secretion via Ca^{2+} entry have been suggested. It was originally proposed that the NSC_{Ca} channel was responsible for Ca^{2+} entry in the phase of sustained secretion. Indeed, although no Ca^{2+} currents were recorded through the NSC_{Ca} channel, a low permeability to Ca^{2+} may be sufficient to account for secretagogue-induced Ca^{2+} entry (PETERSEN and MARUYAMA 1983). However, recent evidence suggests that the NSC_{Ca} channel does not play a major role in Ca^{2+} entry into pancreatic acinar cells (PFEIFFER et al. 1995). The situation is probably different in lacrimal glands, in which a Ca^{2+}-permeable non-selective cation channel is opened by external ATP (SASAKI and GALLACHER 1990). It is not known whether this channel is a true NSC_{Ca} channel.

The standard model of NaCl secretion in exocrine acinar cells is based on the cooperation of the Na^+-K^+ pump, Na^+-K^+-Cl^- cotransport, and Ca^{2+}-activated K^+ channels (maxi K^+) at the basolateral membrane, and Ca^{2+}-dependent Cl^- channels at the apical membrane. The whole system allows the transcellular transfer of Cl^- from the interstitium to the lumen of the acinus whereas Na^+ flows through the paracellular pathway (PETERSEN 1992).

However, this model is not applicable to all exocrine cells and, in particular, not to mouse and rat pancreatic acinar cells which have no maxi K^+ channel. KASAI and AUGUSTINE (1990) have described an ingenious "push-pull" model for mouse and rat pancreatic acinar cells, based a secretagogue-evoked Ca^{2+} signal spreading from the apical to the basolateral pole of the cell. The initial step in electrolyte secretion is the discharge of Cl^- into the lumen due to the opening of Cl^- channels ($V–E_{Cl} < 0$)[5]. The increase in internal Ca^{2+} concentration at the basal pole of the cell then induces the opening of both Cl^- and NSC_{Ca} channels. The activation of the NSC_{Ca} channels induces a membrane depolarization such that $V–E_{Cl} > 0$. Cl^- enters the cell through the Cl^- channels in the basolateral membrane. According to KASAI and AUGUSTINE (1990) and PETERSEN (1992), this model is consistent with all available electrophysiological data. A similar model (except for the inclusion of local Ca^{2+} fluctuations) was originally proposed by MARTY et al. (1984) for rat lacrimal glands, which have maxi K^+ channels. The opening of NSC_{Ca} channels causes membrane depolarization which is amplified by Na^+ blockade of the maxi K^+ channels (due to the entry of Na^+ through the NSC_{Ca} channels) in the basolateral membrane. The depolarization is enough to permit Cl^- entry through Cl^- channels in the basolateral membrane. The electrochemical gradient for Cl^- across the apical membrane is negative (apical K^+ channels are not blocked) so Cl^- diffuses into the lumen of the acinus via apically located Cl^- channels. These two models are remarkable in their simplicity, but they are probably unrealistic in restraining Na^+-K^+-Cl^- cotransport to a marginal role (PETERSEN 1992; PAULAIS and TURNER 1992; TURNER et al. 1993).

The role attributed to the NSC_{Ca} channel in these examples conforms to what we expect from a non-selective cation channel: control of membrane potential and Na^+ entry. The blockade of a K^+ channel, due to Na^+ entry via a cation channel (P_{2X} type), has recently been reported (STRUBING and HESCHELER 1996).

III. Other Epithelia: Speculative Functions

Channel function has not been explored in epithelia. In particular, I am not aware of physiological studies (short-circuit current, microperfused renal tubules . . .) giving evidence that the NSC_{Ca} channel is involved in transepithelial electrolyte transport. Thus the functions attributed to NSC_{Ca} channels in epithelia are almost entirely speculative. Some possible functions are given below.

As for exocrine acinar cells (Sect. I.II), NSC_{Ca} channels in the basolateral membrane of epithelia may participate in Cl^- secretion via two Cl^- channels in series, provided that $V–E_{Cl}$ remains negative at the luminal membrane. The advantage of the arrangement is not clear and, although there is reasonable

[5] V: membrane potential; E_{Cl}: Equilibrium potential for the Cl^- ion

evidence that this transport system functions in mouse pancreas, Ca^{2+}-dependent Cl^- secretion controlled by NSC_{Ca} channels has not been demonstrated for other epithelia. NSC_{Ca} channels on the apical side may be part of a reciprocal system allowing transcellular Cl^- absorption. In both cases, these transport systems may function only if the two membrane potentials vary independently (tight epithelia). In the above examples, the involvement of NSC_{Ca} channels in transcellular ion transport would be indirect. NSC_{Ca} channels may also be directly involved in the transcellular absorption of Na^+ under the control of internal Ca^{2+}concentration. However, the NSC_{Ca} channel has not been identified in apical membranes of native Na^+-absorbing epithelia such as the cortical collecting duct in the kidney, or the distal colon cells. A puzzling case is that of the marginal cell of the stria vascularis, which is in contact with the endolymph, a medium containing more than 100 mmol/l K^+. NSC_{Ca} channels on the apical membrane might theoretically secrete K^+ ions but, according to TAKEUCHI et al. (1992), the channel density is insufficient to account for overall secretion.

With the exception of the acinus, it is unclear whether the NSC_{Ca} channel is involved at all in transcellular ion transport. VOLK et al. (1995), for example, suggested a role in cell volume regulation in medullary renal cells which are in contact with a hypertonic medium. SIEMER and GÖGELEIN (1992, 1993) suggested that the NSC_{Ca} channel is involved in the proliferation of colon cells from the crypt base. NSC_{Ca} channels may also regulate a number of ion transport systems via Na^+ entry (Na^+pump, Na^+/H^+ exchanger, Na^+-K^+-Cl^- cotransport) by affecting chemical gradients. This is perhaps the most striking feature of the channel: we know much about how the channel is regulated in isolated patches, but comparatively less about its functioning in situ. Determining the physiological conditions leading to the opening of these specific channels will be a major aim of future studies, particularly those involving epithelia.

Acknowledgments. I would like to thank all those who, at one time or another, worked with me on these channels: Ahmed Chraïbi, Thierry Van Den Abbeele, Romain Guinamard, Richard Moreau, Pedro Marvao and, primarily, Marc Paulais. I am also greatly indebted to Ian Findlay and David Gallacher for their friendship and for introducing me to the patch-clamp technique, and to professor O.H. Petersen for kindly receiving me in his laboratory, many years ago.

References

Adams DJ, Dwyer TM, Hille B (1980) The permeability of endplate channels to monovalent and divalent metal cations. J Gen Physiol 75:493–510

Barnstable CJ (1993) Cyclic nucleotide-gated nonselective cation channels: a multifunctional gene family. In: Nonselective cation channels:pharmacology, physiology and biophysics. Siemen D, Hescheler J, editors. Basel: Birkhauser Verlag, pp. 121–133

Baron A, Frieden M, Chabaud F, Beny JL (1996) Ca(2+)-dependent non-selective cation and potassium channels activated by bradykinin in pig coronary artery endothelial cells. J Physiol (Lond) 493:691–706

Bevan S, Gray PTA, Ritchie JM (1984) A calcium-activated cation-selective channel in rat cultured Schwann cells. Proc R Soc Lond B 222:349–355

Bleich M, Riedemann N, Warth R, Kerstan D, Leipziger J, Hor M, Driessche WV, Greger R (1996) Ca^{2+} regulated K+ and non-selective cation channels in the basolateral membrane of rat colonic crypt base cells. Pflügers Arch 432:1011–1022

Braun AP, Schulman H (1995) A non-selective cation current activated *via* the multifunctional Ca^{2+}-caldmodulin-dependent protein kinase in human epithelial cells. J Physiol (Lond) 488:37–55

Bregestovski P, Bakhramov A, Danilov S, Moldobaeva A, Takeda K (1988) Histamine-induced inward currents in cultured endothelial cells from human umbilical vein. Brit J Pharmacol 95:429–436

Butt AG, Hamilton KL (1998) Ion channels in isolated mouse jejunal crypts. Pflugers Arch 435:528–538

Cannell MB, Lederer WJ (1986) The arrhythmogenic current I_{TI} in the absence of electrogenic sodium-calcium exchange in sheep cardiac purkinje fibres. J Physiol (Lond) 374:201–219

Caterina MJ, Schumacher MA, Tominaga M, Rosen TA, Levine JD, Julius D (1997) The capsaicin receptor: a heat-activated ion channel in the pain pathway. Nature 389:816–824

Champigny G, Verrier B, Lazdunski M (1991) A voltage, calcium, and ATP sensitive non selective cation channel in human colonic tumor cells. Biochem Biophys Res Comm 176:1196–1203

Chien PY, Farkas RH, Nakajima S, Nakajima Y (1996) Single-channel properties of the nonselective cation conductance induced by neurotensin in dopaminergic neurons. Proc Natl Acad Sci USA 93:14917–14921

Chraibi A, Guinamard R, Teulon J (1995) Effects of internal pH on the nonselective cation channel from the mouse collecting tubule. J Membr Biol 148:83–90

Chraibi A, Van den Abbeele T, Guinamard R, Teulon J (1994) A ubiquitous non-selective cation channel in the mouse renal tubule with variable sensitivity to calcium. Pflügers Arch 429:90–97

Colqhoun D, Neher E, Reuter H, Stevens CF (1981) Inward current channels activated by intracellular Ca in cultured cardiac cells. Nature 294:752–754

Conley EC (1996) The ion channel Facts book. Intracellular ligand-gated channels. Native calcium-activated non-selective cation channels. Academic Press; London pp. 284–304

Cook DI, Poronnik P, Young JA (1990) Characterization of a 25-pS nonselective cation channel in a cultured secretory epithelial cell line. J Membr Biol 114:37–52

Cooper K, Gates P, Rae JL, Dewey J (1990) Electrophysiology of cultured human lens epithelial cells. J Membr Biol 117:285–298

Dinudom A, Young JA, Cook DI (1994) Ion channels in the basolateral membrane of intralobular duct cells of mouse mandibular glands. Pflügers Arch 3–428:202–208

Duszik M, French AS, Man SFP (1991) Cation channels in normal and cystic fibrosis human airway epithelial cells. Biomedical Res 12:17–23

Ecke D, Bleich M, Greger R (1996) Crypt base cells show forskolin-induced Cl- secretion but no cation inward conductance. Pflügers Arch 431:427–434

Ehara T, Noma A, Ono K (1988) Calcium-activated non-selective cation channel in ventricular cells isolated from adult guinea-pig hearts. J Physiol (Lond) 403:117–133

Farkas RH, Chien PY, Nakajima S, Nakajima Y (1996) Properties of a slow nonselective cation conductance modulated by neurotensin and other neurotransmitters in midbrain dopaminergic neurons. J Neurophysiol 76:1968–1981

Feng ZP, Clark RB, Berthiaume Y (1993) Identification of nonselective cation channels in cultured adult rat alveolar type II cells. American Journal of Respiratory Cell & Molecular Biology 9:248–254

Fitz JG, Sostman AH, Middleton JP (1994) Regulation of cation channels in liver cells by intracellular calcium and protein kinase C. Am J Physiol 266:G677–84

Galietta LJV, Mastrocola T, Nobile M (1989) A class of non-selective cation channel in human fibroblasts. FEBS Letters 253:43–46

Garty H, Palmer LG (1997) Epithelial sodium channels: function, structure, and regulation. Physiol Rev 77:359–396

Gauss R, Seifert R, Kaupp UB (1998) Molecular identification of a hyperpolarization-activated channel in sea urchin sperm. Nature 393:583–587

Gögelein H (1990) Ion channels in mammalian proximal renal tubules. Renal Physiol Biochem 13:8–25

Gögelein H, Capek K (1990) Quinine inhibits chloride and nonselective cation channels in isolated rat distal colon cells. Biochim Biophys Acta 1027:191–198

Gögelein H, Dahlem D, Englert HC, Lang HJ (1990) Flufenamic acid, mefenamic acid and niflumic acid inhibit single nonselective cation channels in the rat exocrine pancreas. FEBS Letters 268:79–82

Gögelein H, Greger R (1986) A voltage-dependent ionic channel in the basolateral membrane of late proximal tubules of the rabbit kidney. Pflügers Arch 407:S142–S148.

Gögelein H, Pfannmuller B (1989) The nonselective cation channel in the basolateral membrane of rat exocrine pancreas. Inhibition by 3',5-dichlorodiphenylamine-2-carboxylic acid (DCDPC) and activation by stilbene disulfonates. Pflügers Arch 413:287–298

Gray MA, Argent BE (1990) Non-selective cation channel on pancreatic duct cells. Biochim Biophys Acta 1029:33–42

Haj-Dahmane S, Andrade R (1997) Calcium-activated cation nonselective current contributes to the fast afterdepolarization in rat prefrontal cortex neurons. J Neurophysiol 78:1983–1989

Hamill O, Marty A, Neher E, Sakmann B, Sigworth F (1981) Improved patch-clamp technique for high resolution recording from cells and cell-free membrane patches. Pflugers Archiv Eur 391:85–100

Hamill OP, McBride Jr DW (1996) The pharmacology of mechanogated membrane ion channels. Pharmacol Rev 48:231–252

Hamilton KL, Benos DJ (1990) A non-selective cation channel in the apical membrane of cultured A6 kidney cells. Biochim Biophys Acta 1030:16–23

Herson PS, Dulock KA, Ashford MLJ (1997) Characterization of a nicotinamide-adenine-dinucleotide-dependent cation channel in the CRI-G1 rat insulinoma cell line. J Physiol (Lond) 505:65–76

Hille B (1992) Ionic channels of excitable membranes.Sinauer; Sunderland, USA p. 354

Hofmeier G, Lux HD (1981) The time courses of intracellular free calcium and related electrical effects after injection of $CaCl_2$ into neurons of the snail, *Helix Pomatia*. Pflügers Arch 391:242–251

Isenberg G (1993) Nonselective cation channels in cardiac and smooth muscle cells. In: Nonselective cation channels. Pharmacology, Physiology and Biophysics. Siemen D, Hescheler J, editors. Basel: Birhaüser Verlag, pp. 247–260

Jabr RI, Cole WC (1993) Alterations in electrical activity and membrane currents induced by intracellular oxygen-derived free radical stress in guinea pig ventricular myocytes. Circulation Res 72:1229–1244

Jabr RI, Cole WC (1995) Oxygen-derived free radical stress activates nonselective cation current in guinea pig ventricular myocytes. Circulation Res 76:812–824

Jorissen M, Vereecke J, Carmeliet E, Van den Berghe H, Cassiman JJ (1990) Identification of a voltage- and calcium-dependent non-selective cation channel in cultured adult and fetal human nasal epithelial cells. Pflügers Arch 415:617–623

Kasai K, Augustine GJ (1990) Cytosolic Ca2+ gradients triggering unidirectional fluid secretion from exocrine pancreas. Nature 348:735–738

Kass RS, Lederer WJ, Tsien RW, Weingart R (1978) Role of calcium ions in tansient inward currents and aftercontractions induced by strophantidin in cardiac purkinje fibres. J Physiol (Lond) 281:187–208

Knox RJ, Jonas EA, Kao L-S, Smith PJS, Connor JA, Kaczmarek LK (1996) Ca^{2+} influx and activation of a cation current are coupled to intracellular Ca2+ release in peptidergic neurons of *Aplysia californica*. J Physiol (Lond) 494:627–639
Koch J-P, Korbmacher C (1999) Osmotic shrinkage activates nonselective cation channels (NSC) in various cell types. J Membr Biol 168:131–139
Koivisto A, Dotzler E, Russ U et al. (1993) Nonselective cation channels: Pharmacology, Physiology and biophysics. Nonselective cation channels in brown and white fat cells.; Siemen D, Herscheler J, editors. Birhaüser Verlag; Basel pp. 201–211
Koivisto A, Nedergaard J (1995) Modulation of calcium-activated non-selective cation channel activity by nitric oxide in rat brown adipose tissue. J Physiol (Lond) 486:59–65
Koivisto A, Siemen D, Nedergaard J (1993) Reversible blockade of the calcium-activated nonselective cation channel in brown fat cells by the sulfhydryl reagents mercury and thimerosal. Pflügers Arch 425:549–551
Koliwad SK, Kunze DL, Elliot SJ (1996) Oxydant stress activates a non-selective cation channel responsible for membrane depolarization in calf vascular endothelial cells. J Physiol (Lond) 491:1–12
Korbmacher C, Volk T, Segal AS, Boulpaep EL, Fromter E (1995) A calcium-activated and nucleotide-sensitive nonselective cation channel in M-1 mouse cortical collecting duct cells. J Membr Biol 146:29–45
Kramer RH, Zucker RS (1985) Calcium-dependent inwrd current in *Aplysia* bursting pace-maker neurones. J Physiol (Lond) 362:107–130
Leech CA, Habener JF (1998) A role for Ca2+-sensitive nonselective cation channels in regulating the membrane potential of pancreatic beta-cells. Diabetes 47:1066–1073
Lewis DV (1984) Spike aftercurrents in R15 of *Aplysia*: their relationship to slow inward current and calcium influx. J Neurophysiol 51:387–403
Lidofsky SD, Sostman A, Fitz JG (1997) Regulation of cation-selective channels in liver cells. J Membr Biol 157:231–236
Light DB, McCann FV, Keller TM, Stanton BA (1988) Amiloride-sensitive cation channel in apical membrane of inner medullary collecting duct. Am J Physiol 2552:F278–86
Lindau M, Fernandez JM (1986) A patch-clamp study of histamine-secreting cells. J Gen Physiol 88:349–368
Ling BN, Hinton CF, Eaton DC (1991) Amiloride-sensitive sodium channels in rabbit cortical collecting tubule primary cultures. Am J Physiol 261:F933–F944
Lipton SA (1986) Antibody activates cationic channels via second messenger Ca^{2+}. Biochim Biophys Acta 856:59–67
Loirand G, Pacaud P, Baron A, Mironneau C, Mironneau J (1991) Large conductance calcium-activated non-selective cation channel in smooth muscle cells isolated from rat portal vein. J Physiol (Lond) 437:461–475
Ludwig A, Zong X, Jeglitsch M, Hofman F, Biel M (1998) A family of hyperpolarizatio-activated mammalian cation channels. Nature 393:587–591
Marcus DC, Takeuchi S, Wangemann P (1992) Ca(2+)-activated nonselective cation channel in apical membrane of vestibular dark cells. Am J Physiol 262:C1423–9
Marty A, Tan YP, Trautmann A (1984) Three types of calcium-dependent channel in rat lacrimal glands. J Physiol (Lond) 357:293–325
Marunaka Y (1996) Amiloride-blockable Ca^{2+}-activated N^{a+}-permeant channels in the fetal distal lung epithelium. Pflügers Arch. 431:748–756.
Marunaka Y, Tohda H, Hagiwara N, O'Brodovich H (1992) Cytosolic Ca2+-induced modulation of ion selectivity and amiloride sensitivit of a cation channel and beta agonist action in fetal lung epithelium. Biochem Biophys Res Comm 187:648–656
Maruyama Y, Moore D, Petersen OH (1985) Calcium-activated cation channel in rat thyroid follicular cells. Biochim Biophys Acta 821:229–232
Maruyama Y, Petersen OH (1982) Single-channel currents in isolated patches of plasma membrane from basal surface of pancreatic acini. Nature 299:159–161

Maruyama Y, Petersen OH (1984) Single calcium-dependent cation channels in mouse pancreatic acinar cells. J Membr Biol 81:83–87

Matsunaga H, Yamashita T, Miyajima Y, Okuda T, Chang H, Ogata E, Kurokawa K (1991) Ion channel activities of cultured rat mesangial cells. Am J Physiol 261:F808–F814

Nilius B (1990) Permeation properties of a non-selective cation channel in human vascular endothelial cells. Pflügers Arch 416:609–611

Nilius B, Schwartz G, Oike M, Droogmans G (1993) Histamine-activated, nonselective cation currents and Ca^{2+} transients in endothelial cells from human umbilical vein. Pflügers Arch 424:258–293

Nilius B, Viana F, Droogmans G (1997) Ion channels in vascular endothelium. Annu Rev Physiol 59:145–170

Noma A (1983) ATP-regulated K^+ channels in cardiac muscle. Nature 305:147–148

Nonaka T, Matsuzaki K, Kawahara K, Suzuki K, Hoshino M (1995) Monovalent cation selective channel in the apical membrane of rat inner medullary collecting duct cells in primary culture. Biochim Biophys Acta 1233:163–174

Ono S, Mougouris T, DuBose TD, Jr., Sansom SC (1994) ATP and calcium modulation of nonselective cation channels in IMCD cells. Am J Physiol 267:F558–65

Orser BA, Bertlik M, Fedorko L, O'Brodovitch H (1991) Non-selective cation channel in fetal alveolar type II epithelium. Biochim Biophys Acta 1094:19–26

Palmer LG, Frindt G (1992) Conductance and gating of epithelial Na^+ channels from rat cortical collecting tubule. Effects of luminal Na and Li. J Gen Physiol 92:37–43

Partridge LD, Swandulla D (1987) Single Ca-activated cation channels in bursting neurons of *Helix*. Pflügers Arch 410:627–631

Partridge LD, Swandulla D (1988) Calcium-activated non-specific cation channels. Trends in Neurosciences 11:69–72

Partridge LD, Swandulla D, Müller TH (1990) Modulation of calcium-activated non-specific cation currents by cyclic AMP-dependent phosphorylation in neurones of *Helix*. J Physiol (Lond) 429:131–145

Paulais M, Teulon J (1989) A cation channel in the thick ascending limb of Henle's loop of the mouse kidney: inhibition by adenine nucleotides. J Physiol (Lond) 413:315–327

Paulais M, Turner RJ (1992) β-adrenergic upregulation of the Na^+-K^+-$2Cl^-$ cotransporter in rat parotid acinar cells. J Clin Invest 89:1142–1147

Petersen OH (1992) Simulus-secretion coupling: cytoplasmic calcium signals and the control of ion channels in exocrine acinar cells. J Physiol (Lond) 448:1–51

Petersen OH, Maruyama Y (1983) What is the mechanism of the calcium influx to pancreatic acinar cells evoked by secretagogues? Pflügers Arch 396:82–84

Pfeiffer F, Schmid A, Schulz I (1995) Capacitive Ca^{2+} influx and a Ca^{2+}-dependent non-selective cation pathway are discriminated by genistein in mouse pancreatic acinar cells. Pflügers Arch 430:916–922

Poncet V, Mérot J, Poujeol P (1992) A calcium-permeable channel in the apical membrane of primary cultures of the rabbit distal bright convoluted tubule. Pflügers Arch 422:112–119

Popp R, Gögelein H (1992) A calcium and ATP sensitive nonselective cation channel in the antiluminal membrane of rat cerebral capillary endothelial cells. Biochim Biophys Acta 1108:59–66

Popp R, Hoyer J, Gögelein H (1993) Mechanosensitive nonselective cation channels in the antiluminal membrane of cerebral capillaries (blood-brain barrier). In: Nonselective cation channels. Pharmacology, Physiology and Biophysics pp. 101–105.

Rae JL, Dewey J, Cooper K, Gates P (1990) A non-selective cation channel in rabbit corneal endothelium activated by internal calcium and inhibited by internal ATP. Exp Eye Res 50:373–384

Randriamampita C, Chanson M, Trautmann A (1988) Calcium and secretagogues-induced conductances in rat exocrine pancreas. Pflugers Arch 411:53–57

Razani-Boroujerdi S, Partridge LD (1993) Activation and modulation of calcium-activated non-selective cation channels from embryonic chick sensory neurons. Brain Res 623:195–200

Reale V, Hales CN, Ashford ML (1994a) The effects of pyridine nucleotides on the activity of a calcium-activated nonselective cation channel in the rat insulinoma cell line, CRI-G1. J Membr Biol 142:299–307

Reale V, Hales CN, Ashford ML (1994b) Nucleotide regulation of a calcium-activated cation channel in the rat insulinoma cel line, CRI-G1. J Membr Biol 141:101–112

Reale V, Hales CN, Ashford ML (1995) Regulation of calcium-activated nonselective cation channel activity by cyclic nucleotides in the rat insulinoma cell line, CRI-G1. J Membr Biol 145:267–278

Roe MW, Worley JF 3rd, Qian F, Tamarina N, Mittal AA, Dralyuk F, Blair NT, Mertz RJ, Philipson LH, Dukes ID (1998) Characterization of a Ca^{2+} release-activated nonselective cation current regulating membrane potential and $[Ca^{2+}]_i$ oscillations in transgenically derived beta-cells. J Biol Chem 273:10402–10410

Sasaki T, Gallacher DV (1990) Extracellular ATP activates receptor-operated cation channels in mouse lacrimal acinar cells to promote calcium influx in the absence of phosphoinositide metabolism. FEBS Letters 264:130–134

Sasaki T, Gallacher DV (1992) The ATP-induced inward current in mouse lacrimal acinar cells is potentiated by isoprenaline and GTP. J Physiol (Lond) 447:103–118

Siemen D, Hescheler J (1993) Nonselective cation channels. Pharmacology, Physiology and Biophysics. Birkaüser Verlag; Basel

Siemen D, Reuhl T (1987) Non-selective cationic channel in primary cultured cells of brown adipose tissue. Pflügers Arch 408:534–536

Siemer C, Gögelein H (1992) Activation of nonselective cation channels in the basolateral membrane of rat distal colon crypt cells by prostaglandin E2. Pflügers Arch 420:319–328

Siemer C, Gögelein H (1993) Effects of forskolin on crypt cells of rat distal colon. Activation of nonselective cation channels in the crypt base and of a chloride conductance pathway in other parts of the crypt. Pflügers Arch 424:321–328

Sipido KR, Callewaert G, Carmeliet E (1993) $[Ca^{2+}]_i$ transients and $[Ca^{2+}]_i$-dependent chloride currents in single Purkinje cells from rabbit heart. J Physiol (Lond) 468:641–667

Sipido KR, Callewaert G, Porciatti F, Vereecke J, Carmeliet E (1995) [Ca2+]i-dependent membrane currents in guinea-pig ventricular cells in the absence of Na/Ca exchange. Pflügers Arch 430:871–878

Strubing C, Hescheler J (1996) Potassium current inhibition by nonselective cation channel-mediated sodium entry in rat pheochromocytoma (PC-12) cells. Biophys J 70:1662–1668

Sturgess NC, Hales CN, Ashford ML (1986) Inhibition of a calcium-activated, non-selective cation channel, in a rat insulinoma cell line, by adenine derivatives. FEBS Letters 2:397–400

Sturgess NC, Hales CN, Ashford ML (1987) Calcium and ATP regulate the activity of a non-selective cation channel in a rat insulinoma cell line. Pflügers Arch 409:607–615

Sunose H, Ikeda K, Saito Y, Nishiyama A, Takasaka T (1993) Nonselective cation and Cl channels in luminal membrane of the marginal cell. Am J Physiol 265:C72–8

Suzuki K, Petersen OH (1988) Patch-clamp study of single-channel and whole-cell K^+ currents in guinea pig pancreatic acinar cells. Am J Physiol 255:G275–G285

Suzuki M, Murata M, Ikeda M, Miyoshi T, Imai M (1998). Primary structure and functional expression of a novel non-selective cation channel. Biochem Biophys Res Comm 242:191–196

Swandulla D, Lux HD (1985) Activation of a nonspecific cation conductance by intracellular Ca^{2+} elevation in bursting pacemaker neurons of *Helix Pomatia*. J Neurophysiol 54:1430–1443

Takeuchi S, Ando M, Kozakura K, Saito H, Irimajiri A (1995) Ion channels in basolateral membrane of marginal cells dissociated from gerbil stria vascularis. Hearing Res 83:89–100

Takeuchi S, Marcus DC, Wangemann P (1992) Ca(2+)-activated nonselective cation, maxi K+ and Cl- channels in apical membrane of marginal cells of stria vascularis. Hearing Res 61:86–96

Teulon J, Paulais M, Bouthier M (1987) A Ca2-activated cation-selective channel in the basolateral membrane of the cortical thick ascending limb of Henle's loop of the mouse. Biochim Biophys Acta 905:125–132

Teulon J, Ronco P, Baudoin B, Geniteau-Legendre M, Cassingena R, Vandewalle A (1992) Transformation of renal tubule epithelial cells by simian virus 40 is associated with altered mitogenic sensitivity to renal K^+ channel blockers and emergence of Ca^{2+}-insensitive channels. J Cell Physiol 151:113–125

Thorn P, Petersen OH (1992) Activation of nonselective cation channels by physiological cholecystokinin concentrations in mouse pancreatic acinar cells. J Gen Physiol 100:11–25

Tohda H, Foskett JK, O'Brodovich H, Marunaka Y (1994) Cl- regulation of a Ca(2+)-activated nonselective cation channel in beta-agonist-treated fetal distal lung epithelium. Am J Physiol 266:C104–9

Turner RJ, Paulais M, Manganel M, Lee SI, Moran A, Melvin JE (1993) Ion and water transport mechanisms in salivary glands. Crit Rev Oral Biol Med 4:385–391

Van den Abbeele T, Tran Ba Huy P, Teulon J (1994) A calcium-activated nonselective cationic channel in the basolateral membrane of outer hair cells of the guinea-pig cochlea. Pflügers Arch 417:56–63

Van den Abbeele T, Tran Ba Huy P, Teulon J (1996) Modulation by purines of calcium-activated non-selective cation channels in the outer hair cells of the guinea-pig cochlea. J Physiol (Lond) 494:77–89

Volk T, Frömter E, Korbmacher C (1995) Hypertonicity activates nonselective cation channels in mouse cortical collecting duct cells. Proc Natl Acad Sci USA 92:8478–8842

von Tscharner V, Prod'hom B, Baggiolini M, Reuter H (1986) Ion channels in human neutrophils activated by a rise in free cytosolic calcium concentration. Nature 324:369–372

Wang W, Hebert SC, Giebisch G (1997) Renal K^+ channels: structure and function. Annu Rev Physiol 59:413–436

Wangemann P, Wittner M, Di Stefano A, Englert HC, Lang HJ, Schlatter E, Greger R (1986) Cl- channel blockers in the thick ascending limb of the loop of Henle, structure activity relationship. Pflügers Arch 407:S18–S141

Weber A, Siemen D (1989) Permeability of the non-selective cation channel in brown adipocytes to small cations. Pflügers Arch 414:564–570

Weyand B, Warth R, Bleich M, Kerstan D, Nitschke R, Greger R (1998) Hypertonic cell shrinkage reduces the K+ conductance of rat colonic crypts. Pflügers Archiv 436:227–232

Yang J (1990) Ion permeation through 5-hydroxytryptamine-gated channels in neuroblastoma N18 cells. J Gen Physiol 96:1177–1198

Yeh T-H, Herman P, Tsai M-C, Tran Ba Huy P, Van den Abbeele T (1998) A cationic nonselective stretch-activated channel in the Reissner's membrane of the guinea pig cochlea. Am J Physiol 274:C566–C576

Yellen G (1982) Single Ca^{2+}-activated nonselective cation channels in neuroblastoma. Nature 296:357–359

Zhu X, Jiang M, Peyton M, Boulay G, Hurst R, Stefani E, Birmbaumer L (1996) *Trp*, a novel mammalian gene family essential for agonist-activated capacitive Ca^{2+} entry Cell 85:661–671

Subject Index

abecarnil 508
– structure 507(fig.)
4-acetoamido-4′-isothio-cyanostilbene-2,2′-disulphonic acid (SITS) 593, 630, 634
acetozolamide-sensitive myotonia 9
acetylcholine 297, 637, 639
– induced bradycardia 297
– muscarinic potassium channels, activation 298–305
acidosis, metabolic 252
aconitine 39(table), 42
acromelic acid 431(fig.)
action potential, duration 30–31
adamantane compounds 430–431
ADCI 445(fig.), 446
adenine 584
– compounds 588
adenophostin 606(fig.)
adenophostin A 610
adenosine 588, 637
ADP 588
adrenalectomy 262–263
aequorin 68
Ag 591
agatoxin IVA 69, 88(fig.)
agel-489 430(fig.)
Agelenopsis aperta venom 140–142
Ageltoxin-489 430(fig.)
agitoxin 182–186
AIDS dementia 437
AKAP-15 17
AKT1/KAT1 channels 562
alanine 486
alcohol 511–512
– dependence 545
– 5-HT3R function 554
– potentiation of GlyR function 488–489
alcoholism 437
aldosterone 250–252
α1 subunit, calcium channels (voltage-dependent L-type)
– functional domains 87–91
– – channel activation 95–96
– – channel inactivation 96–98
– – pore and ion selectivity filter 94–95
α2 subunit, calcium channel 91–92
α-chymotrypsin 33
α-dendrotoxin 186
α-scorpion toxins (α-ScTx) 43–44
α-agatoxins 453
alphaxalone 510(fig.), 511
Alzheimer's disease 437
amiloride 133, 572, 629
amino acids 486
1-amino-1-carboxycyclobutane 449(fig.)
1-amino-1-carboxycyclopropane 449(fig.)
4-aminopyridine 629
amiodarone 35
amlodipine 128–129
– structure 123(fig.)
AMPA (α-amino-3-hydroxy-5-methyl-4-isoxazole propionic acid) 393
– autoimmune disease 397
– GluR2 subunit and calcium permeability 396
– grip 397
– phosphorylation 397
– QR site, determinant of channel properties 396–397
– therapeutic potential 434–436
– type subunits 395–396
AMPA (α-amino-3-hydroxy-5-methyl-4-isoxazole propionic acid) receptors
– molecular biology 416–417
– pharmacology
– – agonists 417–418
– – antagonists, competitive 419–427
– – antagonists, non-competitive (benzodiazepine) 427–428

– – channel blockers 429–431
– – positive allosteric modulators 428–429
AMPCPP 588
AMPPCP 588
anchorin protein, potassium channel 162
anesthetics
– general 213
– 5-HT3R function 554–555
– inhalational 590–591
– potentiation of GlyR function 488–489
angina pectoris 119
angiotensin II 210
aniracetam 428–429
ankyrin links 16
anti-inflammatory aromatic compounds (fenamates) 214
anti-inflammatory drugs 629
antiarrhythmic drugs
– sodium channel blockade, mechanism 30–32
– sodium channel models, interaction with 32–38
Arabidopsis thaliana 341
araguspongine B 612
Argiotoxin$_{(636)}$ 453, 454(fig.)
aspirin 629
ataxia, progressive 16
ATP 588, 608, 637
ATP-binding transporters related channels 385
ATP-dependent potassium channels, kidney 243
ATP-sensitivity 285–286
ATPA 432
Avermectin B_{1a} 512
azidobutyl diltiazem, structure 125(fig.)
azidodiltiazem, structure 125(fig.)
azidopine 122(fig.)

barbiturates 509–511
barium ions 587–588
Barter's syndrome 263–264
batrachotoxin (BTX) 39(table), 42–43
Bay K-8644 142–144
benzazepine 134
– structure 126(fig.)
benziazem 135
– structure 126(fig.)
benzil penicillin 499
benzimidazolones 214
benzocaine 37
benzodiazepine 499
– analogues 427–428
– site ligands 506–509
benzolactam (HOE166) 132
benzothiazepines 88(fig.), 100, 121, 131–132
– binding site 103–104
– calcium channel antagonist, structure 125–126(figs)
bepridil 132, 137(fig.)
β spectrin 16
β subunit, calcium channel 92
β-ABA 485(fig.)
β-AIBA 485(fig.)
β-alanine 485(fig.), 486
bi-functional *Kir* subunits channels 384–385
bicuculline 499
– structure 504(fig.), 505
bradykinin 637
brain
– K_{ATP} channels, properties 274
– P2X receptors 534
brevetoxins 44
brucine 486

C-I-2 SDZ-202-791 144
C-I-3 CGP-28392 144
C-II FPL-64176 144
cadaverine 230
Caenorhabditis elegans 340–341
caffeine 588–589, 595(table), 612
calbindin 546
calcicludine 141
calcineurin 6
calciseptrine 140(fig.), 141
calcitonin gene-related peptide (cGRP) 271, 532
calcium 631
– induced calcium release 582–595
– influx, postsynaptic 71–72
– IP3RCa release channel coactivator 607–608
– olfactory channel 569
– release, physiological (PCR) 582–586
"calcium antagonism" 119
"calcium bowl" 202(fig.), 207
calcium channel agonists 142
calcium channel antagonists 119–138
– alkaloids 139–142
– background, historical 119–120
– binding sites 133–138
– – allosteric interaction 121–127
– – reciprocal allosteric modulation,

representation 127(fig.)
– inorganic blockers 138–139
– natural toxins 139–142
– properties, biophysical and pharmacological 127–133
calcium channels (voltage-dependent L-type) 87–107
– α-1 subunit 87–91, 94–98
– auxiliary subunit 91–94
– calcium channel agonists and antagonists, binding sites 100–107
– interaction sites with other proteins 98–100
calcium channels (voltage-gated), classification and function 55–73
– Cav1/L-type calcium channels 59–61
– Cav2 61–64
– Cav3/T-type calcium channels 64–65
– conservation of calcium channel families 68
– functional properties 55–56
– functional roles 66–72
– – excitation-contraction coupling 66–67
– – introduction/subcellular localization 66
– – rhythmic activity 67–72
– gating 55
– inactivation 55
– molecular biological nomenclature 59
– permeation 56
– pharmacology, note 65
– selectivity 55–56
– subunit composition 56–58
calcium-activated non-selective cation channels 623–641
– agonist-mediated control 636–637
– blockers 628–630
– conductive properties 626–628
– hypertonicity, dependence 635–636
– intracellular regulatory elements 631–634
– phosphorylation-dependent regulation 634–635
– physiological role 637–641
– stimulators 630
– tissue distribution 623–626
calcium-calmodulin 568–569
calcium-modulated potassium channels (BKca) 197–215
– auxiliary subunits 204–205
– calcium sensitivity and diversity 205–207
– channel inactivation 209–210
– channel structure 198
– membrane topology 203(fig.)
– metabolic modulation 210–211
– pharmacology
– – channel activators 213–215
– – organic blockers 211–213
– – toxins 211–212
– TEA 203
– voltage dependence, origin 208–209
calmidazolium 572
calmodulin 567, 591–592
calsequestrin 592
capsaicin receptor 380–381
carbamylcholine 637
cardiac action potential, main potassium channels 349(fig.)
cardiac atrial myocytes 319–321
cardiac repolarization, physiological role of Iks 355–356
– determinants 356–357
– pharmacological acquired modulation LQTS 357–358
cardiac/skeletal muscle, type K_{ATP} channel 283
CCNC1 (cone cyclic-nucleotide channel subunit-1) 562
cell adhesion molecule (CAM) 15
CGP-28392 143(fig.), 144
CGS8216, structure 507(fig.)
Charcot-Marie-Tooth syndrome, type-4B 16
charybdotoxin 182–186, 211–212
chloramine T 35
7-chloro-5-iodokynuremic acid 451(fig.)
4-chloro-*m*-cresol 593
cholecystokinin 637, 639
chromaffin 209
chromosome 11g3 16
chymotrypsin 35
ciguatoxin 44
cilnidipine 129
– structure 123(fig.)
cinnarizine 132, 137(fig.)
7-CIQ 485(fig.)
5,7-CIQA 485(fig.)
clentiazem 132
– structure 125(fig.)
clofibric acid 592–593
cloramine T 35
clormethiazole 509, 510(fig.)
CNQX 423, 451
CNS-1102 445(fig.), 446
cocaine 594
concavalin A 428, 434

contactin 15
coronary vasodilators 119
cortical collecting duct
– ATP-sensitive potassium channel
– – function 247–248
– – regulation 250–252
– cell model 251(fig.)
– ion transport, model 247(fig.)
– ROMK density, regulation 262–263
CP-101,606 454(fig.), 455
CREB (nuclear binding protein) 72
cromakalin 271, 273
– structure 272(fig.)
CTB 485(fig.)
CTP 588
curare 553
cyanotriphenylborate 487–488
cyclic ADP-ribose 591, 612–613
cyclic nucleotide-gated channels 379–380, 561–573
– blockers 571–572
– channel gating 565–568
– cyclic-nucleotide binding 565–568
– ion permeation properties 563–565
– modulations 568–571
– structure 562–563
cyclophilin 170
cyclosporin 252
cyclosporin A 170
cys-loop receptors 374–378
cystic fibrosis transmembrane conductance regulator (CFTR) 280, 385
– regulation 260

DZKA 431(fig.)
D-α-aminoadipate 440, 441(fig.)
D-tubocurarine 552
dantrolene 584, 594, 595(table)
DCQX 451
decahydroisoquinolines 423–425, 434
depolarization 585–586, 595(table)
"depolarization"-induced calcium release (DICR) 585
depolarizing phase 3
depression 437
desensitization, "short-term" 298
Desmodium adscendens, medicinal herb 213–214
devapamil 103
dextromethorphan 445(fig.)
diabetes mellitus, non-insulin dependent 271
diacylglycerol, analogues 571
diamide 591
diazepam 506
– structure 507(fig.)
diazoxide 245, 271
– sensitivity 286–287
– structure 272(fig.)
3′,4′-dichlorobenzamil 572
dihydropyridines 3, 88(fig.), 121, 128–130
– binding site 100–103
– calcium channel antagonist 122–123(figs.)
– recemic 142–144
– receptor 584
4,4′-diisothiocyanostilbene-2,2′-disulphonic acid (DIDS) 593, 630
diltiazem 103, 119, 121, 131–132, 629
– analogues 121
– pharmacological effects 128(table)
– structure 125(fig.)
– vs. tetrandrine, differences 140
diphenyl hidantoin 133
disopyramide 34
dithiothreitol (DTT) 261, 591, 634
dizocilpine 445(fig.)
DNQX 423
domoic acid 431(fig.)
doxorubicin 591
DPI-201-106 44–45
DTZ-323 134
dynorphin, release 69
dysiherbaine 431(fig.)

E666-E726 peptide 98
efonidipine 129–130
– structure 123(fig.)
endistomin D 589
endocrine cells, G protein-gated potassium channels 322
endocytosis 69
enflurane 213
enkephalin 532
epilepsy, generalized (with febrile seizures) 15
etazolate 509, 510(fig.)
ethanol *see also* alcohol 214–215, 511, 532
etidocaine 35, 37
etomidate 509, 510(fig.)
excitation-contraction coupling 581, 585
excitatory aminoacid (EAA) receptors,

classification 415–416
exocrine acinar cells 639–641
extracellular matrix molecules 17

falodipine 128
– structure 123(fig.)
fatty acids (unsaturated), TRAAK forms potassium channel activation 340
fendiline 132
FK-506 binding protein 592, 611
FKBP-12 611
flavanoids 507(fig.), 509
flecainide 37
flufenamic acid 637
flunaridine 120(table)
flunarizine 132, 137(fig.)
fluspirilene 132
forskolin 637
FPL-12495 445(fig.), 446
FPL-64176 143(fig.)
FS-2 141
FTX 140–141
fura-2 587–588, 611
furanocumarins 509
– structure 507(fig.)

G $\beta\gamma$ 168–169
G protein 210
– α subunit, possible role in the G protein-gated potassium channel regulation 317–318
– coupled potassium channels, domains 169(fig.)
– cycle, representation 300(fig.)
– receptor reaction, Thomsen's and Mackay's models 305(fig.)
– sensitivity 287–288
– sodium channels, interactions 6
– subunit, effector subtype-specific effect 301(table)
G protein $\beta\gamma$ subunit
– binding domains, GIRK subunits 313–316
– induced activation, G protein-gated potassium channels, putative mechanism 316
G protein-gated potassium channels
– molecular analysis 305–318
– various organs, localization 318–323
GABA site
– agonists 502–503
– antagonists 504–505
– structure 504(fig.)
$GABA_A$ receptors
– activators and inhibitors 502–512
– function molecular structure, subtypes 499–502
– – anesthetics, general 511–512
– – barbiturates and related drugs 509–511
– – benzodiazepine site ligands 506–507
– – GABA site 502–505
– – miscellaneous agents 512
– – neuroactive steroids 511
– – picrotoxin site 505–506
– – structure 501(fig.)
$GABA_B$ receptors 499
gadolinium 629–630
galopamyl 103–104, 119
γ subunit, calcium channels 93–94
ganaxalone 511
gephyrin 484
GIRK 1,2,4 159–160
GIRK channels, expression 309–312
GIRK subfamily 307–309
glibenclamide 271, 273
– structure 272(fig.)
Glu (2,3′,4′)P3 606(fig.)
Glu-758 42
glutamate 377, 435
glutamate receptor (GluR) channels
– additional members 405–406
– AMPA subtype 395–397
– A subfamily 406
– L subfamily 405–406
– kainate subtype 397–398
– molecular diversity 394(table)
– NMDA subtype 398–405
– subunits families and subtypes 393–394
– transmembrane topology model 394–395
glutamate-gated cation channels 378–379
glyburide 248
glycine 448–449
– neurotransmitter 479
– structure 485(fig.)
glycine receptors 479–484
– clustering by the anchoring protein gephyrin 484
– heterogeneity 480–482
– ion channel function, determinants 483–484
– ligand-binding domain 482–483

– pathology 489–491
– – human hereditary disorder hyperkplexia 490–491
– – mouse glycine receptor mutants 489–490
– pharmacology
– – aminoacids 486–487
– – function, potentiation by anesthetic, alcohol and zinc 488–489
– – piperidine carboxylic acid compound 486–487
– – strychnine poison, antagonist 484–486
Goldman-Hodgkin-Katz equation 339
granisetron 552
grayanotoxin 39, 42
GRIP (glutamate receptor interacting protein) 397
GTP 588
GYKI-53655 427, 436

H_2O_2 591
H5 pore region 165
H-8 572
haloperidol 455
halothane 213, 510(fig.), 511–512, 554–555, 590–591
hanatoxin-1 (HaTx1) 190
heart
– K_{ATP} channels, properties 272–273
– pacemaker activity 561–562
HEK-293 cells 105–106, 214
Henle's loop
– ATP-sensitive potassium channels
– – function 245–247
– – regulation 249–250
heparin 611–612
HERG current 351(fig.)
HERG gene 348
HERG (human ether-go-go related gene) 169
20-HETE 250
heteropodatoxins (HpTx1/2/3) 190
Hill's coefficient 547, 567
histamine 637
hNav-2.1 10, 14
hNE-Na 12
HOE-166 132
homoquinolate 439, 440(fig.)
7-HPCE 417(fig.)
5-HT3 receptor 541–555
– allosteric regulation 554–555
– distribution 542
– molecular structure 542–545
– nervous system, function 545–551
– pharmacological properties 552–554
human cardiac sodium channel (hH1), molecular organization 28(fig.)
human hereditary disorder hyperkplexia 489–491
hydrazinium 139
5-hydroxyindole 555
hydroxylammonium 139
hyperinsulinemic hypoglycemia of infancy (PHHI) 288
hyperkalemic periodic paralysis 9
hyperpolarization 32
hypertension 119
hypokaliemia 358
hypomagnesemia 358

iberiotoxin 212
ibuprofen 629
IDRA-21 436
IEM-1460 430(fig.)
IEM-1754 430(fig.)
If channel 561
ifenprondil 437, 453, 454(fig.)
Ih channel 561
Ikr current 348–352
imidazobenzodiazepines 508
imidazopuridines 506, 507(fig.)
immunophilin ligands 611
imperatoxin activator 592
indole diterpenes 213
indomethacin 629
inositol triphosphate (IP3) receptors 380, 582, 605
– analogues 606(fig.), 609–610
– caged 610
– calcium release channels, regulation 603–615
– – activators 609–611
– – agonists 605–609
– – function 603–605
– – inhibitors 611–613
– – modulators 605–609
– – molecular structure 603–605
– – pharmacology 613(table)
– – RyR, comparison between 613
– – spatio-temporal patterns 613–615
– structures 606(fig.)
insulin secretion, β cells 69
"inverse agonists" 499
inward rectification, mechanism 225–236
– Kir channel pore: binding sites for polyamines 231–233

- nature, classical considerations 225–227
- other potassium channels 229
- polyamine-induced rectification 236
- pore block and intrinsic 230–231
- structural requirements, blocking particles 233–235
- two transmembrane domain potassium channels 227–229
ionotropic glutamate receptors (iGluRs) 415
IRK-1 169
IRK-3 162, 170
ischemia 435
IsK, activation 165
isoflurane 213, 554–555
isoproterenol 210
isoxazoles 425–426
isradipine 128, 130
- structure 122(fig.)
ITP 588

Jarvell and Lange-Nielsen syndrome 347, 352–354

κ-conotoxin PVIIA 189–190
KAB-2 162
kainate receptors 415
- agonists 431–433
- antagonists, competitive 424(fig.), 433–434
- ligands, therapeutic potential 434–437
- molecular biology 416–417
kainate subtype, GluR channels 397
kainic acid 431(fig.)
kalichidines 188–189
kaliotoxin 182–186
KCO-1 341
ketamine 512
kidney
- inner medullary collecting duct 562
- K_{ATP} channels, properties 275
kidney, ATP-dependent potassium channels 243–264
- ATP-sensitive potassium channels, function
- - cortical collecting duct 247–248
- - Henle's loop 245–247
- - proximal tubule 243–245
- - thick ascending limb 245–247
Kir channel family 227–229
- Kir-1 subfamily 227–228
- Kir-2 subfamily 228
- Kir-3 subfamily 228
- Kir-4 and 5 subfamily 228
- Kir-6 subfamily 229
- Kir-D subfamily 229
Kir channel pore, binding sites for polyamines 231–233
kurtoxin 64, 102(table), 142
Kv (voltage-gated) channel, dendrotoxin binding site 186
Kv (voltage-gated) potassium channels 157, 159
- activation 162–163
- C-type inactivation 164–165
- functional devices, location 163(fig.)
- N-type inactivation 163–164
- pharmacology 177–191
- - molecular and functional organization 178–182
- - peptide toxin binding sites 182–191
- S4 region, dynamic translocation 164(fig.)
KvLQT-1 353
KvLQT-1/IsK 352

L-689,560 451(fig.)
L-701,324 451(fig.)
L-CCG-IV 440(fig.)
L-*cis*-diltiazem 571–572
lanthanum 630
lidocaine 34–35, 37, 594
local anesthetic drugs 30–1
- frequency-dependent block 31
- tonic block 31
long QT syndrome-3 10
long QT syndromes (LQTS) 347–348, 352–355
- acquired, pharmacological considerations 356–358
long term potentiation/depression 437
lorecelazole 509, 510(fig.)
lorecezole 509
LQT-2 348–352
LU-49888 135
LY-293558 424(fig.)
LY-294486 424(fig.)
LY-302679 424(fig.)
LY-339434 431(fig.)
LY-382884 424(fig.)
LY-83583 572

magnesium 593, 595(table), 611
- cytoplasmic second messenger 170
- sensitivity 287
magnesium ion 591
malignant hyperthermia 590–591

margatoxin 212
MDL-105,519 451(fig.)
mefenamic acid 630
memantine 445(fig.)
metabotropic glutamate receptors (mGluRs) 415
metals, transition 570
methoxyverapamil 134
9-methyl-7-bromoeudistomin D (MBED) 589
methylammonium 139
mexiletine 35
Meyer-Overton correlation 512
MgATP 581
mibefradil 64, 88(fig.), 120(table), 133, 137(fig.)
μ-conotoxins 27, 41
– peptide sodium channel blockers 41–42
midazolam 508
mineralocorticoids 262–263
mitochondria, K_{ATP} channels, properties 276
MK-801 323, 438
mNav-2.3 10, 14
MNQX 451
molecular heterogeneity 3
3-morpholino-sydnonomine 571
multi-drug resistance associated proteins (MRP) 280
muscarinic potassium (KAch) channels 297
– acetylcholine-activation 298–305
– – G protein's cyclic reaction 299–301
– – GTP, positive cooperative effect 302–304
– – receptor-G protein reaction, incorporation 304–305
muscimol 499
– structure 504(fig.), 505
myosin-actin-tropomyosin-troponin system 581
myotonia fluctuans 9
myotoxin 593

N-bromocatemide 33
N-ethylmaleimide 570, 591
N-methyl-D-aspartate (NMDA) receptor channel
– agonist binding 402–403
– channel pore and gating 401–402
– distribution of the subunits, dynamic variations 399–400
– heteromeric nature 398–399
– modulation 403–404
– phosphorylation 403
– post-synaptic proteins, associated 404–405
– splice variants 400–401
– synaptic plasticity, neural development 404
– transmembrane topology 395(fig.)
N-methyl-D-aspartate (NMDA) receptors
– agonists and partial agonists (glycine site) 449(fig.)
– allosteric modulatory sites 452–456
– antagonists (glutamate site) 441(fig.)
– antagonists (glycine site) 451(fig.)
– channel blockers 444–446
– glutamate recognition site 438–444
– glycine recognition site 446–451
– molecular biology 416–417
– therapeutic considerations 436
N-methyl-D-glucamine 549
NaCh6/Scn 8a sodium channel 3
NaNG 13
Nas 12
NBQX 423, 435
nefenamic acid 637
nerve conduction 3
nerve growth factor 532
neurons, G protein gated potassium channels 321–322
neuropeptide Y 70
neurotoxin receptor sites, sodium channels 39(table)
neurotransmission, calcium ion role 68
neurotransmitter transporters related channels 385–388
neurotrophin-3 211
NG108–15 cells 548
nicardipine 128, 130
– structure 122(fig.)
nickel 570
nicotine 571
nifedipine 64, 111
– pharmacological effects 128(table)
– structure 122(fig.)
niflumid acid 214, 593
nimodipine 64, 130
– pharmacological effects 128
– structure 123(fig.)
NIP 163
nipicotic acid 485(fig.)
nisoldipine, structure 122(fig.)
nitrendipine, structure 122(fig.)
nitric oxide 210–211, 571

nitro compounds 210
noxinstoxin 212
NS-102 424(fig.)
NS-256 423
NS-257 422(fig.)
nucleotide disulphate (NDP) sensitivity 286
nucleotide-sensitive potassium channels 383–384
nucleotides
– adenosine-based, inhibition 632(table)
– intracellular cyclic, stimulatory effects 633

OCNC1 (olfactory cyclic-nucleotide channel subunit-1) 562
octanol 120(table), 133
olfactory receptor neurons 561
olfactory transduction 561
ω-agatoxin 120(table)
– IVA 140(fig.), 142
ω-conotoxin 3, 120(table)
– GVIA 140(fig.), 142
ondasetron 552
ouabain 245
oxidizing agents 591
oxytocin 69

P2X receptors
– distribution 523–527
– functional properties 527–535
– – modulation 531–532
– molecular biology 519–523
"P region" 5
P-glycoprotein 280
pancreatic β-cell type K_{ATP} channels 276–282
– model 285(fig.)
– pathophysiology 288
– properties 273–274
– subunits, physical interaction and stoichiometry 284–285
Parabuthus transvaalicus 142
paramyotonia congenita 9
Parkinson's disease 437
patch-clamp technique 243
paxilline 213
PC12 209
PDZ 162
penticainide 34
pentobarbital 509
pentylenetetrazole 499
PEPA 429
peptide toxins 30
peptide-gated channels 383
peripheral neurons, P2X receptors 534
pH, regulation 260–2
phellopterin 507(fig.), 509
phencyclidine 438
phenobarbital 509, 510(fig.)
phentolamine, sensitivity 287
phenylalanine analogues 426–427
phenylalkylamines 88(fig.), 100, 121, 130–131
– binding site 103–104
– calcium channel antagonist, structure 124(fig.)
phenylethanolamine 454
phenylglycine 426–427
phenytoin 37
philanthotoxin 453
phloretin 214
phosphatase 2A 6
phosphatidylinositol 4,5-biphosphate (PIP2) 262
– mediation of G-β-γ-activation of KG channels 316–317
phosphoinositides, regulation 262
phospholipase C-β 603
phospholipase C-γ 603
phosphorylation 569–70, 608–609
photoreceptors, retinal 561
picrotoxin 499
picrotoxinin 485(fig.), 487–488
– site 505–506
– structure 504(fig.), 505
pimozide 132, 572
pinacidil 273
piperidine carboxylic acid compounds 486
pitrazepin 504(fig.), 505
PN4 10
PN-1 12–13
PNQX 435–436
polyamines 452
– induced rectification 236
potassium channel families 157–171
– gating mechanism 162–165
– heteromultimetric assembly 159–162
– inward rectification mechanism 166–167
– ion permation and block 165–168
– perspectives 170–171
– regulation mechanisms 168–170
– 2-repeat type 159
– structure, diversity 158(fig.)
– 2-TM type 158

– 1-TM type 158
– 6-transmembrane (TM) type 157–158
potassium channels
– β subunit 160–161
– Kir inwardly rectifying 305
– – subunits, evolutionary free 306(fig.)
– – tetrameric structure 312
– modulatory, β subunit (voltage gated) 181–182
– one pore domain 333
– protein-gated 297–324
– selective cyclic nucleotide-activated 561
– structural domains, α subunit (voltage gated) 178–181
– two pore domains 334–343
– – channel, structure 337
– – cloning and gene organization 334–335
– – functional expression 335–337
– – mammals, related potassium channels 337–344
procaine 584, 588, 591, 594, 595(table)
pronase 35
propaphenone 34
propofol 510(fig.), 511–512
protein kinase A 634–635
– regulation by phosphorylation 257–258
protein kinase C
– dependent-phosphorylation, expressed L-type calcium channel modulation 106–107
– regulation by phosphorylation 259
proteins, endogenous modulatory 591–592
proton-gated channels 381–383
protons 571
proximal tubule
– ATP-sensitive potassium channels regulation 248–249
– ion transport model 244(fig.)
putrescine 230, 236, 452
pyrazinone 510(fig.)
pyrazolopyridines 509
pyrethrin toxins 33, 35
pyrethroids 44

quazepam 506, 507(fig.)
quinidine 37
quinine 629
quinolonic acid compounds 487–488
quinoxalinediones and related compounds 419–423, 433–434
quisqualate receptors 415
QX-222 35–36
QX-314 34, 35–36, 323

R-(+)-HA966 449(fig.)
R-serine 449(fig.)
Radix stephania tetrandrae 139
rapamycin 611
Rasmussen's encephalitis 397
RCNC1 (rod cyclic-nucleotide channel subunit-1) 562
Reissner membrane 630
"relaxing factor" 581
REST (repressor protein) 17
Ro15-1788 508
Ro-04-5595 454(fig.), 455
Ro-25-6981 455
Ro-8-4304 455
Romano-Ward syndrome 347, 353–354
ROMK (cloned ATP-sensitive potassium channels) 252–263
– channel isoforms and localization 254–255
– channel pore-rectification 256–257
– channel structure 252–254
– native secretory ATP-sensitive potassium channel, comparison 256
– phosphorylation, regulation, protein kinase A 257–259
– topology 253(fig.)
RS-TetGly 440(fig.)
RU-5135, structure 504(fig.), 505
ruthenium red 594
ryanodine 589–590, 595(table)
ryanodine receptor (RyR) 100, 367, 368(table), 380
– activators 586–593
– calcium release channels 581–595
– function 582–583
– inhibitors 593–594
– mammalian 583(table)
– molecular structure 582–583
– opening, different modes 583–586
ryanoid 589–590

S-nitroso-cysteine 571
Saccharomyces cerevisiae 341
sarcosine 485(fig.), 486
saxitoxin 6, 38–41
schizophrenia 437, 545
SCL-11 14
SCN4A 17
SCN5A 10, 17, 28(fig.)
Scn8a 10–11, 17

scorpion toxins 182–186, 212
– α and β 39(table)
Sdz(+)(S)-202791 143(fig.)
sea anemone toxins 39(table), 188
secretagogues 639
semotiadil, structure 126(fig.)
sensory neurons, P2X receptors 533–534
serine 486
serotonin 637
serotonin-gated channel 377
Shaker B peptide 209–210
Shaker potassium channel family 563
skeletal muscles, K_{ATP} channels, properties 273
Slo channels 202(fig.)
Slo protein 199
smooth muscles
– K_{ATP} channels 283–284
– – properties 275
– P2X receptor 532–533
snail toxins 189
snake toxin 186–187
sodium ions, membrane permeability 27
sodium (Na) channels
– activators 42–45
– blocking agents 27–42
– classification and structure 27–30
– voltage-gated
– – α subunit 4–15
– – β-1 subunit 15
– – β-2 subunit 15–16
sodium (Na) channels (voltage dependent) 3–18
– accessory subunits 15–18
– – associated proteins 16–17
– – β-1 subunit 15
– – β-2 subunit 15–16
– α subunit 4–15
– – atypical sodium channels 14
– – brain types I,II and III 7–9
– – heart I/SkM2/hH1/SCN5A 9–10
– – mammalian phylogenetic tree 7(fig.)
– – NaCh6(Rat)/Scn8a(Mouse)/IPN4 10
– – NaN/SNS2 13–14
– – PN1/Na/hNE-Na/Scn9a 12–13
– – skeletal muscle, μ-I/SkM1/SCN4A 9
– – SNS/PN3/Na-NG/Scn10a 13
– – structure 5(fig.)
– architecture 4
– genomic structure 17
spermidine 230, 236, 430(fig.), 452
spermine 230, 236, 430(fig.), 452
– sensitivity 287
spider toxins 190–191, 453
SR-95531, structure 504(fig.), 505
Startle disease 490–491
steroid, neuroactive 511
stiff baby syndrome 490–491
stilbene sulphonates 630
store-operated channels 380–381
strontium 587–588
strychnine 484, 585(fig.)
substance P 532
sulfhydryl reagents 570–571
sulfonylurea
– receptor (SUR2B) 260, 277–281, 283, 383–384
– – membrane topology 28(fig.)
– sensitivity 287
syntrophins 17

T-477 132
– structure 126(fig.)
TASK 338–339, 342–343
taurine 486
TBPS, structure 504(fig.), 505
terbutaline 637
tetracaine 572, 594
tetradoxin 3, 6, 9–10, 27, 38–41
– Lipkind-Fozzard binding site model 40(fig.)
tetraethylammonium 212–213
tetramethrine 133
tetrandine 139
7-TFQ 485(fig.)
7-TFQA 485(fig.)
theobromine 589
theophylline 589
thick ascending limb
– ATP-sensitive potassium channels
– – function 245–247
– – ion transport, model 246(fig.)
– – regulation 249–250
thimerosal 591, 610–611, 634
thiopental 509
thromboxane A2 210
tipE 16
TOK/DUK/YKC/YORK, yeast channel 341
tolbutamide 271, 273
– structure 272(fig.)
torsades de pointes 347–348, 356–358
TRAAK 338(fig.), 340, 342–343
trans-MCG-4 431(fig.)

transmitter-gated channel
– membrane domains, structural elements 366–373
– scope of 365–366
– subclasses 373–386
TREK 338–339, 342
triadin 592
triazolam 508
triazolobenzoxazine, structure 507(fig.)
trifenoperazine 572
trifluoroperazine 572
tropisetron 552
TWIK 334–335, 342

U-101017, structure 507(fig.)
U-97775, structure 507(fig.)
UTP 588

"vagusstoff" 297
vanilloid receptors 380–381
vascular smooth muscle type K_{ATP} channel 284
vasepressin 69
vasoactive intestinal peptide (VIP) 271
ventricular fibrillation 347
verapamil 103–104, 119, 130–131
– pharmacological effect 128(table)
– structure 124(fig.)
veratridine 39(table), 42–43
visual transduction 561
volatile agents 511–512
"voltage signals" 638

W-7 572
W-89 553
wasp toxins 453
Weaver mutant mice 323
willardiine 417(fig.), 418

X-ray crystallography 4
xanthine derivatives 589
xenovulene A 509
xestospongin 612
xestospongin C 606(fig.)

YM90K 423

zacopride 552
ZAPA 503
zinc 455–456, 531–532
– GlyR function, potentiation 488–489
zolpidem 506, 507(fig.)
zopiclone, structure 507(fig.)

Printing (Computer to Film): Saladruck, Berlin
Binding: Stürtz AG, Würzburg